Algorithms and Computation in Mathematics

Volume 31

With this forward-thinking series Springer recognizes that the prevailing trend in mathematical research towards algorithmic and constructive processes is one of long-term importance. This series is intended to further the development of computational and algorithmic mathematics. In particular, Algorithms and Computation in Mathematics emphasizes the computational aspects of algebraic geometry, number theory, combinatorics, commutative, non-commutative and differential algebra, geometric and algebraic topology, group theory, optimization, dynamical systems and Lie theory. Proposals or manuscripts that center on content in non-computational aspects of one of these fields will also be regarded if the presentation gives consideration to the contents' usefulness in algorithmic processes.

Cláudio L. Lucchesi • U. S. R. Murty

Perfect Matchings

A Theory of Matching Covered Graphs

Springer

Cláudio L. Lucchesi
University of Sao Paulo
São Paulo, Brazil

U. S. R. Murty
University of Waterloo
Waterloo, ON, Canada

ISSN 1431-1550
Algorithms and Computation in Mathematics
ISBN 978-3-031-47503-0 ISBN 978-3-031-47504-7 (eBook)
https://doi.org/10.1007/978-3-031-47504-7

This Springer imprint is published by the registered company Springer Nature Switzerland AG
The registered company address is: Gewerbestrasse 11, 6330 Cham, Switzerland

Paper in this product is recyclable.

To the memory of our parents

Foreword

In recent decades the theory of matching covered graphs has blossomed. The concept was introduced by Tutte, who showed that every 2-connected cubic graph is matching covered, thereby strengthening a result of Petersen. The subject was studied in depth by Lovász, beginning in the 1980s. His contribution included the noteworthy theorem that any two tight cut decompositions of a given matching covered graph yield the same list of bricks and braces, up to multiple edges. The numbers of bricks and braces in this list are therefore important invariants of the original matching covered graph. Remarkably, it turns out that the number of bricks whose underlying simple graph is isomorphic to the Petersen graph is also significant.

The authors' interest in this subject was piqued by the early work of Lovász, and they began a close collaboration together that also included other authors. They developed an extensive theory of removable edges in matching covered graphs. These are edges whose removal preserves the matching covered property of the graph. The authors used this theory to unify the subject of matching covered graphs, simplify some proofs and make further significant advances. This volume is the result of that collaboration and presents a unified account of the most important results in this area. Its climax is a remarkable proof of a characterisation of Pfaffian near bipartite graphs.

The authors present a subject of great beauty and depth. I now invite the reader to turn the page and begin to enjoy the treasures that await.

Charles Little

Preface

A graph is *matchable* if it has a perfect matching. A *matching covered graph* is a connected graph on at least two vertices in which each edge is covered by some perfect matching. The theory of matching covered graphs, though of relatively recent vintage, has an array of interesting results with elegant proofs, several surprising applications and challenging unsolved problems.

Lovász and the matching lattice: It was the article *Matching Structure and the Matching Lattice* by László Lovász (1987, [58]) that inspired us to learn the subject. The theorem which states that any two tight cut decompositions of a matching covered graph yield the same list of bricks and braces, up to multiple edges, is an example of an interesting and surprising assertion with an elegant proof. The crowning achievement of that paper was the characterization of the matching lattice. The *matching lattice* $\mathcal{L}at(G)$ of a graph G is the set of all integer linear combinations of incidence vectors of perfect matchings of G. A necessary condition for an integer vector $\mathbf{x}$ in the edge space of G to belong to $\mathcal{L}at(G)$ is that it be *regular*, that is $\sum_{e\in\partial(u)} \mathbf{x}_e = \sum_{e\in\partial(v)} \mathbf{x}_e$, for any two vertices u and v of G.[1] But, in general, a regular integer vector in the edge space of a matching covered graph G need not belong to $\mathcal{L}at(G)$. For example, the all 1's vector $\mathbf{1}$ in the edge space of the Petersen graph is not in its matching lattice. Generalizing a theorem of Seymour (1979, [85]), Lovász proved the stunning statement that if G is any brick whose underlying simple graph is not the Petersen graph, then any regular integer vector in the edge space of G is in $\mathcal{L}at(G)$. One cannot but be impressed by the sheer 'audacity' of a non-obvious assertion like this which says that 'the Petersen graph is the only exception'!

Tutte and the 1-factor theorem: The idea of a matching covered graph first appeared in an article by William Thomas (Bill) Tutte who was one of the preeminent graph theorists and cryptographers of the twentieth century. In his high school days, Bill learnt basic graph theory from Rouse Ball's book *Mathematical Recreations and Essays* and was familiar with Petersen's Theorem, which states that every 2-connected cubic graph is matchable. In 1935 he went to Cambridge to study chemistry. There he made friends with Leonard Brooks, Cedric Smith and Arthur

[1] For any subset S of the vertex set of a graph G, we denote the edge cut consisting of the set of edges of G with one end in S and one end in $\overline{S}$ by $\partial(S)$.

Stone, who were students of mathematics. The four of them worked on the problem of dissecting rectangles into squares no two of which are of the same size. They observed that there is a one-to-one correspondence between dissections of rectangles into squares and flows of currents in a certain associated electrical network. This observation made it possible for them to employ the determinantal formulas which are used for computing currents in an electrical network to find desired dissections of rectangles into squares. Their work was published in 1940 [4].

Bill Tutte switched to mathematics in 1940, but his studies were interrupted by call to wartime duty at Bletchley Park, the fabled organization of code breakers. We refer the reader to Younger (2012, [95]) for a biographical sketch of Bill Tutte and for an account of his pivotal contributions to code breaking when he was at Bletchley.

After returning to Cambridge in 1945, Tutte resumed his research in graph theory. One of the problems he set himself to work on was to find a characterization of matchable graphs. He writes in his mathematical autobiography *Graph Theory as I Have Known It* (1998, [90]) that he was hoping to find a formula for the number of perfect matchings of a graph analogous to Kirchhoff's determinantal formula for the number of spanning trees. He did not succeed in finding such a formula, but using properties of polynomials associated with matrices, called Pfaffians, he did succeed in finding a characterization of matchable graphs by showing that a graph G is matchable if and only if $o(G - S) \leq |S|$, for any subset S of V, where $o(G - S)$ denotes the number of odd components of $G - S$. (Tutte's hope of using Pfaffians to count the number of perfect matchings in a graph was partially fulfilled by Kasteleyn (1963, [44]), who showed how they could be used to count the number of perfect matchings of a planar graph. This is described in Chapter 19.)

Tutte's fundamental result appeared in the paper *The Factorization of Linear Graphs* (1947, [89]). As a corollary, he established a strengthening of Petersen's Theorem by showing that every 2-connected cubic graph is matching covered. Thus the notion of matching covered graphs goes back to this article by Tutte. Since Tutte referred to perfect matchings as 1-factors, some authors called such graphs 1-factor covered graphs. It appears that Lovász (1983, [57]) was the first to refer to them as matching covered graphs.

Lovász has made many important contributions to combinatorics and computer science, but matching theory seems to have been his first love. His book *Matching Theory* with M. D. Plummer (1986, [59]) contains accounts of developments in the theory of matching covered graphs[2] that took place up to the mid 1980's. But that book was written before the appearance of his article *Matching Structure and the Matching Lattice* which established the uniqueness of tight cut decompositions of matching covered graphs, thereby providing an elegant theoretical foundation to this subject.

[2] A matchable graph is *elementary* if the union of perfect matchings in it induces a connected subgraph. It should be noted that some results in [59] about matching covered graphs are presented in the more general context of elementary graphs.

Tight Cut Decompositions and Bricks and Braces: Given any cut $C := \partial(X)$ of a graph G one can obtain two graphs G/X and $G/\overline{X}$ by contracting, respectively, the shores X and $\overline{X}$ of C to single vertices. These two graphs are called the *C-contractions of G*. A cut C of a matching covered graph G is said to be *tight* if $|M \cap C| = 1$ for every perfect matching M of G. It is easy to see that if C is a tight cut of a matching covered graph G, then the two C-contractions of G are also matching covered.

A matching covered graph which is free of nontrivial tight cuts is a *brick* if it is nonbipartite and a *brace* if it is bipartite. If a graph G has a nontrivial tight cut $C := \partial(X)$, then both G/X and $G/\overline{X}$ have fewer vertices than G. Thus, by repeatedly contracting shores of nontrivial tight cuts, any matching covered graph may be reduced to a list of bricks and braces. Depending on the choice of tight cuts, this 'tight cut decomposition procedure' may yield different lists of bricks and braces. However, Lovász [58] showed that, up to multiple edges, any two applications of this procedure result in the same list of bricks and braces. This result allows us to define two invariants $b(G)$ and $p(G)$ associated with matching covered graphs, where $b(G)$ is the total number of bricks in a tight cut decomposition of G and $p(G)$ is the number of bricks in such a decomposition whose underlying simple graphs are isomorphic to the Petersen graph.

Initial focus of our research: An edge e of a matching covered graph G is *removable* if $G-e$ is also matching covered, and a removable edge e is *b-invariant* if $b(G-e) = b(G)$. We started working systematically on the following conjecture: *any brick G, which is different from the complete graph K_4, the triangular prism $\overline{C_6}$, and the Petersen graph $\mathbb{P}$, has a b-invariant edge.* (This conjecture was communicated to us by Lovász with the observation that its validity would simplify his proof of the theorem in [58] which established a characterization of the matching lattice.) As our work progressed, we realized that a proof of the above conjecture would also settle another (folklore!) conjecture which states that every brick whose underlying simple graph is not $\mathbb{P}$ has an ear decomposition which involves just one double ear addition.

What led us to write this book?: We succeeded in confirming the above-stated conjecture of Lovász in collaboration with Marcelo Carvalho, a former graduate student of Cláudio Lucchesi. Our proofs appeared in the three papers [9], [10] and [11], which were published in 2002. (We were actually able to show that every brick other than K_4, $\overline{C_6}$, $\mathbb{P}$ and a brick on eight vertices which we call the bicorn has at least two b-invariant edges and use this result to show that any matching covered graph G has an ear decomposition which involves just $b(G) + p(G)$ double ear additions and that this result is best possible.) We continued to work on other problems in this area and have been able to make further contributions to the subject. Our efforts have led us to discover some interesting new facts and to develop some new methods of proof. Our object in writing this book has been to share with readers what we have learnt about this fascinating subject. We list below some of the many ideas which have enhanced our understanding of this subject.

Barriers and matchable edges: A subset B of the vertex set of a matchable graph G is a *barrier* if $|B| = o(G - B)$. Theorem 2.1, which is a straightforward consequence of Tutte's Theorem 1.3, shows that if u and v are any two vertices of a matchable graph G, then $G - u - v$ is also matchable if and only if there is no barrier of G which contains both u and v. Graphs which do not have any nontrivial barriers are said to be *bicritical graphs*. As we shall see, all bricks are bicritical, but not all bicritical graphs are bricks. (The bicritical graph shown in Figure 3.5 is not a brick because it has nontrivial tight cuts.)

An edge e of a matchable graph G is *matchable* if there is some perfect matching of G which contains e. Theorem 2.1 stated above implies that e is matchable if and only if there is no barrier of G which contains both ends of e. Thus a matching covered graph is a connected matchable graph in which each barrier is a stable set.

A connected bipartite graph $G := G[A, B]$, with $|A| = |B| \geq 2$, is matching covered if and only if $|N(X)| \geq |X| + 1$, for all $X \subset A$ such that $1 \leq |X| \leq |A| - 1$. It is easy to prove this result, by adapting Hall's well-known characterization of matchable bipartite graphs. Corollary 2.10, which follows from this characterization, states that in any matchable bipartite graph $G := G[A, B]$ there exists a subset S of A such that the subgraph of G induced by $S \cup N(S)$ is matching covered. This useful observation is the basis of what is known as the *Dulmage-Mendelsohn decomposition* of matchable bipartite graphs. (See Section 2.3.1.)

Maximal barriers and the Canonical Partition Theorem: A barrier B of a matchable graph G is *maximal* if there is no barrier of G which properly contains it. Theorem 3.2 states that if B is any maximal barrier of a matchable graph then each component of $G - B$ is critical (factor-critical, hypomatchable); in particular, this means that no component of $G - B$ is even. As noted earlier, all barriers in a matching covered graph are stable sets; in particular, all maximal barriers of such a graph are stable sets.

One of the basic results in this theory, known as the *canonical partition theorem* (Lemma 5.2.1 in Lovász and Plummer [59]), states that the maximal barriers of a matching covered graph G constitute a partition of the vertex set of G. In Chapter 3 we give a simple proof of this theorem as described below.

We first consider the case in which G is a bipartite matching covered graph with bipartition $\{A, B\}$. If S is any maximal barrier of G, then each component of $G - S$ has just one vertex because there are no critical bipartite graphs of order greater than one. Consequently $\overline{S} := V - S$ is also a stable set, implying that $\{S, \overline{S}\}$ is a bipartition of G. However, being a matching covered graph, G is connected, and every connected bipartite graph has a unique bipartition. It follows that either $S = A$ or $S = B$. This also implies that if $a \in A$ and $b \in B$, then there is no barrier which includes both a and b and, furthermore, by Theorem 2.1, that $G - a - b$ is matchable.

To deal with the case in which G is not bipartite, we consider a maximal barrier B of G and associate with it the bipartite matching covered graph $\mathbb{H}(B)$ obtained from G by shrinking each component of $G - B$ to a single vertex. It follows from the fact that each component of $G - B$ is critical, and the above stated result concerning bipartite matching covered graphs, that if $b \in B$ and $a \notin B$, then the graph $G - a - b$ is matchable. Thus, if there is a maximal barrier $B' \neq B$ such that $b \in B \cap B'$

and $a \in B' - B$, the graph $G - a - b$ would be matchable. By Theorem 2.1 this is impossible because both a and b belong to the barrier B'.

The core of a matchable graph with respect to a barrier: The bipartite $\mathbb{H}(B)$ associated with a maximal barrier B in a matching covered graph is a special case of a bipartite graph associated with any barrier B of a matchable graph. The *core* of a matchable graph G with respect to a barrier B, denoted by $\mathbb{H}(B)$, is the bipartite graph obtained from G by deleting all edges with both ends in B, all the vertices belonging to even components of $G - B$, and then shrinking each odd component of $G - B$ to a single vertex. If G is matching covered, then, clearly, $\mathbb{H}(B)$ is also matching covered. More generally, every matchable graph G has a barrier B such that $\mathbb{H}(B)$ is matching covered, with the additional property that each odd component of $G - B$ is critical. We refer to such a barrier as a *Dulmage-Mendelsohn barrier* as it is possible to use Corollary 2.10 to show that any maximal barrier of a matchable graph includes such a barrier (see Lemma 5.4). This result plays a crucial role in our proof of an important result in this theory which is described below (and also in the proof of a useful result in Chapter 10).

Edmonds-Lovász-Pulleyblank (ELP) Theorem: Recall that bricks and braces are, respectively, nonbipartite and bipartite matching covered graphs which are free of nontrivial tight cuts. As mentioned earlier, Lovász [58] proved that any matching covered graph may be decomposed into bricks and braces by means of the tight cut decomposition procedure. But how does one recognize whether or not a given matching covered graph is free of nontrivial tight cuts? For example, how can we tell whether or not a given nonbipartite graph is a brick?

ELP-cuts (barrier cuts and 2-separation cuts): To answer this question raised above, it would be convenient to know how tight cuts arise. There is no known simple answer to this question, but two scenarios which yield nontrivial tight cuts of a matching covered graph G were recognized early on: (i) if B is a barrier of G, and K is an odd component of $G - B$, then $\partial(K)$ is a tight cut; such a cut is called a *barrier cut*; (ii) if $\{u, v\}$ is a 2-vertex cut which is not a barrier, then all components of $G - u - v$ are even, and if H is the union of any nonempty proper subset of the components of $G - u - v$, then the cut $\partial(\{u\} \cup V(H))$, and similarly $\partial(\{v\} \cup V(H))$, are tight cuts of G, such cuts are called *2-separation cuts*. We refer to these two types of tight cuts as *ELP-cuts*. A fundamental result of matching theory, due to Edmonds, Lovász, and Pulleyblank (1982, [32]) states that if a matching covered graph has a nontrivial tight cut, then it also has a nontrivial ELP-cut. Their proof of this result was based on linear programming techniques. An easier and purely graph theoretical proof was given by Szigeti (2002, [87]).

We give an alternative proof of the Edmonds-Lovász-Pulleyblank (ELP) Theorem in Chapter 5. Our proof is inspired by the one due to Szigeti mentioned above and relies on certain key properties of (i) Dulmage-Mendelsohn (DM) barriers of matchable graphs, and of (ii) peripheral tight cuts in matching covered graphs. (A nontrivial tight cut $\partial(X)$ of a matching covered graph G is *peripheral* if there is no nontrivial tight cut $\partial(X')$ of G such that $X' \subset X$. Clearly, any matching covered graph which has nontrivial tight cuts must have a peripheral tight cut.)

Lemma 5.5 (key lemma related to the existence of DM-barriers) states that if G is a matchable graph and X is a nonempty proper subset of $V(G)$ such that both subgraphs $G[X]$ and $G[\overline{X}]$ are connected and no edge of $\partial(X)$ is matchable in G, then there exists a DM-barrier B of G such that B and the vertex sets of all the components of $G - B$ are entirely contained in one of the shores of $\partial(X)$.

Lemma 5.1 (key Lemma related to peripheral tight cuts) states that if $C := \partial(X)$ is a peripheral tight cut of a matching covered graph G then there exists an edge uv of G, with $u \in X$ and $v \in \overline{X}$, such that both the subgraphs $G[X - u]$ and $G[\overline{X} - v]$ are connected.

To see how the two lemmas stated above may be used to prove the ELP-Theorem, let us consider (using the notation of Lemma 5.1) the special case in which v is the only neighbour of u in $\overline{X}$ and u is the only neighbour of v in X. In this case, it is easy to see that $G' = G - u - v$ is a matchable graph in which no edge of the cut $\partial(X - u)$ is matchable. By Lemma 5.5, there is a Dulmage-Mendelsohn barrier B' of G' such that B' as well as the vertex sets of all the components of $G' - B'$ are contained either in $X - u$ or $\overline{X} - v$. Assuming, without loss of generality that $B' \subset X - u$, it is not difficult to show that $B' + u$ is a nontrivial barrier of G.

There are some other cases to be examined because v may not be the only neighbour of u in $\overline{X}$ and u may not be the only neighbour of v in X. We analyse all possible cases and deduce that G either (i) has a 2-separation or (ii) a nontrivial barrier, say B. In the first case G has a 2-separation cut. In the second case either $G - B$ has a nontrivial component, say K, implying that $\partial(K)$ is a barrier cut of G; or all components of $G - B$ are trivial, implying that G is a bipartite matching covered graph. By Theorem 4.7, whose proof is fairly simple, all tight cuts in a bipartite matching covered graph are barrier cuts. Thus, in all possible cases, G has an ELP-cut.

As a consequence of the ELP Theorem, it follows that a matching covered graph G is a brick if and only if it is bicritical and 3-connected. Thus, to check that G is a brick, it suffices to verify that, for any two vertices u and v of G the graph $G-u-v$ is connected and is matchable. The ELP Theorem does not provide a characterization of braces. But Theorem 5.17 provides a characterization of braces. According to this theorem, a bipartite graph $G := G[A, B]$ with $|A| = |B| \geq 3$ is a brace if and only if $G - u_1 - u_2 - v_1 - v_2$ is matchable for any two distinct vertices u_1 and u_2 in A, and any two distinct vertices v_1 and v_2 in B.

A strengthening of the ELP Theorem: Using a refinement of the method we have used in our proof of the ELP Theorem, Chen, Feng, Lu, Lucchesi and Zhang (2021, [20]) have been able to establish a strengthening of that result; they showed that, given any tight cut C of a matching covered graph G, there exists an ELP-cut D of G which is laminar to C.

Separating Cuts and Solid Bricks: A cut C of a matching covered graph G is a *separating cut* if both the C-contractions of G are also matching covered. All tight cuts are separating cuts, but not every separating cut is tight. For example, the brick $\mathbb{P}$ (the Petersen graph), being a brick, has no nontrivial tight cuts but, for any pentagon Q of $\mathbb{P}$, the cut $\partial(Q)$ is a separating cut.

A matching covered graph G is *solid* if it has no separating cuts other than those which are tight. All bipartite graphs are solid; a brick is solid if it has no nontrivial separating cuts. We discovered this concept in connection with our work on the conjecture of Lovász concerning the existence of b-invariant edges in bricks; we were able to show that in a solid brick G any edge which is removable is also b-invariant. Our inductive proof of Lovász's conjecture was based on this result.

The *perfect matching polytope* of a graph G is the set of all convex linear combinations of incidence vectors of perfect matchings of G. Edmonds (1965 ,[30]) showed that a vector $\mathbf{x}$ in the edge space $\mathbb{R}^E$ of G is in this polytope if and only if (i) $\mathbf{x}(e) \geq 0$, for each $e \in E$ (non-negativity); (ii) $\mathbf{x}(\partial(v)) = 1$, for all $v \in V$ (degree constraints); and (iii) $\mathbf{x}(\partial(S)) \geq 1$, for each odd subset S of V (odd set constraints). We were pleasantly surprised to discover that the only bricks whose perfect matching polytopes can be described without using the odd set constraints are the solid bricks.

It is not known if the problem of deciding whether or not a given brick is solid is in the complexity class NP. This is one of the challenging unsolved problems of this theory.

The dependence relation: An edge e of a matching covered graph G is said to *depend* on another edge f if every perfect matching of G which contains e also contains f. Two edges e and f are *mutually dependent* if they depend on each other. Mutual dependence is an equivalence relation, and we refer to equivalence classes with respect to this relation as *dependency classes*. The partial order of dependence among the edges could be extended to partial order among dependency classes. This leads to the definition of a *removable class*, which is a generalization of the idea of a *removable edge*.

Lemma 5.4.5 in Lovász and Plummer [59] states that any minimal removable class in a matching covered graph consists of either a single edge or a pair of edges. In Chapter 8 we give a new proof of that Lemma.

(The notion of a removable edge extends naturally to a removable single ear, and that of a removable doubleton to a removable double ear. We have found it convenient to adopt a 'top-down' approach to the theory of ear decompositions of matching covered graphs. Lovász's conjecture is clearly suggestive of this viewpoint!)

Most of the above described material is confined to the first part of the book which deals with what we regard as basic theory. In the remaining two parts, there are many new results whose proofs involve intricate proof techniques. We hope that all readers, regardless of the degree of their familiarity with this subject, will find something of interest among these pages.

Notation and Terminology: General mathematical notation which we use is mostly standard except, when there is no scope for confusion, we write $X + Y$ for the union $X \cup Y$ of two sets X and Y and write $X - Y$ for their difference $X \setminus Y$. Graph theoretical notation and terminology which we use, with minor differences, are mostly the same as the ones used in the book *Graph Theory* by Bondy and Murty (2008, [3]). Whenever in doubt, readers should either refer to the section *Notation and Terminology* which appears in the Front Matter or to the index at the end.

Broad outline: There are twenty-one chapters, divided into three parts. Each chapter begins with a mini-table of contents which lists the titles of all the sections and subsections in that chapter. Throughout the book, there are a number of exercises of varying difficulty. At the end of each chapter there is a Notes section which includes references to useful related work and to statements of unsolved problems. There is an appendix at the end of the book which includes solutions to some of the harder exercises.

How to use the book: The theory of matching covered graphs is a fascinating subject. It is our hope that those who take a course based on this book will come to appreciate the scope and beauty of this area and that some of them may even end up doing research in it. However, a warning is in order. If one tries to learn all the results and their proofs in a chapter before proceeding to the next, one would not get very far with the subject in just one term or two. For this reason, it is better for the instructor to judiciously select topics to be covered in class and to avoid presenting detailed proofs of all the theorems. A better approach would probably be to convey through examples the ideas underlying concepts introduced and statements of some of the harder theorems and their proofs. It would also be good to encourage student participation. This could be done by assigning selected material for classroom presentation. (For this purpose, we have highlighted certain sections of the book and have indicated them by the symbol $\sharp$.) (There are some sections which are marked by $\sharp\sharp$. We recommend that they be skipped on a first reading.)

It might also be a good idea to encourage students to work together on selected exercises, which are marked with $\triangleright$, a small triangle. Some exercises have solutions in the appendix, and are marked with $*$, an asterisk.

Acknowledgements: First and foremost, we would like to express our gratitude to László Lovász whose article *Matching Structure and the Matching Lattice*, published in 1987, served as an inspiration for us to pursue research in this field.

The first proof of Lovász's conjecture concerning the existence of b-invariant edges in bricks was a part of Marcelo Carvalho's PhD thesis, written under the supervision of Cláudio Lucchesi, and finished in 1996. A strengthening of this result, which determined the minimum number of double ear additions required in any ear decomposition of a matching covered graph, appeared in 2002 in a series of papers co-authored by CLM.[3] We are thankful to Marcelo for his many contributions to this subject and his warm friendship.

Charles Little has done pioneering work in the theory of Pfaffian orientations of graphs. We have known him for a number of years and have learned much about this subject from him. When we finished a draft of this book and were looking for knowledgeable readers, he was the first person whose name occurred to us. We were happy that he undertook the arduous task of going through our work. He read the manuscript we sent to him very thoroughly, literally line-by-line, and sent us detailed comments. Apart from the more substantive observations about the mathematical content, he pointed out many misprints, instances of improper punctuation and grammatical errors. We owe a huge debt of gratitude to Charles.

[3] CLM is an abbreviation of Carvalho, Lucchesi and Murty.

Arnaldo Mandel is another friend who agreed to read our manuscript. He has an extensive knowledge of many areas of mathematics and computer science. He is jovial and is fond of cryptic multilingual puns. Luckily, he communicated his comments to us in plain language. We thank him for his pertinent suggestions regarding terminology and style.

Adrian Bondy has been a dear friend to us for many years. Clarity of his thinking and his artistic sensibilities are reflected in his lucid expository writing. We have learned a lot from him. Thank you dear Adrian.

Alberto Miranda, who did his doctorate under the supervision of Cláudio Lucchesi, and Nishad Kothari, who did his under the partial supervision of Rama Murty, have provided valuable input. We appreciate their help.

Gabriel Morete de Azevedo, Sebastião F. Nachtergaele Gomes Navarro, Thiago Lima Oliveira and Ariana Maite Quispe Porras, who are four students in the University of São Paulo, have read the proofs of our book with utmost care and found some errors or misprints and, more importantly, suggested various changes to the text to make it more readable. We appreciate their help and wish them the very best in their academic careers.

Cláudio would like to express his gratitude to Luísa Angélica (in memoriam), to Rosa Marilyn, and to Gabriela, Ricardo and Andrea, for their personal and emotional support.

On his part, Rama has been the grateful beneficiary of affection and unwavering support from his friends Daksha and Kirti Shah and their entire family, and his siblings and their families.

Finally, we wish to acknowledge the partial financial support of the Brazilian CAPES, and of CNPq, grant 309451/2021-3.

Cláudio Lucchesi and Rama Murty
June 2023

Contents

Terminology and Notation

The reader is urged to peruse the glossary, which describes the notation used in the book.

1 Sets and Subsets

Much of the notation we use in dealing with sets and their subsets is standard. However, in some cases, we have found it natural and convenient to depart from the general conventions. For example, if S and T are two sets, we denote the set of all elements in S which are not also elements of T by $S - T$. This subset of S is commonly represented by $S \setminus T$.

If A is any subset of a set S, and x is any element of S that may or may not be an element of A, we write $A - x$ instead of $A \setminus \{x\}$, and $A + x$ instead of $A \cup \{x\}$.

2 Graphs and subgraphs

All graphs considered in this book are loopless, but they may have multiple edges. For graph theoretical notation and terminology, we essentially follow *Graph Theory* (2008, [3]) by Bondy and Murty, or an earlier version *Graph Theory with Applications* (1976, [2]) of that book.

We shall assume that the reader is familiar with the basic notions of graph theory such as subgraphs, induced subgraphs, spanning subgraphs, paths, cycles, trees, connected graphs, cut vertices, cut edges and blocks. However, for the convenience of the reader, we list below some of the graph theoretical terms (and the symbols that pertain to them) that are used in this work. In some cases, our notation and terminology vary from the ones used in the standard texts of graph theory. These differences are also noted below.

Neighbour set: For a vertex v of a graph G, the *neighbour set* of v, denoted by $N_G(v)$, is the set of all vertices of G that are adjacent to v. More generally, for a subset S of the vertex set $V(G)$ of G, the neighbour set $N_G(S)$ of S is the set of all vertices of G in $V(G) - S$ which are adjacent to at least one vertex in S.

When graph G is evident from the context, we simply write $N(v)$ for $N_G(v)$ and $N(S)$ for $N_G(S)$. We use this convention of leaving out the suffix G from graph theoretical parameters whenever there is no scope for ambiguity. For example, we simply write V for $V(G)$ and E for $E(G)$.

ORDER AND SIZE: The *order* of a graph is the number of its vertices, and its *size* is the number of its edges. When the graph under consideration is denoted by G, we denote its order, which is $|V|$, by n, and its size, which is $|E|$, by m.

DELETIONS OF VERTICES AND EDGES: For a subset S of the vertex set V of a graph G, the subgraph obtained by deleting all the vertices in S is denoted by $G - S$. In the special case when S has just one vertex v, we simplify the cumbersome notation $G - \{v\}$ to $G - v$. Similarly, for a subset F of the edge set E of G, we denote the subgraph of G obtained by deleting all edges in F by $G - F$. Again, if F happens to have a single edge e, we write $G - e$ instead of $G - \{e\}$.

CONNECTIVITY: A *vertex cut* of a graph G is a subset S of V such that $G - S$ is disconnected. A k*-vertex-cut* is a vertex cut of cardinality k. A graph that has a complete graph as a spanning subgraph has no vertex cuts, all other graphs do. If a graph G has at least one pair of nonadjacent vertices, the *connectivity* of G, denoted by $\kappa(G)$, is the minimum value of k for which G has a k-vertex-cut; otherwise we define $\kappa(G)$ to be $n - 1$. (This definition is the same as the one given in [2], but for graphs with a complete graph as a spanning subgraph, it differs from the one in [3].) A graph is k*-connected* if $\kappa(G) \geq k$. By Menger's theorem (9.1 in [3]), a graph G of order $k+1$ or more is k-connected if and only if there are at least k internally-disjoint paths between any two nonadjacent vertices of G.

DEFINITIONS AND NOTATION: EDGE CUTS AND SHORES

For a nonempty proper subset S of the set of vertices of a graph G, we follow [3] and denote the set of edges with precisely one end in S by $\partial(S)$, and refer to it as the *edge cut*, or simply the *cut*, associated with S. If v is a vertex, then, we denote $\partial(\{v\})$, for brevity, by $\partial(v)$. In the same spirit, if H is a subgraph of G, we denote the cut $\partial(V(H))$ simply by $\partial(H)$.

The *shores* of a cut $\partial(S)$ of a connected graph G are the two sets S and $\overline{S} := V - S$.

A cut $\partial(S)$ of a graph is *trivial* if either $|S|$ or $|\overline{S}|$ (or both) have exactly one vertex.

A cut $\partial(S)$ in a graph G is a *bond* if both its shores induce connected subgraphs of G.

EDGE CONNECTIVITY: A k*-edge-cut* of a graph G is an edge cut with k edges. All nontrivial graphs do have edge cuts. If a graph G is nontrivial its *edge connectivity*, denoted by $\kappa'(G)$, is the minimum value of k for which G has a k-edge-cut; if G is trivial, then we define $\kappa'(G)$ to be zero. A graph is k*-edge-connected* if its edge-connectivity is at least k. By the edge version of Menger's theorem (9.7 in [3]), a nontrivial graph G is k-edge-connected if and only if there are at least k edge-disjoint paths between any two distinct vertices of G.

For any graph G, the inequality $\kappa' \geq \kappa$ holds. There are graphs for which this inequality is strict. However, if G is a cubic graph, then $\kappa' = \kappa$.

Following Bondy and Murty [3], the minimum and maximum degrees of the vertices of G are denoted respectively $\delta(G)$ and $\Delta(G)$.

3 Operations

Contraction of a Set of Vertices: Let X be a set of vertices of a graph G. The graph obtained from G by contracting X to a single new vertex x (and removing all resulting loops) is denoted by $G/(X \to x)$. If the name of the *contraction vertex* x is irrelevant we just write G/X.

Part I
Basic Theory

Chapter 1
Perfect Matchings

Contents

1.1 Matchings and Perfect Matchings

The objects of principal interest in this book are perfect matchings in graphs. A *matching* in a graph G is a subset M of $E(G)$ such that no vertex of G is incident with more than one edge in M. A *perfect matching* in G is a matching such that every vertex of G is incident with precisely one edge in it. In other words, a subset M of $E(G)$ is a perfect matching in G if the subgraph $G[M]$ of G induced by M is a spanning 1-regular subgraph of G. For this reason, perfect matchings in a graph are also known as its *1-factors*. The Petersen graph (Figure 1.1(a)), which plays a prominent role in our subject, has six perfect matchings. On the other hand, the Sylvester graph (Figure 1.1(b)) has no perfect matchings.

Perfect matchings in graphs have been of interest ever since Tait showed that the four colour conjecture was equivalent to the statement that every 3-connected cubic planar graph is 3-edge-colourable or, equivalently, that such a graph has three pairwise edge-disjoint perfect matchings (see Bondy and Murty [3]). In fact, the four colour conjecture was later shown to be equivalent to the seemingly stronger

C. L. Lucchesi, U. S. R. Murty, *Perfect Matchings*, Algorithms and Computation in Mathematics 31, https://doi.org/10.1007/978-3-031-47504-7_1

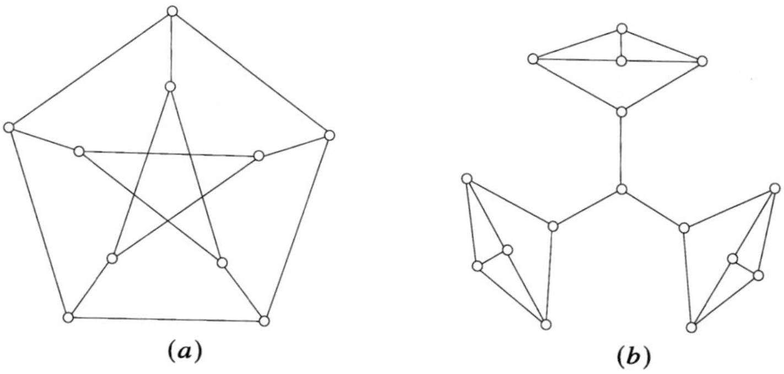

Fig. 1.1 (a) The Petersen graph; (b) The Sylvester graph

statement that every 2-connected cubic planar graph is 3-edge-colourable (see [3]). The Sylvester graph shows that a cubic planar graph need not have even a single perfect matching if there were no restriction on its connectivity. As we shall see (Theorem 1.4), Petersen [80] showed that any 2-connected cubic graph (planar or not) has at least one perfect matching. However, a nonplanar 2-connected cubic graph need not have two edge-disjoint perfect matchings; the Petersen graph is an example of such a graph (Exercise 1.1.2).

This book is concerned mainly with graphs which have perfect matchings. For brevity, we shall refer to such graphs by the term defined in the following inset.

MATCHABLE GRAPHS
A graph is *matchable* if it has a perfect matching.

Before working on the exercises below, the readers should familiarize themselves with the notation we use for edge cuts in graphs. (If X is a subset of the vertex set V of a graph G, then $\partial(X)$ is the set of edges of G with one end in X and one end in $V - X$. For a subgraph H of G, we simply write $\partial(H)$ for $\partial(V(H))$; and for a vertex v, we write $\partial(v)$ for $\partial(\{v\})$.)

The following proposition plays a useful role in arguments involving perfect matchings and cuts in a graph.

Proposition 1.1 *For any perfect matching M and for any subset X of the vertex set V of a graph G the following identity holds:*

$$|\partial(X) \cap M| \equiv |X| \pmod 2. \qquad \square$$

We leave the proof of the above proposition as Exercise 1.1.3(ii).

Exercises

1.1.1 *Show that the Sylvester graph is not matchable.*

1.1.2

(i) *Show that the Petersen graph is the unique cubic graph of girth five on ten vertices.*
(ii) *For any pentagon Q and any perfect matching M of the Petersen graph, show that $|\partial(Q) \cap M|$ is either one or five.*
(iii) *Show that every perfect matching of the Petersen graph is one of its 5-edge cuts.*
(iv) *Deduce that every edge of the Petersen graph is in exactly two perfect matchings, and that any two perfect matchings have exactly one edge in common. (This also shows that the Petersen graph is nonhamiltonian.)*
(v) *List all the perfect matchings of the Petersen graph.*

▷**1.1.3** *Let G be a graph.*

(i) *For any subset X of V, show that $|\partial(X)| \equiv r \pmod 2$, where r is the number of vertices in X that have odd degree in G.*
(ii) *Deduce Proposition 1.1 from part (i).*

1.2 The Advent of Tutte

Frobenius (1917, [35]) and Hall (1935, [40]) established necessary and sufficient conditions for the existence of perfect matchings in bipartite graphs. But the more general problem of characterizing graphs which have a perfect matching remained unsolved until W. T. Tutte entered the field. Tutte started working in graph theory during his undergraduate days at Cambridge, but his studies were interrupted by the call to wartime duty at Bletchley Park, the fabled organization of codebreakers. During his tenure there (1941-45), he made pivotal contributions to cryptography (see Younger (2012, [95]). He returned to Cambridge after the war to resume his research in graph theory. Within a short period, he produced a number of brilliant results, one of them is a solution to the problem mentioned above. In his mathematical autobiography "Graph Theory As I Have Known It" (1998, [90]), he describes in a delightfully engaging manner how he was led to his solution of that problem.

Obviously, in order for a graph G to have a perfect matching, it must have an even number of vertices. Thus, for any odd subset X of V, the subgraph $G[X]$ induced by X cannot have a perfect matching. Consequently, if M is any perfect matching of G, and X is any odd subset of V, then $|\partial(X) \cap M| \geq 1$. (In fact $|\partial(X) \cap M|$ must be odd.) This simple observation is the basis of a necessary condition for the existence of a perfect matching in a graph G. The notation described in the following inset is convenient for stating that condition.

NOTATION RELATED TO ODD COMPONENTS: $O(G)$ AND $o(G)$

We denote by $O(G)$ the set of odd components of a graph G and by $o(G)$ the number $|O(G)|$ of odd components of G.

Suppose that G has a perfect matching M and let $S \subseteq V$. Then, by the previous observation, for any odd component K of $G - S$, there must be at least one edge of M in $\partial(K)$. As K is a component of $G - S$, every edge in $\partial(K)$ has one end in K and one end in S.

A NECESSARY CONDITION FOR THE EXISTENCE OF A PERFECT MATCHING IN G

Since no two distinct edges of a perfect matching of G can have a common end vertex, the following condition

$$o(G - S) \leq |S|, \quad \text{for any subset } S \text{ of } V \tag{1.1}$$

must be satisfied in order for G to have a perfect matching.

Thus, if there exists a subset S of V such that $o(G - S) > |S|$, then G cannot be matchable. See Example 1.2.

Example 1.2 Figure 1.2 depicts a subset S of the vertex set of a graph G (indicated by solid dots) which violates condition 1.1. Thus G is not matchable.

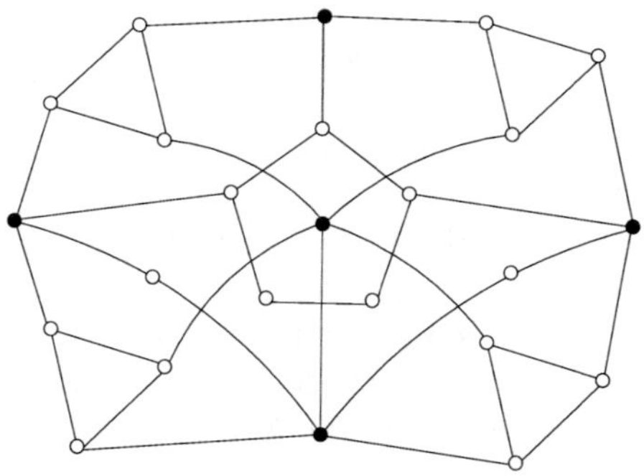

Fig. 1.2 $|S| = 5$, and $o(G - S) = 7$

Tutte (1947, [89]) showed that the necessary condition (1.1) for the existence of a perfect matching in a graph G is also sufficient. Tutte's proof is based on properties of invariants, known as Pfaffians, associated with skew-symmetric matrices. (They are described in Section 1.5.) As this proof involves many ideas that may be unfamiliar to readers who are more accustomed to combinatorial methods than algebraic ones, we have chosen to present, in the next section, a purely graph theoretic proof due to Lovász (1975, [56]) which essentially follows the same line of reasoning used by

Tutte but which avoids the use of Pfaffians. Part-III of the book (Pfaffian Orientations) is partly dedicated to questions arising from Tutte's seminal work.

Exercises

▷**1.2.1** *Let G be a graph, and let X be a subset of V. Show that:*

$$|X| - o(G - X) \equiv |V| \pmod 2.$$

> DEFICIENCY OF A MATCHING
>
> The *deficiency* of a matching M in a graph G, denoted $\mathrm{def}(M)$, is the number of vertices of G that are not covered by M.

∗**1.2.2** *Let M be a matching of a graph G.*

(i) *Show that for any subset S of V,*

$$\mathrm{def}(M) \geq o(G - S) - |S|. \tag{1.2}$$

(ii) *Conclude that if equality holds in* (1.2) *for some set S then M is a maximum matching of G (that is, a matching of largest possible cardinality).*

1.2.3 *Find a maximum matching in the graph shown in Figure 1.2 and justify your answer.*

1.3 Theorems of Tutte and Petersen

We now proceed to describe Tutte's characterization of graphs which have a perfect matching.

1.3.1 Matchable graphs

> TUTTE'S THEOREM [89]
>
> **1.3** *A graph G has a perfect matching if and only if*
>
> $$o(G - S) \leq |S|, \text{ for every subset } S \text{ of } V. \tag{1.3}$$

Proof The proof we give here is essentially the same as it is in Tutte's original paper, but we use the approach of Lovász (1975, [56]). See Bondy and Murty [2] or Lovász and Plummer (1986, [59]).

We have already seen that if G has a perfect matching, then (1.3) holds. Conversely, suppose that G is a graph that has no perfect matchings. It is incumbent on us to establish the existence of a subset S of V for which

$$o(G - S) > |S|.$$

We shall refer to such a set as a *Tutte set*.

Clearly, we may assume without loss of generality that G is simple. If $|V|$ is odd, then the empty set is a Tutte set. So, we may assume that $|V|$ is even.

Let G^* be a maximal simple graph, with $V(G^*) = V(G)$ and $E(G^*) \supseteq E(G)$, such that G^* also has no perfect matchings. (If u and v are two nonadjacent vertices of G such that $G - u - v$ has no perfect matchings, then $G + uv$, obtained from G by adding the edge uv, also has no perfect matchings. A graph G^* satisfying the required property may be obtained by repeatedly adding edges in this manner.) We note that, for any subset S of V, $o(G - S) \geq o(G^* - S)$. Thus, any Tutte set S of G^* would also be a Tutte set of G.

Let S denote the set of vertices of degree $n - 1$ in G^*, where n is the number of vertices of G. (Thus, any vertex in S is joined in G^* to all other vertices in V.) If $S = V$, then G^* is a complete graph and hence would be matchable. Therefore S is a proper subset of V (it is possible for S to be empty). Clearly the subgraph of G^* induced by S is a complete graph.

1.3.1 *Every component of $G^* - S$ is also complete.*

Proof Consider any component J of $G^* - S$. If it is not complete, it would have two nonadjacent vertices x and y with a common neighbour z (see Exercise 1.3.1). Since z is in $V(J)$, and not in S, the degree of z is less than $n - 1$. Let w be a vertex that is not a neighbour of z. Clearly w is not in S, but it may not be in $V(J)$ either. See Figure 1.3.

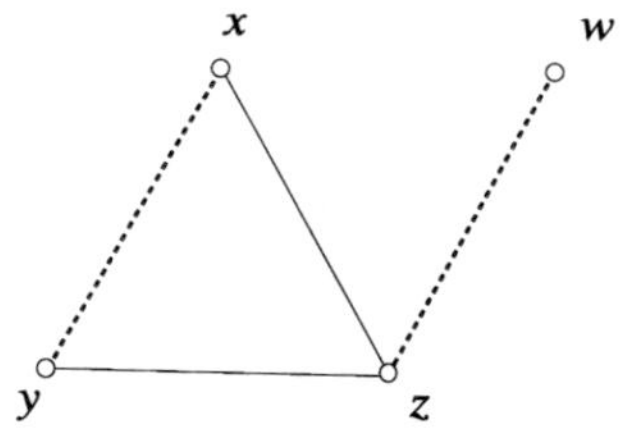

Fig. 1.3 The case in which J is not complete

Since G^* is a maximal simple graph that is not matchable, the graph $G^* + uv$ is matchable, for any two nonadjacent vertices u and v of G^*. Let M_1 and M_2 denote perfect matchings in $G^* + xy$ and $G^* + wz$, respectively, and denote by H the subgraph of $G^* + xy + wz$ induced by the symmetric difference $M_1 \triangle M_2$. Since each vertex of H has degree two, H is a disjoint union of cycles. Furthermore, all these cycles are even, since edges of M_1 alternate with edges of M_2 around them. We distinguish two cases:

Case 1 *The edges xy and wz are in the same cycle C of H (Figure 1.4(a)).*

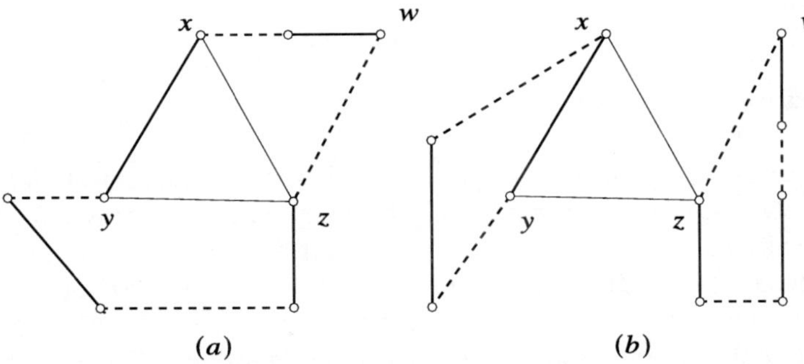

Fig. 1.4 Perfect matching M_1 is indicated by heavy lines, and M_2 by dashed lines

By the symmetry of x and y, we may assume that the vertices z, w, x and y occur in C in that cyclic order. The edges of M_1 in the segment zx of C, together with the edge xz and the edges of M_2 not in the segment zx constitute a perfect matching of G^*. This is a contradiction to the definition of G^*.

Case 2 *The edges xy and wz are in distinct components of H (Figure 1.4(b)).*

Let C denote the cycle of H that contains the edge wz. The edges of M_1 in C, together with the edges of M_2 not in C, constitute a perfect matching of G^*. This is a contradiction to the definition of G^*. □

We now use the fact that every component of $G^* - S$ is complete. For brevity, let us write $t := o(G^* - S)$. Our objective is to show that $t > |S|$. Assume to the contrary that $t \leq |S|$. In this case, let T be a subset of S of cardinality t, and let T' be a set of t vertices consisting of distinct representatives of the vertex sets of the odd components of $G^* - S$. Since all vertices in S have degree $n - 1$, there exists a matching, say M, which matches the vertices in T with vertices in T'. As $|V|$ is even, $|S|$ and $o(G^* - S)$ have the same parity (see Exercise 1.2.1). Thus $S - T$ is even, and the subgraph of G^* induced by this set is a complete graph. Furthermore, each component of $G^* - (S \cup T')$ is a complete graph on an even number of vertices. Hence the matching M can be extended to a perfect matching of G^*. This contradicts the assumption that G^* has no perfect matchings. It follows that $t = o(G^* - S) > |S|$. □

The problem of deciding whether or not a given graph G has a perfect matching is obviously in NP; if G has a perfect matching, that fact can be demonstrated by displaying a perfect matching in G. The import of Tutte's theorem is that the above problem is also in co-NP; if G does not have perfect matchings, that fact can be demonstrated by displaying a Tutte set in G. In fact, the literature has several polynomial-time algorithms for determining maximum perfect matchings. The first such algorithm was given by Edmonds (1965, [31]). That article, and another article published by Cobham (1965, [24]), are regarded by computer scientists as fundamental contributions to the theory of Computational Complexity, by the introduction of the class P, the class of problems for which there are polynomial-time algorithms.

Complexity of Algorithms

In a pioneering paper called "Paths, Trees and Flowers" (1965, [31]), Jack Edmonds persuasively argued the importance of polynomial-time algorithms and, as an example, presented an algorithm which finds a maximum matching in a graph with running time $O(n^2m)$. Hopcroft and Karp (1973, [43]) discovered an algorithm to determine a maximum matching of a bipartite graph in time $O(\sqrt{n}m)$. Micali and Vazirani (1980, [70]) discovered an $O(\sqrt{n}m)$-time algorithm for finding a maximum matching of a graph.

1.3.2 Cubic matchable graphs

Tutte had known Petersen's Theorem on cubic graphs as a schoolboy. (A very interesting account of his encounter with a published proof of this theorem, in his own words, can be found in Chapter 6 of McCarthy and Stanton's book (1979, [67]).) With the aid of his characterization of graphs which have a perfect matching, he was able to present a simple proof of that result.

PETERSEN'S THEOREM (1891, [80])

1.4 *Every 2-connected cubic graph is matchable.*

Proof Let G be a 2-connected cubic graph. As all vertices of G have degree three, $|V|$ is even and $o(G) = 0$. Thus, $o(G - \emptyset) = 0$. Let S be any nonempty proper subset of V. For every odd component K of $G - S$, $|\partial(K)|$ is odd (Exercise 1.1.3). On the other hand, by hypothesis, G is 2-connected. Hence $|\partial(K)| \geq 3$, for each odd component K of $G - S$. We thus have:

$$3|\mathcal{O}(G - S)| \leq \sum_{K \in \mathcal{O}(G-S)} |\partial(K)| \leq |\partial(S)| \leq 3|S|, \tag{1.4}$$

implying that $o(G - S) \leq |S|$, for all subsets S of V. Hence, by Theorem 1.3, G has a perfect matching. □

Tutte not only deduced from his Theorem 1.3 that every 2-connected cubic graph is matchable, but also showed that, in fact, every such graph has the property that every edge is in some perfect matching. This stronger assertion follows from a more general result about matchable graphs which will be discussed in Chapter 2 (see Theorem 2.8). But a keen student should be able to work out its proof without much difficulty.

The above-mentioned property of 2-connected cubic graphs suggests the definition of the following more general class of graphs whose name appears as the subtitle of this book: a nontrivial connected graph is *matching covered* if, given any edge of the graph, there is some perfect matching which contains that edge. Basic terminology related to this class and several interesting families of examples will be described in the next chapter.

Exercises

▷**1.3.1** *Show that if a connected graph G is not a complete graph, then it would have two nonadjacent vertices x and y with a common neighbour z.*

> ALTERNATING PATHS AND CYCLES, AUGMENTING PATHS
>
> Let M and N be matchings in a graph G. A path or a cycle of G is *M-alternating* if its edges are alternately in M and in $E - M$, and is *(M, N)-alternating* if its edges are alternately in $M - N$ and in $N - M$.
>
> A nontrivial M-alternating path in G is *M-augmenting* if neither of its ends is covered by M.

▷**1.3.2**

(i) *Let M and N be two matchings in a graph G. Show that each component of the subgraph $G[M \triangle N]$ of G induced by $M \triangle N$ is either an (M, N)-alternating path or an (M, N)-alternating cycle. Moreover, if $|N| > |M|$, show that some component of $G[M \triangle N]$ is an (M, N)-alternating path whose first and last edges are in N.*

(ii) *Deduce from part (i) that a matching M in a graph G is maximum (that is, one of maximum size) if and only if G has no M-augmenting paths. (Berge, 1962)*

∗**1.3.3** *Let M_0 be a maximal matching of G (that is, one which is not a proper subset of any other matching of G). Show that $|M| \leq 2|M_0|$, for every matching M of G. In particular, if M^* denotes a maximum matching of G, deduce that $|M^*| \leq 2|M_0|$.*

1.3.4 *Let G be a cubic graph that has a path which contains all cut edges of G. Show that G has a perfect matching. (The Sylvester graph shows that this is the best possible statement one can make about cubic graphs which have a perfect matching.)*

∗**1.3.5** *For each $r \geq 3$, construct an $(r - 2)$-edge-connected r-regular graph on an even number of vertices which is not matchable.*

r-GRAPHS

For an integer $r \geq 2$, an *r-graph* is an $(r-1)$-edge-connected r-regular graph on an even number of vertices.

1.3.6 *Show that every r-graph is matchable.*

▷**1.3.7** *Show that the graph depicted in Figure 1.5 is a 4-edge connected 4-graph and that it is not 4-edge-colourable. (For an example of a simple graph with these properties, see Exercise 2.4.7.)*

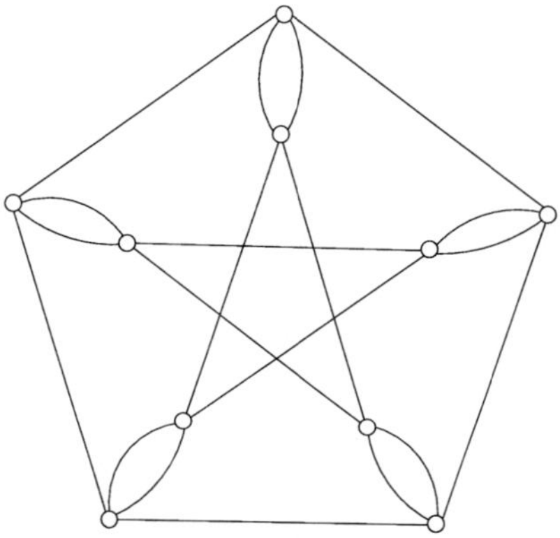

Fig. 1.5 A 4-graph which is not 4-edge-colourable

CRITICAL GRAPHS

A graph G is *factor-critical*, or simply *critical*, if, for any $v \in V$, the graph $G - v$ has a perfect matching. (For example, the odd cycle C_{2k+1} is critical for all $k \geq 1$.) Critical graphs are also known as *hypomatchable* graphs. Clearly, a critical graph has an odd number of vertices.

1.3.8 *Let G be a critical graph on three or more vertices. Show that G is connected and contains an odd cycle.*

1.3.9 *Using Tutte's Theorem 1.3 show that a graph G is critical if and only if:*

$$o(G - S) \leq |S| - 1$$

for every nonempty subset S of V.

1.4 Hall's Theorem ♯

We now present an attractive characterization of bipartite matchable graphs which is usually attributed to Hall [40]. We shall denote by $G[A, B]$ a bipartite graph G with bipartition (A, B). Hall established necessary and sufficient conditions under which a bipartite graph $G[A, B]$ has a matching which matches each vertex in A with some vertex in B (see Exercise 1.4.1). When $|A| = |B|$, such a matching is precisely a perfect matching of G, and this special case is what is relevant to our purposes.

Hall's Theorem

1.5 *A bipartite graph* $G := G[A, B]$, *with* $|A| = |B|$, *has a perfect matching if and only if*

$$|N(X)| \geq |X|, \tag{1.5}$$

for all $X \subseteq A$.

Proof Suppose first that G has a perfect matching M. Then M matches distinct vertices in A with distinct vertices of B. This shows that (1.5) must hold for each $X \subseteq A$.

Conversely, suppose that G has no perfect matching. In this case, we must establish the existence of a subset X of A such that $|X| > |N(X)|$. Towards this end, let G' be the spanning supergraph of G obtained by replacing the stable set B in G by a complete graph on B. (See Figure 1.6 for an illustration.) Clearly, as G has no perfect matching, the graph G' does not have one either. Therefore, by Tutte's Theorem 1.3 and Exercise 1.2.1, applied to G', there must exist a subset S of $V(G')$ such that

$$o(G' - S) \geq |S| + 2. \tag{1.6}$$

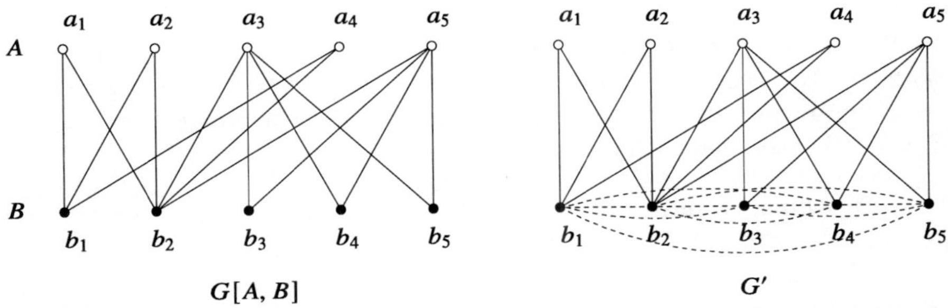

Fig. 1.6 Graph G' is obtained from the bipartite graph $G[A, B]$ by joining vertices in B to each other. The set $S := \{b_1, b_2\}$ is a Tutte set of G' as $o(G' - S)$ has four odd components; a_1, a_2 and a_4 are the isolated vertices of $G' - S$. If we set $X := \{a_1, a_2, a_4\}$, then $|X| = 3$, whereas $|N(X)| = 2$.

As $G'[B]$ is complete, it follows that all vertices in $B - S$ will be in the same component of $G' - S$. Every other component of $G' - S$ is of order one and is the subgraph induced by an isolated vertex of $G' - S$ which belongs to A. Let X denote the subset of A consisting of the isolated vertices of $G' - S$. Then it follows from (1.6) that

$$|X| \geq o(G' - S) - 1 \geq |S| + 1 \geq |B \cap S| + 1. \tag{1.7}$$

Now note that a vertex in A is isolated in $G' - S$ only if all its neighbours in G are in $B \cap S$; hence, $B \cap S \supseteq N(X)$. This observation combined with (1.7) implies that $|X| > |N(X)|$. □

There are many simple and elegant proofs of Hall's Theorem which do not rely on Tutte's Theorem. A constructive proof, based on the notion of alternating paths and cycles, is sketched in Exercises 1.3.2 and 1.4.3. For more details, see Bondy and Murty [3].

Exercises

A matching M in a graph G is said to *cover* a vertex v if v is incident with some edge in M. The assertion that is generally known as Hall's Theorem is the following:

> A bipartite graph $G[A, B]$ has a matching which covers all vertices in A if and only if $|N(X)| \geq |X|$, for all $X \subseteq A$.

▷**1.4.1** *Show that the above assertion is equivalent to Theorem 1.5.*

1.4.2

(i) *Show that, for any $k > 0$, every k-regular bipartite graph has a perfect matching.*

(ii) *Deduce that every k-regular bipartite graph is k-edge-colourable.*

▷**1.4.3** *Let $G[A, B]$ be a bipartite graph, and let M be a maximum matching in G. Suppose that u is a vertex in A which is not covered by M, and that Z denotes the set of all vertices in G which are reachable from u by M-alternating paths, and let $X := A \cap Z$. Show that $|N(X)| < |X|$.*

1.4.4 (Deduction of Tutte's Theorem from Hall's Theorem) *Let G be a graph and suppose that G satisfies the Tutte condition $o(G - S) \leq |S|$, for each $S \subseteq V$, and let us refer to a subset B of V as a barrier of G if $o(G - B) = |B|$. Clearly, the empty set is a barrier of G. Let B denote a maximal barrier of G. Assume, as inductive hypothesis, the sufficiency of Tutte's condition for graphs having fewer vertices than G.*

(i) *Prove that every component of $G-B$ is odd. Hint: if $G-B$ has an even component K, add a vertex of K to B.*
(ii) *Let H be the bipartite graph obtained from G by removing all the edges having both ends in B and by contracting each component of $G-B$ to a single vertex. Prove that H has a perfect matching. Hint: apply Hall's Theorem to H, by proving that for each set X of contracted vertices of H, the set of neighbours of X (in B) is at least as large as X.*
(iii) *Prove that if K is a component of $G-B$ that has a nonempty set X of vertices such that $o(K-X) \geq |X|+1$, then the set $B \cup X$ is a barrier of G.*
(iv) *Prove that every component K is critical. Hint: use the inductive hypothesis and the result in Exercise 1.3.9.*
(v) *Prove that G is matchable. Hint: use the fact that every component is critical to extend the perfect matching of H to a perfect matching of G.*

(Younger, 1969)

▷**1.4.5 (Analogue of Exercise 1.2.2)** *We denote by $I(G)$ the set of isolated vertices of a graph G. Let M be a matching of a bipartite graph G.*

(i) *Prove the following inequality involving the deficiency of M, for every set $S \subset V$.*

$$\mathrm{def}(M) \geq |I(G-S)| - |S|. \tag{1.8}$$

(ii) *Prove that if equality holds in* (1.8) *for some set S then M is a maximum matching of G.*

1.5 The Magic of Pfaffians

An $n \times n$ matrix $\mathbf{A} := (a_{ij})$ is *skew-symmetric* if $a_{ij} = -a_{ji}$, for $1 \leq i, j \leq n$. An example of a skew-symmetric matrix that is appropriate to the present context is the adjacency matrix of an orientation D of a simple graph G. In fact, every skew-symmetric matrix $\mathbf{A} := (a_{ij})$ of order n may be viewed as the 'adjacency matrix' of an arc-weighted tournament D which is an orientation of an edge-weighted complete graph G on $\{1, 2, \ldots, n\}$. The weight assigned an edge ij in G is $|a_{ij}|$ and, in D, edge ij is oriented from i to j if $a_{ij} \geq 0$, and from j to i if $a_{ij} < 0$. Note that each edge in G corresponds to two entries of the matrix; when $a_{ij} \neq 0$, one of those entries is positive and the other is negative. The role of the orientation is to indicate which entry is positive (and, implicitly, which one is negative). See Figure 1.7 for an example. If all the entries a_{ij} of $\mathbf{A}$ are integers, one may view $\mathbf{A}$ as the incidence matrix of an orientation of a graph with multiple edges (see Exercise 1.5.2).

This process can be reversed in an obvious manner; given any arc-weighted digraph, one may associate a skew-symmetric matrix with it.

$$\begin{bmatrix} 0 & 2 & -2 & -\pi \\ -2 & 0 & 0 & -\pi \\ 2 & 0 & 0 & -3 \\ \pi & \pi & 3 & 0 \end{bmatrix}$$

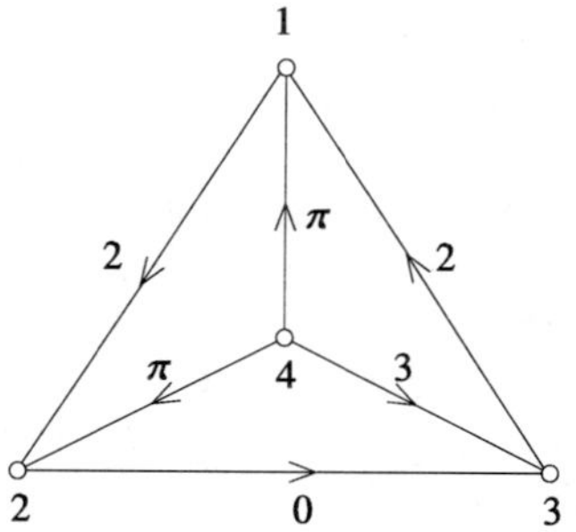

Fig. 1.7 A skew-symmetric matrix and the corresponding weighted digraph

1.5.1 Pfaffians

It is easy to show that when n is odd, the determinant $\det(\mathbf{A})$ of a skew-symmetric matrix $\mathbf{A}$ of order n is zero (Exercise 1.5.1). When n is even, say $n = 2k$, then $\det(\mathbf{A})$ is the square of a polynomial $\mathrm{Pf}(\mathbf{A})$ in the entries of $\mathbf{A}$ known as the Pfaffian of $\mathbf{A}$. In this polynomial, there is one term for each perfect matching of the graph G associated with $\mathbf{A}$. Let $M := \{e_1, e_2, \ldots, e_k\}$ be a perfect matching of G, and suppose that in the orientation D of G, the tail and the head of the edge e_i, for $1 \le i \le k$, are u_i and v_i, respectively. We may then associate with M the permutation $\pi(M)$ as defined below:

$$\pi(M) := \begin{pmatrix} 1 & 2 & 3 & 4 & \ldots & 2k-1 & 2k \\ u_1 & v_1 & u_2 & v_2 & \ldots & u_k & v_k \end{pmatrix}.$$

The *sign* of M, denoted by $\mathrm{sign}(M)$, is +1 if $\pi(M)$ is an even permutation, and is -1 if $\pi(M)$ is an odd permutation. It can be seen that $\mathrm{sign}(M)$ is independent of the order in which the edges of M are enumerated. However, the sign does depend on the orientation D of G. Thus, the same perfect matching of G may have different signs relative to different orientations of G.

Example 1.6 Consider the transitive tournament on $\{1, 2, 3, 4\}$ (an orientation of K_4 in which i precedes j whenever $1 \le i < j \le 4$). The three perfect matchings of this tournament, listed according to the convention mentioned above, are: $M_1 := \{(1,2),(3,4)\}$, $M_2 := \{(1,3),(2,4)\}$, $M_3 := \{(1,4),(2,3)\}$. The permutations $(1,2,3,4)$ and $(1,4,2,3)$ are even permutations of $(1,2,3,4)$; and the permutation $(1,3,2,4)$ is odd. Thus, $\mathrm{sign}(M_1) = +1$, $\mathrm{sign}(M_2) = -1$, and $\mathrm{sign}(M_3) = +1$.

Example 1.7 Figure 1.8 shows the signs of the three perfect matchings in two orientations of K_4.

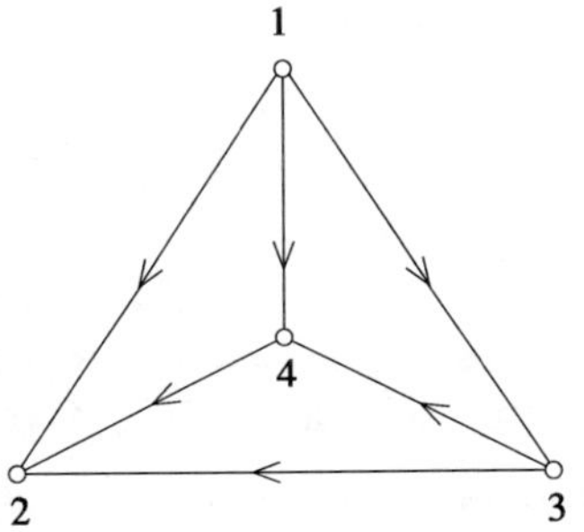

(1, 2) (3, 4) +
(1, 3) (4, 2) +
(1, 4) (3, 2) −

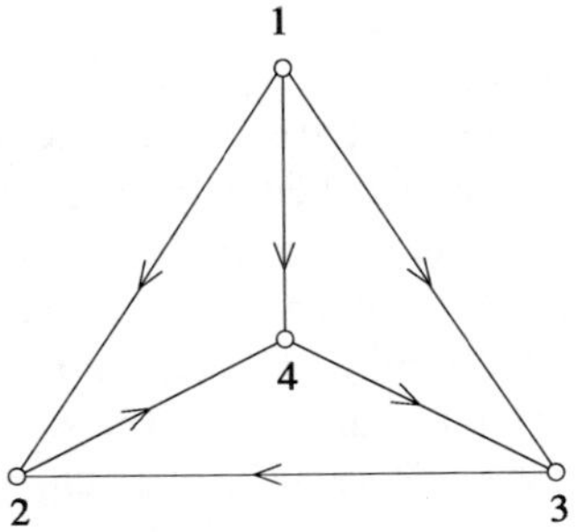

(1, 2) (4, 3) −
(1, 3) (2, 4) −
(1, 4) (3, 2) −

Fig. 1.8 The signs of the perfect matchings in two orientations of K_4

PFAFFIAN OF A SKEW-SYMMETRIC MATRIX

Let **A** be a skew-symmetric matrix of order $n = 2k$, and let G be the associated weighted graph, and its orientation D as defined in the first paragraph of this section. Let $\mathcal{M}$ denote the set of all perfect matchings of G. For a perfect matching $M := \{u_1v_1, u_2v_2, ..., u_kv_k\}$ of G, where, for each edge in M, its tail in D is listed first, let sign(M) denote the sign of M. The *Pfaffian* of **A**, denoted by Pf$(\mathbf{A})$, is the following polynomial in the entries of **A**:

$$\mathrm{Pf}(\mathbf{A}) := \sum \mathrm{sign}(M)\ a_{u_1v_1}a_{u_2v_2}\cdots a_{u_kv_k}, \tag{1.9}$$

where the sum is taken over all M in $\mathcal{M}$, the set of perfect matchings of G.

Example 1.8 Consider the incidence matrix $\mathbf{A} = (a_{ij})$ of the transitive tournament described in Example 1.6. Then,

$$\mathrm{Pf}(\mathbf{A}) = a_{12}a_{34} - a_{13}a_{24} + a_{14}a_{23} = 1 - 1 + 1 = 1.$$

The following classical theorem, due to Cayley, expresses the magical relationship between the determinant and the Pfaffian of a skew-symmetric matrix.

CAYLEY'S THEOREM

1.9 *For any skew-symmetric matrix* **A**, *its determinant* det(**A**) *is the square of its Pfaffian* Pf(**A**). □

For a readable account of Pfaffians, in particular for a proof of Theorem 1.9, we refer the reader to Godsil (1993, [38]). See also Halton (1966, [41]).

When **A** is the adjacency matrix of an orientation D of a graph G, the Pfaffian of **A** is $\sum_{M \in \mathcal{M}} \text{sign}(M)$, where $\mathcal{M}$ is the set of all perfect matchings of G. Thus, each nonzero term in the expansion of Pf(**A**) is either +1 or −1. If all the perfect matchings happen to have the same sign, then the absolute value of Pf(**A**) would be equal to the number of perfect matchings of G, and that number could be computed by evaluating the determinant of A and taking its square root. However, there is no guarantee that all perfect matchings have the same sign. For some graphs, it is possible to choose an orientation which would make the signs of all perfect matchings the same. Such an orientation is called a *Pfaffian orientation* of the graph. The smallest graph that does not admit a Pfaffian orientation is $K_{3,3}$. (The study of Pfaffian orientations is the subject of this chapter and also of Chapters 19, 20 and 21).

Using Theorem 1.9, Tutte was hoping to find a determinantal formula for the number of perfect matchings. But he soon realized the difficulties involved in this approach, as explained in the previous paragraph. However, he found an ingenious way of circumventing this difficulty when the objective is just to determine whether or not the number of perfect matchings of G is positive.

1.5.2 The matrix of indeterminates

Let **A** be the adjacency matrix of an orientation D of a simple graph G. Replace all the nonzero entries above the diagonal by algebraically independent indeterminates, and each nonzero entry below the diagonal by the negative of the corresponding entry above the diagonal. We shall denote the resulting skew-symmetric matrix also by **A** (so as not to proliferate notation), and refer to it as the *matrix of indeterminates* associated with G. Since, corresponding to each edge of G, there is one nonzero entry above the diagonal of the incidence matrix, the total number of indeterminates above the diagonal of **A** is $|E|$. The Pfaffian of **A** is then a polynomial in these $|E|$ indeterminates.

Proposition 1.10 *The Pfaffian* Pf(**A**) *(of the matrix* **A** *of indeterminates) vanishes if and only if G has no perfect matchings.*

Proof By the assumption that the $|E|$ upper diagonal indeterminate entries of **A** are algebraically independent, it follows that Pf(**A**), which is a polynomial in those $|E|$ indeterminates, would vanish only if each term in that polynomial is zero. This would happen precisely when G has no perfect matchings, regardless of the orientation chosen. (For concreteness, a student might wish to take the vertex set of

G to be $\{1, 2, \ldots, n\}$, and the orientation to be the transitive orientation where, for $1 \leq i < j \leq n$, vertex i precedes vertex j.) □

The above observation is the basis of Tutte's characterization of graphs which have a perfect matching.

For any pair x and y of vertices, we shall denote the minor of the matrix $\mathbf{A}$ obtained by deleting the two rows and two columns corresponding to x and y by $\mathbf{A}_{xy}$, and the Pfaffian of $\mathbf{A}_{xy}$ by P_{xy}. (Thus $\mathbf{A}_{xy}$ is the matrix of indeterminates associated with the subgraph $G - x - y$, and P_{xy} is its Pfaffian.) Similar notation applies to any even subset of vertices; however, we shall simply write P for the Pfaffian of $\mathbf{A}$ itself. As an example, consider the 6×6 matrix $\mathbf{A}$ shown in equation (1.10).

$$\mathbf{A} = \begin{pmatrix} 0 & a_{12} & a_{13} & a_{14} & a_{15} & a_{16} \\ -a_{12} & 0 & a_{23} & a_{24} & a_{25} & a_{26} \\ -a_{13} & -a_{23} & 0 & a_{34} & a_{35} & a_{36} \\ -a_{14} & -a_{24} & -a_{34} & 0 & a_{45} & a_{46} \\ -a_{15} & -a_{25} & -a_{35} & -a_{45} & 0 & a_{56} \\ -a_{16} & -a_{26} & -a_{36} & -a_{46} & -a_{56} & 0 \end{pmatrix}. \tag{1.10}$$

Then

$$\mathbf{A}_{13} = \begin{pmatrix} 0 & a_{24} & a_{25} & a_{26} \\ -a_{24} & 0 & a_{45} & a_{46} \\ -a_{25} & -a_{45} & 0 & a_{56} \\ -a_{26} & -a_{46} & -a_{56} & 0 \end{pmatrix} \text{ and } \mathbf{A}_{1234} = \begin{pmatrix} 0 & a_{56} \\ -a_{56} & 0 \end{pmatrix} \tag{1.11}$$

with $P_{13} = a_{24}a_{56} - a_{25}a_{46} + a_{26}a_{45}$, and $P_{1234} = a_{56}$.

An important ingredient in Tutte's proof is an identity involving Pfaffians of minors of $\mathbf{A}$ of the above type. To get a feel for its statement, consider the skew-symmetric matrix of order six, shown in equation (1.10). Then the following identity in the fifteen variables a_{ij} may be verified by routine calculations (see Exercise 1.5.3).

$$P_{12}\, P_{34} - P_{13}\, P_{24} + P_{14}\, P_{23} = P\, P_{1234}. \tag{1.12}$$

If we set some of the a_{ij}'s to be zero, this equation would clearly still hold. Thus equation (1.12) is valid for any graph on six vertices.

Using a classical determinantal identity due to Jacobi, Tutte proved the following Pfaffian identity for any graph G, and any four vertices x, y, z, w of G:

$$\pm\, P_{xy}\, P_{zw} \pm\, P_{xz}\, P_{yw} \pm P_{xw}\, P_{yz} = P\, P_{xyzw}. \tag{1.13}$$

Because of the ambiguity of the values of signs, some explanation is in order. What equation (1.13) means is that $P\, P_{xyzw}$ is equal to one of the eight polynomials obtained from the expression on the left hand side by specific valuations of the three signs. (See Exercise 1.5.5 for a proof of this identity.) The ambiguities of signs arise because this Pfaffian identity is obtained from three determinantal identities by taking square roots, and the Pfaffian of a skew-symmetric matrix is one of the two square roots of its determinant. The precise values of the signs turned out to be irrelevant for the application Tutte had in his mind. The ambiguity of the values of the signs was resolved in 1966 by Halton (1966, [42]).

The following lemma, which plays a crucial role in Tutte's proof of Theorem 1.3, is a consequence of the above identity. (Recall that a graph is matchable if it has a perfect matching.)

Lemma 1.11 *Let G be a graph which is not matchable, and let x, y, z and w be any four vertices of G. If neither $G - x - z$ nor $G - y - z$ is matchable, then at least one of $G - x - y$ and $G - z - w$ is also not matchable.*

Proof Let $\mathbf{A}$ be the matrix of indeterminates associated with G. Since G is not matchable, the Pfaffian P of G is zero. Hence the right hand side of equation (1.13) is zero. Also, as $G - x - z$ and $G - y - z$ are not matchable, both P_{xz} and P_{yz} are zero. By equation (1.13) it now follows that at least one of P_{xy} and P_{zw} must be zero. Hence, by Proposition 1.10, at least one of $G - x - y$ and $G - z - w$ is not matchable. □

1.5.3 Tutte's original proof of his theorem

In Chapter 1, Tutte's Theorem 1.3, in the proof of statement 1.3.1, we adopted the approach of Lovász (1975, [56]). In that proof, we consider a graph G^*, and the four vertices w, x, y and z of G^*. It is assumed that G^* is not matchable, and that xz and yz are edges of G^*, but neither xy nor wz are edges of G^*. It is then shown that at least one of the graphs $G^* + xy$ and $G^* + wz$ is not matchable. This result is equivalent to the statement of Lemma 1.11.

Exercises

1.5.1 *Show that the determinant of an odd order skew-symmetric matrix is zero. (Hint: For any skew-symmetric matrix* $\mathbf{A}$*, the matrix obtained from* $\mathbf{A}$ *by multiplying each of its columns by* -1 *is the transpose of* $\mathbf{A}$*.)*

1.5.2 *Let G be a graph which is not necessarily simple, and let D be an orientation of G such that, for any adjacent pair $\{u, v\}$ of vertices, all edges joining u and v are oriented the same way, that is, either u is the tail of all the corresponding arcs in D, or u is their head. In this case, in the adjacency matrix* $\mathbf{A} = (a_{uv})$ *of D, we set $a_{uv} = 0$, if u and v are not adjacent in G, and a_{uv} is the number of edges joining u to v, or is the negative of that number, according to whether u is the tail, or is the head, of the corresponding arcs in D. The digraph shown in Figure 1.9 is such an orientation of a graph on four vertices. (It happens to be a Pfaffian orientation of its underlying graph.)*

Compute the determinant and Pfaffian of the adjacency matrix shown in Figure 1.9 and verify that the number of perfect matchings in the graph is the absolute value of the Pfaffian.

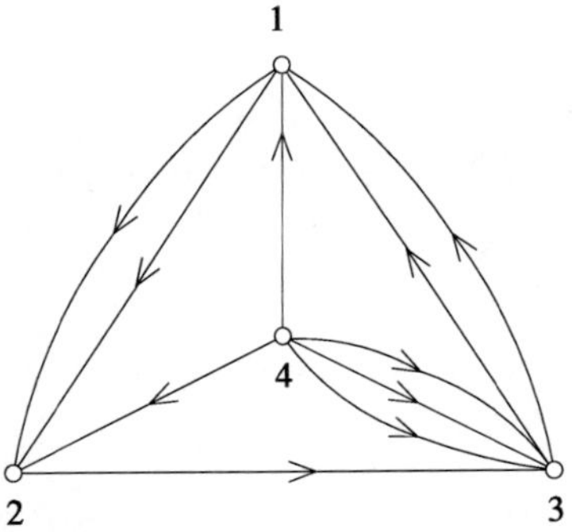

$$\begin{bmatrix} 0 & 2 & -2 & -1 \\ -2 & 0 & 1 & -1 \\ 2 & -1 & 0 & -3 \\ 1 & 1 & 3 & 0 \end{bmatrix}$$

Fig. 1.9 A digraph and its adjacency matrix

1.5.3 *Write down the expansions of* P, P_{12}, P_{34}, P_{13}, P_{24}, P_{14}, P_{23}, *and* P_{1234} *corresponding to the* 6×6 *matrix* $\mathbf{A}$ *shown in equation (1.10), and verify the identity equation (1.12).*

1.5.4 (The cofactor matrix) *Let* $\mathbf{A} = (a_{ij})$ *be an* $n \times n$ *matrix. For* $1 \leq i, j \leq n$, *let* $\mathbf{A}[i, j]$ *denote the matrix obtained from* $\mathbf{A}$ *by deleting the* i^{th} *row and* j^{th} *column. (This is not to be confused with* $\mathbf{A}_{ij}$, *which is the matrix obtained from* $\mathbf{A}$ *by deleting the* i^{th} *and* j^{th} *rows and the* i^{th} *and* j^{th} *columns.) The cofactor matrix* $\mathbf{C} = (c_{ij})$ *is defined by* $c_{ij} := (-1)^{i+j} \det \mathbf{A}[i, j]$. *(Thus the adjugate* $\mathbf{Adj}(\mathbf{A})$ *of* $\mathbf{A}$ *is the transpose of the* $n \times n$ *matrix* $\mathbf{C} := (c_{ij})$.*) Verify that if* $\mathbf{A}$ *is a skew-symmetric matrix of even order then* $\mathbf{C}$ *is a skew-symmetric matrix by showing that:*

(i) $c_{ii} = 0$, *for* $1 \leq i \leq n$, *and*
(ii) $c_{ij} = -c_{ji}$, *for* $1 \leq i < j \leq n$.

1.5.5 (Jacobi's Identity) *Let* $\mathbf{A} = (a_{ij})$ *be a skew-symmetric matrix of even order and let* $\mathbf{C}$ *be its cofactor matrix (defined in Exercise 1.5.4). Suppose that* $R = \{1, 2, \ldots, r\}$, *where* r *is even, and let* $\mathbf{B}$ *denote the matrix obtained from the identity matrix* $\mathbf{I}_n$ *by replacing its first* r *columns by the first* r *rows of* $\mathbf{C}$.

(i) *By considering the matrix* $\mathbf{AB}$, *and using the fact that* $\mathbf{AC}^t = \det(\mathbf{A})\mathbf{I}_n$, *show that*

$$\det(\mathbf{A}) \det(\mathbf{C}_{\overline{R}}) = (\det(\mathbf{A}))^r \det(\mathbf{A}_R), \tag{1.14}$$

where $\overline{R} = \{1, 2, \ldots, n\} - R$.
(For clarity of visualization, we have stated this identity taking R *to be* $\{1, 2, \ldots, r\}$, *but it can be seen that it holds for any* r*-subset* R *of* $\{1, 2, \ldots, n\}$.*)*

(ii) *Considering the special case in which* $\mathbf{A}$ *is the matrix of indeterminates associated with an orientation of the complete graph* K_n, *and* R *is equal to* $\{x, y, z, w\}$, *deduce that:*

$$(c_{xy}c_{zw} - c_{xz}c_{yw} + c_{xw}c_{yz})^2 = (\det(\mathbf{A}))^3 \det(\mathbf{A}_{xyzw}).$$

∗ **1.5.6**

(i) *By taking R to be $\{i, j\}$ in equation (1.14), deduce that* $c_{ij} = \pm P\, P_{ij}$.
(ii) *Now taking R to be $\{x, y, z, w\}$, and applying equation* (1.14) *deduce the identity stated in equation (1.13).*

1.6 Notes

We defer to the treatises by Lovász and Plummer [59], and Schrijver (2003, [84]) for an account of the history of matching theory. Efficient algorithms related to matchings can be found in the books by Schrijver [84], by Cormen, Leiserson, Rivest and Stein (2009, [25]) and also in the book by Bondy and Murty [3].

Tutte's Theorem 1.3 was clearly a leap from Petersen's Theorem 1.4 on cubic graphs and Hall's Theorem 1.5 on bipartite graphs. As such it was received enthusiastically. As his original proof used Pfaffian identities, some authors felt that it was unnatural to use such algebraic machinery to prove a graph theoretical statement; they argued that graphs were simple 'geometric' objects, and it ought to be possible to prove statements concerning them by manipulations of diagrams that could be visualized. Alternative proofs of Theorem 1.3 which did not depend on properties of Pfaffians were discovered. But, as it turned out, Pfaffians continued, and continue, to play an important role in matching theory. We will present an account of some these developments in Part III.

As noted in the introduction to this chapter, a perfect matching of a graph G is the edge set of a spanning 1-regular subgraph of G, or, to use Tutte's terminology, a 1-factor of G. Having been able to characterize graphs which have 1-factors, Tutte was led to the more general problem of characterizing graphs which have spanning subgraphs with specified degree sequences, which he called f-factors. An account of Tutte's work, and of subsequent developments in this area, can be found in Lovász and Plummer [59].

Chapter 2
Matching Covered Graphs

Contents

2.1 Matchable Edges in Matchable Graphs

This book is primarily concerned with matching covered graphs which were briefly introduced at the end of Section 1.3. We begin our exploration of their rich theory with a study of notions and statements applicable to the more general class of matchable graphs. By definition, a graph is matchable if it has a perfect matching. In any such graph G, by Tutte's Theorem 1.3, $o(G - S) \leq |S|$, for all $S \subseteq V$.

Barriers in matchable graphs

A subset B of the vertex set of a matchable graph G is a *barrier* of G if $o(G - B) = |B|$. The empty set and all singletons are barriers in any matchable graph. We shall refer to a barrier of cardinality at most one as a *trivial barrier*. The study of properties of barriers plays an important role in this theory.

C. L. Lucchesi, U. S. R. Murty, *Perfect Matchings*, Algorithms and Computation in Mathematics 31, https://doi.org/10.1007/978-3-031-47504-7_2

The graph G shown in Figure 2.1 has several nontrivial barriers. For example the set $B := \{u_1, u_2, u_3\}$ is a barrier of cardinality three as $G - B$ has three odd components (and one even component). The two sets $\{u_1, u_2, u_3, u_4, u_5\}$ and $\{v_7, v_8, v_9\}$ are also nontrivial barriers.

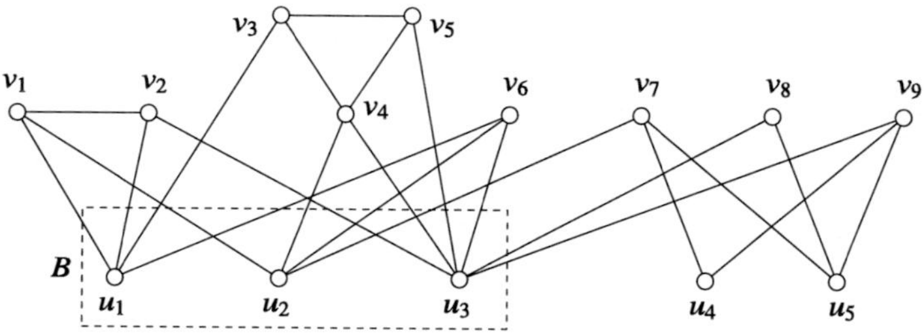

Fig. 2.1 Barriers in a matchable graph

MATCHABLE EDGES

An edge e of a graph G is *matchable* in G if there is some perfect matching of G that contains it, and is *unmatchable* otherwise.

Corollary 2.2 provides a characterization of matchable edges in a matchable graph. We derive that corollary from the following more general statement which will be of use in the next chapter.

Theorem 2.1 *Let u and v be any two vertices in a matchable graph G. Then the graph $G - u - v$ is matchable if and only if there is no barrier of G which contains both u and v.*

Proof Firstly suppose that $G - u - v$ is matchable. We wish to show that no subset of V containing both u and v is a barrier of G. Towards this end, let B be any subset of V which includes both u and v. By setting $S := B - u - v$, we have:

$$\begin{aligned} o(G - B) &= o((G - u - v) - S) \\ &\le |S| \qquad \text{(because } G - u - v \text{ is matchable)} \\ &= |B| - 2, \end{aligned}$$

implying that $o(G - B) < |B|$, and hence that B is not a barrier of G.

Conversely, suppose that $G - u - v$ is not matchable. Then, by Theorem 1.3, and Exercise 1.2.1, there is a subset S of $V(G - u - v)$ such that

$$o(G - u - v - S) \ge |S| + 2. \tag{2.1}$$

Now, letting $B := S + u + v$, we have:

$$o(G - B) \ge |B|. \tag{2.2}$$

By hypothesis, G itself is matchable. Thus, by Theorem 1.3, strict inequality cannot hold in (2.2). Therefore $o(G - B) = |B|$, implying that B is a barrier of G which contains both u and v. □

If $e = uv$ is an edge of G, then e is matchable in G if and only if $G - u - v$ is matchable. Thus, as a consequence of the above theorem, we have:

MATCHABLE EDGES AND BARRIERS

Corollary 2.2 *An edge of a matchable graph is matchable if and only if there is no barrier of the graph which contains both ends of that edge.* □

The following proposition is a simple consequence of the above corollary.

Proposition 2.3 *Suppose that B is a barrier of a matchable graph G and that K is an even component of $G - B$. Then any edge uv of G with $u \in B$ and $v \in V(K)$ is unmatchable in G.*

Proof Consider the set $B' := B + v$. Clearly, each odd component of $G - B$ is also an odd component of $G - B'$. In addition, any odd component of $K - v$ is also an odd component of $G - B'$. There must be at least one odd component of $K - v$ because K is an even component of $G - B$. Thus $o(G - B') \geq |B| + 1 = |B'|$. However, as G is a matchable graph, equality must hold implying that B' is a barrier of G. Since both ends of the edge uv belong to this barrier, it follows from Corollary 2.2 that uv is unmatchable. □

Example 2.4 Consider the barrier $B := \{u_1, u_2, u_3\}$ of the matchable graph G shown in Figure 2.1. The subgraph of order two induced by $\{v_1, v_2\}$ is an even component of $G - B$. Adding any vertex of this even component of $G - B$ to B yields another barrier of G. The two edges u_1v_1, u_2v_1 have both their ends in the barrier $B + v_1$, and the two edges u_1v_2 and u_3v_2 have both their ends in the barrier $B + v_2$. Thus the four edges u_1v_1, u_2v_1, u_1v_2, and u_3v_2 are unmatchable in G.

Exercises

2.1.1 *Find all the matchable edges in the graph shown in Figure 2.1.*

▷**2.1.2** *Describe a polynomial-time algorithm to determine whether an edge vw of a graph G is matchable. Hint: determine whether the graph $H := G - v - w$ is matchable; if a perfect matching M of G is known, the complexity is reduced by considering the restriction of M to H.*

SPECIAL BARRIERS

A barrier B in a matchable graph G is a *special barrier* if $G - B$ has precisely one nontrivial odd component.

2.1.3 *Show that in the graph G obtained by deleting an edge from the Petersen graph every nontrivial barrier is a special barrier.*

2.2 Matching Covered Graphs

Adopting the terminology introduced in Section 2.1 we may now define a matching covered graph as follows:

> **MATCHING COVERED GRAPHS**
>
> A *matching covered graph* is a nontrivial connected graph in which every edge is matchable. The complete graph K_2 is the smallest matching covered graph. A plethora of examples of graphs and families of graphs which play important roles in this theory can be found at the end of this section and in Sections 2.4 and 2.5.

Thus matching covered graphs are special types of matchable graphs. We make use of this fact implicitly in establishing statements concerning matching covered graphs.

A subset of the edge set of a graph G is a perfect matching of G if and only it is a union of perfect matchings of the components of G. Also, clearly, the set of perfect matchings of G is the same as the set of perfect matchings of the graph obtained from G by deleting all unmatchable edges. For this reason, the theory of matching covered graphs is the natural setting for studying problems such as counting or packing or covering problems involving perfect matchings.

Corollary 2.2 implies the following distinguishing property of the class of matching covered graphs. (Recall that a subset S of the vertex set of a graph G is stable if no edge of G has both its ends in S.)

Theorem 2.5 *A connected matchable graph is matching covered if and only if each of its barriers is a stable set.* □

Hall's Theorem 1.5 can be used to derive a characterization of matching covered graph that is specific to bipartite graphs, see Section 2.3.

2.2.1 Connectivity of matching covered graphs

If B is a barrier of a matchable graph G then, by definition, the subgraph $G - B$ has $|B|$ odd components. The following proposition asserts that, if G is matching covered, then $G - B$ cannot have any even components.

Proposition 2.6 *For each nonempty barrier B of a matching covered graph G, each component of $G - B$ is odd.*

Proof Assume that $G - B$ has an even component, say K. The graph G, being matching covered, is connected. Thus, G has an edge uv with $u \in B$ and $v \in V(K)$. By Proposition 2.3, edge uv is unmatchable in G. This is absurd because G is matching covered. We conclude that every component of $G - B$ is odd. □

The following assertion is a basic property of matching covered graphs:

Corollary 2.7 *Every matching covered graph on four or more vertices is 2-connected.*

Proof Let v be any vertex of a matching covered graph G. Then $G - v$ has odd components. As G is matchable, it has precisely one odd component. By Proposition 2.6, $G - v$ has no even component. Therefore $G - v$ has just one component, implying that v is not a cut vertex of G. It follows that if the order of G is at least four, then G is 2-connected. □

2.2.2 Cubic matching covered graphs

As noted at the end of Section 1.3, Tutte [89] deduced from his Theorem 1.3 that every 2-connected cubic graph is matching covered. Here we present a proof which is based on the characterization of matching covered graphs given by Theorem 2.5.

CUBIC MATCHING COVERED GRAPHS

Theorem 2.8 *Every 2-connected cubic graph is matching covered.*

Proof Let G be a 2-connected cubic graph and let S be any subset of V. Then, as in the proof of Theorem 1.4, the inequality $3 \leq |\partial(K)|$ holds for each odd component K of $G - S$. We thus have:

$$3\,o(G - S) \leq |\partial(S)| = 3|S| - 2|E(G[S])| \leq 3|S|. \tag{2.3}$$

We deduce that $o(G - S) \leq |S|$. This conclusion holds for each subset S of V. By Tutte's Theorem 1.3, G is matchable. Moreover, from equation (2.3), we deduce that $o(G - S) = |S|$ only if S is stable. This conclusion holds for each subset S of V. Thus, every barrier of G is stable. The assertion now follows from Theorem 2.5. □

A similar argument shows that, for $r \geq 2$, any r-graph (as defined in Exercise 1.3.6) is matching covered (Exercise 2.2.3).

2.2.3 Three special matching covered graphs

Even cycles and wheels (Exercise 2.2.4) are some examples of matching covered graphs which shall come across frequently. However, most of the graphs which play prominent roles in this theory are 3-connected cubic graphs. Figure 2.2 shows three such graphs (K_4, the complete graph on four vertices, $\overline{C_6}$, the triangular prism, and $\mathbb{P}$, the Petersen graph) which play special roles in our theory. (These three graphs feature in the conjecture of Lovász mentioned in the preface.) The operation of

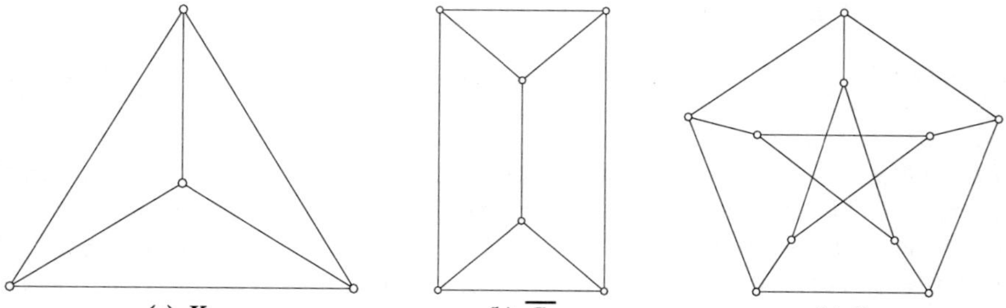

(a) K_4 (b) $\overline{C_6}$ (c) $\mathbb{P}$

Fig. 2.2 The trio of important cubic matching covered graphs

splicing which can be used to 'combine' two matching covered graphs to obtain a larger matching covered graph is described in Section 2.4.

Exercises

▷**2.2.1** *Let R be the set of unmatchable edges of a graph G. Show that every nontrivial component of $G - R$ is matching covered and show that G is matchable if and only if each component of $G - R$ is nontrivial.*

▷**2.2.2** *Describe a polynomial-time algorithm, which, given a graph G, determines whether G is matching covered. Hint: check in linear-time whether G is connected, determine in polynomial-time a maximum matching of G and, if G is matchable then, for each edge $vw \in E$, determine in polynomial-time whether vw is matchable (Exercise 2.1.2.).*

2.2.3 *Show that any r-graph (defined in Exercise 1.3.6), with $r \geq 2$, is matching covered.*

WHEELS

For an integer $k \geq 3$, the *wheel* W_k is the graph obtained from a cycle R of length k by adding a new vertex h and joining it to all vertices of R.

Thus W_k is a graph on $k + 1$ vertices. (When a graph is denoted by a symbol with a subscript, that subscript usually indicates the order of the graph. However, here we are deviating from this general convention so as to keep our notation for wheels consistent with the one in standard use.)

The cycle R is the *rim* of W_k, the vertex h is its *hub*, and the edges incident with h are its *spokes*. Clearly, the degree of the hub of W_k is k. We note that W_3 is isomorphic to the complete graph K_4, and any one of its vertices may be regarded as its hub. For all $k \geq 4$, the wheel W_k has a unique hub. An *odd wheel* is a wheel with an odd number of spokes.

2.2.4 *Consider now a wheel W_k, where k is odd.*

(i) *Show that every spoke of W_k lies in precisely one perfect matching of W_k.*
(ii) *Show that W_k has precisely k perfect matchings.*
(iii) *Show that every edge in the rim of W_k lies in precisely $(k-1)/2$ perfect matchings of W_k. Conclude that W_k is matching covered.*
(iv) *Prove that, if $k \geq 5$, and e is any edge of W_k, then $W_k - e$ is matching covered if and only if e is a spoke of W_k.*

$*$ **2.2.5** *Prove that if $C := \partial(X)$ is a cut of a matching covered graph G and $G[X]$ has an odd cycle that is hamiltonian then the graph G_1, obtained from G by contracting $\overline{X}$ to a single vertex $\overline{x}$ and removing all resulting loops, is matching covered.*

$*$ **2.2.6** *Let $G[A, B]$ be a bipartite graph and let M be a perfect matching of G. Let H be the graph obtained from G by the addition of a parallel edge to each edge of M. Let D be the directed graph obtained from H by directing every edge of $E(H) - M$ from A to B and by directing every edge of M from B to A.*

(i) *Show that G is matching covered if and only if D is strongly connected. Hint: a digraph is strongly connected if and only if it is connected and each one of its edges is in a directed cycle.*
(ii) *Design a linear-time algorithm to decide if a given bipartite graph is matching covered, assuming as input the bipartite graph G and a perfect matching M of G.*

2.3 Bipartite Matching Covered Graphs

A HALL-TYPE CHARACTERIZATION OF BIPARTITE MATCHING COVERED GRAPHS

Theorem 2.9 *Let $G := G[A, B]$ be a bipartite graph of order four or more, such that $|A| = |B|$. The graph G is matching covered if and only if $|N(S)| \geq |S| + 1$, for all subsets S of A such that $1 \leq |S| \leq |A| - 1$.*

Proof Suppose that there exists a nonempty proper subset S of A such that $|N(S)| \leq |S|$. Then, every edge (if any) joining $N(S)$ with $A - S$ is not matchable. If there were no such edge then G would not be connected, and consequently, not matching covered. In any case, G is not matching covered.

Now, suppose that $|N(S)| \geq |S| + 1$, for all nonempty proper subsets S of A. Let ab be an edge of G, where $a \in A$ and $b \in B$. Then, in $G - a - b$, we have $|N(X)| \geq |X|$, for all $X \subseteq A - a$. By Hall's Theorem, $G - a - b$ is matchable. It follows that ab is matchable in G. This conclusion holds for each edge ab of G. Moreover, G is matchable.

Finally, for any connected component K of G, as G is matchable we have that $|A \cap V(K)| = |B \cap V(K)|$. As $|N(S)| \geq |S| + 1$ for every nonempty proper subset S of A, it follows that $A \cap V(K) = A$, hence $B \cap V(K) = B$. Thus, $G = K$, and G is connected. We conclude that G is matching covered. □

We leave the proof of the next assertion as Exercise 2.3.1.

Corollary 2.10 *A matchable bipartite graph $G := G[A, B]$ is not matching covered if and only if there exists a subset S of A such that $1 \leq |S| \leq |A|-1$ and $|N(S)| = |S|$.* □

2.3.1 Dulmage-Mendelsohn decomposition

In analyzing the properties of a given matching covered graph we are sometimes led to considering the properties of a related matchable graph. For example, if G is a matching covered graph with at least two edges, and e is an edge of G, then the graph $G - e$ may not be matching covered, but it is matchable because any perfect matching of G that does not include the edge e is a perfect matching of $G - e$. If B is a barrier of $G - e$ then either B is also a barrier of G or the two ends of the edge e lie in different odd components of $G - e - B$.

We conclude this section by noting a fundamental property of matchable bipartite graphs which we shall have occasion to use in future chapters. For this, we need the following auxiliary result, whose proof is left as Exercise 2.3.2.

Proposition 2.11 *In any matchable bipartite graph $G[A, B]$ there exists a nonempty subset S of A, possibly equal to A, such that the subgraph of G induced by $S \cup N(S)$ is matching covered. (Likewise, there is a nonempty subset T of B such that the subgraph induced by $N(T) \cup T$ is matching covered.) See Figure 2.3 for an illustration.* □

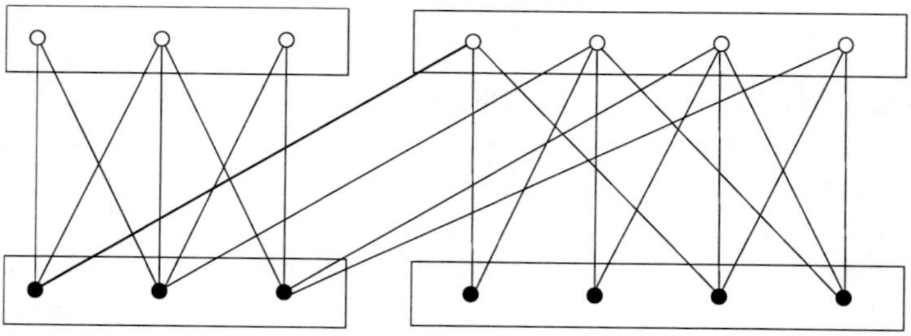

Fig. 2.3 The subgraph of $G[A, B]$ induced by $S \cup N(S)$ is matching covered

THE DULMAGE-MENDELSOHN DECOMPOSITION

A *Dulmage-Mendelsohn decomposition of a matchable bipartite graph* $G := G[A, B]$ consists of a partition $(A_1, \ldots, A_r)$ $(r \geq 1)$ of A and a partition $(B_1, \ldots, B_r)$ of B such that

(i) for $i = 1, 2, \ldots, r$, the subgraph $G[A_i \cup B_i]$ of G induced by $A_i \cup B_i$ is matching covered, and
(ii) if $a_i b_j$, with $a_i \in A_i$ and $b_j \in B_j$, is an edge of G, then $i \geq j$.

Proposition 2.11 is a special case of the following more general assertion.

Theorem 2.12 (Dulmage and Mendelsohn (1958, [28]))
Every matchable bipartite graph has a Dulmage-Mendelsohn decomposition.

We leave the proof of Theorem 2.12 as Exercise 2.3.3.

Exercises

∗ **2.3.1** *Prove Corollary 2.10.*

∗ **2.3.2** *Deduce Proposition 2.11.*

∗ **2.3.3** *Prove Theorem 2.12. Hint: use Proposition 2.11 and apply induction.*

2.4 The Operation of Splicing

Let G and H be two vertex-disjoint graphs and let u and v be vertices of G and H, respectively, such that $d_G(u) = d_H(v)$. In addition, let θ be a given bijection between the set $\partial_G(u)$ of edges incident with u in G and the set $\partial_H(v)$ of edges incident with v in H. We define $(G \odot H)_{u,v,\theta}$ to be the graph obtained from the union of $G - u$ and $H - v$ by joining, for each edge e in $\partial_G(u)$, the end of e in G belonging to $V(G) - u$ to the end of $\theta(e)$ in H belonging to $V(H) - v$; and refer to $(G \odot H)_{u,v,\theta}$ as the graph obtained by *splicing G at u, with H at v, with respect to the bijection θ.* Figure 2.4 illustrates this operation, with the bijection θ between $\partial_G(u)$ and $\partial_H(v)$, where $\theta(e_i) = f_i$, for $i = 1, 2, \ldots, 5$.

It is not always necessary to use the cumbersome notation $(G \odot H)_{u,v,\theta}$ for describing a splicing of two graphs G and H; in most cases of interest either the parameters (u, v, and θ) are implicit, or the statements made about the graph resulting from splicing does not depend on the parameters. We leave verification of the following assertion as Exercise 2.4.1.

When both G and H happen to be simple graphs, the bijection θ in the definition of splicing may be regarded just as a bijection between the neighbour set $N_G(u)$ of u in G and the neighbour set $N_H(v)$ of v in H, and $(G \odot H)_{u,v,\theta}$ may be defined as the graph obtained from the union of $G - u$ and $H - v$ by joining each vertex x in $N_G(u)$ to the vertex $\theta(x)$ in $N_H(v)$. Both the pentagonal prism and the Petersen graph may be obtained by splicing two copies of the 5-wheel at their hubs under suitable bijections between the vertex sets of their rims (see Exercise 2.4.2).

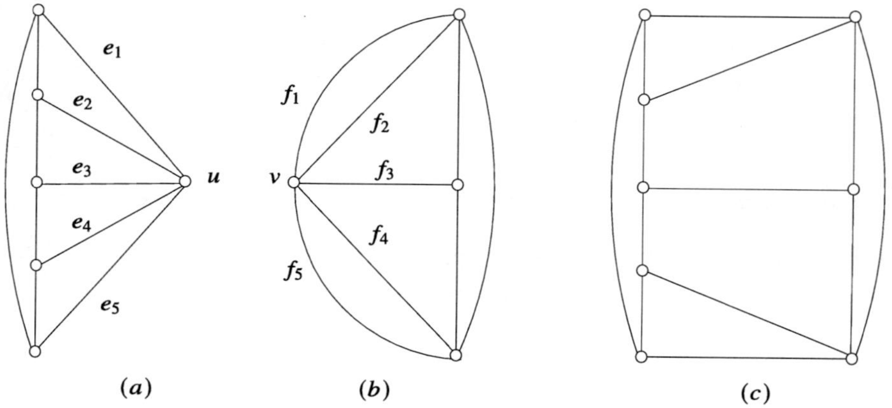

Fig. 2.4 (a) G; (b) H; (c) $(G \odot H)_{u,v,\theta}$

As the example mentioned above shows, in general the result of splicing two graphs G and H depends on the choices of u, v, θ. However, if G is a vertex-transitive cubic graph, then the result of splicing G with $H = K_4$ does not depend, up to isomorphism, on the choices of u, v, and θ, and we denote it simply by $G \odot K_4$. More generally, for any cubic graph G, the result of splicing G and K_4 depends, up to isomorphism, only on the orbit of the automorphism group of G to which u belongs (and the choices of v and θ are immaterial); and we denote it simply by $(G \odot K_4)_u$.

For example, since both K_4 and the triangular prism $\overline{C_6}$ are vertex-transitive, there is only one way of splicing K_4 with itself or with $\overline{C_6}$. The result of splicing K_4 with itself is clearly $\overline{C_6}$. The result of splicing $\overline{C_6}$ with K_4 is shown in Figure 2.5(c). The automorphism group of this graph has three different orbits. The result of splicing a K_4 with it at this vertex is shown in Figure 2.5(d). We refer to the two graphs in Figure 2.5(c) and (d), respectively, as the *bicorn* and the *tricorn* as they resemble, in our imagination, to the two-cornered and three-cornered hats worn by pirates. We denote the bicorn by $\mathbb{H}_8$ and the tricorn by $\mathbb{H}_{10}$.

> SPLICING TWO MATCHING COVERED GRAPHS
>
> **2.13** *Any graph obtained by splicing together two matching covered graphs is also matching covered.* □

Exercises

∗ **2.4.1** *Give a proof of statement 2.13.*

∗ **2.4.2** *Consider two copies G and H of the 5-wheel W_5, where u and v denote, respectively, the hubs in the two copies. Describe bijections θ and ϕ between $N_G(u)$ and $N_H(v)$ such that $(G \odot H)_{u,v,\theta}$ is the pentagonal prism, and $(G \odot H)_{u,v,\phi}$ is*

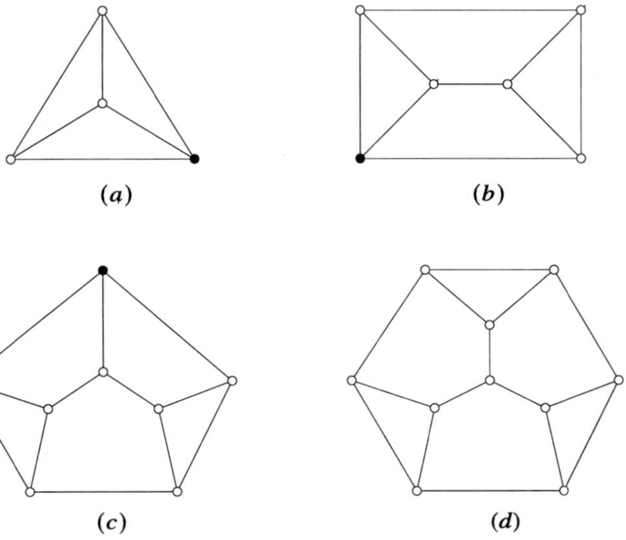

Fig. 2.5 (a) K_4; (b) $\overline{C_6}$; (c) $\mathbb{H}_8$, the bicorn; (d) $\mathbb{H}_{10}$, the tricorn

the Petersen graph. Show that all such splicings are 3-edge-colourable, with the exception of the Petersen graph.

▷**2.4.3** *Show that, up to isomorphism, there are four graphs that are obtainable by the splicing of two 5-wheels at their hubs.*

▷**2.4.4** *Let G be a graph, let u and w be two vertices of G, both having degree three. Show that if there is an automorphism θ of G that maps u to w then the graph obtained by splicing G and K_4 at u is isomorphic to the one obtained by splicing G and K_4 at w. Conclude that, up to isomorphism, there is only one graph that can be obtained by splicing a vertex-transitive graph G to K_4.*

▷**2.4.5** *Find all the graphs, up to isomorphism, that can be obtained by splicing the bicorn with K_4.*

∗**2.4.6** *Given any family $\mathcal{G}_n$ of cubic graphs of order n, let $\mathcal{G}_{n+2}$ denote the family of cubic graphs of order $n+2$ (up to isomorphism) that can be obtained by splicing members of $\mathcal{G}_n$ with K_4. Taking $\mathcal{G}_4$ to be $\{K_4\}$, determine all members of $\mathcal{G}_6$, $\mathcal{G}_8$, and $\mathcal{G}_{10}$.*

▷**2.4.7** *The Meredith graph, shown in Figure 2.6, is obtained from the 4-graph in Figure 1.5 by splicing a $K_{4,4}$ at each of its vertices as indicated. Show that this graph is not 4-edge-colourable. (For $r \geq 4$, similar ideas can be used to construct r-connected r-graphs which are not r-edge-colourable.)*

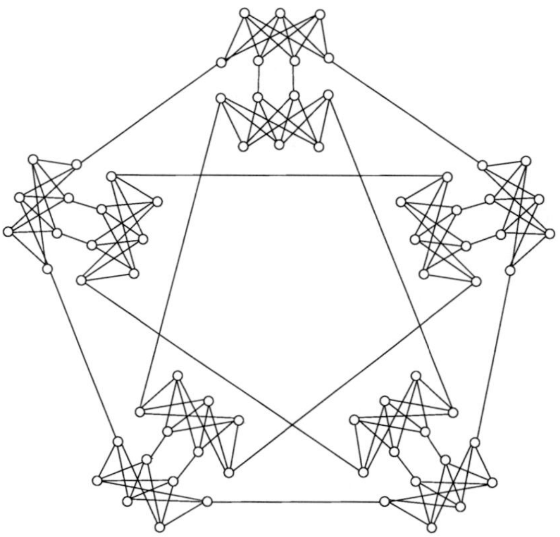

Fig. 2.6 The Meredith graph

2.5 Families of Matching Covered Graphs ♯

In addition to wheels (see Exercise 2.2.4), there are several other families of matching covered graphs which play important roles in this theory. We introduce some of them in this section. These families play significant roles in the works of McCuaig [68] and Norine and Thomas [78], which are described in Chapter 18.

> **PRISMS**
>
> The *prism* $\mathbb{P}_{2n}$, $n \geq 3$, is the graph obtained from two disjoint cycles
>
> $$u_1u_2 \ldots u_nu_1 \text{ and } v_1v_2 \ldots v_nv_1$$
>
> of length n by the addition of the n edges u_iv_i, $i = 1, 2, \ldots, n$.

Prisms are so named because they are the graphs associated with the familiar geometrical objects known as 'prisms'. For this reason, we have special names for those of small orders. Thus $\mathbb{P}_6$, which is the same as $\overline{C_6}$, is the *triangular prism*; $\mathbb{P}_8$ is the *cube*; and $\mathbb{P}_{10}$, shown in Figure 2.7(a), is the *pentagonal prism*. The family of all prisms is denoted by $\mathcal{P}$.

> **MÖBIUS LADDERS**
>
> The *Möbius ladder* $\mathbb{M}_{2n}$, $n \geq 2$, is the graph obtained from a cycle
>
> $$v_1v_2 \ldots v_{2n}v_1$$

of length $2n$ by the addition of the n *chords* $v_i v_{i+n}$, $1 \le i \le n$, joining antipodal pairs of vertices of the cycle.

All Möbius ladders are embeddable in the projective plane. The two graphs $\mathbb{M}_4$, and $\mathbb{M}_6$ are isomorphic to K_4 and $K_{3,3}$, respectively. Figure 2.7(b) shows the Möbius ladder $\mathbb{M}_{10}$. The family of all Möbius ladders is denoted $\mathcal{ML}$. (We are using $\mathcal{ML}$ to denote the family of Möbius ladders since $\mathcal{M}$ is reserved for the set of all perfect matchings of a graph.)

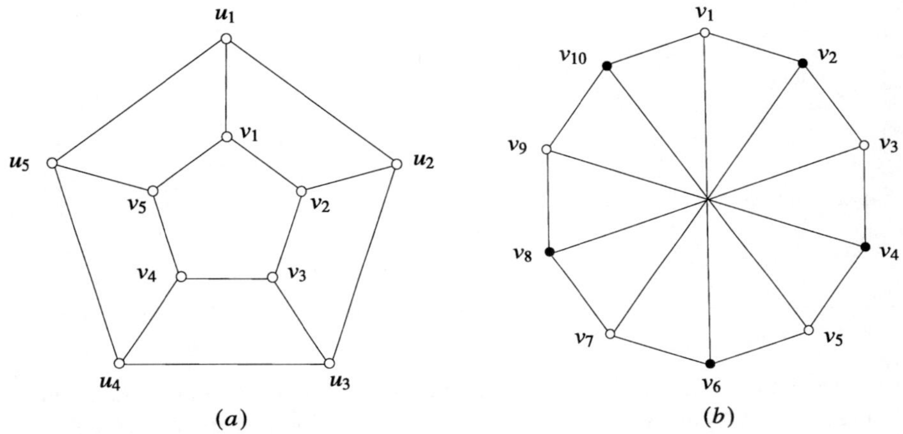

Fig. 2.7 (a) prism $\mathbb{P}_{10}$; (b) Möbius ladder $\mathbb{M}_{10}$

Prisms and Möbius ladders may be defined alternatively as supergraphs of ladders, which are defined below.

LADDERS

For $n \ge 2$, the *ladder* $\mathbb{L}_{2n}$ is the graph obtained from two disjoint paths

$$u_1 u_2 \ldots u_n \text{ and } v_1 v_2 \ldots v_n$$

of length $n-1$ by the addition of the n *rungs* $u_i v_i$, $i = 1, 2, \ldots, n$.

Figure 2.8(a) shows the ladder $\mathbb{L}_8$. Clearly, the prism $\mathbb{P}_{2n}$, $n \ge 3$, is just the planar graph obtained from $\mathbb{L}_{2n}$ by adding the two edges $u_1 u_n$ and $v_1 v_n$. Although it is less obvious, it can be shown that the Möbius ladder $\mathbb{M}_{2n}$, $n \ge 3$ is isomorphic to the graph obtained from $\mathbb{L}_{2n}$ by adding the two edges $u_1 v_n$ and $v_1 u_n$ (see Exercise 2.5.1). This explains the 'ladder' part in their names.

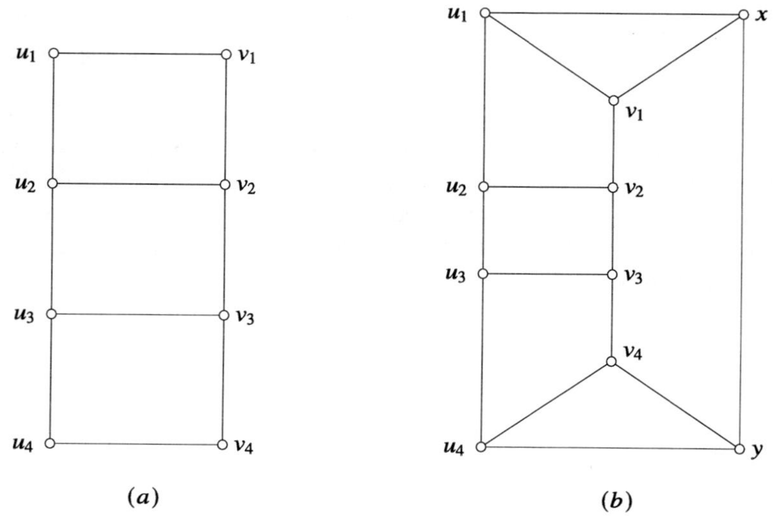

Fig. 2.8 (a) Ladder $\mathbb{L}_8$; (b) Staircase $\mathbb{S}_{10}$

STAIRCASES

Consider the ladder $\mathbb{L}_{2n-2}$ obtained from two disjoint paths $u_1u_2\ldots u_{n-1}$ and $v_1v_2\ldots v_{n-1}$ by adding, for $1 \le i \le n-1$, an edge joining u_i and v_i. For $n \ge 3$, the *staircase* $\mathbb{S}_{2n}$ is the graph obtained from $\mathbb{L}_{2n-2}$ by adding two new vertices x and y, and then joining x to u_1 and v_1, y to u_{n-1} and v_{n-1}, and x and y to each other.

See Figure 2.8(b) for a drawing of $\mathbb{S}_{10}$. The staircase $\mathbb{S}_6$ on six vertices is isomorphic to the triangular prism $\overline{C_6}$. The staircase $\mathbb{S}_8$ on eight vertices is the bicorn $\mathbb{H}_8$. The family of all staircases is denoted $\mathcal{ST}$.

BIWHEELS

The *biwheel* $\mathbb{B}_{2n}$, $n \ge 4$, is the bipartite graph obtained from a cycle $v_1v_2\ldots v_{2n-2}v_1$ of length $2n-2$, called the *rim* of $\mathbb{B}_{2n}$, by the addition of two vertices, h_1 and h_2, called the *hubs* of $\mathbb{B}_{2n}$, and by the addition of edges $h_1v_1, h_1v_3, \ldots, h_1v_{2n-3}$ and edges $h_2v_2, h_2v_4, \ldots, h_2v_{2n-2}$.

The family of biwheels is denoted $\mathcal{B}$. Figure 2.9 shows two drawings of the biwheel $\mathbb{B}_{10}$ where the second drawing depicts it as a 'wheel' with two hubs. We note that the cube, apart from being the prism $\mathbb{P}_8$, is also the biwheel $\mathbb{B}_8$.

TRUNCATED BIWHEELS

The *truncated biwheel* $\mathbb{T}_{2n}$, $n \ge 3$, is the graph obtained from a path $v_1v_2\ldots v_{2n-2}$ of length $2n-3$, by the addition of two vertices, h_1 and h_2, and

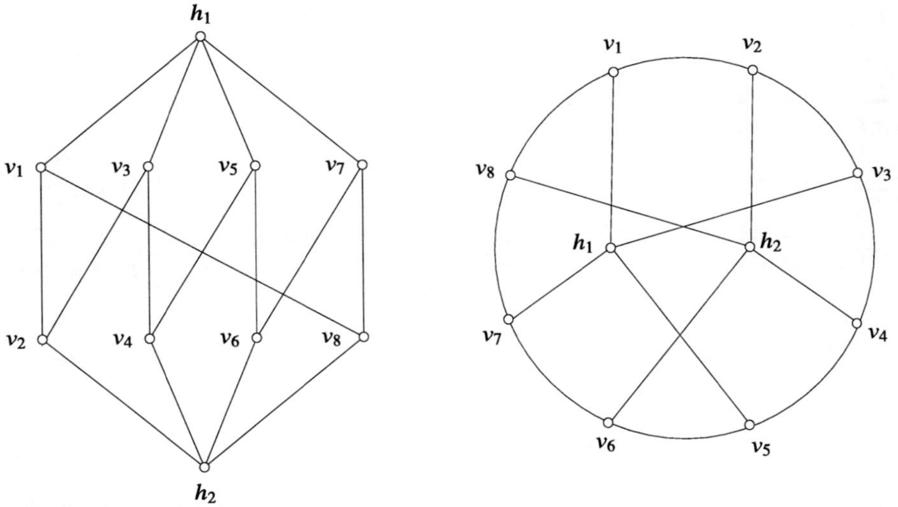

Fig. 2.9 Two drawings of the biwheel $\mathbb{B}_{10}$

by the addition of edges $h_1v_1, h_1v_3, \ldots, h_1v_{2n-3}$, edges $h_2v_2, h_2v_4, \ldots, h_2v_{2n-2}$ and edges h_1v_{2n-2} and h_2v_1. Two drawings of $\mathbb{T}_{10}$ are shown in Figure 2.10.

Note that for $n \geq 4$, the truncated biwheel $\mathbb{T}_{2n}$ is isomorphic to the graph obtained from the biwheel $\mathbb{B}_{2n}$, by deleting the edge v_1v_{2n-2} and adding the two edges h_1v_{2n-2} and h_2v_1. We remark that the graph $\overline{C_6}$ is the truncated biwheel $\mathbb{T}_6$, the triangular prism $\mathbb{P}_6$, as well as the staircase $\mathbb{S}_6$. We denote the family of truncated biwheels by $\mathcal{TB}$. Norine and Thomas [79] refer to truncated biwheels as *prismoids*.

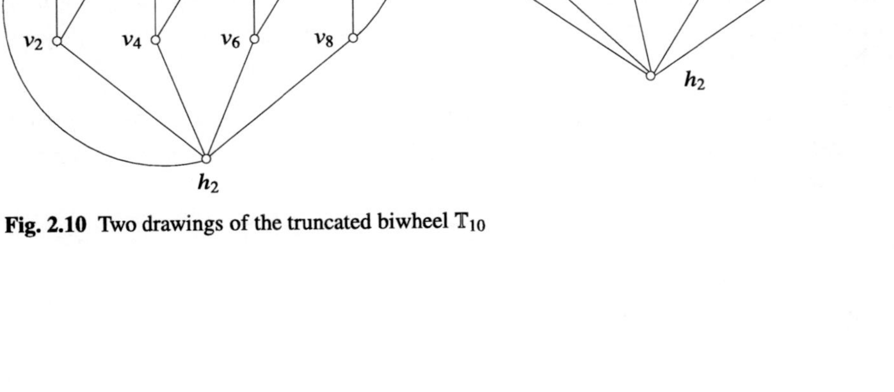

Fig. 2.10 Two drawings of the truncated biwheel $\mathbb{T}_{10}$

Exercises

2.5.1 *Show that the two definitions of the Möbius ladder $\mathbb{M}_{2n}$ given above result in the same graph (up to isomorphism).*

2.5.2 *Show that:*

(i) *the prism $\mathbb{P}_{2n}$ is bipartite when n is even and nonbipartite when n is odd,*
(ii) *the Möbius ladder $\mathbb{M}_{2n}$ is bipartite when n is odd, and nonbipartite when n is even.*

2.5.3 (Ladders and Fibonacci Numbers) *We denoted by $\Phi(G)$ the number of perfect matchings of a graph G. Show that:*

$$\Phi(\mathbb{L}_{2n}) = F(n+1), \text{ for } n \geq 2. \tag{2.4}$$

(The Fibonacci numbers $F(n)$ are the numbers in the sequence defined by the recursion $F(n) = F(n-1)+F(n-2)$, for $n \geq 3$, with initial values $F(1) = F(2) = 1$. Thus, $F(1) = 1$, $F(2) = 1$, $F(3) = 2$, $F(4) = 3$, $F(5) = 5$, $F(6) = 8$, $F(7) = 13$, etc.)

∗ **2.5.4** *Using (2.4), deduce the following:*

(i) *for $n \geq 3$,*

$$\Phi(\mathbb{P}_{2n}) = \begin{cases} F(n+1) + F(n-1) & \text{if } n \text{ is odd} \\ F(n+1) + F(n-1) + 2 & \text{if } n \text{ is even} \end{cases}$$

(ii) *for $n \geq 2$,*

$$\Phi(\mathbb{M}_{2n}) = \begin{cases} F(n+1) + F(n-1) + 2 & \text{if } n \text{ is odd} \\ F(n+1) + F(n-1) & \text{if } n \text{ is even} \end{cases}$$

(iii) *for $n \geq 2$,*

$$\Phi(\mathbb{S}_{2n}) = F(n) + 2.$$

2.5.5 *Contraction of a triangle in the staircase $\mathbb{S}_{2n}$ results in the staircase $\mathbb{S}_{2(n-1)}$. Using this argument, write a proof that $\Phi(\mathbb{S}_{2n}) = F(n) + 2$, for $n \geq 2$, by induction on the order of $\mathbb{S}_{2n}$.*

2.5.6 *Show that $\Phi(\mathbb{B}_{2n}) = (n-1)^2$. (Hint: Show that any edge incident with hub h_1, and any edge incident with hub h_2 are contained together in a unique perfect matching.*

2.5.7 *Contraction of a triangle in the truncated biwheel $\mathbb{T}_{2n}$ results in the truncated biwheel $\mathbb{T}_{2(n-1)}$ plus an edge joining the two hubs. Use this argument and induction to deduce that*

$$\Phi(\mathbb{T}_{2n}) = \frac{n(n-1)}{2} + 1, \quad n \geq 3.$$

2.5.8 *For each even integer $n \geq 6$, let A_n be the graph obtained from a matching $\{u_1u_2, u_3u_4, \ldots, u_{n-3}u_{n-2}\}$, by adjoining two vertices, h_1 and h_2, and then joining h_1 to each of $u_1, u_3, \ldots, u_{n-3}$ by a pair of multiple edges, and h_2 to each of $u_2, u_4, \ldots, u_{n-2}$ by a pair of multiple edges. Figure 2.11 shows a drawing of A_{10}.*

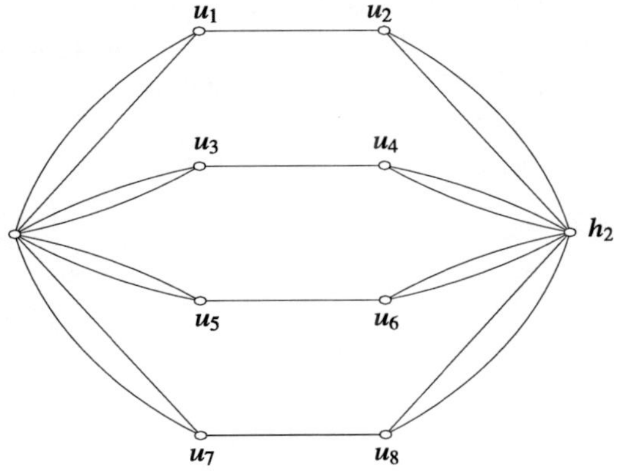

Fig. 2.11 The graph A_{10}

Show that:

(i) *the graph A_n is a bipartite matching covered graph for each even integer $n \geq 6$, and*

(ii) $\Phi(A_n) = 2n - 4$.

2.6 Notes

In the treatise by Lovász and Plummer [59], matching covered graphs are referred to as 1-*extendable graphs* as any edge of such a graph can be 'extended' to a perfect matching of the graph. (More generally, a connected graph G is k-*extendable* if any set of k edges which constitutes a matching can be extended to a perfect matching of G.) The term 'matching covered graphs' has been widely accepted since the appearance of Lovász's seminal paper (1987, [58]).

A matchable graph is *elementary* if the set of matchable edges in it induces a connected subgraph. Every matching covered graph is elementary, but not every elementary graph is matching covered. However a connected bipartite graph is matching covered if and only if it is elementary. In Chapters 4 and 5 in Lovász and Plummer's book [59] many results about matching covered graphs are derived as special cases of results about elementary graphs. In this work, we have chosen to confine ourselves to matching covered graphs.

CLM (2013, [17]) showed that, for each even integer $n \geq 14$, any bipartite matching covered graph G_n on n vertices with $\delta(G_n) \geq 3$ has at least $2n-4$ perfect matchings and that the graph A_n described in Exercise 2.5.8 achieves this lower bound. We do not know the best possible lower bound to $\Phi(G_n)$ if we require, as an additional hypothesis, that G_n be simple. In particular, we do not know any infinite family $\{G_n\}$ of simple bipartite matching covered graphs of minimum degree at least three such that $\Phi(G_n)$ grows linearly. The number of perfect matchings of biwheels grows quadratically.

For a special class of bipartite graphs known as braces, which will be introduced later, we have shown that the number of perfect matchings grows quadratically with the number of vertices. This result also appears in the same paper mentioned above. See the Notes Section of Chapter 6 for a related conjecture.

Chapter 3
Canonical Partitions

Contents

3.1 Barriers in Matchable Graphs

A barrier B in a matchable graph G is *maximal* if there is no other barrier of G that properly contains B. The barrier $B = \{u_1, u_2, u_3\}$ in the graph shown in Figure 2.1 is not maximal because there are barriers of the graph which properly contain B. It can be verified that $\{v_1, u_1, u_2, u_3, u_4, u_5\}$ is a maximal barrier of that graph.

The objective of this chapter is to establish one of the basic results of our subject, which states that the maximal barriers of a matching covered graph constitute a partition of its vertex set[1]. Our strategy is to first verify its validity for bipartite graphs, and then deduce the general result from this special instance.

[1] According to Lovász and Plummer [59, Chapter 5], this result seems to have originated in the works of Kotzig (that appeared in a Slovenian journal in 1959 and 1960) and further developed by Lovász (in a paper that appeared in 1972).

C. L. Lucchesi, U. S. R. Murty, *Perfect Matchings*, Algorithms and Computation in Mathematics 31, https://doi.org/10.1007/978-3-031-47504-7_3

3.1.1 Properties of barriers of matchable graphs

We record below two properties of barriers of matchable graphs which are frequently used. We leave the proof of their validity as Exercise 3.1.1.

Proposition 3.1 *Let B be a barrier in a matchable graph G, and let M be any perfect matching of G. Then:*

(i) *if K is an odd component of $G - B$, then $M \cap \partial(K)$ has precisely one edge; and if v is the end of that edge in $V(K)$, then $M \cap E(K)$ is a perfect matching of $K - v$, and*

(ii) *if L is an even component of $G - B$, then $M \cap E(L)$ is a perfect matching of L and no edge in $\partial(L)$ is matchable in G.* □

3.1.2 Maximal barriers in matchable graphs

Theorem 3.2 establishes an interesting and useful property of maximal barriers. (Recall that a graph G is critical if $G - v$ is matchable for every vertex v of G. See Exercises 1.3.8 and 1.3.9.)

MAXIMAL BARRIERS AND CRITICAL SUBGRAPHS

Theorem 3.2 *Let B be a maximal barrier in a matchable graph G. Then each component of $G - B$ is a critical graph. (In particular, $G - B$ has no even components.)*

Proof Let J be any component of $G - B$. We consider separately the cases (i) J is even and (ii) J is odd but not critical. If J is even, then let $B' := B + v$, where v is a vertex of J; it follows that $o(G - B') \geq |B'|$. Alternatively, if J is odd but not critical, there exists a nonempty subset S of $V(J)$ such that $o(J - S) \geq |S| + 1$ (by Exercises 1.2.1 and 1.3.9). Now consider the set $B' = B \cup S$. It can be verified that $o(G - B') \geq |B'|$. In both alternatives, $o(G - B') \geq |B'|$. However, as G is matchable, we must in fact have $o(G - B') = |B'|$, implying that B' is a barrier of G, and contradicting the maximality of B. The assertion follows. □

In fact, the converse of Theorem 3.2 is true (Exercise 3.1.4):

Theorem 3.3 *Let B denote a barrier of a matchable graph G. If every component of $G - B$ is critical then B is maximal.* □

3.1.3 Maximal barriers in bipartite matching covered graphs

Let S be any maximal barrier of a matching covered graph G. Then each component of $G - S$ is a critical graph by Theorem 3.2. In the special case in which S is a maximal barrier in a bipartite graph $G[A, B]$, this means that all components of $G - S$ are trivial, because no nontrivial bipartite graph is critical. As a consequence we have the following simple result.

Theorem 3.4 *The only two maximal barriers in a bipartite matching covered graph $G := G[A, B]$ are its two parts A and B.*

Proof Let S be any maximal barrier of G. Since G is matching covered, S is a stable set. By the observation made above, each component of $G - S$ is trivial. Thus, $\overline{S} := V(G) - S$ is another stable set, and $(S, \overline{S})$ is a bipartition of G. However, the graph G, being a nontrivial connected bipartite graph, has a unique bipartition (see Exercise 3.1.5), implying that S is equal to either A or to B. □

As any barrier is a subset of a maximal barrier, we have:

Corollary 3.5 *Every barrier in a bipartite matching covered graph $G[A, B]$ is a subset of one of the two sets A and B.* □

Another consequence of Theorem 3.4 is the following result which will prove to be very useful in the proof of Theorem 3.11.

Corollary 3.6 *Let $G := G[A, B]$ be a bipartite matching covered graph, and let a and b any two vertices of G with $a \in A$ and $b \in B$. Then $G - a - b$ is matchable.*

Proof Suppose that $G - a - b$ is not matchable. Then, by Theorem 2.1, there exists a barrier of G which contains both a and b. By Corollary 3.5, this is impossible. □

As a straightforward consequence of Corollary 3.6, we have the following useful result:

Corollary 3.7 *Let $G := G[A, B]$ be a bipartite matching covered graph, and let a and b be any two vertices of G with $a \in A$ and $b \in B$. Then the graph $G + ab$, obtained from G by adding an edge joining a and b is also a bipartite matching covered graph with bipartition $[A, B]$.* □

In fact, the converse of Corollary 3.6 also holds. In addition to the Hall-type characterization of bipartite matching covered graphs (Theorem 2.9), there is the following additional characterization:

A SECOND CHARACTERIZATION OF BIPARTITE MATCHING COVERED GRAPHS

Theorem 3.8 *A bipartite graph $G = G[A, B]$ which has at least one edge is matching covered if and only if $G - a - b$ is matchable, for every $a \in A$ and every $b \in B$.*

The proof of Theorem 3.8 is left as Exercise 3.1.7.

Exercises

3.1.1 *Prove Proposition 3.1.*

3.1.2 *Starting with the barrier B in the graph shown in Figure 2.1, and applying the idea of the proof of Theorem 3.2, find all the maximal barriers containing B.*

$*$**3.1.3** *Draw the graph $K_{3,3} \odot K_{3,3}$ and list all the barriers in it.*

$*$**3.1.4** *Prove Theorem 3.3. Hint: use the statement of Exercise 1.3.9 to prove that no proper superset of B is a barrier of G.*

$\triangleright$**3.1.5** *A graph G is said to have a unique bipartition if its vertex set V has a unique partition into two stable sets. Show that a nonempty graph is a connected bipartite graph if and only if it has a unique bipartition.*

$\triangleright$**3.1.6** *Prove Corollary 3.7.*

$*$**3.1.7** *Prove Theorem 3.8.*

3.2 Barriers and Cores

Although Theorem 3.4 is specific to bipartite graphs, it can be applied to prove that, in any matching covered graph, the maximal barriers of the graph partition its vertex set. We now proceed to introduce the device that makes this generalization possible.

> CORE WITH RESPECT TO A BARRIER
>
> For a barrier B of a matchable graph G, the bipartite graph $\mathbb{H}(B)$ is defined as the graph obtained by deleting all edges with both ends in B, deleting all vertices in the even components of $G - B$, and then shrinking the vertex set of each odd component of $G - B$ to a single vertex. We shall refer to this bipartite graph $\mathbb{H}(B)$ as the *core of G* with respect to the barrier B. See Example 3.9 for an illustration.

Example 3.9 Let G be the graph shown in Figure 3.1(a), and let B be the barrier $\{u_1, u_2, u_3\}$. Then $G - B$ has one even component, and three odd components. The core $\mathbb{H}(B)$ with respect to the barrier B is shown in Figure 3.1(b), where a_1 is the vertex resulting from shrinking $\{v_3, v_4, v_5\}$, a_2 is the vertex resulting from shrinking $\{v_6\}$, and a_3 is the vertex resulting from shrinking $\{u_4, u_5, v_7, v_8, v_9\}$.

It is an easy matter to verify that $\mathbb{H}(B)$ is matchable, and that it is matching covered when G is matching covered (see Exercise 3.2.1). This bipartite graph will play a crucial role in our approach to the subject. The proof of Theorem 3.11 provides a good illustration of its usefulness!

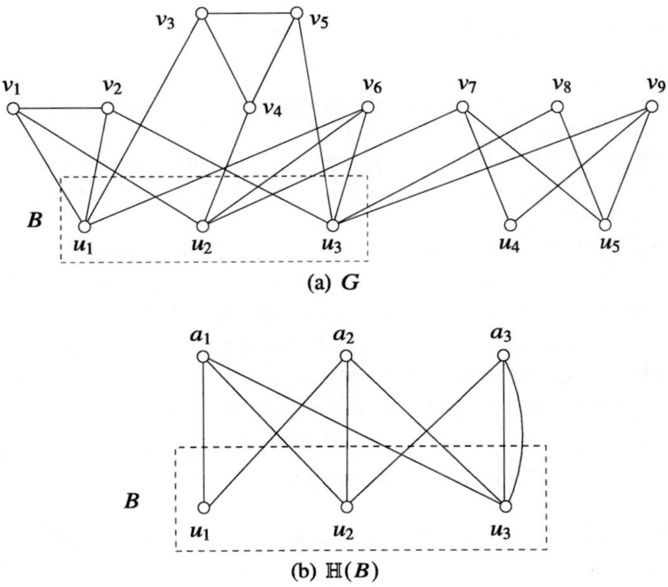

Fig. 3.1 The core $\mathbb{H}(B)$ of G with respect to the barrier B

Exercises

3.2.1 *Let G be a matchable graph and let B denote a barrier of G. Show that the core $\mathbb{H}(B)$ with respect to B is matchable, and that if G is matching covered then $\mathbb{H}(B)$ is a bipartite matching covered graph. Hint: use Proposition 3.1.*

∗**3.2.2**

(i) *Show that if a bipartite nontrivial graph $H := H[A, B]$ has a unique perfect matching, then at least one vertex in A, and at least one in B, have degree one in H. (Hint: Let M be the only perfect matching in H, and let P be a maximal M-alternating path. Show that the ends of P have degree one in H and show also that P starts and ends with an edge in M.)*

(ii) *Deduce that a 2-edge-connected matchable graph G has at least two perfect matchings. (Hint: Consider the core $\mathbb{H}(B)$ of G with respect to any maximal barrier B of G.)* *(Kotzig)*

(iii) *Show that if a bipartite graph $H := H[A, B]$ has a unique perfect matching M, then there is an enumeration $a_1, a_2, \ldots, a_k$ of the vertices of A and an enumeration $b_1, b_2, \ldots, b_k$ of the vertices of B such that (i) $a_i b_i \in M$, for $i = 1, 2, \ldots, k$ and (ii) $a_i b_j \notin E(H)$ for $1 \le i < j \le k$.*

3.3 Canonical Partitions

We are now ready to prove the result mentioned in the beginning of this chapter.

Lemma 3.10 *Let B be a maximal barrier of a matching covered graph G, and let $a \notin B$, and $b \in B$ be any two vertices of G. Then $G - a - b$ is matchable.*

Proof Since $a \notin B$, it follows that it belongs to an odd component, say K, of $G - B$. Now consider the core $\mathbb{H}(B)$ of G relative to B. Let a_K denote the vertex in $\mathbb{H}(B)$ resulting from the shrinking of the odd component K. It follows from Corollary 3.6 that $\mathbb{H} - a_K - b$ is matchable. Let N be a perfect matching in $\mathbb{H} - a_K - b$. Using the fact that each component of $G - B$ is critical, it is now easy to show that N can be extended to a perfect matching of $G - a - b$ (Exercise 3.3.1). □

THE CANONICAL PARTITION THEOREM

Theorem 3.11 *The maximal barriers of a matching covered graph G partition its vertex set.*

Proof Since every singleton is a barrier, every vertex of G is contained in some maximal barrier of G. So, to establish the assertion, it suffices to show that maximal barriers of G are pairwise disjoint. With this objective in mind, let B and B' be any two distinct maximal barriers of G, and suppose that $b \in B \cap B'$. The fact that B' is maximal and is distinct from B implies that there must exist a vertex a in $B' - B$. By Lemma 3.10, it follows that the graph $G - a - b$ is matchable. However, because a and b both belong to the barrier B', this is impossible by Theorem 2.1. □

CANONICAL PARTITION

The partition of the vertex set V of a matching covered graph G into maximal barriers is called its *canonical partition.*

It follows from Corollary 3.5 that the canonical partitions of bipartite matching covered graphs have just two parts. Curiously, the canonical partition of any nonbipartite matching covered graph has at least four parts (see Exercise 3.3.5).

3.3.1 Bicritical graphs

A nontrivial graph G is *bicritical* if $G - u - v$ is matchable for any two vertices u and v of G. Bicritical graphs are a special type of matching covered graphs (see Exercise 3.3.6). By Theorem 2.1 this is the case precisely when all maximal barriers of G are singletons. The three graphs shown in Figure 2.2 are bicritical; another example is shown in Figure 3.5. Three-connected bicritical graphs, which are known as *bricks*, will play an important role in our theory.

3.3.2 Bisubdivisions

A *subdivision* of an edge $e = uv$ of a graph G consists of replacing the edge e by an uv-path so that all the internal vertices (if any) of that path have degree two in the resulting graph. As the name suggests, a subdivision of e consists of dividing e by the insertion of a certain number (possibly zero) of vertices of degree two. If the number of vertices inserted is even, then we refer to the subdivision as a *bisubdivision*. Thus, in case of a bisubdivision, we replace e by a path of odd length (with an even number of added vertices). For this reason, some authors refer to a bisubdivision of an edge as an *odd subdivision* (and some as an *even subdivision*). The following proposition is easy to verify (see Exercise 3.3.9).

Proposition 3.12 *Suppose that a graph $\tilde{G}$ is obtained by bisubdividing an edge $e = uv$ of a matching covered graph G distinct from K_2, effected by replacing the edge uv by the path $P := us_1s_2 \ldots s_{2k}v$, where $k > 0$. Then the graph $\tilde{G}$ is also matching covered. And if B_u and B_v are the maximal barriers of G containing u and v, respectively (see Figure 3.2), then:*

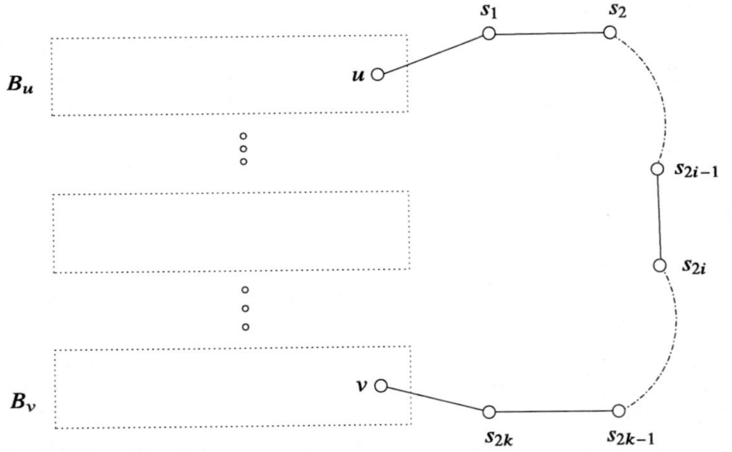

Fig. 3.2 Bisubdivision of an edge $e = uv$

(i) *all parts of the canonical partition of G, except B_u and B_v, are also parts in the canonical partition of $\tilde{G}$, and*
(ii) *$\tilde{B}_u = B_u \cup \{s_2, s_4, \ldots, s_{2k}\}$ and $\tilde{B}_v = B_v \cup \{s_1, s_3, \ldots, s_{2k-1}\}$ are the maximal barriers containing u and v, respectively, in the canonical partition of $\tilde{G}$.* □

A *bisubdivision* of a graph G is a graph obtained from G by bisubdividing each edge in a subset of E. (We regard a graph as a bisubdivision of itself.) Figure 3.3 shows a spanning subgraph of the Petersen graph which is a bisubdivision of K_4. (The edges of this subgraph are indicated by solid lines.)

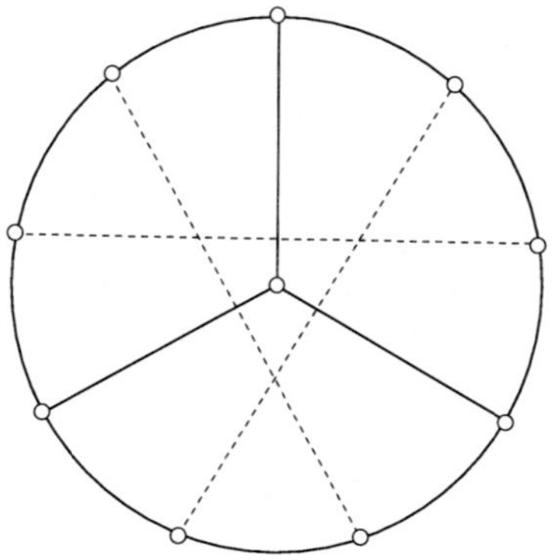

Fig. 3.3 A bisubdivision of K_4 contained in the Petersen graph

If $\tilde{G}$ denotes a bisubdivision of a matching covered graph G, then by the above proposition, the graph $\tilde{G}$ is also matching covered. We refer to the vertices in $\tilde{G}$ which are also vertices of G as the *branch vertices* of $\tilde{G}$, and those vertices of $\tilde{G}$ which are not vertices of G as the *subdivision* vertices.

We conclude this subsection by noting that the graph obtained by subdividing an edge e in a graph G may be constructed as the result of splicing G with a graph H whose underlying simple graph is an even cycle (see Exercise 3.3.11).

Exercises

$*$**3.3.1** *In the proof of Lemma 3.10, supply the missing details in showing that N can be extended to a perfect matching of $G - a - b$.*

3.3.2 *Let G be a matching covered graph and let u and v be two vertices of G such that $G + uv$ is matching covered. Prove that the canonical partition of $G + uv$ is a refinement of the canonical partition of G in the sense that every maximal barrier of $G + uv$ is a subset of a maximal barrier of G.*

3.3.3 *Determine the canonical partition of the matching covered graph shown in Figure 3.4 (which is obtained by splicing $K_{3,3}$ with K_4).*

3.3.4 *Determine the canonical partitions of the following graphs:*

(i) *the wheel W_{2k+1}, $k \geq 1$,*
(ii) *$W_{2k+1} - e$, where $k \geq 2$, and e is a spoke of W_{2k+1}.*

$*$**3.3.5** *Show that the canonical partition of a nonbipartite matching covered graph has at least four parts. For each integer $r \geq 4$, give an example of a matching covered graph G whose canonical partition consists of precisely r parts.* *(Carvalho)*

$\triangleright$**3.3.6** *Prove that every bicritical graph is matching covered.*

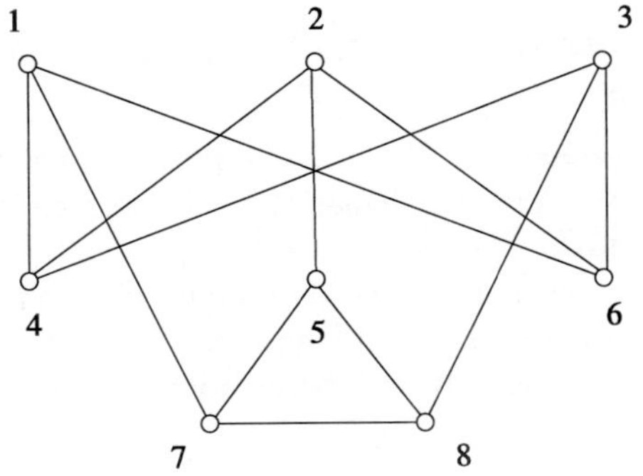

Fig. 3.4 Figure for Exercise 3.3.3

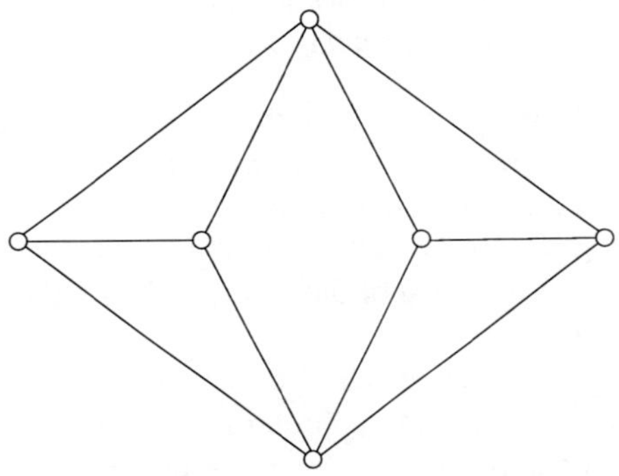

Fig. 3.5 A bicritical graph of order six

3.3.7 *Show that the graph shown in Figure 3.5 is bicritical.*

3.3.8 *Show that a graph obtained by splicing together two bicritical graphs is also bicritical.*

▷**3.3.9** *Prove Proposition 3.12.*

∗**3.3.10** *Let G be a matching covered graph, and let $\{u_1, v_1\}$ and $\{u_2, v_2\}$ be any two pairs of nonadjacent vertices of G. State and prove the necessary and sufficient conditions under which it is possible to add two new edges $e_1 = u_1v_1$ and $e_2 = u_2v_2$ to G such that neither $G + e_1$ nor $G + e_2$ is matching covered, but $G + e_1 + e_2$ is.*

3.3.11 *Let e be an edge of a graph G incident with a vertex u of degree $k > 2$, and let H be a graph, disjoint from G, which is obtained from a $2r$-cycle, $r \geq 2$, by replacing one of its edges vv' by $k - 1$ multiple edges. (Thus, in H, the vertices v and v' are adjacent and have degree k; the rest of the vertices have degree two.) Find a suitable bijection θ between $\partial_G(u)$ and $\partial_H(v)$ so that $(G \odot H)_{u,v,\theta}$ is isomorphic to the graph obtained from G by subdividing e by the insertion of $2r - 2$ vertices.*

3.4 Ear Decompositions of Bipartite Graphs ♯

Given a matching covered graph H, when is it possible to add an edge joining two vertices u and v of H by a new edge so that the resulting graph $H+uv$ is also matching covered? The answer to this question is provided by Lemma 3.10 and Theorem 3.11: the graph $H+uv$ is matching covered if and only if u and v belong to different maximal barriers in the canonical partition of H.

Corollary 3.7 implies the following observation concerning bipartite matching covered graphs:

Proposition 3.13 *Let $H[A, B]$ be a spanning bipartite matching covered subgraph of a bipartite matching covered graph $G[A, B]$. Then G can be obtained from H by adding the edges in $E(G)-E(H)$, one at a time, in any order, so that all intermediate graphs in the sequence are also matching covered.* □

The above result does not extend to nonbipartite graphs. Consider for example the spanning subgraph of the Petersen graph shown in Figure 3.3. There are three edges of the Petersen graph (indicated by dotted lines) which do not belong to this subgraph. None of those three edges can be added to the subgraph to obtain a matching covered graph. (We have to add at least two of those three edges simultaneously to obtain a matching covered graph.)

Now let us suppose that H is a matching covered subgraph of a matching covered graph G, with $V(H) \subset V(G)$, and suppose that we wish to obtain G from H by means of elementary operations which preserve matching coveredness. Since the addition of one or more edges to H does not result in a graph of order larger than that of H, it is also necessary to employ an operation that augments the number of vertices. An obvious candidate for such an operation is bisubdivision.

> **EAR**
>
> A (single) *ear* in a matching covered graph is a path of odd length[2] all of whose internal vertices have degree two.

The idea of adding an ear to a matching covered graph combines the ideas of edge additions and bisubdivisions.

> **EAR PATH**
>
> Suppose H is a subgraph of a matching covered graph G. An odd path $P := us_1s_2\cdots s_{2k}v$ in G connecting two vertices u and v of H is called an *ear path of H in G* if it is internally disjoint from H. If P is an ear path of H in G, then it is an an ear in the graph $H \cup P$. For this reason, we say that $H + P := H \cup P$ is obtained from H by *adding the ear P* to H.

[2] On occasion, we shall use the term ‘even ear’ in a graph to mean a path of even length. Unless otherwise specifically stated, the term ‘ear’ should be taken to mean a path of odd length.

Clearly, the above-defined operation is tantamount to first adding to H an edge joining u and v, and then bisubdividing that edge by inserting new vertices $s_1, s_2, \ldots, s_{2k}$. When H is matching covered, any graph obtained from H by adding an ear connecting two vertices in different maximal barriers of H is also matching covered.

A natural question that arises is if it is possible to generate all matching covered graphs by starting with the smallest matching covered graph, namely K_2, and sequentially adding ears (with the proviso that all intermediate graphs in the sequence are also matching covered). However, it follows from Proposition 3.12 that adding an ear to a bipartite matching covered graph results in another bipartite matching covered graph. Thus it is not possible to generate nonbipartite matching covered graphs in the above described manner. We shall however see that this is possible in case of bipartite graphs.

Before formally defining what an ear decomposition of a bipartite graph is, let us consider the example shown in Figure 3.6. The paths

$$P_1 := 12, \qquad P_2 := 1342, \qquad P_3 := 1562,$$
$$P_4 := 3874, \qquad P_5 := 58, \quad \text{and} \qquad P_6 := 67,$$

are paths of odd length in the cube and

$$G_i := P_1 \cup P_2 \cup \cdots \cup P_i, \quad \text{for } 1 \leq i \leq 6.$$

The graph G_1 is K_2 and, for $2 \leq i \leq 6$, the graph G_i is obtained from G_{i-1} by adding the ear P_i.

EAR DECOMPOSITION OF A BIPARTITE GRAPH

A *bottom-up ear decomposition* of a bipartite matching covered graph $G[A, B]$ or, simply, an *ear decomposition* of $G[A, B]$), is a sequence

$$G_1 \subset G_2 \subset \cdots \subset G_r$$

of matching covered subgraphs of G, where

(i) $G_1 = K_2$, $G_r = G$ and,
(ii) for $2 \leq i \leq r$, $G_i = G_{i-1} + P_i$, where P_i is an ear of G_i in G.

Note that if $P := us_1s_2 \cdots s_{2k}v$ is a path of odd length, then the set $E_{even}(P) := \{s_1s_2, s_3s_4, \ldots, s_{2k-1}s_{2k}\}$ of even numbered edges of P is a perfect matching of $P - \{u, v\}$ (when P is a path of length one, then $E_{even}(P)$ is empty). It follows that if a bipartite matching covered graph G has an ear decomposition as described above, where e is the unique edge of G_1, then

$$M_e := \{e\} \cup E_{even}(P_2) \cup \cdots \cup E_{even}(P_r)$$

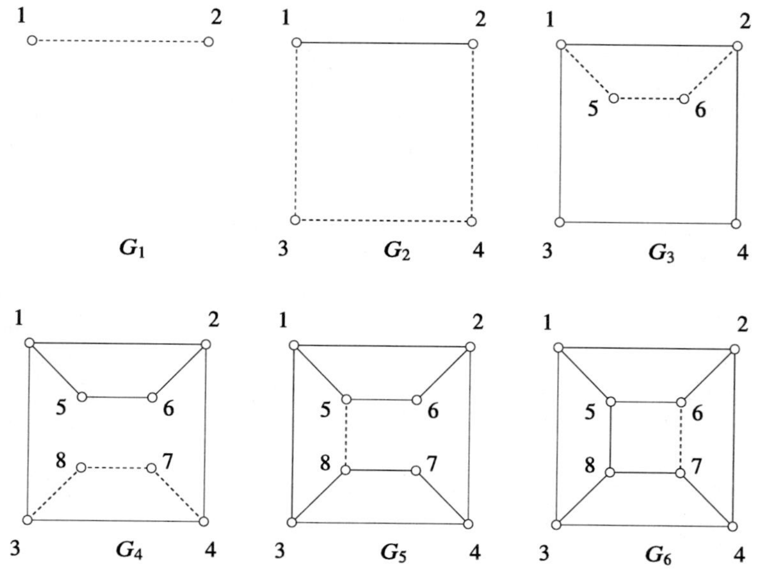

Fig. 3.6 An ear decomposition of the cube

is a perfect matching of G such that, for $1 \leq i \leq r$, the set $M_e \cap E(G_i)$ is a perfect matching of G_i. This observation is the basis of the proof of the following fundamental theorem.

THEOREM: EAR DECOMPOSITION OF A BIPARTITE GRAPH

Theorem 3.14 *Every bipartite matching covered graph $G[A, B]$ has an ear decomposition.*

Proof Let e be any edge of G, and let M_e be any fixed perfect matching containing e. Take G_1 to be the subgraph of order two whose only edge is e. Assume, inductively, that a partial ear decomposition $G_1 \subset G_2 \subset \cdots \subset G_i$ of G has been constructed for some $i \geq 1$, with the property that the restriction of M_e to $E(G_i)$ is a perfect matching of G_i.

If $G_i = G$, we have the required ear decomposition of G with $i = r$. So, suppose that $G_i \subset G$. In this case, since G is connected, there must exist an edge f belonging to $E(G) - E(G_i)$ which has at least one end in $V(G_i)$. Now let M_f be any perfect matching of G containing f. Such a perfect matching must exist because G is matching covered. Furthermore, $M_f \neq M_e$ because $M_e \cap E(G_i)$ is a perfect matching of G_i, and f has an end in $V(G_i)$ but is not in $E(G_i)$.

Consider the (M_f, M_e)-alternating cycle starting with the edge f, and let g denote the edge with which this cycle re-enters $V(G_i)$ for the first time. (If f has both its ends in $V(G_i)$, then $g = f$.) This edge g must also be in the perfect matching M_f. Thus the segment of the (M_f, M_e)-alternating cycle starting with f and ending with g is a path of odd length which has its ends in $V(G_i)$, but which is internally disjoint from G_i.

Now, by taking this path to be P_{i+1}, we obtain the (matching covered) subgraph $G_{i+1} := G_i + P_{i+1}$ of G such that $M_e \cap E(G_{i+1})$ is a perfect matching of G_{i+1}. Thus we can extend the given partial decomposition of G by appending G_{i+1}. This process can clearly be continued until we obtain an ear decomposition of G. □

The above proof implicitly contains an algorithm for finding an ear decomposition of G. The fact that, for $1 \leq i < r$, G_i can be extended to G_{i+1} by adding an ear is guaranteed by the fact that M_e is a perfect matching of G such that its restriction to G_i is a perfect matching of G_i. For this reason we refer to M_e as the *reference matching*; it may be chosen arbitrarily but, once chosen, it guides the process of finding an ear decomposition.

A bipartite graph need not have a unique ear decomposition. (For example, it is easy to find an ear decomposition of the cube in which the second graph G_2 is a hamiltonian cycle in it.) However, it is easy to see that the number of subgraphs in an ear decomposition is independent of the choices of ears in obtaining the decomposition (Exercise 3.4.2).

Theorem 3.15 *In any ear decomposition $(G_1, G_2, \ldots, G_r)$ of a bipartite matching covered graph G,*

$$r = m - n + 2, \quad \text{where } m = |E| \text{ and } n = |V|.$$

We conclude this chapter with a simple application of the above theorem.

Corollary 3.16 *Any bipartite matching covered graph G has at least*

$$m - n + 2 \qquad \text{distinct perfect matchings.}$$

Proof Let $K_2 = G_1, G_2, \ldots, G_r = G$ be an ear decomposition of G. We prove the assertion by induction on r. The assertion holds if $G = K_2$. Assume that $G \neq K_2$. Let P_r denote the ear added to G_{r-1} to obtain $G_r = G$. By induction, G_{r-1} has at least

$$m_{r-1} - n_{r-1} + 2 \quad \text{distinct perfect matchings,}$$

where $m_{r-1} = |E(G_{r-1})|$ and $n_{r-1} = |V(G_{r-1})|$. Each of these perfect matchings can be extended to a perfect matching of G by adding the edges in the set of even numbered edges of P_r. In addition, the set of odd numbered edges of P_r are contained in at least one perfect matching of G. (The restriction of such a perfect matching of G to $E(G_{r-1})$ cannot be a perfect matching of G_{r-1}.) Thus, altogether, G has at least

$$m_{r-1} - n_{r-1} + 2 + 1 = m - n + 2$$

distinct perfect matchings. □

The proof of the above theorem illustrates how the ear decomposition procedure for bipartite graphs can be used as an inductive tool. There is an ear decomposition procedure which is applicable to all matching covered graphs where simultaneous addition of two disjoint ears is permitted. Establishing the existence of such a procedure is much harder and will be discussed in Chapter 11. In the next chapter we turn to a different type of decomposition which is of central importance in this theory.

Exercises

3.4.1 *Show that any bipartite matching covered graph G with minimum degree at least three has an edge e such that $G - e$ is matching covered.*

Give an example to show that the above statement is invalid without the assumption that G is bipartite.

3.4.2 *Give a proof of Theorem 3.15.*

> **CONFORMAL SUBGRAPHS**
>
> A matchable subgraph H of a matchable graph G is *conformal* if the graph $G - V(H)$ has a perfect matching (equivalently, if any perfect matching of H may be extended to a perfect matching of G).

▷**3.4.3**

(i) *Prove that if G_1 is a conformal subgraph of G_2 and if G_2 is a conformal subgraph of G_3 then G_1 is also a conformal subgraph of G_3.*
(ii) *Find all the conformal cycles in the graph shown in Figure 3.7.*

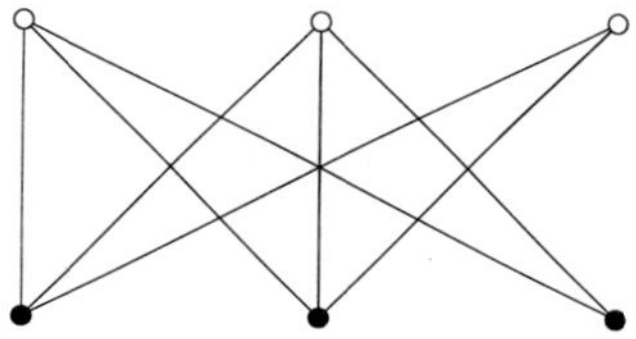

Fig. 3.7 Graph for Exercise 3.4.3.

(iii) *Find all conformal cycles in the Petersen graph.*

∗**3.4.4** *Let H be a matching covered subgraph of a bipartite matching covered graph G. Show that any ear decomposition of H may be extended to an ear decomposition of G if and only if H is a conformal subgraph of G.*

▷**3.4.5** *Let H be any conformal matching covered subgraph of a matching covered graph G (which may or may not be bipartite), and let g be an edge in $E(G) - E(H)$ which has an end in $V(H)$. Suppose that M is a perfect matching of G such that $M \cap E(H)$ is a perfect matching of H, let N be a perfect matching of G containing the edge g, and let C be an (N, M)-alternating cycle containing the edge g. Show that $H \cup C$ is a conformal matching covered subgraph of G.*

> **CONFORMAL MINORS**
>
> A matching covered graph J is a *conformal minor* of another matching covered graph G if some bisubdivision H of J is a conformal subgraph of G. (Figure 3.3 shows that K_4 is a conformal minor of the Petersen graph.)

▷**3.4.6** *Show that the triangular prism (the graph $\overline{C_6}$) is not a conformal minor of the Petersen graph $\mathbb{P}$.*

3.5 Notes

We shall see in Chapter 6 that the set of perfect matchings obtained as in the proof of Corollary 3.16 have a special significance, namely that their incidence vectors form a basis for the vector space generated by the set of incidence vectors of perfect matchings of G (see Exercise 6.3.3).

Conformal subgraphs are known by various names in the literature. Lovász and Plummer [59] referred to them as 'nice' subgraphs, and Robertson, Seymour and Thomas [82] called them 'central' subgraphs. McCuaig [68] had the more descriptive name 'well-fitted' subgraphs. The term 'conformal' subgraphs that we are using (and have used in several of our papers) is in the same spirit as McCuaig's terminology.

The notion of a conformal minor of a matching covered graph (and related notions which will be introduced in due course) plays a role in this theory that is akin to that of the notion of a minor in the theory of graphs. For example, the structure of bipartite matching covered graphs which do not have $K_{3,3}$ as a conformal minor plays a pivotal role in the theory of Pfaffian orientations which will be discussed in Part III.

Chapter 4
Tight Cuts

Contents

4.1 Separating Cuts

In Chapter 2 we introduced the operation of splicing which can be used to 'combine' two matching covered graphs to obtain another matching covered graph. Here we introduce the related notion of a separating cut which leads to a way of 'decomposing' a matching covered graph into two matching covered graphs. Tight cuts are a special type of separating cuts. They play a pivotal role in many aspects of this theory, including the study of questions concerning the existence of an edge in a matching covered graph whose deletion results in another matching covered graph.

Before we proceed further, it is necessary to establish the notation and terminology we use in connection with cuts in graphs. Familiarity with these is crucial for understanding much of the rest of the book.

C. L. Lucchesi, U. S. R. Murty, *Perfect Matchings*, Algorithms and Computation in Mathematics 31, https://doi.org/10.1007/978-3-031-47504-7_4

Given a nonempty set X of vertices of a graph G, we denote by $G/(X \to x)$ the graph obtained from G by contracting X to a single new vertex x (and removing all resulting loops). If the name of the contraction vertex x is not important, we simply write G/X.

CUTS AND CUT-CONTRACTIONS

Given any cut $C := \partial(X)$ of a connected graph G, where X is a nonempty proper subset of V, we refer to the two graphs $G/(X \to x)$ and $G/(\overline{X} \to \overline{x})$ as the *C-contractions of G*, where $\overline{X} = V - X$.

With the aid of the above notation, we are now in a position to introduce the important notion of a separating cut in a matching covered graph.

SEPARATING CUT

A cut $C := \partial(X)$ in a matching covered graph G is a *separating cut* if both C-contractions G/X and $G/\overline{X}$ are also matching covered.

Figure 4.1 shows two examples of separating cuts which are indicated by dotted lines. Figure 4.1(a) depicts a separating cut in the Petersen graph, both the contractions with respect to this cut are 5-wheels. Referring to Figure 4.1(b), one of the contractions with respect to the indicated separating cut is $K_{3,3}$, and the other is K_4.

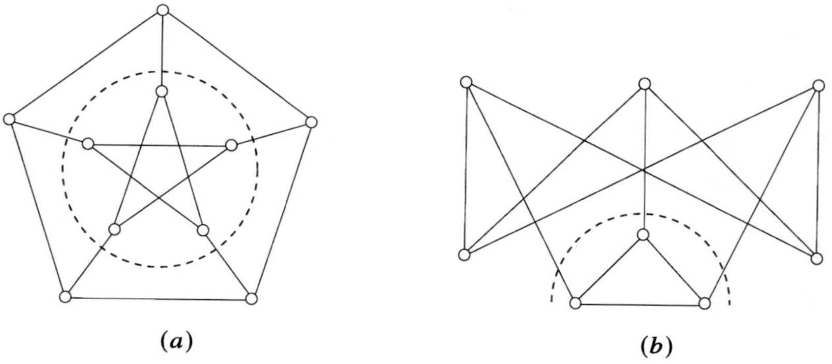

Fig. 4.1 Examples of separating cuts

Note that if G_1 and G_2 are the two C-contractions of a matching covered graph G with respect to a separating cut C, then the sets of edges incident with the contraction vertices in G_1 and G_2 are both equal to C. Thus, G can be recovered from G_1 and G_2 by splicing them together in the obvious manner.

The following assertion can be deduced from the definition of a separating cut and the fact that no matching covered graph can have a cut vertex (see Exercise 4.1.1):

Proposition 4.1 *For any separating cut C of a matching covered graph G, the subgraphs of G induced by the shores of C are both connected. In other words, C is a bond.* □

The ensuing result provides a useful characterization of separating cuts.

CHARACTERIZATION OF SEPARATING CUTS

Theorem 4.2 *A cut $C = \partial(X)$ of a matching covered graph G is separating if and only if each edge of G lies in a perfect matching that contains precisely one edge in C.*

Proof Let $G_1 := G/(\overline{X} \to \overline{x})$ and $G_2 := G/(X \to x)$ be the two C-contractions of G.

Suppose first that any edge e of G lies in a perfect matching M_e of G which meets C in exactly one edge. The graph G, a matching covered graph, is connected, hence every contraction of G is connected. In particular, G_1 is connected. Furthermore, for any edge e of G_1, the fact that $|M_e \cap C| = 1$ implies that $M_e \cap E(G_1)$ is a perfect matching of G_1 containing e. We conclude that G_1 is matching covered. Likewise, G_2 is also matching covered. Hence C is a separating cut.

Conversely, suppose that C is a separating cut. Then, by definition, both G_1 and G_2 are matching covered. Let e be any edge of G. Adjust notation so that e is an edge of G_1. Then, since G_1 is matching covered, there is a perfect matching of G_1, say M_1, that contains e. Let f (possibly equal to e) be the edge of M_1 incident with the contraction vertex $\overline{x}$ of G_1. Then f belongs to C, and thus, is an edge of G_2 incident with the contraction vertex x. Now, since G_2 is matching covered, there is a perfect matching of G_2, say M_2, that contains f. We conclude that $M_e := M_1 \cup M_2$ is a perfect matching of G that contains the edge e and has just one edge, namely f, in cut C. □

Exercises

▷**4.1.1** *Prove Proposition 4.1.*

4.1.2 *Show that in a cubic matching covered graph every 3-edge cut is a separating cut.*

4.2 Tight Cuts

TIGHT CUT

A cut C of a matching covered graph G is *tight* if $|M \cap C| = 1$, for every perfect matching M of G.

Simplest examples of tight cuts are the trivial cuts. The cut shown in Figure 4.1(b) is a nontrivial tight cut. By Theorem 4.2, every tight cut of a matching covered graph is also a separating cut. But, in general, a separating cut need not be tight. (For example, the separating cut in the Petersen graph shown in Figure 4.1(a) is not tight!) We describe below two special types of tight cuts in matching covered graphs.

4.2.1 Barrier cuts

Let B be any barrier of a matching covered graph G. It follows from Proposition 3.1 that, for each (odd) component K of $G-B$, the cut $\partial(K)$ is a tight cut in G. Tight cuts which arise in this manner are known as *barrier cuts*. Every trivial cut is a barrier cut. The cut shown in Figure 4.1(b) is an example of a nontrivial barrier cut. Since bicritical graphs (defined in Section 3.3) do not have nontrivial barriers, the trivial cuts are the only barrier cuts in a bicritical graph.

Special barrier cuts

Recall that a barrier B in a matching covered graph G is special if $G - B$ has precisely one nontrivial odd component. A tight cut C of G is a *special barrier cut* if there is some special barrier B of G such that $C = \partial(K)$, where K is the unique nontrivial component of $G - B$. The cut C shown in Figure 4.2 is the special barrier cut associated with the barrier $\{u, v, w, x\}$.

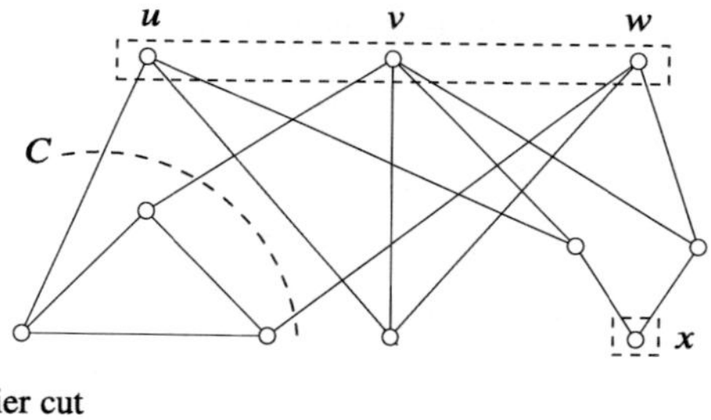

Fig. 4.2 A special barrier cut

We leave the proof of the following simple proposition as Exercise 4.2.1.

Proposition 4.3 *A tight cut C in a matching covered graph of order four or more is a special barrier cut if and only if at least one of the two shores of C is bipartite.* □

4.2.2 2-Separation cuts

A *2-separation* of a matching covered graph G is a 2-vertex cut $\{u, v\}$ of G that is not a barrier. If $\{u, v\}$ is a 2-separation of G, then each component of $G - u - v$ is even. Suppose that H_1 is the union of a nonempty proper subset of the components of $G - u - v$, and H_2 is the union of the remaining components of $G - u - v$. Then H_1 and H_2 are two nonempty vertex-disjoint even order subgraphs whose union is $G - u - v$.

With any such expression of $G-u-v$ as the union of H_1 and H_2, we may associate the cuts $C := \partial(V(H_1) + v)$ and $D := \partial(V(H_2) + v)$, and it is easy to see that:

4.4 *The two cuts C and D defined above are tight (Exercise 4.2.2).* □

Tight cuts of a graph which arise in this manner are known as *2-separation cuts* of G. Figure 4.3 depicts a 2-separation $\{u, v\}$ in a matching covered graph G, a partition $\{H_1, H_2\}$ of $G - u - v$ and the pair $\{C, D\}$ of 2-separation cuts associated with the partition $\{H_1, H_2\}$.

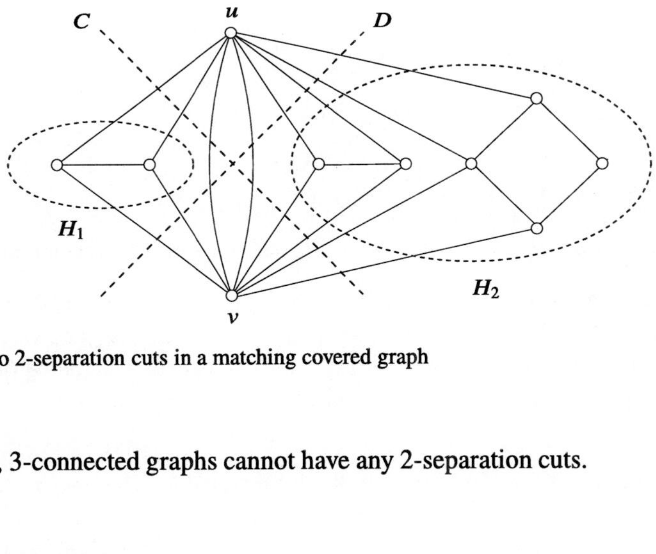

Fig. 4.3 Two 2-separation cuts in a matching covered graph

Clearly, 3-connected graphs cannot have any 2-separation cuts.

4.2.3 Tight 3-edge cuts ♯

A 3-edge cut of a graph is an edge cut with three edges. A 3-edge cut in a matching covered graph need not be tight. (For example, the graph $\overline{C_6}$ has a 3-edge cut that is not tight.) However, those 3-edge cuts which happen to be tight satisfy the following interesting and useful property.

Theorem 4.5 *In a matching covered graph, every 3-edge cut that is tight is a barrier cut.*

Proof Let $C := \partial(X)$ be a tight 3-edge cut of a matching covered graph G. Let

$$C := \{u_i v_i : 1 \leq i \leq 3, u_i \in X, v_i \in \overline{X}\}.$$

Case 1 *The cut C is not a matching.*

Without loss of generality, we may assume that the vertices u_i, $i = 1, 2, 3$ are not pairwise distinct. If the three vertices coincide then C is trivial, hence a barrier cut. If two of them are the same, adjust notation so that $u_1 = u_2 \neq u_3$. The shore X of C, being a shore of a tight cut, is odd. Thus $\{u_1, u_3\}$ is a barrier of G where $X - \{u_1, u_3\}$ and $\overline{X}$ are the sets of vertices of the two components of $G - u_1 - u_3$. In both alternatives, the ends of the edges of C in X constitute a barrier of G and C is a barrier cut associated with that barrier.

Case 2 *The cut C is a matching. See Figure 4.4.*

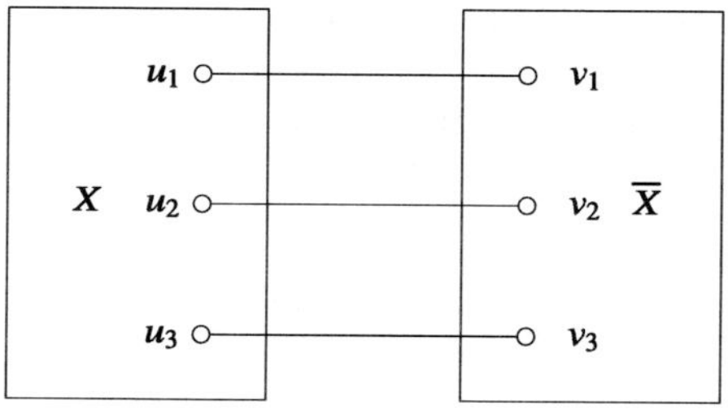

Fig. 4.4 The case in which the 3-cut C is a matching

Let $G_1 := G/(\overline{X} \to \overline{x})$ and let $G_2 := G/(X \to x)$.

4.5.1 *At least one of the graphs $G_1 - u_1 - u_2$ and $G_2 - v_1 - v_2$ is not matchable.*

Proof Assume, to the contrary, that both graphs $G_1 - u_1 - u_2$ and $G_2 - v_1 - v_2$ are matchable. Let M_1 and M_2 be perfect matchings of $G_1 - u_1 - u_2$ and $G_2 - v_1 - v_2$, respectively. Then, $M := M_1 \cup M_2 \cup \{u_1v_1, u_2v_2\}$ is a perfect matching of G which contains all the three edges of C, a contradiction to the hypothesis that C is tight in G. □

Adjust notation so that $G_1 - u_1 - u_2$ is not matchable. The graph G_1, being a C-contraction of G, is matching covered. Thus, G_1 has a barrier that contains both u_1 and u_2. Let B denote the maximal barrier of G_1 that contains both u_1 and u_2. As G_1 is matching covered, the set B is independent, and thus the contraction vertex $\overline{x}$ does not lie in B. We deduce that $G_1 - B$ has a component, say K, which contains the vertex $\overline{x}$.

4.5.2 *Vertex u_3 also lies in B.*

Proof Assume the contrary. As u_3v_3 is an edge and $\overline{x}$ belongs to $V(K)$, it follows that u_3 also belongs to $V(K)$. In fact, u_3 is the only vertex of K that is adjacent to $\overline{x}$. Thus, $K - u_3$ is not matchable, hence K is not critical. This is a contradiction, as B is a maximal barrier of G_1. □

In sum, the vertex $\overline{x}$ is isolated in the graph $G_1 - B$, therefore C is a barrier cut of G, associated with B, which is also a barrier of G. □

Tight cuts may be neither barrier nor 2-separation cuts

A matching covered graph may have a tight cut which is neither a barrier cut, nor a 2-separation cut; Figure 4.5 depicts such a cut (see Exercise 4.2.6). However, as we shall now proceed to show, every separating cut in a bipartite matching covered graph is (or may be constructed as) a barrier cut. We shall in fact see that this statement is more generally true for a separating cut in any matching covered graph if that cut happens to have a shore that is bipartite.

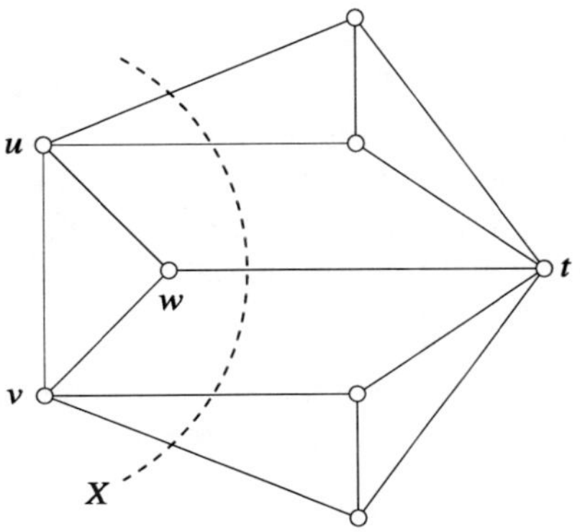

Fig. 4.5 A tight cut that is neither a barrier cut nor a 2-separation cut

4.2.4 Separating cuts with bipartite shores

> MAJORITY AND MINORITY PARTS
>
> When the set X of vertices of a connected bipartite graph H is odd, the two parts of the bipartition of H have distinct cardinalities; the larger part is called *the majority part*, the other *the minority part*; we denote the majority part of X by X_+, and the minority part by X_-.

Lemma 4.6 *Suppose that G is a matching covered graph and that v is a vertex of G. If the subgraph $G - v$ is bipartite, then G is also bipartite. Furthermore, the majority part of $G - v$ has just one more vertex than its minority part.*

Proof Suppose that $G - v$ is a bipartite graph with bipartition (U, W). Assume without loss of generality that $|U| > |W|$. Since G is connected, the vertex v is joined to at least one vertex in $G - v$. If v is joined in G to a vertex w in W, then no matching containing the edge vw can cover all the vertices in U, and hence vw would not be matchable in G. This is impossible because G is matching covered. Thus all neighbours of v in G are in U, implying that G is a bipartite graph with bipartition $(U, W + v)$. As G is matching covered, it must be the case that $|U| = |W| + 1$. □

SEPARATING CUTS WITH A BIPARTITE SHORE

Theorem 4.7 *Let $C := \partial(X)$ be a separating cut in a matching covered graph G. If the subgraph $G[X]$ of G induced by X is bipartite, and if the subgraph $G[\overline{X}]$ of G induced by $\overline{X}$ is nontrivial then the majority part X_+ of $G[X]$ is a barrier of G and C is a special barrier cut of G.*

Proof Consider the C-contraction $G' := G/(\overline{X} \to \overline{x})$. Then $G' - \overline{x} = G[X]$, which is bipartite by the hypothesis. Thus, by Lemma 4.6, G' is also bipartite, and the vertex $\overline{x}$ is adjacent in G' only to the vertices in X_+. Furthermore, $|X_+| = |X_-| + 1$. Thus, $G - X_+$ has precisely $|X_+|$ odd components, one of them being the nontrivial graph $G[\overline{X}]$, and the remaining being the trivial components corresponding to the vertices in the minority part X_- of X. □

It is easy to deduce from the above theorem the following useful characterization of separating cuts in bipartite matching covered graphs. We leave its proof as Exercise 4.2.8.

TIGHT CUTS IN BIPARTITE GRAPHS

Theorem 4.8 *Let $C := \partial(X)$ be a separating cut in a bipartite matching covered graph $G[A, B]$. Then:*

(i) $|X_+| = |X_-| + 1$, *(and $|\overline{X}_+| = |\overline{X}_-| + 1$),*
(ii) *the cut C is tight,*
(iii) *X_+ and $\overline{X}_+$ are subsets of different parts of the bipartition of G, and*
(iv) *every edge in the cut C has one end in X_+ and one end in $\overline{X}_+$. (In other words, all edges in C join vertices in the majority part of one shore of C to vertices in the majority part of the other shore. See Figure 4.6.)*

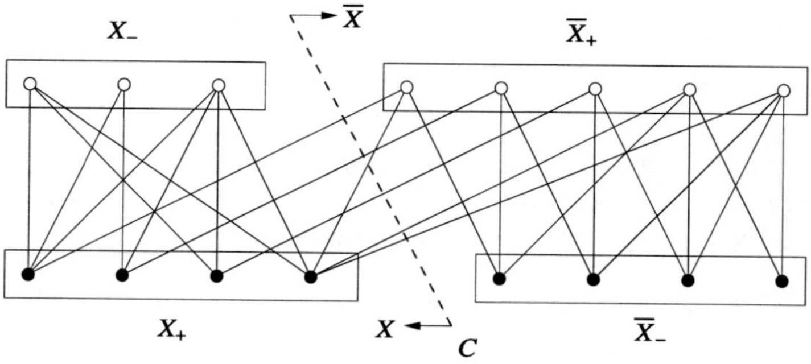

Fig. 4.6 A tight cut in a bipartite matching covered graph

The fact that all tight cuts in bipartite matching covered graphs conform to the pattern described in the statement of Theorem 4.8 will be found to be very useful in many contexts.

As immediate consequences of Theorem 4.7, we have:

Corollary 4.9 *Every separating cut in a bipartite matching covered graph is tight.* □

Corollary 4.10 *If a separating cut of a matching covered graph is not tight then both its shores induce graphs that have odd cycles.* □

Corollary 4.11 *Both shores of any nontrivial separating cut of a bicritical graph induce graphs that have odd cycles. Consequently, if there do not exist two vertex-disjoint odd cycles in a bicritical graph then the graph is free of nontrivial separating cuts.* □

Exercises

4.2.1 *Prove Proposition 4.3*

$*$**4.2.2** *Give a proof of statement 4.4.*

By Theorem 4.2, a cut C of a matching covered graph G is separating if and only if each edge of G is in a perfect matching which contains only one edge in C. We now extend the notion of separating cut to a collection of cuts.

Cohesive collection of cuts
A collection $\mathcal{C}$ of cuts of a matching covered graph G is *cohesive* if every edge of G lies in a perfect matching that contains precisely one edge in each cut in $\mathcal{C}$.

Consequently, each cut of a cohesive collection of cuts is separating.

▷**4.2.3** *Let G be a matching covered graph and let C be a separating cut of G.*

(i) *Prove that every collection of tight cuts of G is cohesive.*
(ii) *Prove that the addition of C to a collection of tight cuts of G is cohesive.*

$*$**4.2.4** *Let $C := \partial(X)$ be a separating cut of a matching covered graph G, and let D be a separating cut of a C-contraction $H := G/X$ of G.*

(i) *Prove that $\{C, D\}$ is cohesive.*
(ii) *Prove that if C is tight in G and D is tight in G/X then D is tight in G.*
(iii) *Give an example where D is tight in G/X but not tight in G.*
(iv) *Give an example of a graph and a pair of separating cuts which is not cohesive. Hint: consider the Petersen graph.*

▷**4.2.5** *Let G be a cubic matching covered graph. Show that:*

(i) *every 2-separation cut of G is also a barrier cut of G, and that*
(ii) *every barrier cut of G is a 3-edge-cut.*

▷**4.2.6** *Show that the cut shown in Figure 4.5 is tight, but it is neither a barrier cut, nor a 2-separation cut.*

4.2.7

(i) *Verify that the Petersen graph $\mathbb{P}$ is bicritical. (Hint: use the 4-transitivity of the Petersen graph and the fact that its diameter is equal to two, in order to reduce to two the number of pairs of vertices $\{v, w\}$ to check that $\mathbb{P} - v - w$ has a perfect matching.)*
(ii) *List all the nontrivial separating cuts of the Petersen graph. (Hint: use the result of Corollary 4.10.)*
(iii) *Show that the Petersen graph is free of nontrivial tight cuts.*

▷**4.2.8** *Give a proof of Theorem 4.8.*

▷**4.2.9** *Use Theorem 4.8 to show that all tight cuts in a bipartite cubic matching covered graph are 3-edge cuts.*

4.3 Tight Cut Decompositions

4.3.1 Bricks and braces

Some matching covered graphs do not have nontrivial tight cuts. The reader would be able to verify easily that K_2, C_4, K_4, $\overline{C_6}$, $K_{3,3}$ and the Petersen graph are examples of such graphs. We refer to a matching covered graph which is free of nontrivial tight cuts as a *brace* if it is bipartite, and as a *brick* if it is nonbipartite. Among the graphs listed above, K_2, C_4, and $K_{3,3}$ are braces, and K_4, $\overline{C_6}$ and the Petersen graph are bricks. Simple ways of recognizing bricks and braces will be described in Chapter 5. (Bricks turn out to be 3-connected bicritical graphs, and braces turn out to be connected 2-extendable bipartite graphs.)

A 3-connected cubic graph is *essentially 4-edge-connected* if it has no 3-edge cuts other than those which are trivial. It follows from Exercise 4.2.9 that every essentially 4-edge-connected bipartite cubic graph is a brace. It is easy to verify that prisms $\mathbb{P}_{2n}$, for even $n \geq 4$, and Möbius ladders $\mathbb{M}_{2n}$, for odd $n \geq 3$, are essentially 4-edge-connected bipartite cubic graphs and, as such, they are braces.

Every prism $\mathbb{P}_{2n}$, for odd $n \geq 3$, is bicritical and has a unique nontrivial separating cut and that cut is not tight (Exercise 4.3.1). Möbius ladders $\mathbb{M}_{2n}$, for even $n \geq 2$, are also bicritical and have no nontrivial separating cuts at all, and hence are free of nontrivial tight cuts (Exercise 4.3.2). It follows that prisms $\mathbb{P}_{2n}$, for odd $n \geq 3$, and Möbius ladders $\mathbb{M}_{2n}$, for even $n \geq 2$, are bricks.

TIGHT CUT DECOMPOSITION

Given any matching covered graph G, we may apply to it a procedure, called a *tight cut decomposition* of G, which produces a list of bricks and braces. If G itself is a brick or a brace then the list consists of just G. Otherwise, let C be any nontrivial tight cut of G. Then, both C-contractions of G are matching covered. One may recursively apply the tight cut decomposition procedure to each C-contraction of G, and then combine the resulting lists to produce a tight cut decomposition of G itself.

As one might expect, the list of bricks and braces obtained by an application of this procedure might depend on the choice of tight cuts. Consider, for example, the graph G shown in Figure 4.7. The cut $\partial(X)$ is a tight cut in G and the corresponding contractions of G are $G_1 := G/(\overline{X} \to \overline{x})$, and $G_2 := G/(X \to x)$.

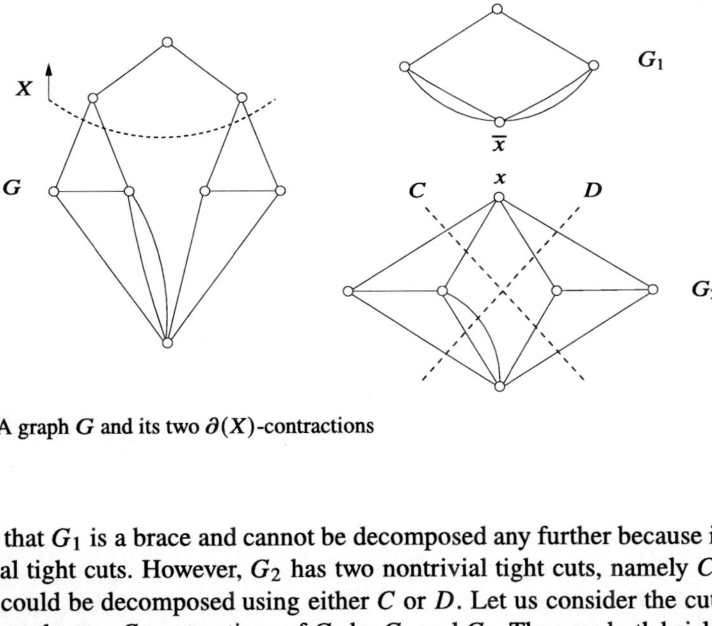

Fig. 4.7 A graph G and its two $\partial(X)$-contractions

Note that G_1 is a brace and cannot be decomposed any further because it has no nontrivial tight cuts. However, G_2 has two nontrivial tight cuts, namely C and D, and G_2 could be decomposed using either C or D. Let us consider the cut C first, and denote the two C-contractions of G_2 by G_3 and G_4. They are both bricks whose underlying simple graphs are K_4's. Thus $\{G_1, G_3, G_4\}$ is a tight cut decomposition of G. Considering the second option D, let G_5 and G_6 denote the two D-contractions of G_2. They are also bricks whose underlying simple graphs are K_4's, and thus $\{G_1, G_5, G_6\}$ is another tight cut decomposition of G. The reader will be able to check that although the underlying simple graphs of $\{G_1, G_3, G_4\}$ are the same, in some order, as the underlying simple graphs of $\{G_1, G_5, G_6\}$, the two lists of graphs are not exactly the same (Exercise 4.3.3).

The above example shows that different applications of the tight cut decomposition on a matching covered graph may yield different lists of bricks and braces. However, a remarkable result of Lovász (1987, [58]) states that any two applications of the tight cut decomposition procedure on a given graph G yield the same list of bricks and braces, up to multiple edges. A proof of this fundamental result will be presented in the next section. It relies on the notion of uncrossing cuts which we define below.

4.3.2 Uncrossing cuts

Let G be a matching covered graph. Let $C := \partial(X)$ and $D := \partial(Y)$ be two odd cuts of G. The four sets $X \cap Y$, $X \cap \overline{Y}$, $\overline{X} \cap Y$ and $\overline{X} \cap \overline{Y}$ are the *quadrants* defined by C and D. The cuts C and D *cross* if each of these four quadrants is nonnull. (The cuts C and D shown in Figure 4.7 are two tight cuts in G_2 that cross each other.) A collection $\mathcal{C}$ of cuts of G is *laminar* if no two of its members cross. We note that if two cuts C and D are laminar, then one of the shores of C is a subset of a shore of D. Figure 4.8 shows two crossing cuts, C and D.

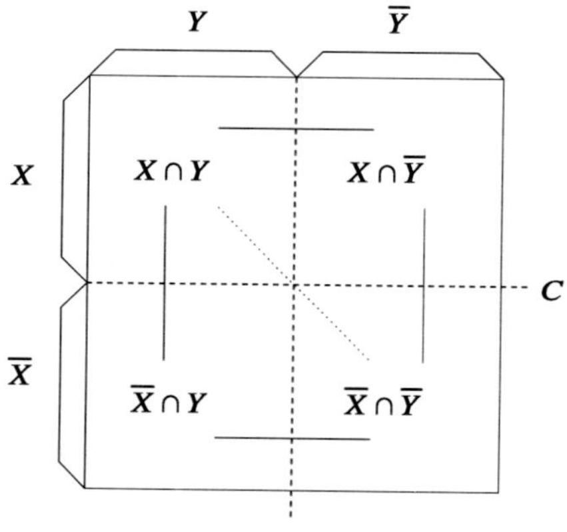

Fig. 4.8 Crossing cuts $C := \partial(X)$, $D := \partial(Y)$, $I := \partial(X \cap Y)$, $U := \partial(\overline{X} \cap \overline{Y})$ and the four quadrants

If $C := \partial(X)$ and $D := \partial(Y)$ are crossing odd cuts, then one of the two quadrants $X \cap Y$ and $X \cap \overline{Y}$ is odd and the other is even (and nonempty). We shall often find it convenient to adjust notation so that the quadrant $X \cap Y$ is odd, which implies that $X \cap \overline{Y}$ and $\overline{X} \cap Y$ are even and $\overline{X} \cap \overline{Y}$ is odd. In addition, if C and D are both tight, pictorially we display the four quadrants as in Figure 4.8. An unbroken line between two quadrants indicates the presence of at least one edge between them; a broken line between the quadrants $X \cap Y$ and $\overline{X} \cap \overline{Y}$ indicates the possible presence of edges between them; Figure 4.8 depicts the situation in which the four subgraphs $G[X]$, $G[\overline{X}]$, $G[Y]$, and $G[\overline{Y}]$ are connected, and all the four quadrants are nonempty.

The following result provides a crucial step in the proof of Lovász's fundamental result mentioned at the beginning of this section.

UNCROSSING TIGHT CUTS

Theorem 4.12 *Let $C := \partial(X)$ and $D := \partial(Y)$ be two tight cuts of a matching covered graph G, where $|X \cap Y|$ is odd. Then:*

(i) *no edge of G joins a vertex in $X \cap \overline{Y}$ to a vertex in $\overline{X} \cap Y$, and*
(ii) *the cuts $I := \partial(X \cap Y)$ and $U := \partial(\overline{X} \cap \overline{Y})$ are both tight in G.*

Proof Suppose first that C and D do not cross and, without loss of generality, that $X \cap \overline{Y} = \emptyset$. In this case $X \subset Y$. Thus $I = \partial(X) = C$ and $U = \partial(Y) = D$, and hence the assertion holds.

So, we may assume that C and D do cross. Let M be any perfect matching of G. Denote by F the set of edges of G that have an end in $X \cap \overline{Y}$ and an end in $\overline{X} \cap Y$. Clearly,

$$|M \cap C| + |M \cap D| = |M \cap I| + |M \cap U| + 2|M \cap F|. \tag{4.1}$$

Let e be any edge of G, let M_e be a perfect matching of G that contains the edge e. By hypothesis, C and D are tight, thus M_e contains just one edge in each of C and D. The cuts I and U are both odd, hence M_e contains at least one edge in each cut in $\{I, U\}$. From (4.1) we have that

$$2 = |M_e \cap C| + |M_e \cap D| = |M_e \cap I| + |M_e \cap U| + 2|M_e \cap F| \geq 2 + 2|M_e \cap F|. \tag{4.2}$$

We conclude that for each edge e of G and any perfect matching M_e that contains edge e we have

(a) $e \in M_e$, $|M_e \cap F| = 0$,
(b) $|M_e \cap C| = 1$, $|M_e \cap D| = 1$, $|M_e \cap I| = 1$ and $|M_e \cap U| = 1$.

From (a) we deduce that $e \notin F$. This conclusion holds for each edge e of G, hence F is empty. That is, part (i) of the assertion holds. From (b) we conclude that every perfect matching that contains edge e contains just one edge in I and just one edge in U. This conclusion holds for each edge e of G, hence I and U are both tight cuts. That is, part (ii) of the assertion holds. □

The following corollary will prove to be very useful in the proof of Theorem 4.17.

Corollary 4.13 *Let G be a matching covered graph, and let $C := \partial(X)$ and $D := \partial(Y)$ be two tight cuts of G that cross, where $|X \cap Y|$ is odd. Then the underlying simple graphs of the two graphs $G/X/\overline{X} \cap \overline{Y}$ and $G/\overline{Y}/X \cap Y$ are isomorphic.*

Proof For the convenience of referring to the two graphs, let us write:

$$G_1 := G/(X \to x)/(\overline{X} \cap \overline{Y} \to s), \text{ and } G_2 := G/(\overline{Y} \to \overline{y})/(X \cap Y \to t)$$

See Figure 4.9.

Observe that $V(G_1) = (\overline{X} \cap Y) \cup \{x, s\}$, and $V(G_2) = (\overline{X} \cap Y) \cup \{\overline{y}, t\}$. As D is a tight cut, $G[\overline{Y}]$ is connected. This implies that x and s are adjacent in G_1. Similarly, since C is a tight cut, $G[X]$ is connected, implying that t and $\overline{y}$ are adjacent in G_2.

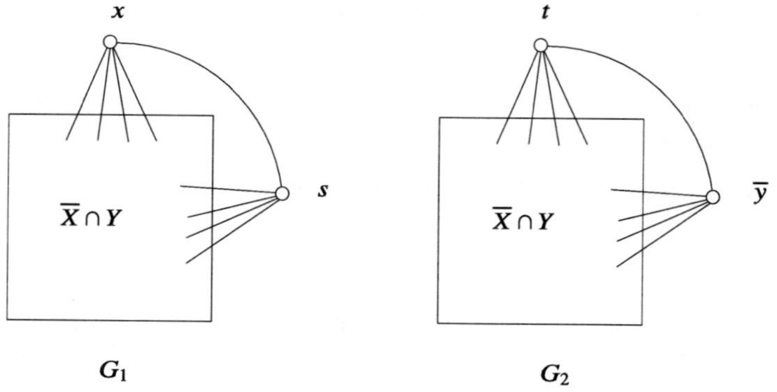

Fig. 4.9 Isomorphic contractions

Furthermore, as there are no edges between $\overline{X} \cap Y$ and $X \cap \overline{Y}$ by Theorem 4.12, it follows that the mapping θ, where

$$\theta(v) := v, \text{ for each } v \in \overline{X} \cap Y, \quad \theta(x) := t \quad \text{and} \quad \theta(s) := \overline{y}$$

is an isomorphism between the underlying simple graphs of G_1 and G_2. We leave the details as Exercise 4.3.4. □

Although the underlying simple graphs of G_1 and G_2 are isomorphic, the graphs G_1 and G_2 may not be isomorphic, as the number of edges joining x and s in G_1 may not coincide with the number of edges joining t and $\overline{y}$ in G_2. In fact, the number of edges between x and s in G_1 is the number of edges between X and $\overline{X} \cap \overline{Y}$ in G, whereas the number of edges between t and $\overline{y}$ in G_2 is the number of edges between $X \cap Y$ and $\overline{Y}$ in G. Clearly, there is no reason for these two numbers to be the same. Thus G_1 and G_2 may not be isomorphic, but they are isomorphic up to multiple edges!

Uncrossing separating cuts

Since tight cuts are special types of separating cuts, it is natural to wonder if the assertion of Theorem 4.12 concerning tight cuts holds more generally for separating cuts. Alas, this is not the case; the cuts $I := \partial(X \cap Y)$ and $U := \partial(\overline{X} \cap \overline{Y})$ obtained by uncrossing two separating cuts $C := \partial(X)$ and $D := \partial(Y)$ that cross need not be separating cuts (see Exercise 4.3.7).

As stated in the next result, what is really required of separating cuts C and D for I and U to be also separating is that each edge of the graph must be in some perfect matching that contains just one edge in C and just one edge in D. In other words, it is required that the pair $\{C, D\}$ be cohesive. (Recall that a collection $\mathcal{C}$ of cuts of a matching covered graph G is cohesive if each edge of G is in a perfect matching that contains precisely one edge in each cut in $\mathcal{C}$; see Exercise 4.2.4).

The proof of Theorem 4.14 requires quite minor modifications in the proof of Theorem 4.12 and is left as Exercise 4.3.8.

UNCROSSING COHESIVE PAIRS

Theorem 4.14 *Let $C := \partial(X)$ and $D := \partial(Y)$ be two separating cuts of a matching covered graph G, where $|X \cap Y|$ is odd. If $\{C, D\}$ is cohesive then:*

(i) *no edge of G joins a vertex in $X \cap \overline{Y}$ to a vertex in $\overline{X} \cap Y$,*
(ii) *for each perfect matching M of G,*

$$|M \cap C| + |M \cap D| = |M \cap I| + |M \cap U|, \tag{4.3}$$

(iii) *the collection $\{C, D, I, U\}$ is cohesive, and*
(iv) *the cuts I and U are separating.* □

If a cut C is separating and a cut D is tight then the pair $\{C, D\}$ is cohesive (Exercise 4.2.3). Thus, Corollary 4.15, below, is a consequence of Theorem 4.14.

Corollary 4.15 *Let $C := \partial(X)$ be a separating cut of G, and let $D := \partial(Y)$ be a tight cut of G. Suppose that $|X \cap Y|$ is odd. Then:*

(i) *no edge of G joins a vertex in $X \cap \overline{Y}$ to a vertex in $Y \cap \overline{X}$,*
(ii) *both $I := \partial(X \cap Y)$ and $U := \partial(\overline{X} \cap \overline{Y})$ are separating cuts of G, and*
(iii) *if C is tight then both I and U are tight whereas if C is not tight, then at least one of I and U is not tight.* □

Note that Corollary 4.15 implies immediately Theorem 4.12. We leave the proof of Corollary 4.15 as Exercise 4.3.9.

Corollary 4.16 *Let G be a matching covered graph, and let D be a tight cut of G. If G has a separating cut that is not tight, then one of the D-contractions of G also has a separating cut that is not tight.* □

The above corollary is used in the proof an important result (Theorem 7.1) in Chapter 7.

Exercises

4.3.1 *Consider a prism $G := \mathbb{P}_{2n}$, where n is odd.*

(i) *Prove that G is bicritical.*
(ii) *Prove that G has a unique pair of disjoint odd cycles, both of length n.*
(iii) *Prove that G is a brick. (Hint: use the result of Corollary 4.11.)*

4.3.2 *Consider a Möbius ladder* $G := \mathbb{M}_{2n}$, *where n is even.*

(i) *Prove that G does not have two disjoint odd cycles.*
(ii) *Prove that G is a brick. (Hint: use the result of Corollary 4.11.)*

▷**4.3.3** *Show that the graph G in Figure 4.7 has two tight cut decompositions which are not the same if we take multiplicities of edges into consideration.*

▷**4.3.4** *Supply the missing details in the proof of Corollary 4.13.*

▷**4.3.5** *Let G be a matching covered graph of order* $n \geq 4$, *let* $\mathcal{L}$ *and* $\mathcal{C}$ *be, respectively, a maximal laminar collection of nontrivial tight cuts of G and the collection of braces and bricks of the corresponding tight cut decomposition of G.*

(i) *Prove that* $|\mathcal{L}| \leq n/2 - 2$ *and* $|\mathcal{C}| \leq n/2 - 1$. *Hint: use induction on n.*
(ii) *Prove that for each even integer* $n \geq 4$ *there exists a bipartite matching covered graph of order n such that both upper bounds are attained exactly.*
(iii) *Similarly, prove that for each even integer* $n \geq 4$ *there exists a bicritical matching covered graph of order n such that both upper bounds are attained exactly.*

4.3.6 *Consider the two cuts C and D of the graph G depicted in Figure 4.10. Show that C is a separating cut which is not tight, that D is a tight cut, and that C and D cross.*

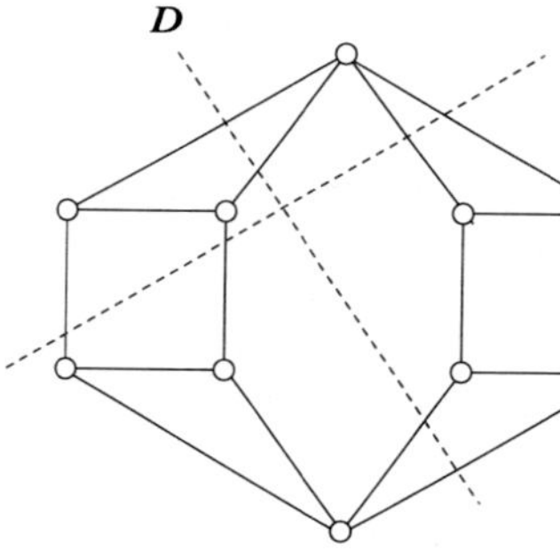

Fig. 4.10 Two cuts that cross (Exercise 4.3.6)

▷**4.3.7** *Give an example to show that the cuts I and U obtained by uncrossing two crossing separating cuts C and D of a matching covered graph need not be separating cuts. (Hint: Try the Petersen graph. See also Exercise 4.2.4.)*

∗**4.3.8** *Make minor modifications to the proof of Theorem 4.12 in order to obtain a proof of Theorem 4.14.*

∗**4.3.9** *Deduce Corollary 4.15 from Theorem 4.14.*

Characteristic of a separating cut

Let C be a separating cut of a matching covered graph G. If C is tight, then $|M \cap C| = 1$, for every perfect matching M of G; otherwise, there is some perfect matching M of G such that $|M \cap C| \geq 3$. We define the *characteristic* $\lambda(C)$ of C to be ∞ if C is tight and, otherwise, to be the minimum of $|M \cap C|$, where the minimum is taken over the set of all those perfect matchings M of G which have at least three edges in C.

▷4.3.10

(i) *Show that the Petersen graph has precisely six nontrivial separating cuts (each consisting, coincidentally, of a perfect matching).*
(ii) *Find the characteristics of all the nontrivial separating cuts in the Petersen graph.*
(iii) *Find the characteristics of the four separating cuts C_1, C_2, C_3 and C_4 in the graph shown in Figure 4.11.*

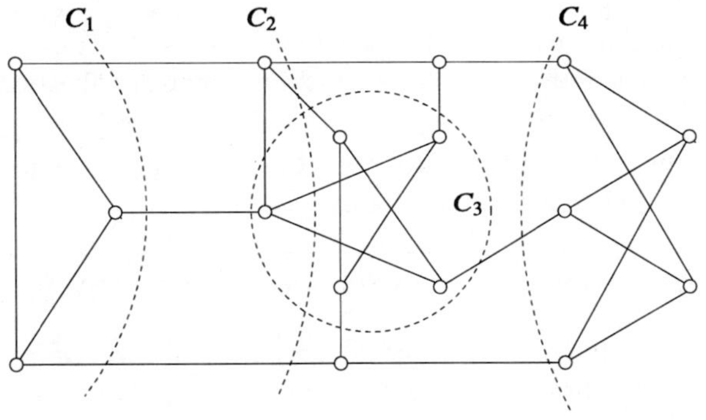

Fig. 4.11 Graph for Exercise 4.3.10

∗4.3.11 (Uncrossing Separating Cuts) *Let C, D, I and U be as in the statement of Corollary 4.15. Prove that*

$$\lambda(C) = \min\{\lambda(I), \lambda(U)\}.$$

4.3.12 *Let D be a tight cut of a matching covered graph G, and let C be a cut of a D-contraction H of G.*

(i) *Prove that C is separating in G if and only if it is separating in H.*
(ii) *Prove that if C is separating in G then its characteristics in G and in H coincide.*

> CHARACTERISTIC OF A MATCHING COVERED GRAPH
>
> The characteristic of a matching covered graph G, denoted by $\lambda(G)$, is the minimum of the characteristics of its separating cuts.

4.3.13

(i) *Determine the characteristics of the Petersen graph and the graph in Figure 4.11.*
(ii) *Using the assertion of Exercise 4.3.11, show that the characteristic of a matching covered graph G is the minimum of the characteristics of its bricks and braces.*

4.3.14 *Let G be a matching covered graph, let $C := \partial(X)$ be a separating cut of G which is not tight. Let H be the underlying simple graph of $G/(X \to x)$, let d denote the degree of x in H. Show that $\lambda(G) \leq d$. Deduce that if $d = 3$ then $\lambda(G) = 3$.*

4.4 Uniqueness of Tight Cut Decompositions

Suppose that C is a tight cut of a matching covered graph G. Then any tight cut of a C-contraction of G is a tight cut of G that does not cross C. Thus, associated with any tight cut decomposition of G there is a maximal laminar collection of nontrivial tight cuts of G, and *vice versa*.

Using Theorem 4.12, and Corollary 4.13, Lovász [58] proved the following remarkable result on tight cut decompositions.

> THE UNIQUE DECOMPOSITION THEOREM (1987, [58])
>
> **4.17** *Any two applications of the tight cut decomposition procedure to a matching covered graph G produce the same list of bricks and braces, up to multiple edges.*

Proof We shall refer to two maximal laminar families $\mathcal{C}$ and $\mathcal{D}$ of nontrivial tight cuts of G as *equivalent*, and write $\mathcal{C} \equiv \mathcal{D}$, if they produce the same list of bricks and braces, up to multiple edges. We prove that any two maximal laminar collections of nontrivial tight cuts of G are equivalent, by induction on the number of vertices.

For $i = 1, 2$, let $\mathcal{C}_i$ be two maximal laminar collections of nontrivial tight cuts of G. We consider various cases.

Case 1 *Collections $\mathcal{C}_1$ and $\mathcal{C}_2$ contain a common cut C.*

Let G_1 and G_2 denote the two C-contractions of G. For $i, j \in \{1, 2\}$, let $\mathcal{C}_{ij}$ denote the restriction of $\mathcal{C}_i - C$ to G_j. For $i = 1, 2$, any tight cut of G_i is also a tight cut of G, it follows that $\mathcal{C}_{ij}$ is a maximal laminar collection of nontrivial tight cuts of G_j. By induction, for $j = 1, 2$, $\mathcal{C}_{1j}$ and $\mathcal{C}_{2j}$ are equivalent. Thus, $\mathcal{C}_1$ and $\mathcal{C}_2$ are also equivalent. The assertion holds in this case.

Case 2 *There are cuts $C_1 \in \mathcal{C}_1$ and $C_2 \in \mathcal{C}_2$ that do not cross.*

By hypothesis, $\{C_1, C_2\}$ is laminar. Let $\mathcal{C}_3$ denote a maximal laminar collection of nontrivial tight cuts of G that includes $\{C_1, C_2\}$. For $i = 1, 2$, $\mathcal{C}_i$ and $\mathcal{C}_3$ have cut C_i in common. By Case 1,

$$\mathcal{C}_1 \equiv \mathcal{C}_3 \equiv \mathcal{C}_2.$$

Thus, $\mathcal{C}_1$ and $\mathcal{C}_2$ are equivalent. The assertion holds in this case.

Case 3 *There are cuts $C_1 := \partial(X_1) \in \mathcal{C}_1$ and $C_2 := \partial(X_2) \in \mathcal{C}_2$ such that $|X_1 \cap X_2|$ is odd and nontrivial.*

If C_1 and C_2 do not cross then Case 2 is applicable. We may thus assume that C_1 and C_2 cross. Let $C_3 := \partial(X_1 \cap X_2)$. Then, C_3 is nontrivial. By Theorem 4.12, cut C_3 is tight in G.

Let $\mathcal{C}_4$ denote a maximal laminar collection of nontrivial tight cuts of G that contains C_3. Cuts C_1 and C_3 do not cross. By Case 2, collections $\mathcal{C}_1$ and $\mathcal{C}_4$ are equivalent. Likewise, cuts C_2 and C_3 do not cross. By Case 2, collections $\mathcal{C}_2$ and $\mathcal{C}_4$ are equivalent. In sum,

$$\mathcal{C}_1 \equiv \mathcal{C}_4 \equiv \mathcal{C}_2.$$

Thus, $\mathcal{C}_1$ and $\mathcal{C}_2$ are equivalent. The assertion holds in this case.

Case 4 *None of the previous cases is applicable.*

If G is a brick or a brace then the assertion holds trivially. We may thus assume that G has nontrivial tight cuts. Then, $\mathcal{C}_1$ and $\mathcal{C}_2$ are both nonempty. For $i = 1, 2$, let $C_i := \partial(X_i)$ denote a cut in $\mathcal{C}_i$. If C_1 and C_2 do not cross then Case 2 applies. We may thus assume that C_1 and C_2 cross.

Adjust notation so that $|X_1 \cap X_2|$ is odd, whereupon $|\overline{X_1} \cap \overline{X_2}|$ is also odd. If $|X_1 \cap X_2| > 1$ or if $|\overline{X_1} \cap \overline{X_2}| > 1$ then Case 3 applies. We may thus assume that $X_1 \cap X_2$ and $\overline{X_1} \cap \overline{X_2}$ are both singletons. Let u denote the only vertex of $X_1 \cap X_2$, let v denote the only vertex of $\overline{X_1} \cap \overline{X_2}$.

Assume that $\mathcal{C}_i = \{C_i\}$, for $i = 1, 2$. By Corollary 4.13, G/X_1 and $G/\overline{X_2}$ are isomorphic, up to multiple edges. Likewise, $G/\overline{X_1}$ and G/X_2 are also isomorphic, up to multiple edges. Thus, $\mathcal{C}_1 \equiv \mathcal{C}_2$. The assertion holds in this case.

We now prove that $\mathcal{C}_i = \{C_i\}$, thereby completing the proof. Assume, to the contrary, that one of $\mathcal{C}_1$ and $\mathcal{C}_2$ contains two or more cuts. Adjust notation so that $\mathcal{C}_1 - C_1$ contains a cut $D := \partial(Y)$. By hypothesis, $\mathcal{C}_1$ is laminar, therefore C_1 and D do not cross. Adjust notation so that one of Y and X_1 is a subset of the other. Adjust notation, by complementing the three sets X_1, X_2 and Y if necessary, so that $Y \subset X_1$. Then,

$$Y \cap X_2 \subseteq X_1 \cap X_2 = \{u\} \quad \text{and} \quad \{v\} = \overline{X_1} \cap \overline{X_2} \subseteq \overline{Y} \cap \overline{X_2}. \tag{4.4}$$

Cuts C_2 and D cross, otherwise Case 2 applies. Thus, $Y \cap X_2$ is nonempty. From (4.4) we deduce that $Y \cap X_2 = X_1 \cap X_2 = \{u\}$. Then, $|\overline{Y} \cap \overline{X_2}|$ is odd. If $|\overline{Y} \cap \overline{X_2}| > 1$ then Case 3 applies. We may thus assume that $|\overline{Y} \cap \overline{X_2}| = 1$. From (4.4), we deduce that $\overline{Y} \cap \overline{X_2} = \overline{X_1} \cap \overline{X_2}$, whence $Y \cup X_2 = X_1 \cup X_2$. In sum,

$$Y \cap X_2 = X_1 \cap X_2 \quad \text{and} \quad Y \cup X_2 = X_1 \cup X_2.$$

We conclude that Y and X_1 coincide, a contradiction. As assumed, $\mathcal{C}_i = \{C_i\}$, for $i = 1, 2$. □

4.4.1 The number of bricks

One particular consequence of Lovász's Theorem is that any two tight cut decompositions of a given matching covered graph G yield the same numbers of bricks and braces. We refer to these two invariants as the numbers of bricks and braces of G, respectively. For example, the graph G shown in Figure 4.7 has two bricks and one brace. We denote the number of bricks of a matching covered graph G by $b(G)$, or simply b if G is understood.

Let H be any given simple brick or brace, such as the Petersen graph $\mathbb{P}$ or $K_{3,3}$. A specific implication of Theorem 4.17 is that the number of graphs in a tight cut decomposition of G whose underlying simple graphs are isomorphic to H does not depend on the choice of tight cuts used in the decomposition. In a later chapter, the number of 'Petersen bricks' will play an important role.

4.4.2 Near-bricks

A *near-brick* is a matching covered graph with a single brick. Near-bricks feature prominently in many fundamental results which will be described in Part II. The following result is a simple consequence of Proposition 4.3.

Proposition 4.18 *Every tight cut of a near-brick is a special-barrier cut.*

Proof Let $C := \partial(X)$ be a tight cut of a near-brick G. By definition, $b(G) = 1$. It follows that precisely one of the C-contractions of G is bipartite. Adjusting notation, if necessary, we may suppose that $G_1 := G/(\overline{X} \to \overline{x})$ is not bipartite and $G_2 := G/(X \to x)$ is bipartite, with bipartition, say (A, B), with x belonging to A. Then B is a special barrier of G, $G[\overline{X}]$ is the nontrivial component of $G - B$ and C is a special barrier cut associated with B. □

As bicritical graphs do not have nontrivial barriers, we have the following obvious corollary to the above assertion:

Corollary 4.19 *Every bicritical near-brick is a brick.* □

If B is any barrier of a near-brick G, then one of the components of $G - B$ is nonbipartite. But $G - B$ may also have nontrivial bipartite components (see, for example, the near-brick shown in Figure 4.2). Thus, not every barrier of a near-brick is special. However, every maximal barrier in a near-brick is special (Exercise 4.4.5).

Exercises

4.4.1 *Construct a matching covered graph with two bricks and three braces such that (i) the underlying simple graphs of the bricks are $\overline{C_6}$ and $\mathbb{P}$, and (ii) the underlying simple graphs of the braces are $K_{3,3}$ and two copies of $K_{2,2}$.*

▷**4.4.2** *Show that if G is bipartite, then $b(G) = 0$. (Hint: Use Theorem 4.7.)*

▷**4.4.3** *For any matching covered graph G, let $b_0(G)$ denote the number of braces of G.*

(i) *Prove that if G has order $n \geq 4$ then $b(G) + b_0(G) \leq n/2 - 1$.*
(ii) *Give an infinite family of bipartite matching covered graphs G of order n such that $b_0(G) = n/2 - 1$.*
(iii) *Give an infinite family of matching covered graphs G of order n such that $b(G) = n/2 - 1$.*

4.4.4 *Let G be a nonbipartite matching covered graph. Prove that G has a laminar collection $\mathcal{C}$ of $b - 1$ nontrivial tight cuts such that the (possibly partial) tight cut decomposition produced by these $b - 1$ cuts consists of b near-bricks.*

4.4.5 *Prove that every maximal barrier in a near-brick is special.*

4.5 Subadditivity of the Number of Bricks

The number of bricks $b(G)$ of a matching covered graph G is additive over the tight cuts of G in the sense that if G_1 and G_2 are the two cut-contractions with respect to a tight cut of G, then $b(G) = b(G_1) + b(G_2)$. This property, as we shall see, distinguishes tight cuts from separating cuts which are not tight; the function $b(G)$ is subadditive over separating cuts of G. This property will only be used in the second part of the book, in the proof of a conjecture of Lovász concerning bricks.

SUBADDITIVITY OF FUNCTION $b(G)$

Theorem 4.20 *Let G be a matching covered graph. Let C be a separating cut of G, and let G_1 and G_2 be the two C-contractions of G. Then*

$$b(G) \leq b(G_1) + b(G_2), \tag{4.5}$$

with equality if, and only if, C is a tight cut.

Proof If C is a tight cut of G then, by Theorem 4.17, it follows that (4.5) holds with equality. So, suppose that C is a separating cut that is not tight. Let X be a shore of C, and take G_1 to be $G/\overline{X}$ and G_2 to be G/X. We shall prove the desired inequality by induction on $|V(G)|$.

Firstly observe that if either G_1 or G_2 is bipartite then C would be a barrier cut of G, implying that C is tight in G. We may therefore assume that $b(G_1) \geq 1$ and $b(G_2) \geq 1$. Now, if G is free of nontrivial tight cuts then the desired inequality is satisfied. Therefore we may assume that G has a nontrivial tight cut, say $D := \partial(Y)$. Let $H_1 = G/\overline{Y}$ and $H_2 = G/Y$ denote the two D-contractions of G. The case in which C and D do not cross is easier and is left as Exercise 4.5.1. See Figure 4.12.

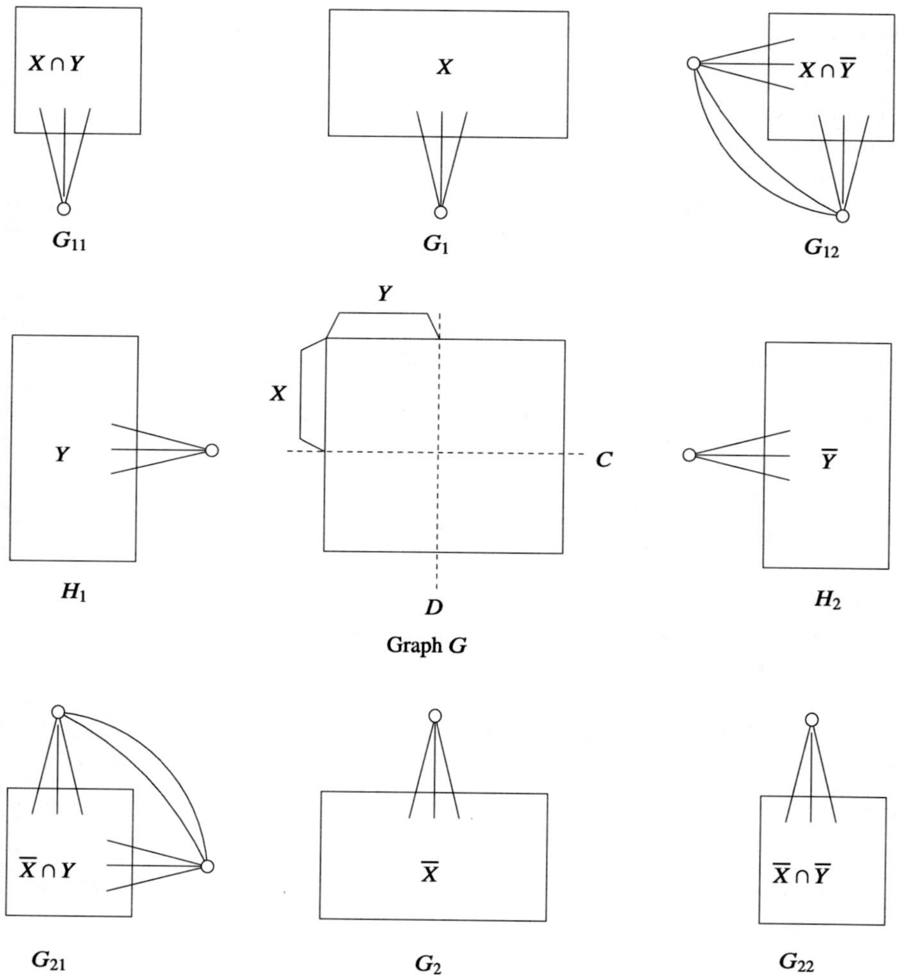

Fig. 4.12 Illustration for the proof of Theorem 4.20

Adjust notation, by interchanging Y with $\overline{Y}$ if necessary, so that $|X \cap Y|$ is odd. Let $I := \partial(X \cap Y)$ and $U := \partial(\overline{X} \cap \overline{Y})$. The cut C is separating and the cut D is tight. By Corollary 4.15, (i) no edge of G joins a vertex in $X \cap \overline{Y}$ to a vertex in $\overline{X} \cap Y$, (ii) each one of the cuts I and U is separating in G, and (iii) at least one of I and U is not tight. Let M be any perfect matching of G. From (i) and recalling that D is tight we infer that

$$|M \cap I| + |M \cap U| = |M \cap C| + 1. \tag{4.6}$$

Every perfect matching of G_1 can be extended to a perfect matching of G that meets C in just one edge. Thus, it follows from (4.6) that I is a tight cut of G_1. Similarly, U is a tight cut of G_2.

Let $G_{11} := G/(\overline{X} \cup \overline{Y})$, and observe that G_{11} is one of the I-contractions of the graph G_1. Now let G_{12} denote the other I-contraction of G_1. Similarly, let $G_{22} := G/(X \cup Y)$, and observe that G_{22} is one of the U-contractions of G_2. Now let G_{21} denote the other U-contraction of G_2. Finally, note that G_{11} and G_{21} are the I-contractions of H_1; and similarly, G_{12} and G_{22} are the U-contractions of H_2. See Figure 4.12.

We then have

$$b(G) = b(H_1) + b(H_2), \tag{4.7}$$

$$b(H_1) \leq b(G_{11}) + b(G_{21}), \tag{4.8}$$

$$b(H_2) \leq b(G_{12}) + b(G_{22}), \tag{4.9}$$

$$b(G_{11}) + b(G_{12}) = b(G_1), \tag{4.10}$$

$$b(G_{21}) + b(G_{22}) = b(G_2). \tag{4.11}$$

The validity of (4.7) follows from the fact that D is tight in G. Since the cut I is a separating cut in G and D is tight in G, it follows that I is a separating cut in H_1. Likewise, the cut U is a separating cut in H_2. The validity of the inequalities (4.8) and (4.9) follows from the induction hypothesis; moreover, since at least one of I and U is not tight in G it follows that either I is not tight in H_1 or U is not tight in H_2. Therefore, at least one of (4.8) and (4.9) is strict. Finally, the facts that I is tight in G_1 and that U is tight in G_2, imply the two equations (4.10) and (4.11). By adding up inequalities (4.7)-(4.11), and simplifying, we obtain the asserted (strict) inequality. □

ROBUST CUTS

A separating cut C of a matching covered graph G is *robust* if C is not tight and both C-contractions of G are near-bricks.

From the subadditivity of function b we now derive the following consequence which indicates the importance of robust cuts.

4.21 *If a matching covered graph has a robust cut then it is a near-brick.*

Proof Let G be a matching covered graph and assume that G has a robust cut C. Let G_1 and G_2 denote the two C-contractions of G. By definition, C is not tight, and both G_1 and G_2 are near-bricks. By the subadditivity of function b, we have that $b(G) < b(G_1) + b(G_2) = 2$. Thus, $b(G) \in \{0, 1\}$. The cut C is separating but not tight. In a bipartite matching covered graph every separating cut is tight (Corollary 4.9). Thus, G is not bipartite. We conclude that $b(G) = 1$. As asserted, G is a near-brick. □

Exercises

▷**4.5.1** *Complete the proof of Theorem 4.20 in the case where C and the nontrivial tight cut D of G do not cross.*

∗**4.5.2** *For any integer $k \geq 2$, show that there exists a brick G, with a separating cut C, such that $b(G') + b(G'') = k$, where G' and G'' are the two C-contractions of G.*

4.6 Notes

Tight cut decompositions and the elegant Uniqueness Theorem 4.17 were first described in the landmark paper by Lovász (1987, [58]). But a related procedure known as a 'brick decomposition procedure' predates this paper. This procedure is essentially the tight cut decomposition procedure where one restricts oneself to cut-contractions with respect to barrier cuts and 2-separation cuts. The term 'brace' appeared for the first time in [58], but the notion of a 'brick' appeared in earlier papers (see [59]). In those works, a brick was defined as a 3-connected bicritical graph. It turns out that this is equivalent to the definition that a brick (as defined in this chapter) is a nonbipartite matching covered graph which is free of nontrivial tight cuts. (This is an important result due to Edmonds, Lovász and Pulleyblank (1982, [32]) which, together with a new proof, will be described in the next chapter.)

In Chapter 5 of their book entitled *Matching Theory* (1982, [59]), Lovász and Plummer remark that the collection of bricks determined by the brick decomposition procedure is not unique. However, they present, in Chapter 7 of their book, a proof of the fact that any two applications of the brick decomposition procedure on a matching covered graph G yield the same number of bricks, implying that this number $b(G)$ is an invariant of G. This is established, indirectly, by relating it to the dimension of the perfect matching polytope of G. (This connection will be described in Chapter 6.) The power and beauty of the Unique Decomposition Theorem 4.17 is that any two applications of the brick decomposition procedure (more generally, of the tight cut decomposition procedure) on a matching covered graph G yield the same collection of bricks up to multiple edges. This directly implies that $b(G)$ is an invariant of G.

Chapter 5
Characterizations of Bricks and Braces

Contents

5.1 Characterization of Bricks

Recall that a matching covered graph which is free of nontrivial tight cuts is a brace if it is bipartite, and is a brick if it is nonbipartite. In Chapter 4 we presented a proof of Lovász's Uniqueness Theorem (4.17) for tight cut decompositions, which is one of the cornerstones of this theory. According to that theorem, every matching covered graph has a unique decomposition into bricks and braces (up to multiple edges). The proof of this fundamental result simply relied on the fact that bricks and braces are free of nontrivial tight cuts. But this fact by itself is inadequate to suggest efficient procedures for recognizing these basic objects. In this chapter, we address these issues by establishing characterizations of bricks and braces which lead to polynomial-time recognition algorithms. We deal with bricks in this section and braces in Section 5.4.

C. L. Lucchesi, U. S. R. Murty, *Perfect Matchings*, Algorithms and Computation in Mathematics 31, https://doi.org/10.1007/978-3-031-47504-7_5

Let us begin by recalling the definitions of barrier cuts and 2-separation cuts. For any barrier B of a matching covered graph G, and any odd component K of $G - B$, the cut $\partial(K)$ is a tight cut of G (see Figure 4.1(b) for an example). Cuts which arise in this manner are barrier cuts of G. (The set $\{\partial(K), K \in \mathcal{O}(G - B)\}$ is the set of barrier cuts *associated with* the barrier B.) As bicritical graphs do not have nontrivial barriers, they do not have nontrivial barrier cuts.

Recall that a 2-separation of a matching covered graph G is a 2-vertex cut of G which is not a barrier. Thus, if $\{u, v\}$ is a 2-separation of G, the graph $G - u - v$ has no odd components. Suppose that H_1 is the union of a nonempty proper subset of the components of $G - u - v$, and H_2 is the union of the remaining components of $G - u - v$. Then H_1 and H_2 are two vertex-disjoint even order subgraphs whose union is $G - u - v$. The cuts $\partial(V(H_1) + v)$ and $\partial(V(H_2) + v)$ are two tight cuts of G which cross each other (see Figure 4.3 for an example). Cuts which arise in this manner are 2-separation cuts. Clearly, 3-connected graphs cannot have 2-separation cuts.

Edmonds, Lovász and Pulleyblank (1982, [32]) proved that if a matching covered graph has a nontrivial tight cut then it has a nontrivial barrier cut or a (nontrivial) 2-separation cut. We refer to this assertion as the *ELP Theorem* and use the term ELP cut to mean either a barrier cut or a 2-separation cut. As we noted in Chapter 4, not all tight cuts in a matching covered graph need be ELP cuts. For example, the cut C in the matching covered graph shown in Figure 5.1 is not an ELP cut; however, the cut D is.

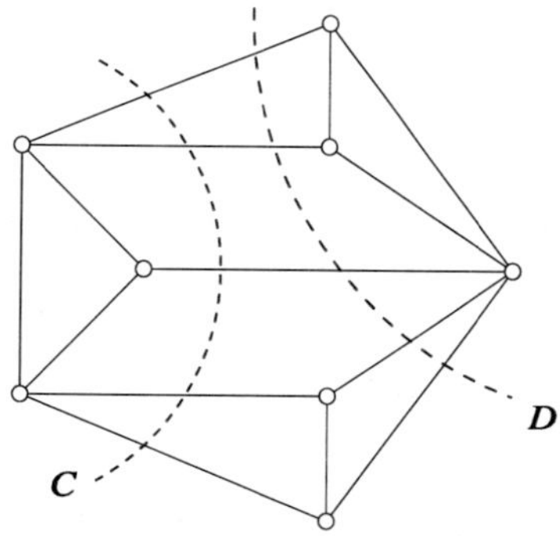

Fig. 5.1 The cut C is not an ELP cut, but D is a 2-separation cut

The proof of the above mentioned theorem by its three authors was based on linear programming duality. About twenty years later, Szigeti (2002, [87]) presented a simpler and purely graph theoretical proof. The proof we give here is inspired by the one given by Szigeti. It relies on the properties of special types of barriers in matchable graphs, which we call Dulmage-Mendelsohn barriers. But before we introduce them, we describe a property that is shared by all matching covered graphs with nontrivial tight cuts.

5.2 Proof of the ELP Theorem

5.2.1 Peripheral tight cuts

A nontrivial tight cut C of a matching covered graph G is *peripheral* if one of its shores, say X, has the property that there is no nontrivial proper subset X' of X such that $\partial(X')$ is a tight cut of G. In other words, C is peripheral if at least one of the C-contractions of G is free of nontrivial tight cuts. Clearly, any matching covered graph which has nontrivial tight cuts has a peripheral tight cut. We shall establish a property which all such tight cuts have. This will prove to be a key step in our proof of the ELP Theorem.

A PROPERTY OF PERIPHERAL TIGHT CUTS

Lemma 5.1 *Let $C := \partial(X)$ be a peripheral tight cut of G. Then there exists an edge $e := uv \in C$, with $u \in X$ and $v \in \overline{X}$, such that both the subgraphs $G[X-u]$ and $G[\overline{X}-v]$ are connected.*

Proof By hypothesis, C is peripheral. Adjust notation so that no cut $\partial(X')$ is tight, for any nontrivial proper subset X' of X. Consider the two C-contractions

$$G/\overline{X} := G/(\overline{X} \to \overline{x}) \text{ and } G/X := G/(X \to x)$$

of G, and note that

$$G[X] = (G/\overline{X}) - \overline{x} \text{ and } G[\overline{X}] = (G/X) - x.$$

As C is a tight cut of G, both $G/\overline{X}$ and G/X are matching covered. Furthermore, by the minimality of X, the graph $G/\overline{X}$ is free of nontrivial tight cuts.

Let us first show that C contains an edge $e := uv$, where v lies in $\overline{X}$, such that $G[\overline{X}-v]$ is connected. The graph G/X, being matching covered, is 2-connected, and therefore the graph $G[\overline{X}] = (G/X) - x$ is connected. If $G[\overline{X}]$ happens to be 2-connected then $G[\overline{X}-v]$ would be connected, for every vertex v of $\overline{X}$. We may thus assume that $G[\overline{X}]$ has two or more blocks. Then it has an end block F that contains precisely one cut vertex, say w, of $G[\overline{X}]$. Now, as G/X is 2-connected, it follows that F contains a vertex v, different from w, which is incident with an edge $e = uv$ of C (otherwise, w would be a cut vertex of G). Since $v \neq w$, it is not a cut vertex of $G[\overline{X}]$, and hence $G[\overline{X}-v]$ is connected (see Figure 5.2).

To complete the proof, we now proceed to show that the graph $G[X-u]$ is connected. For this, assume the contrary, that $G[X-u]$ is disconnected. As the graph $G/\overline{X}$ is matching covered, and $G[X-u] = (G/\overline{X}) - \overline{x} - u$, it follows that $\{u, \overline{x}\}$ is either a barrier or a 2-separation of $G/\overline{X}$. But the ends of e in $G/\overline{X}$ are u and $\overline{x}$. Moreover, as $G/\overline{X}$ is matching covered, the edge e is matchable in $G/\overline{X}$. Consequently, $\{u, \overline{x}\}$ is not a barrier of $G/\overline{X}$. It follows that $\{u, \overline{x}\}$ is a 2-separation of $G/\overline{X}$. Thus, $G/\overline{X}$ has nontrivial tight cuts, contradicting the minimality of X. This proves the assertion. □

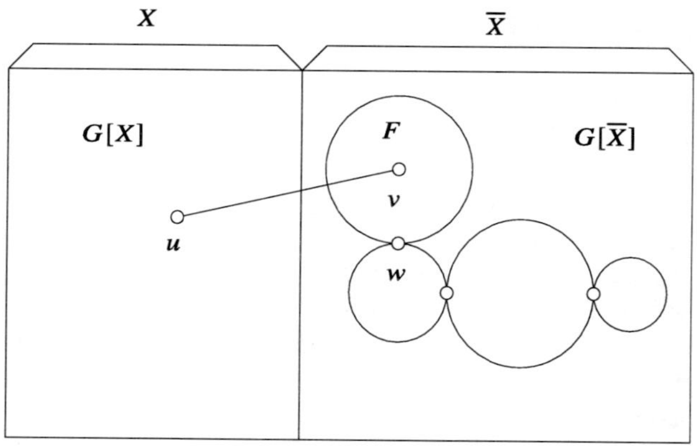

Fig. 5.2 An edge uv in C such that $G[\overline{X} - v]$ is connected

It should be noted that the property of peripheral tight cuts established above is not valid in general for all nontrivial tight cuts in matching covered graphs. For example, the cut C which is a nonperipheral tight cut in the matching covered graph shown in Figure 5.3 does not have the desired property (see Exercise 5.2.2).

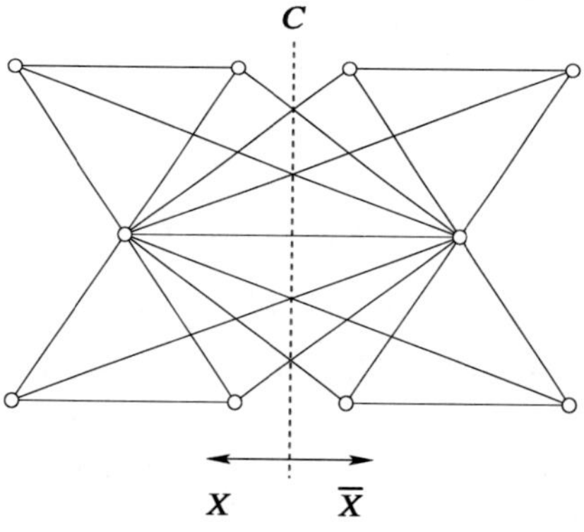

Fig. 5.3 A nonperipheral cut in a matching covered graph

We shall use Lemma 5.1 to show that any matching covered graph G which has nontrivial tight cuts either has a nontrivial barrier (associated with a nontrivial barrier cut), or it has a 2-separation (in which case, it has a 2-separation cut). We are able to do this by considering matchable graphs which are closely related to the graph G. For example, with the notation used in the statement of Lemma 5.1, the graph $G - u - v$ is a matchable graph in which no edge in the cut $\partial(X - u)$ is matchable (Exercise 5.2.1). Under certain conditions, it turns out that this graph $G - u - v$ has a barrier which can be extended to a barrier of G by adding to it either u or v. This line of inquiry has led us to study the existence of a special type of barriers in matchable graphs which we now proceed to introduce.

5.2.2 Dulmage-Mendelsohn barriers

Recall that if B is a barrier of a matchable graph G, the core of G with respect to B, denoted by $\mathbb{H}(B)$, is the bipartite graph obtained from G by deleting edges with both ends in B, deleting the vertices in the even components of $G - B$, and then shrinking each odd component of $G - B$ to a single vertex (see Section 3.2).

DULMAGE-MENDELSOHN BARRIERS

A nonempty barrier B of a matchable graph G is a *Dulmage-Mendelsohn barrier* if:

- the core $\mathbb{H}(B)$ of G with respect to the barrier B is matching covered, and
- each odd component of $G - B$ is critical.

For brevity, we shall refer to a Dulmage-Mendelsohn barrier as a *DM-barrier.*

DM-barriers have a fundamental property that will be used in our proof of the ELP Theorem (Exercise 5.2.3).

Lemma 5.2 *Let B denote a DM-barrier of a matchable graph G. Then every edge of $\mathbb{H}(B)$ is matchable in G.* □

An essential ingredient in our proof of the ELP Theorem is the assertion that every maximal barrier of a matchable graph contains a DM-barrier as a subset. Thus, let G be any matchable graph, and let B^* be a maximal barrier of G. Then B^* is one of the parts of the bipartition of the core $\mathbb{H}(B^*)$ of G with respect to B^*. The other part of the bipartition of $\mathbb{H}(B^*)$, which we denote by A^*, consists of vertices resulting from shrinking the odd components of $G - B^*$. Proposition 2.11, applied to the bipartite graph $\mathbb{H} := \mathbb{H}[A^*, B^*]$, yields the following result:

5.3 *There exists a nonempty subset S of A^* such that the subgraph of $\mathbb{H}$ induced by the set $S \cup N(S)$, where $N(S)$ is the neighbour set of S in $\mathbb{H}$, is matching covered.* □

MAXIMAL BARRIERS AND DM-BARRIERS

Lemma 5.4 *Every maximal barrier B^* of a matchable graph G includes a Dulmage-Mendelsohn barrier B. Moreover, every odd component of $G - B$ is an odd component of $G - B^*$.*

Proof As described above, let $\mathbb{H} := \mathbb{H}[A^*, B^*]$ denote the core of G with respect to the barrier B^*, where A^* is the set of vertices resulting from shrinking odd components of $G - B^*$. By Theorem 3.2 every component of $G - B^*$ is odd and critical. Furthermore, for any perfect matching of G, the set $M \cap E(\mathbb{H})$ is a perfect matching of the core. Thus, $\mathbb{H}[A^*, B^*]$ is a matchable bipartite graph. Hence, by (5.3), there exists a nonempty subset S of A^* such that the subgraph of $\mathbb{H}$ induced by $S \cup N(S)$ is matching covered. It follows that $B := N(S)$ is a nonempty barrier of

$\mathbb{H}$. Moreover, the core $\mathbb{H}(B)$ of G with respect to B is the same as the subgraph of $\mathbb{H}[A^*, B^*]$ induced by $S \cup B$, implying that $\mathbb{H}(B)$ is matching covered. As the barrier B^* is maximal, each odd component of $G - B^*$ is critical by Theorem 3.2. Thus, each odd component of $G - B$, which is also an odd component of $G - B^*$, is critical. We conclude that B is a DM-barrier of G. □

5.2.3 A key lemma

There is one more intermediate result that is needed before we are able to present a proof of the ELP Theorem. It asserts the existence of special types of DM-barriers in matchable graphs with certain very specific properties.

DM-BARRIERS INCLUDED IN SHORES OF CUTS

Lemma 5.5 *Let G be a matchable graph, and let X be a nonempty proper subset of $V(G)$ such that:*

1. *both the subgraphs $G[X]$ and $G[\overline{X}]$ are connected, and*
2. *no edge in the cut $\partial(X)$ is matchable in G.*

Then G has a DM-barrier B which is a subset of one of the shores of $\partial(X)$ (that is, a subset of X or of $\overline{X}$). Furthermore, the vertex sets of all the odd components of $G - B$ are also subsets of that same shore.

Proof By hypothesis, G has perfect matchings but no edge of $\partial(X)$ is matchable. It follows that the graphs $G[X]$ and $G[\overline{X}]$ both have perfect matchings.

Consider first the case in which $\partial(X)$ is empty. Let $B^\star$ be any maximal barrier of $G[X]$. By Lemma 5.4, $G[X]$ has a DM-barrier B that is a subset of $B^\star$. This barrier B of $G[X]$ is also a DM-barrier of G. The assertion holds in this case.

We may thus assume that $\partial(X)$ is nonempty. Let $e := uv$ denote an edge of $\partial(X)$, where $u \in X$ and $v \in \overline{X}$. By hypothesis, e is not matchable. By Corollary 2.2, G has a barrier that contains both u and v. Let $B^\star$ be a maximal barrier of G that contains both u and v. By Lemma 5.4, some subset B of $B^\star$ is a DM-barrier of G.

Let us first show that if K is any odd component of $G - B$ then its vertex set $V(K)$ is a subset of one of the shores of the cut $\partial(X)$. Suppose that this is not the case. Then, both $V(K) \cap X$ and $V(K) \cap \overline{X}$ are nonempty. One of these sets has to be even and the other odd because $V(K)$ is an odd set. Without loss of generality, assume that $|V(K) \cap X|$ is even. Clearly, K is also an odd component of $G - B^*$. Since $B^* \cap X$ is not empty, $V(K) \cap X$ is a nonempty proper subset of X. As $G[X]$ is connected by the hypothesis, and K is a component of $G - B$, it follows that there is some edge, say e_1, which joins a vertex in $V(K) \cap X$ to a vertex in $B \cap X$. That edge e_1, being an edge of the graph $\mathbb{H}(B)$, is matchable in G, by Lemma 5.2.

Let M_1 be a perfect matching of G containing e_1. Since K is an odd component of $G - B$, the edge e_1 is the only edge of M_1 in $\partial(V(K))$. However, since $V(K) \cap X$ is an even set, $|M_1 \cap \partial(X \cap V(K))|$ is even. This implies that some edge with one end in $V(K) \cap X$ and one end in $V(K) \cap \overline{X}$ is in M_1. This is impossible because, by hypothesis, no edge in $\partial(X)$ is matchable. We conclude that $V(K)$ is a subset of one of the shores of $\partial(X)$.

Now observe that since B is a DM-barrier, graph $\mathbb{H}(B)$ is matching covered. Thus $\mathbb{H}(B)$ is connected. Moreover, by Lemma 5.2, each edge of $\mathbb{H}(B)$ is matchable in G. As no edge of $\partial(X)$ is matchable in G, it follows that B and the vertex sets of all the odd components of $G - B$ are all subsets of one and the same shore of $\partial(X)$. □

5.2.4 The ELP Theorem

EDMONDS-LOVÁSZ-PULLEYBLANK (1982, [32])

Theorem 5.6 *If a matching covered graph has a nontrivial tight cut, then it has a nontrivial tight cut which is an ELP cut.*

Proof We begin the proof by making the following observation:

5.6.1 *If G is not bicritical then it has a nontrivial barrier cut.*

Proof Suppose that G is not bicritical; let B denote a nontrivial barrier of G. As G is matching covered, it follows that every component of $G - B$ is odd. For each component K of $G - B$, the cut $\partial(K)$ is a barrier cut. If some component K of $G - B$ is nontrivial then $\partial(K)$ is also nontrivial, because B is nontrivial.

Suppose then that every component of $G - B$ is trivial. In that case, the graph G is bipartite, where B is one of the parts of the bipartition of G. By Theorem 4.8, every tight cut of G is a barrier cut. By hypothesis, G has nontrivial tight cuts. □

We shall now analyse several possibilities, all but one of which leads to the conclusion that G is not bicritical, a case already considered. Finally, one last case will lead to the conclusion that G has a 2-separation.

Let $C := \partial(X)$ be a nontrivial tight cut of G, and let $e := uv$, with $u \in X$ and $v \in \overline{X}$, be an edge of C such that both the subgraphs graphs $G[X - u]$ and $G[\overline{X} - v]$ are connected. (The existence of such a cut C, and edge e, is guaranteed by Lemma 5.1.) As C is tight, both C-contractions of G are matching covered, hence 2-connected. Thus, $G[X]$ and $G[\overline{X}]$ are also connected. In sum, the four graphs $G[X]$, $G[X - u]$, $G[\overline{X}]$ and $G[\overline{X} - v]$ are connected.

The rest of the proof consists of analysing various cases depending on the neighbourhoods of u and v.

Case 1 *The vertex v is the only neighbour of u in $\overline{X}$ and u is the only neighbour of v in X.*

Let $G' := G - u - v$. As G is matching covered, it has perfect matchings that contain e, and the restriction of any such perfect matching of G to $E(G')$ is a perfect matching of G'. Therefore, G' is matchable. Moreover,

$$C - e = \partial_{G'}(X - u) = \partial_{G'}(\overline{X} - v)$$

is a cut of G'. For every perfect matching M' of G', the set $M' + e$ is a perfect matching of G. As C is tight in G, the edge e has to be the only edge of $M' + e$ in C. It follows that no edge of $C - e$ is matchable in G'. By Lemma 5.5, the graph G' has a DM-barrier B' such that B', as well as the vertex sets of all the odd components of $G' - B'$, are subsets of one of $X - u$ and $\overline{X} - v$. Adjust notation so that $B' \subseteq X - u$. (See Figure 5.4 for an illustration, and Exercise 5.2.4 for a concrete example.)

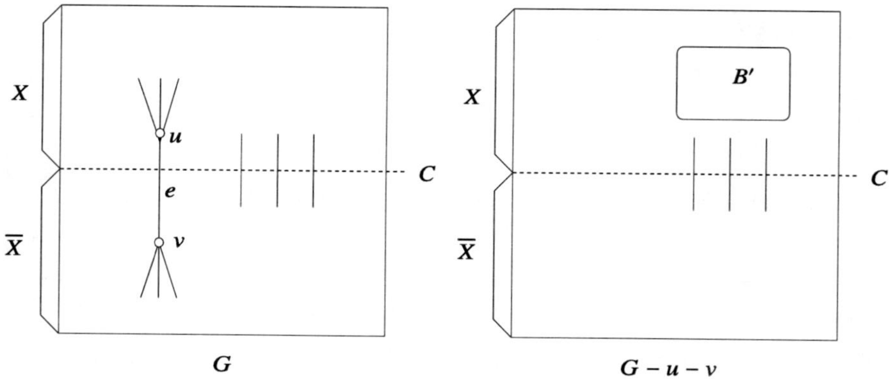

Fig. 5.4 $\partial(u) \cap \overline{X} = \{v\}$ and $\partial(v) \cap X = \{u\}$

Let $B := B' + u$. By hypothesis, u is the only vertex of X adjacent to v. Thus, v is not adjacent to any vertex in any of the $|B| - 1$ odd components of $G' - B'$. It follows that all the $|B| - 1$ odd components of $G' - B'$ are also odd components of $G - B$. By parity, B is a nontrivial barrier of G. We deduce that G is not bicritical. By statement 5.6.1, G has a nontrivial barrier cut.

Case 2 *Either u has two or more neighbours in $\overline{X}$, or v has two or more neighbours in X.*

We may adjust notation so that u has at least two neighbours in $\overline{X}$, and let $R := \partial(u) - C$. As $G[X]$ is connected and nontrivial, the set R is not empty. Now consider the graph $G'' := G - R$, together with the cut $D := \partial(X - u) - R = \partial(\overline{X} + u) - R$. (See Figure 5.5.)

5.6.2 *The graphs $G''[X - u]$ and $G''[\overline{X} + u]$ are both connected. Furthermore, the graph G'' is matchable and no edge in the cut D is matchable in G''.*

Proof Note that the graph $G''[X - u]$ is the same as the graph $G[X - u]$, which is connected by the choice of the cut C and the edge e.

The graph $G[\overline{X}]$ is connected and the vertex u is adjacent to vertices of $\overline{X}$ (v is one such vertex). Thus, the second graph $G''[\overline{X} + u]$ is connected as well.

Every perfect matching of G that contains edge e is also a perfect matching of G''. Therefore, the graph G'' is matchable. If possible, let M be a perfect matching of G'' which has an edge in the cut D. Since D is an even cut of G'', this perfect

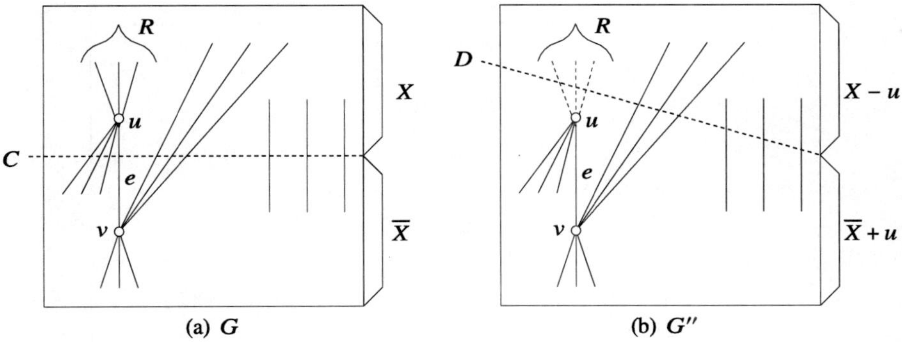

Fig. 5.5 The graphs G and G''; dashed lines indicate removed edges

matching M would have to meet D in at least two edges, and we would have $|M \cap C| > |M \cap D| \geq 2$. As M is also a perfect matching of G, and C is a tight cut of G, this is impossible. We conclude that no edge of D is matchable in G''. □

By Lemma 5.5, it now follows that G'' has a DM-barrier B'' such that B'' and the vertex sets of all the odd components of $G'' - B''$ are subsets of one of $X - u$ and $\overline{X} + u$. The rest of the analysis depends on where the vertex u is in relation to the barrier B''.

Case 2.1 Vertex u lies in an even component of $G'' - B''$.

In this case, $B := B'' + u$ is a nontrivial barrier of G''. But $G'' - B = G - B$, and therefore B is a nontrivial barrier of G. We conclude that G is not bicritical. By statement 5.6.1, G has a nontrivial barrier cut.

Suppose now that u is either in B'' or in the vertex set of one of the odd components of $G'' - B''$. Since u is in $\overline{X} + u$, it follows that B'' and the vertex sets of all the odd components of $G'' - B''$ are subsets of $\overline{X} + u$. As $G''[X - u]$ is connected, we conclude that $X - u$ is a subset of the vertex set of an even component, say L, of the graph $G'' - B''$.

Case 2.2 Vertex u lies in B''.

In this case, $G'' - B'' = G - B''$. Consequently, L is an even component of $G - B''$. Let w be any vertex of L. Then, $B'' + w$ is a nontrivial barrier of G and we conclude that G is not bicritical. By statement 5.6.1, G has a nontrivial barrier cut.

Now there is just one possibility that remains to be analysed, namely the case in which u lies in some odd component of $G'' - B''$.

Case 2.3 Vertex u lies in some odd component, K, of $G'' - B''$.

In this case, the even component L of $G'' - B''$ contains the graph $G[X - u]$. See Figure 5.6.

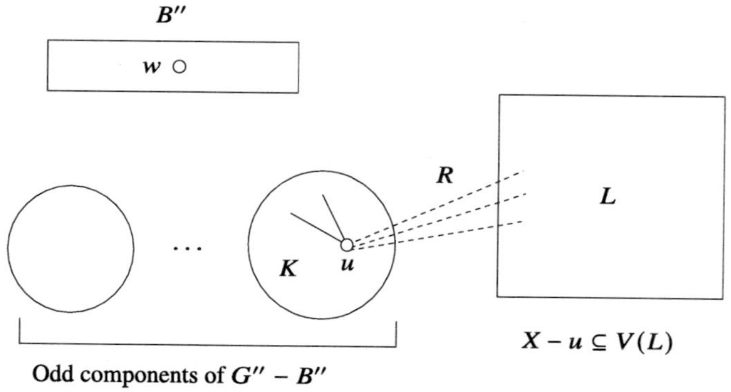

Fig. 5.6 Vertex u is in an odd component K of $G'' - B''$

Clearly, no edge of G'' joins vertices in two different components of $G'' - B''$. But the edges in $R = E(G) - E(G'')$ may join vertices belonging to different components of $G'' - B''$. However, each edge in R joins u to a vertex in $X - u$. Thus, each odd component of $G'' - B''$ other than K is also an odd component of $G - B''$. We may thus conclude that $o(G - B'') \geq |B''| - 1$, and hence, by parity, that B'' is in fact a barrier of G itself.

If B'' is nontrivial, then G is not bicritical. Assume that B'' has just one vertex, say w. By the hypothesis of the case under consideration, the vertex u has at least two neighbours in $\overline{X}$. Thus u has a neighbour in $\overline{X}$ that is different from w, and that neighbour has to be a vertex in K, implying that K is nontrivial. In this case, $G - u - w$ is not connected; hence either $\{u, w\}$ is a nontrivial barrier of G or $\{u, w\}$ is a 2-separation of G, implying that $\partial(V(L) + u)$ is a 2-separation cut of G. If G is not bicritical then, by statement 5.6.1, G has a nontrivial barrier cut. We conclude that G has a nontrivial ELP cut.

In all cases considered, we deduced that the existence of a nontrivial tight cut in G implies that G has a nontrivial ELP cut. The proof of the ELP Theorem is complete. □

The ELP Theorem implies the following characterization of bricks, as discovered by Edmonds, Lovász and Pulleyblank [32].

CHARACTERIZATION OF BRICKS

Theorem 5.7 *A nonbipartite matching covered graph is a brick if and only if it is 3-connected and bicritical.* □

The above result suggests an obvious polynomial-time algorithm for recognizing bricks. (Exercise 5.2.8).

The ELP Theorem simply says that every matching covered graph that has a nontrivial tight cut also has an ELP cut. Theorem 4.8 provides an explicit characterization of tight cuts in bipartite matching covered graphs. But the nature of tight cuts in nonbipartite matching covered graphs remains a mystery. In Section 5.5, we shall establish a precise description of tight cuts in matching covered graphs with at most two bricks and present an extension of that result, concerning tight cuts in arbitrary matching covered graphs.

Exercises

▷**5.2.1** *Let G be a matching covered graph, let $\partial(X)$ be a tight cut of G, and let uv be an edge of G, with $u \in X$ and $v \in \overline{X}$ (as in Lemma 5.1). Show that $G - u - v$ is a matchable graph in which no edge in the cut $\partial(X - u)$ is matchable.*

▷**5.2.2** *Show that the cut $C = \partial(X)$ in the matching covered graph G shown in Figure 5.3 is a nonperipheral tight cut and, furthermore, that for any edge uv of the cut with $u \in X$ and $v \in \overline{X}$, either $G[X - u]$ or $G[\overline{X} - v]$ is disconnected.*

▷**5.2.3** *Prove Lemma 5.2. (Hint: prove that every perfect matching of $\mathbb{H}(B)$ is extendable to a perfect matching of G.)*

5.2.4 *Consider the graph G shown in Figure 5.7 with the indicated cut C. Use the method of the proof of Theorem 5.6, Case 1 to find a barrier in G.*

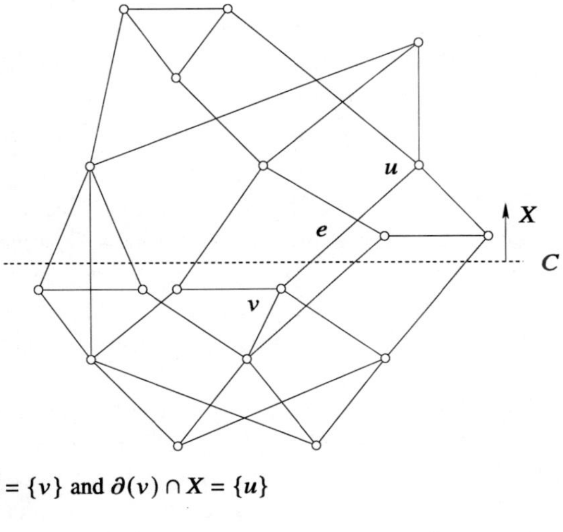

Fig. 5.7 $\partial(u) \cap \overline{X} = \{v\}$ and $\partial(v) \cap X = \{u\}$

▷**5.2.5** *Use Theorem 5.6 to show that the following five families of matching covered graphs defined in Section 2.5 are bricks:*

(i) *odd wheels (W_{2k+1}, for $k \geq 1$),*
(ii) *prisms of order* $2p \pmod 4$ *($\mathbb{P}_{2n}$, for odd $n \geq 3$),*
(iii) *Möbius ladders of order* $0 \pmod 4$ *($\mathbb{M}_{2n}$, for even $n \geq 2$),*
(iv) *staircases ($\mathbb{S}_{2n}$, $n \geq 3$), and*
(v) *truncated biwheels ($\mathbb{T}_{2n}$, $n \geq 3$).*

The bricks belonging to the above five families are known as Norine-Thomas bricks *and will play an important role in Part II.*

5.2.6 *Let G be a brick and let $G + e$ be a graph obtained from G by adding an edge e joining two vertices of G. Show that $G + e$ is also a brick.*

$*$ **5.2.7** *Let G be a brick and let $C := \partial(X)$ be a 3-cut of G. Prove that if the subgraph $G[X]$ of G induced by X is bipartite then X is trivial.*

$*$ **5.2.8** *Based on Theorem 5.7, describe a polynomial-time algorithm, which, given a nonbipartite matching covered graph G, either attests that G is a brick or finds a nontrivial tight cut of G.*

5.3 Some Applications of the ELP Theorem

In this section we note two interesting consequences of the ELP Theorem.

Tight cuts in cubic graphs

Theorem 5.8 *In a cubic matching covered graph every tight cut is a 3-edge-cut.*

Proof The smallest cubic matching covered graph is the θ-graph (two vertices joined by three parallel edges), and the statement clearly holds for this graph. We shall prove the validity of the statement in general by induction on the number of vertices. Thus let G be any cubic matching covered graph on four or more vertices.

If G is either a brick or a brace, then the only tight cuts of G are the trivial cuts which are obviously 3-edge-cuts. If not, by the ELP Theorem, G has either a nontrivial barrier cut or a 2-separation cut. However, by Exercise 4.2.5(i), every 2-separation cut in G is also a barrier cut. Also, by the second part of the same exercise, every barrier cut in G is a 3-edge-cut. Let $D := \partial(Y)$ denote any nontrivial barrier cut of G. Let $C = \partial(X)$ be any nontrivial tight cut of G. Our objective is to show that C is a 3-edge cut.

Since D is a 3-edge-cut, both D-contractions of the cubic graph G are also cubic. If C does not cross D, then it is a tight cut in one of the D-contractions of G and hence, by induction, is a 3-edge of that D-contraction which clearly is also a 3-edge-cut of G itself. So, we may assume that C and D cross. As usual, fix notation so that $X \cap Y$ is odd. Then, by Theorem 4.12, the two cuts $I := \partial(X \cap Y)$ and $U := \partial(\overline{X} \cap \overline{Y})$ are tight, and there are no edges joining vertices in $X \cap \overline{Y}$ to vertices in $\overline{X} \cap Y$, implying the following equality:

$$|C| + |D| = |I| + |U|.$$

The cut D, being a barrier cut of the cubic graph G, is a 3-edge-cut. The two cuts I and U, being tight cuts of the D-contractions of G, are (by induction) 3-edge-cuts. The above equation now implies the stated result. □

Corollary 5.9 *In a cubic matching covered graph, every tight cut is a barrier cut.*

Proof Let C be any tight cut in a cubic matching covered graph G. By the above theorem, C is a 3-edge cut, and by Theorem 4.5, C is a barrier cut. □

ESSENTIALLY 4-EDGE-CONNECTED GRAPHS

We recall this definition, given in Section 4.3. A 3-edge-connected graph is *essentially* 4-*edge-connected* if all its 3-edge cuts are trivial. Thus, a 3-edge-connected graph G is essentially 4-edge-connected if $|C| \geq 4$, for any nontrivial cut C of G.

For future use, we note the following consequences of Theorem 5.8.

Corollary 5.10 *Every essentially* 4*-edge-connected cubic graph is either a brick or a brace.* □

Corollary 5.11 *Every nonbipartite essentially* 4*-edge-connected cubic graph is a brick.* □

It should be noted that the above-stated condition is not a necessary condition for a cubic graph to be a brick. This is because nonbipartite cubic graphs, such as $\overline{C_6}$, may have 3-edge cuts which are separating but not tight. However, any 3-edge cut in a cubic graph is a separating cut (Exercise 4.1.2), and every separating cut in a bipartite graph is tight (Corollary 4.9). These two facts, together with Theorem 5.8 now imply the following characterization of cubic braces of order six or more.

Corollary 5.12 *A bipartite cubic graph of order six or more is a brace if and only if it is essentially* 4*-edge-connected.* □

Tight cuts in bicritical graphs

Theorem 5.13 *Any tight cut decomposition of a bicritical matching covered graph consists entirely of bricks.*

Proof Let G be a bicritical graph. If G is itself a brick, then there is nothing more to prove. We shall establish the assertion in general by induction on $|V|$.

We first note that since G is bicritical, it follows that it cannot have any nontrivial barriers and, consequently, it cannot have nontrivial barrier cuts. Thus, by the ELP Theorem, G must have a 2-separation cut. Let C be such a cut and let G_1 and G_2 denote the two C-contractions of G. Using the fact that G is bicritical and C is a 2-separation cut, it is easy to show that both G_1 and G_2 are also bicritical (Exercise 5.3.1). By the induction hypothesis, it follows that any tight cut decompositions of G_1 and of G_2 consists entirely of bricks, implying that any tight cut decomposition of G starting with C consists just of bricks. Now Lovász's Unique Decomposition Theorem 4.17, implies that every tight cut decomposition of G consists only of bricks. □

We note that the analogue of the above statement for separating cut decompositions does not hold, in the sense that a separating cut decomposition of a bicritical graph may not consist exclusively of bicritical graphs (see Exercise 5.3.2). However, interestingly, the converse of the above statement holds not only for tight cut decompositions, but more generally for separating cut decompositions, in the sense that any graph which has a separating cut decomposition consisting exclusively of bicritical graphs is itself bicritical. This is a consequence of the following proposition:

Proposition 5.14 *Let C be a separating cut of a matching covered graph G. If both C-contractions of G are bicritical, then G is also bicritical.* □

The above assertion is equivalent to the statement of Exercise 3.3.8 which says that any matching covered graph obtained by splicing two bicritical graphs is also bicritical. The following is an interesting and useful result concerning graphs obtained by splicing two bricks.

Proposition 5.15 *Let C be a separating cut of a matching covered graph G. If both C-contractions of G are bricks, then G is also a brick unless the cut C is a 2-separation cut of G.*

Proof Let $G_1 := G/(\overline{X} \to \overline{x})$ and $G_2 := G/(X \to x)$ be the two C-contractions of G. By the hypothesis, both G_1 and G_2 are bricks which, by the ELP Theorem, are bicritical and also 3-connected. Using the fact that both G_1 and G_2 are bicritical and Proposition 5.14 we deduce that G is bicritical. If G is also 3-connected, then it would follow from the ELP Theorem that G is a brick. Being matching covered, G is 2-connected and hence has no cut vertices. Thus, if G is not 3-connected, it must have a 2-vertex cut. Since G_1 is 3-connected, it follows that for any 2-subset S of $V(G_1) - \overline{x}$ the graph $G - S$ is connected. Similarly, if S is any 2-subset of $V(G_2) - x$, the graph $G - S$ is connected. Thus, if G does have a 2-vertex cut S, then S has one vertex, say u in $V(G_1) - \overline{x}$, and one vertex, say v in $V(G_2) - x$. As G_1 is 3-connected, the graph $G[X - u] = G_1 - \overline{x} - u$ is connected. Likewise, $G[\overline{X} - v] = G_2 - x - v$ is connected. As $G - u - v$ is not connected, we conclude that every edge of C is incident with a vertex in $\{u, v\}$. In this case, $G - u - v$ has two even components, one with the vertex set $V(G_1) - \overline{x} - u$ and one with the vertex set $V(G_2) - x - v$, and C is a 2-separation cut of G associated with the 2-separation $\{u, v\}$. □

Exercises

5.3.1 *Let C be a 2-separation of a bicritical graph G, and let G_1 and G_2 be the two C-contractions of G. Show that both G_1 and G_2 are also bicritical.*

5.3.2 *Consider the graph G in Figure 5.8. Show that G is bicritical and that C is a separating cut in G such that one of the C-contractions of G is not bicritical.*

▷**5.3.3** *Prove that a matching covered graph of order four or more is bicritical if and only if its tight cut decomposition consists solely of bricks.*

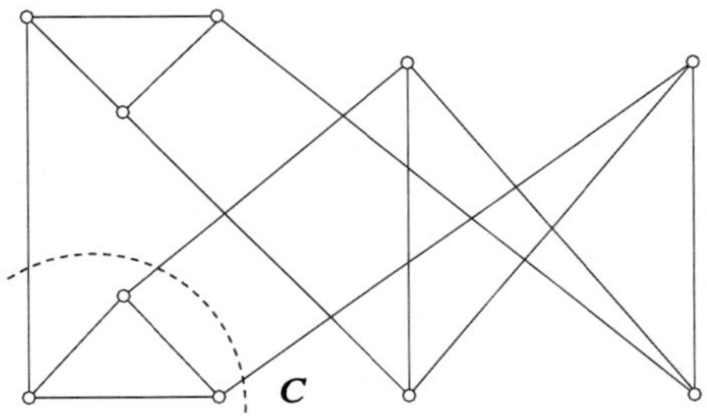

Fig. 5.8 The graph G is bicritical, but one of its C-contractions is not

5.3.4 *Let G be a cubic matching covered graph of order n.*

(i) *Show that if G is not a brick and $n \geq 4$ then G is not bicritical.*
(ii) *Using the previous part and Theorem 5.8, prove that $b(G) \leq n/4$, (and show that this bound is best possible by giving for each $k \geq 1$, an example of a cubic graph on $4k$ vertices with exactly k bricks).*

EDGE-EXTENSIONS

Let H be a cubic graph, and let u_1v_1 and u_2v_2 be two disjoint edges of H. The cubic graph obtained from H by (i) subdividing the edge u_1v_1 by inserting a vertex w_1 between u_1 and v_1, (ii) subdividing the edge u_2v_2 by inserting a vertex w_2 between u_2 and v_2, and (iii) adding a new edge w_1w_2 joining w_1 and w_2, is said to be obtained from H by an *edge-extension based on* $\{u_1v_1, u_2v_2\}$. (The graph in Figure 5.9 is a graph obtained from the cube by a sequence of two edge-extensions.)

▷**5.3.5**

(i) *Let H be an essentially 4-edge-connected cubic graph. Show that any graph G obtained from H by an edge-extension is also an essentially 4-edge-connected cubic graph.*
(ii) *Wormald (1979, [94]) showed that any essentially 4-edge-connected cubic graph of order four or more can be obtained from either K_4 or from the cube by a sequence of edge-extensions. Show how $K_{3,3}$, $\mathbb{M}_8$, and the Petersen graph $\mathbb{P}$ can be obtained from K_4 by means of edge-extensions.*

▷**5.3.6** *Let G be a matching covered graph, let $C := \partial(X)$ be an odd cut of G, and let $G[X]$ denote the subgraph of G induced by X. Prove the following statements:*

(i) *if G is bicritical then $G/\overline{X}$ is bicritical if and only if $G[X]$ is critical,*
(ii) *if G is 3-connected then $G/\overline{X}$ is 3-connected if and only if $G[X]$ is 2-connected,*
(iii) *if G is a brick then $G/\overline{X}$ is a brick if and only if $G[X]$ is critical and 2-connected.*

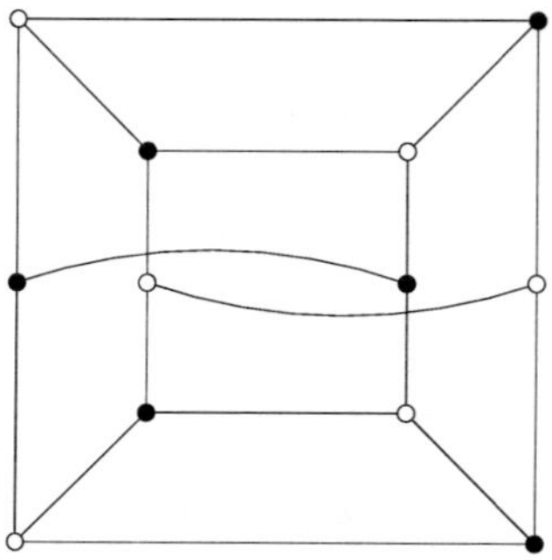

Fig. 5.9 A graph obtained from the cube by two edge-extensions

5.4 Characterizations of Braces

We now turn to the problem of characterizing braces. Any bipartite matching covered graph of order at most four is a brace. This is not the case with graphs of order six or more. In the ensuing discussion, we shall restrict ourselves to graphs with at least six vertices. We present two characterizations of braces; the first is a characterization in terms of a 'Hall-type' property, and the second implies a polynomial-time recognition algorithm for braces (and also shows that braces are the same as 2-extendable bipartite graphs). Both these characterizations rely on the fact that all tight cuts in bipartite matching covered graphs are barrier cuts, and conform to the pattern described in Theorem 4.8. For the convenience of the reader, a profile of a tight cut $C := \partial(X)$ in a bipartite matching covered graph is displayed in Figure 5.10.

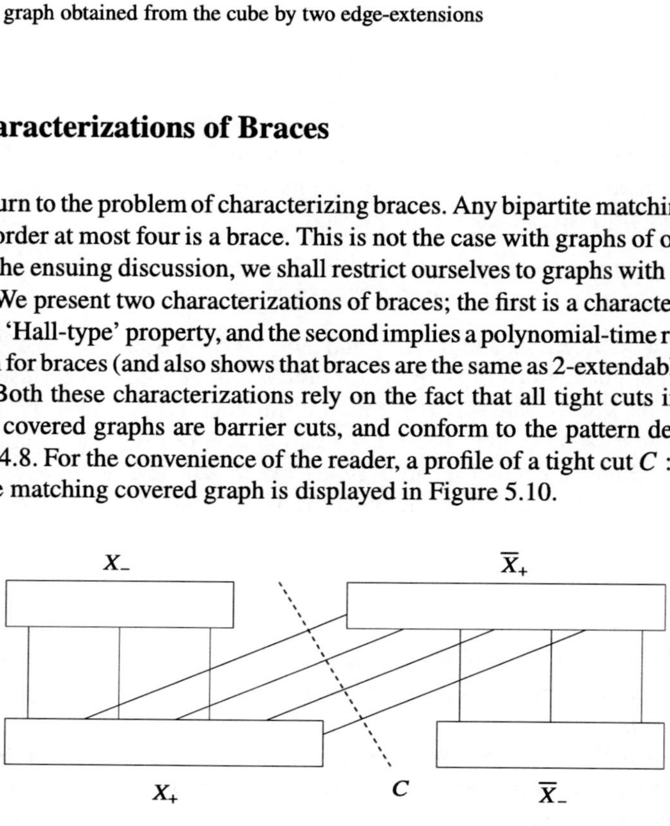

$N(X_-) = X_+$ and $|X_+| = |X_-| + 1$

Fig. 5.10 Profile of a tight cut in a bipartite matching covered graph

The next result is similar to Theorem 2.9.

A HALL-TYPE CHARACTERIZATION OF BRACES

Theorem 5.16 *A bipartite graph $G[A, B]$, with $|A| = |B| \geq 3$, is a brace if and only if*

$$|N(X)| \geq |X| + 2, \quad \textit{for all } X \subset A \textit{ such that } 0 < |X| < |A| - 1. \tag{5.1}$$

Proof Let us first suppose that G is a brace. Then G is, in particular, a matching covered graph. By Theorem 2.9, $|N(S)| \geq |S| + 1$ for each nonempty proper subset S of A. Suppose that for some subset S of A, with $0 < |S| < |A| - 1$, the equality $|N(S)| = |S| + 1$ holds. Then it is easy to see that $\partial(X)$, where $X := S \cup N(S)$, is a nontrivial tight cut of G (Exercise 5.4.1). We conclude that the property (5.1) must hold.

Conversely, suppose that the property (5.1) holds. Then it is straightforward to verify (Exercise 5.4.1) that G is matching covered. If G were not a brace, it would have a nontrivial tight cut $\partial(X)$. By interchanging X and $\overline{X}$, if necessary, we may assume that X_- is a subset of A. By Theorem 4.8, the set X_- violates (5.1). We conclude that G is a brace. □

The next result is similar to Theorem 3.8.

A SECOND CHARACTERIZATION OF BRACES

Theorem 5.17 *A bipartite graph $G[A, B]$, with $|A| = |B| \geq 3$, is a brace if and only if*

$$G - u_1 - u_2 - v_1 - v_2 \textit{ is matchable,} \tag{5.2}$$

for any pair $\{u_1, u_2\}$ of distinct vertices in A and any pair $\{v_1, v_2\}$ of distinct vertices in B.

Proof Let us first suppose that G is a brace. Then, by Theorem 5.16, the property (5.1) must hold. Now let u_1 and u_2 be any two distinct vertices of A, and let v_1 and v_2 be any two distinct vertices of B. Then $H := G - u_1 - u_2 - v_1 - v_2$ is a bipartite graph with bipartition $(A - u_1 - u_2, B - v_1 - v_2)$. For any subset S of $A - u_1 - u_2$, clearly, $N_H(S) = N_G(S) - v_1 - v_2$. Thus, it follows from inequality (5.1) that $|N_H(S)| \geq |S|$ for any subset S of $A - u_1 - u_2$ and, hence, by Hall's Theorem 1.5, that H is matchable.

Conversely, suppose that the property (5.2) holds. Let X be a proper subset of A such that $0 < |X| < |A| - 1$. Then $|A - X| \geq 2$; let u_1 and u_2 be two distinct vertices of $A - X$. Let v_1 and v_2 be two distinct vertices of B such that $I := \{v_1, v_2\} \cap N_G(X)$ is as large as possible. Let $H := G - u_1 - u_2 - v_1 - v_2$. By hypothesis, H is matchable. By Hall's Theorem,

$$|N_G(X)| = |N_H(X)| + |I| \geq |X| + |I|. \tag{5.3}$$

We now show that $|I| = 2$. By definition, X is nonempty. From (5.3) we infer that

$$|N_G(X)| \geq |I| + 1. \tag{5.4}$$

From inequality (5.4) we deduce that $N_G(X)$ is nonempty. By the maximality of I, it follows that $|I| \geq 1$. Again, from inequality (5.4) we infer that $|N_G(X)| \geq 2$. By the maximality of I, we conclude that $|I| = 2$.

From (5.3) we infer that (5.1) holds for X. This conclusion holds for each proper subset X of A such that $0 < |X| < |A| - 1$. By Theorem 5.16, G is a brace. □

Braces are 2-extendable

Recall that a graph G is said to be 2-extendable if, given any two disjoint edges e and f of G, there exists a perfect matching of G which contains $\{e, f\}$. We leave the proof of the following theorem as an exercise.

Theorem 5.18 *A connected bipartite graph of order six or more is a brace if and only if it is 2-extendable.* □

The above theorem suggests an obvious polynomial-time algorithm for determining whether or not a given bipartite graph is a brace. Let $G[A, B]$ be a bipartite graph on n vertices and m edges, and assume that a perfect matching M has been determined. It can be checked in linear time $O(m)$ if G is matching covered in the following way: let D be a directed graph obtained from G by directing every edge of $G - M$ from A to B and directing every edge in M from B to A. Then G is matching covered if and only if D is strongly connected (see Exercise 2.2.6). The strongly connected property is well known and an $O(m)$-time procedure is described, for instance, in Section 22.5 of [25]. Now, to check if a bipartite graph G is a brace, it suffices to remove the ends of every edge of G and decide if the resulting graph is matching covered. This procedure can be computed in polynomial-time. A more mindful reader may consider a faster algorithm (Exercise 5.4.4).

The following three factors imply the existence of a polynomial-time algorithm to determine a tight cut decomposition of a matching covered graph (Exercise 5.4.5):

(i) the polynomial-time algorithm for recognition of braces (Exercise 5.4.4),
(ii) the polynomial-time algorithm for recognition of bricks (Exercise 5.2.8), and
(iii) the fact that the number of bricks and braces in a tight cut decomposition of a matching covered graph of order n is bounded above by $n/2 - 1$ (Exercise 4.3.5).

Exercises

▷**5.4.1** *Supply the missing details in the proof of Theorem 5.16.*

5.4.2 *Let $H[A, B]$ be a brace on four or more vertices. Let a_1, a_2 be any two vertices in A, and b_1, b_2 be any two vertices in B. Show that the graph G obtained from H by adding the two edges a_1a_2 and b_1b_2 is a brick.*

∗ **5.4.3** *Prove Theorem 5.18.*

∗ **5.4.4** *Let G be a connected bipartite graph of order six or more and let M be a perfect matching of G. Show that G is a brace if and only if $G - u - v$ is matching covered, for every edge uv of M. Based on this property, design a polynomial-time algorithm, which, given G and a perfect matching of G, either attests that G is a brace, or determines a nontrivial tight cut of G. Hint: Exercise 2.2.6.*

∗ **5.4.5** *Describe a polynomial-time algorithm to determine a tight cut decomposition of a given matching covered graph.*

∗ **5.4.6** *Let $G[A, B]$ be any brace of order at least six. Show that, for any two vertices u and v of G, the graph $G - u - v$ is connected (and thus, that G is 3-connected). Furthermore show that if S is any subset of V of cardinality three which is neither a subset of A, nor a subset of B, then $G - S$ is connected.*

5.5 Tight Cuts in Graphs with Two Bricks

Barrier cuts and essentially 2-separation cuts

Every tight cut in a matching covered graph with at most one brick is an ELP cut. More precisely, if C is a tight cut in a matching covered graph G with $b(G) \leq 1$ then one of the two shores of C is bipartite, and it follows from Theorem 4.7 that C is a special barrier cut. But, as we have seen, a matching covered graph with more than two bricks may have tight cuts which are not ELP cuts. In their attempts to confirm a conjecture of Lovász (Chapter 15), CLM found it necessary to have a precise description of all tight cuts in matching covered graphs with two bricks. In this special case, they found that every tight cut is either an ELP cut or is close to being one in the sense to be described below[1].

In Chapter 4, we defined a barrier B of a matching covered graph G to be *special* if $G - B$ has precisely one nontrivial component. Thus, if G is nonbipartite, and B is a special barrier of G, then $G - B$ has one nonbipartite component, say K, while the remaining $|B| - 1$ components are trivial. We also referred to $\partial(K)$ as the *special barrier cut* associated with B.

ESSENTIALLY 2-SEPARATION CUTS

Let G be a nonbipartite matching covered graph. An *essential* 2*-separation* of G is a pair $\{S, T\}$ of two disjoint odd subsets S and T of $V(G)$ such that:

- both $G[S]$ and $G[T]$ are bipartite subgraphs of G such that their majority parts S_+ and T_+ are special barriers of G,

[1] We recommend to the first-time readers who wish to gain an overview of the basic results of the subject to skip this section and return to it at a later stage.

- the cut $\partial(S)$ is the special barrier cut associated with the barrier S_+ and the cut $\partial(T)$ is the special barrier cut associated with the barrier T_+, and
- the pair $\{s, t\}$ is a 2-separation of the graph $G/(S \to s)/(T \to t)$.

Clearly, if $\{S, T\}$ is an essential 2-separation of G, than any 2-separation cut of $G/(S \to s)/(T \to t)$ associated with the 2-separation $\{s, t\}$ corresponds naturally to a tight cut of G itself, and we refer to it as an *essentially 2-separation cut* of G.

See Figure 5.11 for an illustration.

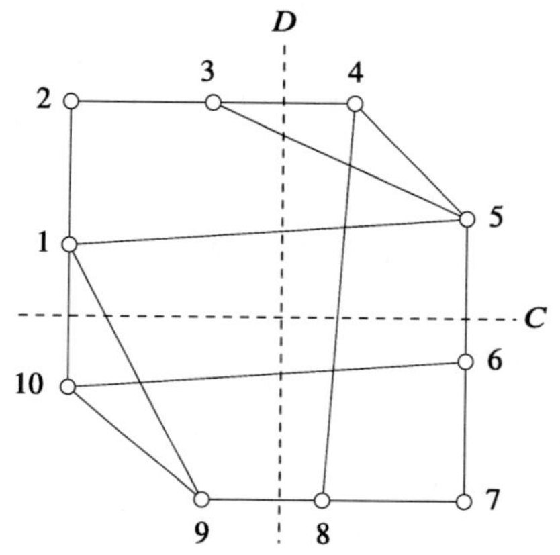

Fig. 5.11 Essentially 2-separation cuts in a graph

In Figure 5.11, the pair $\{S, T\}$ of subsets of V, where $S := \{1, 2, 3\}$ and $T = \{6, 7, 8\}$, constitutes an essential 2-separation of G and the indicated tight cuts C and D of G are the two essentially 2-separation tight cuts of G associated with $\{S, T\}$. A second example of an essential 2-separation of G is the pair $\{S', T\}$, where $S' = \{1\}$ and $T = \{6, 7, 8\}$. Note that one of the two essentially 2-separation cuts associated with $\{S', T\}$ is the same cut C while the other cut, $D' := \partial(\{1, 9, 10\})$, is different from D.

The above-given definition may be phrased more briefly as follows: a tight cut $C := \partial(X)$ of a matching covered graph G is an essentially 2-separation cut if there is a *companion* tight cut $D := \partial(Y)$ which crosses C such that $X \cap Y$ is odd and both $G[X \cap Y]$ and $G[\overline{X} \cap \overline{Y}]$ are bipartite subgraphs of G. In the above example, C is an essentially 2-separation cut with either D or D' as its companion.

TYPES OF NONTRIVIAL TIGHT CUTS IN GRAPHS WITH TWO BRICKS

Theorem 5.19 *In a matching covered graph with just two bricks, every nontrivial tight cut is either a barrier cut or is an essentially 2-separation cut.*

Proof Let G be a matching covered graph such that $b(G) = 2$, and let $C := \partial(X)$ be a nontrivial tight cut of G. All matching covered graphs on fewer than six vertices are either bricks or braces. Thus, a graph with two bricks has to have at least six vertices. It can be easily checked that if $n = 6$ and $b = 2$, then C is a 2-separation cut (Exercise 5.5.1). We prove, by induction on $|V|$, that C is either a barrier cut or is an essentially 2-separation cut. To begin with, we may assume that both C-contractions of G are nonbipartite; otherwise, by Theorem 4.7, the cut C would be a barrier cut. Thus, since $b(G) = 2$, we may assume that both C-contractions of G are near-bricks.

Lemma 5.20 *If the graph G is bicritical, then C is a 2-separation cut.*

Proof Suppose that G is bicritical. Theorem 5.13 and the fact that $b(G) = 2$ together imply that both C-contractions of G are bricks. As G itself is not a brick, it follows from Proposition 5.15 that C is a 2-separation cut and the assertion holds. □

When G is not bicritical, it has nontrivial barriers. The proof of the theorem then turns out to be relatively simple if the graph G happens to have a barrier with a special relationship with the cut C.

C-AVOIDING BARRIER

We define a nontrivial barrier B of G to be *C-avoiding* if C does not cross any of the barrier cuts associated with B. Figure 5.12 shows two examples of such barriers.

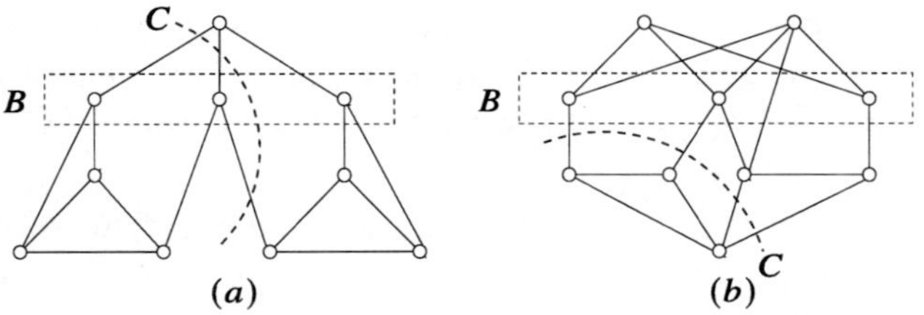

Fig. 5.12 C-avoiding barriers

We note that if a nontrivial barrier B is C-avoiding, then either C is a tight cut of the bipartite graph $H := \mathbb{H}(B)$ obtained by shrinking each odd component of $G - B$ to a single vertex (as in the example shown in Figure 5.12(a)) or there is a nontrivial odd component K of $G - B$ such that C is a tight cut of the graph $G/\overline{(V(K))}$ (as in the example shown in Figure 5.12(b)). See Exercise 5.5.2.

CASE: THERE IS A C-AVOIDING BARRIER B

Lemma 5.21 *If the graph G has a C-avoiding barrier B, then C is either a barrier cut or is an essentially 2-separation cut of G.*

Proof We consider the two alternatives mentioned above.

Firstly, suppose that C is a cut of the bipartite graph H, and let X be the shore of C in H such that $X_+ \subset B$. Then, X_+ is a barrier of G, and C is a barrier cut of G associated with X_+.

If C is not a cut of the bipartite graph H, the fact that B is C-avoiding implies that there an odd component, say K, of $G - B$, such that C is a nontrivial tight cut of the graph $J := G/(\overline{V(K)} \to \overline{k})$.

5.21.1 *If C is a barrier cut of J, then it is also a barrier cut of G.*

Proof Suppose that the cut C is a barrier cut associated with a barrier B_J of J. If the contraction vertex $\overline{k}$ lies in B_J then $B \cup (B_J - \overline{k})$ is a barrier of G and the cut C is associated with that barrier. Alternatively, if the contraction vertex $\overline{k}$ of J does not lie in B_J then B_J is a barrier of G and cut C is associated with B_J in G. In both alternatives, C is a barrier cut of G. □

In view of the above statement, we may assume that C is not a barrier cut of J. As any tight cut of a graph with at most one brick is a barrier cut, it follows that $b(J) \geq 2$. However, as $b(G) = 2$, it follows that $b(J) = 2$ and hence that $G/V(K)$ is bipartite. As J clearly has fewer vertices than G, the induction hypothesis implies that C is either a barrier cut of J or is an essentially 2-separation cut of J. In the former case, by statement 5.21.1, the cut C is a barrier cut of G. The following statement deals with the latter case.

5.21.2 *If C is an essentially 2-separation cut of J then it is also an essentially 2-separation cut of G.*

Proof Let $\{S, T\}$ be an essential 2-separation of J such that C is a 2-separation cut of J associated with it. If the contraction vertex $\overline{k}$ does not lie in either S or T, then $\{S, T\}$ is an essential 2-separation of G itself, and C is an essentially 2-separation cut of G.

So we may assume that $\overline{k}$ is in one of the sets S and T. Adjust notation so that $\overline{k} \in S$. In this case, the pair $\{S', T\}$, where $S' := (S - \overline{k}) \cup \overline{V(K)}$, is an essential 2-separation of G and C is also an essentially 2-separation cut of G associated with $\{S', T\}$. □

This completes the proof of the assertion of the lemma. □

How does one go about finding a C-avoiding nontrivial barrier of G? The proof of the following lemma shows that given any maximal nontrivial barrier B^* of G one can either find a C-avoiding barrier or deduce that C is a cut of the required type.

CASE: A MAXIMAL BARRIER B^* IS NOT C-AVOIDING

Lemma 5.22 *If the graph G has a maximal nontrivial barrier B^* which is not C-avoiding, then:*

- *either G has another nontrivial barrier B which is C-avoiding,*
- *or C is a 2-separation cut of G.*

Proof Since B^* is not C-avoiding, there exists a barrier cut $D := \partial(Y)$ associated with B^* such that C and D cross, where Y is the vertex set of an odd component of $G - B^*$ (and thus, the barrier B^* is a subset of $\overline{Y}$.) Adjust notation so that $X \cap Y$ and $\overline{X} \cap \overline{Y}$ are odd and, as usual, let $I := \partial(X \cap Y)$ and $U := \partial(\overline{X} \cap \overline{Y})$. By Theorem 4.12, both I and U are tight cuts and there are no edges of G between $X \cap \overline{Y}$ and $\overline{X} \cap Y$. We shall use the following notation to denote the various graphs obtained from G by tight cut contractions (see Figure 5.13):

$$\begin{aligned} G_1 &:= G/(\overline{X} \to \overline{x}), & G_2 &:= G/(X \to x), \\ H_1 &:= G/(\overline{Y} \to \overline{y}), & H_2 &:= G/(Y \to y), \\ G_{11} &:= G/(\overline{X} \cup \overline{Y}), & G_{22} &:= G/(X \cup Y), \\ G_{12} &:= G_1/(X \cap Y), & G_{21} &:= G_2/(\overline{X} \cap \overline{Y}). \end{aligned}$$

(G_{12} is the same as $H_2/(\overline{X} \cap \overline{Y})$ up to multiple edges.)
(G_{21} is the same as $H_1/(X \cap Y)$ up to multiple edges.)

If either of the two shores of C is bipartite, then the majority part of that shore would be a C-avoiding barrier of G and there would be nothing more to prove. So, we may assume that both shores of C are nonbipartite. Since $b(G) = 2$, this implies that both G_1 and G_2 are near-bricks. Keeping this in mind, we now proceed to draw conclusions about the various graphs displayed in Figure 5.13.

5.22.1 *The vertex $\overline{y}$ does not lie in any nontrivial barrier of H_1.*

Proof Let B' denote a barrier of H_1 that contains vertex $\overline{y}$. Then the set $B'' := (B' - \overline{y}) \cup B^*$ is a barrier of G. By the maximality of B^*, it follows that $B' = \{\overline{y}\}$. Indeed, the only barrier of H_1 that contains the vertex $\overline{y}$ is trivial. □

5.22.2 *The graph $G_{21} = G/(\overline{Y} \to \overline{y})/(X \cap Y \to s)$ is not bipartite.*

Proof Assume the contrary. Let (A', B') denote the bipartition of G_{21} and adjust notation so that s lies in A'. By Proposition 4.1, the subgraph $G[X]$ induced by X is connected. This implies that $\overline{y}$ and s are adjacent in G_{21} implying that $\overline{y}$ lies in B'. Then, B' is a nontrivial barrier of H_1 that contains vertex $\overline{y}$, which contradicts statement 5.22.1. □

5.22.3 *The graph G_{22} is bipartite. Moreover, $|\overline{X} \cap \overline{Y}| = 1$.*

Proof The graph G_2, a C-contraction of G, is a near-brick. Moreover, the cut $U := \partial(\overline{X} \cap \overline{Y})$ is tight in G_2, and the U-contraction G_{21} of G_2 is not bipartite. Therefore, G_{22}, the other U-contraction of G_2, is bipartite implying that the graph $G[\overline{X} \cap \overline{Y}]$ is bipartite.

If $|\overline{X} \cap \overline{Y}| \geq 3$, the majority part of $\overline{X} \cap \overline{Y}$ would be a nontrivial special barrier of G which is C-avoiding and there would be nothing more to prove. Therefore, we may assume that $|\overline{X} \cap \overline{Y}| = 1$. □

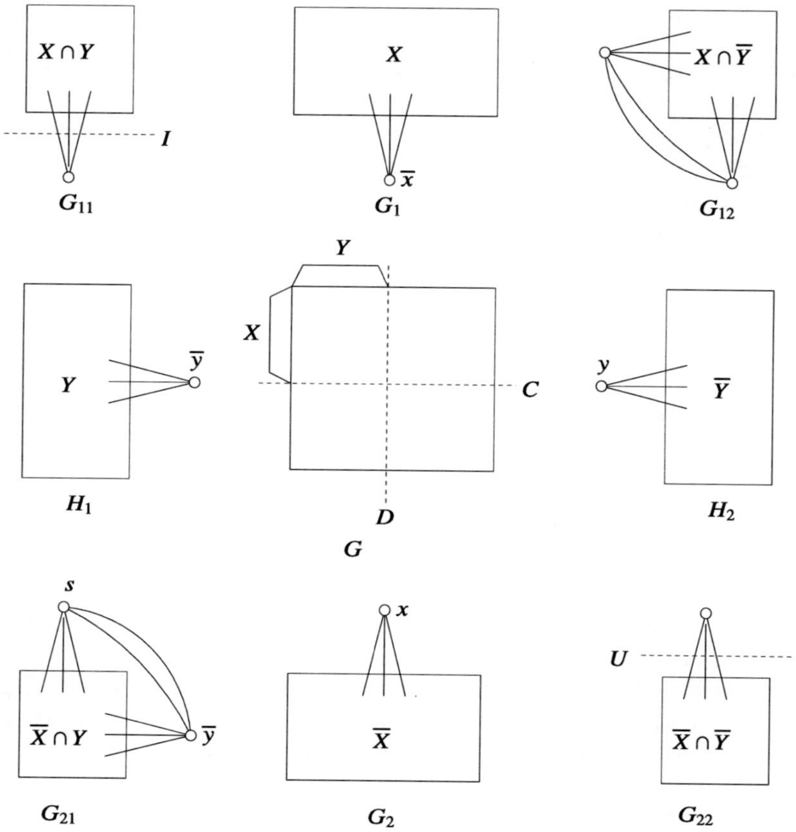

Fig. 5.13 The graph G and its cut contractions

We shall let t denote the unique vertex in $\overline{X} \cap \overline{Y}$. Now we consider two cases depending on whether or not the graph G_{11} is bipartite.

Case 1 *The graph G_{11} is bipartite.*

In this case, $G[X \cap Y]$ is the bipartite shore of the cut $I = \partial(X \cap Y)$ in G. If $|X \cap Y| \geq 3$, then the majority part of $X \cap Y$ would be a nontrivial special barrier of G which is C-avoiding. On the other hand, if $|X \cap Y| = 1$, say $X \cap Y = \{s\}$, then $\{s, t\}$ would be a 2-separation of G, and C would be a 2-separation cut of G. Both possibilities lead to a confirmation of the assertion of the lemma.

Case 2 *The graph G_{11} is not bipartite.*

In this case, neither of the two I-contractions of H_1 is bipartite. As H_1 is a D-contraction of G which has two bricks, it follows that G_{11} and G_{21} are near-bricks and, furthermore, that $b(H_1) = 2$ and that H_2 is bipartite.

Since $b(H_1) = 2$ and I is a tight cut of H_1, by the induction hypothesis, I is either a barrier cut of H_1, or is an essentially 2-separation cut of H_1.

Case 2.1 I is a barrier cut of H_1.

Let B_1 denote a barrier of H_1 such that the cut $I = \partial(X \cap Y)$ is associated with B_1. Since G_{11} and G_{21} are the two I-contractions of H_1, it follows that either (i) B_1 is a subset of $(\overline{X} \cap Y) + \overline{y}$ and $X \cap Y$ is the vertex set of an odd component of $H_1 - B_1$, or (ii) B_1 is a subset of $X \cap Y$ and $V(G_{21}) - s$ is the vertex set of an odd component of $H_1 - B_1$.

If B_1 were a subset of $(\overline{X} \cap Y) + \overline{y}$ and $X \cap Y$ the vertex set of an odd component of $H_1 - B_1$, the contraction vertex $\overline{y}$ would have to belong to B_1 because there are edges in G between $X \cap Y$ and $\overline{Y}$. Likewise, some vertex of $\overline{X} \cap Y$ would also have to belong to B_1, because there are edges that join vertices in $X \cap Y$ to vertices in $\overline{X} \cap Y$. But, in this case, B_1 would be a nontrivial barrier of H_1 that contains vertex $\overline{y}$, in contradiction to statement 5.22.1. So, this situation cannot arise.

Thus, we may suppose that B_1 is a subset of $X \cap Y$ and $V(G_{21}) - s$ is the vertex set of an odd component of $H_1 - B_1$. In this case, B_1 is a barrier of G such that the shore $\overline{X}$ of C is a subset of the vertex set of one of the odd components of $G - B_1$, implying that B_1 is a C-avoiding barrier of G and hence the assertion of the lemma.

Case 2.2 I is an essentially 2-separation cut of H_1.

Let $\{X_1, X_2\}$ denote an essential 2-separation of H_1 such that I is a 2-separation cut of $H_1/(X_1 \to x_1)/(X_2 \to x_2)$ associated with the pair $\{x_1, x_2\}$. For $i = 1, 2$, let (A_i, B_i) denote the bipartition of $H_1[X_i]$, where B_i is the majority part.

One of X_1 and X_2 is a subset of $X \cap Y$, the other a subset of $(\overline{X} \cap Y) + \overline{y}$. Adjust notation so that X_1 is a subset of $X \cap Y$.

If X_1 is not a singleton, the majority part B_1 of X_1, being a subset of $X \cap Y$, does not contain the contraction vertex $\overline{y}$. Hence B_1 would be a C-avoiding nontrivial barrier of G itself, and the first alternative in the assertion of the lemma follows. So, we may assume that $|X_1| = 1$. Let x_1 be the vertex of X_1.

5.22.4 *Every edge of $\partial(X \cap \overline{Y})$ is incident in G with a vertex in $\{x_1, t\}$.*

Proof Let e be an edge of $\partial(X \cap \overline{Y})$. The cuts C and D are tight in G; therefore, no edge of G joins a vertex of $X \cap \overline{Y}$ to a vertex of $\overline{X} \cap Y$. As t is the only vertex of $\overline{X} \cap \overline{Y}$, we infer that e is incident in G with a vertex in $(X \cap Y) + t$. Suppose that e is not incident in G with vertex t. In that case, e is incident with a vertex of $X \cap Y$. Thus, e is in I and has both ends in X. Being an edge of I, the edge e is incident in H_1 with a vertex in $X_2 + x_1$. The set X_2 is disjoint with X. The edge e has both ends in X. Thus, e is incident with x_1. This conclusion holds for each edge e of $\partial(X \cap \overline{Y})$ which is not incident with t. We deduce that every edge of $\partial(X \cap \overline{Y})$ is incident with a vertex in $\{x_1, t\}$, as asserted. □

Graph H_2 is bipartite; therefore $G[\overline{Y}]$ is also bipartite. The barrier B^* is one of the parts in the bipartition of $G[\overline{Y}]$. Let A^* denote the other part. The vertex t is the only vertex of $\overline{X} \cap \overline{Y}$, and it lies in B^*. See Figure 5.14.

It follows that $A^* \cup (B^* - t) = X \cap \overline{Y}$. By statement (5.22.4), every edge of $\partial(A^* \cup (B^* - t))$ is incident with a vertex in $\{x_1, t\}$. The vertices of $B^* - t$ are not adjacent to vertex t. Thus, every edge of $\partial(B^* - t)$ is incident with $A^* + x_1$. It

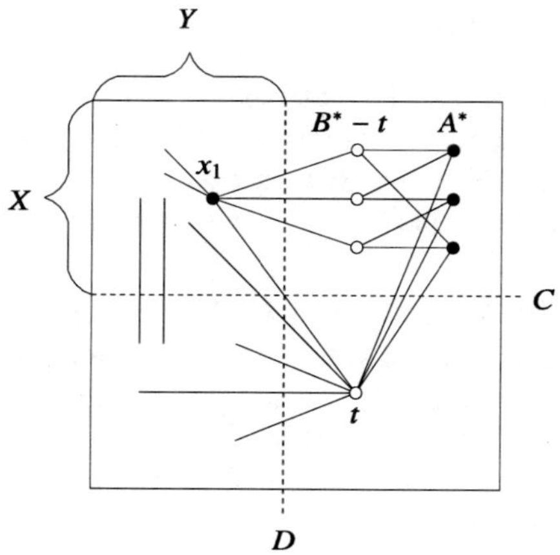

Fig. 5.14 The bipartition (A^*, B^*) of $G[\overline{Y}]$

follows that the $|A^*|$ vertices of $B^* - t$ are isolated in $G - A^* - x_1$. We conclude that $B := A^* + x_1$ is a special barrier of G, and the only nontrivial component of $G - B$ contains all the vertices of $\overline{X}$. In sum, B is a nontrivial C-avoiding barrier of G. □

It is now straightforward to deduce the assertion of the theorem from Lemmas 5.20, 5.21, and 5.22. □

We leave the proof of the following corollary as Exercise 5.5.3.

Corollary 5.23 *Let G be a matching covered graph such that $b(G) = 2$. Let C be a tight cut of G. Then, one of the following alternatives holds:*

(i) *one of the shores of C is bipartite,*
(ii) *the cut C is a member of a pair of nontrivial barrier cuts,*
(iii) *the cut C is a member of an essentially 2-separation cut pair.* □

5.5.1 The laminar ELP Theorem

As mentioned before, the need for a precise characterization of all tight cuts in matching covered graphs with two bricks arose in connection with the work of CLM on a conjecture of Lovász which will be described in Chapter 15. This naturally led them to the question of whether a similar description of tight cuts in arbitrary matching covered graphs was possible. For many years they were not able to find an answer to this question, and were led to conjecture that, if C is any tight cut in a matching covered graph G, then there exists an ELP cut of G which is laminar to C (CLM (2018, [19])). This conjecture was confirmed in 2021 by Guantao Chen, Xing Feng, Fuliang Lu, Cláudio L. Lucchesi and Lianzhu Zhang (2021, [20]). They were able to establish this assertion by showing that a statement which is in the same spirit as Theorem 5.19 is true for tight cuts in all matching covered graphs. We briefly sketch below the main ingredients in their proof.

C-SHELTERED AND C-AVOIDING SETS

Let G be a matching covered graph, let C be a tight cut of G and let S be a set of vertices of G which is either a barrier or a 2-separation. The set S is *C-sheltered* if S is a subset of a shore of C, and is *C-avoiding* if each ELP cut associated with S is laminar with C. If S is C-sheltered then, as each shore of C induces a connected subgraph of G, one of the components of $G - S$ contains all the vertices of the shore of C disjoint with S. It follows that if S is C-sheltered then S is C-avoiding. We record this result for later reference.

Proposition 5.24 *Let S be either a 2-separation or a barrier of a matching covered graph G and let C be a tight cut of G. If S is C-sheltered then some cut associated with S has a shore that is a superset of a shore of C, say, X, and all the other cuts associated with S have a shore that is a subset of $\overline{X}$. Consequently, if S is C-sheltered then it is C-avoiding.* □

Theorem 5.25 (The Laminar ELP Theorem) *Let C be a nontrivial tight cut of a matching covered graph G. Then either G has a C-sheltered nontrivial barrier or a 2-separation cut which is laminar with C (see Example 5.26).* □

Example 5.26 Consider the graph depicted in Figure 5.15. The tight cut C is laminar with the cut D, which is a 2-separation cut associated with the pair $\{u_1, u_2\}$. The cut C is also laminar with the 2-separation cut F, which is associated with the pair $\{1, b_3\}$. The barriers $\{b_1, b_2, b_3\}$ and $\{2, 3\}$ are C-sheltered, whereas the barrier $\{1, 2, 3\}$ is not C-avoiding.

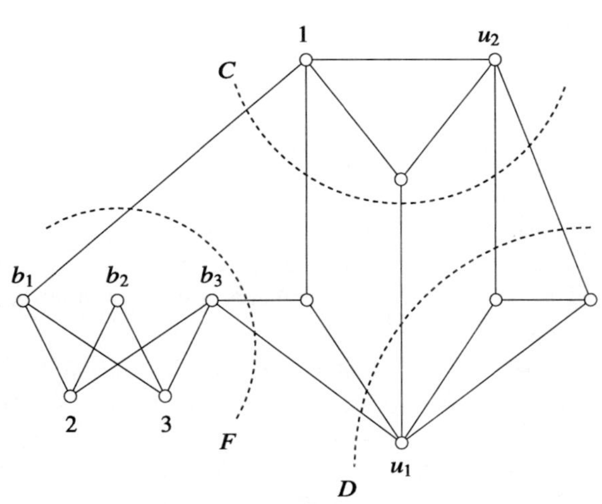

Fig. 5.15 The graph of Example 5.26

5.5.2 ELP tight cut decompositions

Given any matching covered graph G, we may apply to it a procedure, called an *ELP tight cut decomposition* of G, which produces a list of bricks and braces. If G itself is a brick or a brace then the list consists of just G. Otherwise, let C be any nontrivial ELP cut of G. Then, both C-contractions of G are matching covered. One may recursively apply the ELP tight cut decomposition procedure to each C-contraction of G, and then combine the resulting lists to produce an ELP tight cut decomposition of G itself.

An interesting consequence of Theorem 5.25 is that any nontrivial tight cut C of a matching covered graph G participates in some ELP tight cut decomposition of G. In this sense, every tight cut in a matching covered graph is ultimately an ELP cut.

It is our hope that Theorem 5.25 which provides a characterization of all tight cuts could be used as a tool for establishing interesting results concerning the structure of matching covered graphs!

Exercises

5.5.1 *Show that any nontrivial tight cut in a matching covered graph G with $n = 6$ and $b = 2$ is a 2-separation cut.*

5.5.2 *Show that if B is a C-avoiding barrier, then either C is a tight cut of the core H of G with respect to B, or that it is a nontrivial tight cut of $G/\overline{V(K)}$ for some odd component K of $G - B$.*

▷**5.5.3** *Prove Corollary 5.23.*

5.6 Notes

Recall that a matching covered graph G is k-extendable if any matching of size k in G can be extended to a perfect matching of G. Every matching covered graph is 1-extendable and, by Theorem 5.18, a bipartite matching covered graph is 2-extendable if and only if it is a brace.

The graph in Figure 5.16 is 2-extendable, but is not 3-extendable.

For $k \geq 3$, the graph obtained from K_{2k} by deleting the edges of a perfect matching is known as a *cocktail party graph* as it can be viewed as representing the pattern of intermingling among n couples at a cocktail party where each guest socializes with all people except with his or her own spouse.

It is fairly easy to show that if a matching covered graph is 2-extendable, then it has to be either a brick or a brace. The complete graph K_4 is a 2-extendable brick. Prisms of order n, where $n \equiv 2 \pmod 4$, Möbius ladders of order $n \geq 8$, where $n \equiv 0 \pmod 4$, and the Petersen graph are all examples of bricks which are not 2-extendable.

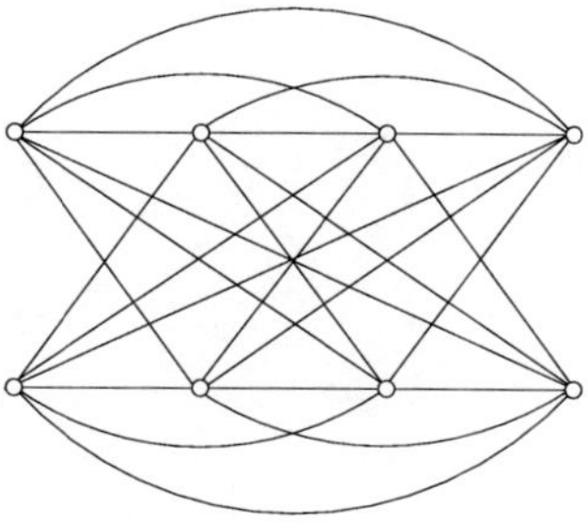

Fig. 5.16 The cocktail party graph of order eight

Figure 5.16 depicts the cocktail party graph of order eight. For $k \geq 3$, the cocktail party graph of order $2k$ is $(k-2)$-extendable, but not $(k-1)$-extendable. We refer the reader to the appendix in the second edition of Lovász and Plummer (1989,[60]) for further information on k-extendable graphs.

Chapter 6
The Perfect Matching Polytope

Contents

6.1 Edmonds' Theorem

A geometrical object of great interest in the theory of combinatorial optimization is the perfect matching polytope of a graph. It turns out that tight cut decompositions of a matching covered graph, which were described in Chapter 4, are very relevant to understanding the properties of its perfect matching polytope. The purpose of this chapter is to explore some of these connections.

> THE EDGE SPACE $\mathbb{R}^E$ OF A GRAPH G
>
> The *edge space* $\mathbb{R}^E$ of a graph G is the set of all real-valued functions on the edge set E of G.

It is convenient to view each member $\mathbf{x}$ of $\mathbb{R}^E$ as vector whose coordinates are indexed by the edges of G, with $\mathbf{x}(e)$ representing the value of $\mathbf{x}$ in the position corresponding to edge e.

C. L. Lucchesi, U. S. R. Murty, *Perfect Matchings*, Algorithms and Computation in Mathematics 31, https://doi.org/10.1007/978-3-031-47504-7_6

Thus, any vector $\mathbf{x}$ in $\mathbb{R}^E$ may be specified by a drawing of G in which, for each edge e, the value of $\mathbf{x}(e)$ is indicated by juxtaposition. Figure 6.1 displays two members of the edge space of $\overline{C_6}$ (the missing entries are to be taken as zeros). The *support* of a vector $\mathbf{x}$ in $\mathbb{R}^E$ is the subset $\{f \in E : \mathbf{x}(f) \neq 0\}$. For example, the support of the vector displayed in Figure 6.1(a) has six edges.

INCIDENCE VECTORS AND A USEFUL NOTATION

With each subset F of the edge set E of a graph G we associate a vector χ^F in $\mathbb{R}^E$ called its *incidence vector*, where $\chi^F(e) = 1$ if $e \in F$, and $\chi^F(e) = 0$ if $e \notin F$. The vector displayed in Figure 6.1(b) is the incidence vector of the indicated perfect matching in $\overline{C_6}$.

Given any vector $\mathbf{x}$ in $\mathbb{R}^E$, and a subset F of E, we write $\mathbf{x}(F)$ to denote $\sum\{\mathbf{x}(e) : e \in F\}$, which is the same as the dot product of χ^F and $\mathbf{x}$. As an example, take $\mathbf{x}$ to be the vector in the edge space of $\overline{C_6}$ shown in Figure 6.1(a) and M to be the perfect matching indicated by dotted lines in Figure 6.1(b); then $\mathbf{x}(M) = 1$.

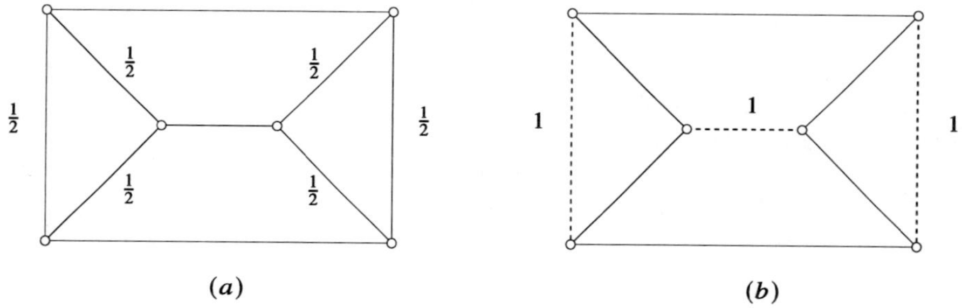

Fig. 6.1 Two vectors in the edge space of $\overline{C_6}$ (zero entries not shown)

6.1.1 Convex polytopes

The perfect matching polytope of a graph G is a convex polytope in its edge space. Pertinent material relating to these objects can be found in various books on combinatorial optimization such as the treatises by Lovász and Plummer [59] and Schrijver [84]. For the convenience of the readers, we provide here a brief account of the relevant definitions and the notation we use. (Readers who are familiar with the basic theory of convex polytopes may skip this subsection.)

Convex Linear Combinations: Recall that a linear combination of r vectors $\mathbf{x}_1$, $\mathbf{x}_2, \ldots, \mathbf{x}_r$ in a Euclidean space $\mathbb{R}^n$ over the field $\mathbb{R}$ of real numbers is an expression of the form

$$\sum_{i=1}^{r} \alpha_i \mathbf{x}_i, \text{ where } \alpha_i \in \mathbb{R}, \text{ for } 1 \le i \le r. \tag{6.1}$$

A linear combination $\sum_{i=1}^{r} \alpha_i \mathbf{x}_i$ of vectors $\mathbf{x}_1, \mathbf{x}_2, \ldots, \mathbf{x}_r$ is a *convex linear combination* if (i) $0 \le \alpha_i \le 1$, for $1 \le i \le r$, and (ii) $\sum_{i=1}^{r} \alpha_i = 1$. (To visualize these notions geometrically, let $\mathbf{x}_1$ and $\mathbf{x}_2$ be two distinct points in a Euclidean space $\mathbb{R}^n$. Then a point $\mathbf{x}$ in $\mathbb{R}^n$ is a convex linear combination of $\mathbf{x}_1$ and $\mathbf{x}_2$ if and only if it is on the line segment joining $\mathbf{x}_1$ and $\mathbf{x}_2$; in particular the vector $\frac{1}{2}\mathbf{x}_1 + \frac{1}{2}\mathbf{x}_2$ represents the midpoint of that line segment.)

Convex Sets and Convex hulls: A subset S of $\mathbb{R}^n$ is *convex* if it is closed under the operation of taking convex linear combinations. It is easy to show that the intersection of any two convex sets is a convex set.

The *convex hull* of $\mathbf{x}_1, \mathbf{x}_2, \ldots, \mathbf{x}_r$ is the set of all convex linear combinations of these vectors. Convex hulls of finite sets are examples of convex sets. Not all convex sets are convex hulls of finite sets. For example, a disc is a convex set but it is not the convex hull of a finite set of points.

Extreme Points of Convex Sets: A point $\mathbf{x}$ in a convex set S is an *extreme point* of S if it cannot be expressed as a convex combination of two distinct points in S. For example, the line segment joining two distinct points $\mathbf{x}$ and $\mathbf{y}$ in $\mathbb{R}^n$ has precisely two extreme points, namely $\mathbf{x}$ and $\mathbf{y}$. Not all convex sets have extreme points. For example, an open disc (that is, a disc without its boundary) is such a convex set.

Convex Polytopes: A subset P of a Euclidean space $\mathbb{R}^n$ is a *convex polytope* (or, simply, a *polytope*) if it is the convex hull of some finite subset of $\mathbb{R}^n$. It should be noted that a polytope may happen to be the convex hull of two different finite subsets of $\mathbb{R}^n$. For example, the convex hulls of the two sets $\{(1, 0, 0), (0, 1, 0), (0, 0, 1)\}$ and $\{(1, 0, 0), (0, 1, 0), (0, 0, 1), (\frac{1}{2}, \frac{1}{6}, \frac{1}{3})\}$ are the same, namely the triangle in $\mathbb{R}^3$ with the three unit vectors as its vertices. Clearly, every polytope is uniquely defined by the set of its extreme points; it is the convex hull of that set.

Many important problems in combinatorial optimization may be expressed as a problem of optimizing a linear function over the set of points in a polytope. For example, the assignment problem [59] may be expressed as the problem of maximizing a linear function over the polytope which is the convex hull of the set of incidence vectors of perfect matchings of a bipartite graph. For applying linear programming techniques to solve such problems, it is necessary to be able to describe polytopes in terms of linear inequalities. All polytopes afford such descriptions as we now proceed to describe.

Hyperplanes and half-spaces: A *hyperplane* in $\mathbb{R}^n$ is the set of all solutions to a single linear equation

$$\sum_{i=1}^{n} a_i x_i = b,$$

where not all the coefficients a_i are zero. (In $\mathbb{R}^2$, hyperplanes are lines, in $\mathbb{R}^3$, hyperplanes are planes, etc.) A *half-space* in $\mathbb{R}^n$ is the set of all solutions to a single linear inequality

$$\sum_{i=1}^{n} a_i x_i \geq b.$$

(If the inequality is strict, then the set of its solutions is called an 'open half-space'. Here we restrict ourselves to 'closed half-spaces'.) Note that a hyperplane $\sum_{i=1}^{n} a_i x_i = b$ is the intersection of the two half-spaces $\sum_{i=1}^{n} a_i x_i \geq b$ and $\sum_{i=1}^{n} a_i x_i \leq b$.

It is easy to see that hyperplanes and half-spaces are convex sets. Hence the set of all solutions to a system of linear inequalities, being the intersection of a finite number of half-spaces, is a convex set. A classical result due to Minkowski and Weyl (see [84]) states that every convex polytope is the intersection of a finite number of half-spaces. Equivalently, every convex polytope is the set of solutions to a system of linear inequalities. For example, the convex polytope in $\mathbb{R}^3$ whose extreme points are the three unit vectors $(1, 0, 0)$, $(0, 1, 0)$ and $(0, 0, 1)$ is the set of solutions to the following system:

$$x_1, x_2, x_3 \geq 0 \tag{6.2a}$$

$$x_1 + x_2 + x_3 = 1 \tag{6.2b}$$

This establishes the alluded connection between convex polytopes and systems of linear inequalities.

Optimizing a linear function over a convex polytope: A fundamental theorem in the theory of convex polytopes states that a the optimal value of a linear function $\sum_i^n c_j x_j$ over a convex polytope in $\mathbb{R}^n$ occurs at an extreme point of that polytope. Thus, if one could find a system of linear inequalities which describe that polytope, then one would be able to apply linear programming techniques to find the desired optimal solution. For example, suppose that with each edge e of a graph G there is an associated cost c_e and our aim is to find a minimum cost perfect matching of G. This amounts to finding the minimum value of the function $\sum_{e \in E} c_e x_e$ over the set of incidence vectors of perfect matchings of G. Thus, if one could find a linear inequality description of the polytope whose extreme points are the incidence vectors of perfect matchings of G, then it would be possible to find a minimum cost perfect matching of G by using linear programming techniques. Such a description is given in the following subsection.

6.1.2 The perfect matching polytope of a graph

Recall that the set of all perfect matchings of a graph G is denoted by $\mathcal{M}(G)$, or simply by $\mathcal{M}$ when the graph G is understood from the context.

THE PERFECT MATCHING POLYTOPE

The *perfect matching polytope* of G, denoted by $\mathcal{P}oly(G)$, is the set of all convex linear combinations of incidence vectors of perfect matchings of G. In other words:

$$\mathcal{P}oly(G) = \left\{\mathbf{x} \in \mathbb{R}^E \;:\; \mathbf{x} = \sum_{M \in \mathcal{M}} \alpha_M \chi^M,\; 0 \le \alpha_M \le 1,\; \sum_{M \in \mathcal{M}} \alpha_M = 1\right\}.$$

Jack Edmonds (1965, [30]) established the following description of the perfect matching polytope of a graph in terms of linear inequalities.

EDMONDS' CHARACTERIZATION OF $\mathcal{P}OLY(G)$

Theorem 6.1 *The perfect matching polytope $\mathcal{P}oly(G)$ of a graph G is the set of solutions to the following system of linear inequalities:*

$$\mathbf{x}(e) \ge 0, \quad \textit{for each } e \in E, \tag{6.3a}$$

$$\mathbf{x}(\partial(v)) = 1, \quad \textit{for each } v \in V, \tag{6.3b}$$

$$\mathbf{x}(\partial(S)) \ge 1, \quad \textit{for each odd subset } S \textit{ of } V. \tag{6.3c}$$

We refer to the constraints in (6.3a) as *non-negativity constraints*, those in (6.3b) as the *degree constraints*, and those in (6.3c) as the *odd set constraints*. In case of bipartite graphs, (6.3a) and (6.3b) together imply (6.3c); see Exercise 6.1.1. However, in general, this is not the case. For example, the vector $\mathbf{x}$ in the edge space of $\overline{C_6}$ which is displayed in Figure 6.1(a) satisfies both the non-negativity and degree constraints, but it fails to satisfy (6.3c) for the set S consisting of the three vertices of a triangle in $\overline{C_6}$.

For each edge e of a graph G, let $\mathcal{M}^e$ denote the set of perfect matchings of G that contain the edge e. The following observation is a straightforward consequence of the definition of the characteristic vector χ^M:

Proposition 6.2 *If* $\mathbf{x} := \sum_{M \in \mathcal{M}} \alpha_M \chi^M$ *then* $\mathbf{x}(e) = \sum_{M \in \mathcal{M}^e} \alpha_M$. □

Proof (of Theorem 6.1) Let $\mathbf{P}(G)$ denote the set of all solutions to the system of linear inequalities consisting of (6.3a), (6.3b) and (6.3c). Our task is to show that the two convex polytopes $\mathbf{P}(G)$ and $\mathcal{P}oly(G)$ are one and the same.

It is easy to see that, for any perfect matching M of G, its incidence vector χ^M is in $\mathbf{P}(G)$. This implies that $\mathcal{P}oly(G)$ is a subset of $\mathbf{P}(G)$ (Exercise 6.1.2). Our objective is to show that $\mathbf{P}(G) \subseteq \mathcal{P}oly(G)$, and deduce that the two polytopes $\mathbf{P}(G)$ and $\mathcal{P}oly(G)$ are the same. This is clearly true if $\mathbf{P}(G) = \emptyset$. On the other hand, if $\mathbf{P}(G)$ is not empty, it can be shown that G is matchable (Exercise 6.1.3).

Since $\mathbf{P}(G)$ is a convex polytope, every vector in it is a convex linear combination of its extreme points. Thus, to show that $\mathbf{P}(G) \subseteq \mathcal{P}oly(G)$, it suffices to show that every extreme point of $\mathbf{P}(G)$ is in $\mathcal{P}oly(G)$. Towards this end, let $\mathbf{w}$ be an extreme point of $\mathbf{P}(G)$. Our proof is divided into two main cases. In the first case, we use induction on the number of vertices to show that $\mathbf{w}$ is in $\mathcal{Poly}(G)$. In the second case, we show directly that $\mathbf{w}$ is the incidence vector of a perfect matching of G.

Case 1 *There exists a nontrivial odd cut* $C := \partial(S)$ *of* G *such that* $\mathbf{w}(C) = 1$.

Let $G_1 := G/(\overline{S} \to \overline{s})$ and $G_2 := G/(S \to s)$ denote the two C-contractions of G. Also let $\mathcal{M}_1$ and $\mathcal{M}_2$ denote, respectively, the sets of perfect matchings of G_1 and G_2 and define, for $i = 1, 2$, the vector $\mathbf{w}_i$ as the restriction of $\mathbf{w}$ to the edge set $E(G_i)$ of G_i.

(To make sense of the notation used below, it might be helpful to the reader to work through the details using the graph shown in Figure 6.2 as an example. See Exercise 6.1.4.)

Now, using the fact that $\mathbf{w} \in \mathbf{P}(G)$ and that $\mathbf{w}(C) = 1$, it is easy to check that $\mathbf{w}_1$ and $\mathbf{w}_2$ satisfy the non-negativity, degree, and odd-set constraints in graphs G_1 and G_2, respectively (Exercise 6.1.5). Therefore, by the induction hypothesis, $\mathbf{w}_1$ and $\mathbf{w}_2$ are in $\mathcal{Poly}(G_1)$ and $\mathcal{Poly}(G_2)$, respectively. Thus, there exist coefficients β_{M_1}, for $M_1 \in \mathcal{M}_1$ and λ_{M_2}, for $M_2 \in \mathcal{M}_2$, such that

$$\mathbf{w}_1 = \sum_{M_1 \in \mathcal{M}_1} \beta_{M_1} \chi^{M_1}, \quad \beta_{M_1} \geq 0, \quad \sum \beta_{M_1} = 1,$$
$$\mathbf{w}_2 = \sum_{M_2 \in \mathcal{M}_2} \lambda_{M_2} \chi^{M_2}, \quad \lambda_{M_2} \geq 0, \quad \sum \lambda_{M_2} = 1.$$

Whenever M_1 and M_2 are perfect matchings of G_1 and G_2, respectively, such that $M_1 \cap C = M_2 \cap C$, the union of M_1 and M_2 is a perfect matching of G which meets C in just one edge. Consider the set $\mathcal{N}$ of all perfect matchings of G which have exactly one edge in common with C, and define the coefficient α_M for χ^M, for each $M \in \mathcal{N}$, as follows:

$$\alpha_M := \begin{cases} \dfrac{\beta_{M_1}\lambda_{M_2}}{\mathbf{w}(e)}, & \text{if } \mathbf{w}(e) > 0 \\ 0 & \text{otherwise} \end{cases} \tag{6.4}$$

where (i) $M_1 = M \cap E(G_1)$, (ii) $M_2 = M \cap E(G_2)$, and (iii) e is the unique edge of $M \cap C$. There may be perfect matchings of G with more than one edge in the cut C. If M is any such perfect matching of G, we take α_M to be zero.

Using (i) the fact that $\mathbf{w}_1$ and $\mathbf{w}_2$ belong, respectively, to $\mathcal{Poly}(G_1)$ and $\mathcal{Poly}(G_2)$, (ii) the definition of the coefficients α_M, and (iii) the validity of the equality $\sum_{e \in C} w(e) = 1$, we now proceed to verify that:

$$\text{(a) } \alpha_M \geq 0, \quad \text{(b) } \sum_{M \in \mathcal{N}} \alpha_M = 1, \quad \text{and} \quad \text{(c) } \mathbf{w} = \sum_{M \in \mathcal{N}} \alpha_M \chi^M.$$

(a) It is easy to see $\alpha_M \geq 0$, for each $M \in \mathcal{N}$.

(b) We shall use the following notation in our verification of parts (b) and (c). For each $e \in C$ and for $i = 1, 2$, we let $\mathcal{M}_i^e$ denote the set of all perfect matchings of G_i whose intersection with C is $\{e\}$. Every perfect matching in $\mathcal{N}$ is of the form $M_1 \cup M_2$, where M_1 belongs to $\mathcal{M}_1^e$ and M_2 belongs to $\mathcal{M}_2^e$. Thus, for any $e \in C$

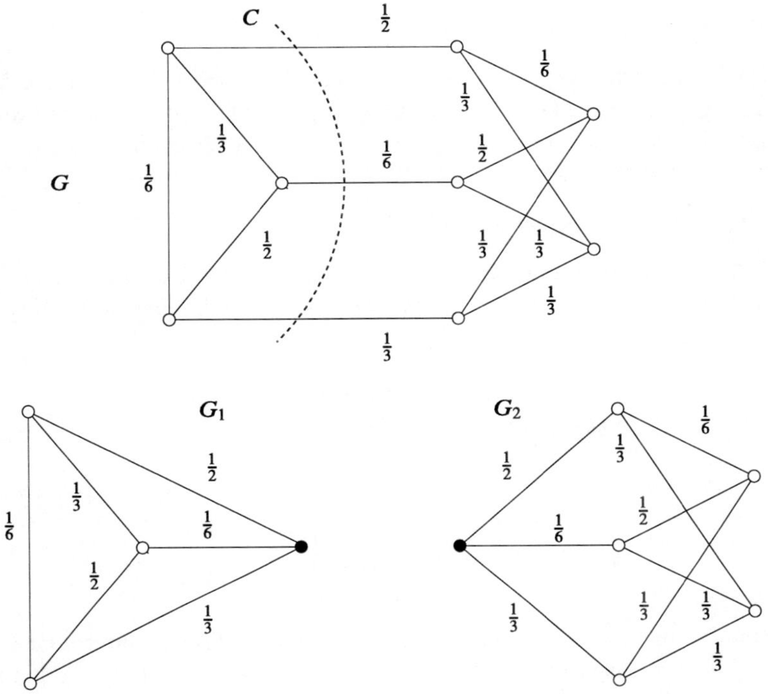

Fig. 6.2 Proof of Edmonds' Theorem, Case 1

with $\mathbf{w}(e) \neq 0$, using the definitions of the α, β, and λ-coefficients (6.4), it can be seen (Exercise 6.1.6) that:

$$\sum_{M_1 \in \mathcal{M}_1^e,\ M_2 \in \mathcal{M}_2^e} \alpha_{M_1 \cup M_2} = \sum_{M_1 \in \mathcal{M}_1^e,\ M_2 \in \mathcal{M}_2^e} \frac{\beta_{M_1} \lambda_{M_2}}{\mathbf{w}(e)} = \mathbf{w}(e). \tag{6.5}$$

By the hypothesis of the case under consideration, $\sum_{e \in C} \mathbf{w}(e) = 1$. Thus $\sum_{M \in \mathcal{N}} \alpha_M = 1$.

(c) It remains to be seen that $\mathbf{w} = \sum_{M \in \mathcal{N}} \alpha_M \chi^M$.

The proof of this relies on Proposition 6.2. Equation 6.5 and Proposition 6.2 show that the two vectors $\sum_{M \in \mathcal{N}} \alpha_M \chi^M$ and $\mathbf{w}$ agree in the coordinates corresponding to edges in the cut C. To complete the proof of (c), it is necessary to show that this agreement also holds for edges of G which are not in C. We leave the details as Exercise 6.1.7.

The three observations made above imply that $\mathbf{w}$ is a convex linear combination of the incidence vectors of perfect matchings of G; in other words that $\mathbf{w}$ is in $\mathcal{Poly}(G)$. In fact, as $\mathbf{w}$ is an extreme point of the polytope $\mathbf{P}(G)$, it follows that $\mathbf{w}$ is itself the incidence vector of a perfect matching of G.

Case 2 $\mathbf{w}(C) > 1$ *for every nontrivial odd cut* $C := \partial(S)$ *of* G.

Let $G[\mathbf{w}]$ denote the subgraph of G induced by the support of $\mathbf{w}$. The constraints 6.3a and 6.3b ensure that $G[\mathbf{w}]$ is a spanning subgraph of G with no isolated vertices. Our objective is to show that the edge set of $G[\mathbf{w}]$ is a perfect matching of G. We start with the following simple observation.

6.2.1 *The graph* $G[\mathbf{w}]$ *has no even cycles.*

Proof Assume to the contrary that $G[\mathbf{w}]$ has an even cycle Q whose edges in cyclic order are $e_1, e_2, \ldots, e_{2k}$. Now define two vectors $\mathbf{x}$ and $\mathbf{y}$ in 'close-proximity' of $\mathbf{w}$ as follows:

$$\mathbf{x}(e) = \begin{cases} \mathbf{w}(e) + \epsilon & \text{if } e = e_1, e_3, \ldots, e_{2k-1} \\ \mathbf{w}(e) - \epsilon & \text{if } e = e_2, e_4, \ldots, e_{2k} \\ \mathbf{w}(e) & \text{if } e \notin E(Q) \end{cases}$$

and

$$\mathbf{y}(e) = \begin{cases} \mathbf{w}(e) - \epsilon & \text{if } e = e_1, e_3, \ldots, e_{2k-1} \\ \mathbf{w}(e) + \epsilon & \text{if } e = e_2, e_4, \ldots, e_{2k} \\ \mathbf{w}(e) & \text{if } e \notin E(Q). \end{cases}$$

Using the fact that $\mathbf{w}$ belongs to $\mathbf{P}(G)$ and that $\mathbf{w}(C) > 1$ for each nontrivial odd cut C of G, it is now easy to verify that there exists a positive ϵ such that both $\mathbf{x}$ and $\mathbf{y}$ are also in $\mathbf{P}(G)$. Also, $\mathbf{w} = \frac{1}{2}\mathbf{x} + \frac{1}{2}\mathbf{y}$, for any value of ϵ. This is impossible because $\mathbf{w}$ is an extreme point of $\mathbf{P}(G)$. □

It is an elementary fact that if a graph is free of even cycles then each of its blocks is either K_1, or K_2, or is an odd cycle (Exercise 6.1.8). It follows that every block of $G[\mathbf{w}]$ of order three or more is an odd cycle. Suppose that an odd cycle Q is an end block of a component of $G[\mathbf{w}]$. Then either Q by itself is that component, or it is attached to the rest of that component at a single cut vertex v. In the former case $\partial(Q)$ is empty and $\mathbf{w}(\partial(Q)) = 0 < 1$, and in the latter case, $\partial(Q)$ is a proper subset of $\partial(v)$ and $\mathbf{w}(\partial(Q)) < \partial(v) = 1$ (see Figure 6.3). Thus, in either case, $\mathbf{w}(\partial(Q)) < 1$ which violates (6.3c). It follows that no end block of $G[\mathbf{w}]$ is an odd cycle.

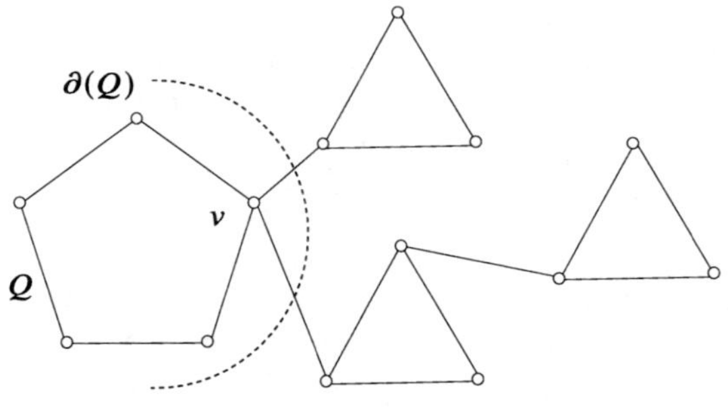

Fig. 6.3 If an odd cycle Q is an end block of $G[\mathbf{w}]$, then $\mathbf{w}(\partial(Q)) < 1$

On the other hand, it follows from 6.3a and 6.3b that the graph $G[\mathbf{w}]$ has no isolated vertices, and if it has a vertex of degree one, then the edge incident with that vertex is the only edge of a component of $G[\mathbf{w}]$. We conclude that $G[\mathbf{w}]$ is a spanning 1-regular subgraph of G or, equivalently, that $\mathbf{w}$ is the incidence vector of a perfect matching of G. □

Edmonds' own proof of the above theorem was based on a polynomial-time algorithm for finding a maximum weight perfect matching in a weighted graph (1965, [30]). For a description of that algorithm, and other proofs of this theorem, we refer the reader to Schrijver's treatise [84].

The following result concerning perfect matching polytopes of matching covered graphs may be proved by adapting the arguments used in Case 1 of the proof of Theorem 6.1.

Theorem 6.3 *Let C be a tight cut of a matching covered graph G and let G_1 and G_2 be the two C-contractions of G. A vector $\mathbf{x}$ in $\mathbb{R}^E$ is in $\mathcal{P}oly(G)$ if and only if the restrictions of $\mathbf{x}$ to $E(G_1)$ and $E(G_2)$ are in $\mathcal{P}oly(G_1)$ and $\mathcal{P}oly(G_2)$, respectively.* □

Thus, when G is a matching covered graph, to check if a vector $\mathbf{x}$ is in the perfect matching polytope of G it suffices to check if the restrictions of $\mathbf{x}$ to the edge sets of the bricks and braces of G are in their respective perfect matching polytopes.

Exercises

▷**6.1.1** *Let G be a bipartite graph, and let $\mathbf{x}$ be any vector in its edge space. Show that if $\mathbf{x}$ satisfies (6.3a) and (6.3b), then it also satisfies (6.3c).*
Hint: Use the fact that for any subset S of V:

$$\mathbf{x}(\partial(S)) \geq \mathbf{x}(\partial(S_+)) - \mathbf{x}(\partial(S_-)) = \mathbf{x}(S_+) - \mathbf{x}(S_-) = |S_+| - |S_-|.$$

∗**6.1.2** *In the proof of Theorem 6.1, explain why $\mathcal{P}oly(G)$ is a subset of $\mathbf{P}(G)$.*

▷**6.1.3** *Show that if the polytope $\mathbf{P}(G)$ defined in the proof of Theorem 6.1 is nonempty, then G is matchable. Hint: let $\mathbf{x} \in \mathbf{P}(G)$ and show that for each subset S of V,*

$$o(G-S) \leq \sum_{K \in \mathcal{O}(G-S)} \mathbf{x}(\partial(K)) \leq \mathbf{x}(\partial(S)) \leq \mathbf{x}(S) = |S|.$$

∗**6.1.4** *Consider the graph G and its two C-contractions G_1 and G_2 shown in Figure 6.2. The weights attached to the edges of G are the components of a vector $\mathbf{w}$ which satisfies the conditions necessary for it to belong to $\mathcal{P}oly(G)$.*

(i) *For $i = 1, 2$, express the restriction $\mathbf{w}_i$ of $\mathbf{w}$ to $E(G_i)$ as a convex combination of incidence vectors of perfect matchings of G_i.*
(ii) *Then, using the method of proof of Theorem 6.1 find a convex linear combination of incidence vectors of perfect matchings of G which is equal to the vector $\mathbf{w}$.*

▷**6.1.5** *In Case 1 of the proof of Theorem 6.1, show that, for* $i = 1, 2$, *the vector* $\mathbf{w}_i$ *is in* $\mathbf{P}(G_i)$.

▷**6.1.6** *Verify the validity of Equation* (6.5).

Hint: Note that

$$\sum_{M_1 \in \mathcal{M}_1^e,\ M_2 \in \mathcal{M}_2^e} \frac{\beta_{M_1} \lambda_{M_2}}{\mathbf{w}(e)} = \sum_{M_1 \in \mathcal{M}_1^e} \frac{\beta_{M_1}}{\mathbf{w}(e)} \sum_{M_2 \in \mathcal{M}_2^e} \lambda_{M_2}$$

and use the fact that $\sum_{M_1 \in \mathcal{M}_1^e} \beta_{M_1} = \sum_{M_2 \in \mathcal{M}_2^e} \lambda_{M_2} = \mathbf{w}(e)$.

∗**6.1.7** *In Case 1 of the proof of Theorem 6.1, verify that* $\mathbf{w} = \sum_{M \in \mathcal{M}} \alpha_M$.

6.1.8 *Show that if a graph G has no even cycles, then each block of G is either* K_1, K_2, *or an odd cycle.*

6.1.9 *Give a proof of Theorem 6.3.*

6.1.10 *Let G be an r-graph (as defined in Exercise 1.3.6). Show that the vector* $(\frac{1}{r}, \frac{1}{r}, \cdots, \frac{1}{r})$ *is in* $\mathcal{P}oly(G)$.

THE PERFECT MATCHING CONE

The *Perfect Matching Cone* of a graph G, denoted $\mathcal{C}one(G)$, is the set of all non-negative linear combinations of incidence vectors of perfect matchings of G. In other words:

$$\mathcal{C}one(G) := \{\mathbf{x} \in \mathbb{R}^E \ :\ \mathbf{x} = \sum_{M \in \mathcal{M}} \alpha_M \chi^M,\ \ \alpha_M \geq 0\}.$$

Theorem 6.4 (A characterization of $\mathcal{C}one(G)$) *Let G be a graph. A vector* $\mathbf{x} \in \mathbb{R}^E$ *is in* $\mathcal{C}one(G)$ *iff it satisfies the three restraints below, for some* $r \in \mathbb{R}$:

$$\mathbf{x}(e) \geq 0, \quad \text{for each } e \in E, \tag{6.6a}$$
$$\mathbf{x}(\partial(v)) = r, \quad \text{for each } v \in V, \tag{6.6b}$$
$$\mathbf{x}(\partial(S)) \geq r, \quad \text{for each odd subset } S \text{ of } V. \tag{6.6c}$$

▷**6.1.11** *Prove that Theorem 6.1 implies Theorem 6.4 and, conversely, that Theorem 6.4 implies Theorem 6.1.*
(Hint: let $\mathbf{C}(G)$ *denote the set of all solutions to the system of linear inequalities consisting of (6.6a), (6.6b) and (6.6c), let* $\mathbf{x} \in \mathbb{R}_+^E - \mathbf{0}$, *let* $r := \mathbf{x}(\partial(v))$ *for some vertex v and consider the equivalence of the following statements if* $r > 0$:

$$\mathbf{x} \in \mathcal{C}one(G), \quad \mathbf{y} := \frac{\mathbf{x}}{r} \in \mathcal{P}oly(G), \quad \mathbf{y} \in \mathbf{P}(G), \quad \textit{and} \quad \mathbf{x} \in \mathbf{C}(G).)$$

∗**6.1.12** *The objective of this exercise is to give a proof of Theorem 6.4, due to Seymour (1979, [85]). For this, adopt the definition of* $\mathbf{C}(G)$ *given in Exercise 6.1.11. Let* $\mathbf{x} \in \mathbf{C}(G)$, *let* $r := \mathbf{x}(\partial(v))$ *for some vertex v and let S be the support of* $\mathbf{x}$. *Prove that* $\mathbf{x} \in \mathcal{C}one(G)$ *by induction on* $|V| + |S|$.

(i) *Prove that if S is nonempty then the graph $G[S]$ of G induced by S is a spanning matchable subgraph of G. Thus, let M be a perfect matching of $G[S]$ (and of G).*

(ii) *Suppose that G has a nontrivial odd cut C such that $\mathbf{x}(C) = r$; apply the inductive hypothesis to both C-contractions of G, as in the first part of the proof of Theorem 6.1.*

(iii) *Suppose that $\mathbf{x}(C) > r > 0$ for each nontrivial odd cut of G. Let*

$$\epsilon := \min\left[\{\mathbf{x}(e) : e \in M\} \cup \{\frac{\mathbf{x}(C) - r}{|M \cap C| - 1} : |M \cap C| > 1,\ C \text{ odd}\}\right].$$

Let $\mathbf{y} := \mathbf{x} - \epsilon \cdot \chi^M$ and prove that (i) $\mathbf{y} \in \mathbf{C}(G)$ and (ii) either the support of $\mathbf{y}$ is a proper subset of S or the previous case is applicable with $\mathbf{y}$ playing the role of $\mathbf{x}$.

6.2 The Matching Space ♯

THE PERFECT MATCHING SPACE

The *perfect matching space* of a graph G is the linear subspace of $\mathbb{R}^E$ generated by the set

$$\{\chi^M\} := \{\chi^M : M \in \mathcal{M}\}$$

of incidence vectors of perfect matchings of G. We denote this space by $\mathcal{L}in(G)$ and, for brevity, refer to it simply as the *matching space*.

The purpose of this section is to present a characterization of the matching space of a matching covered graph G, and to determine a formula for its dimension (which, interestingly, involves the parameter $b(G)$, the number of bricks of the graph G.)

REGULAR VECTORS

We denote the set of all tight cuts in a matching covered graph G by $\mathcal{T}(G)$. A vector $\mathbf{x}$ in the edge space $\mathbb{R}^E$ of G is *r-regular* if

$$\mathbf{x}(C) = r, \text{ for each } C \in \mathcal{T}(G). \tag{6.7}$$

A *regular vector* is one which is r-regular for some real number r.

The following result establishes a necessary condition for a vector to be in the matching space of a matching covered graph G.

Proposition 6.5 *Every vector $\mathbf{x}$ in $\mathcal{L}in(G)$ is regular.*

Proof Since $\mathbf{x}$ is in $\mathcal{L}in(G)$, by definition, it may be expressed as a linear combination of incidence vectors of perfect matchings of G. Thus, there exist coefficients α_M, $M \in \mathcal{M}$, such that $\mathbf{x} = \sum_{M\in\mathcal{M}} \alpha_M \chi^M$. Now let C be any tight cut of G. Then, using the fact that $|M \cap C| = 1$, for each $M \in \mathcal{M}$, we have:

$$\mathbf{x}(C) = \sum_{M\in\mathcal{M}} \alpha_M \chi^M(C) = \sum_{M\in\mathcal{M}} \alpha_M. \tag{6.8}$$

It follows that $\mathbf{x}$ is r-regular, with $r = \sum_{M\in\mathcal{M}} \alpha_M$. □

Every vector in $\mathcal{P}oly(G)$ is a non-negative 1-regular vector, by Proposition 6.2; but not every 1-regular vector is in $\mathcal{P}oly(G)$. For example, the vectors $\mathbf{x}$ and $\mathbf{y}$ displayed in Figure 6.4 are 1-regular vectors in the edge space of $G := \overline{C_6}$. But, only one of them, namely $\mathbf{y}$, is in $\mathcal{P}oly(G)$ (see Exercise 6.2.1). However, as the following Lemma shows, both $\mathbf{x}$ and $\mathbf{y}$ and, in fact, all 1-regular vectors are in $\mathcal{L}in(G)$.

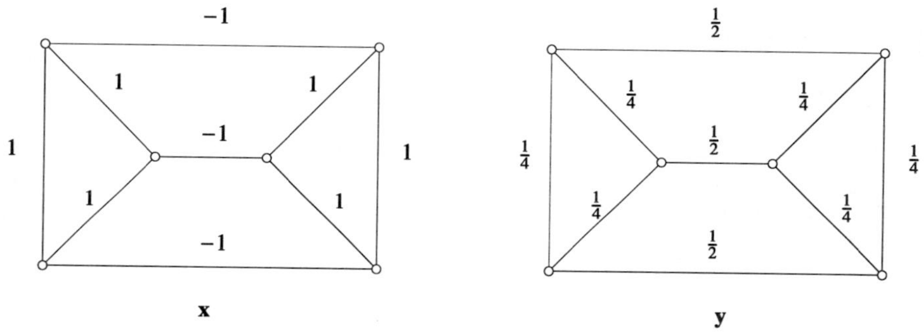

Fig. 6.4 Two 1-regular vectors in the edge space of $\overline{C_6}$

Lemma 6.6 *Every* 1*-regular vector* $\mathbf{x}$ *in* $\mathbb{R}^E$ *belongs to* $\mathcal{L}in(G)$.

Proof Every 1-regular vector in $\mathcal{P}oly(G)$ is clearly in $\mathcal{L}in(G)$. Suppose that $\mathbf{x}$ is a 1-regular vector that is not in $\mathcal{P}oly(G)$. To show that $\mathbf{x}$ is in $\mathcal{L}in(G)$, it suffices to show that it can be expressed as a linear combination of two vectors $\mathbf{y}$ and $\mathbf{z}$ in $\mathcal{P}oly(G)$.

Let $\mathbf{y} = \sum_{M\in\mathcal{M}} \alpha_M \chi^M$ be a vector in $\mathcal{P}oly(G)$ in which each α_M is strictly positive. (For example, $\sum_{M\in\mathcal{M}} |\mathcal{M}|^{-1}\chi^M$ is such a vector.) Then, $\mathbf{y}$ is 1-regular and strictly positive. Furthermore, for each odd cut C that is not tight,

$$\mathbf{y}(C) = \sum_{M\in\mathcal{M}} |M \cap C|\, \alpha_M > \sum_{M\in\mathcal{M}} \alpha_M = 1,$$

because $|M \cap C| \geq 1$ for all perfect matchings M, with strict inequality for at least one perfect matching. Now it can be verified that, for any 'sufficiently' small positive real number ϵ, the vector

$$\mathbf{z} := (1 - \epsilon)\mathbf{y} + \epsilon\mathbf{x}$$

is also a vector in $\mathcal{Poly}(G)$ (see Exercise 6.2.1). Since both $\mathbf{y}$ and $\mathbf{z}$ are in $\mathcal{Poly}(G)$, they are in $\mathcal{Lin}(G)$. We conclude that

$$\mathbf{x} = \left(1 - \frac{1}{\epsilon}\right)\mathbf{y} + \frac{1}{\epsilon}\mathbf{z}$$

is also in $\mathcal{Lin}(G)$. □

CHARACTERIZATION OF THE MATCHING SPACE

Theorem 6.7 *A vector* $\mathbf{x}$ *in the edge space of a matching covered graph* G *belongs to* $\mathcal{Lin}(G)$ *if and only if it is regular.*

Proof If $\mathbf{x}$ is in $\mathcal{Lin}(G)$ then, by Proposition 6.5, it is regular. To prove the converse, we must show that every regular vector $\mathbf{x}$ in $\mathbb{R}^E$ is in $\mathcal{Lin}(G)$.

By Lemma 6.6, every 1-regular vector is in $\mathcal{Lin}(G)$. So, suppose that $\mathbf{x}$ is an r-regular vector, where $r \neq 1$. Let $\mathbf{y}$ be any 1-regular vector, and consider the vector $\mathbf{w}$ defined as follows:

$$\mathbf{w} := \frac{1}{r-1}(\mathbf{x} - \mathbf{y}) \tag{6.9}$$

It is easy to check that this vector $\mathbf{w}$ is 1-regular (Exercise 6.2.2). Therefore, by Lemma 6.6, the vector $\mathbf{w}$ is in $\mathcal{Lin}(G)$; hence the vector $\mathbf{x} = (r-1)\mathbf{w} + \mathbf{y}$ is also in $\mathcal{Lin}(G)$. □

We shall now proceed to derive a formula for the dimension of the matching space of a matching covered graph. It turns out that in the case of braces (and more generally, of all bipartite graphs), and in the case of bricks, this formula may be derived using elementary linear algebra. After dealing with these two special cases, we shall use the tight cut decomposition procedure to deduce the formula for the dimension of the matching space of an arbitrary matching covered graph by induction.

6.2.1 Linear equality description

As $\mathcal{Lin}(G)$ is a subspace of $\mathbb{R}^E$, it is the set of solutions to a system of linear homogeneous equations. We begin by introducing the notation necessary for describing such a system of equations. (We shall adopt the convention that all vectors in $\mathbb{R}^E$ are row vectors.)

Let G be any matching covered graph. Recall that $\mathcal{T} := \mathcal{T}(G)$ denotes the set of all tight cuts (including the trivial cuts) of G. We shall denote the $|\mathcal{T}| \times m$ matrix whose rows are the incidence vectors of the tight cuts of G by $\mathbf{T}$. Figure 6.5 shows an example of a graph G and the matrix $\mathbf{T}$, where $C = \partial(\{w,x,y\})$, and $D = \partial(\{u,w,x\})$. We note that the rows of the $n \times m$ incidence matrix $\mathbf{A}$ of G are just the incidence vectors of the trivial tight cuts of G. Thus $\mathbf{A}$ is an $n \times m$ submatrix of $\mathbf{T}$, and if G is free of nontrivial tight cuts, then $\mathbf{T} = \mathbf{A}$.

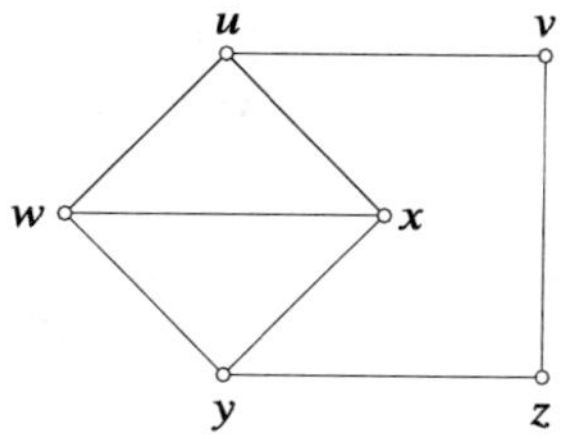

	uv	uw	ux	vz	wx	wy	xy	yz
$\partial(u)$:	1	1	1	0	0	0	0	0
$\partial(v)$:	1	0	0	1	0	0	0	0
$\partial(w)$:	0	1	0	0	1	1	0	0
$\partial(x)$:	0	0	1	0	1	0	1	0
$\partial(y)$:	0	0	0	0	0	1	1	1
$\partial(z)$:	0	0	0	1	0	0	0	1
C :	0	1	1	0	0	0	0	1
D :	1	0	0	0	1	1	1	0

Fig. 6.5 The matrix $\mathbf{T}(G)$ (the first six rows constitute $\mathbf{A}(G)$)

Suppose that r is a fixed real number. By definition, a vector $\mathbf{x}$ in $\mathbb{R}^E$ is r-regular if and only $\mathbf{x}(C) = r$, for each $C \in \mathcal{T}$, which amounts to saying that, for each $C \in \mathcal{T}$, the dot product of χ^C and $\mathbf{x}$ is r. Thus, in matrix notation, the set of all r-regular vectors is the set of all solutions to the system

$$\mathbf{T}\mathbf{x}^t = \mathbf{r}, \tag{6.10}$$

of linear equations where (i) $\mathbf{T}$ is the $|\mathcal{T}| \times m$ matrix defined above, (ii) $\mathbf{x}$ is the $1 \times m$ vector of variables, one per each edge, and (iii) $\mathbf{r}$ is the $|\mathcal{T}| \times 1$ column vector each of whose entries is the real number r. (By our convention, all vectors in $\mathbb{R}^E$ are row vectors. Thus, for the sake of consistency of matrix multiplication, it is necessary to take the transpose of $\mathbf{x}$.)

Let $\mathcal{L}in_0(G)$ denote the subset of $\mathcal{L}in(G)$ consisting of all 0-regular vectors. It follows from the above observation that $\mathcal{L}in_0(G)$ is the subspace of $\mathbb{R}^E$ consisting of all solutions to the system $\mathbf{T}\mathbf{x}^t = \mathbf{0}$ of linear homogeneous equations. Since the number of variables is m (one per each edge), it follows that

$$\dim(\mathcal{L}in_0(G)) = m - \operatorname{rank}(\mathbf{T}).$$

A basis of $\mathcal{L}in_0(G)$ together with any 1-regular vector clearly constitutes a basis of $\mathcal{L}in(G)$ (Exercise 6.2.5). We therefore have:

Theorem 6.8 *The dimension of the matching space of G is:*

$$\dim(\mathcal{L}in(G)) = m + 1 - \mathrm{rank}(\mathbf{T}). \tag{6.11}$$

In view of the above theorem, determining a formula for the dimension of $\mathcal{L}in(G)$ reduces to determining a formula for the rank of the matrix $\mathbf{T}$. (This was the approach used by Naddef (1982, [76]) to determine the dimension of $\mathcal{L}in(G)$.)

Let us first consider the case in which G is a graph which is free of nontrivial tight cuts. In this case, the matrix $\mathbf{T}$ is identical to the incidence matrix $\mathbf{A}$ of G. Using the fact that G is connected, it is not difficult to show (Exercise 6.2.7) the following:

$$\mathrm{rank}(\mathbf{A}) = \begin{cases} n-1 & \text{if } G \text{ is bipartite,} \\ n & \text{otherwise.} \end{cases} \tag{6.12}$$

This, combined with Equation (6.11), immediately yields the following result:

DIMENSION FORMULAS FOR BRICKS AND BRACES

Theorem 6.9 *The dimension of $\mathcal{L}in(G)$ is $m-n+2$ if G is a brace, and $m-n+1$ if G is a brick.* □

For example the dimension of the matching space of $K_{3,3}$ is $9-6+2=5$, and the dimension of the matching space of the pentagonal prism is $15-10+1=6$.

Let $G(A,B)$ be a bipartite matching covered graph, and let $C := \partial(X)$ be a tight cut of G. Then, by Theorem 4.8, there exist partitions of A into (A_1, A_2) and B into (B_1, B_2) such that (i) $|A_1| = |B_1| + 1$ and $|A_2| = |B_2| - 1$, and (ii) every edge of the cut C has one end in A_1 and one end in B_2.

It follows that the incidence vector of C is the sum of the incidence vectors of the trivial cuts associated with vertices in A_1 minus the sum of the trivial cuts of the vertices in B_1. Hence the incidence vector of any tight cut of G is in the row space of the incidence matrix of G. Since G is bipartite, the rank of the incidence matrix of G is $n-1$. It now follows from Theorem 6.8 that the dimension of the matching space of G is $m-n+2$. Similarly, when G is a near-brick (a graph with $b(G) = 1$), the incidence vector of any tight cut may be expressed as an affine linear combination of incidence vectors of trivial cuts, and the dimension of $\mathcal{L}in(G)$ is $m-n+1$ (Exercise 6.2.3). The next section is dedicated to finding a formula for the dimension of the matching space of a matching covered graph that takes into account the number of its bricks.

Exercises

∗**6.2.1**

(i) *Consider the two vectors* $\mathbf{y}$ *and* $\mathbf{x}$ *in the edge space of* $G := \overline{C_6}$ *shown in Figure 6.4. Show that both* $\mathbf{y}$ *and* $\mathbf{x}$ *are* 1*-regular vectors, where* $\mathbf{y}$ *is in* $\mathcal{Poly}(G)$ *and* $\mathbf{x}$ *is not. Furthermore, find a value of* ϵ *such that* $\mathbf{z} := (1-\epsilon)\mathbf{y} + \epsilon\mathbf{x}$ *is in* $\mathcal{Poly}(G)$.

(ii) *In the proof of Lemma 6.6, show that* $\mathbf{z}$ *is in* $\mathcal{Poly}(G)$ *for all small enough positive values of* ϵ.

6.2.2 *Show that the vector* $\mathbf{w}$ *defined in Equation (6.9) is a* 1*-regular vector.*

▷**6.2.3** *Let* G *be a matching covered graph with at most one brick.*

(i) *Show that the incidence vector of any tight cut can be expressed as an affine linear combination of incidence vectors of trivial tight cuts.*

(ii) *Deduce that a vector* $\mathbf{x}$ *in* $\mathbb{Z}^E$ *is regular if and only if* $\mathbf{x}(\partial(u)) = \mathbf{x}(\partial(v))$*, for any two vertices* u *and* v *of* G.

▷**6.2.4** *Consider the matching covered graph* G *with two bricks which is shown in Figure 6.6.*

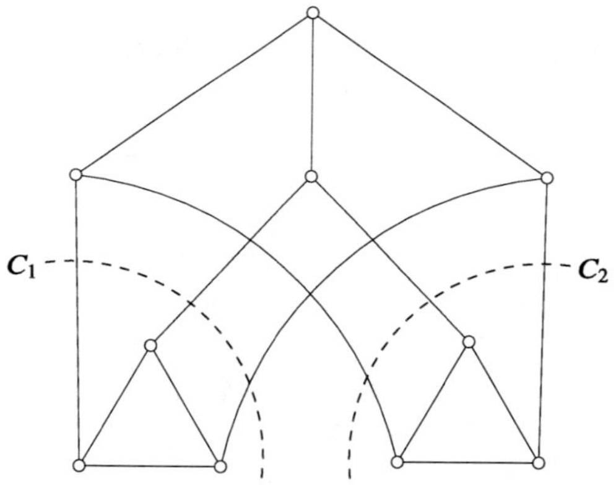

Fig. 6.6 A matching covered graph G with two bricks

(i) *Find a non-negative vector* $\mathbf{x}$ *in* $\mathbb{R}^E$ *such that* $\mathbf{x}(\partial(v)) = 1$*, for all* $v \in V$*, with the property that neither* $\mathbf{x}(C_1)$ *nor* $\mathbf{x}(C_2)$ *is equal to* 1*. (Such a vector satisfies the non-negativity and degree constraints stipulated in the statement of Edmonds' Theorem 6.1 without being a* 1*-regular vector.)*

(ii) *For* $i = 1, 2$*, show that the incidence vector of the cut* C_i *can be expressed as an affine linear combination of incidence vectors of trivial tight cuts and the cut* C_{3-i}.

∗**6.2.5** *Show that a basis for* $\mathcal{Lin}_0(G)$ *together with any* 1*-regular vector* $\mathbf{y}$ *constitutes a basis for* $\mathcal{Lin}(G)$*. (Hint: Let* $\mathbf{x}$ *be any* r*-regular vector, where* $r \neq 0$*; then* $\frac{1}{r}\mathbf{x} - \mathbf{y}$ *is a* 0*-regular vector.)*

> **Matching orthogonal vectors**
>
> A vector $\mathbf{x}$ in the edge space of a matching covered graph G is *matching orthogonal* if it belongs to $(\mathcal{L}in(G))^{\perp}$; equivalently, if $\mathbf{x}(M) = 0$, for each perfect matching M of G.

$*$ **6.2.6** *Let G be a matching covered graph and let* $\mathbf{T}$ *be the matrix whose rows are the incidence vectors of the tight cuts of G. Let D be a tight cut of G and let* $\mathbf{B}$ *denote the matrix obtained from* $\mathbf{T}$ *by replacing each row* χ^C, $C \neq D$, *by* $\chi^C - \chi^D$.

(i) *Prove that* $\mathbf{Bx} = \mathbf{0}$ *if and only if* $\mathbf{x} \in \mathcal{L}in(G)$ *and deduce that the rows of* $\mathbf{B}$ *span* $\mathcal{L}in(G)^{\perp}$.
(ii) *Let* $\mathbf{p}$ *be any vector in* $\mathbb{R}^V$ *such that* $\sum_{v \in V} p(v) = 0$. *Then show that the vector* $\mathbf{x}$ *in* $\mathbb{R}^E$ *induced by* $\mathbf{p}$ *where, for each edge uv of G,* $\mathbf{x}(uv) = \mathbf{p}(u) + \mathbf{p}(v)$, *is a matching orthogonal vector.*
(iii) *Now suppose that G is a brick. Using the characterization of the matching space of a brick and part (i), show that every matching orthogonal vector in* $\mathbb{R}^E$ *is a function as described in part (ii).*

$*$ **6.2.7** *Give a proof of the formula (6.12) for the rank of* $\mathbf{A}$. *(Hint: First show that* $\text{rank}(\mathbf{A}) \geq n - 1$ *by showing that the columns corresponding to the edges of a spanning tree of G are linearly independent. Then note that any linear combination of the rows of* $\mathbf{A}$ *is of the form* $\sum_{v \in V} p(v)\partial(v)$. *In such a linear combination, the component corresponding to an edge uw is* $p(u) + p(w)$. *Hence the rows of* $\mathbf{A}$ *are linearly dependent if and only if there is an assignment of weights* $p(v)$ *to* $v \in V$ *such that the 'induced' weights* $(p(u) + p(w))$ *on the edges of G are all zero.)*

6.2.8 *Find two different bases for the row space of the matrix* $\mathbf{T}(G)$, *where G is the graph shown in Figure 6.5.*

6.2.9

(i) *Express the dimension of the matching space of a cubic brick of order n as a function of n.*
(ii) *Find bases for the matching spaces of the Petersen graph and the pentagonal prism.*

6.3 The Dimension Formula

If a matching covered G is neither a brace nor a brick, then it has a nontrivial tight cut C. In this case, it is possible to obtain a basis for the matching space of G by combining in a certain manner bases for the matching spaces of the two C-contractions of G. This involves the operation defined below.

6.3.1 The merger operation

Let C be a separating cut of a matching covered graph G, and let G_1 and G_2 be the two C-contractions of G. For $i = 1, 2$, let $\mathcal{B}_i$ denote a set of perfect matchings of G_i such that $\{\chi^M : M \in \mathcal{B}_i\}$ is a basis of $\mathcal{L}in(G_i)$. We shall now describe a procedure which may be used to obtain from $\mathcal{B}_1$ and $\mathcal{B}_2$ a set $\mathcal{B} := \mathcal{B}_1 \vee \mathcal{B}_2$ of $|\mathcal{B}_i| + |\mathcal{B}_2| - |C|$ perfect matchings of G whose incidence vectors are linearly independent and, when C is a tight cut, constitute a basis of $\mathcal{L}in(G)$. (Although our interest here is only in tight cuts, with future applications in mind, we allow the possibility of C being a separating cut which may not be tight.)

In describing the above-mentioned procedure and its properties, for brevity, we shall sometimes informally refer to a set of perfect matchings as being independent if their incidence vectors are independent. Similarly, we shall refer to a set of perfect matchings as a basis of the matching space of a graph to mean that the set of their incidence vectors is a basis of that space.

For $i = 1, 2$, each perfect matching of G_i clearly contains precisely one edge incident with the contraction vertex in G_i, and the edges of G_i incident with this vertex are in one-to-one correspondence with the edges of C. We obtain $\mathcal{B}_1 \vee \mathcal{B}_2$ as follows:

- For $i = 1, 2$, group the members of $\mathcal{B}_i$ according to the edge of C they contain. For each $e \in C$, denote the subset of perfect matchings in $\mathcal{B}_i$ containing e by $\mathcal{B}_i^e$.
- For each $e \in C$, and $i = 1, 2$, fix (select) a perfect matching F_i^e from $\mathcal{B}_i^e$.
- For any perfect matching M_1 of G_1 in $\mathcal{B}_1^e$, and the fixed perfect matching F_2^e of G_2, the set $M_1 \cup F_2^e$ is a perfect matching of G. Similarly, for the fixed perfect matching F_1^e of G_1 and for any perfect matching M_2 of G_2 in $\mathcal{B}_2^e$, the set $F_1^e \cup M_2$ is a perfect matching of G.
  Define the set $\mathcal{B}^e$ of perfect matchings and the perfect matching F^e as below:
  $$\mathcal{B}^e := \{M_1 \cup F_2^e : M_1 \in \mathcal{B}_1^e\} \cup \{F_1^e \cup M_2 : M_2 \in \mathcal{B}_2^e\} \quad \text{and} \quad F^e := F_1^e \cup F_2^e.$$
- Now let $\mathcal{B} := \cup_{e \in C} \mathcal{B}^e$.
  We denote this set of perfect matchings of G by $\mathcal{B}_1 \vee \mathcal{B}_2$ and refer to it as the *merger* of $\mathcal{B}_1$ and $\mathcal{B}_2$.

It can be seen that $|\mathcal{B}^e| = |\mathcal{B}_1^e| + |\mathcal{B}_2^e| - 1$, since $F_1^e \cup F_2^e$ is counted twice in the sum $|\mathcal{B}_1^e| + |\mathcal{B}_2^e|$. Consequently,

$$|\mathcal{B}_1 \vee \mathcal{B}_2| = |\mathcal{B}_1| + |\mathcal{B}_2| - |C|. \tag{6.13}$$

Example 6.10 Consider the graph G shown in Figure 6.7(a) in which the cut C is a tight cut. The graph G_1, one of the two C-contractions of G, is the pentagonal prism; it has a basis $\mathcal{B}_1$ consisting of six perfect matchings of which two, M_1 and M_2, contain the edge e. In Figure 6.7(b), the edges labeled i are the edges in M_i, $i = 1, 2$. The graph G_2, the other C-contraction of G, is the graph obtained by splicing $K_{3,3}$ and K_4; it has a basis $\mathcal{B}_2$ consisting of five perfect matchings of which two, N_1 and N_2, contain e. In Figure 6.7(c), the edges labeled i are the edges in N_i, $i = 1, 2$.

The result of the merger operation on edge e results in three perfect matchings of G, $M_1 \cup N_1$, $M_1 \cup N_2$ and $M_2 \cup N_1$, labeled respectively 11, 12 and 21 in Figure 6.7(a); the perfect matchings M_1 of G_1 and the perfect matching N_1 of G_2 play, respectively, the roles of F_1^e and F_2^e.

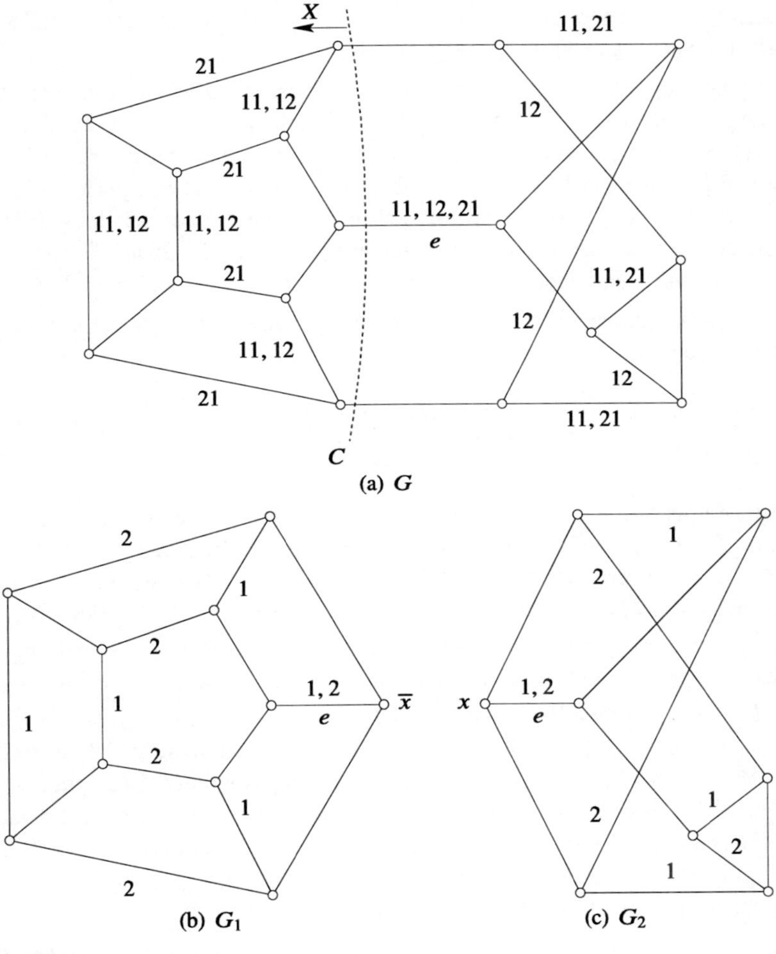

Fig. 6.7 Merger of bases of tight cut-contractions

Theorem 6.11 *Let G_1 and G_2 be the two C-contractions of a matching covered graph G with respect to a separating cut C, and, for $i = 1, 2$, let $\mathcal{B}_i$ denote a set of perfect matchings of G_i whose incidence vectors constitute a basis of $\mathcal{L}in(G_i)$. Then:*

(i) *the set of incidence vectors of $\mathcal{B} := \mathcal{B}_1 \vee \mathcal{B}_2$ is an independent subset of $\mathcal{L}in(G)$;*
(ii) *furthermore, if C is a tight cut, then this set is a basis of $\mathcal{L}in(G)$.*

Proof Let us first show that $\{\chi^M : M \in \mathcal{B}\}$ is a linearly independent set. This requires us to show $\alpha_M = 0$, for all $M \in \mathcal{B}$, is the only solution to the equation

$$\sum_{M \in \mathcal{B}} \alpha_M \chi^M = \mathbf{0}, \tag{6.14}$$

where $\mathbf{0}$ is the vector of all zeros. We shall establish this by considering the restrictions of the above equation to $E(G_1)$ and to $E(G_2)$ and using the fact that $\mathcal{B}_1$ and $\mathcal{B}_2$ are 'bases' of $\mathcal{L}in(G_1)$ and $\mathcal{L}in(G_2)$, respectively.

Every perfect matching in $\mathcal{B}$ is either of the form $M_1^e \cup F_2^e$, for some $M_1^e \in \mathcal{B}_1^e$, or of the form $F_1^e \cup M_2^e$ for some $M_2^e \in \mathcal{B}_2^e$. Thus, the restrictions of members of $\mathcal{B}$ to $E(G_1)$ are perfect matchings in $\mathcal{B}_1$. For each $e \in C$, a perfect matching M_1^e in $\mathcal{B}_1^e$ which is different from F_1^e, appears as the restriction of just one perfect matching in $\mathcal{B}$, namely $M_1^e \cup F_2^e$. However, for each $e \in C$, the perfect matching F_1^e arises as the restriction of each of the perfect matchings in $\mathcal{B}$ of the form $F_1^e \cup M_2^e$ where $M_2^e \in \mathcal{B}_2^e$. (Similar statements apply to restrictions of members of $\mathcal{B}$ to $E(G_2)$.)

In view of the above observation, by restricting the equation (6.14) to $E(G_1)$ we obtain the equation:

$$\sum_{e \in C} \left(\sum_{M_1^e \in \mathcal{B}_1^e - F_1^e} \left(\alpha_{M_1^e \cup F_2^e}\right) \chi^{M_1^e} + \sum_{M_2^e \in \mathcal{B}_2^e} \left(\alpha_{F_1^e \cup M_2^e}\right) \chi^{F_1^e} \right) = \mathbf{0}. \tag{6.15}$$

Since, by hypothesis, the set of incidence vectors of $\mathcal{B}_1$ is a linearly independent subset of $\mathcal{L}in(G_1)$, it follows that $\alpha_M = 0$ for all perfect matchings M in $\mathcal{B}$ of the form $M_1^e \cup F_2^e$, where $M_1^e \in \mathcal{B}_1^e - F_1^e$, for any $e \in C$. Likewise, $\alpha_M = 0$ for all perfect matchings M in $\mathcal{B}$ of the form $F_1^e \cup M_2^e$, where $M_2^e \in \mathcal{B}_2^e - F_2^e$, for any $e \in C$. When we take these two facts into account, the coefficient of $\chi^{F_1^e}$ in equation (6.15) reduces to $\alpha_{F_1^e \cup F_2^e}$. Now we can infer that $\alpha_M = 0$, for all $M \in \mathcal{B}$, establishing that $\{\chi^M : M \in \mathcal{B}\}$ is a linearly independent set.

Now we turn to the proof of the second part of the statement which asserts that $\{\chi^M : M \in \mathcal{B}\}$ is a basis of $\mathcal{L}in(G)$ under the assumption that C is a tight cut.

To prove this, it suffices to show that the incidence vector of any perfect matching of G is in the span of $\{\chi^M : M \in \mathcal{B}\}$. So, let N be a perfect matching of G. Since C is a tight cut, N has precisely one edge in C. Let $N \cap C = \{e\}$, and let N_1 and N_2 denote the restrictions of N to $E(G_1)$ and $E(G_2)$, respectively.

Since, by hypothesis, $\{\chi^{M_1} : M_1 \in \mathcal{B}_1\}$ is a basis of $\mathcal{L}in(G_1)$, it follows that there exist coefficients α_{M_1} such that

$$\chi^{N_1} = \sum_{M_1 \in \mathcal{B}_1} \alpha_{M_1} \chi^{M_1}. \tag{6.16}$$

The set N_1 is a perfect matching of G_1. Thus, χ^{N_1} is 1-regular in G_1. By Proposition 6.5,

$$\sum_{M_1 \in \mathcal{B}_1} \alpha_{M_1} = 1.$$

Now, using the above α_{M_1}'s, define the vector $\mathbf{x}$ in $\mathbb{R}^E$ as follows:

$$\mathbf{x} := \sum_{M_1 \in \mathcal{B}_1} \alpha_{M_1} \chi^{M_1 \cup F_2^e}. \tag{6.17}$$

Clearly, $\mathbf{x}$ is in the span of $\{\chi^M : M \in \mathcal{B}\}$. By (6.16), the restriction of $\mathbf{x}$ to $E(G_1)$ is χ^{N_1}. And, by (6.17), the restriction of $\mathbf{x}$ to $E(G_2)$ is $\chi^{F_2^e}$.

By an analogous argument, it follows that there exists a vector

$$\mathbf{y} := \sum_{M_2 \in \mathcal{B}_2} \beta_{M_2} \chi^{F_1^e \cup M_2}$$

in the span of $\{\chi^M : M \in \mathcal{B}\}$ such that the restriction of $\mathbf{y}$ to $E(G_2)$ is χ^{N_2}, and the restriction of $\mathbf{y}$ to $E(G_1)$ is $\chi^{F_1^e}$.

To complete the proof, we note the easily verified identity (Exercise 6.3.1):

$$\chi^N = \chi^{N_1 \cup F_2^e} + \chi^{F_1^e \cup N_2} - \chi^{F_1^e \cup F_2^e}. \tag{6.18}$$

Hence

$$\chi^N = \mathbf{x} + \mathbf{y} - \chi^{F_1^e \cup F_2^e}$$

and it follows that χ^N is in the span of $\{\chi^M : M \in \mathcal{B}\}$. We conclude that $\{\chi^M : M \in \mathcal{B}\}$ is a basis of $\mathcal{L}in(G)$. □

When C is a separating cut that is not tight, the above arguments do not apply, and $\mathcal{B}$ is not a basis for $\mathcal{L}in(G)$. However, the following statement may be easily deduced (Exercise 6.3.4).

Theorem 6.12 *Let G be a matching covered graph, and let C be a separating cut of G which is not tight, and let G_1, G_2 and $\mathcal{B}$ be as in the statement of Theorem 6.11. Then $\mathcal{B}$ is a basis for the subspace $\mathcal{L}in(G, C)$ of $\mathcal{L}in(G)$ generated by the set of incidence vectors of those perfect matchings of G that have just one edge in common with C.* □

THE DIMENSION FORMULA

The next result establishes the dimension of the matching space of a matching covered graph, in terms of its number of vertices, edges and bricks (Edmonds, Lovász and Pulleyblank [32], see also Naddef (1982, [76])).

Theorem 6.13 *The dimension of the matching space $\mathcal{L}in(G)$ of a matching covered graph G with m edges, n vertices and b bricks is $m - n + 2 - b$.*

Proof By induction on the number of edges of G. We have already seen that if G is bipartite (that is, if $b = 0$), $\dim(\mathcal{L}in(G)) = m - n + 2$, and that if G is a near-brick (that is, if $b = 1$), then $\dim(\mathcal{L}in(G)) = m - n + 1$. Suppose that $b \geq 2$, and that C is a nontrivial tight cut of G. Let G_1 and G_2 denote the two C-contractions of G and, for $i = 1, 2$, let m_i, n_i, and b_i denote the numbers of edges, vertices and bricks of G_i, respectively. By the induction hypothesis,

$$\dim(\mathcal{L}in(G_1)) = m_1 - n_1 + 2 - b_1, \quad \text{and} \quad \dim(\mathcal{L}in(G_2)) = m_2 - n_2 + 2 - b_2.$$

Furthermore, by Theorem 6.11 and equation (6.13),

$$\dim(\mathcal{L}in(G)) = \dim(\mathcal{L}in(G_1)) + \dim(\mathcal{L}in(G_2)) - |C|.$$

We thus have

$$\begin{aligned}\dim(\mathcal{L}in(G)) &= (m_1 - n_1 + 2 - b_1) + (m_2 - n_2 + 2 - b_2) - |C| \\ &= (m_1 + m_2 - |C|) - (n_1 + n_2 - 2) + 2 - (b_1 + b_2) \\ &= m - n + 2 - b,\end{aligned}$$

where the last equality follows from Theorem 4.17. □

Since $\mathcal{L}in(G)$ is the subspace of $\mathbb{R}^E$ spanned by $\{\chi^M : M \in \mathcal{M}\}$, the set of incidence vectors of perfect matchings of G, there clearly exist bases of $\mathcal{L}in(G)$ which are subsets of $\{\chi^M : M \in \mathcal{M}\}$. Thus, if $\dim(\mathcal{L}in(G)) = r$, there exists an $r \times m$ matrix of rank r whose rows are incidence vectors of perfect matchings of G. We refer to any such matrix $\mathbf{M}$ as a *basis matrix* of $\mathcal{L}in(G)$.

The problem of finding a basis for $\mathcal{L}in(G)$ from among the incidence vectors of perfect matchings of G is related to the theory of ear decompositions. In case of bipartite graphs, this is straightforward. In fact, when G is bipartite, using the method of proof of Corollary 3.16, one can find a basis for $\mathcal{L}in(G)$ such that any integer vector in $\mathcal{L}in(G)$ may be expressed as an integer linear combination of vectors in that basis (see Exercise 6.3.3). However, dealing with nonbipartite graphs is much harder and will require several tools which are yet to be developed.

6.3.2 Matching lattice and the seminal work of Lovász

The *perfect matching integer cone* of a graph G, (or simply the *matching integer cone*), denoted by $\mathcal{I}nt\mathcal{C}one(G)$, is the set of all non-negative integer linear combinations of incidence vectors of perfect matchings of G. It is easy to observe that, for any integer $r \geq 1$, an r-graph G is r-edge-colourable if and only if the vector $\mathbf{1}$ of all 1's is in $\mathcal{I}nt\mathcal{C}one(G)$. Thus, in particular, the 4-colour theorem is equivalent to the statement that if G is any planar 3-graph then $\mathbf{1} \in \mathcal{I}nt\mathcal{C}one(G)$. This equivalent formulation of the the 4-colour theorem does not make it any easier to prove it, but it allows us to consider relaxations of this problem, and more generally, of the problem of deciding which r-graphs are r-edge-colourable. We have already come across one such relaxation; Exercise 6.1.10 shows that if G is any r-graph then $\mathbf{1} \in \mathcal{C}one(G)$. In an article with many novel ideas, Seymour (1979,[85]) considered a more challenging relaxation which he referred to as 'subtractive edge-colourings'. This led to the idea of the matching lattice described below.

THE PERFECT MATCHING LATTICE

The *perfect matching lattice* (or simply the *matching lattice*) of a matching covered graph G, denoted by $\mathcal{L}at(G)$, is the set of all integer linear combinations of the incidence vectors of perfect matchings of G. Clearly, $\mathcal{L}at(G)$ is a subset of $\mathcal{L}in(G) \cap \mathbb{Z}^E$, for any graph G. Thus, every vector in $\mathcal{L}at(G)$ is a regular integer vector.

It follows from the observation made in the previous subsection that when G is bipartite, every regular integer vector in $\mathbb{Z}^E$ is in $\mathcal{L}at(G)$. However, in general, not every regular integer vector in $\mathbb{Z}^E$ is in $\mathcal{L}at(G)$. For example, if G is the Petersen graph, then $\mathbf{1}$ is a 3-regular integer vector but it cannot be expressed as an integer linear combination of incidence vectors of perfect matchings of G (see Exercise 6.3.5). One of the many interesting results proved by Seymour in the paper cited above implies that if G is any planar 3-graph then $\mathbf{1} \in \mathcal{L}at(G)$.

In what may be regarded as the watershed paper of this entire subject, Lovász (1987, [58]) established a characterization of the matching lattice of a matching covered graph. In particular, he proved:

LOVÁSZ'S THEOREM

Theorem 6.14 *If G is any brick whose underlying simple graph is not the Petersen graph, then $\mathcal{L}at(G) = \mathcal{L}in(G) \cap \mathbb{Z}^E$.*

We shall present a proof of this remarkable result in Chapter 16. Many of the ideas introduced in Section 6.3.1 will play a crucial role there. For example, the proof of Theorem 6.11 can be adapted to show that if C is a tight cut of G, and $\mathcal{B}_1$ and $\mathcal{B}_2$ are 'bases' of $\mathcal{L}at(G_1)$ and $\mathcal{L}at(G_2)$, respectively, then $\mathcal{B}$ is a 'basis' of $\mathcal{L}at(G)$.

Exercises

6.3.1 *Give a proof of the identity* (6.18).

▷**6.3.2 (Merger of bases of separating cut-contractions)** *The graph G shown in Figure 6.8 is a brick, and the cut C is a separating cut in G. One of the C-contractions, G_1, is the pentagonal prism with one edge duplicated, and the other C-contraction, G_2, is the brick S_8 (shown in Figure 7.2). The dimensions of the matching spaces of G_1 and G_2 are 7 and 6.*

(i) *Find a basis $\mathcal{B}_1$ of $\mathcal{L}in(G_1)$ and a basis $\mathcal{B}_2$ of G_2. (Such bases can be found by 'extending' the bases of the pentagonal prism and $K_{3,3} \odot K_4$).*
(ii) *Determine the merger $\mathcal{B} := \mathbf{B}_1 \vee \mathbf{B}_2$.*
(iii) *Show that $\mathcal{B}$ does not constitute a basis of $\mathcal{L}in(G)$.*
(iv) *Extend $\mathcal{B}$ to a basis of $\mathcal{L}in(G)$. (This would require adding to $\mathcal{B}$ one perfect matching which has more than one edge in the cut C.)*

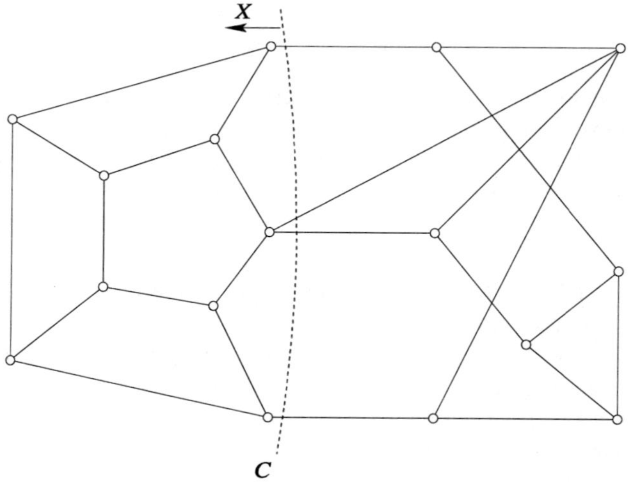

Fig. 6.8 Merger across a separating cut

$*$ **6.3.3** *Let G be a bipartite matching covered graph. Using the method of the proof of Corollary 3.16 show that there exist perfect matchings $M_1, M_2, \ldots, M_r$ in G, where $r = m - n + 2$, such that every integer vector in $\mathcal{L}in(G)$ may be expressed as an integer linear combination of $\{\chi^{M_1}, \chi^{M_2}, \ldots, \chi^{M_r}\}$.*

$\triangleright$**6.3.4** *Give a proof of Theorem 6.12.*

6.3.5 *Show that the vector **1** cannot be expressed as an integer linear combination of incidence vectors of perfect matchings of the Petersen graph. (Hint: Observe that the set $\mathcal{M}$ of the six perfect matchings of the Petersen graph is a basis for its matching space, and that $\mathbf{1} = \sum_{M \in \mathcal{M}} \frac{1}{2}\, \chi^M$.)*

6.3.6 *Using Theorem 6.13 and Exercise 5.3.4, show that the number of perfect matchings in a cubic matching covered graph G is at least $n/4 + 2$.*

$\triangleright$**6.3.7 (A basis for the matching space of a complete graph)** *Let G be the complete graph on the vertex set $\{h, v_1, v_2, \ldots, v_{2r-1}\}$, $r \geq 2$.*

(i) *Show that G can be expressed as the union of $r - 1$ spanning subgraphs $G_1, G_2, \ldots, G_{r-1}$ where, for $1 \leq i \leq r - 1$, the graph G_i is a wheel with the vertex h as its hub and a cycle of length $2r-1$ on the vertex set $\{v_1, v_2, \ldots, v_{2r-1}\}$ as its rim. (Thus, for $1 \leq i < j \leq r - 1$, $E(G_i) \cap E(G_j)$ consists of the $2r - 1$ edges incident with h, but the rims of G_i and G_j are edge-disjoint $(2r-1)$-cycles on $\{v_1, v_2, \ldots, v_{2r-1}\}$.)*

(ii) *For $1 \leq i \leq r - 1$, let $\mathcal{B}_i$ denote the set of all perfect matchings of the wheel G_i, and let $\mathcal{B} := \mathcal{B}_1 \cup \mathcal{B}_2 \cup \cdots \cup \mathcal{B}_{r-1}$. Show that the set of the incidence vectors of the perfect matchings in $\mathcal{B}$ constitutes a basis for the matching space of G.*

$\triangleright$**6.3.8** *Consider the brick in Figure 6.9. Show that the number of perfect matchings in this brick is precisely the same as the dimension of its matching space.*

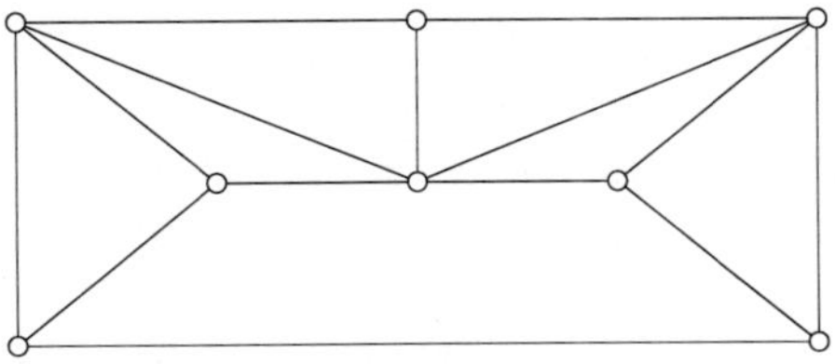

Fig. 6.9 The brick of Exercise 6.3.8.

6.4 Matching Spaces over Finite Fields ♯

The matching space $\mathcal{L}in(G)$ associated with a matching covered graph G, as defined in the previous section, is the vector space over the field $\mathbb{R}$ of real numbers generated by the set of incidence vectors of perfect matchings of G. Clearly, one may define matching spaces over any field $\mathbb{F}$, where one regards the incidence vector χ^M of a perfect matching M as a $(0, 1)$-vector with its entries in $\mathbb{F}$. We denote by $\mathcal{L}in(G, \mathbb{F})$ the linear space over $\mathbb{F}$ generated by the set of incidence vectors of perfect matchings of G, and refer to it as the *matching space of G over $\mathbb{F}$*.

It is a routine exercise to show that Theorem 6.11 extends to matching spaces over arbitrary fields (Exercise 6.4.1). However, the dimension formula (Theorem 6.13) does not extend to fields of characteristic 2. Once again, it is the Petersen graph $\mathbb{P}$ that is the exception. As we have seen, over the reals, the dimension of $\mathcal{L}in(\mathbb{P})$ is six. However, the dimension of $\mathcal{L}in(\mathbb{P}, \mathbb{Z}_2)$, where $\mathbb{Z}_2$ is the field of integers modulo 2, is only five (Exercise 6.4.2). If G is any brick other than a Petersen brick, we shall see in a Chapter 16 that the dimension of $\mathcal{L}in(G, \mathbb{Z}_2)$ is $m - n + 1$. The techniques for proving this result turn out to be the same as the one we use to prove Theorem 6.14!

Exercises

▷**6.4.1** *Prove that Theorem 6.11 extends to matching spaces over arbitrary fields.*

6.4.2 *Show that the dimension of $\mathcal{L}in(\mathbb{P}, \mathbb{Z}_2)$ is equal to five. (Hint: Use the fact that each edge of the Petersen graph is in precisely two of its perfect matchings, and that any two of its perfect matchings have exactly one edge in common.)*

6.5 Notes

A good reference for useful proof techniques and interesting unsolved problems related to the perfect matching polytope and edge colourings of graphs is the paper by Seymour (1979, [85]). In that paper Seymour asked if it is possible to use the four colour theorem to show that every planar 4-graph is 4-edge-colourable. This leads to the more general question: For each $r \geq 4$, does the four colour theorem imply that every planar r-graph is r-edge-colourable?

Guenin proved that this is the case for $r = 4, 5$ (2003, [39]). Dvořák et al. proved the case $r = 6$ (2016, [29]). The cases $r = 7, 8$ were proved by Chudnovsky et al. (2016, [22, 23]).

In that same paper [85], Seymour proposed several conjectures. We state two of them now.

6.15 (Conjecture 3.3 in [85]) *Any r-graph is $(r + 1)$-edge-colourable.*

6.16 (Conjecture 3.5 in [85]) *If $r \geq 4$ then every r-graph has a perfect matching whose deletion gives an $(r - 1)$-graph.*

To the best of our knowledge, Conjecture 6.15 is open. Conjecture 6.16 implies Conjecture 6.15, because every 3-graph has a perfect matching and every graph G having $\Delta(G) = 2$ has a 3-edge colouring.

However, Conjecture 6.16 would imply that if $r \geq 4$ then every r-graph has two disjoint perfect matchings. The latter is false: for each $r \geq 4$ there exist r-graphs which do not have two disjoint perfect matchings, as proved by Rizzi (1999, [81]).

The number of perfect matchings of a graph G is denoted by $\Phi(G)$. Suppose that G is a 2-connected cubic graph. The fact that every such graph G is matching covered (Theorem 2.8) implies that $\Phi(G) \geq 3$. Since, clearly, $\Phi(G) \geq \dim(\mathcal{L}in(G))$, the dimension formula (Theorem 6.13) implies a lower bound for $\Phi(G)$ that is a linear function of n (see Exercise 6.3.6).

CLM (2005, [13]) defined a matching covered graph G to be *extremal* if $\Phi(G) = \dim(\mathcal{L}in(G))$. Figure 6.9 depicts an example of an extremal brick. It turned out that the only extremal bipartite cubic graph is the θ-graph on three edges. All other extremal cubic graphs are bricks, and they were able to show that there are precisely ten extremal cubic bricks. One of them is the Petersen graph. Of the remaining nine, the complete graph K_4, the triangular prism $\overline{C_6}$, the bicorn $\mathbb{H}_8$, and the tricorn $\mathbb{H}_{10}$ are the ones of order four, six, eight and ten. The largest extremal cubic graph is the cubic brick of order sixteen shown in Figure 6.10.

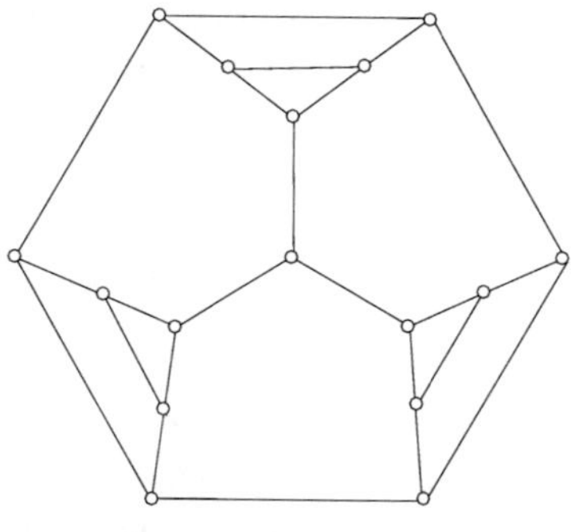

Fig. 6.10 Extremal cubic brick of order 16

Some years later, confirming a conjecture of Lovász and Plummer, Esperet, Kardoš, King, Král' and Norine (2011, [33]) showed that any 2-connected cubic graph on n vertices has at least $2^{n/3656}$ perfect matchings.

The work on the number of perfect matchings in bipartite matching covered graphs [17], mentioned in the Notes section of Chapter 2, led CLM to the following conjecture:

THE NUMBER OF PERFECT MATCHINGS IN A BRACE

Conjecture 6.17 *There exists a positive integer N such that*

$$\Phi(G) \geq \frac{(n-2)^2}{4},$$

for any brace G of order $n \geq N$.

The biwheel on n vertices has $(n-2)^2/4$ perfect matchings. Using a Theorem of McCuaig [68], which will be described in Part II, CLM were able to show in [17] that any brace on n vertices has at least $(n-2)^2/8$ perfect matchings.

An odd wheel of order n has $n-1$ perfect matchings. It may be true that there exists a positive integer N such that $\Phi(G) \geq n-1$, for any brick G on $n \geq N$ vertices!

Chapter 7
Solid Bricks

Contents

7.1 Solid Matching Covered Graphs

The fact that bipartite graphs are free of odd cycles often makes the task of analyzing their properties easier, and makes it possible to state assertions concerning them more simply, and prove them more easily, than the corresponding assertions that are valid for all graphs. For example, the description of the perfect matching polytope of a bipartite graph is simpler, and its proof easier, than the one that applies to all graphs (see Exercise 6.1.1). Solid matching covered graphs defined below have many properties akin to those of bipartite matching covered graphs.

> SOLID MATCHING COVERED GRAPHS
>
> A matching covered graph is *solid* if every separating cut in it is a tight cut.

By Corollary 4.9, every bipartite matching covered graph is solid. But not all solid matching covered graphs are bipartite. (The smallest solid matching covered graph that is not bipartite is K_4.)

C. L. Lucchesi, U. S. R. Murty, *Perfect Matchings*, Algorithms and Computation in Mathematics 31, https://doi.org/10.1007/978-3-031-47504-7_7

> **SOLID BRICKS**
>
> By definition, bricks are free of nontrivial tight cuts. Thus a brick is *solid* if it is free of nontrivial separating cuts. In other words, a brick is solid if it cannot be obtained by splicing together two smaller matching covered graphs. (In this sense, solid bricks are 'unbreakable'!) The brick $\overline{C_6}$ is not solid because it is equal to $K_4 \odot K_4$.

According to Corollary 4.16, if D is a tight cut of a matching covered graph G, and G has a separating cut C that is not tight, then one of the D-contractions of G also has a separating cut which is not tight. It follows that G is not solid if and only if some member in a tight cut decomposition of G is not solid. The assertion stated below follows on noting that braces, being bipartite, are solid (see Exercise 7.1.1).

Theorem 7.1 *A matching covered graph is solid if and only if each of its bricks is solid.* □

Thus, the problem of recognizing whether or not a given matching covered graph is solid reduces to the problem of recognizing whether or not a given brick is solid.

7.1.1 Odd-intercyclic bricks

By Corollary 4.10, if a separating cut in a matching covered graph is not tight, then both shores of that cut induce subgraphs which are not bipartite. This observation leads to the following definition.

> **ODD-INTERCYCLIC GRAPHS**
>
> A graph is said to be *odd-intercyclic* if any two odd cycles in it have at least one vertex in common.
>
> It follows from Corollary 4.10 that every odd-intercyclic brick is solid. Odd wheels, which are easily seen to be odd-intercyclic bricks, will play a special role in future chapters.

Example 7.2 (Odd-intercyclic projective planar graphs) Let H be a 2-connected planar bipartite graph and let $Q := v_1v_2 \dots v_{2k}v_1$ be a facial cycle of H. Obtain G from H by joining, for $1 \le i \le k$, the vertices v_i and v_{i+k} by a new edge which we shall refer to as a chord of the cycle Q. For $1 \le i < j \le k$, the two chords v_iv_{i+k} and v_jv_{j+k} cross in the sense that Q traverses the ends of the these chords in the cyclic order $v_iv_jv_{i+k}v_{j+k}$. (See Figure 7.1 for an illustration.) Each graph G obtained in this way has an embedding in the projective plane so that all faces are even. (When H is just an even cycle of length $2k$, the resulting graph is the Möbius ladder $\mathbb{M}_{2k}$.) We leave it to the reader as an exercise to show that all graphs that can be obtained by this construction are odd-intercyclic (Exercise 7.1.5).

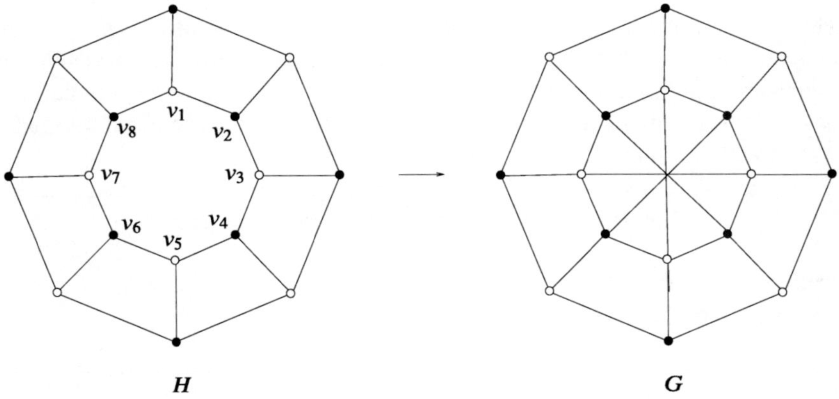

Fig. 7.1 A projective planar odd-intercyclic graph

Subject to some connectivity conditions, every odd-intercyclic graph turns out to be a projective planar graph of the type described in Example 7.2. This had been known for a long time to many prominent mathematicians, but no proof was ever published. Eventually, a paper by Kawarabayashi and Ozeki [45], which appeared in 2013, gave an elegant proof of this result. They showed that if G is any internally 4-connected graph, then G has no two vertex-disjoint odd cycles if and only if G satisfies one of the following: (i) $G-v$ is bipartite for some $v \in V$, (ii) $G-\{e_1, e_2, e_3\}$ is bipartite for some three edges which constitute the edges of a triangle of G, (iii) $|V| \leq 5$, or (iv) G can be embedded in the projective plane so that each face boundary has even length.

The above-mentioned result leads to a polynomial-time algorithm for recognizing solid bricks which are odd-intercyclic bricks. But, alas, not all solid bricks are odd-intercyclic! The brick depicted in Figure 7.2 is a member of an infinite family of non-odd-intercyclic solid bricks which will be described in the next section.

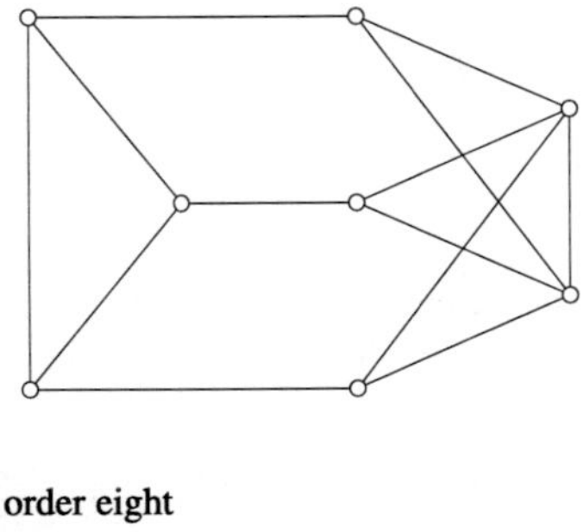

Fig. 7.2 Σ_8, a solid brick of order eight

The problem of recognizing whether a given brick is solid is clearly in the complexity class co-NP; if a brick happens to be nonsolid, to certify that it is nonsolid it suffices to display a nontrivial separating cut in it. However, no polynomial-time algorithm for recognizing solid bricks is yet known. It is not even known if the related decision problem is in NP!

UNSOLVED PROBLEM: RECOGNIZING SOLID BRICKS

Problem 7.3 *Is the problem of recognizing whether a given brick is solid in the complexity class NP? Is it in P?*

7.1.2 Separating cut decompositions

Analogous to a tight cut decomposition of a matching covered graph, one may define a separating cut decomposition by considering contractions with respect to nontrivial separating cuts, instead of the more restrictive tight cuts. Using such a decomposition, any matching covered graph may be reduced to a list of graphs which are free of nontrivial separating cuts.

Since any separating cut in a bipartite graph is also tight, such a list would consist of braces and solid bricks. However, no statement similar to Theorem 4.17 holds for separating cuts: distinct separating cut decompositions may even yield distinct numbers of graphs. Consider, for example, the brick shown in Figure 7.3.

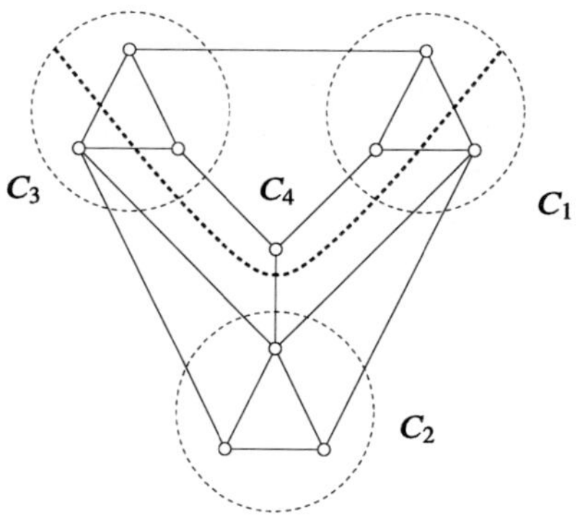

Fig. 7.3 A brick which does not admit a unique separating cut decomposition

One separating cut decomposition of this brick, using the three cuts C_1, C_2, and C_3 yields a collection of four K_4's, up to multiple edges, whereas another separating cut decomposition that uses only the cut C_4 produces two W_5's, up to multiple edges.

Do separating cut decompositions of a matching covered graph shed any light on the structure of that graph? There is no known evidence to suggest that they do. But, as we shall see in future chapters, properties of solid bricks do have many uses. It is conceivable that a parameter associated with separating cut decompositions, such as the minimum possible number of solid bricks in a decomposition of a graph, might prove to be relevant in some context!

Exercises

▷**7.1.1** *Give a proof of Theorem 7.1.*

7.1.2 *Show that the brick shown in Figure 7.2 is not odd-intercyclic, but is solid.*

▷**7.1.3** *Prove that the odd wheel* W_{2k+1} *is an odd-intercyclic brick for all* $k \geq 1$.

▷**7.1.4** *Prove that the Möbius ladder* $\mathbb{M}_{2n}$ *(defined in Section 2.5) is an odd-intercyclic brick, for all even* $n \geq 2$.

▷**7.1.5** *Let* G *be a graph obtained by the procedure described in Example 7.2. Show that:*

(i) *every odd cycle of* G *must include an odd number of chords of* Q, *and*
(ii) *if* C *and* D *are two odd cycles in* G, *then there exist chords of* Q *belonging to* C *and* D, *respectively, which cross.*

Now, using the planarity of H, *deduce that* G *is odd-intercyclic.*

7.1.6 *Determine an infinite family of odd-intercyclic cubic graphs which are not Möbius ladders. (Hint: Note that in the construction described in Example 7.2, it is not necessary to add all the* k *chords of the cycle* Q *to obtain an odd-intercyclic graph. Consider a suitable 2-connected planar bipartite graph* H *in which there is a facial cycle* Q *with four vertices of degree two and add two crossing chords of* Q *to obtain a cubic graph. A Möbius ladder is obtained by adding two edges to a ladder!)*

7.1.7 *Consider the graph* $\Sigma_8 \odot \Sigma_8$ *obtained by a splicing two copies of the brick* Σ_8 *(Figure 7.2). Show that it is a brick and show that it has two distinct separating cut decompositions.*

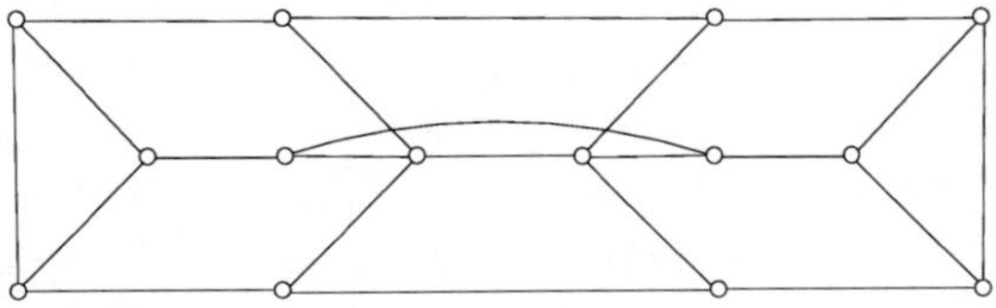

Fig. 7.4 $\Sigma_8 \odot \Sigma_8$

7.2 The Perfect Matching Polytope and Solid Matching Covered Graphs

By Edmonds' Theorem 6.1 there are three types of constraints that every vector $\mathbf{x}$ in the perfect matching polytope $\mathcal{P}oly(G)$ of a matching covered graph G must satisfy. We consider two other polytopes associated with G which are obtained by relaxing the odd set constraints.

7.2.1 Polytopes $\mathcal{P}oly^t(G)$ and $\mathcal{P}oly^*(G)$

The first polytope, which we denote by $\mathcal{P}oly^t(G)$, is obtained by restricting the odd set constraints to equalities corresponding to the tight cuts of G, and the second one, which we denote by $\mathcal{P}oly^*(G)$, is obtained by dropping all the odd set constraints. Thus $\mathcal{P}oly^t(G)$ is the set of solutions to the following system of linear inequalities:

$$\mathbf{x}(e) \geq 0, \quad \text{for each } e \in E \tag{7.1a}$$
$$\mathbf{x}(\partial(S)) = 1, \quad \text{for each tight cut } \partial(S) \text{ of } G, \tag{7.1b}$$

and $\mathcal{P}oly^*(G)$, which is known as the *fractional perfect matching polytope*, is the set of solutions to:

$$\mathbf{x}(e) \geq 0, \quad \text{for each } e \in E \tag{7.2a}$$
$$\mathbf{x}(\partial(v)) = 1, \quad \text{for each vertex } v \text{ of } G. \tag{7.2b}$$

For want of a better name, we shall refer to $\mathcal{P}oly^t(G)$ as *the tight-cut perfect matching polytope*.

The following assertion is an easy consequence of the above definitions:

Proposition 7.4 *For any matching covered graph G*

$$\mathcal{P}oly(G) \subseteq \mathcal{P}oly^t(G) \subseteq \mathcal{P}oly^*(G). \tag{7.3}$$

It is easy to see that for any bipartite graph G, both inclusions in (7.3) hold as equalities. There are examples of nonbipartite matching covered graphs which show that either one or both inclusions in (7.3) may hold strictly (Exercise 7.2.6). We are thus led to the natural questions of characterizing matching covered graphs for which these inclusions hold as equalities. Interestingly, the notion of solid matching covered graphs plays a crucial role in the answers to these questions. We shall see, for example, that $\mathcal{P}oly(G) = \mathcal{P}oly^t(G)$ if and only if G is solid, and that $\mathcal{P}oly(G) = \mathcal{P}oly^*(G)$ if and only if G is a solid matching covered graph with at most one brick.

We shall first deal with bricks and then return to the general case later on. Thus let G be a brick. As bricks do not have nontrivial tight cuts it follows that the equality $\mathcal{P}oly^t(G) = \mathcal{P}oly^*(G)$ holds. Thus, every nonnegative vector in $\mathbb{R}^E$ which satisfies the degree constraints is in $\mathcal{P}oly^t(G)$; it follows that $\mathcal{P}oly(G) = \mathcal{P}oly^*(G)$ if and only if every such vector is also in $\mathcal{P}oly(G)$. This happens to be the case if and only if G is a solid brick. The proof of this surprising fact requires the notion of a certain partial order on the set of odd cuts of a matching covered graph. (This notion will also feature as a crucial tool in the proofs of several important theorems in future chapters.)

7.2.2 Precedence relation on cuts

PRECEDENCE RELATION

A cut C of a matching covered graph G *precedes* a cut D (written $C \preceq D$) if

$$|M \cap C| \leq |M \cap D|, \quad \text{for all } M \in \mathcal{M}.$$

If strict inequality holds for at least one perfect matching M, then we say that C *strictly precedes* D and write $C \prec D$. On the other hand, if $|M \cap C| = |M \cap D|$ for all $M \in \mathcal{M}$, then C and D are *matching equivalent*.

For example, in the brick G shown in Figure 7.5, the cut C strictly precedes the cut D; whereas, in the graph $G - e$, the two cuts C and D are matching equivalent (Exercise 7.2.1).

We say that a cut C is *minimal* with respect to the relation $\preceq$ if there is no cut that strictly precedes C.

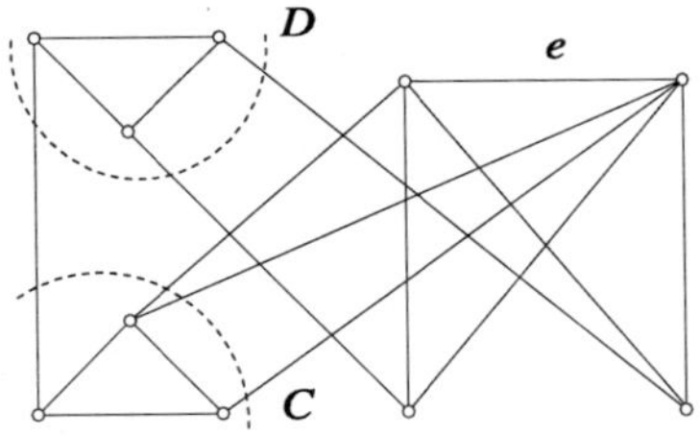

Fig. 7.5 C precedes D

Theorem 7.5 *The perfect matching polytope $\mathcal{P}oly(G)$ of a brick G consists of all non-negative 1-regular vectors if and only if G is solid. (In other words, the equality $\mathcal{P}oly(G) = \mathcal{P}oly^*(G)$ holds for a brick G if and only G is solid.)*

Proof Firstly suppose that G is not solid. We wish to show that there is some non-negative 1-regular vector in $\mathbb{R}^E$ that does not belong to $\mathcal{P}oly(G)$. Since G is nonsolid, we first note that it has a nontrivial separating cut, say C. Let M_0 be a perfect matching of G such that $|M_0 \cap C| > 1$. (Such a perfect matching must exist; otherwise C would be a tight cut in a brick which is absurd.) Also, since C is separating, for every edge e of G, there is a perfect matching M_e of G that contains edge e and just one edge in C, by Theorem 4.2. Now let

$$\mathbf{x} := \frac{1}{|E|-1}\left(\left(\sum_{e\in E}\chi^{M_e}\right) - \chi^{M_0}\right) \tag{7.4}$$

It can be verified that $\mathbf{x}$ is non-negative 1-regular with

$$\mathbf{x}(C) < 1$$

(Exercise 7.2.2).

Conversely, suppose that G is solid. We wish to prove that every non-negative 1-regular vector in $\mathbb{R}^E$ belongs to $\mathcal{Poly}(G)$. Assume to the contrary that there is a non-negative 1-regular vector $\mathbf{x}$ that does not belong to $\mathcal{Poly}(G)$. Then, by Theorem 6.1 there must exist at least one odd cut C such that $\mathbf{x}(C)$ is strictly less than one. Let $\mathcal{C}$ denote the set of all odd cuts C for which $\mathbf{x}(C) < 1$ and let $D := \partial(Y)$ be a cut in $\mathcal{C}$ that is minimal with respect to the precedence relation $\preceq$ defined above. We shall show that D is a separating cut and thereby obtain a contradiction to the hypothesis that G is solid.

Let $G_1 := G/(\overline{Y} \to \overline{y})$ and $G_2 := G/(Y \to y)$ denote the two D-contractions of G. We wish to show that both G_1 and G_2 are matching covered. Let us first consider G_1. Clearly it is connected. If it is not matching covered, then it is either not matchable, or is matchable but has an unmatchable edge. In the former case, by Tutte's Theorem 1.3, there exists a subset S of $V(G_1)$ such that $o(G_1 - S) > |S|$ (Figure 7.6(a)). In the latter case, by Theorem 2.1, there exists a subset S of $V(G_1)$ such that $o(G_1 - S) = |S|$, but there is an edge e of G_1 with both its ends in S (Figure 7.6(b)). As G is matching covered, on both alternatives the contraction vertex $\overline{y}$ is in S (see Figure 7.6).

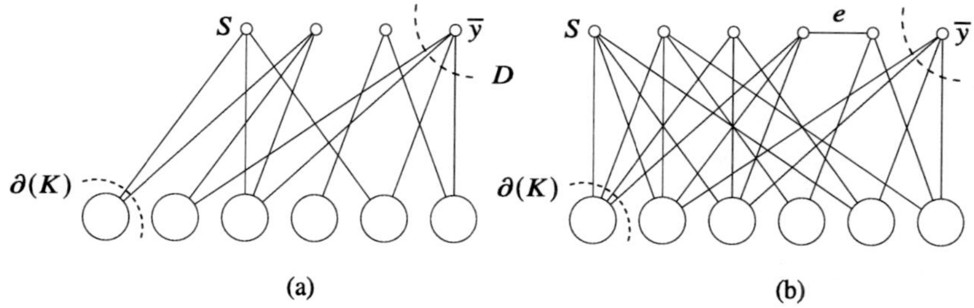

Fig. 7.6 Finding a separating cut using precedence

In either case, a simple counting argument shows that there must be an odd component K of $G_1 - S$ for which $\mathbf{x}(\partial(K)) < 1$ (Exercise 7.2.3(i)). One may now verify that $\partial(K) \prec D$ (Exercise 7.2.3(ii)), contradicting the choice of D. Therefore G_1 is matching covered. Similarly, G_2 is also matching covered; hence D is a separating cut. As $\mathbf{x}(D) < 1$, the cut D is nontrivial. In sum, D is a nontrivial separating cut of a brick. Hence D is not tight, in contradiction to the hypothesis that G is solid. □

The following statements related to the perfect matching polytope of a matching covered graph are easy consequences of the above theorem. We leave their proofs as Exercise 7.2.5.

Corollary 7.6 *For every matching covered graph G, the following properties hold:*

(i) $\mathcal{P}oly(G) = \mathcal{P}oly^t(G)$ *if and only if G is solid,*
(ii) $\mathcal{P}oly(G) = \mathcal{P}oly^*(G)$ *if and only if G is a solid matching covered graph with at most one brick.* □

Exercises

▷**7.2.1** *Verify that in the brick shown in Figure 7.5 the cut C strictly precedes the cut D.*

▷**7.2.2** *Verify that $\mathbf{x}(C) < 1$ for the vector $\mathbf{x}$ defined in Equation (7.4).*

∗**7.2.3** *Consider the graph $G_1 := G/\overline{Y}$ defined in the proof of Theorem 7.5, and a subset S of $V(G_1)$ with the properties specified there. Show that:*

(i) *the graph $G_1 - S$ has an odd component K such that $\mathbf{x}(\partial(K)) < 1$, and*
(ii) *the cut $\partial(K)$ strictly precedes D.*

7.2.4 *Describe the perfect matching polytope of the Petersen graph using as few odd set constraints as possible. Justify your answer.*

7.2.5 *Give a proof of Corollary 7.6.*

7.2.6 *Give examples of matching covered graphs to show that each of the following instances may occur:*

(i) $\mathcal{P}oly(G) \subset \mathcal{P}oly^t(G) = \mathcal{P}oly^*(G)$,
(ii) $\mathcal{P}oly(G) = \mathcal{P}oly^t(G) \subset \mathcal{P}oly^*(G)$,
(iii) $\mathcal{P}oly(G) \subset \mathcal{P}oly^t(G) \subset \mathcal{P}oly^*(G)$.

7.3 A Characterization of Solid Bricks ♯

The characterization of solid bricks presented in this section does not seem to lead in any obvious way to a polynomial-time algorithm for recognizing solid bricks. However, it provides a convenient means for showing that certain bricks are solid.

It is straightforward to verify that the incidence vector of any perfect matching of any matching covered graph G is an extreme point of the polytope $\mathcal{P}oly^*(G)$ (Exercise 7.3.1). If all the extreme points of $\mathcal{P}oly^*(G)$ were incidence vectors of perfect matchings of G, then $\mathcal{P}oly(G)$ and $\mathcal{P}oly^*(G)$ would be the same polytope. Thus if G is a brick that is not solid, the polytope $\mathcal{P}oly^*(G)$ must have an extreme point which is not the incidence vector of a perfect matching. The following result concerning extreme points of $\mathcal{P}oly^*(G)$ may be proved by means of arguments analogous to the ones used in our proof of Theorem 6.1 (Exercise 7.3.2).

EXTREME POINTS OF $\mathcal{P}OLY^*(G)$

Theorem 7.7 *Every extreme point of $\mathcal{P}oly^*(G)$ is a $(0, \frac{1}{2}, 1)$-vector. Furthermore, a vector $\mathbf{w}$ is an extreme point of $\mathcal{P}oly^*(G)$ if and only if the subgraph $G[\mathbf{w}]$ of G induced by the support of $\mathbf{w}$ is a spanning subgraph of G in which each component is either an odd cycle or has just one edge, and*

$$\mathbf{w}(e) = \begin{cases} 0 & \text{if } e \text{ is not an edge of } G[\mathbf{w}] \\ \frac{1}{2} & \text{if } e \text{ is an edge of an odd cycle in } G[\mathbf{w}] \\ 1 & \text{if } e \text{ is the only edge of a component of } G[\mathbf{w}]. \end{cases}$$

□

For example, the polytope $\mathcal{P}oly^*(\overline{C_6})$ has five extreme points; four of them are the incidence vectors of the four perfect matchings of $\overline{C_6}$, and the fifth is the $(0, \frac{1}{2})$-valued vector displayed in Figure 6.1(a).

With the help of the above theorem we can now present the result mentioned at the beginning of this section. It was communicated to us in 2003 by four friends from the University of São Paulo: de Pina, Fernandes, Reed and Wakabayashi (2003, [27]). It first appeared in a paper by CLM (2004, [12]), which presented a purely graph theoretic proof. That proof relies on results concerning robust cuts in nonsolid bricks that will be described in a later chapter. Here we shall derive it as a consequence of Theorem 7.7. It may also be deduced from more general results by Balas (1981, [1]).

A CHARACTERIZATION OF SOLID BRICKS

Theorem 7.8 *A brick G is nonsolid if and only if it has two disjoint odd cycles C_1 and C_2 such that $G - (V(C_1) \cup V(C_2))$ has a perfect matching.*

Proof By Theorem 7.5, G is nonsolid if and only if $\mathcal{P}oly(G)$ is a proper subset of $\mathcal{P}oly^*(G)$. As observed earlier this is the case if and only if $\mathcal{P}oly^*(G)$ has an extreme point $\mathbf{w}$ which is not the incidence vector of a perfect matching of G. It follows from Theorem 7.7 that each component of the subgraph $G_{\mathbf{w}}$ of G induced by the support of $\mathbf{w}$ is either an odd cycle or has just two vertices. Let $C_1, C_2, \ldots, C_k$ denote the components of $G_{\mathbf{w}}$ which are odd cycles. Since $\mathbf{w}$ is not the incidence vector of a perfect matching of G, and since $|V|$ is even, it follows that k is greater than zero and that it is even. Since all components of G_w other than $C_1, C_2, \ldots, C_k$ have just one edge each, it follows that $G - (V(C_1) \cup V(C_2) \cup \cdots \cup V(C_k))$ has a perfect matching. If $k = 2$, there is nothing more to prove. So, suppose that $k \geq 4$. Let u be any vertex of C_k, let M_k be a maximum matching of $C_k - u$, and let L denote a perfect matching of $G - (V(C_1) \cup V(C_2) \cup \cdots \cup V(C_k))$. Then $M = M_k \cup L$ is a matching of G which leaves all vertices of $C_1, C_2, \ldots, C_{k-1}$ and just one vertex of C_k, namely u, exposed. Since G is matchable, there exists a M-augmenting path, say P, starting at u. Let v denote the other end of P and adjust notation, if necessary, so that v is a vertex of C_{k-1}. Denote by M_{k-1} the perfect matching of $C_{k-1} - v$, and let $M' := M_{k-1} \cup (M \triangle P)$. This is a perfect matching of $G - (V(C_1) \cup V(C_2) \cup \cdots \cup V(C_{k-2}))$. See Figure 7.7.

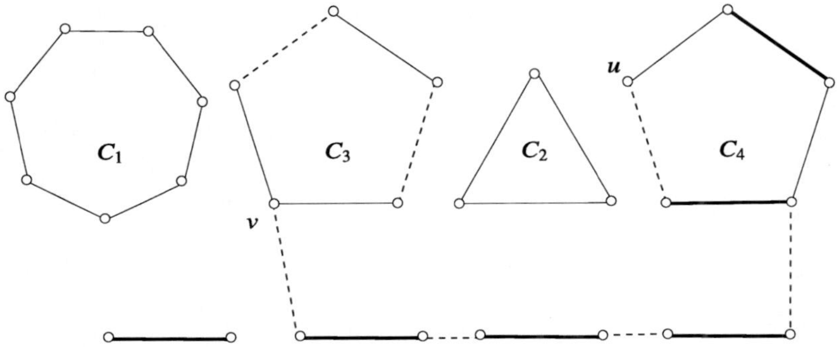

Fig. 7.7 Proof of Theorem 7.8. The edges of M are indicated by thick lines. The edges of $E(P) - M$ and those of the perfect matching M_3 of $C_3 - v$, indicated by dotted lines, constitute the perfect matching M' of $G - (V(C_1) \cup V(C_2))$.

By applying this procedure $(k-2)/2$ times we can reduce the list of k odd cycles to just two odd cycles C_1 and C_2 such that $G - (V(C_1) \cup V(C_2))$ has a perfect matching. □

7.3.1 An infinite family of solid bricks that are not odd-intercyclic

The graph shown in Figure 7.2 is the smallest solid brick that is not odd-intercyclic. A description of an infinite family of such bricks, which includes the one in Figure 7.2 can be found in the article by Lucchesi, Carvalho, Kothari and Murty (2018, [64]). At the time of writing that paper, the authors were unaware of any examples of cubic solid bricks with disjoint odd cycles. In a recent manuscript, Chen, Feng, Lu and Zhang described an infinite family of non-odd-intercyclic cubic solid bricks (2019, [21]). The graph shown in Figure 5.9 is the first member of that infinite family. Using the idea of edge-extensions, defined in Exercise 5.3.5, we are now able to present a simpler description of another infinite family of non-odd-intercyclic cubic solid bricks. Each graph in our family is obtained from a bipartite prism by means of two edge-extensions.

Let $k \geq 3$ be any integer and let $\mathbb{P}_{4k}$ be the prism obtained from two disjoint even cycles $u_1u_2 \cdots u_{2k}u_1$ and $v_1v_2 \cdots v_{2k}v_1$ by the addition of the $2k$ edges u_iv_i, $i = 1, 2, \ldots, 2k$. Now obtain G_{4k} from P_{4k} by deleting the two edges u_3v_3 and $u_{2k}v_{2k}$, and then adding an edge joining u_3 to v_{2k} and an edge joining v_3 to u_{2k}. Figure 7.8 depicts the graph G_{20} obtained from the prism $\mathbb{P}_{20}$. (The graph shown in Figure 5.9 is isomorphic to G_{12}.)

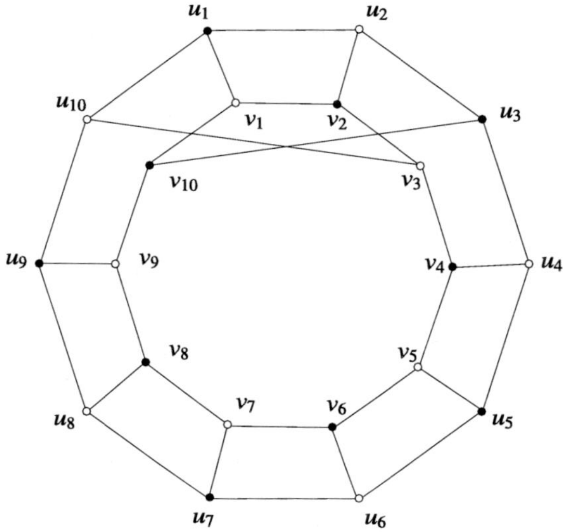

Fig. 7.8 G_{20}: a cubic solid brick with disjoint odd cycles

Lemma 7.9 *For each $k \geq 3$, the graph G_{4k} obtained from the prism $\mathbb{P}_{4k}$ as described above is a brick with disjoint odd cycles.*

Proof Let us first note that G_{4k} is cubic and that it is not bipartite. (The 5-cycle $u_1v_1v_{2k}u_3u_2u_1$ and the $(2k-1)$-cycle $u_{2k}v_3v_4 \cdots v_{2k-1}u_{2k-1}u_{2k}$ are two disjoint odd cycles in G_{4k}.)

Since G_{4k} is cubic and nonbipartite, to conclude that it is a brick it suffices to show, by Corollary 5.11, that G_{4k} is essentially 4-edge-connected. This is indeed the case. Consider the prism H obtained from the two $(2k-2)$-cycles $u_1u_2u_4u_5 \cdots u_{2k-1}u_1$ and $v_1v_2v_4v_5 \cdots v_{2k-1}v_1$ by adding the $2k-2$ edges u_iv_i, for $i = 1, 2, 4, 5, \ldots, 2k-1$. Clearly, H is isomorphic to the prism $\mathbb{P}_{4k-4}$. (It is important to note that neither u_3 nor v_3 are vertices of H.) It can be seen that the graph G_{4k} is the graph obtained from H by two edge-extensions; one based on the pair $\{u_2u_4, v_{2k-1}v_1\}$, and the other based on pair $\{v_2v_4, u_{2k-1}u_1\}$. As $k \geq 3$, the prism $\mathbb{P}_{4k-4}$ is a (cubic) brace; hence H is a cubic brace. By Corollary 5.12, H is essentially 4-edge connected. As G_{4k} is obtained from H by two edge extensions, it is essentially 4-edge connected (Exercise 5.3.5(i)). Indeed, G_{4k} is a brick. □

Theorem 7.10 *For each $k \geq 3$, the graph $G := G_{4k}$ obtained from the prism $\mathbb{P}_{4k}$ is a cubic solid brick with disjoint odd cycles.*

Proof It follows from Lemma 7.9 that G is a brick with disjoint odd cycles. To show that G is solid we use Theorem 7.8. Thus let C_1 and C_2 be any two disjoint odd cycles in G. Note that the graph $G - \{u_3v_{2k}, v_3u_{2k}\}$ is bipartite. Since C_1 and C_2 are disjoint odd cycles, it follows that one of these two cycles must pass through u_3v_{2k} and the other through v_3u_{2k}. The set of the vertices of the 4-cycle $Q := u_1u_2v_2v_1u_1$ is a 4-vertex cut of G. Furthermore, it is easy to see that the graph $G - V(Q)$ is isomorphic to a bisubdivision of the Möbius ladder $\mathbb{M}_{4k-8}$ (Exercise 7.3.4). Since $\mathbb{M}_{4k-8}$ is odd-intercyclic, not both C_1 and C_2, which are disjoint odd cycles, can be confined to the graph $G - V(Q)$. Without loss of generality, we may assume that the cycle C_1 passes through the edge u_3v_{2k} and that it has edges in common with the 4-cycle Q. Since C_1, being disjoint from C_2, cannot pass through either v_3 or u_{2k}, it follows that C_1 is either $u_3v_{2k}v_1u_1u_2u_3$ or $u_3v_{2k}v_1v_2u_2u_3$. Thus, in the graph $G - V(C_1 \cup C_2)$, one of the two vertices u_1 and v_2 is an isolated vertex, implying that $G - V(C_1 \cup C_2)$ is not matchable. Hence, by Theorem 7.8, G is solid. □

Exercises

▷**7.3.1** *For any perfect matching M of a graph G, show that χ^M is an extreme point of $\mathcal{P}oly^*(G)$.*

▷**7.3.2** *Let G be any graph, let $\mathbf{w}$ be any extreme point of its fractional perfect matching polytope $\mathcal{P}oly^*(G)$, and let $G[\mathbf{w}]$ be the subgraph of G induced by the support of $\mathbf{w}$. Give a proof of Theorem 7.7 along the following lines.*

(i) *The graph $G[\mathbf{w}]$ is free of even cycles (see proof of statement 6.2.1).*
(ii) *Each block of $G[\mathbf{w}]$ is either an odd cycle or is K_2.*
(iii) *If a component of $G[\mathbf{w}]$ has a vertex v of degree one, then the edge incident with v is the only edge of that component (implying that if a component of $G[\mathbf{w}]$ has more than one edge, then all end blocks of that component are odd cycles).*
(iv) *Every component of $G[\mathbf{w}]$ is either K_2 or an odd cycle. (Suppose that, to the contrary, there is a component of $G[\mathbf{w}]$ with more than one end block. Such a component will contain two odd cycles, say Q_1 and Q_2, and a path P which connects a vertex u of Q_1 with a vertex v of Q_2, but which is internally-disjoint from both Q_1 and Q_4. Now obtain a contradiction to the assumption that $\mathbf{w}$ is an extreme point of $\mathcal{P}oly^*(G)$ by showing that there exist two vectors $\mathbf{x}$ and $\mathbf{y}$ in $\mathcal{P}oly^*(G)$ such $\mathbf{w} = \frac{1}{2}\mathbf{x} + \frac{1}{2}\mathbf{y}$. It may be convenient to consider two cases depending on the parity of the length of P.)*
(v) *Now show that $\mathbf{w}$ is $(0, \frac{1}{2}, 1)$-valued and is as described in the statement of Theorem 7.7.*

∗**7.3.3** *Describe all the extreme points of the fractional perfect matching polytope $\mathcal{P}oly^*(\mathbb{P})$ of the Petersen graph.*

▷**7.3.4** *In the proof of Theorem 7.10 it was claimed that the graph $G - V(Q)$, where $Q = u_1u_2v_2v_1u_1$, is a bisubdivision of the Möbius ladder $\mathbb{M}_{4k-4}$. Justify this claim.*

∗**7.3.5** *Show that the brick G depicted in Figure 7.9 does not have two disjoint odd cycles C_1 and C_2 such that $G - (V(C_1) \cup V(C_2))$ has a perfect matching, and deduce that it is solid. (This is one of an infinite family of solid bricks, discovered by Nishad Kothari, that are not odd-intercyclic which will be described in a later chapter.)*

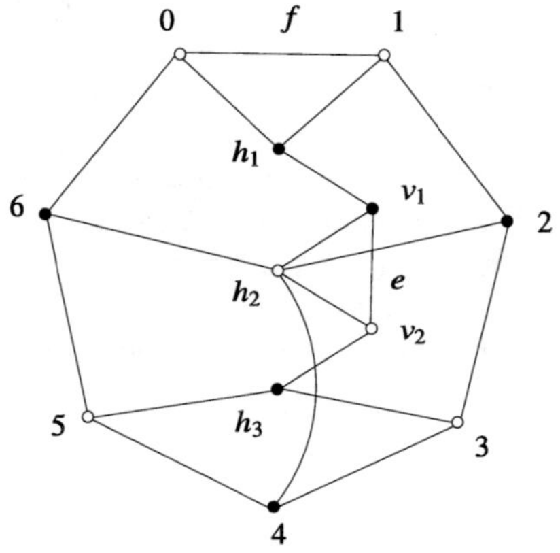

Fig. 7.9 The solid brick H_{12}

7.3.6 *All the members of the family $\{G_{4k} : k \geq 3\}$ described in Section 7.3.1 are near-bipartite, that is, have a pair of edges whose deletion results in a bipartite matching covered graph. The graph shown in Figure 7.10 was found by Nishad Kothari by exhaustive computer search. It is not near-bipartite. Show that it is solid.*

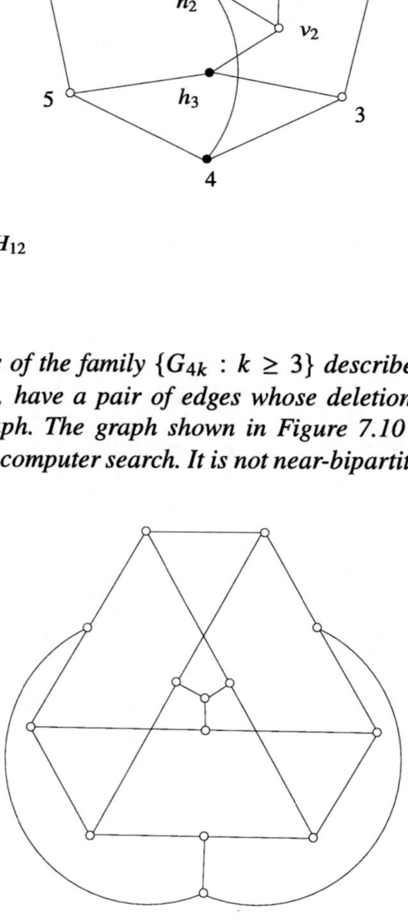

Fig. 7.10 A cubic solid brick of order 14

7.4 Notes

CLM discovered solid matching covered graphs in connection with their work on a conjecture of Lovász, and have since found them to have many interesting and useful properties (see (2002, [9]), (2005, [13]) and (2012, [16])).

The most puzzling unsolved problem concerning solid matching covered graphs is of course Problem 7.3. But there are several other intriguing questions related to them.

Let $f(2n)$ denote the maximum number of edges a simple solid matching covered graph on $2n$ vertices can have. It is easy to see that $f(4) = 6$, and not too difficult to show that $f(6) = 11$. A simple bipartite graph of order $2n$ has at most n^2 edges, which is the number of edges of $K_{n,n}$. It seems highly plausible that the following conjecture is true:

Conjecture 7.11 *There exists a positive integer N such that $f(2n) = n^2$, for all $n \geq N$.*

We do not know any solid bricks which are 'dense'. In particular, we do not know any solid bricks whose connectivity is greater than three.

Problem 7.12 *Do there exist solid bricks which are 4-connected?*

Apart from the fact that there are many intriguing problems concerning solid matching covered graphs, their properties will prove to be very useful in inductive proofs of several important results of the subject. In particular, solid matching covered graphs play a fundamental role in Chapter 15, where it is proved Lovász's conjecture, on the existence of b-invariant edges in every brick distinct from K_4, $\overline{C_6}$ and the Petersen graph (an edge e of a matching covered graph G is b-invariant if $G - e$ is matching covered and $b(G - e) = b(G)$).

Chapter 8
Dependence Relation and Removable Classes

Contents

8.1 Deletions of Edges

Deletions and contractions of edges are two common inductive tools in graph theory. For example, the famous theorem of Kuratowski states that a graph is nonplanar if and only if it can be reduced to either K_5 or to $K_{3,3}$ by means of deletions and contractions of edges. There are several theorems of similar flavour in the theory of matching covered graphs. But here the deletion and contraction operations used are, of necessity, more restrictive as the deletion of an arbitrary edge from a matching covered graph need not result in a matching covered graph, and the contraction of a single edge does not even preserve the parity of the number of vertices.

Recall that the contraction of the shore X of a separating cut $\partial(X)$ of a matching covered graph G amounts to contracting each edge of G with both its ends in X and then deleting all the resulting loops. The graph G/X thus obtained is also matching covered. We have effectively used tight-cut contractions as inductive tools for establishing properties of matching covered graphs such as the formula for the

C. L. Lucchesi, U. S. R. Murty, *Perfect Matchings*, Algorithms and Computation in Mathematics 31, https://doi.org/10.1007/978-3-031-47504-7_8

dimension of $\mathcal{L}in(G)$ (Theorem 6.13). Separating cut-contractions and tight-cut contractions of a very special type, known as 'bicontractions', will play prominent roles in future chapters. In this chapter and the next two, our focus will be on deletions of edges.

REMOVABLE EDGES

An edge e of a matching covered graph G is *removable* if the graph $G - e$ is also matching covered.

For example, every edge of $K_{3,3}$ is removable, but no edge of K_4 or of $\overline{C_6}$ is. In the context of matching covered graphs, by the *deletion of an edge*, we always mean the deletion of a removable edge.

More generally, a nonempty (proper) subset R of edges is *removable* if $G - R$ is matching covered. Bipartite matching covered graphs have the following special property:

Theorem 8.1 *In a bipartite matching covered graph, every minimal removable set of edges is a singleton.*

Proof Let $G := G[A, B]$ be a bipartite matching covered graph with bipartition (A, B). Let R be a removable set of edges of G. Suppose that $|R| \geq 2$, and let $e := ab$, with $a \in A$ and $b \in B$, be an edge in the set R. By hypothesis $G' := G - R$ is a bipartite matching covered graph, and a and b are two vertices which belong to different parts of the bipartition of G'. It follows from Corollary 3.7 (with G' playing the role of G) that $G' + e$ is matching covered. This implies that R is not minimal because $G' + e = (G - R) + e = G - (R - e)$. We conclude that every minimal removable set of edges of G is a singleton. □

A nonbipartite graph G may contain removable pairs of edges of the following type:

REMOVABLE DOUBLETONS

A pair $\{e, f\}$ of edges of a matching covered graph G is a *removable doubleton* if $G - e - f$ is matching covered, but neither $G - e$ nor $G - f$ is.

For example, each of K_4 and $\overline{C_6}$ have three removable doubletons, and the bicorn $\mathbb{H}_8$ has two.

In the above definition, the fact that neither $G - e$ nor $G - f$ is matching covered implies that every perfect matching of G which contains one of the two edges e and f also contains the other.

In the next section we shall introduce an equivalence relation on the edge set of a matching covered graph that is suggested by the above observation, and a partial order on the set of equivalence classes with respect to this relation. Using the properties of these equivalence classes, we shall establish the striking fact (Corollary 8.18) that if a matching covered graph G has a removable set R of edges, with the proviso that no proper subset of R is removable, then R is either a removable edge or is a removable doubleton of G. In other words, every minimal removable set of a matching covered graph contains just one or two edges.

It should be noted that, in general, a matching covered graph G need not have any removable sets of edges at all. For example, if each edge of G is incident with a vertex of degree two and $|E| > 2$, then the deletion of any nonempty subset of $E(G)$ results in a graph which is not matching covered. However, it follows from the theory of ear decompositions to be described in Chapter 11 that every matching covered graph of minimum degree at least three has either a removable edge or a removable doubleton (Corollary 11.9). (In case of bipartite graphs, this is just the statement of Exercise 3.4.1.)

8.1.1 Removable edges in bipartite graphs

Let $G := G[A, B]$ be a bipartite matching covered graph. Suppose that there exist partitions (A_0, A_1) of A and (B_0, B_1) of B into nonempty subsets such that $|A_0| = |B_0|$, $|A_1| = |B_1|$, and ab, with $a \in A_0$ and $b \in B_1$, is the only edge of G with one end in A_0 and one end in B_1. (See Figure 8.1 for an illustration.)

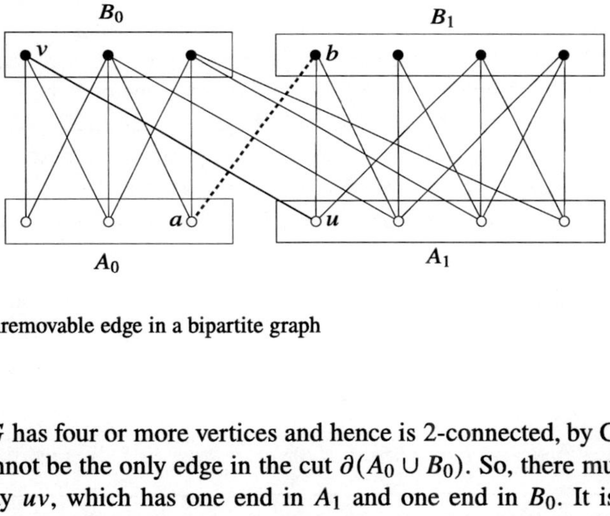

Fig. 8.1 A nonremovable edge in a bipartite graph

Clearly, G has four or more vertices and hence is 2-connected, by Corollary 2.7. Thus, ab cannot be the only edge in the cut $\partial(A_0 \cup B_0)$. So, there must be at least one edge, say uv, which has one end in A_1 and one end in B_0. It is easy to see that every perfect matching of G which contains the edge uv also contains ab. (In the terminology to be introduced in the next section, the edge uv 'depends' on the edge ab.) Thus, the deletion of ab would result in a graph in which uv would be unmatchable, implying that ab is not removable.

The following lemma establishes necessary and sufficient conditions under which an edge in a bipartite matching covered graph is not removable.

NONREMOVABLE EDGES IN BIPARTITE MATCHING COVERED GRAPHS

Lemma 8.2 *Let $G := G[A, B]$ be a bipartite matching covered graph with $|E| \geq 2$. An edge ab of G, with $a \in A$ and $b \in B$, is not removable in G if and only if there exist nonempty proper subsets A_1 and B_1 of A and B, respectively, such that:*

(i) *the subgraph $G[A_1 \cup B_1]$ induced by $A_1 \cup B_1$ is matching covered, and*
(ii) *$a \in A - A_1$ and $b \in B_1$, and ab is the only edge joining a vertex in $A - A_1$ to a vertex in B_1.*

We shall refer to the cut $\partial(A_1 \cup B_1)$ as a *certificate for the nonremovability of the edge ab.*

Proof Suppose first that there exist nonempty proper subsets A_1 and B_1 of A and B, respectively, satisfying the required properties. Then, as explained above, there must exist an edge, say uv, with $u \in A_1$ and $v \in B - B_1$, and any perfect matching of G which contains edge uv must also contain the edge ab, implying that ab is not removable in G.

Suppose now that ab is not removable. Then, by definition, the subgraph $H := G - ab$ is not matching covered. But it is matchable because the hypothesis that $|E| \geq 2$ implies that the vertex a has degree at least two, and any perfect matching of G which does not contain the edge ab is also a perfect matching of H. Therefore, by Theorem 2.12, there exists a nonempty proper subset T of B containing b such that the subgraph of H induced by $N_H(T) \cup T$ is matching covered. Since G itself is matching covered, it must be the case that ab has one end in $A - N_H(T)$ and one end in T, for, otherwise, $|N_G(T)| = |T|$ which is impossible because G is matching covered (Theorem 2.9). Thus, if we let $A_1 := N_H(T)$ and $B_1 := T$, we have the two subsets A_1 and B_1 of A and B, respectively, satisfying both the two required conditions. □

As a consequence of the above results, we have the following special property of braces.

REMOVABLE EDGES IN BRACES

Theorem 8.3 *Every edge in a brace of order at least six is removable.*

Proof Let $G[A, B]$ be a brace of order at least six. If G has a nonremovable edge ab then it follows from Lemma 8.2 that there exist subsets A_1 and B_1 of A and B, respectively, satisfying the two conditions specified in its statement. If we set $A_0 := A - A_1$ and $B_0 := B - B_1$, it is easy to see that both $\partial(A_0 \cup (B_0 + b))$ and $\partial((A_1 + a) \cup B_1)$ are tight cuts of G and that at least one of them is nontrivial (Exercise 8.1.6). But this contradicts the hypothesis that G is a brace. □

In an unpublished paper, Lovász and Vempala [61] proved the following useful generalization of Lemma 8.2 which provides a way of simultaneously certifying the nonremovability of each of the edges in a specified subset of $\partial(a)$.

NONREMOVABLE EDGES IN BIPARTITE MATCHING COVERED GRAPHS

Theorem 8.4 *Let $G[A, B]$ be a matching covered bipartite graph and let a be a vertex in A with $b_1, b_2, \ldots, b_d$, $d \geq 3$, as its neighbours. Suppose that the edges $ab_1, ab_2, \ldots, ab_r$, $r \leq d$, incident with the vertex a, are not removable in G. Then there exist partitions $(A_0, A_1, \ldots, A_r)$ of A and $(B_0, B_1, \ldots, B_r)$ of B such that, for $1 \leq i \leq r$,*

(i) *the subgraph $G[A_i \cup B_i]$ induced by $A_i \cup B_i$ is matching covered, and*
(ii) *$a \in A_0$, $b_i \in B_i$ and ab_i is the only edge that joins a vertex of $A - A_i$ to a vertex in B_i.*

Consequently, $|A_i| = |B_i|$, for $i = 0, 1, 2, \ldots, r$ and $b_{r+1}, b_{r+2}, \ldots, b_d \in B_0$. (Figure 8.2 illustrates the case in which $r = 2$ and $d = 3$.)

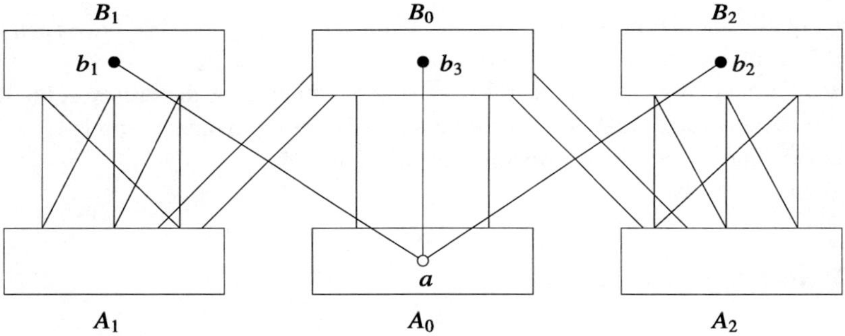

Fig. 8.2 Illustration for the case $r = 2$ and $d = 3$ (Theorem 8.4)

Proof The edge ab_i is not removable in G, for $1 \leq i \leq r$. Therefore, by Lemma 8.2, there are nonempty sets $A_i \subset A$ and $B_i \subset B$ satisfying the two properties described in the statement of the theorem, with b_i, A_i and B_i playing respectively the roles of b, A_1 and B_1. So, for any index i, where $1 \leq i \leq r$, the subgraph $G[A_i \cup B_i]$ is matching covered, $a \in A - A_i$, $b_i \in B_i$ and ab_i is the only edge that joins a vertex of $A - A_i$ to a vertex in B_i. Let

$$A_0 := A - \bigcup_{i=1}^{r} A_i \quad \text{and} \quad B_0 := B - \bigcup_{i=1}^{r} B_i.$$

The vertex a does not belong to any of the sets $A_1, A_2, \ldots, A_r$. Thus, $a \in A_0$. Of the neighbours of the vertex a, for $1 \leq i \leq r$, the vertex b_i is the only vertex of B_i adjacent to a; hence it belongs only to B_i (and not to B_j for $j \neq i$).

To establish the desired conclusion, it thus suffices to show that the B_i, as well as the A_i, are disjoint. Consider, say, B_1 and B_2, and let $S := B_1 \cap B_2$. The vertex b_1 is in $B_1 - B_2$; hence $N(S) \subseteq A_1$. Likewise, $N(S) \subseteq A_2$. Thus, $N(S) \subseteq A_1 \cap A_2$.

As $d \geq 3$, the graph G has a perfect matching, M, that contains the edge ab_3. As ab_1 is the only edge that joins a vertex in $A - A_1$ to a vertex in B_1, it follows that the vertices of B_1 are matched by M to vertices of A_1. Likewise, the vertices of B_2 are matched by M with vertices of A_2.

The vertices of $B_1 \cup B_2$ are matched by M with vertices of $A_1 \cup A_2$. Thus, $|B_1 \cup B_2| \leq |A_1 \cup A_2|$. Likewise, every vertex of S is matched by M with a vertex of $A_1 \cap A_2$. Thus, $|S| \leq |A_1 \cap A_2|$. As $G[A_1 \cup B_1]$ is matching covered, we deduce that $|B_1| = |A_1|$. Likewise, $|B_2| = |A_2|$. In sum,

$$\begin{aligned} |B_1| &= |A_1|, \\ |B_2| &= |A_2|, \\ |B_1 \cup B_2| &\leq |A_1 \cup A_2| \text{ and} \\ |S| &\leq |A_1 \cap A_2|. \end{aligned}$$

It follows that

$$|B_1| + |B_2| = |B_1 \cup B_2| + |S| \leq |A_1 \cup A_2| + |A_1 \cap A_2| = |A_1| + |A_2| = |B_1| + |B_2|;$$

hence equality holds throughout. In particular, $|S| = |A_1 \cap A_2|$. We have seen that $N(S) \subseteq A_1 \cap A_2$; hence $|N(S)| = |S|$.

As G is matching covered, it follows that S is either B or the empty set (Theorem 2.9). As b_1 is not in S, we conclude that S is empty; hence so too is $A_1 \cap A_2$. We deduce that B_1 and B_2 are disjoint, as are A_1 and A_2. □

8.1.2 Near-bipartite graphs

The deletion of a removable edge from a nonbipartite matching covered graph never results in a bipartite matching covered graph (Exercise 8.1.3), but the deletion of a removable doubleton may.

> NEAR-BIPARTITE GRAPHS
>
> A matching covered graph G is *near-bipartite* if it has a removable doubleton R such that the subgraph $G - R$ obtained by the deletion of R is a bipartite (matching covered) graph.
>
> Many prominent classes of bricks, such as prisms of order 2 (mod 4) and Möbius ladders of order 0 (mod 4), are near-bipartite.

Proposition 8.5 *Let G be a near-bipartite matching covered graph and let R be a removable doubleton of G such that $G - R$ is bipartite. Let (A, B) denote the bipartition of $G - R$. Then, one of the edges of R has both ends in A and the other edge of R has both ends in B.*

Proof Let e and f denote the two edges of R. Assume that one of the edges in R, say e, has one end in A and one end in B. As $G - e - f$ is a matching covered graph with bipartition (A, B), and e is an edge that has one end in A and one end in B, it follows from Corollary 3.7 that $(G - e - f) + e = G - f$ is matching covered, implying that f is removable in G. This is contrary to the definition of a removable doubleton. We deduce that e has both ends in the same part of the bipartition of $G - e - f$, say A.

Likewise, edge f has both its ends in the same part of $G - e - f$. It cannot have both ends also in A because, in that case, both e and f would be unmatchable in G. We deduce that f has both ends in B. □

By an ingenious use of elementary linear algebra, Lovász (1987, [58]) showed that every brick with a removable doubleton is near-bipartite; see Exercise 8.1.7. We shall present an alternative proof of this result in Section 9.1.1 in a more general context. But, it should be noted that not every matching covered graph with a removable doubleton is near-bipartite. For example, the two edges 26 and 59 constitute a removable doubleton in the graph shown in Figure 8.4, but that graph is not near-bipartite.

We conclude this section with a result which will be found to be useful in Section 9.3. Its proof makes use of Theorem 8.4.

Lemma 8.6 *Let G be a near-bipartite brick, where R is a removable doubleton of G such that $H := G - R$ is bipartite. Let v be a vertex of G that has degree three or more in H. Then, at most two edges incident with v are not removable in H.*

Proof Let e and f denote the two edges of R, and let (A, B) denote the bipartition of H. Adjust notation so that e has both ends in A and f has both ends in B. Let a denote a vertex in A having degree $d \geq 3$ in H. Let $ab_1, ab_2, \ldots, ab_r, r \leq d$, denote r edges of H incident with a that are not removable in H. Let $(A_0, A_1, \ldots, A_r)$ and $(B_0, B_1, \ldots, B_r)$ be partitions of A and B, respectively, as in the statement of Theorem 8.4.

For $i = 1, 2, \ldots, r$, the set $A_i \cup \{a\}$ is a nontrivial barrier of H, where the $|A_i|$ vertices of B_i are isolated in $G - A_i - a$ (See Figure 8.2). By hypothesis, G is bicritical; therefore edge f must have an end in the set B_i. (Otherwise, $A_i \cup \{a\}$ would be a nontrivial barrier of G itself which is absurd because G is bicritical.) This conclusion holds for each i such that $1 \leq i \leq r$. Since f has just two ends, it follows that $r \leq 2$. □

Making use of their Theorem 8.4, and the results in Exercises 8.1.8 and 3.2.2, Lovász and Vempala [61] proved that every near-bipartite brick other than K_4, $\overline{C_6}$ and the bicorn $\mathbb{H}_8$ has at least two removable edges. In fact, every simple near-bipartite brick other than the three bricks mentioned above has at least two nonadjacent removable edges. We shall present a proof of this interesting result in the next chapter.

8.1.3 Removable sets and separating cuts

The following useful assertion, showing a connection between removable sets of edges and separating cuts, is easy to verify.

Proposition 8.7 *Let G be a matching covered graph, let C be a separating cut of G, let G_1 and G_2 be the two C-contractions of G and let R be a set of edges of G. If $G_i - R$ is matching covered, for $i = 1, 2$, then $G - R$ is also matching covered and $C - R$ is a separating cut of $G - R$.* □

The converse of the above assertion does not, in general, hold. For instance, consider the cut C of the bicorn $\mathbb{H}_8$ shown in Figure 8.3. The edge e is removable in $\mathbb{H}_8$, but it is not removable in $\overline{C_6}$, which is one of its C-contractions.

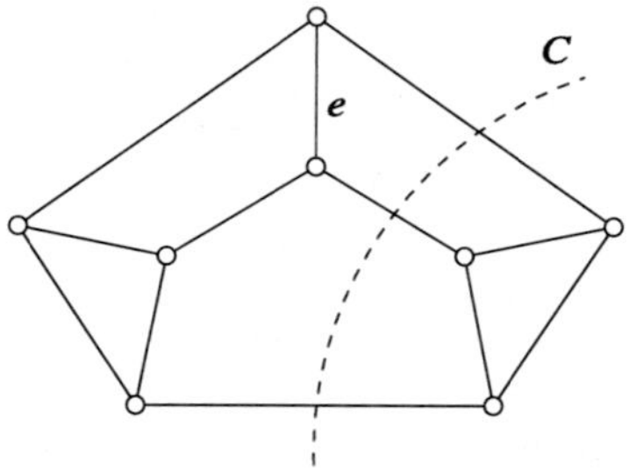

Fig. 8.3 Edge e is removable in $\mathbb{H}_8$ but not in one of its C-contraction

If C is a tight cut of G, and R is a removable set of edges of G, then $C - R$ is a tight cut of $G - R$, because every perfect matching of $G - R$ is also a perfect matching of G. This observation implies that the converse of Proposition 8.7 does hold when C is a tight cut. Thus:

Proposition 8.8 *Let G be a matching covered graph, let C be a tight cut of G, let G_1 and G_2 be the two C-contractions of G and let R be a set of edges of G. The graph $G - R$ is matching covered if and only if, for $i = 1, 2$, the graph $G_i - R$ is matching covered.* □

Let C be a separating cut of a matching covered graph G. We shall adopt the convention that if an edge e does not belong to the edge set of a C-contraction H of G, then e is removable in H.

Corollary 8.9 *Let G be a matching covered graph, and let C be a separating cut of G. If an edge e is removable in both C-contractions of G then $C - e$ is a separating cut of $G - e$, and e is removable in G.* □

Exercises

8.1.1 *Find all the removable edges and doubletons in the graph shown in Figure 8.4.*

8.1.2 *Find all the removable edges and doubletons in each of the families of matching covered graphs described in Section 2.5.*

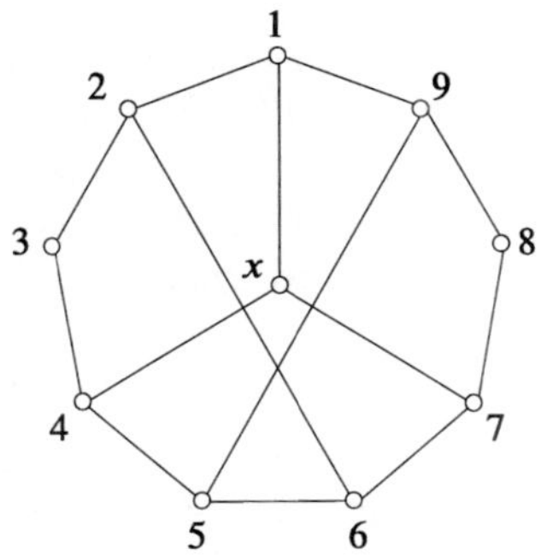

Fig. 8.4 Graph obtained by deleting an edge from the Petersen graph

▷**8.1.3** *Let G be a matching covered graph and let e be a removable edge of G. Show that G is bipartite if and only if G − e is bipartite. (Compare with Lemma 4.6 in Chapter 4.)*

▷**8.1.4** *Prove Propositions 8.7 and 8.8.*

8.1.5 *Let $G[A, B]$ be a bipartite matching covered graph on four or more vertices, and let x be a vertex of G. Show that either there is an edge of G, not incident with x, which is removable in G; or there is a subset Y of $V - x$, $|Y| \geq 3$, such that $\partial(Y)$ is a tight cut of G.*

▷**8.1.6** *Supply the missing details in the proof of Theorem 8.3.*

∗**8.1.7 (Graphs with a removable doubleton, Lovász [58])** *Let e and f be two edges of a matching covered graph G such that $G - e - f$ is matching covered, but neither $G - e$ nor $G - f$ is. Show that:*

(i) *every perfect matching of G which contains one of e and f also contains the other,*
(ii) *deduce that if $\mathbf{x} = \chi^e - \chi^f$, where χ^e and χ^f are the incidence vectors of the singleton sets $\{e\}$ and $\{f\}$, respectively, then $\mathbf{x} \circ \chi^M = 0$, for any perfect matching M of G, or, in other words, **x** is matching orthogonal (this is just a translation of the statement in the first item),*
(iii) *now, using the result in Exercise 6.2.6, show that if G is a brick, then $G - e - f$ is bipartite.*

An alternative proof of (iii) in a more general context will be presented in Section 9.1.

▷**8.1.8** *Let $H[A, B]$ be a bipartite matching covered graph, and let v be a vertex of degree three or more in H. Suppose that H has a 4-cycle Q that contains vertex v. Using Theorem 8.4, show that some edge of $\partial(v) \cap E(Q)$ is removable in H.*

▷**8.1.9 (The Exchange Property)** *We say that a matching covered graph G has the* exchange property with respect to the deletion of edges *if e is any removable edge of G, and if f is any removable edge of $G - e$, then the edge f is removable in G itself (and e is removable in $G - f$).*

(i) *Show that all bipartite matching covered graphs have the exchange property. (We shall show in Chapter 10 that all solid matching covered graphs have the exchange property.)*
(ii) *Find two suitable edges e and f of the tricorn $\mathbb{H}_{10}$ (Figure 2.5(d)) to show that the tricorn does not have the exchange property.*

$*$ **8.1.10** *Let G be a brace of order at least six, and let $e := ab$ be any edge of G such that $G - e$ is not a brace.*

(i) *Let $C := \partial(X)$ and $D := \partial(Y)$ be nontrivial cuts of G such that the cuts $C - e$ and $D - e$ are tight in $G - e$ and $a \in X \cap Y$. Let $I := \partial(X \cap Y)$ and let $U := \partial(\overline{X} \cap \overline{Y})$. Show that $I - e$ and $U - e$ are both nontrivial tight cuts in $G - e$.*
(ii) *Show that there exists a nested family $X_1 \subset X_2 \subset \cdots \subset X_k$ of subsets of V, each containing a, but not b, such that*

$$\left.\begin{array}{l}(G-e)/\overline{X_1} \text{ is a brace,}\\ (G-e)/X_i/\overline{X_{i+1}} \text{ is a brace for } 2 \le i < k,\\ (G-e)/X_k \text{ is a brace.}\end{array}\right\} \qquad (8.1)$$

(iii) *In the above decomposition, show that X_1 and X_k are uniquely determined and deduce that if $G - e$ has at most three braces, then $G - e$ has a unique tight cut decomposition, that is, $G - e$ has a unique maximal laminar family of nontrivial tight cuts.* (CLM *(2015, [18]))*

8.2 The Dependence Relation

In this section we introduce an equivalence relation on the edge set of a matching covered graph G which was first defined and used in a paper by Carvalho and Lucchesi (1996, [7]) and later studied more extensively in a paper by CLM (1999, [8]). It turns out that every minimal nonempty removable set of edges of G is an equivalence class with respect to this relation.

Let G be a matching covered graph, and let e and f be any two edges of G. Then we say that e *depends* on f, or e *implies* f, if every perfect matching that contains e also contains f. Equivalently, e depends on f if and only if e is not matchable in $G - f$. (In the graph depicted in Figure 8.1, the edge uv depends on the edge ab. In fact, every edge between A_1 and B_0 depends on the edge ab.)

We write $e \Rightarrow f$ to indicate that e depends on f. Clearly, $\Rightarrow$ is a reflexive and transitive relation on the edge set $E(G)$. Two edges e and f are *mutually dependent* if $e \Rightarrow f$ and $f \Rightarrow e$. In this case we write $e \Leftrightarrow f$. Clearly mutual dependence is an equivalence relation on $E(G)$. We refer to the equivalence classes with respect to this relation as *dependence classes* .

It is convenient to visualize the relation $\Rightarrow$ by means of the digraph it defines on the set of edges of G, called the *dependence digraph*, and denoted by $D(G)$. Figure 8.5 depicts the dependence digraph $D(\mathbb{H}_8)$ of the bicorn $\mathbb{H}_8$, which is a brick of order eight. For every edge e of $\mathbb{H}_8$ distinct from the edge 27 there is an edge distinct from e that depends on e. In other words, if e is any edge other than 27, then some edge of $\mathbb{H}_8 - e$ is unmatchable in that subgraph. But $\mathbb{H}_8 - 27$ has no unmatchable edges. Thus 27 is the only removable edge of $\mathbb{H}_8$.

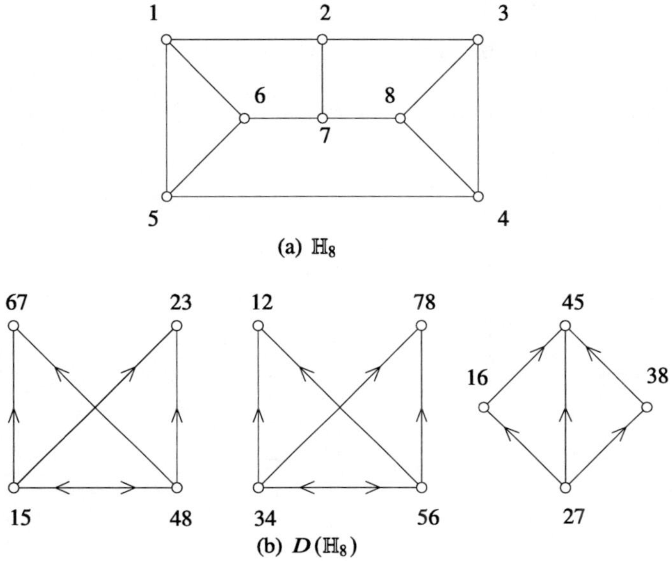

Fig. 8.5 The bicorn $\mathbb{H}_8$ and the associated dependence digraph $D(\mathbb{H}_8)$

As already noted, an edge e depends on another edge f if and only if e is not matchable in $G - f$. Thus, by Corollary 2.2:

Proposition 8.10 *An edge e of a matching covered graph G depends on another edge f of G if and only if $G - f$ has a barrier B that contains both ends of e.* □

Proposition 8.11 *Let e and f be distinct edges of a matching covered graph G such that $G - f$ has a barrier B that contains both ends of e. Then, the edge f has its ends in distinct odd components of $G - f - B$.*

Proof The graph G is matching covered, hence e is matchable in G. As e has both ends in B, then B is not a barrier of G, hence f has its ends in distinct components of $G - f - B$. If f has an end, v, in an even component of $G - f - B$ then $B + v$ is a barrier of G that contains both ends of e. This is not possible, because e is matchable in G. We deduce that f has its ends in distinct odd components of $G - f - B$. □

The next assertion is rather technical, but will play a fundamental role in several proofs in the book.

Lemma 8.12 *Let f be an edge of a matching covered graph G, let B denote a barrier of $G - f$ such that f has its ends in distinct odd components of $G - f - B$. Let E_B denote the set of edges of G that have both ends in B and let C be the set of edges that join a vertex of B to a vertex of some even component of $G - f - B$, as illustrated in Figure 8.6. For each perfect matching M of G,*

$$2|M \cap E_B| + |M \cap C| \leq 2|M \cap \{f\}|, \tag{8.2}$$

with equality if and only if $|M \cap \partial(K)| = 1$, for each odd component K of $G - f - B$.

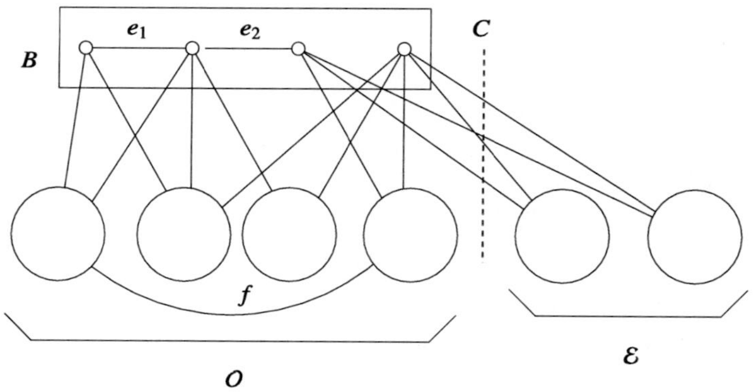

Fig. 8.6 Illustration for Lemma 8.12, where $E_B = \{e_1, e_2\}$, and $\mathcal{O}$ and $\mathcal{E}$ denote respectively the collection of odd and even components of $G - f - B$

Proof (of Lemma 8.12) Let $\mathcal{O}$ denote the collection of odd components of the graph $G - f - B$ and let F denote the set of edges that join a vertex of B to a vertex of some odd component of $G - f - B$. Then,

$$|B| \leq \sum_{K \in \mathcal{O}} |M \cap \partial(K)|, \tag{8.3}$$

$$\sum_{K \in \mathcal{O}} |M \cap \partial(K)| = 2|M \cap \{f\}| + |M \cap F|$$

$$|M \cap F| + |M \cap C| = |M \cap \partial(B)|$$

$$|M \cap \partial(B)| + 2|M \cap E_B| = |B|.$$

Note that equality holds in (8.3) if and only if $|M \cap \partial(K)| = 1$ for each $K \in \mathcal{O}$. Addition of the inequalities and simplification yields the asserted inequality (8.2). Moreover, equality holds in (8.2) if and only if $|M \cap \partial(K)| = 1$ for each $K \in \mathcal{O}$. □

8.2.1 Dependence classes

The following theorem describes properties of a matching covered graph which are implied by the existence of two mutually dependent edges.

Theorem 8.13 *Let e and f be two distinct mutually dependent edges of a matching covered graph G, and let B be a barrier of $G - f$ that contains both ends of e. Then, the following properties hold:*

(i) *edge f has its ends in distinct odd components of $G - f - B$,*
(ii) *for each component K of $G - f - B$, the cut $\partial(K)$ is tight in G,*
(iii) *edge e is the only edge of G that has both ends in B, and*
(iv) *every component of $G - f - B$ is odd.*

Proof Let $\mathcal{O}$ denote the collection of odd components of $G - f - B$. The validity of part (i) follows immediately from Proposition 8.11. To prove the other parts, let M be a perfect matching of G. By hypothesis, the edges e and f are mutually dependent, hence $|M \cap \{e\}| = |M \cap \{f\}|$. Let E_B and C be as in the statement of Lemma 8.12. Then,

$$2|M \cap \{e\}| + 2|M \cap (E_B - e)| + |M \cap C| \leq 2|M \cap \{f\}| = 2|M \cap \{e\}|, \quad (8.4)$$

with equality if and only if $|M \cap \partial(K)| = 1$, for each component $K \in \mathcal{O}$.

Equality holds throughout in (8.4). Thus, $2|M \cap (E_B - e)| + |M \cap C| = 0$ and $|M \cap \partial(K)| = 1$, for each $K \in \mathcal{O}$. These equalities hold for each perfect matching M of G. Thus, $\partial(K)$ is tight, for each $K \in \mathcal{O}$. This proves part (ii).

As G is matching covered, every edge of G is matchable. We deduce that $E_B - e$ and C are both empty. By hypothesis, $e \in E_B$, thus $E_B = \{e\}$. This proves part (iii). As G is matching covered, it is connected. As C is empty, we conclude that $G - f - B$ does not have any even components. This proves part (iv). □

Exercises

8.2.1 *Draw the dependence digraphs of K_4, $\overline{C_6}$, $K_{3,3}$, and the graph shown in Figure 8.4 which is obtained by deleting an edge from the Petersen graph.*

EVEN AND ODD 2-EDGE CUTS

▷**8.2.2** *Let e and f be two edges of a matching covered graph G such that $\{e, f\}$ is a cut of G. Show that e and f depend on each other if and only if $\{e, f\}$ is an even cut of G. (Hint: Use the parity condition stated in Exercise 1.2.1.)*

▷**8.2.3** *Let G_1 and G_2 denote the two C-contractions with respect to a tight cut C of a matching covered graph G. Show that:*

(i) *for $i = 1, 2$, the dependence relation in G_i is the restriction to $E(G_i)$ of the dependence relation in G, and*
(ii) *if Q is any equivalence class in G, then $|Q \cap C| \leq 1$.*

8.2.4 *Let u, v, and w be the vertices of a triangle in a matching covered graph G, and suppose that the degree of u in G is three and that uv, uw, and uu', where $u' \neq v, w$, are the three edges incident with u (see Figure 8.7).*

(i) *Show that the edge vw depends on the edge uu'.*
(ii) *Deduce that uu' is not removable in G.*

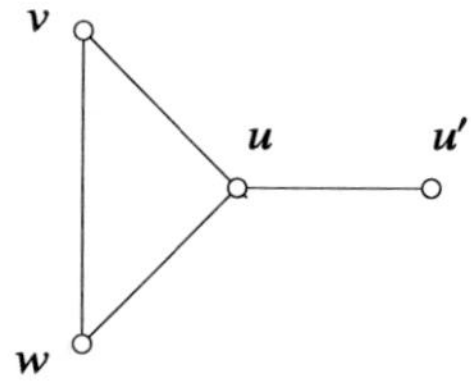

Fig. 8.7 Edge vw depends on the edge uu'

8.2.5 *Consider the wheel W_{2k+1}, where $k \geq 2$.*

(i) *Show that every spoke of W_{2k+1} is removable.*
(ii) *Using Exercise 8.2.4, show that no edge in the rim of W_{2k+1} is removable.*

8.3 Removable Classes

8.3.1 Minimal classes

By identifying the vertices in the equivalence classes in the dependence digraph, we obtain the *reduced dependence digraph* $\widehat{D}(G)$. The reduced dependence digraph of the bicorn $\mathbb{H}_8$ shown in Figure 8.5 is the digraph shown in Figure 8.8.

Clearly $\widehat{D}(G)$ is acyclic. The sources in $\widehat{D}(G)$ are called *minimal dependence classes*, or simply *minimal classes*[1]. For an edge e of G, any source Q of $\widehat{D}(G)$ that contains an edge f of G that depends on e is said to be a *minimal dependence class induced by* e. For example, in $\widehat{D}(\mathbb{H}_8)$ shown in Figure 8.8, the minimal class induced by the edges 67, 34 and 45 are, respectively, $\{15, 48\}$, $\{34, 56\}$ and $\{27\}$.

[1] Minimality here is with respect to the partial order $\Rightarrow$ and it should not be confused with minimality with respect to set containment.

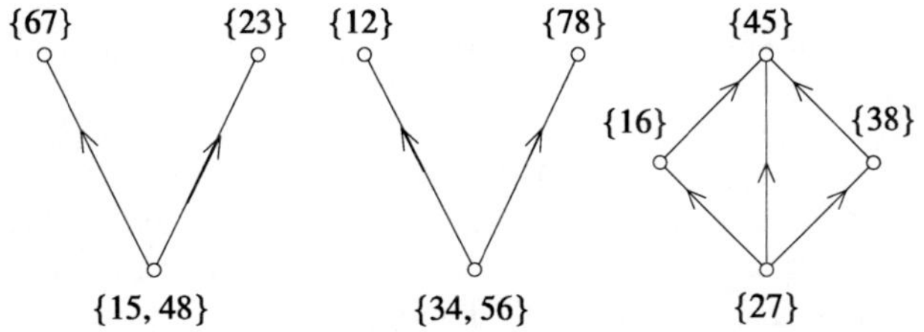

Fig. 8.8 The digraph $\widehat{D}(\mathbb{H}_8)$ corresponding to the graph $\mathbb{H}_8$ in Figure 8.5

Proposition 8.14 *Every edge of a matching covered graph induces at least one minimal dependence class.* □

In general, an edge may induce several minimal classes. For example, the graph shown in Figure 8.9 illustrates this phenomenon.

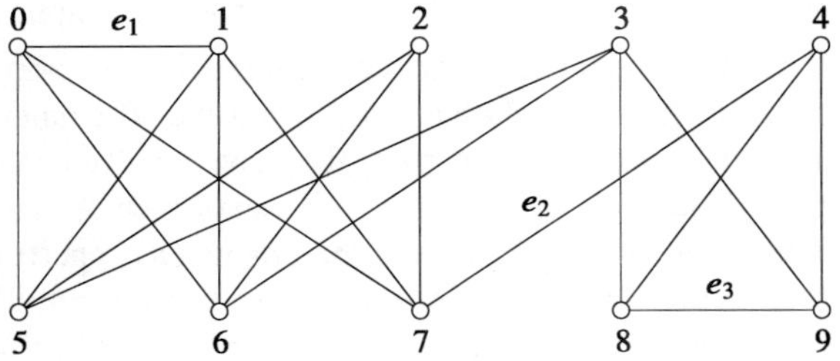

Fig. 8.9 The set $Q := \{e_1, e_2, e_3\}$ is a dependence class and 35 and 36 are removable edges. Any edge in Q induces both $\{35\}$ and $\{36\}$.

Removable classes

An equivalence class R of the dependence relation of a matching covered graph G is a *removable class* if the graph $G - R$ is matching covered. Clearly, every removable class is a minimal class. By the definition of the dependence relation, a minimal class Q of a matching covered graph G is a minimal nonempty subset of E such that every edge in $E - Q$ is matchable in $G - Q$. Thus, if $G - Q$ happens to be connected, then it is also matching covered; in this case, Q is a removable class. For example, if G has at least two edges and $|Q| = 1$, then Q is a removable class of G because any matching covered graph G with at least two edges is 2-connected.

The bicorn $\mathbb{H}_8$ shown in Figure 8.5 has three minimal classes, and each of them is also a removable class. Although, in general, every removable class is minimal, a minimal class need not always be a removable class. As an example, consider the cycle C_{2k}, where $k \geq 2$. It has precisely two minimal classes, namely its perfect matchings, but neither of them is a removable class. As another example, consider the graph shown in Figure 8.9. In this graph $\{e_1, e_2, e_3\}$ is a minimal class, but it is not removable!

When a removable class R of a matching covered graph G consists of just one edge, that edge is a removable edge of G, and when R consists of two edges, then those two edges form a removable doubleton of G. The bicorn $\mathbb{H}_8$ in Figure 8.5 has one removable edge and two removable doubletons.

8.3.2 Sizes of removable classes

All removable classes in a bipartite matching covered graph are singletons (Theorem 8.1). But, as we have seen, nonbipartite matching covered graphs may have removable classes which have more than one edge. However, as we shall demonstrate in this section, all removable classes in any matching covered graph are either removable edges or removable doubletons. This result (Corollary 8.18) plays an important role in the theory of ear decompositions of matching covered graphs (Chapter 11).

The above-mentioned fundamental fact concerning sizes of removable classes may be easily deduced from Lemma 5.4.5 in Lovász and Plummer [59]. But the statement of that lemma in [59] and the proof they give involve concepts and techniques that are not included in this book. Here we shall deduce that result by first establishing, entirely within the framework of this chapter, a property of all minimal classes of size three or more. Our proof of this more general result (Theorem 8.17) is inductive and depends crucially on Theorem 8.13 and the following lemma.

Lemma 8.15 *Let G_1 and G_2 denote the two C-contractions with respect to a tight cut C of a matching covered graph G, and let Q be a minimal class in G. Then:*

(i) *for $i = 1, 2$, the set $Q \cap E(G_i)$ is either empty or a minimal class in G_i,*
(ii) *$|Q \cap C| \leq 1$, and*
(iii) *if Q and C are disjoint then Q is entirely contained in one of $E(G_1) - C$ and $E(G_2) - C$.*

Proof The first two parts of the assertion are covered by Exercise 8.2.3. Turning to the last part, suppose that Q and C are disjoint. If possible, let e_1 and e_2 be two edges of Q such that $e_1 \in E(G_1) - C$ and $e_2 \in E(G_2) - C$.

Let M_1 be a perfect matching of G containing e_1, and let f be the edge of M_1 in C. As Q and C are disjoint, f is not in Q. Thus, as Q is a minimal class, edge f is matchable in $G - Q$. If M_f is any perfect matching of G in $G - Q$ that contains the edge f, the matching $(M_1 \cap E(G_1)) \cup (M_f \cap E(G_2))$ would be a perfect matching of G that contains e_1 but not e_2. This would imply that e_1 and e_2 are not equivalent, which is absurd because e_1 and e_2 are both edges of the minimal class Q. □

We have noted that minimal classes of matching covered graphs can be arbitrarily large. This is the case even when the graph under consideration has no vertices of degree two. For example, in the graph on 16 vertices shown in Figure 8.10, the set of the eight edges indicated by dotted lines is a minimal class.

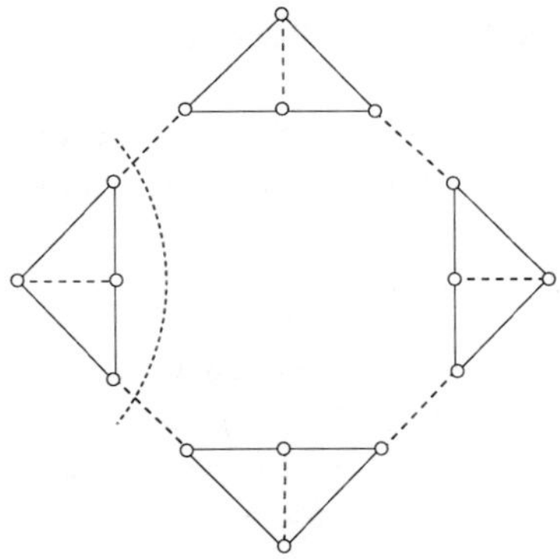

Fig. 8.10 A minimal class with eight edges (indicated by dotted lines)

Any two edges in a matching covered graph which constitute an even 2-cut are equivalent; see Exercise 8.2.2. (The cut shown in Figure 8.10 is an even 2-cut.) The following result indicates a very important property of minimal classes that are doubletons.

Proposition 8.16 *Let Q be a minimal class of a matching covered graph G. If $|Q| = 2$ then either Q is removable in G or Q is an even 2-cut of G.*

Proof Suppose that $|Q| = 2$. As Q is minimal, every edge of the graph $G - Q$ is matchable in $G - Q$. If $G - Q$ is connected then the graph $G - Q$ is matching covered; hence Q is a removable class. Alternatively, suppose that $G - Q$ is not connected. Then, Q includes some nonempty cut, C, of G. As $|Q| = 2$, the (matching covered) graph G has two or more edges; hence it is 2-connected. Thus, $|C| \geq 2$. We deduce that $C = Q$. That is, Q is a 2-cut of G. It is now easy to conclude that Q is an even cut of G. See Exercise 8.2.2. □

Let Q be a minimal class of a matching covered graph G with at least two edges. We have seen that if $|Q| = 1$ then Q is a removable class of G, and that if $|Q| = 2$ then either Q is a removable class or it is an even 2-cut of G. We now turn our attention to the case in which $|Q| \geq 3$.

MINIMAL CLASSES AND EVEN 2-CUTS

Theorem 8.17 *Any minimal class Q of a matching covered graph G, with $|Q| \geq 3$, includes an even 2-cut of G.*

Proof Suppose that $|Q| \geq 3$. We shall prove, by induction on $|V|$, that Q includes an even 2-cut. Towards this end, let e_1 and e_2 be two edges which belong to Q. As this means that e_1 and e_2 are mutually dependent, Proposition 8.10 implies that there exists a barrier B of $G - e_2$ which contains both ends of e_1. By Theorem 8.13, the only edge of G with both ends in B is e_1, for each component K of $G - e_2 - B$, the cut $\partial(K)$ is tight in G, and the edge e_2 has its ends in two different components of $G - e_2 - B$.

We first of all consider the case in which all components of $G - e_2 - B$ are trivial. It turns out that it is possible to deal with this case without recourse to induction.

Case 1 *Every component of $G - e_2 - B$ is trivial.*

In this case, the graph $H := G - e_1 - e_2$ is bipartite with bipartition (A, B), where A is the set of isolated vertices of $G - e_2 - B$. Since Q is a dependence class, any perfect matching which contains one of the edges of Q contains all edges of Q. Thus, since every perfect matching of H is also a perfect matching of G, it follows that no edge of $Q - e_1 - e_2$ is matchable in H. As $|Q| \geq 3$, we conclude that H is not matching covered.

On the other hand, any perfect matching of G that does not include the edges in Q is a perfect matching of H. So, H is matchable. Moreover, since Q is a minimal class, all edges in $E - Q$ are matchable in $G - Q$ and hence also in the graph H.

As H is matchable but not matching covered, it follows (from Theorem 2.9) that there exists a nonempty proper subset S of B such that $|N(S)| = |S|$, where $N(S)$ is the neighbour set of S in H. Take X to be the set $S \cup N(S)$ and consider the cut $\partial_H(X)$. All the edges of H in this cut have one end in $N(S)$ and one end in $B - S$. As $|N(S)| = |S|$, no such edge is matchable in H. By the above observation, it follows that all edges in the cut $\partial_H(X)$ are in Q.

Now let M denote a perfect matching of G which includes all the edges of Q. A simple counting argument implies that $|(M - e_1 - e_2) \cap \partial_G(X)|$ is equal to the number of ends of e_1 in S minus the number of ends of e_2 in $N(S)$. Since, as noted above, all edges in $\partial_G(X)$ are in Q, it follows that $|Q \cap \partial_G(X)| \leq 2$. Furthermore $|Q \cap \partial_G(X)|$ has to be even because $\partial_G(X)$ is an even cut, and it cannot be zero because if that were the case graph G would be disconnected. It is now easy to conclude that $Q \cap \partial_G(X)$ is an even 2-cut of G. The four possible scenarios which give rise to even 2-cuts are illustrated in Figure 8.11. (We leave the details as Exercise 8.3.5.)

Case 2 *The graph $G - e_2 - B$ has a nontrivial component, K.*

By Theorem 8.13, $C := \partial(K)$ is a tight cut of G. Now let

$$G_1 := G/(V(K) \to k), \quad \text{and} \quad G_2 := G/(\overline{V(K)} \to \overline{k})$$

be the two C-contractions of G, and, for $i = 1, 2$, let $Q_i := Q \cap E(G_i)$ denote the restrictions of the minimal class Q of G to the edge set of G_i.

Case 2.1 Either Q_1 or Q_2 has three or more edges.

Let $i \in \{1, 2\}$ be such that $|Q_i| \geq 3$. By Lemma 8.15(i) the set Q_i is a minimal class of G_i. Since $|V(G_i)| < |V|$ and $|Q_i| \geq 3$, it follows by induction that Q_i includes an even 2-cut of G_i. This validates the assertion of the Theorem, because an even 2-cut of G_i is also an even 2-cut of G itself.

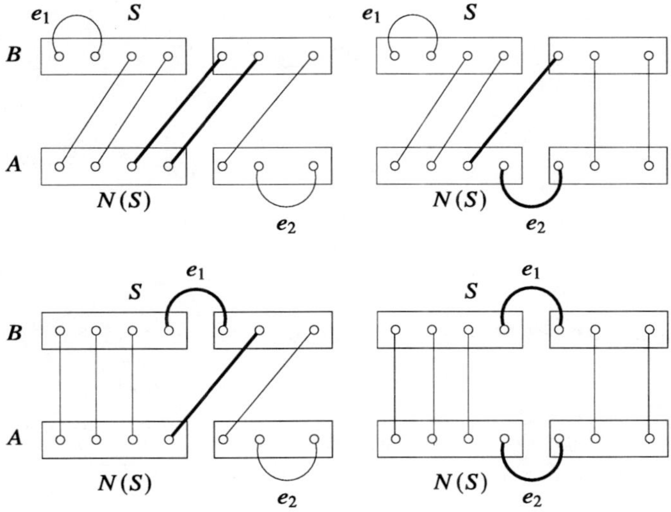

Fig. 8.11 The four scenarios which give rise to even 2-cuts (Case 1)

Case 2.2 $|Q_1| \leq 2$ and $|Q_2| \leq 2$.

In this case, the following assertion holds.

8.17.1 *The edge e_2 belongs to the cut C, $|Q| = 3$, and $|Q_1| = |Q_2| = 2$.*

Proof Since $|Q| \geq 3$, it follows that both Q_1 and Q_2 are strict subsets of Q. Lemma 8.15 now implies that $|Q \cap C| = 1$. Let f be the edge of Q in C. If $f \neq e_2$, the three edges e_1, e_2 and f belong to Q_1, contradicting the hypothesis that $|Q_1| \leq 2$. Hence e_2 must be the edge of Q in C, and it is the only edge in $Q_1 \cap Q_2$. Since $Q_1 \cup Q_2 = Q$, it follows that $|Q_1| + |Q_2| = |Q| + 1 \geq 4$. Since both Q_1 and Q_2 have fewer than three edges, this implies that $|Q| = 3$ and $|Q_1| = |Q_2| = 2$. □

The minimal class Q then has just one edge other than e_1 and e_2. Let e_3 be that edge. This edge has both its ends in $V(K)$, the set Q_1 is equal to $\{e_1, e_2\}$ and the set Q_2 is equal to $\{e_2, e_3\}$. Let x be the end of e_2 in $V(K)$, let y be the end of e_2 in $\overline{V(K)}$. See Figure 8.12.

As Q_2 is a minimal dependence class in G_2 there exists a barrier B' in $G_2 - e_3$ such that (i) the ends of e_2 in G_2, namely x and the contraction vertex $\overline{k}$, belong to B', and (ii) edge e_3 joins vertices in different components of $G_2 - e_3 - B'$. Now let $B'' := (B' - \overline{k}) \cup B$. It can be verified (Exercise 8.3.6) that (i) B'' is a barrier of $G - e_3$ which contains both ends of e_1 and (ii) the vertex x is the only end of e_2 in B''. The components of $G - e_3 - B''$ are all those of $G_2 - e_3 - B'$ and those of $G - e_2 - B$ other than K. See Figure 8.13.

If all components of $G - e_3 - B''$ are trivial, then Case 1 applies, with e_3 and B'' playing respectively the roles of e_2 and B. Alternatively, if the graph $G - e_3 - B''$ has a nontrivial component, L, then $G/V(L)$ contains all the three edges of Q and Case 2.1 applies, with e_3, B'' and L playing respectively the roles of e_2, B and K. In all alternatives considered, we conclude that Q includes an even 2-cut. □

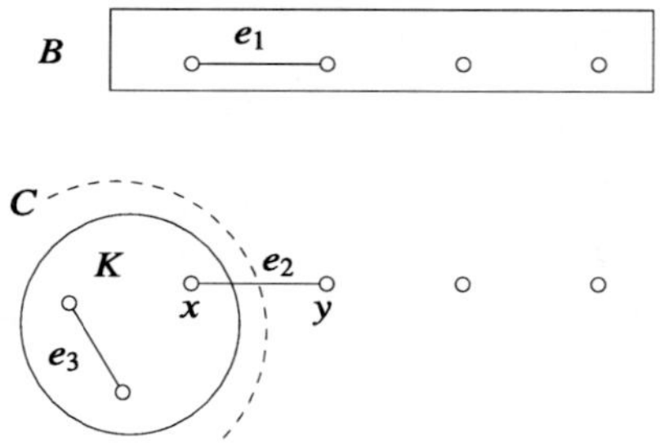

Fig. 8.12 The case in which $Q_1 = \{e_1, e_2\}$ and $Q_2 = \{e_2, e_3\}$

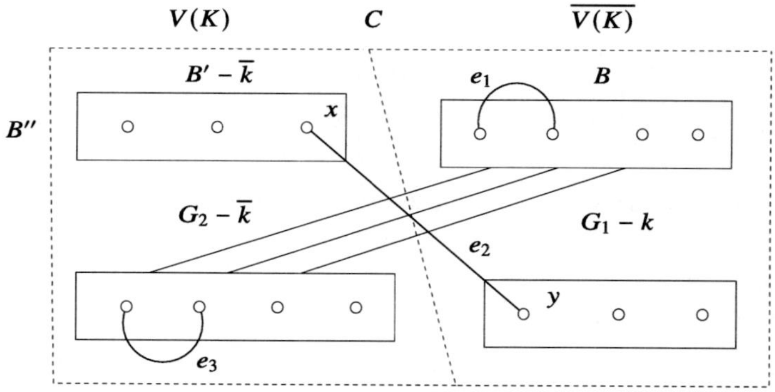

Fig. 8.13 Barrier B'' of $G - e_3$ containing the two ends of e_1

Size of removable classes

Corollary 8.18 *Every removable class of a matching covered graph is either a singleton or a doubleton.*

Proof Let R be a removable class of a matching covered graph G. As R is a removable class, it is necessarily a minimal class with respect to the dependence relation. The graph $G - R$ is matching covered, hence connected. Thus, R does not include any nonempty cut of G. In particular, R does not include any even 2-cut. By Theorem 8.17, it follows that $|R| \leq 2$. □

From now on, for the above reason, we shall employ the term removable class to mean either a removable edge or a removable doubleton.

A matching covered graph G need not have any removable classes. For example, as noted in the first section of this chapter, if $|E| \geq 3$ and every edge of G is incident with a vertex of degree two, then G cannot have removable classes. However, as also mentioned in the first section, we shall see in Chapter 11 that if the minimum degree of a matching covered graph G is at least three, then G does have at least one removable class.

Exercises

▷**8.3.1** *Draw the reduced dependence digraph of the matching covered graph shown in Figure 8.4, and determine all the minimal classes of that graph.*

▷**8.3.2** *Show that in a wheel of order 10 every edge in the rim induces four minimal classes, each of which consists of a single spoke.*

▷**8.3.3** *The objective of this exercise is to prove that every near-bipartite matching covered graph has exactly one brick. (In other words, every near-bipartite graph is a near-brick.) For this, let G be a near-bipartite graph, let $R := \{e, f\}$ be a removable doubleton of G such that $G - R$ is bipartite (and matching covered).*

(i) *Prove that G is not bipartite. Hint: otherwise the set $E \triangle (E - R) = R$ is a cut of G.*
(ii) *Suppose that G is not a brick. Let $C := \partial(X)$ be a nontrivial tight cut of G, note that $C - R$ is tight in $G - R$ and consider all the possible scenarios for the edges e and f in a way similar to that in Case 1 of the proof of Theorem 8.17. Conclude that one C-contraction of G is bipartite and the other is near-bipartite.*

▷**8.3.4** *Show that, in the graph shown in Figure 8.14:*

(i) *the set $Q := \{e_1, e_2, e_3\}$ of edges is a dependence class and*
(ii) *both $\{f_1\}$ and $\{f_2\}$ are minimal classes induced by e_i, for $i = 1, 2, 3$.*

Hint: Observe that $\{e_1, f_1, f_2\}$ is an even cut, and use the fact that every perfect matching meets an even cut in an even number of edges.

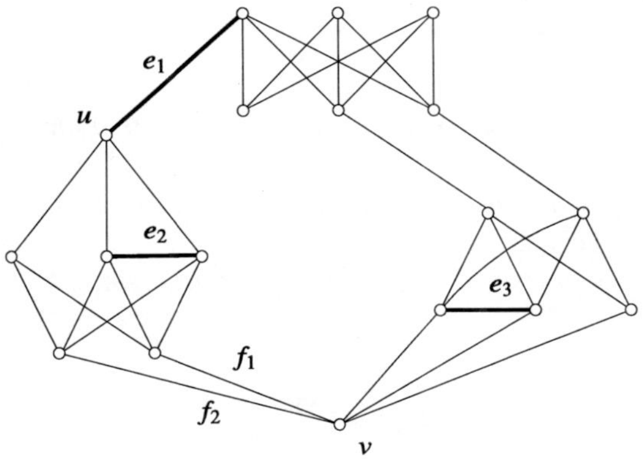

Fig. 8.14 The set $\{e_1, e_2, e_3\}$ is a dependence class which is not minimal. For any $i \in \{1, 2, 3\}$, both $\{f_1\}$ and $\{f_2\}$ are minimal classes induced by the edge e_i.

▷**8.3.5** *In Case 1 of the proof of Theorem 8.17, supply all the details that are required to show that $\partial_G(X)$ is an even 2-cut which is a subset of Q.*

▷**8.3.6** *In the proof of Theorem 8.17, verify the fact that* $B'' = (B' - \overline{k}) \cup B$ *is a barrier of* $G - e_3$ *which contains the two ends of* e_1 *and that* e_2 *has precisely one end in* B''.

Hint: Use the fact that B *is a barrier of* $G - e_2$ *containing both ends of* e_1 *and that* B' *is a barrier of* G_2 *containing the ends* x *and* $\overline{k}$ *of the edge* e_2 *in* G_2.

▷**8.3.7** *Let* B *denote a barrier of a matching covered graph and let* H *be the bipartite graph obtained from* G *by shrinking each component of* $G - B$ *to a single vertex. Let* R *be a removable class of* G. *Prove that* $|R \cap E(H)| \leq 1$ *and also that* $G - B$ *has precisely one component* K *such that* $R \subset E(G_1)$, *where* $G_1 := G/\overline{V(K)}$.

∗**8.3.8** *Let* G *be a matching covered graph such that* $\delta(G) \geq 3$. *Let* $\mathcal{C}$ *be a (possibly empty) maximal laminar collection of even 2-cuts of* G. *Prove that* G *has at least* $\Delta(G) + |\mathcal{C}|$ *removable classes.*

8.3.9 *Consider the brick* G *depicted in Figure 8.15.*

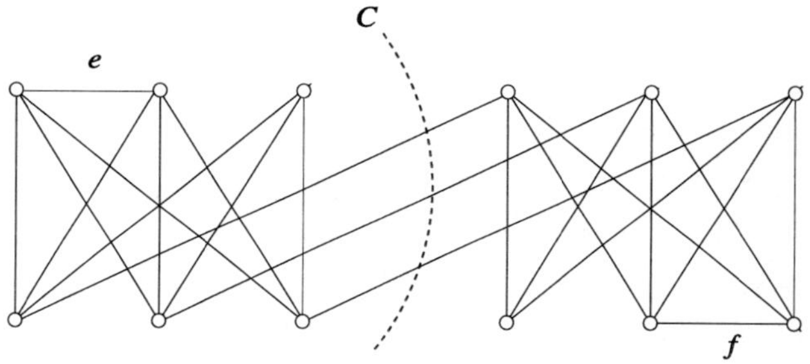

Fig. 8.15 The doubleton $\{e, f\}$ is a nonremovable dependence class

(i) *Show that* $\{e, f\}$ *is a dependence class in* G *and that any edge of the cut* C *depends on* $\{e, f\}$.
(ii) *Deduce that* $\{e, f\}$ *is not minimal (with respect to the partial order* $\Rightarrow$*) and hence not removable.*
(iii) *Show that every edge of* G *distinct from* e *and* f *is removable.*

8.4 Two Lemmas for Future Use

When C is the barrier cut associated with a special barrier of a matching covered graph, one of the C-contractions of that graph is bipartite. Using Lemma 8.2 and Theorem 8.4 we now derive a lemma concerning removable edges in bipartite graphs which arise in this way in a very particular context. It will play a crucial role in the proofs of several important results in Part II, but is not needed for an understanding of the basic theory.

Lemma 8.19 *Let G be a bicritical graph. Let e be a removable edge of G and let $C := \partial(X)$ be a cut of G. Suppose that the graph $H := (G - e)/(\overline{X} \to \overline{x})$ is bipartite and matching covered. The following properties hold:*

(i) *every edge of $C - e$ is removable in H, and*
(ii) *for every vertex v of the minority part X_- of X, if v has degree three or more in $G - e$ then at most one edge of $\partial(v) - e$ is not removable in $G - e$.*

Proof Let (A, B) denote the bipartition of H, where $B := X_+$ and $A := X_- + \overline{x}$.

8.19.1 *Let (A', A'') be a partition of A and let (B', B'') be a partition of B, such that $|A'| = |B'|$ and H has just one edge, say f, that joins a vertex in A' to a vertex in B''. Then, the contraction vertex $\overline{x}$ is in A''. (See Figure 8.16(a).)*

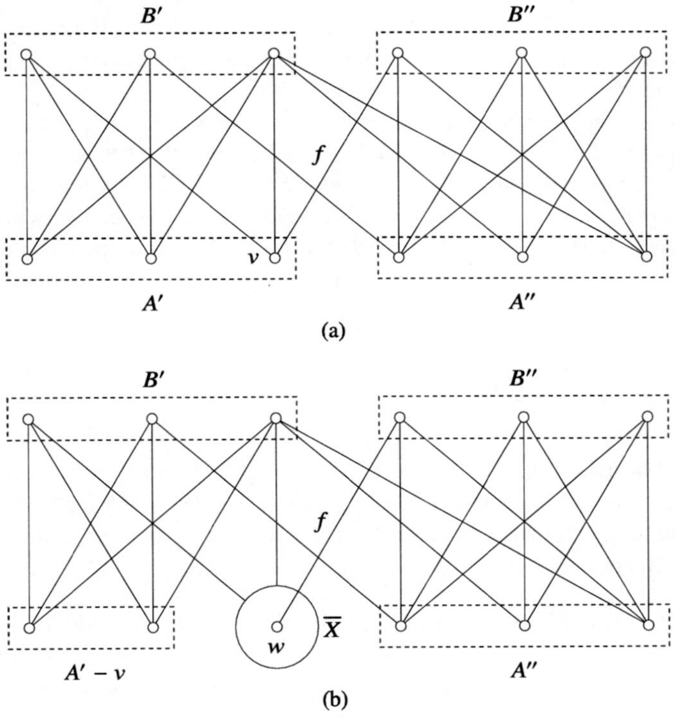

Fig. 8.16 Illustration for Lemma 8.19

Proof Suppose that $\overline{x} \in A'$. Let v denote the end of f in A. If $f \in C$, let w denote the end of f in $\overline{X}$ (see Figure 8.16(b)). If f is not in C, then v is an end of f in G, let $w := v$. In either case, the set $A'' + w$ is a nontrivial barrier of G. This conclusion is a contradiction to the hypothesis that G is bicritical. □

Let f be any edge of $C - e$, and assume that f is not removable in H. By Lemma 8.2, there exists a partition (A', A'') of A and a partition (B', B'') of B such that $|A'| = |B'|$ and f is the only edge of H that joins a vertex in A' to a vertex in B''. In that case, as $f \in C$, the end of f in A is $\overline{x}$; hence $\overline{x}$ is not in A'', a contradiction to statement (8.19.1). We conclude that f is removable in H. This conclusion holds for each edge f in $C - e$.

Let v be a vertex of $A - \overline{x}$ that has degree three or more in $G - e$. Then, v has degree three or more in H. Suppose that $\partial(v) - e$ contains two edges, say vb_1 and vb_2, that are not removable in H. By Theorem 8.4, with v playing the role of a, there exist a partition (A_0, A_1, A_2) of A and a partition (B_0, B_1, B_2) of B such that, for $1 \leq i \leq 2$, $|A_i| = |B_i|$ and $v \in A_0$, $b_i \in B_i$ and vb_i is the only edge that joins a vertex of $A - A_i$ to a vertex in B_i.

The sets A_1 and A_2 are disjoint; therefore one of them does not contain the contraction vertex $\overline{x}$. Adjust notation so that $\overline{x} \notin A_2$. We then have a contradiction to statement (8.19.1), with $A_0 \cup A_1$, A_2, $B_0 \cup B_1$ and B_2 playing respectively the roles of A', A'', B' and B''. We conclude that at most one edge of $\partial(v) - e$ is not removable in H.

The vertex v and the contraction vertex $\overline{x}$ are in the same part of the bipartition of H. Thus, no edge of $\partial(v) - e$ is in $C - e$. Moreover, as H is matching covered, the cut $C - e$, a barrier cut of $G - e$, is tight in $G - e$. It follows that every edge of $\partial(v) - e$ which is removable in H is also removable in $G - e$. We deduce that at most one edge of $\partial(v) - e$ is not removable in $G - e$. This conclusion holds for each vertex v in $A - \overline{x}$ that has degree three or more in $G - e$. □

8.4.1 The Zigzag Lemma

We shall now make use of Theorem 8.4 and Lemma 8.2 to derive another lemma concerning removable edges in bipartite graphs. This result will have a crucial role in our proof of an important result in Part III, but is not needed for an understanding of the basic theory.

Lemma 8.20 (The Zigzag Lemma) *Let $G := G[A, B]$ be a bipartite matching covered graph and let $P := (v_1, v_2, v_3, v_4)$, $v_1 \in A$, be a path in G such that no edge of P is removable in G and the vertices v_2 and v_3 are adjacent to three or more vertices. See Figure 8.17. Then, A has a partition (A_1, A_2, A_3, A_4) and B has a partition (B_1, B_2, B_3, B_4) such that*

(i) $v_1 \in A_1$, $v_2 \in B_2$, $v_3 \in A_3$, $v_4 \in B_4$,
(ii) *for* $i = 1, 2, 3, 4$, $|A_i| = |B_i|$,
(iii) v_1v_2 *is the only edge that joins a vertex in* A_1 *to a vertex of* $B - B_1$,
(iv) v_3v_4 *is the only edge that joins a vertex of* $A - A_4$ *to a vertex of* B_4, *and*
(v) v_2v_3 *is the only edge that joins a vertex of* $B_1 \cup B_2$ *to a vertex of* $A_3 \cup A_4$.

Proof The edges v_2v_3 and v_3v_4 are not removable in G and the vertex v_3 is adjacent to three or more vertices in G. From Theorem 8.4, with v_3, v_2 and v_4 playing

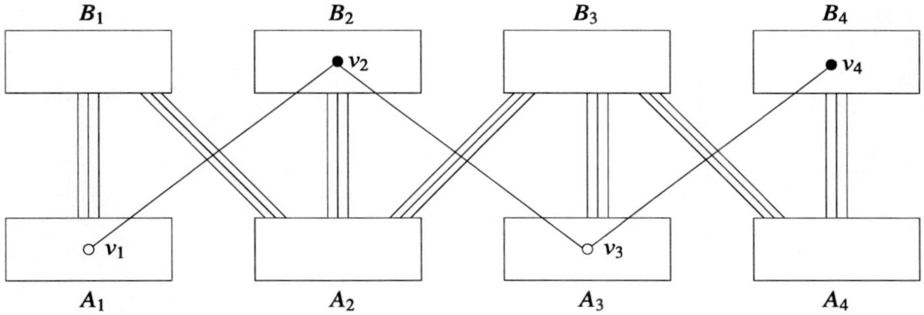

Fig. 8.17 The path P and the partitions of A and B

respectively the role of a, b_1 and b_2, we infer that A has a partition (A_{12}, A_3, A_4) and B has a partition (B_{12}, B_3, B_4) such that

(i) $v_2 \in B_{12}$, $v_3 \in A_3$ and $v_4 \in B_4$,
(ii) $|A_{12}| = |B_{12}|$, $|A_3| = |B_3|$ and $|A_4| = |B_4|$,
(iii) v_3v_4 is the only edge that joins a vertex in $A - A_4$ to a vertex in B_4, and
(iv) v_2v_3 is the only edge that joins a vertex in B_{12} to a vertex in $A - A_{12}$.

See Figure 8.18.

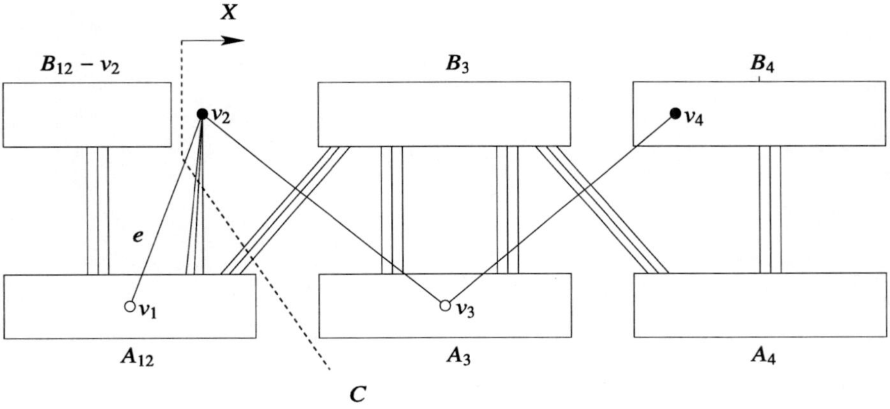

Fig. 8.18 The partitions (A_{12}, A_3, A_4) and (B_{12}, B_3, B_4)

Let $e := v_1v_2$ in G. Note that $v_1 \in A_{12}$ and $v_2 \in B_{12}$. Let $C := \partial(X)$, where $X := A_3 \cup A_4 \cup B_3 \cup B_4 \cup \{v_2\}$. The cut C is tight in G. Let $H := G/(X \to x)$. Then, $(A_{12}, (B_{12} - v_2) + x)$ is the bipartition of H. See Figure 8.19(a).

By hypothesis, the vertex v_2 is adjacent to three or more vertices in G. Thus, v_2 is joined to one or more vertices in $A_{12} - v_1$; hence e is a multiple edge in $H' := G/(\overline{X} \to \overline{x})$. As e is not removable in G, we conclude that e is not removable in H. From Lemma 8.2, with A_{12}, $(B_{12} - v_2) + x$, v_1 and x playing respectively the roles of A, B, a and b, we infer that there is a partition (A_1, A_2') of A_{12} and a partition

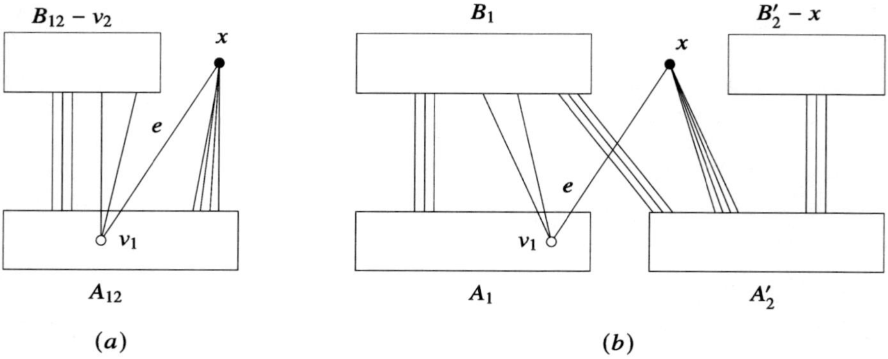

Fig. 8.19 The graph H and the partitions of A_H and B_H

(B_1, B_2') of $(B_{12} - v_2 + x)$ such that e is the only edge that joins a vertex of A_1 to a vertex in B_2', $|A_1| = |B_1|$ and $|A_2'| = |B_2'|$. See Figure 8.19(b).

Let us now expand vertex x. Let f be an edge of $C - e$. As can be seen in Figure 8.19(b), f is incident with a vertex in A_2' and as can be seen in Figure 8.18, f is in $\partial(B_3 + v_2)$. See Figure 8.20.

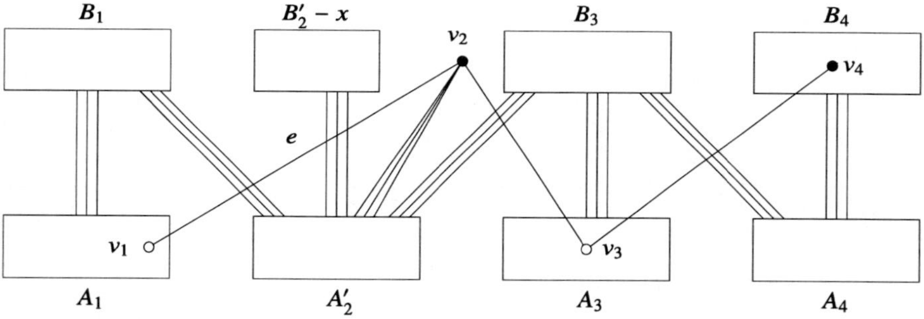

Fig. 8.20 The expansion of vertex x

The assertion holds, with $A_2 := A_2'$ and $B_2 := (B_2' - x) + v_2$. □

Recall the concept of certificate for nonremovability of an edge in a bipartite graph, defined in the statement of Lemma 8.2. In a paper submitted for publication, Lucchesi [63, 2023] proved the following generalization of that Lemma.

Theorem 8.21 *Let $G[A, B]$ be a bipartite matching covered graph of order four or more and let T be a nontrivial tree which is a subgraph of G. If (i) no edge of T is removable in G and (ii) each internal vertex of T has degree three or more in G, then there is a laminar collection C of cuts, each of which is a certificate of an edge of T.* □

Example 8.22 In Figure 8.21, the tree T consists of the four edges $e_1 := v_1v_2$, $e_2 := v_2v_3$, $e_3 := v_3v_4$, and $e_4 := v_3v_5$. For each e_i, $i = 1, 2, 3, 4$, the corresponding certificate cut C_i is defined as follows:

$$C_1 := \partial(S_1), \quad C_2 := \partial(S_1 \cup S_2), \quad C_3 := \partial(S_4) \quad \text{and} \quad C_4 := \partial(S_5).$$

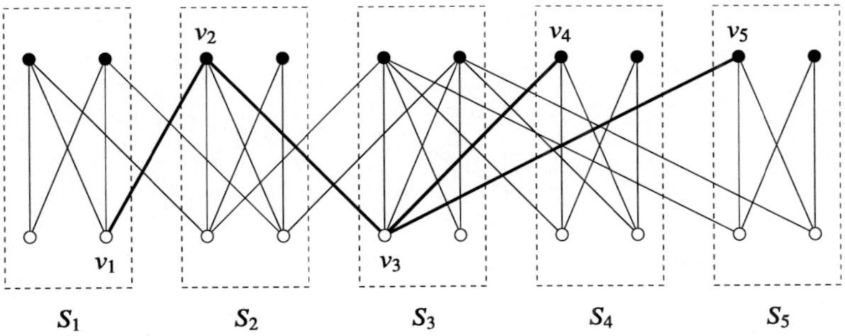

Fig. 8.21 An illustration for Theorem 8.21.

Theorem 8.21 may be proved by induction on the number of edges of the tree (Exercise 8.4.2). Interestingly enough, Theorem 8.4 and Lemma 8.20 my be easily derived from Theorem 8.21 (Exercises 8.4.3 and 8.4.4).

Exercises

$*$**8.4.1** *Let G be brace of order at least six and let e be an edge of G. Show that:*

(i) *for any vertex v that has degree at least three in $G - e$, at most one edge of $G - e$ incident with v is not removable in $G - e$, and*

(ii) *if $C := \partial(X) - e$ is a nontrivial tight cut of $G - e$ then each edge of C is removable in $G - e$.* *(CLM [18])*

▷**8.4.2** *Prove Theorem 8.21 by induction on the number of edges of G.*

▷**8.4.3** *Deduce Theorem 8.4 from Theorem 8.21.*

▷**8.4.4** *Deduce Lemma 8.20 from Theorem 8.21.*

8.5 Notes

The paper written by CLM (1999, [8]) is a self-contained introduction to the theory of removable classes in matching covered graphs. That paper can be thought of as a useful supplement to the material in this chapter and in Chapters 9 and 11.

The next chapter is devoted to the study of removable classes in bricks. We shall see that in a brick every dependence class, minimal or not, has at most two edges. In 2014, CLM found the graph in Figure 8.9 to show that 3-connected matching covered graphs which are not bricks could have dependence classes of size more than two. But both of us forgot about it. Recently, Lu, Kothari, Feng and Zhang (2020, [62]) showed that, given any $k \geq 2$ and $\ell \geq 1$, there exists a k-connected matching covered graph which has a dependence class of size ℓ. The graph depicted in Figure 8.22 is a 3-connected matching covered graph in which $\{e_1, e_2, e_3\}$ is a dependence class of size three.

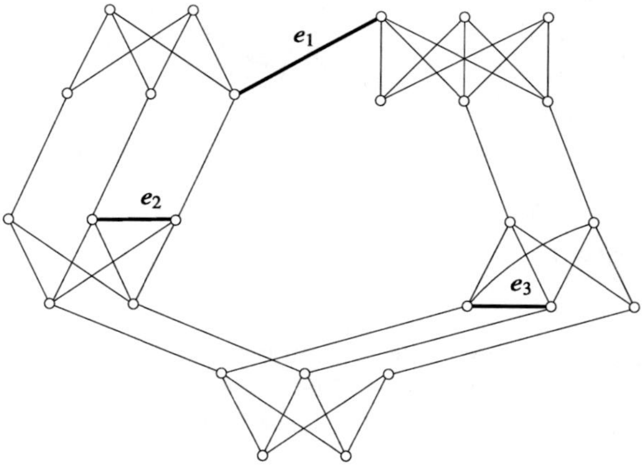

Fig. 8.22 A 3-connected matching covered graph with a dependence class of size three

The above graph is obtained from the one shown in Figure 8.14 by splicing bipartite graphs at vertices u and v. Interestingly, this trick preserves the dependence class $\{e_1, e_2, e_3\}$ while boosting the connectivity up by one! Lu, Kothari, Feng and Zhang [62] use similar ideas to construct, for any given $k \geq 2$ and $\ell \geq 1$, a k-connected matching covered graph which has a dependence class of specified size ℓ.

Chapter 9
Dependence Classes in Bricks

Contents

9.1 Dependence Classes in Bricks

We continue our study of dependence classes in matching covered graphs. But here our attention is mainly devoted to bricks.

We have seen that, in general, dependence classes in matching covered graphs can be arbitrarily large (see Figure 8.10), although no removable class can have more than two edges (Corollary 8.18).

In case of bricks, as we shall see, all dependence classes, regardless of whether they are removable or not, are either singletons or doubletons.

This is a consequence of the following theorem, which deals with mutually dependent edges in bricks, and is a generalization of a result proved by Lovász [58].

MUTUALLY DEPENDENT EDGES IN BRICKS

Theorem 9.1 (CLM(1999, [8])) *Let G be a brick and let e and f be two mutually dependent edges of G. Then $G - e - f$ is a bipartite graph, with parts of equal cardinality, such that both ends of e lie in one part of its bipartition and both ends of f lie in the other.*

C. L. Lucchesi, U. S. R. Murty, *Perfect Matchings*, Algorithms and Computation in Mathematics 31, https://doi.org/10.1007/978-3-031-47504-7_9

Proof As e and f are mutually dependent by the hypothesis, it follows from Proposition 8.10 that there exists a barrier, say B, of $G - f$ which contains both ends of the edge e. Furthermore, by Theorem 8.13,

- edge f has its ends in different components of $G - f - B$,
- edge e is the only edge of G having both ends in B,
- for each component K of $G - f - B$, the cut $\partial(K)$ is tight in G.

But the graph G, being a brick, is free of nontrivial tight cuts. We conclude that each component of $G - f - B$ is in fact trivial. Let A denote the set $V - B$. Then, $|A| = |B|$ and $G - e - f$ is a bipartite graph with bipartition (A, B). Hence the assertion. □

SIZES OF DEPENDENCE CLASSES IN BRICKS

Corollary 9.2 *Any dependence class of a brick G has at most two edges.*

Proof Assume to the contrary that G has a dependence class, Q, such that $|Q| \geq 3$, and let e, f and g denote three distinct edges of Q. Since $e \Leftrightarrow f$, it follows from Theorem 9.1 that $E - e - f$ is a cut of G. Likewise, $E - e - g$ is also a cut of G. As the symmetric difference of any two cuts of G is a cut of G, it follows that $(E - e - f) \triangle (E - e - g) = \{f, g\}$ is a cut of G. This is absurd because G, being a brick, is 3-edge-connected. □

The brick shown in Figure 8.15 (Exercise 8.3.9) shows that, in general, a dependence class in a brick need not be minimal and hence may not be removable. However, using the fact all bricks are 3-connected, it is easy to deduce the following corollary, whose proof is left as Exercise 9.1.2.

Corollary 9.3 *Every minimal dependence class in a brick is removable.* □

The following is another simple consequence of Theorem 9.1.

Corollary 9.4 *The union $Q_1 \cup Q_2$ of any two dependence classes Q_1 and Q_2 of size two of a brick G is a 4-cut of G.*

Proof Clearly, the two doubletons Q_1 and Q_2 are disjoint; hence $|Q_1 \cup Q_2| = 4$. As in the proof of Corollary 9.2, $E - Q_1$ and $E - Q_2$ are cuts; hence so is their symmetric difference $Q_1 \cup Q_2$. We conclude that $Q_1 \cup Q_2$ is a 4-cut of G. □

For example, the prism $\mathbb{P}_{2n}$, where $n \geq 3$ is odd, and the Möbius ladder $\mathbb{M}_{2n}$, where $n \geq 4$ is even, are bricks. Each of these bricks has n dependence classes of size two, and the union of any two of these classes is a 4-cut. The brick K_4 is a Möbius ladder; it has three dependence classes of size two and the union of any two of them is a 4-cut of K_4.

Theorem 9.10 describes the structure of bricks that have three or more removable doubletons.

In all the examples mentioned above, each dependence class of size two is also a removable class. But, as the example in Figure 8.15 shows, a dependence class R of size two in a brick G need not be removable in that brick. However, if R is removable, that is, if $G - R$ is matching covered, then we obtain the following result, due to Lovász [58], as a consequence of Theorem 9.1. (See Exercise 8.1.7 for Lovász's original proof.)

REMOVABLE DOUBLETONS IN BRICKS

Theorem 9.5 *If R is a removable doubleton of a brick G, then the graph $G - R$ is a bipartite matching covered graph.* □

Thus bricks with removable doubletons are the same as near-bipartite bricks, implying that if R is a removable doubleton in a brick G, then $b(G - R) = 0$. A more general result to be proved later (Theorem 9.9) shows that if R is a removable doubleton in any (nonbipartite) matching covered graph G, then $b(G - R) = b(G) - 1$.

Recall that a minimal class R of a matching covered graph G is said to be induced by an edge e of G if R contains an edge f such that f depends on e, that is, if every perfect matching of G containing f also contains e. Clearly minimal classes which are induced by two adjacent edges have to be distinct. This observation leads to the following assertion.

A LOWER BOUND ON THE NUMBER OF REMOVABLE CLASSES IN BRICKS

Theorem 9.6 *Every brick G has at least $\Delta(G)$ removable classes, where $\Delta(G)$ is the maximum of the degrees of vertices in G.*

Proof Let $e_1, e_2, \ldots, e_\Delta$ be the edges of G incident with a vertex, say v, of degree Δ, and, for $1 \le i \le \Delta$, let R_i denote a minimal class of G induced by e_i. Since any two edges incident with v are adjacent, it follows that $R_1, R_2, \ldots, R_\Delta$ are distinct minimal classes of G. Moreover, since G, being a brick, is 3-connected, it follows from Corollary 9.2 that each R_i is either a singleton or a doubleton, and hence that $G - R_i$ is connected. We conclude that $R_1, R_2, \ldots, R_\Delta$ are distinct removable classes. □

If a brick has no removable doubletons, then all its removable classes are removable edges. Thus:

Corollary 9.7 *Any brick G that is not near-bipartite has at least $\Delta(G)$ removable edges.* □

It follows from the above corollary that a cubic brick which is not near-bipartite has at least three removable edges. The tricorn $\mathbb{H}_{10}$ (Figure 2.5(d)) is an example of such a cubic brick with precisely three removable edges (Exercise 9.1.1). We do not know if there exist arbitrarily large cubic bricks which are free of removable doubletons but have just three removable edges. See Conjecture 9.19 in the Notes Section at the end of this chapter.

This result clearly does not, in general, hold for near-bipartite bricks. (For example K_4 and $\overline{C_6}$ are cubic bricks with no removable edges, and the bicorn $\mathbb{H}_8$ is a cubic brick with just one removable edge. See Figure 2.5.) In the next section we shall study the structure of near-bipartite bricks which will help us in deducing lower bounds on the number of removable edges in near-bipartite bricks. We shall see in particular that any simple near-bipartite brick distinct from the three mentioned above has at least two disjoint removable edges.

9.1.1 Monotonicity of functions b and $b + p$

The invariant $b(G)$ of a matching covered graph G, which denotes the number of bricks of G, was introduced in Chapter 4, and its significance in determining the dimension of the matching space was described in Chapter 6. There is another invariant of G which will prove to be of special interest in later chapters: a *Petersen brick* is any graph whose underlying simple graph is the Petersen graph. Theorem 4.17 implies that the number of Petersen bricks in a tight cut decomposition of G is independent of the tight-cut-contractions employed in the decomposition. We denote this number by $p(G)$. We may consider counting the number of bricks, Petersen bricks counted twice. This is the invariant $(b + p)(G) := b(G) + p(G)$. As we shall see in later chapters, it plays a special role in various aspects of this theory, especially in the theory of ear decompositions. (See the Notes Section at the end of this chapter for an example of an assertion about matching covered graphs which involves the parameter $b + p$.)

When the graph G under consideration is clear from the context, we shall simply write b and p for $b(G)$ and $p(G)$. The purpose of this subsection is to study the effect of deletions of removable classes from a matching covered graph on the two parameters b and $b + p$.

MONOTONICITY WITH RESPECT TO DELETIONS OF REMOVABLE EDGES

Theorem 9.8 (CLM (2002, [9])) *Let G be a matching covered graph and let e be a removable edge of G. Then,*

$$b(G - e) \geq b(G) \text{ and } (b + p)(G - e) \geq (b + p)(G). \tag{9.1}$$

Proof By induction on $|V|$. Consider first the case in which G has a nontrivial tight cut, say C. Let G' and G'' be the two C-contractions of G. By Proposition 8.8, both $G' - e$ and $G'' - e$ are matching covered. Moreover, every perfect matching of $G - e$ is a perfect matching of G, whence $C - e$ is tight in $G - e$. By induction, using the uniqueness of the tight cut decomposition, we then have

$$b(G - e) = b(G' - e) + b(G'' - e) \geq b(G') + b(G'') = b(G).$$

Likewise,

$$(b+p)(G-e) = (b+p)(G'-e) + (b+p)(G''-e) \\ \geq (b+p)(G') + (b+p)(G'') = (b+p)(G).$$

The assertion holds in this case.

We may thus assume that G is free of nontrivial tight cuts. That is, G is either a brick or a brace. If G is a brace then clearly $G-e$ is bipartite, whence $b(G-e) = (b+p)(G-e) = b(G) = (b+p)(G) = 0$ and the assertion holds. We may thus assume that G is a brick.

If the underlying simple graph of G is the Petersen graph then $p(G) = 1$. In that case, for every edge e of G, if e is a multiple edge then $b(G-e) = p(G-e) = 1$, whereas if e is not a multiple edge then $b(G-e) = 2$. In both alternatives, the assertion holds.

Finally, we may assume that G is a brick and $p(G) = 0$. It now suffices to prove that $G-e$ is not bipartite. This follows from Exercise 8.1.3.

In all cases considered, the asserted inequalities hold. □

The inequality established in the above theorem need not be tight; by deleting an edge from a brick it is possible to obtain a matching covered graph with an arbitrarily large number of bricks (see Exercise 9.1.3). In contrast, as is shown by the following assertion, the changes in the values of the functions b and $b+p$ effected by the deletion of a removable doubleton are predictable. We note that, by definition, a removable doubleton is a minimal removable class. By Theorem 8.1, if a matching covered graph has a removable doubleton then it is nonbipartite.

MONOTONICITY WITH RESPECT TO DELETIONS OF REMOVABLE DOUBLETONS

Theorem 9.9 (CLM (2002, [11])) *Let G be a (nonbipartite) matching covered graph, and let R be a removable doubleton of G. Then,*

$$b(G-R) = b(G) - 1 \text{ and } (b+p)(G-R) = (b+p)(G) - 1. \tag{9.2}$$

Proof By induction on $|V|$.

Case 1 *Graph G is a brick.*

As G is a brick, $b(G) = 1$. Furthermore, since G has a removable doubleton, it cannot be a Petersen brick because such bricks do not have removable doubletons, and thus $(b+p)(G) = 1$.

Since R is a removable doubleton of the brick G, by Theorem 9.5, the graph $G-R$ is a bipartite matching covered graph. Hence $b(G-R) = 0 = b(G) - 1$, and $(b+p)(G-R) = 0 = (b+p)(G) - 1$.

Case 2 *Graph G has a nontrivial tight cut C such that $|C \cap R| = 0$.*

Let G_1 and G_2 denote the two C-contractions of G. As R and C are disjoint, it follows, from Lemma 8.15, that R is is entirely contained in one of $E(G_1) - C$ and $E(G_2) - C$, and also that if $R \subset E(G_i) - C$, then R is a removable doubleton of G_i. Assume without loss of generality that $R \subset E(G_2) - C$. Clearly every perfect matching of $G - R$ is a perfect matching of G, and therefore C is a tight cut of $G - R$. By induction and Theorem 4.20, we have:

$$b(G - R) = b(G_1) + b(G_2 - R) = b(G_1) + b(G_2) - 1 = b(G) - 1.$$

Likewise,

$$\begin{aligned}(b + p)(G - R) &= (b + p)(G_1) + (b + p)(G_2 - R) \\ &= (b + p)(G_1) + (b + p)(G_2) - 1 = (b + p)(G) - 1.\end{aligned}$$

Hence the required assertion holds in this case.

Case 3 *There is no nontrivial tight cut C of G such that $|C \cap R| = 0$.*

As R is a removable doubleton, every perfect matching which contains one of the edges of R also contains the other. Hence no tight cut of G can contain both the edges of R. Thus, if C is a tight cut such that $|C \cap R| \neq 0$, then $|C \cap R| = 1$. Let $R := \{e, f\}$ and suppose that e is the edge of R in C. Let G_1 and G_2 be the two C-contractions of G, and assume without loss of generality that $E(G_1) \cap R = \{e\}$ and $E(G_2) \cap R = R$. Since R is a removable class in G it follows that e is a removable edge of G_1 and R is a removable doubleton of G_2.

At this stage one might think that the desired identities can be obtained by induction as in the case above. But this is not quite so straightforward. Although one can assert, using induction, that $b(G_2 - R) = b(G_2) - 1$, there is no obvious way to assert that $b(G_1 - e) = b(G_1)$. For this reason, rather than considering arbitrary tight cuts of G, we restrict our attention to barrier cuts and 2-separation cuts. (By the ELP Theorem 5.6, any matching covered graph that has a nontrivial tight cut either has a nontrivial barrier cut or a 2-separation cut.)

Case 3.1 Graph G has a nontrivial barrier B.

Let G_1 denote the bipartite matching covered graph obtained by shrinking the vertex set of each odd component of $G - B$ to a single vertex.

The graph $G - B$ must have one component K such that $R \subset E(G_2)$, where $G_2 := G/\overline{V(K)}$ (see Exercise 8.3.7). By the hypothesis of the case, the cut $\partial(K)$ contains an edge of R, say e. Thus, f has both ends in $V(K)$. For any component L of $G - B$ distinct from K, the (tight) cuts $\partial(K)$ and $\partial(L)$ are disjoint; therefore $\partial(L)$ and R are disjoint. By the hypothesis of the case, L is trivial. It follows that K is the unique nontrivial odd component of $G - B$. Consequently, the bipartite graph G_1 is equal to $G/V(K)$.

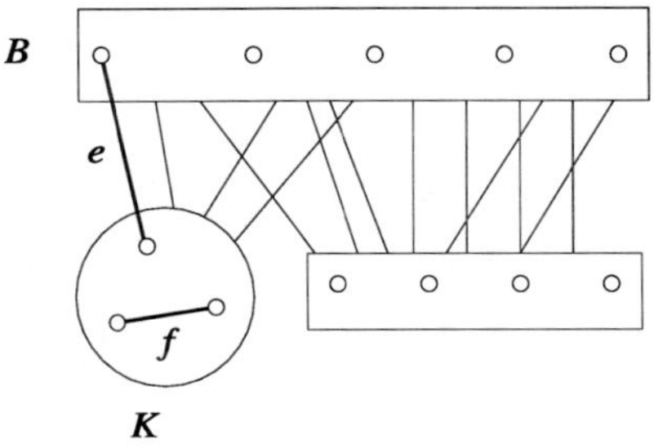

Fig. 9.1 $G - B$ has one nontrivial component K and $C_K \cap R = \{e\}$

Since R is removable in G, it follows that $E(G_1) \cap R = \{e\}$ is a removable singleton of G_1, and $E(G_2) \cap R = R$ is a removable doubleton of G_2. Since G_1 is bipartite, $b(G_1 - e) = b(G_1) = 0$. By induction, $b(G_2 - R) = b(G_2) - 1$ and $(b+p)(G_2 - e) = (b+p)(G_2) - 1$. The asserted equalities may now be deduced as before.

Case 3.2 Graph G is bicritical but it has a 2-separation $\{u, v\}$.

Let H_1 and H_2 be two vertex disjoint subgraphs of G such that $G - u - v$ is the union of H_1 and H_2. Then, as described in Chapter 4, the two cuts $C := \partial(V(H_1) + v)$ and $D := \partial(V(H_2) + v)$ are the two 2-separation cuts associated with $\{u, v\}$.

We shall first show that both the vertices u and v must be adjacent to at least two distinct vertices in each of the even sets $V(H_1)$ and $V(H_2)$. Suppose that this is not the case and assume that u is adjacent to just one vertex w in $V(H_1)$. In this case the graph $G - v - w$ is disconnected and has an odd component whose vertex set is contained in $V(H_1) - w$. It follows that $\{v, w\}$ is a nontrivial barrier of G. This is not possible because, by hypothesis of the subcase, G is bicritical. Thus each of u and v must be adjacent to at least two distinct vertices in each of the sets $V(H_1)$ and $V(H_2)$. A consequence of this observation is that each edge of C is a multiple edge in one of the C-contractions, and an analogous statement holds for D. Furthermore, since Case 2 does not apply, $|C \cap R| = |D \cap R| = 1$. (It is easy to see that any edge joining u and v is removable in G and hence cannot be in the doubleton R (Exercise 9.1.4). So, $C \cap R \neq D \cap R$.)

Adjust notation so that $C \cap R = \{e\}$. In one of the two C-contractions of G, say G_1, the edge e is a multiple edge, and in the other C-contraction of G, say G_2, the set R is a removable doubleton. In the illustration depicted in Figure 9.2 edge e of the doubleton R is in $C := \partial(X)$, where $X := V(H_1) + v$, and the second edge f of R has both its ends in $\overline{X} := V(H_2) + u$.

As the edge e is multiple in G_1, it is removable in G_1. Moreover, the underlying simple graphs of G_1 and of $G_1 - e$ are the same. Therefore, $b(G_1 - e) = b(G_1)$ and $(b+p)(G_1 - e) = (b+p)(G_1)$. On the other hand, as R is a subset of $E(G_2)$, it is a removable doubleton of G_2. Therefore, by induction, $b(G_2 - R) = b(G_2) - 1$ and $(b+p)(G_2 - R) = (b+p)(G_2) - 1$. The desired equalities in (9.2) now follow, as in the previous case. □

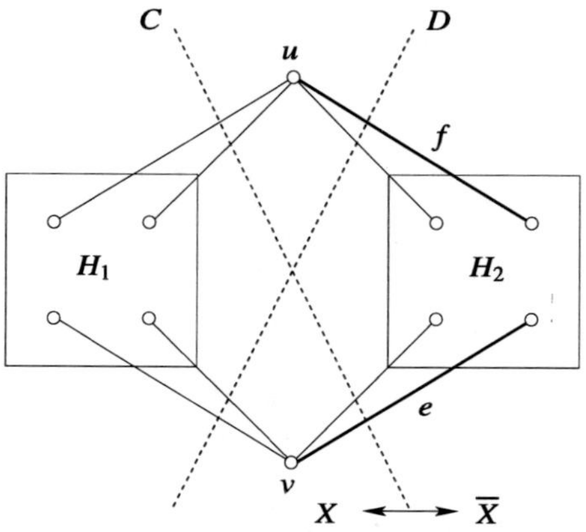

Fig. 9.2 Edge e is a multiple edge in one of the C-contractions of G, and R is a removable doubleton in the other C-contraction of G

9.1.2 Bricks with more than two removable doubletons

The next assertion describes the structure of bricks that have three or more removable doubletons. Its proof is left as Exercise 9.1.6.

Theorem 9.10 (Carvalho, Kothari, Lucchesi, Murty (2015, [26])) *Let G be a brick distinct from K_4 which has precisely $k \geq 3$ removable doubletons, and let R be the union of the k doubletons. Then, $G - R$ has precisely k components, each of which is bipartite and matchable. Moreover, the k components may be enumerated as $H_1, H_2, \ldots, H_k$ and the k removable doubletons may be enumerated as $R_1, R_2, \ldots, R_k$ such that H_i is joined to H_{i+1} by the two edges of R_i, as illustrated in Figure 9.3 (indices are taken modulo k).* □

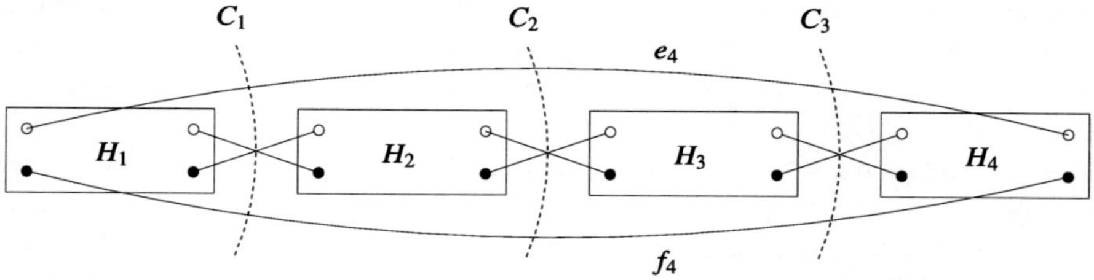

Fig. 9.3 The general structure of G, where $k = 4$, $R_4 = \{e_4, f_4\}$ and $C_i := R_i \cup R_4$, for $i = 1, 2, 3$

A brick need not have any removable doubletons. But it may have as many as $n/2$ removable doubletons. Moreover, if a simple brick has $n/2$ removable doubletons, then it is either a prism or a Möbius ladder, as stated below.

Corollary 9.11 *Every brick of order $n \geq 6$ has at most $n/2$ removable doubletons. Moreover, a simple brick of order n has $n/2$ removable edges if and only if it is a prism or a Möbius ladder.* □

Exercises

9.1.1 *Show that the tricorn has no removable doubletons and that it has precisely three removable edges.*

▷**9.1.2** *Prove Corollary 9.3.*

▷**9.1.3** *For each positive integer k, find a brick G with a removable edge e such that $b(G - e) = k$.*

▷**9.1.4** *Let $\{u, v\}$ be a separation of a matching covered graph G as in Case 3.2 of the proof of Theorem 9.9. Show that any edge joining u and v is removable in G and hence cannot be in a removable doubleton of G.*

▷**9.1.5** *For each positive integer k, find a matching covered graph G with a removable doubleton R such that $b(G) = k$ and $b(G - R) = k - 1$.*

∗**9.1.6** *The objective of this exercise is to obtain a proof of Theorem 9.10. Let G be a brick, let $\mathcal{R}$ be the collection of the $k \geq 3$ removable doubletons of G, let R be the union of the k doubletons, let R_k be one of the doubletons, and let $\mathcal{C} := \{R_k \cup S : S \in \mathcal{R} - R_k\}$. Note that each one of the $k - 1$ sets in $\mathcal{C}$ is a 4-cut of G, by Corollary 9.4.*

(i) *Prove that if $\mathcal{C}$ is not laminar then G is K_4. Hint: suppose that $\mathcal{C}$ contains two cuts, $C := \partial(X)$ and $D := \partial(Y)$, that cross.*

  (a) *Consider the four quadrants $X \cap Y$, $X \cap \overline{Y}$, $\overline{X} \cap Y$ and $\overline{X} \cap \overline{Y}$ and the corresponding collection $\mathcal{F}$ of four cuts each of which has one of the quadrants as a shore.*

  (b) *Each edge of $C \cup D$ is in precisely two cuts in $\mathcal{F}$ and each edge in R_k is in $C \cap D$.*

  (c) *Deduce that the sum of the number of edges of each one of the four cuts in $\mathcal{F}$ is equal to 12 and conclude that each cut in $\mathcal{F}$ is a 3-cut.*

  (d) *Apply the result of Exercise 5.2.7 and conclude that each quadrant is a singleton.*

(ii) *Let $e_k := v_k w_k$ be an edge of R_k. For each cut C in $\mathcal{C}$, let X be the shore of C that contains the vertex v_k. Let $\mathcal{X}$ be the collection of the $k - 1$ shores thereby obtained. Prove that if G is not K_4 then there is an enumeration $X_1, X_2, \ldots, X_{k-1}$ of the shores in $\mathcal{X}$ such that $X_1 \subset X_2 \subset \cdots \subset X_{k-1}$. Hint:use the laminarity of $\mathcal{C}$.*

(iii) *Suppose that G is not K_4. Complete the proof of the Theorem. Hint: see the general structure of G in Figure 9.3.*

9.2 Mutually Exclusive Classes

By definition, if Q is an equivalence class of the dependence relation on the edge set of a matching covered graph G, every perfect matching of G which contains one edge of Q contains all edges of Q. Two distinct equivalence classes Q_1 and Q_2 are *mutually exclusive* if no perfect matching of G contains all the edges of the set $Q_1 \cup Q_2$. In other words, Q_1 and Q_2 are mutually exclusive if any perfect matching of G which contains all edges of one of Q_1 and Q_2 contains none of the edges of the other.

Two examples should serve to clarify the above definition. In the Petersen graph (Figure 9.4(a)), all equivalence classes are singletons, as each edge is removable. There is no perfect matching containing both e and f, but there is a perfect matching containing both e and g (Exercise 9.2.1). Therefore, $\{e\}$ and $\{f\}$ are mutually exclusive but $\{e\}$ and $\{g\}$ are not. The pentagonal prism (Figure 9.4(b)) has five removable edges; no two of these are mutually exclusive because there is a perfect matching containing all of them. But it also has five pairs of equivalence classes corresponding to the five removable doubletons. It is easy to see that the two equivalence classes $\{e_1, f_1\}$ and $\{e_2, f_2\}$ are mutually exclusive whereas $\{e_1, f_1\}$ and $\{e_3, f_3\}$ are not.

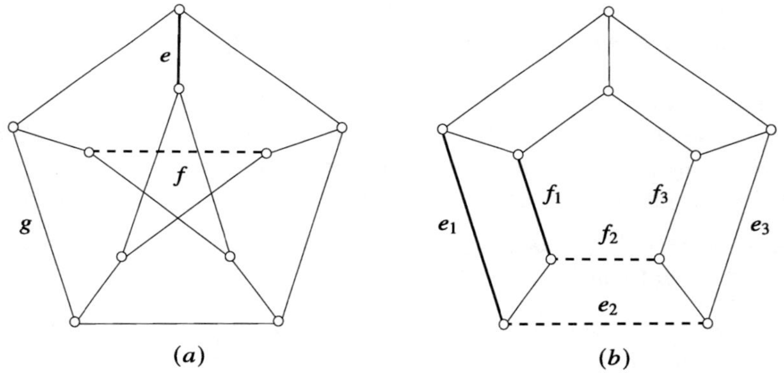

Fig. 9.4 Mutually exclusive equivalence classes–examples

If some edge of Q_1 is adjacent to some edge of Q_2 then, obviously, Q_1 and Q_2 are mutually exclusive. As the example of the Petersen graph shows, this condition of adjacency is not necessary for two singleton equivalence classes to be mutually exclusive. However, if two doubleton equivalence classes Q_1 and Q_2 of a brick G are mutually exclusive then the edges of Q_1 and Q_2 lie in close proximity to each other in a special way. To make this precise, we need the definition given below.

AITCH CONFIGURATION

Let $Q_1 := \{e_1, f_1\}$ and $Q_2 := \{e_2, f_2\}$ be two pairs of edges of a graph G such that $Q_1 \cap Q_2 = \emptyset$. We say that Q_1 and Q_2 *induce an aitch configuration* in G

if G has a subgraph H, where $V(H)$ is the set of ends of edges in $Q_1 \cup Q_2$, $E(H) = Q_1 \cup Q_2 \cup \{h\}$, and h is an edge not in $Q_1 \cup Q_2$, satisfying the following properties:

(i) the edges e_1 and e_2 share precisely one common end, say u_{12},
(ii) the edges f_1 and f_2 share precisely one common end, say v_{12},
(iii) the vertices u_{12} and v_{12} both have degree three and are joined by edge h.

Figure 9.5 illustrates an aitch configuration. For specific examples, consider the pentagonal prism which is shown in Figure 9.4(b), and, for $i = 1, 2, 3$, let $Q_i := \{e_i, f_i\}$. Then Q_1 and Q_2 induce an aitch configuration, and so do Q_2 and Q_3.

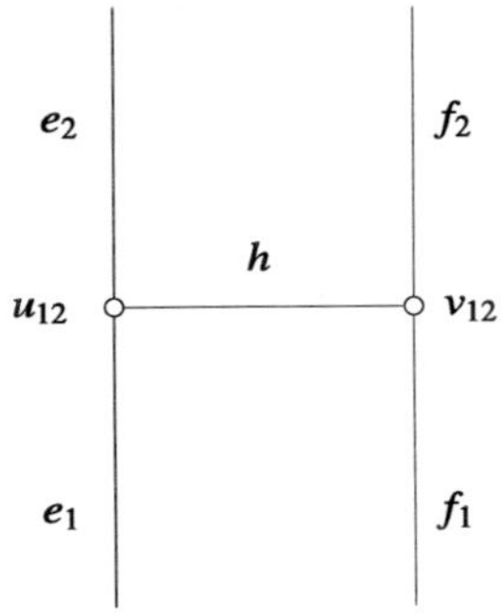

Fig. 9.5 The doubletons $\{e_1, f_1\}$ and $\{e_2, f_2\}$ induce an aitch configuration

The brick K_4 induces an extreme form of an aitch graph, in which Q_1 and Q_2 are two distinct perfect matchings of K_4. See Figure 9.6.

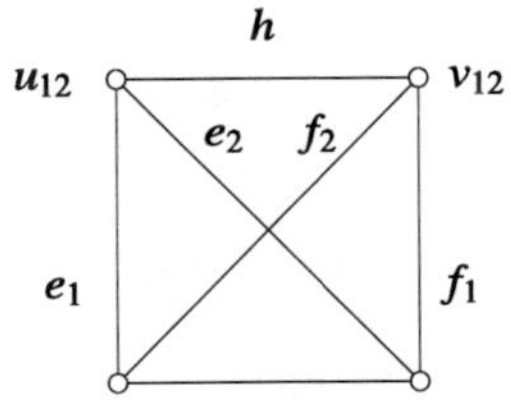

Fig. 9.6 An extreme form of aitch graph induced by K_4

Lemma 9.12 *Let Q_1 and Q_2 be two (not necessarily removable) doubletons that are equivalence classes of the dependence relation of a simple brick G. If Q_1 and Q_2 are mutually exclusive then they induce an aitch configuration in G.*

Proof Let $Q_i := \{e_i := u_i w_i, f_i := v_i x_i\}$, $i = 1, 2$, be two mutually exclusive doubletons of G. By Theorem 9.1, $E(G) - Q_i$ is a cut of G; it is the set of edges of a bipartite graph $H_i(A_i, B_i)$, with $|A_i| = |B_i|$. By Corollary 9.4, $Q_1 \cup Q_2 = \partial(A_1 \Delta A_2)$ is a 4-cut of G.

Adjust notation so that e_1 has both ends in A_1, implying that f_1 has both ends in B_1. The ends of e_1 are u_1 and w_1. One of them is in A_2, the other in B_2. Adjust notation so that $u_1 \in A_2$, implying that $w_1 \in B_2$. Likewise, note that the ends of f_1 are v_1 and x_1, and adjust notation so that $x_1 \in A_2$ and $v_1 \in B_2$. Adopt similar notational adjustments for the ends of the two edges e_2 and f_2 in Q_2 (Figure 9.7). Any edge of G which does not belong to the cut $Q_1 \cup Q_2$ belongs to both $\partial(A_1)$ and $\partial(A_2)$. Therefore, any such edge either joins a vertex in $A_1 \cap A_2$ to a vertex in $B_1 \cap B_2$, or a vertex in $A_1 \cap B_2$ to a vertex in $B_1 \cap A_2$, as indicated by dotted lines in Figure 9.7. Note that, up to this point, we have not used the hypothesis that Q_1 and Q_2 are mutually exclusive. In what follows, we shall use this hypothesis to show that $Q_1 \cup Q_2$ is not a matching. (In other words, not all the eight ends of the four edges in $Q_1 \cup Q_2$ are distinct as Figure 9.7 might suggest.)

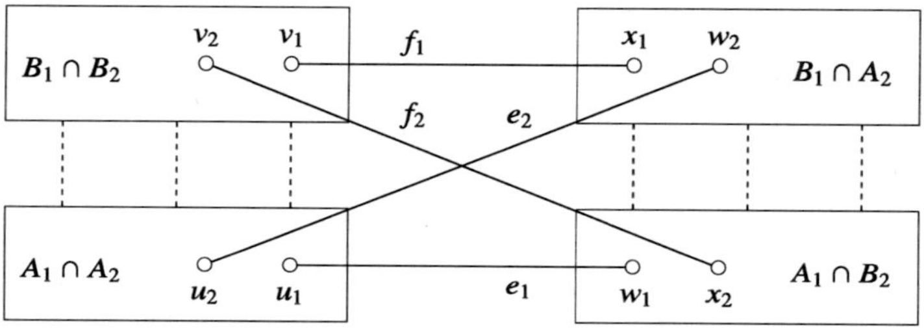

Fig. 9.7 The cut $Q_1 \cup Q_2$

9.12.1 *The following equalities hold:*

$$|A_1 \cap A_2| = |B_1 \cap B_2| \text{ and } |A_1 \cap B_2| = |B_1 \cap A_2|. \tag{9.3}$$

Moreover, the set $Q_1 \cup Q_2$ is not a matching.

Proof Let M_1 be a perfect matching of G that contains e_1. The doubleton Q_1 is an equivalence class and therefore M_1 contains f_1 as well. The two doubletons Q_1 and Q_2 are mutually exclusive. Thus M_1 and Q_2 are disjoint. A simple counting argument then establishes the asserted equalities in (9.3).

Now let X and Y denote the two subsets of $V(G)$ as defined below:

$$X := \left[(A_1 \cap A_2) \cup (B_1 \cap B_2)\right] - \{u_1, u_2, v_1, v_2\},$$
$$Y := \left[(A_1 \cap B_2) \cup (B_1 \cap A_2)\right] - \{w_1, w_2, x_1, x_2\}.$$

Assume, contrary to the second part of the assertion, that $Q_1 \cup Q_2$ is a matching. Graph G, being a brick, is bicritical. Therefore $G - u_1 - u_2$ has a perfect matching, say F. As $|A_1 \cap A_2| = |B_1 \cap B_2|$, the matching F must contain both f_1 and f_2. Hence F includes a matching, say M, that saturates precisely the vertices of X. Likewise, $G - x_1 - w_2$ has a perfect matching that contains both e_1 and f_2, whence it includes a matching, say N, that saturates precisely the vertices of Y. It now follows that $M \cup N \cup Q_1 \cup Q_2$ is a perfect matching of G that includes $Q_1 \cup Q_2$. This is absurd because, by hypothesis, Q_1 and Q_2 are mutually exclusive. Hence $Q_1 \cup Q_2$ cannot be a matching. □

9.12.2 *Suppose that the edge e_1 of Q_1 is adjacent to the edge e_2 of Q_2. Then:*

- *the two edges f_1 of Q_1 and f_2 of Q_2 are also adjacent, and*
- *the common end $u_1 = u_2$ of e_1 and e_2 is adjacent to the common end $v_1 = v_2$ of f_1 and f_2. Moreover, as illustrated in Figure 9.8,*

$$|A_1 \cap A_2| = 1 = |B_1 \cap B_2|, \ N(u_1) = \{v_1, w_1, w_2\} \text{ and } N(v_1) = \{u_1, x_1, x_2\}.$$

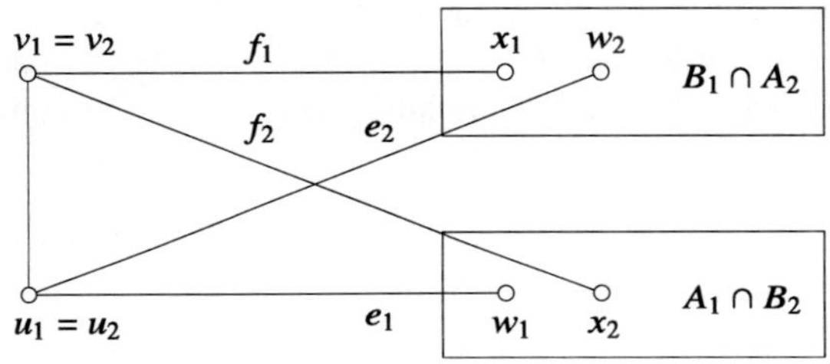

Fig. 9.8 Edges f_1 and f_2 adjacent

Proof Assume to the contrary that $v_1 \neq v_2$. In this case, since $u_1 = u_2$ by the hypothesis, and since $|A_1 \cap A_2| = |B_1 \cap B_2|$ by (9.12.1), it follows that $G - v_1 - v_2$ cannot have a perfect matching. This cannot be the case because G is bicritical. Therefore $v_1 = v_2$ must be the common end of f_1 and f_2.

If $A_1 \cap A_2$ had any vertex other than u_1, the pair $\{u_1, v_1\}$ would be a 2-vertex-cut of G. This cannot be the case because G, being a brick, is 3-connected. We conclude that $A_1 \cap A_2 = \{u_1\}$ and $B_1 \cap B_2 = \{v_1\}$. The last part of the assertion, that u_1 and v_1 must be adjacent follows from the fact that in a brick every vertex has at least three distinct neighbours. □

Each of the doubletons Q_1 and Q_2 is a matching. Thus, as $Q_1 \cup Q_2$ is not a matching, some edge of $Q_1 = \{e_1, f_1\}$ must be adjacent to some edge of $Q_2 = \{e_2, f_2\}$. By renaming the edges of Q_1 and Q_2, if necessary, we may assume that e_1 and e_2 are adjacent. Then, by the above result (9.12.2), the edges e_1 and e_2 share a common end $u_{12} := u_1 = u_2$, the edges f_1 and f_2 share a common end $v_{12} := v_1 = v_2$, and the vertices u_{12} and v_{12} both have degree three and are joined by an edge, say h, which is not in $Q_1 \cup Q_2$. Indeed, Q_1 and Q_2 induce an aitch configuration. □

As an application of Lemma 9.12, we now derive the following consequence.

BRICKS WITH THREE MUTUALLY EXCLUSIVE DOUBLETONS

Theorem 9.13 *If a simple brick G has three pairwise mutually exclusive doubletons then G is one of K_4 and $\overline{C_6}$.*

Proof Let $Q_i := \{e_i, f_i\}$, $i = 1, 2, 3$ be three mutually exclusive doubletons of a simple brick G. By Lemma 9.12 the pair Q_1, Q_2 induces an aitch configuration. Adjust notation so that the edges e_1 and e_2 share a common end labelled u_{12}, and the edges f_1 and f_2 share a common end labelled v_{12}. The vertices u_{12} and v_{12} have degree three and are adjacent in G. Let h denote the edge that joins u_{12} and v_{12}. See Figure 9.5.

Case 1 *The edge h is in* Q_3.

Adjust notation so that $h = e_3$. Then, the vertex u_{12} is the common end of the edges e_1 and e_3. By Lemma 9.12, with Q_3 playing the role of Q_2, the edge e_2 joins the common end u_{12} of e_1 and e_3 to the common end of f_1 and f_3, say v_{13}. See Figure 9.9(a). Also, the vertex v_{12} is the common end of f_1 and e_3. Thus, f_2 joins v_{12} to the common end of e_1 and f_3, say, u_{13}. Moreover, the vertices u_{13} and v_{13} are adjacent and have degree three. We conclude that G is K_4. See Figure 9.9(b).

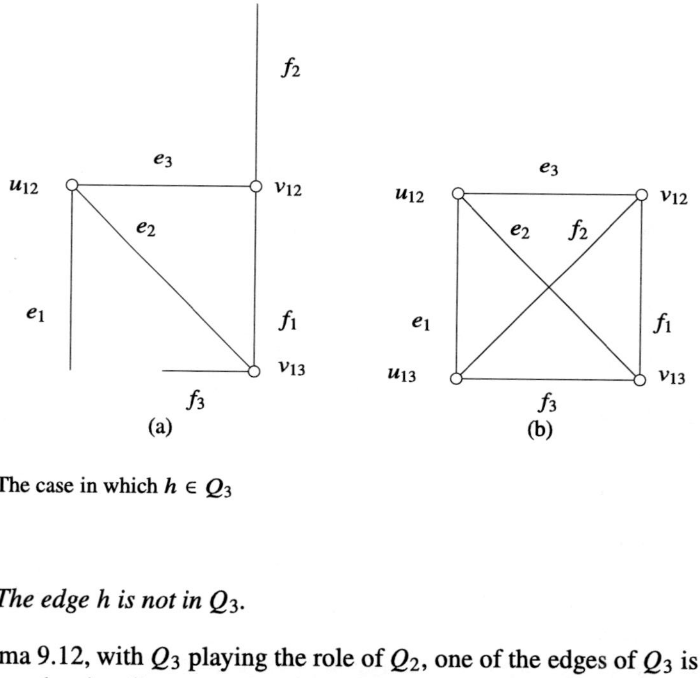

Fig. 9.9 The case in which $h \in Q_3$

Case 2 *The edge h is not in* Q_3.

By Lemma 9.12, with Q_3 playing the role of Q_2, one of the edges of Q_3 is adjacent to e_1, the other is adjacent to f_1. Adjust notation so that e_3 is adjacent to e_1 and f_3 is adjacent to f_1. As $h \notin Q_3$, the edges e_1 and e_3 share a common end, say u_{13}, distinct from u_{12}. Likewise, f_1 and f_3 share a common end, say v_{13}, distinct from v_{12}. Moreover, the vertices u_{13} and v_{13} are adjacent and have degree three. See Figure 9.10(a).

Again, by Lemma 9.12, with Q_3 playing the role of Q_1, we deduce that one of e_3 and f_3 shares an end, say u_{23}, with e_2 and the other shares an end, say v_{23}, with f_2. Moreover, u_{23} and v_{23} are adjacent and have degree three. It follows that the six vertices $u_{12}, v_{12}, u_{13}, v_{13}, u_{23}$, and v_{23} are all the vertices of G and that G is a cubic

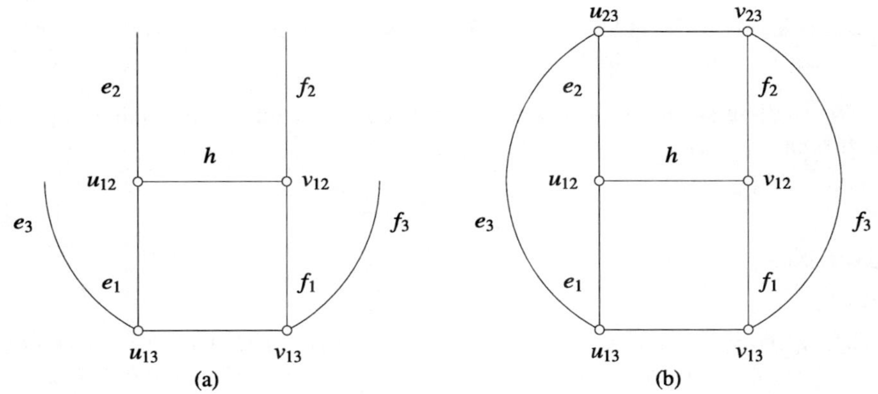

Fig. 9.10 The case in which $h \notin Q_3$

graph. Also, as G is a brick, it is not bipartite. Therefore u_{23} is not an end of f_2 and v_{23} is not an end of e_2; otherwise G would be the complete bipartite graph $K_{3,3}$. We conclude that u_{23} is an end of e_2 and v_{23} is an end of f_2, and hence that G is $\overline{C_6}$. See Figure 9.10(b). □

Prisms of order $2n$, for each odd $n \geq 3$, Möbius ladders of order $2n$, for each even $n \geq 4$, and staircases of order $2n$, for $n \geq 4$ are examples of bricks which have two, but not three, pairwise mutually exclusive removable doubletons. But one can construct other examples (Exercise 9.2.3).

9.2.1 Removable edges in bricks

Theorem 9.13 proved above has the following interesting implication:

A LOWER BOUND ON THE NUMBER OF REMOVABLE EDGES IN BRICKS

Theorem 9.14 *Every simple brick G distinct from K_4 and $\overline{C_6}$ has at least $\Delta(G) - 2$ removable edges, where $\Delta(G)$ is the maximum of the degrees of vertices of G.*

Proof Let G be a simple brick distinct from K_4 and $\overline{C_6}$. Let v be a vertex of G of degree $\Delta(G)$. For each edge e of $\partial(v)$, let $R(e)$ denote a minimal class induced by e. Class $R(e)$ has at most two edges. Moreover, as G is a brick, it is 3-edge-connected. Thus, $R(e)$ is a removable class. It is easy to see that, for any two distinct edges e and f in $\partial(v)$, the classes $R(e)$ and $R(f)$ are mutually exclusive (Exercise 9.2.2). Thus, G has $\Delta(G)$ mutually exclusive removable classes. By hypothesis, G is neither K_4 nor $\overline{C_6}$, implying that at most two of the removable classes induced by edges in $\partial(v)$

are doubletons. Thus, at least $\Delta(G) - 2$ removable classes among those induced by the Δ edges in $\partial(v)$ are singletons. □

We shall pursue in the next section a more detailed study of removable edges in near-bipartite bricks.

Exercises

▷**9.2.1** *Referring to the Figure 9.4(a) of the Petersen graph, find a perfect matching containing e and g, and show that there is no perfect matching containing both e and f.*

9.2.2 *For any vertex v of a matching covered graph G, show that any two minimal classes of G induced by different edges e and f in $\partial(v)$ are mutually exclusive.*

∗**9.2.3** *Find an example of a brick which is not a prism or a Möbius ladder or a staircase which has two, but not three mutually exclusive doubletons.*

STAIRCASES

Theorem 9.15 *Let G be a brick that has two removable, mutually exclusive doubletons $R_1 := \{e_1, f_1\}$ and $R_2 := \{e_2, f_2\}$. For $i = 1, 2$, let M_i be a perfect matching of G that includes the edges of R_i. If M_1 and M_2 are unique and disjoint then either G is K_4 or G is a staircase.*

∗**9.2.4** *Prove Theorem 9.15, following the steps listed below.*

(i) *Prove that R_1 and R_2 induce an aitch configuration of G; adjust notation so that (i) e_1 and e_2 share a common end, u_{12}, (ii) f_1 and f_2 share a common end, v_{12}, and (iii) the vertices u_{12} and v_{12} both have degree three in G and are joined by an edge h not in $R_1 \cup R_2$.*

(ii) *Let C be an (M_1, M_2)-alternating cycle in G that contains the edge e_1. Prove that C is a hamiltonian cycle of G, where h is a chord of C. Hint: use the fact that M_1 and M_2 are unique and disjoint.*

(iii) *Let P and Q denote the two subpaths of C, both from u_{12} to v_{12}, denoted as follows:*

$$P := u_{12}w_1w_2\cdots w_sv_{12},\ s \geq 1 \quad \textit{and} \quad Q := u_{12}x_1x_2\cdots x_tv_{12},\ t \geq 1.$$

Prove that every chord of C distinct from h joins an internal vertex of P to an internal vertex of Q. Hint: use the uniqueness of M_i and the fact that $G - R_i$ is bipartite, for $i = 1, 2$.

(iv) *Prove that for $i = 1, 2, \ldots, \min\{s, t\}$, w_i and x_i both have degree three and are adjacent. Hint: use induction on i.*

(v) *Conclude that $s = t$ and also that G is either K_4 or is a staircase.*

BRICKS WITH AT MOST ONE REMOVABLE EDGE

Theorem 9.16 *Let G be a brick that has at most one removable edge. Then, G is one of the following three bricks: K_4, $\overline{C_6}$ or $\mathbb{H}_8$.*

$*$**9.2.5** *Using Theorem 9.15, prove Theorem 9.16, following the steps listed below.*

(i) *Prove that G has a pair of mutually exclusive removable doubletons, R_1 and R_2. Hint: for any vertex v, consider the minimal dependence classes induced by the edges of $\partial(v)$.*
(ii) *For $i = 1, 2$, let M_i be a perfect matching of the brick G that includes R_i and let H_i be the (bipartite) graph $G - R_i$. Prove that every removable edge of H_i which is not in M_i is removable in G.*
(iii) *Prove that every edge of $M_i - R_i$ is removable in H_i. Hint: suppose the contrary, let $e := vw$ be an edge in $M_i - R_i$ which is not removable in H_i and apply Lemma 8.6 to vertices v and w, to obtain two edges of H_i which are not in M_i but are removable in H_i and then deduce a contradiction, using part (ii).*
(iv) *Prove that M_1 and M_2 are unique. Hint: suppose that M_i is not unique, let C be an M_i-alternating cycle that does not contain any edge in R_i, and use parts (iii) and (ii) to prove that every edge of $E(C) - M_i$ is removable in G.*
(v) *Prove that M_1 and M_2 are disjoint. Hint: suppose that $M_1 \cap M_2$ contains an edge, e, and use part (iii) to prove that e is a removable edge of H_1 and conclude that G has a perfect matching that contains both edges of R_2 but does not contain e, contradicting the uniqueness of M_2.*
(vi) *Apply Theorem 9.15 to prove that G is either K_4 or a staircase.*
(vii) *Conclude that G has order eight or less and is thus K_4, $\overline{C_6}$ or $\mathbb{H}_8$.*

9.2.6 *Prove the converse of Theorem 9.15.*

9.3 Removable Edges in Near-Bipartite Bricks

The bricks K_4 and $\overline{C_6}$ have no removable edges. The bicorn $\mathbb{H}_8$ has only one removable edge. In fact, they are the only three bricks that have fewer than two removable edges (Exercise 9.2.5). The following theorem in fact shows that every simple near-bipartite brick distinct from these three bricks has two nonadjacent removable edges. The proof we present here is also from the unpublished paper [61] by Lovász and Vempala.

REMOVABLE EDGES IN NEAR-BIPARTITE BRICKS

Theorem 9.17 *Every simple near-bipartite brick distinct from K_4, $\overline{C_6}$ and the bicorn $\mathbb{H}_8$ has two nonadjacent removable edges.*

Proof Let G be a simple near-bipartite brick. We shall prove that either G has two nonadjacent removable edges or G is either K_4, $\overline{C_6}$ or the bicorn. By hypothesis, G is near-bipartite. Let $R := \{e, f\}$ denote a removable doubleton of G such that $H := G - R$ is bipartite. The next assertion is immediate.

9.17.1 *Let M be a perfect matching of G that contains both edges of R. Every edge of $E - M$ which is removable in H is also removable in G.* □

As R is a dependence class, every perfect matching of G that contains one edge of R contains both edges of R. Let M be a perfect matching of G that contains both edges of R. We now consider several cases. In each case we assume that the previous cases do not apply.

Case 1 *The matching M is not unique.*

The analysis of this case is quite similar to that used in the proof required in Exercise 9.2.5(iii). Assume that G has a perfect matching M', distinct from M, that contains the edges e and f. Let C be an (M, M')-alternating cycle. Consider first the case in which every edge in $C \cap M$ is removable in H. As G is simple, the cycle C has length four or more; hence $C \cap M$ contains two or more edges. The edges of $C \cap M$ are not in M'; hence they are removable in G, by statement 9.17.1. The assertion holds in this case, as the edges of $C \cap M$ are not adjacent.

We may thus assume that some edge of $C \cap M$, say $h := vw$, is not removable in H. Clearly, the edges e and f are not adjacent to edge h. Thus, v has degree three or more in the bipartite graph H. By Lemma 8.6, $\partial(v)$ contains an edge, say e', that is removable in H. The edge e' is not the edge h, because h is not removable in H. Thus, $e' \notin M$. By statement 9.17.1, e' is removable in G. Likewise, $\partial(w) - h$ contains an edge, e'', which is removable in G. As G is simple and since h is neither e' nor e'', we conclude that e' and e'' are two removable edges of G that are not adjacent. The assertion holds in this case.

$$\text{We may thus assume that the matching } M \text{ is unique.} \tag{9.4}$$

Let (A, B) denote the bipartition of H. One of e and f has both ends in A, the other has both ends in B. Adjust notation so that e has both ends in A and f has both ends in B. Let $e := u_1u_2$ and $f := v_1v_2$. Consider the subgraph $H' := G - \{u_1, u_2, v_1, v_2\}$ obtained by deleting the ends of the two edges e and f. Now let $A' := A - \{u_1, u_2\}$, let $B' := B - \{v_1, v_2\}$, and let $M' := M - \{e, f\}$. The graph H' is bipartite; the pair (A', B') is a bipartition of H'. The set M' is a perfect matching of H'. From equation 9.4, M is the only perfect matching of G that contains the edges e and f. Thus, M' is the only perfect matching of H'. This implies that there is an enumeration $a_1, a_2, \ldots, a_k$ of the vertices of A' and an enumeration $b_1, b_2, \ldots .b_k$ of the vertices of B' such that (i) $a_ib_i \in M$, for $i = 1, 2, \ldots, k$ and (ii) $a_ib_j \notin E(H)$ for $1 \le i < j \le k$ (Exercise 3.2.2). If $k = 0$ then G has order four; hence $G = K_4$. We may thus assume that $k \ge 1$.

9.17.2 $N(a_1) = \{v_1, v_2, b_1\}$ *and* $N(b_k) = \{u_1, u_2, a_k\}$. *See Figure 9.11.*

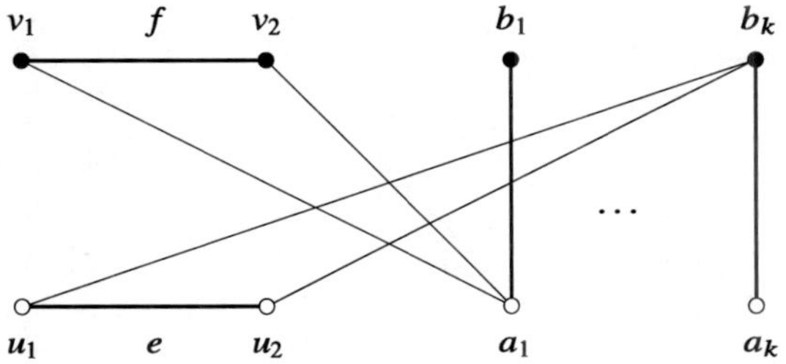

Fig. 9.11 The adjacencies of a_1 and b_k.

Proof In H', the vertex a_1 has degree one. Thus, in G, a_1 is adjacent to both v_1 and v_2; hence $N(a_1) = \{v_1, v_2, b_1\}$. Likewise, $N(b_k) = \{u_1, u_2, a_k\}$. □

Case 2 $k = 1$.

If $k = 1$ then G has $\overline{C_6}$ as a spanning subgraph. In that case, either $G = \overline{C_6}$ or it has two nonadjacent removable edges. (Exercise 9.3.1.)

$$\text{We may thus assume that } k \geq 2. \tag{9.5}$$

In H', the vertex a_2 is adjacent to b_2 and possibly to b_1, but it is not adjacent to any vertex in $B' - \{b_1, b_2\}$. Thus, $N(a_2) \subseteq \{b_1, b_2, v_1, v_2\}$. As a_2 has degree three or more, it is adjacent to a vertex in $\{v_1, v_2\}$. Adjust notation so that a_2 and v_2 are adjacent. The vertices a_2 and b_2 are also adjacent. Thus, $\{b_2, v_2\} \subset N(a_2)$. It follows that $N(a_2) - \{b_2, v_2\}$ is a nonempty subset of $\{b_1, v_1\}$. See Figure 9.12.

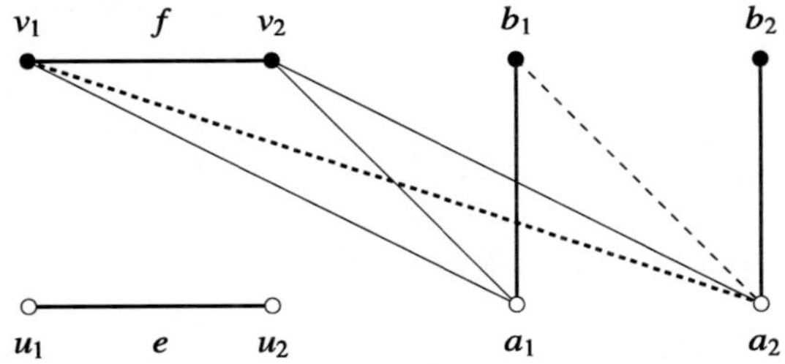

Fig. 9.12 The adjacencies of a_1 and a_2. At least one of the two dashed lines represents an edge of G.

Likewise the vertex b_{k-1} is adjacent to a_{k-1} and possibly to a_k, but it is not adjacent to any vertex in $A' - \{a_{k-1}, a_k\}$. Thus, $N(b_{k-1}) \subseteq \{a_k, a_{k-1}, u_1, u_2\}$. As b_{k-1} has degree three or more, it is adjacent to a vertex in $\{u_1, u_2\}$. Adjust notation so that b_{k-1} and u_2 are adjacent. The vertices b_{k-1} and a_{k-1} are also adjacent. Thus, $\{a_{k-1}, u_2\} \subset N(b_{k-1})$. It follows that $N(b_{k-1}) - \{a_{k-1}, u_2\}$ is a nonempty subset of $\{a_k, u_1\}$. See Figure 9.13.

Exercise 8.1.8 and statement 9.17.1 immediately imply the next assertion (Exercise 9.3.2).

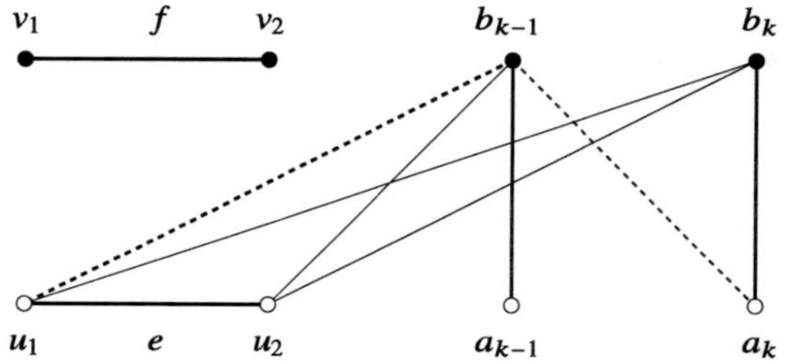

Fig. 9.13 The adjacencies of b_{k-1} and b_k. At least one of the two dashed lines represents an edge of G.

9.17.3 *Let C be a quadrilateral of H and let v be a vertex of C. If v has degree three or more in H and the sets $\partial(v) \cap E(C)$ and M are disjoint then at least one edge of $\partial(v) \cap E(C)$ is removable in G.* □

9.17.4 *If a_2 and v_1 are adjacent then one of the edges a_1v_1 and a_1v_2 is removable in G.*

Proof Suppose that a_2 and v_1 are adjacent. Then H has the quadrilateral $C := a_2v_1a_1v_2a_2$. Neither a_1v_1 nor a_1v_2 is in M; thus one of them is removable in G, by statement 9.17.3. □

9.17.5 *If a_2 and b_1 are adjacent then one of the edges a_2b_1 and a_2v_2 is removable in G.*

Proof Suppose that a_2 and b_1 are adjacent. Then H has the quadrilateral $C := a_2b_1a_1v_2a_2$. Neither a_2b_1 nor a_2v_2 is in M, thus one of them is removable in G, by statement 9.17.3. □

A reasoning similar to that used in the proofs of statements 9.17.4 and 9.17.5 may be used to prove the next two assertions (Exercise 9.3.3).

9.17.6 *If b_{k-1} and u_1 are adjacent then one of the edges b_ku_1 and b_ku_2 is removable in G.* □

9.17.7 *If b_{k-1} and a_k are adjacent then one of the edges $b_{k-1}a_k$ and $b_{k-1}u_2$ is removable in G.* □

Case 3 *Either a_2 is adjacent to v_1 or b_{k-1} is adjacent to u_1.*

Suppose that a_2 is adjacent to v_1. From statement 9.17.4 we infer that one of the edges a_1v_1 and a_1v_2, say, e', is removable in G. We have seen that b_{k-1} is adjacent to a vertex in $\{a_k, u_1\}$. Thus, statements 9.17.6 and 9.17.7 imply that at least one of the edges b_ku_1, b_ku_2, $b_{k-1}a_k$ and $b_{k-1}u_2$, say, e'', is removable in G. The edges e' and e'' are distinct and nonadjacent. Likewise, if b_{k-1} and u_1 are adjacent then one of the edges b_ku_1 and b_ku_2, say, f', is removable in G and one of the edges a_1v_1, a_1v_2, a_2b_1 and a_2v_2, say, f'', is removable in G. The edges f' and f'' are distinct and nonadjacent. In both alternatives, we concluded that G has two nonadjacent removable edges. The assertion holds in this case.

We may thus assume that

$$N(a_2) = \{b_1, b_2, v_2\} \quad \text{and} \quad N(b_{k-1}) = \{a_k, a_{k-1}, u_2\}. \tag{9.6}$$

Case 4 $k \geq 3$.

From statement 9.17.5 we infer that one of the edges a_2b_1 and a_2v_2, say, e', is removable in G. Likewise, statement 9.17.7 implies that one of the edges $b_{k-1}a_k$ and $b_{k-1}u_2$, say, e'', is removable in G. As $k \geq 3$, the edges e' and e'' are distinct and nonadjacent. We conclude that G has two nonadjacent removable edges. The assertion holds in this case.

Case 5 $k = 2$.

In this case, in view of statement 9.17.2 and equation (9.6), we infer that either $u_1v_1 \in E$ or $\{u_1v_2, u_2v_1\} \subset E$. See Figure 9.14.

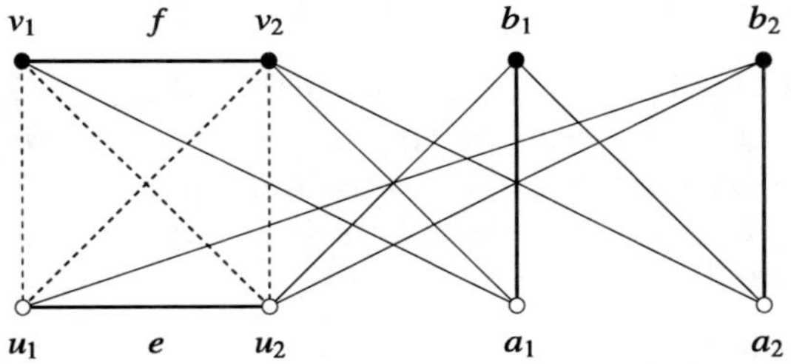

Fig. 9.14 The adjacencies of u_1, u_2, v_1 and v_2. Either $u_1v_1 \in E$ or $\{u_1v_2, u_2v_1\} \subset E$. Vertices u_2 and v_2 may also be adjacent.

Suppose that $u_1v_1 \notin E$. Then, $u_1v_2, u_2v_1 \in E$. In that case, H has two vertex-disjoint quadrilaterals, $C_1 := a_1v_1u_2b_1a_1$ and $C_2 = a_2v_2u_1b_2a_2$. By statement 9.17.3, one of the edges u_2b_1 and u_2v_1 is removable in G. Likewise, one of the edges v_2a_2 and v_2u_1 is removable in G. (In fact, u_2b_1 and v_2a_2 are both removable in G.)

Alternatively, if $u_1v_1 \in E$ then $\mathbb{H}_8$ is a spanning subgraph of G and the edge a_2b_1 is removable in $\mathbb{H}_8$ and in G. Thus, either G is the bicorn or any edge of $G - E(\mathbb{H}_8)$ is an edge that joins a vertex in $\{u_1, u_2\}$ to a vertex in $\{v_1, v_2\}$ and is thus removable in G and not adjacent to a_2b_1.

In all cases considered, we concluded that either $G \in \{K_4, \overline{C_6}, \mathbb{H}_8\}$ or G has two nonadjacent removable edges. □

We note that the conclusion of Theorem 9.17 does not hold without the hypothesis that the brick G is near-bipartite. There exist infinite families of bricks in which all removable edges are incident with a single vertex. One such family consists of the odd wheels $W_n, n \geq 5$. A particular example is shown in Figure 9.15 (see Exercise 9.3.5). In Chapter 11 we shall present a construction of another infinite family of bricks which do not have nonadjacent removable edges.

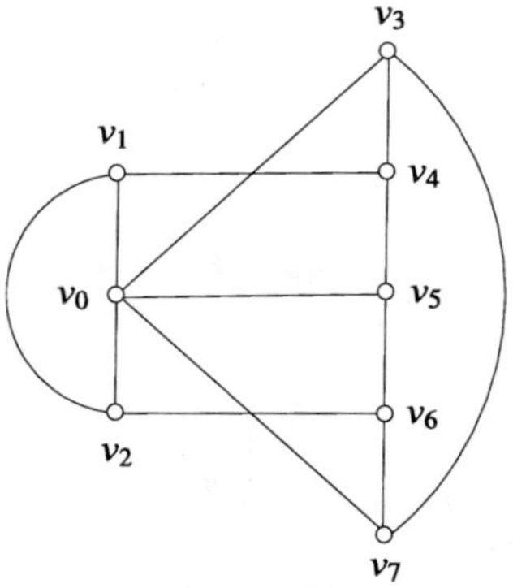

Fig. 9.15 A brick in which all removable edges are incident with one vertex.

9.3.1 Nonremovable edges in near-bipartite bricks

We now consider near-bipartite bricks G, with a removable doubleton R, in which an edge is removable in $G - R$ but is not removable in G.

Theorem 9.18 (CLM [8]) *Let G be a brick and let $R := \{e, f\}$ be a removable doubleton of G. Let (A, B) denote the bipartition of $G - R$, where e has both ends in A and f has both ends in B. Let h be a removable edge of $G - R$. Edge h is not removable in G if and only if A and B have partitions (A_1, A_2) and (B_1, B_2), as illustrated in Figure 9.16, such that:*

(i) $|A_1| = |B_1| + 1$ *(and $|B_2| = |A_2| + 1$),*
(ii) *edge e has both ends in A_1,*
(iii) *edge f has both ends in B_2, and*
(iv) *edge h is the only edge of G that joins a vertex of B_1 to a vertex of A_2.*

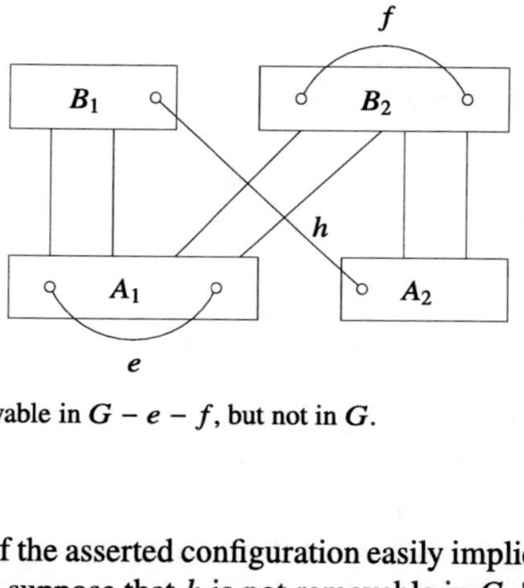

Fig. 9.16 Edge h is removable in $G - e - f$, but not in G.

Proof The existence of the asserted configuration easily implies that h is not removable in G. Conversely, suppose that h is not removable in G. Then, e and f are the only edges of G that depend on h (in addition to h itself). Let B_2 denote a maximal barrier of $G - h$ that contains both ends of edge f.

The graph $G - f - h$ is not bipartite. To see this, suppose the contrary. Then, both pairs $\{e, f\}$ and $\{f, h\}$ would be complements of cuts of G; hence $\{e, h\} = (E - \{e, f\}) \triangle (E - \{f, h\})$ would be a cut of G, a contradiction, as G, a brick, is 3-connected. Thus, $J := G - h - B_2$ has nontrivial components.

Let K be a nontrivial component of J. The maximality of B_2 implies that K is critical; thus, K is not bipartite. As $G - R$ is bipartite, it follows that $K - e$ is bipartite. We deduce that $e \in E(K)$. This conclusion implies that K is the only nontrivial component of J. Let A_2 denote the set of isolated vertices of J. Then, $|B_2| = |A_2| + 1$. Let $A_1 := A - A_2$ and let $B_1 := B - B_2$. Then, $|A_1| = |B_1| + 1$. The edge e has both ends in K and also both ends in A. Thus, e has both ends in A_1.

The edge h has one end in A and one end in B. But h has at least one end in A_2 and no end in B_2. We conclude that h joins a vertex in B_1 to a vertex in A_2. Finally, each edge of $\partial_G(K) - h$ is incident with a vertex in B_2 and has one end in $A - A_2$; hence each edge of $\partial_G(K) - h$ joins a vertex of A_1 to a vertex of B_2. □

Exercises

▷**9.3.1** *In the proof of Theorem 9.17, in the case where M is unique and $k = 1$, prove that (i) G has $\overline{C_6}$ as a spanning subgraph and (ii) if $\overline{C_6}$ is a proper subgraph of G then G has two nonadjacent removable edges.*

9.3.2 *Prove statement 9.17.3.*

9.3.3 *Prove statements 9.17.6 and 9.17.7.*

▷**9.3.4** *Let G be a matching covered graph distinct from K_4, $\overline{C_6}$ and $\mathbb{H}_8$, and such that $\delta(G) \geq 3$. Prove the following properties:*

(i) *if G is a near-brick then it has at least two removable edges, and*
(ii) *if G is a simple near-bipartite graph then G has at least two nonadjacent removable edges.*

▷**9.3.5** *The graph shown in Figure 9.15 is a brick. Show that all the removable edges in it are incident with the vertex v_0.*

9.4 Notes

By Theorem 9.13, every brick G which is distinct from K_4 and $\overline{C_6}$ has a removable edge, that is, an edge e such that $G - e$ is matching covered. This result by itself is not sufficiently strong to be useful as an inductive tool for establishing properties of bricks because the deletion of an edge from a brick may result in a matching covered graph with more than one brick. A removable edge e in a brick G is *b-invariant* if $b(G - e) = 1$. (Recall that a matching covered graph with one brick is called a near-brick. Most properties of bricks that are of interest to us in this theory are also shared by near-bricks.) Motivated by his work on the matching lattice, Lovász conjectured in 1987 that every brick distinct from K_4, $\overline{C_6}$ and the Petersen graph has a b-invariant edge. We shall present a proof of this result and several of its implications in a future chapter. One of the interesting implications is the fact that the dimension of the matching space $\mathcal{L}in(G)$ of a matching covered graph G over the field GF(2) is $m - n + 2 - (b(G) + p(G))$. (This formula is an example of an assertion about matching covered graphs in which the parameter $b(G) + p(G)$ plays an essential role.)

The lower bounds for the number of removable edges in a brick established in Corollary 9.7 and Theorem 9.14 are both obtained by just considering the minimal classes induced by the Δ edges incident with one vertex of maximum degree. We find it difficult to believe that such a 'local argument' could provide a tight bound for the number of removable edges in bricks of large orders!

Conjecture 9.19 *There exist a positive real number c and an integer N such that any brick G of order $n \geq N$ has at least cn removable edges.*

It would be of interest to characterize bricks in which all removable edges are incident with a single vertex.

Chapter 10
Removable Classes in Solid Bricks

Contents

10.1 Separating Cuts and Dulmage-Mendelsohn Barriers

Recall that a matching covered graph is solid if every separating cut in it is a tight cut and, in particular, that a brick is solid if it is free of nontrivial separating cuts. If a brick G is not solid, then it is easy to see that it has a nontrivial separating cut C such that at least one of the two C-contractions of G is a solid brick (Exercise 10.1.4). Because of this, properties of bricks which are solid may sometimes be used to analyse properties of those which are not. It is this idea that first led CLM (2002, [9]) to the study of solid bricks. There was one particular result concerning removable edges in solid bricks (called the Lemma on Odd Wheels (2002, [10])) which served as an essential tool in our resolution of a conjecture of Lovász in the above cited papers. According to that result, if G is a simple solid brick in which all removable edges are incident with a vertex v_0, then G is a wheel with v_0 as its hub. In this section, we shall derive a stronger version of that result (called here the Lemma on Wheels) from a general assertion related to nonremovable edges in bicritical graphs. The proof of this assertion makes crucial use of Dulmage-Mendelsohn barriers which were introduced in Chapter 5.

C. L. Lucchesi, U. S. R. Murty, *Perfect Matchings*, Algorithms and Computation in Mathematics 31, https://doi.org/10.1007/978-3-031-47504-7_10

10.1.1 Nonremovable edges in bicritical graphs

If an edge f of a nontrivial matching covered graph G is not removable, by definition, there must be unmatchable edges in $G - f$. Thus, $G - f$ must necessarily contain a barrier which includes both ends of some edge. All the known criteria for deciding whether or not a given edge of a matching covered graph is removable are based on the above observation.

Let B denote a barrier of a matchable graph G. In Section 3.2 we defined the core of G associated with B to be the bipartite graph $\mathbb{H}(B)$ obtained by deleting all edges with both ends in B, deleting all vertices in the even components of $G - B$, and then shrinking the vertex set of each odd component of $G - B$ to a single vertex. In Section 5.1 we defined a barrier B to be a Dulmage-Mendelsohn barrier of G if

- the core $\mathbb{H}(B)$ of G with respect to B is matching covered and
- each odd component of $G - B$ is critical.

NONREMOVABLE EDGES AND DULMAGE-MENDELSOHN BARRIERS

Theorem 10.1 (CLM (2012, [16])) *Let G be a bicritical graph of order at least four, and let f be a nonremovable edge of G. Then, $G - f$ contains a Dulmage-Mendelsohn barrier B that satisfies the following properties:*

(i) *edge f has its ends in distinct odd components of $G - f - B$, and*
(ii) *for each odd component K of $G - f - B$, the cut $C_K := \partial_G(K)$ is separating in G.*

Proof Since G has order four or more and as f is not removable, the graph $G - f$ has nonmatchable edges. Let e be an edge of $G - f$ that depends on f in G. By Proposition 8.10, $G - f$ has barriers that contain both ends of e. Let $B^\star$ be a maximal barrier of $G - f$ that contains both ends of e. By Lemma 5.4, $B^\star$ includes a Dulmage-Mendelsohn barrier, say B. Let $\mathbb{H} := \mathbb{H}(B)$ denote the core of G associated with B.

By Proposition 8.11, f has its ends in distinct odd components of $G - f - B^\star$. If f has an end v in an odd component of $G - f - B^\star$ which is not also an odd component of $G - f - B$, then $B + v$ would be a nontrivial barrier of G, in contradiction to the hypothesis that G is bicritical. Thus, f has its two ends in distinct odd components of $G - f - B$. This proves part (i). See Figure 10.1.

The proof of part (ii) is less straightforward. We begin by establishing the notation which is convenient for explaining the steps involved in its proof:

- E_B denotes the set of edges that have both ends in B,
- C denotes the set of edges that join a vertex in B to a vertex in some even component of $G - f - B$,
- $\mathcal{O}$ denotes the set of odd components of $G - f - B$,

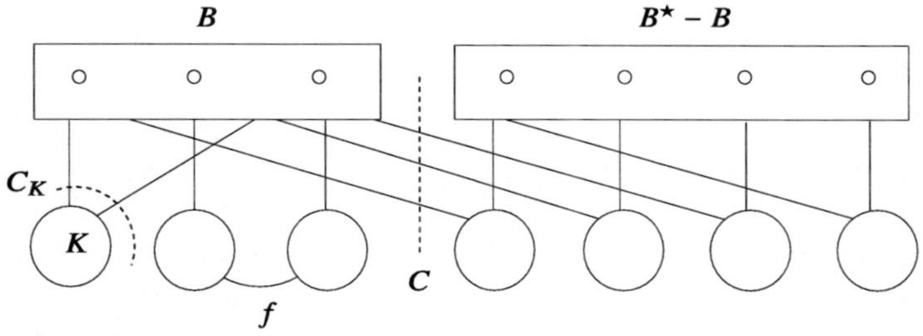

Fig. 10.1 A DM barrier B in $G - f$

- for each $K \in \mathcal{O}$,
  - C_K denotes the cut $\partial_G(K)$ of G,
  - G_K denotes the C_K-contraction $G/(\overline{V(K)} \to \overline{v_K})$ of G,
  - G'_K denotes the C_K-contraction $G/(V(K) \to v_K)$ of G, and
- $\mathbb{L}$ denotes the graph obtained from G by contracting the vertex set of each component in $\mathcal{O}$ to a single vertex.

10.1.1 *The set $E_B \cup C$ is nonempty.*

Proof If $B = B^\star$ then e, an edge having both ends in $B^\star$, has both ends in B. In that case, $e \in E_B$. Alternatively, if $B \subset B^\star$ then, as G is connected, the cut C is nonempty. □

In order to show that C_K is a separating cut of G for each $K \in \mathcal{O}$, we show that both C_K-contractions of G are matching covered. The following lemma shows that G_K is bicritical, which clearly implies that it is matching covered.

Lemma 10.2 *For any component K in $\mathcal{O}$, the graph G_K is bicritical.*

Proof Let K be any component in $\mathcal{O}$. As B is a DM-barrier, by definition of DM-barrier we infer that K is critical. The vertex set of G_K is $V(K) + \overline{v_K}$. So, for any vertex $v \in V(K)$, the graph $G_K - \{v, \overline{v_K}\}$ has a perfect matching. Hence $\overline{v_K}$ is not contained in any nontrivial barrier of G_K. On the other hand, any subset of $V(K)$ that is a nontrivial barrier of G_K is also a nontrivial barrier of G. This is impossible because, by hypothesis, G is bicritical. We conclude that G_K is bicritical. □

For any particular $K \in \mathcal{O}$, the C_K-contraction G'_K of G is obtained by shrinking $V(K)$ to a single vertex. The following lemma shows that $\mathbb{L}$, which is obtained from G by shrinking each $V(K)$, for $K \in \mathcal{O}$, to a single vertex, is matching covered. We shall see then that this, together with the above lemma, implies that G'_K is matching covered.

Lemma 10.3 *The graph $\mathbb{L}$ is matching covered.*

Proof The graph $\mathbb{L}$ is obtained from G by the contraction of some sets of vertices. Thus, $\mathbb{L}$ is connected. Let us prove that every edge of $\mathbb{L}$ is matchable. Towards this end, let g be an edge of $\mathbb{L} - f$, and let M_g be a perfect matching of G that contains it.

Case 1 *M_g contains an edge in $E_B \cup C$.*

By Lemma 8.12,

$$2|M_g \cap E_B| + |M_g \cap C| \leq 2|M_g \cap \{f\}|, \tag{10.1}$$

with equality if and only if $|M_g \cap \partial_G(K)| = 1$ for each component K in $\mathcal{O}$. If M_g contains an edge in C, in turn an even cut, then in fact it contains two edges in C. It follows that equality holds in (10.1); hence $M_g \cap E(\mathbb{L})$ is a perfect matching of $\mathbb{L}$ that contains edges f and g.

Case 2 *M_g and $E_B \cup C$ are disjoint.*

Let J be the graph $\mathbb{L} - V(\mathbb{H})$. As M_g and C are disjoint, the set $M_J := M_g \cap E(J)$ is a perfect matching of J. Suppose that g is an edge of J. The core $\mathbb{H}$ is matching covered. Let M be any perfect matching of $\mathbb{H}$. The set $M \cup M_J$ is a perfect matching of $\mathbb{L}$ that contains the edge g.

Alternatively, suppose that g is not an edge of J. As M_g and $E_B \cup C$ are disjoint, we infer that g is an edge of $\mathbb{H}$. As $\mathbb{H}$ is matching covered, let N be a perfect matching of $\mathbb{H}$ that contains the edge g. Then, $N \cup M_J$ is a perfect matching of $\mathbb{L}$ that contains the edge g.

In all cases considered, every edge of $\mathbb{L} - f$ is matchable in $\mathbb{L}$. Moreover, as noted before, any perfect matching of $\mathbb{L}$ that contains an edge in the nonempty set $E_B \cup C$ also contains the edge f. Every edge of the connected graph $\mathbb{L}$ is matchable, hence $\mathbb{L}$ is matching covered. □

Using Lemma 10.3, we now proceed to prove that the C_K-contraction G'_K of G is matching covered for any component $K \in \mathcal{O}$. For every $K \in \mathcal{O}$, the graph G_K is matching covered, by Lemma 10.2. Thus, splicing $\mathbb{L}$ with G_J, for each $J \in \mathcal{O} - K$, yields the matching covered graph $G'_K = G/(V(K) \rightarrow v_K)$. The other C_K-contraction $G_K = G/(\overline{V(K)} \rightarrow \overline{v_K})$ is, as noted earlier, also a matching covered graph. We conclude that C_K is a separating cut. As this is true for any $K \in \mathcal{O}$, it establishes the validity of part (ii) of the assertion of the theorem. □

We note that part (ii) of the above theorem is not valid in general for components of $G - f - B^*$. In Figure 10.2, we give an example of a bicritical graph G, a nonremovable edge f of G, a maximal barrier $B^\star$ of $G - f$, a component K of $G - f - B^\star$ and the unique perfect matching of G that contains edge g. One of the C_K-contractions of G is K_4, up to multiple edges, but the other C_K-contraction $G/(V(K) \rightarrow k)$ of G is not matching covered, and hence C_K is not a separating cut of G (Exercise 10.1.1).

10.1.2 Removable edges in solid bricks

We shall now use Theorem 10.1 to derive useful results concerning removable edges in solid bricks. We begin with a simple consequence of that theorem, with edge e playing the role of f. (We leave its proof as Exercise 10.1.5.)

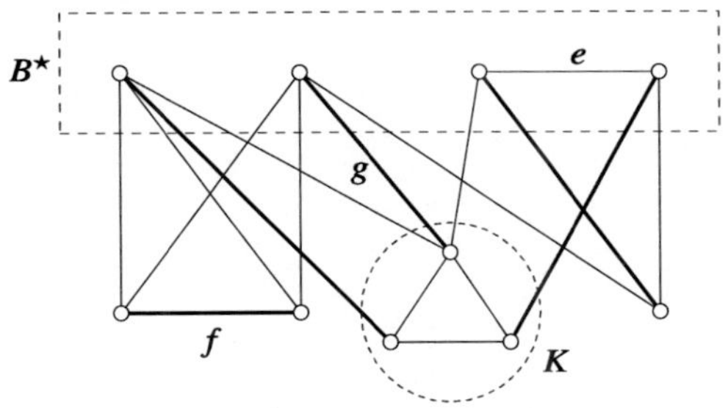

Fig. 10.2 A nonremovable edge in a bicritical graph

Corollary 10.4 *Let G be a solid brick, and let e be a nonremovable edge of G. Then, $G - e$ has a Dulmage-Mendelsohn barrier B such that every odd component of $G - e - B$ is trivial and edge e has its ends in distinct odd components of $G - e - B$.* □

Theorem 10.5 *Let G be a solid brick and let v be a vertex of G. Suppose that vv_1 and vv_2 are distinct nonremovable edges of G. Then, v has exactly three neighbours, v_1, v_2 and v_3. Moreover, for $i = 1, 2$, there exists an equipartition (B_i, I_i) of $V(G)$ such that*

(i) *vv_i is the only edge of G that has both ends in I_i,*
(ii) *every edge that has both ends in B_i is incident with v_3,*
(iii) *the subgraph H_i of G, obtained by the removal of e_i and each edge having both ends in B_i, is matching covered and bipartite, with bipartition $\{B_i, I_i\}$, and*
(iv) *$B_1 = I_2 - v + v_3$ and $B_2 = I_1 - v + v_3$.*

(See Figure 10.3 for an illustration.)

Proof Let d be the degree of v. Enumerate the d edges of $\partial(v)$ as $e_1, e_2, \ldots, e_d$, where e_i joins v to v_i, for $i = 1, 2, \ldots, d$. By hypothesis, neither e_1 nor e_2 is removable in G. By Corollary 10.4, for $i = 1, 2$, graph $G - e_i$ has a Dulmage-Mendelsohn barrier B_i such that (i) each odd component of $G - e_i - B_i$ is trivial and (ii) edge e_i has its ends in distinct odd components of $G - e_i - B_i$.

The following result will be used in this proof. (Exercise 10.1.3).

Lemma 10.6 *Let G be a bicritical graph, let e_1 and e_2 be two adjacent edges of G and let B_i be a barrier of $G - e_i$, for $i = 1, 2$. Then, $|B_1 \cap B_2| \leq 1$.* □

Let us now prove that v_1, v_2 and v_3 are the only three neighbours of v. Let us also prove that $B_1 \cap B_2 = \{v_3\}$. As e_i has both ends in odd components of $G - e_i - B_i$, and since each such odd component is trivial, it follows that v is the vertex of a trivial component of $G - e_i - B_i$. Clearly, e_i is the only edge of G that has its ends in distinct components of $G - e_i - B_i$. It follows that $\{v_3, v_4, \ldots, v_d\}$ is a subset of $B_1 \cap B_2$. By Lemma 10.6, $B_1 \cap B_2$ is a singleton. We deduce that v has only three neighbours and $B_1 \cap B_2 = \{v_3\}$.

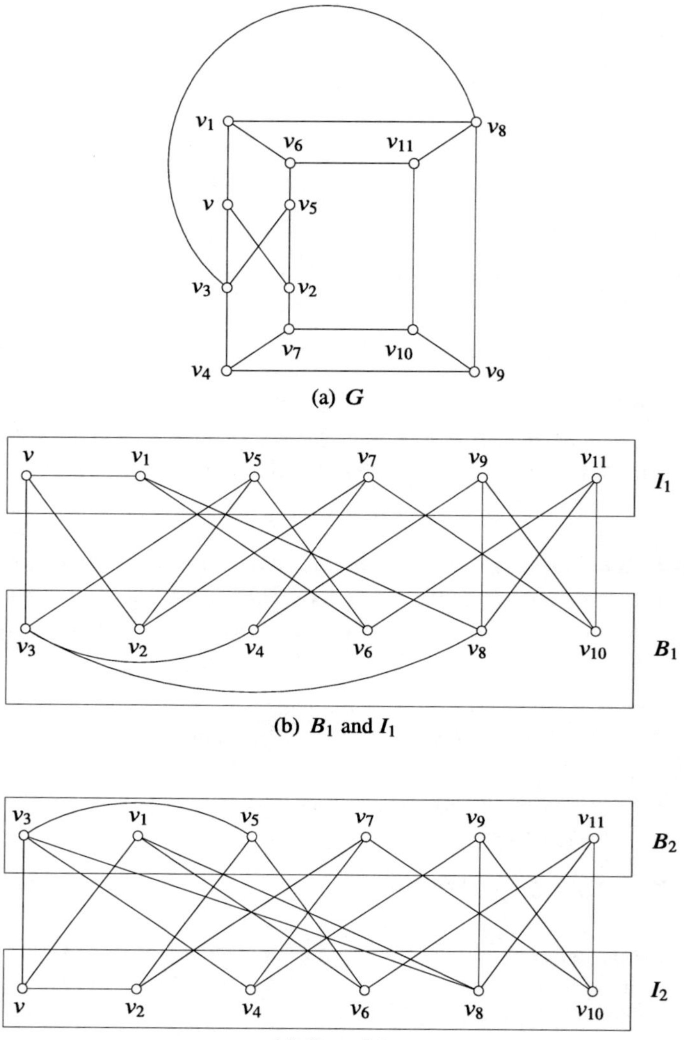

(a) G

(b) B_1 and I_1

(c) B_2 and I_2

Fig. 10.3 The solid brick G and the pairs B_1, I_1 and B_2, I_2

For $i = 1, 2$, if x is a vertex in an even component of $G - e_i - B_i$, then $B_i + x$ is a barrier of $G - e_i$. Using this fact, we now proceed to show that $G - e_i - B_i$ has no even components.

Suppose that $G - e_1 - B_1$ has even components and that xy is an edge of such a component. If x is in an even component of $G - e_2 - B_2$, then $B_1 + x$ and $B_2 + x$ are barriers of $G - e_1$ and $G - e_2$, respectively, and $(B_1 + x) \cap (B_2 + x) = \{v_3, x\}$. Similarly, if x is in B_2, $B_1 + x$ and B_2 are barriers of $G - e_1$ and $G - e_2$, respectively, and $(B_1 + x) \cap B_2 = \{v_3, x\}$. In either case, we have a contradiction to Lemma 10.6.

It follows that x is neither in an even component of $G - e_2 - B_2$, nor in B_2; hence x is in an odd component of $G - e_2 - B_2$. The odd components of $G - e_2 - B_2$ are trivial; hence x is the only vertex of an odd component of $G - e_2 - B_2$. Likewise, y is also the only vertex of an odd component of $G - e_2 - B_2$. As the vertices x and y are adjacent, we derive a contradiction. We conclude that $G - e_1 - B_1$ and, similarly, $G - e_2 - B_2$, have no even components.

For $i = 1, 2$, now let $I_i := V(G) - B_i$. Then, I_i is the set of isolated vertices of $G - e_i - B_i$ and edge e_i is the only edge of G that has both ends in I_i. Moreover, H_i is the core of G associated with B_i, in turn a Dulmage-Mendelsohn barrier; therefore H_i is matching covered. As $B_1 \cap B_2 = \{v_3\}$, and since v lies in $I_1 \cap I_2$, it follows that $B_1 = I_2 - v + v_3$ and $B_2 = I_1 - v + v_3$. As e_1 is the only edge of G having both ends in I_1, it follows that every edge having both ends in B_2 is incident with v_3. Likewise, every edge having both ends in B_1 is incident with v_3. □

We leave the proofs of the following two corollaries, proved in [16], as Exercise 10.1.6.

Corollary 10.7 *If G is a solid brick with six vertices or more then every vertex of G is incident with at most two nonremovable edges of G.* □

Corollary 10.8 *If G is a solid brick of maximum degree three or four, then, for every vertex v of G at most one edge incident with v does not lie in a removable class of G.* □

10.1.3 The Lemma on wheels

For a fixed vertex v_0 of a graph G, a subset M of the edges of G is a *v_0-matching* if $|M \cap \partial(v)| = 1$ for each vertex v distinct from v_0, and if $|M \cap \partial(v_0)| > 1$. We say that G is a *v_0-wheel* if G is a wheel having v_0 as a hub. The following is easily verified using a counting argument:

Proposition 10.9 *If a matching covered graph G has a v_0-matching for some vertex v_0 then G is not bipartite.* □

LEMMA ON WHEELS

Lemma 10.10 *Let v_0 be a vertex of a simple solid brick G, and let M_0 be a v_0-matching of G. Then either G is a v_0-wheel or G has two removable edges $e_1, e_2 \notin M_0 \cup \partial(v_0)$.*

Proof Let $H := G - v_0$. For each vertex v of H, let $d_H(v)$ denote the degree of v in H. As G is a brick, every vertex of G has degree three or more. Thus, for every vertex v of H, $d_H(v) \geq 2$, with equality only if v is adjacent to v_0.

Case 1 *The graph H is 2-regular.*

Each component of H is then a cycle. As G is matching covered, it is 2-connected; hence H is connected. It follows that H is a cycle. Every vertex of H is adjacent to v_0; hence G is a v_0-wheel.

10.10.1 *Let v be a vertex of H such that $d_H(v) \geq 3$. Then, at most one edge of $\partial_H(v) - M_0$ is not removable in G.*

Proof Let d be the degree of v in G and let $e_1, e_2, \ldots, e_d$ be the edges of $\partial_G(v)$, where each e_i joins v to a vertex v_i.

Assume, to the contrary, that $\partial_H(v) - M_0$ contains two edges, say e_1 and e_2, that are not removable in G. As G is simple, from Theorem 10.5 we infer that $d = 3$ and G has an equipartition (B_1, I_1) such that e_1 is the only edge of G having both ends in I_1 and every edge of G that has both ends in B_1 is incident with v_3. Moreover, the graph H_1, obtained from G by removing e_1 and each edge having both ends in B_1, is matching covered.

The edges e_1 and e_2 are not in M_0; hence e_3 is the edge of M_0 incident with v. The edge e_3 is an edge of H; hence $v_3 \neq v_0$. It follows that vv_3 is the only edge of M_0 incident with v_3. Thus, no edge of M_0 has both ends in B_1. We conclude that M_0 is a v_0-matching of the bipartite matching covered graph H_1, in contradiction to Proposition 10.9. □

Case 2 *The graph H has a vertex v such that $d_H(v) \geq 4$.*

The set $\partial_H(v) - M_0$ contains at least three edges, of which at least two are removable in G, by statement 10.10.1. The assertion holds in this case.

We have seen that $d_H(v) \geq 2$, for each vertex v of H. We may assume that the previous cases do not apply; therefore $d_H(v) \in \{2, 3\}$ for each vertex v of H. Moreover, at least one vertex v has degree three in H.

Case 3 *The graph H has a vertex v such that $d_H(v) = 3$.*

We shall now prove that G has two removable edges which are not in $M_0 \cup \partial(v_0)$. We have seen that $d_H(u) \in \{2, 3\}$, for each vertex u in H. The number of vertices of odd degree in H is even. Thus, H has a vertex v' distinct from v such that $d_H(v') = 3$.

At least two edges of $\partial_H(v)$ are not in M_0. By statement 10.10.1, $\partial_H(v) - M_0$ contains an edge, e, which is removable in G. Likewise, $\partial(v') - M_0$ contains an edge, e', which is removable in G. If e and e' are distinct then the assertion holds.

We may thus assume that $e = e'$. If H has a third vertex, v'', such that $d_H(v'') = 3$ then again $\partial_H(v'') - M_0$ has an edge, e'', which is removable in G, and distinct from e. In that case, G has at least two removable edges not in $M_0 \cup \partial(v_0)$.

The assertion holds in every case considered, except the case in which v and v' are the only two vertices of H that have degree three in H; all the other vertices of H have degree two. Moreover, v and v' are joined by an edge, $e = e'$, which is removable in G.

As $G - e$ is matching covered, it is 2-connected; hence $H - e$ is connected. Every vertex of $H - e$ has degree two. We deduce that $H - e$ is a cycle, say

$$Q := v_1 v_2 \cdots v_s v_1, \quad s \geq 3, \quad s \text{ is odd.}$$

Adjust notation so that $v = v_1$, let t be such that $v' = v_t$. We may consider the reverse of Q, if necessary, so that t is odd. Thus, s and t are both odd. As G is simple, the vertices v and v' are not adjacent in $H - e$. Thus, $3 \leq t \leq s - 2$, hence $s \geq 5$. Figure 10.4 illustrates the case in which $t = 5$ and $s = 9$.

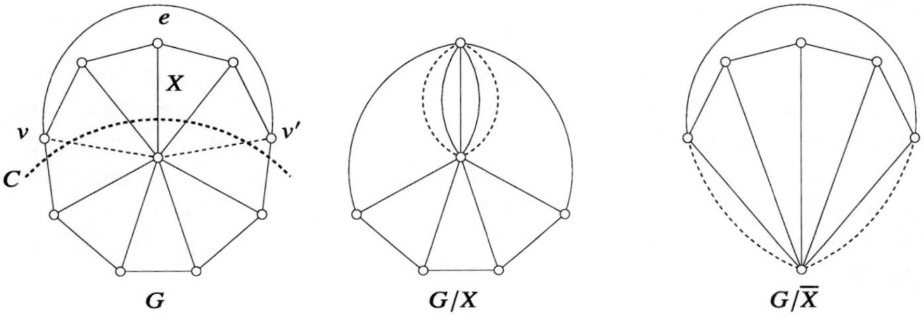

Fig. 10.4 The case $t = 5$ and $s = 9$. Dashed lines indicate the possibility of an edge.

Let $X := \{v_1, v_2, \ldots, v_t\}$ and let $C := \partial(X)$. As $3 \leq t \leq s - 2$, the cut C is nontrivial. We now show that C is separating, therefore obtaining a contradiction to the hypothesis that G is solid. The underlying simple graph of the C-contraction G/X is clearly a wheel of order $s - t + 2 \geq 4$. The underlying simple graph of the C-contraction $G/\overline{X}$ is also a wheel of order $t + 1 \geq 4$. We deduce that G is not solid, a contradiction. □

Figure 10.5 illustrates the necessity for G to be simple.

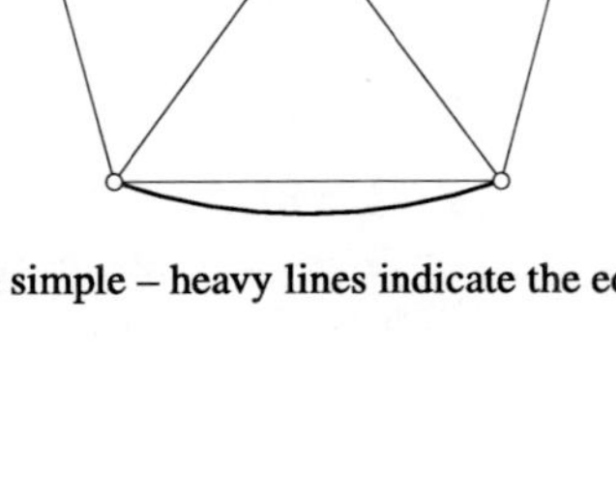

Fig. 10.5 The brick G must be simple – heavy lines indicate the edges of M_0

Corollary 10.11 *Let G be a solid brick and let v_0 be a vertex of G. Either G is a wheel having v_0 as a hub, up to multiple spokes, or G has two removable edges not incident with v_0.*

Proof Suppose that G is simple. Let v_0w and v_0x be two edges incident with v_0. The graph G, a brick, is bicritical. Thus, $G - w - x$ has a perfect matching, M. The set $M_0 := M \cup \{v_0w, v_0x\}$ is a v_0-matching of G. By Lemma 10.10, either G is a v_0-wheel or G has two removable edges not incident with v_0. The assertion holds if G is simple.

We may thus assume that G has multiple edges. If G has a pair of parallel edges not incident with v_0 then each one of these edges is removable in G and not incident with v_0. In that case, the assertion holds.

We may thus assume that every pair of parallel edges of G is incident with v_0. Let H be the underlying simple graph of G. Clearly, H is a simple solid brick. We have seen that the assertion holds for simple solid bricks. Thus, either H is a v_0-wheel or H has two removable edges not incident with v_0. If H is a v_0-wheel then G is a wheel having v_0 as a hub, up to multiple spokes. Alternatively, H has two removable edges not incident with v_0. These two edges are removable in G. The assertion holds in both alternatives. □

Exercises

▷**10.1.1** *Show that the cut $C_K := \partial(K)$ is not a separating cut in the bicritical graph shown in Figure 10.2.*

▷**10.1.2** *Verify that the graph G in Figure 10.3 is solid. Hint: see Example 7.2 and Exercise 7.1.6 to prove that G is odd-intercyclic.*

▷**10.1.3** *Prove Lemma 10.6. Hint: for any two vertices w and x of B_i, the graph $G - w - x$ has a perfect matching that contains the edge e_i.*

▷**10.1.4** *Show that every nonsolid brick G has a nontrivial separating cut C such that at least one of the two C-contractions of G is a solid brick.*

▷**10.1.5** *Give a proof of Corollary 10.4.*

∗**10.1.6** *Give proofs of Corollaries 10.7 and 10.8.*

WHEEL-LIKE BRICKS

A graph G is a *wheel-like-brick* if G is a brick and has a vertex h, called its *hub*, such that every removable class of G has an edge in $\partial(h)$. By Corollary 10.11 every solid wheel-like brick having h as a hub is an odd h-wheel. Example 10.12 shows a wheel-like brick which is not solid.

We do not know any characterization of wheel-like bricks.

Example 10.12 Figure 10.6 depicts an $\overline{x}$-wheel G_1, an x-wheel G_2 and the graph G, which is a splicing of G_1 at $\overline{x}$ with G_2 at vertex x. The graph G is a wheel-like brick having h as its hub. Thus, the converse of Corollary 10.11 does not hold in general.

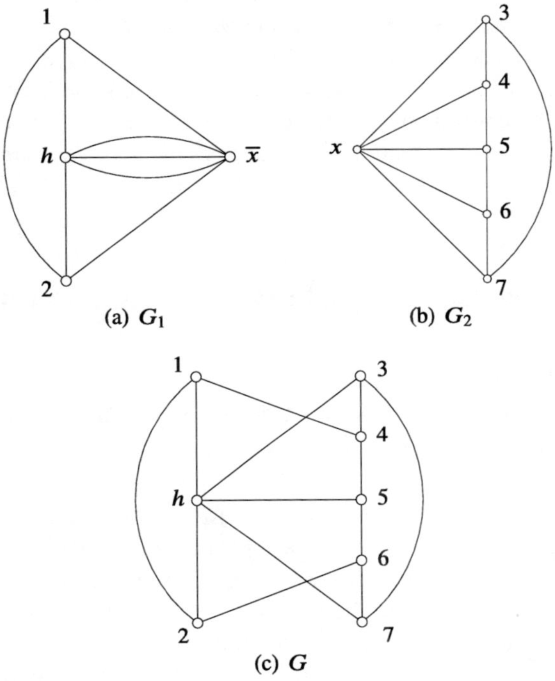

Fig. 10.6 The wheel-like brick G.

▷**10.1.7** *Prove that the graph G depicted in Figure 10.6 is a wheel-like brick having h as its hub.*

10.2 Some Implications of Being Solid

All bipartite matching covered graphs are solid, but not all solid matching covered graphs are bipartite. However, solid matching covered graphs have many properties which are akin to those of bipartite graphs. For example, perfect matching polytopes of solid bricks, as those of bipartite graphs, afford linear inequality descriptions which do not involve odd set constraints (see Theorem 7.5). In this section we shall establish three other properties of solid bricks which exemplify this phenomenon.

The proofs of the first two of these three theorems will follow from the assertion in the next section, whose proof, in turn, makes use of three results:

- Separating cuts with a bipartite shore. Theorem 4.7 asserts that if $C := \partial(X)$ is a separating cut of a matching covered graph G and the subgraph $G[X]$ of G induced by X is bipartite then C is a barrier cut of G; hence C is tight.
- The precedence relation defined in Section 7.2. Recall that a cut C of a matching covered graph G is said to precede another cut D of G if $|M \cap C| \leq |M \cap D|$, for each perfect matching M of G. If, in addition, $|N \cap C| < |N \cap D|$, for some perfect matching N of G, then C strictly precedes D. We write $C \preceq D$ to indicate that C precedes D, and $C \prec D$ to indicate that C strictly precedes D.
- Theorem 4.2. According to that theorem, a cut C of a matching covered graph G is a separating cut of G if and only if, for any edge e of G, there exists a perfect matching M_e of G such that $e \in M_e$ and $|M_e \cap C| = 1$.

10.2.1 Removable edges in solid bricks and separating cuts

Lemma 10.13 *Let G be a solid brick, let e be a removable edge of G and let D be a cut of G. If $D - e$ is a separating cut in $G - e$ then (precisely) one of the $(D - e)$-contractions of $G - e$ is bipartite.*

Proof Let us first prove that at least one $(D-e)$-contraction of $G-e$ is nonbipartite. According to Theorem 9.8, if e is any removable edge of a matching covered graph G, then $b(G-e) \geq b(G)$. In our case, as G is a brick, we must have that $b(G-e) \geq 1$. By Theorem 4.20, at least one of the $(D - e)$-contractions of $G - e$ is nonbipartite.

Having established that at least one of the $(D - e)$-contractions of $G - e$ is not bipartite, assume, to the contrary, that both $(D - e)$-contractions of $G - e$ are nonbipartite. Let $\mathcal{C}$ denote the collection of those odd cuts C of G that have the following properties:

(i) both $(C - e)$-contractions of $G - e$ are nonbipartite, and
(ii) $C \preceq D$.

10.13.1 *For every cut C in $\mathcal{C}$, the cut $C - e$ is separating in $G - e$.*

Proof By hypothesis, $D - e$ is separating in $G - e$. Let f be any edge of $G - e$. By Theorem 4.2, $G - e$ has a perfect matching M_f such that $f \in M_f$ and $|M_f \cap D| = 1$. As $C \preceq D$, and since C is odd, we deduce that $|M_f \cap C| = 1$. This conclusion holds for each edge f of $G - e$. By Theorem 4.2, $C - e$ is separating in $G - e$. □

Let $C := \partial(X)$ be a minimal cut in $\mathcal{C}$, relative to the precedence relation. Let $G_1 := G/(\overline{X} \to \overline{x})$ and $G_2 := G/(X \to x)$ be the two C-contractions of G. As both $(C-e)$-contractions of $G - e$ are nonbipartite, the cut C of G is nontrivial. As G is a solid brick by hypothesis, the cut C cannot be a separating cut of G. Thus the edge e must be unmatchable in one of the two C-contractions of G. (All other edges are matchable in both C-contractions of G because $C-e$ is separating in $G-e$.) Assume without loss of generality that the edge e is unmatchable in G_1. By Corollary 2.2, there must exist a maximal barrier of G_1, say B, which contains both ends of e.

Since G is a brick, B cannot be a barrier of G itself. This means that the contraction vertex $\overline{x}$ (which may or may not be an end of the edge e in G_1) is in B.

Let O denote the set of all odd components of $G_1 - B$ and, for $K \in O$, let C_K denote the cut $\partial(K)$. See Figure 10.7.

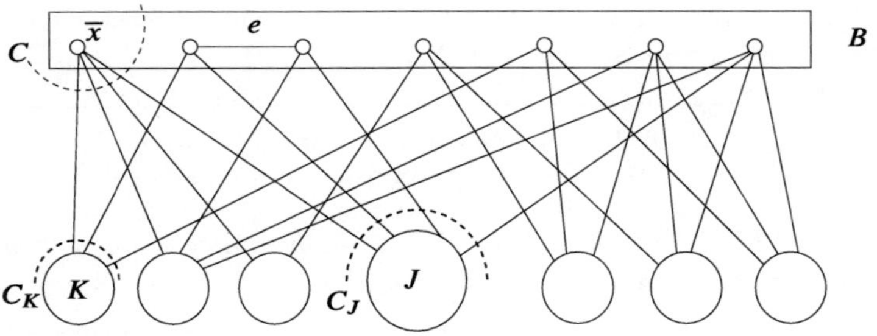

Fig. 10.7 Illustration for Lemma 10.13: B is a maximal barrier of G_1 which includes both ends of the edge e.

It can be verified by simple counting arguments that, for any perfect matching M of G, the following identity holds:

$$\sum_{K \in O} |M \cap C_K| = |M \cap C| + (|O| - 1) - 2|M \cap \{e\}|. \tag{10.2}$$

The above identity has the following implications:

10.13.2 *For each $K \in O$, the cut C_K strictly precedes C.*

Proof Let K be a component in O. From (10.2) we infer that $|M \cap C_K| \leq |M \cap C|$, for each perfect matching M of G. Moreover, equality holds only if $e \notin M$. Thus, for any perfect matching M_e of G that contains the edge e, $|M_e \cap C_K| < |M_e \cap C|$. We conclude that $C_K \prec C$. This conclusion holds for each $K \in O$. □

10.13.3 *There exists a nontrivial component $J \in O$ such that both $(C_J - e)$-contractions of $G - e$ are nonbipartite.*

Proof Let us first observe that, by hypothesis, $G_1 - e$ is not bipartite; hence $G_1 - B$ has a nontrivial component, J. The maximality of B implies that J is critical, hence nonbipartite. Thus, the C_J-contraction $(G - e)/\overline{V(J)}$ is nonbipartite. By hypothesis, the $(C - e)$-contraction $G_2 - e$ of $G - e$ is nonbipartite. By Lemma 4.6, the shore $\overline{X}$ of C induces a nonbipartite graph in $G - e$. Thus, the C_J-contraction $(G - e)/V(J)$ is nonbipartite. □

By statement 10.13.2, $C_J \prec C \preceq D$, and by statement 10.13.3, both $(C_J - e)$-contractions of $G - e$ are nonbipartite. This is in contradiction to the criteria used for choosing C. Thus, (precisely) one of the $(D - e)$-contractions of $G - e$ is nonbipartite. □

10.2.2 Propagation

If G is any bipartite matching covered graph and e is a removable edge of G then, clearly, $G - e$ is also a bipartite matching covered graph. Analogously, a solid matching covered graph bequeaths the property of being solid to any graph obtained from it by the deletion of a removable class. We first consider the case which involves the deletion of removable edges of a brick.

PROPERTY OF BEING SOLID IS HEREDITARY:
DELETIONS OF REMOVABLE EDGES OF A SOLID BRICK

Theorem 10.14 (CLM (2005, [13])) *Every graph obtained from a solid brick by the deletion of a removable edge is also solid.*

Proof Let e be a removable edge of a solid brick G. Let C be a cut of G such that $C - e$ is a separating cut of $G - e$. By Lemma 10.13, one of the $(C - e)$-contractions of $G - e$ is bipartite. By Theorem 4.7, $C - e$ is tight in $G - e$. This conclusion holds for each cut C of G such that $C - e$ is separating in $G - e$. We deduce that $G - e$ is solid. □

The following result can be proved fairly easily by induction on the number of edges of G. We leave it to the reader as Exercise 10.2.2.

PROPERTY OF BEING SOLID IS HEREDITARY:
DELETIONS OF REMOVABLE CLASSES

Theorem 10.15 *If G is a solid matching covered graph and R is a removable class of G, then $G - R$ is solid.* □

Theorem 10.15 may also be deduced from the result stated in Exercise 10.2.6.

10.2.3 b-Invariance (monotonicity revisited)

According to Theorem 9.8, if e is any removable edge of a matching covered graph G, then $b(G - e) \geq b(G)$. Recall that if equality holds, that is, if $b(G - e) = b(G)$, then we refer to e as a b-invariant edge. (See the Notes Section of Chapter 9 for an account of the origin of this notion.) Every brick distinct from K_4 and $\overline{C_6}$ has removable edges, but not every removable edge in a such a brick need be b-invariant. In fact, given any positive integer k, one can find an example of a brick G and a removable edge e of G such that $b(G - e) = k$ (see Exercise 9.1.3). However, a striking property of solid matching covered graphs is that every removable edge in such a graph is b-invariant. This can be deduced from the following assertion concerning solid bricks.

b-INVARIANCE OF REMOVABLE EDGES IN SOLID BRICKS

Theorem 10.16 (CLM (2002, [9])) *In a solid brick, every removable edge is b-invariant.*

Proof Let G be a solid brick, and let e be a removable edge of G. Assume that $b(G - e) \geq 2$. Since e is a removable edge of G, the graph $G - e$ is matching covered, and since $b(G - e) \geq 2$, the graph $G - e$ must have a nontrivial tight cut such that both cut contractions with respect to it are nonbipartite matching covered graphs. This is a contradiction to Lemma 10.13. □

The following result can be proved fairly easily by induction on the number of edges of G. We leave it to the reader as Exercise 10.2.3.

Theorem 10.17 *Let G be a solid matching covered graph. For each removable edge e of G,*

$$b(G - e) = b(G) \text{ and } (b + p)(G - e) = (b + p)(G). \qquad \square$$

Corollary 10.18 *Let G be a solid matching covered graph and let R be a nonempty set of edges of G such that $G - R$ is matching covered. Then,*

$$b(G - R) \geq b(G) + 1 - |R|,$$

with equality if and only if the set R is a removable class of G.

Proof by induction on $|R|$. If R is a removable class of G then equality holds, by Theorems 10.17 and 9.9. Suppose then that R is not a removable class of G. As $G - R$ is matching covered, the set R includes a removable class, S, of G. Let $H := G - S$ and let $T := R - S$. The graph $H - T$ is equal to the graph $G - R$, in turn a matching covered graph. Thus, the set T is a nonempty subset of $E(H)$ such that $H - T$ is matching covered. Moreover, by Theorem 10.15, the graph H is solid. By induction,

$$b(G - R) = b(H - T) \geq b(H) + 1 - |T| = b(G) + 2 - |S| - |T| = b(G) + 2 - |R|.$$

The asserted inequality is strict if R is not a removable class. □

10.2.4 The exchange property

Let G be a matching covered graph, let R be a removable class of G and let S be a removable class of $G - R$. The pair (R, S) has the *exchange property* if G has a removable class T and $G - T$ has a removable class W such that $T \subseteq S$ and $G - T - W = G - R - S$.

A matching covered graph G is said to have the *exchange property with respect to deletions of removable classes* if, for any given removable class R of G, and any removable class S of $G - R$, the pair (R, S) has the exchange property. A particular case of the exchange property holds for bipartite graphs (Exercise 8.1.9): if G is a bipartite matching covered graph, if an edge e is removable in G and an edge f is removable in $G - e$ then f is removable in G itself (and e is removable in $G - f$). We shall prove (Theorem 10.23) that, more generally, all solid matching covered graphs have the exchange property.

Example 10.19 Figure 10.8(a) depicts the graph $G := \Sigma_8$, and $R := \{e\}$ is a removable class of G; the graph $G - e$ is shown in Figure 10.8(b). The pair $S := \{f_1, f_2\}$ is a removable doubleton of $G - e$. Figure 10.8(c) depicts the graph $G - e - S$. The set $T := \{f_1\}$ is a removable class of G. The graph $G - f_1$ is shown in Figure 10.8(d). The set $W := \{e, f_2\}$, which is equal to $R \cup (S - T)$, is a removable doubleton of $G - f_1$. Indeed, (R, S) has the exchange property.

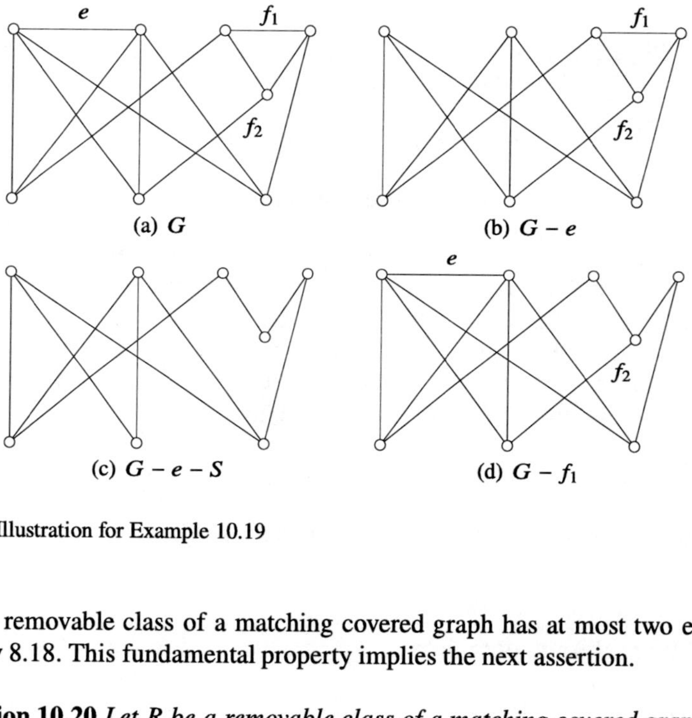

Fig. 10.8 Illustration for Example 10.19

Every removable class of a matching covered graph has at most two edges, by Corollary 8.18. This fundamental property implies the next assertion.

Proposition 10.20 *Let R be a removable class of a matching covered graph G and let S be a removable class of $G - R$. Suppose that (R, S) has the exchange property, let $T \subseteq S$ be a removable class of G and let $W := R \cup (S - T)$ be a removable class of $G - T$. The number of doubletons in $\{R, S\}$ is equal to the number of doubletons in $\{T, W\}$.*

Proof Each one of R and S has either one or two edges. Thus, the number of doubletons in $\{R, S\}$ is equal to $|R| + |S| - 2$. Likewise, the number of doubletons in $\{T, W\}$ is equal to $|W| + |T| - 2$. Moreover, $|R| + |S| = |W| + |T|$. The assertion holds. □

Let us now make a comment about requiring that W be a removable class, in the above definition of exchange property. One might ask that if T is a removable class of G then $G - T - W = G - R - S$, and $G - R - S$ is matching covered. The problem is that W may not be a removable class of $G - T$. This is illustrated in the next example.

Example 10.21 Consider the graph G depicted in Figure 10.9. The edge e is removable in G and the doubleton $\{f_1, f_2\}$ is removable in $G - e$. The edge f_1 is removable in G and the edge f_2 is removable in $G - f_1$ and, finally, the edge e is removable in $G - f_1 - f_2$. Thus, $(\{e\}, \{f_1, f_2\})$ does not have the exchange property, because $G - f_1 - \{e, f_2\}$ is matching covered, but $\{e, f_2\}$ is not a removable doubleton of $G - f_1$.

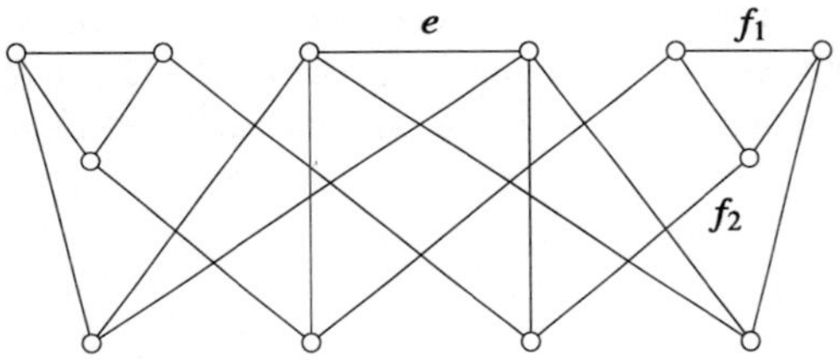

Fig. 10.9 Illustration for Example 10.21

The graph in Example 10.21 is not solid. In fact, it is not possible to exhibit a solid graph which does not have the exchange property, as the next assertion shows.

Proposition 10.22 *Let G be a solid matching covered graph, let R be a removable class of G and let S be a removable class of $G - R$. If S includes a removable class of G then (R, S) has the exchange property.*

Proof Assume that S includes a removable class, T, of G, let $H := G - T$ and let $W := R \cup (S - T)$. The graph $H - W$ is equal to the graph $G - R - S$, in turn a matching covered graph. Thus, W is a nonempty subset of $E(H)$ such that $H - W$ is matching covered. By Theorem 10.15, the graphs $G - R$ and H are solid. By Corollary 10.18,

$$\begin{aligned} b(H - W) &= b(G - R - S) = b(G - R) + 1 - |S| = b(G) + 2 - |R| - |S| \\ &= [b(H) - 1 + |T|] + 2 - |R| - |S| = b(H) + 1 - |W|. \end{aligned}$$

Thus, W is a removable class of H. We deduce that (R, S) has the exchange property.□

EXCHANGE PROPERTY OF SOLID MATCHING COVERED GRAPHS

Theorem 10.23 *Every solid matching covered graph has the exchange property with respect to deletions of removable classes.*

Proof Let G be a solid matching covered graph, let R be a removable class of G and let S be a removable class of $G - R$. We shall prove that (R, S) has the exchange property. In view of Proposition 10.22, it suffices to prove that S includes a removable class of G. We shall prove this by induction on $|V|$. The analysis of the basis of the induction hypothesis is done if G is free of nontrivial tight cuts.

Case 1 *The set S is a singleton, $\{f\}$.*

10.24 *No edge of R depends in G on edge f.*

Proof Assume, to the contrary, that R contains an edge, e, that depends on f in G. The graph $G - R - S$, which is matching covered, is a spanning subgraph of $G - f$; hence $G - f$ is matchable. By Proposition 8.10, $G - f$ has barriers that contain both ends of e. Let B denote a maximal barrier of $G - f$ containing both ends of e. Let E_B denote the set of edges that contain both ends in B. By Proposition 8.11, the edge f joins vertices in two different odd components of $G - f - B$. By Theorem 3.2, the maximality of B implies that each component of $G - f - B$ is odd. Let O denote the set of all (odd) components of $G - f - B$, and for each $K \in O$ let $C_K := \partial(K)$. See Figure 10.10.

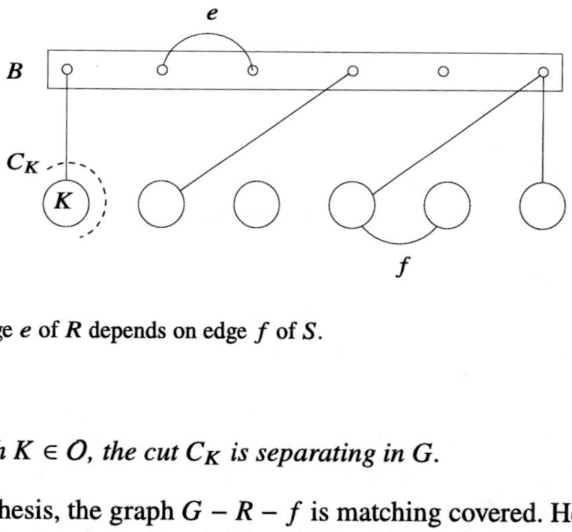

Fig. 10.10 The edge e of R depends on edge f of S.

10.24.1 *For each $K \in O$, the cut C_K is separating in G.*

Proof By hypothesis, the graph $G - R - f$ is matching covered. Hence every edge of $G - R - f$ is matchable in $G - f$. Thus, B is stable in $G - R - f$. The edge f has both ends in distinct components in O. Thus, $E_B \subseteq R$.

Let M be a perfect matching of G. The set $[\cup_{K \in O}(M \cap C_K)] - (M \cap \{f\})$ is equal to $M \cap \partial(B)$. Moreover, $|B| = |O|$. Thus,

$$\sum_{K \in O} |M \cap C_K| - 2|M \cap \{f\}| = |O| - 2|M \cap E_B|. \tag{10.3}$$

To prove the assertion, let h be an edge of G; let us prove that G has a perfect matching that contains edge h and just one edge in C_K, for each $K \in O$. For this, it suffices to prove that h is in a perfect matching M of G such that

$$\sum_{K \in O} |M \cap C_K| = |O|. \tag{10.4}$$

Consider first the case in which h is an edge of $G - R - f$. The graph $G - R - f$ is matching covered; let M be a perfect matching of $G - R - f$ that contains the edge h. As $E_B \subseteq R$, we conclude that M and E_B are disjoint. From equation (10.3) we deduce the validity of equality (10.4).

Suppose then that $h \in R + f$. Let M_e be a perfect matching of G that contains the edge e. As e depends on f, $f \in M_e$. Moreover, $e \in E_B$. From equation (10.3) we infer that $E_B = \{e\}$ and also the validity of equality (10.4). As $e \in R$, and since R is a removable class of G that contains edge e, it follows that $R \subseteq M_e$. In sum, $R + f \subseteq M_e$.

Indeed, every edge of G is in a perfect matching M that satisfies equality (10.4).□

As $G - R$ is matching covered, the edge f is in a perfect matching, M_f, of $G - R$. As $E_B = \{e\}$, no edge of M_f has both ends in E_B. From equation (10.3) we conclude that M_f has precisely three edges in C_L, for some $L \in O$. Thus, C_L is separating but not tight in G. This is not possible, because G is solid. □

The graph $G - R - f$ is matching covered. Thus, each edge of $G - R - f$ is matchable in $G - f$. Let e be an edge of R. We have seen that e does not depend on f; hence G has a perfect matching, N_e, that contains the edge e but does not contain the edge f. As R is an equivalence class of the dependence relation of the edges in G, $R \subset N_e$. Thus, every edge of R is matchable in $G - f$. In sum, every edge of $G - f$ is matchable in $G - f$. We conclude f is removable in G. Indeed, the only edge f of S is removable in G.

We may thus assume that S is a doubleton.

Case 2 *The graph G is free of nontrivial tight cuts.*

From Corollary 10.18 we deduce that $b(G) = b(G - R) - 1 + |R|$. By Theorem 9.9, $b(G - R - S) = b(G - R) - 1$. Moreover, as G is free of nontrivial tight cuts, $b(G) \leq 1$. Thus,

$$1 \geq b(G) = b(G - R) - 1 + |R| = b(G - R - S) + |R| \geq |R| \geq 1.$$

Equality holds throughout, hence G is a brick, R is a singleton and $G - R - S$ is bipartite.

Let e be the edge of R, let f_1 and f_2 be the two edges of S and let (A, B) denote the bipartition of $G - R - S$. The bipartite graph $G - R - S$ is matching covered. Addition to $G - R - S$ of an edge having one end in A and the other in B yields a matching covered graph, by Corollary 3.7. As S is a removable doubleton of $G - R$, no edge of S has one end in A, the other in B. For if, say, f_1 has an end in A and the other in B then the edge f_2 is removable in $G - R$, a contradiction. As $G - R$ is matching covered, one of f_1 and f_2 has both ends in A and the other has its ends in B. If e has one end in A and the other in B, the graph $G - S$ is matching covered. We may thus assume that e has both ends in the same part of (A, B). Adjust notation so that f_1 and e have both ends in A, whereas f_2 has both ends in B. A simple counting argument then shows that every perfect matching of G that contains the edge e contains also the edge f_2, but not the edge f_1. As $G - e - S$ is matching covered, we deduce that every edge of $G - f_1$ is matchable in $G - f_1$. Thus, $G - f_1$ is matching covered. In both alternatives, S includes a removable class of G.

Case 3 *The graph G has a nontrivial tight cut, $C := \partial(X)$.*

Let $G_1 := G/(\overline{X} \to \overline{x})$ and let $G_2 := G/(X \to x)$ be the two C-contractions of G. As G is solid, each brick of G is solid, by Theorem 7.1. Each brick of G_1 is a brick of G, hence solid. By Theorem 7.1, G_1 is solid. Likewise, G_2 is solid.

For $i = 1, 2$, let $R_i := R \cap E(G_i)$ and let $S_i := S \cap E(G_i)$. The cut $C - R$ is tight in $G - R$, because every perfect matching of $G - R$ is a perfect matching of G and $C - R$ is odd in $G - R$. The graph $G_i - R_i$ is a $(C - R)$-contraction of $G - R$; hence $G_i - R_i$ is matching covered. By Lemma 8.15, (i) R_i is either empty or a removable class of G_i, (ii) R has at most one edge in C and (iii) if R and C are disjoint then R is entirely contained in one of $E(G_1) - C$ and $E(G_2) - C$. Likewise, the cut $C - R - S$ is tight in $G - R - S$; hence $G_i - R_i - S_i$, a $(C - R - S)$-contraction of $G - R - S$, is matching covered. By Lemma 8.15, (i) S_i is either empty or is a removable class of $G_i - R_i$, (ii) S has at most one edge in C and (iii) if S and C are disjoint then S is entirely contained in one of $E(G_1) - C$ and $E(G_2) - C$.

The set S has at most two edges; hence it does not have edges in both $E(G_1) - C$ and $E(G_2) - C$. Adjust notation so that S and $E(G_2) - C$ are disjoint. If R_1 is empty then S is a removable class of G_1. Alternatively, if R_1 is not empty then R_1 is a removable class of G_1 and S is a removable class of $G_1 - R_1$. As G_1 is solid, we infer, by induction, that S includes a removable class of G_1. In both alternatives, S includes a removable class, T, of G_1.

If T and C are disjoint then $G - T$ is the splicing of $G_1 - T$ and G_2, both matching covered graphs; hence $G - T$ is matching covered and therefore S includes a removable class of G. Let us thus assume that T and C are not disjoint. The doubleton S contains at most one edge in C and $T \subseteq S$. Then, $T \cap C$ is a singleton, $\{f\}$. Note that $S_2 = \{f\}$. If R_2 is empty then f is a removable edge of G_2. Alternatively, if R_2 is not empty then R_2 is a removable class of G_2 and f is a removable edge of $G_2 - R_2$. As G_2 is solid, we infer, by induction, that f is removable in G_2. In both alternatives, f is a removable edge of G_2. Thus, the graph $G - T$ is matching covered, because it is the splicing of $G_1 - T$ and $G_2 - f$, both matching covered graphs. The assertion holds by induction, if G has nontrivial tight cuts. □

There are essentially two reasons that justify the fact that a nonsolid graph G may not have the exchange property. One reason is given in Example 10.21. A second reason is the fact that given a removable class R of G, the graph $G - R$ may have a removable class which does not include any removable class of G. This is illustrated by the tricorn $\mathbb{H}_{10}$ (Exercise 10.2.7). There are, however, simple nonsolid graphs that have the exchange property; see Exercise 10.2.8.

Exercises

10.2.1

(i) *Show that every near-bipartite graph is a near-brick.*
(ii) *Deduce from Theorem 9.17 that every near-bipartite brick distinct from* K_4, $\overline{C_6}$ *and* $\mathbb{H}_8$ *has at least two nonadjacent b-invariant removable edges.*

∗ **10.2.2** *Give a proof of Theorem 10.15. In fact, prove that if G is a solid matching covered graph and R is a set of edges of G such that* $G - R$ *is matching covered, then* $G - R$ *is solid.*

∗ **10.2.3** *Give a proof of Theorem 10.17.*

∗ **10.2.4** *Let R be a removable class in a matching covered graph G, and suppose that D is a cut in G such that* $D - R$ *is a separating cut in* $G - R$*. Now let* $\mathcal{C}$ *denote the family of all cuts C in G satisfying the following properties:*

(i) $C \preceq D$,
(ii) $C - R$ *is a separating cut in* $G - R$,
(iii) $\lambda_{G-R}(C - R) \leq \lambda_{G-R}(D - R)$,
(iv) *subject to (i), (ii) and (iii), C is minimal.*

Show that every member of $\mathcal{C}$ *is a separating cut of G.*

∗ **10.2.5** *Let G be a solid matching covered graph, let R be a removable class of G and let S be a removable class of* $G - R$ *such that* $G - S$ *is matching covered. The objective of this exercise is to prove that S is a removable class of G. This is certainly the case if S is a singleton. So we may suppose that S is a doubleton. We then prove that S is removable in G by proving that* $b(G - S) \leq b(G) - 1$ *and then applying Theorems 9.8 and 9.9. In fact, these two theorems are used several times in the proof.*

(i) *Prove that in the particular case in which R is a singleton,* $\{e\}$*, then*

$$b(G - S) \leq b(G - e - S) = b(G - e) - 1 = b(G) - 1.$$

Hint: use the fact that G is solid and apply Theorem 10.17.
(ii) *Prove that in the particular case in which R is a doubleton,*

$$b(G - S) \leq b(G - R - S) + 1 = b(G - R) = b(G) - 1.$$

Monotonicity of the function λ [9]

∗ **10.2.6** *Let R be a removable class in a matching covered graph G.*

(i) *Using the result in Exercise 10.2.4, show that* $\lambda(G) \leq \lambda(G - R)$.
(ii) *Deduce that if G is solid, then so is* $G - R$.

▷**10.2.7** *Define a k-*corn*, $k \geq 3$, k odd, to be the graph obtained from the odd wheel W_k by splicing each vertex in the rim with a K_4. The 3-corn is the tricorn. Prove that the collection R_k of removable classes of the k-corn consists of k removable edges and no removable doubletons. Prove that, for $k \geq 5$, removal of any edge in R_k yields a graph with k removable edges. Verify that in the case of the 3-corn, the removal of any edge of R_3 produces a graph with 5 removable edges. Conclude that the k-corn does not have the exchange property.*

▷**10.2.8** *Prove that every prism of order $n \equiv 2 \pmod 4$, $n \geq 10$, has the exchange property.*

▷**10.2.9** *Let G be a near-bipartite brick, let R be a removable doubleton of G and let h be a removable edge of $G - R$. Suppose that h is not removable in G. By Theorem 10.23, G is not solid. Apply Theorem 9.18 to determine a nontrivial separating cut of G.*

10.3 Notes

Using Theorem 10.23, it can be shown that any solid matching covered graph G, with $\delta \geq 3$, has at least $m - n + 1$ removable classes. Proving such a result by induction on the number of edges poses difficulties because a graph obtained from G by the deletion of removable classes may have vertices of degree two and may not have any removable classes at all. Thus, the theory of ear decompositions, which is the subject matter of the next chapter, turns out to be the appropriate context for proving such results by induction. (Ear decompositions of bipartite graphs were introduced in Section 3.4.) This theory incorporates all the ideas related to removable classes, but is unfettered by constraints on degrees!

Chapter 11
Ear Decompositions

Contents

11.1 Additions of Ears

11.1.1 Single and double ears

We now return to the topic of ear decompositions of matching covered graphs which was initiated in Chapter 3.

Recall that a *single ear* in a graph G is a path of odd length all of whose internal vertices have degree two in G. Suppose that H is a subgraph of G, and that P is an odd path in G, which is edge-disjoint from H, and is such that its ends are in $V(H)$ but its internal vertices, if any, are not. Then P is an *ear path* of H and, noting that P

C. L. Lucchesi, U. S. R. Murty, *Perfect Matchings*, Algorithms and Computation in Mathematics 31, https://doi.org/10.1007/978-3-031-47504-7_11

is a single ear in $H \cup P$, we say that the graph $H + P := H \cup P$ is obtained by *adding the ear P to H*.

One convenient way of remembering 'ears' is by thinking of them as bisubdivided 'edges'. Thus adding an ear to a graph amounts to first adding an edge and then bisubdividing that edge.

We showed in Chapter 3 that if H is a matching covered subgraph of a bipartite matching covered graph G, then any bipartite graph obtained by adding a single ear P to H is also a matching covered subgraph of G. This statement does not extend to nonbipartite graphs. Consider, for example, the case in which G is the complete graph K_4, and H is a 4-cycle in G. There are two ear paths of H in G (both being of length one), but adding any one of them singly to H results in a graph which is not matching covered; it is necessary to add both of them simultaneously to obtain G from H.

DOUBLE EARS

A *double ear* in a graph G is a set of two vertex-disjoint single ears in G. When H is a matching covered subgraph of G, and P and Q are two vertex-disjoint ear paths of H in G, the graph $H + \{P \cup Q\} := H \cup (P \cup Q)$ is said to be obtained from H by *adding the double ear* $\{P, Q\}$ to H.

The principal purpose of this chapter is to show that any matching covered graph may be obtained from K_2 by a sequence of (single or double) ear additions such that all intermediate graphs are also matching covered. Before formally stating and proving this result, let us consider the example shown in Figure 11.1. The paths $P_1 := 12$, $P_2 := 157862$, $P_3 := 1342$, $P_4 := 38$, $P'_4 := 47$, and $P_5 := 56$ are paths of odd length in the Möbius ladder $\mathbb{M}_8$ and the graphs $G_1 := P_1$, $G_2 := G_1 + P_2$, $G_3 := G_2 + P_3$, $G_4 := G_3 + \{P_4 \cup P'_4\}$, and $G_5 := G_4 + P_5$ are matching covered. The graph G_1 is K_2 and, for $2 \le i \le 5$, $i \ne 4$, the graph G_i is obtained from G_{i-1} by adding the single ear P_i, whereas G_4 is obtained from G_3 by adding the double ear $\{P_4, P'_4\}$. The graph G_5 is the Möbius ladder $\mathbb{M}_8$.

As a preliminary to proving the above-mentioned principal result, we shall first establish the following weaker statement:

Theorem 11.1 *Given any matching covered graph G, there exists a sequence*

$$(K_2 = G_1 \subset G_2 \subset \cdots \subset G_r = G) \tag{11.1}$$

of matching covered subgraphs of G such that, for $2 \le i \le r$, the graph G_i is obtained from G_{i-1} by simultaneously adding a set of vertex-disjoint ears.

The proof of this result that we shall give is based on the notion of conformal subgraphs introduced in Exercise 3.4.3.

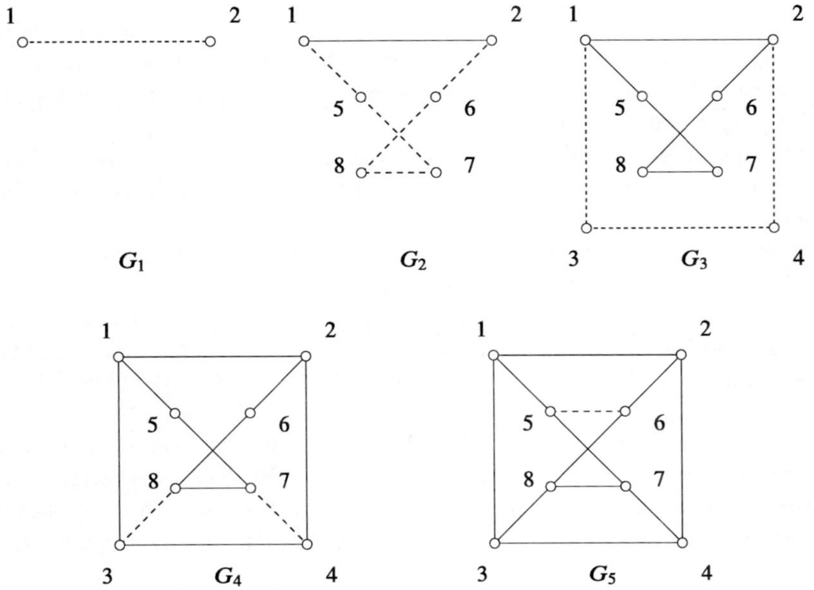

Fig. 11.1 A bottom-up ear decomposition of the Möbius ladder $\mathbb{M}_8$

11.1.2 Conformal subgraphs

> CONFORMAL SUBGRAPHS
>
> Recall that a matchable subgraph H of a matchable graph G is a *conformal subgraph* of G if the graph $G - V(H)$ has a perfect matching. In other words, H is a conformal subgraph of G if any perfect matching of H may be extended to a perfect matching of G.

Suppose that H is a matchable subgraph of a graph G, let $P := us_1s_2 \ldots s_{2k}v$ be an ear path of H in G, and let $E_{\text{even}}(P) := \{s_1s_2, s_3s_4, \ldots, s_{2k-1}s_{2k}\}$ be the set of even numbered edges of P. Likewise, $E_{\text{odd}}(P)$ is the set of odd numbered edges of P. We extend these notations to a pair R of vertex disjoint ear paths of H, by defining $E_{\text{even}}(R)$ and $E_{\text{odd}}(R)$ to be respectively the set of even numbered and odd numbered edges of each path in R.

As we noted in Chapter 3, for any perfect matching M_H of H, the set $M_H \cup E_{\text{even}}(P)$ is a perfect matching of $H + P$. An easy consequence of this observation is the following:

Proposition 11.2 *If $(G_1, G_2, \ldots, G_r)$ is any sequence of subgraphs of a matching covered graph G as described in the statement of Theorem 11.1, then G_i is a conformal subgraph of G_j for $1 \leq i \leq j \leq r$.* □

Let H be a conformal matching covered proper subgraph of a matching covered graph G. We shall see that by adding a suitable set of vertex-disjoint single ears simultaneously to H, we can obtain a conformal matching covered subgraph of G which properly contains H.

Towards this end, let M be any perfect matching of G whose restriction to $E(H)$ is a perfect matching of H (such a perfect matching must exist because H is a conformal subgraph of G).

Now let f be any edge in $E(G) - E(H)$ which has at least one end in $V(H)$ (such an edge must exist because G is connected).

Let M_f be any perfect matching of G containing the edge f, and let C denote the (M_f, M)-alternating cycle which includes the edge f (see Figure 11.2, where edges of M are indicated by dotted lines, and those of $M_f \cap E(C)$ are indicated by solid lines).

It is not difficult to see that $H \cup C$ is a conformal matching covered subgraph of G (see Exercise 11.1.1). But the subgraph of G induced by the edge set $E(C) - E(H)$ consists of vertex-disjoint ear paths of H, and $H \cup C$ is obtained by simultaneously adding those paths as ears to H. (In the Example shown in Figure 11.2, these are the ears P_1, P_2, and P_3.)

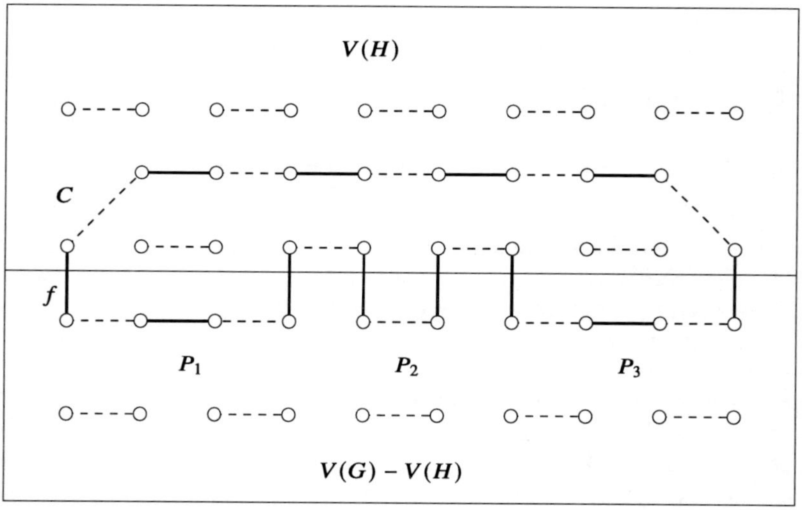

Fig. 11.2 Simultaneous addition of vertex-disjoint single ears

Starting with any given conformal subgraph F of G, one may recursively apply the procedure described above to obtain a sequence of conformal matching covered subgraphs where each member of the sequence, subsequent to F, is obtained from the previous one by simultaneous addition of a set of ears. We thus have the following assertion: (We leave its proof as Exercise 11.1.2.)

Theorem 11.3 *Given any matching covered graph G and a conformal matching covered subgraph F of G, there exists a sequence*

$$(F = G_1 \subset G_2 \subset \cdots \subset G_r = G) \tag{11.2}$$

of conformal matching covered subgraphs of G such that, for $2 \leq i \leq r$, the graph G_i is obtained from G_{i-1} by simultaneously adding a set of vertex-disjoint ears. □

Observe that Theorem 11.1 is a just special case of the above theorem in which the first graph F is K_2.

Our aim is to show that a decomposition of the type described in Theorem 11.3 can be achieved by never having to add more than two single ears simultaneously. For establishing this stronger result, we shall find it convenient to view an ear decomposition as a way of reducing a given matching covered graph G to K_2 by 'deleting' ears. This approach makes it possible for us to use the theory of removable classes described in the previous chapter.

Exercises

▷**11.1.1** *Let H be a conformal matching covered subgraph of a matching covered graph G, and suppose that M is a perfect matching of G whose restriction to $E(H)$ is a perfect matching of H. Let f be an edge in $E(G) - E(H)$ which has at least one end in $V(H)$, and let M_f be a perfect matching of G containing the edge f. Show that:*

(i) *the graph $H \cup C$, where C is the (M_f, M)-alternating cycle containing the edge f, is a a conformal matching covered subgraph of G,*
(ii) *the subgraph of G induced by $E(C) - E(H)$ consists of vertex disjoint ear paths of H, and*
(iii) *the graph $H \cup C$ is the same as the graph obtained from H by adding those ear paths to it.*

▷**11.1.2** *Give a proof of Theorem 11.3.*

11.2 Deletions of Ears

11.2.1 Removable ears

Deletion of an Ear
When P is a single ear in a graph G, we denote by $G - P$ the graph obtained from G by deleting all the edges and internal vertices of P.

The following fact, which will be found to be very useful, is a consequence of Theorem 11.3.

Corollary 11.4 *Let G be a matching covered graph, and let F be a proper conformal matching covered subgraph of G. Then G has a set $\{P_1, P_2, \ldots, P_k\}$ of vertex-disjoint single ears such that the graph $G - P_1 - P_2 - \cdots - P_k$ is matching covered and contains F as a conformal subgraph.*

Proof By Theorem 11.3 there exists a sequence

$$(F = G_1 \subset G_2 \subset \cdots \subset G_{r-1} \subset G_r = G)$$

of matching covered subgraphs of G such that, for $2 \leq i \leq r$, the graph G_i is obtained from G_{i-1} by adding a set of vertex-disjoint single ears. In particular, $G_r = G$ is obtained from G_{r-1} by adding a set of vertex-disjoint single ears, say $P_1, P_2, \ldots, P_k$. By deleting these k ears from G we obtain G_{r-1}, which is matching covered. By Proposition 11.2, F is a conformal subgraph of G_{r-1}. The assertion follows. □

REMOVABLE EARS

A single ear P in a matching covered graph G is *removable* if $G - P$ is matching covered. In the particular case in which a removable single ear P is a path of length one, the edge of P is removable in G.

A double ear $R := \{P, Q\}$ of a matching covered graph G is *removable* if $G - R$ is matching covered, but neither $G - P$ nor $G - Q$ is matching covered. In the particular case in which both the single ears P and Q of a removable double ear $R = \{P, Q\}$ are paths of length one, the two edges of P and Q together constitute a removable doubleton of G.

For example, in the graph G depicted in Figure 11.3, the ear indicated by dotted lines constitutes a removable single ear, and the two paths indicated by solid lines together constitute the two ears of a removable double ear.

Henceforth, we shall use the term 'ear' to designate either a single or double ear. Accordingly, a *removable ear* of G is a removable single ear or a removable double ear of G. For a removable ear R in a matching covered graph G, we say that the graph $G - R$ is obtained by *deleting* R.

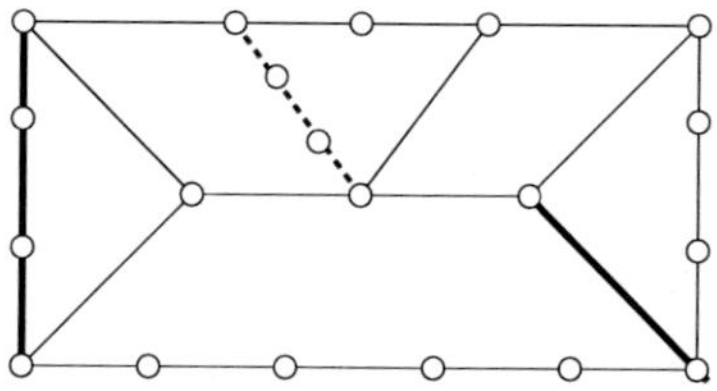

Fig. 11.3 Examples of removable ears

Many statements that apply to a path which is a single ear also apply to each of the two paths that are the constituent single ears of a double ear. For this reason, we shall find it convenient to refer to a path P as a *constituent path of an ear R* if either R is a single ear and $R = P$, or R is a double ear and P is one of the two paths that constitute R.

For future reference, we note two simple properties of removable ears in a matching covered graph.

Proposition 11.5 *Let G be a matching covered graph. Suppose that P is a constituent path of a removable ear R. Then, either G is an even cycle (and $R = P$ is a removable single ear), or both ends of P have degree at least three in G (Exercise 11.2.1).* □

Proposition 11.6 *Suppose that u and v are two vertices of degree at least three in a matching covered graph G, and that Q is an uv-path in G all of whose internal vertices have degree two. Then, for any removable ear R of G, either Q is one of the constituent paths of R, or $E(Q) \cap E(R) = \emptyset$ (Exercise 11.2.1).* □

11.2.2 Series reductions of ears

In Chapter 3 we noted that the operation of adding a single ear $us_1s_2 \ldots s_{2k}v$ to a graph H may be viewed as the composite operation of adding an edge joining u and v first, and then bisubdividing that edge by inserting the vertices $s_1, s_2, \ldots, s_{2k}$. In much the same way, the operation of deleting a removable single ear $P := us_1s_2 \ldots s_{2k}v$ from a matching covered graph G may be construed as the composite operation involving the $\partial(X)$-contraction $H := G/(X \to x)$, where $X := \{s_1, s_2, \ldots, s_{2k}, v\}$, followed by the deletion of the edge ux. Note that $\partial(X)$ is a tight cut of G; hence H is matching covered. Moreover, $H - ux$ is isomorphic to $G - P$; hence ux is a removable edge of H. The above idea may also be described as consisting of a *series reduction* of the path P to a path of length one, and then deleting the edge $e(P)$ of that path. (Similarly, the deletion of a removable double ear $R = \{P, Q\}$ from G may be effected by first reducing P and Q to single edges by series reductions, and then deleting the removable doubleton to which R is reduced.)

11.2.3 The two ear theorem

We present a proof of the following basic result due to Lovász and Plummer [59].

THE TWO EAR THEOREM

Theorem 11.7 *Given a matching covered graph G and a proper conformal matching covered subgraph F of G, there exists a removable (single or double) ear R of G such that F is a conformal subgraph of $G - R$.*

Proof This assertion is clearly valid if G is an even cycle. So, suppose that it is not an even cycle. By Corollary 11.4, G has a set $\{P_1, P_2, \ldots, P_k\}$ of vertex-disjoint single ears such that $G - P_1 - P_2 - \cdots - P_k$ is matching covered and contains F as a conformal subgraph. Let H be the graph obtained from G by the application of series reductions to each ear P_i, thereby obtaining the edges e_i, $i = 1, 2, \ldots, k$. Since the graph $G - P_1 - P_2 - \cdots - P_k$ is matching covered, it follows that the graph $H - e_1 - e_2 - \cdots - e_k$ is matching covered. The set $\{e_1, e_2, \cdots, e_k\}$ must include a removable class, which, by Corollary 8.18, is a removable edge or a removable doubleton. If e_i is a removable edge of H, then $R := P_i$ is a removable single ear in G, because $G - P_i$ is obtained from $H - e_i$ by the bisubdivisions of the ears P_j, $j \neq i$. Likewise, if $\{e_i, e_j\}$ is a removable doubleton of H, then $R := \{P_i, P_j\}$ is a removable double ear in G. In both cases, clearly, F is a conformal subgraph of $G - R$. The assertion holds. □

By taking F to be the subgraph spanned by a single edge of G, and applying the above theorem, we have:

EXISTENCE OF REMOVABLE EARS

Theorem 11.8 *Every matching covered graph $G \neq K_2$ has a removable ear.* □

Corollary 11.9 *Every matching covered graph G whose minimum degree is at least three has a removable class.* □

Exercises

11.2.1 *Give proofs of Propositions 11.5 and 11.6.*

▷**11.2.2 (Monotonicity of functions b and $b + p$)** *Apply Theorems 9.8 and 9.9 to deduce the following inequalities:*

MONOTONICITY WITH RESPECT TO DELETIONS OF REMOVABLE EARS

(i) *for any removable single ear R of G:*

$$b(G - R) \geq b(G) \text{ and } (b + p)(G - R) \geq (b + p)(G); \tag{11.3}$$

(ii) *for any removable double ear R of G:*

$$b(G - R) = b(G) - 1 \text{ and } (b + p)(G - R) = (b + p)(G) - 1. \tag{11.4}$$

▷**11.2.3** *Let H be a conformal, nonbipartite, matching covered subgraph of a matching covered graph G, and let R be a removable single ear of H. Show that $H - R$ is also a conformal nonbipartite matching covered graph of G.*

▷**11.2.4** *Let R be a removable single ear of a solid matching covered graph G. Using Theorem 10.17, show that* $b(G - R) = b(G)$ *and* $(b + p)(G - R) = (b + p)(G)$.

▷**11.2.5** *Let G be any matching covered graph and let R be either a removable double ear of G, or a removable single ear such that* $b(G - R) = b(G)$. *Show that the dimension of* $\mathcal{L}in(G)$ *is just one more than the dimension of* $\mathcal{L}in(G - R)$.

11.3 Ear Decompositions

11.3.1 Two equivalent definitions

Theorem 11.8 immediately implies, by induction, the following assertion:

Theorem 11.10 *Given any matching covered graph G there exists a sequence*

$$\mathcal{G} := (G_1 = G \supset G_2 \cdots \supset G_r = K_2) \tag{11.5}$$

of subgraphs of G such that, for $1 \leq i \leq r - 1$,

$$G_{i+1} = G_i - R_i, \text{ where } R_i \text{ is a removable ear of } G_i. \tag{11.6}$$

We define a *top-down ear decomposition* of a matching covered graph G as a sequence (11.5) of subgraphs of G which satisfies the property (11.6).

Reversing the order of the sequence we obtain the more traditional definition of an ear decomposition of a matching covered graph G as a sequence

$$\mathcal{G} := (G_1 = K_2 \subset G_2 \cdots \subset G_r = G) \tag{11.7}$$

of matching covered subgraphs of G such that, for $2 \leq i \leq r$,

$$G_i = G_{i-1} + R_i, \text{ where } R_i \text{ is an ear of } G_i. \tag{11.8}$$

We define a *bottom-up ear decomposition* (or, simply, *ear decomposition*) of a matching covered graph G as a sequence (11.7) of subgraphs of G which satisfies the property (11.8).

Note that the stipulation that in a top-down ear decomposition each R_i is a removable ear ensures that all subgraphs in the sequence are matching covered. Conversely, in a bottom-up ear decomposition, the stipulation that all the graphs in the sequence are matching covered ensures that each ear R_i is removable. So, the two types of decompositions differ only on a viewpoint.

Suppose that F is a matching covered subgraph of G such that every ear decomposition of F may be extended to an ear decomposition of G. In this case, there exists a sequence

$$\mathcal{G} := (F = G_1 \subset G_2 \cdots \subset G_r = G) \tag{11.9}$$

of subgraphs of G satisfying the property (11.6). We refer to such a sequence as a *partial* ear decomposition of G *starting with* F. The following result is an immediate consequence of Theorem 11.7 and Proposition 11.2:

Theorem 11.11 *A matching covered graph G has an ear decomposition starting with a given graph F if and only if F is a conformal matching covered subgraph of G.* □

An immediate corollary of the above theorem is that an ear decomposition of a matching covered graph G may start with any edge e of G. More precisely, for any edge e of G, the subgraph induced by e could be the first graph G_1 in an ear decomposition of G. The second graph G_2 in any ear decomposition of G is an even cycle. The above theorem implies that any conformal cycle in G could be the second graph G_2 in an ear decomposition of G.

If G has at least two edges, and M is any perfect matching of G, it is easy to see that any given edge of G is contained in some M-alternating cycle, and hence in a conformal cycle of G (see Exercise 11.3.3). Little (1974, [53]) has strengthened this statement to show that any two distinct edges of G are contained together in a conformal cycle of G (see Exercise 11.3.5). Thus, given any two edges e and f of G, there exists an ear decomposition $(G_1, G_2, \ldots, G_r)$ of G such that (i) $E(G_1) = \{e\}$, and (ii) G_2 is a conformal cycle containing both e and f.

11.3.2 An ear decomposition of the Petersen graph

We have already seen an ear decomposition of the Möbius ladder $\mathbb{M}_8$ (Figure 11.1). An ear decomposition of the Petersen graph is shown in Figure 11.4. For $2 \leq i \leq 5$, the ear added to G_{i-1} to obtain G_i (or, equivalently, the ear deleted from G_i to obtain G_{i-1}) is indicated by broken lines. The graph G_3, which is a bisubdivision of K_4, is obtained by adding a double ear to G_2, a conformal 8-cycle of the Petersen graph. The graph G_4 is obtained from G_3 by the addition of a double ear. All other ear additions involved are single ear additions.

One of the many distinguishing properties of the Petersen graph is that, up to isomorphism, it has a unique ear decomposition (Exercise 11.3.2).

Exercises

▷**11.3.1** *Show that every nonbipartite matching covered graph G contains a bisubdivision of a near-bipartite graph as a conformal subgraph.*

11.3.2 *Show that, up to isomorphism, the Petersen graph has a unique ear decomposition.*

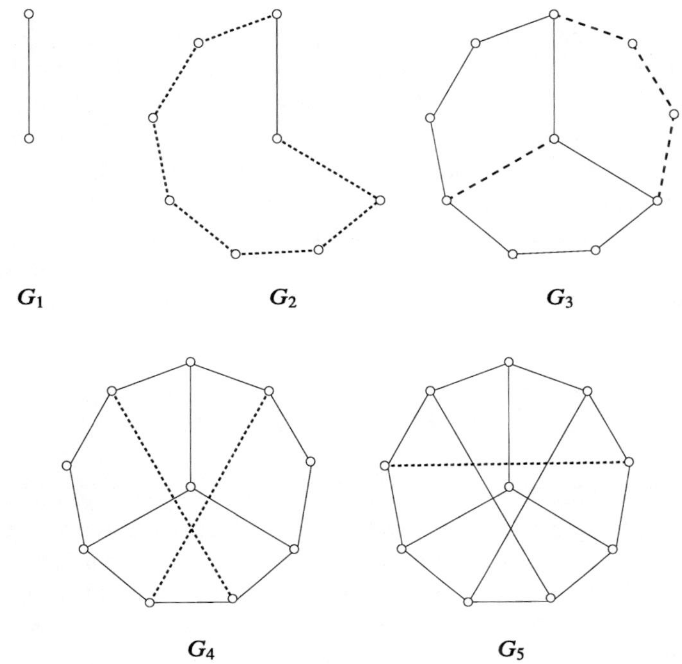

Fig. 11.4 An ear decomposition of the Petersen graph

11.3.3 *Let G be a matching covered graph on at least two edges, and let M be any perfect matching of G. Given any edge e of G, show that there is some M-alternating cycle that contains e.*

11.3.4 *Let H be a proper conformal matching covered subgraph of a matching covered graph G, and let M be a perfect matching of G whose restriction to $E(H)$ is a perfect matching of H. Show that G has a removable (single or double) ear R such that (i) $H \subseteq G - R$, and (ii) if f is any edge of M in $E(R)$, then both ends of f have degree two in G.*

11.3.5 *Let e and f be two edges of a matching covered graph G, and let M_f be a perfect matching of G containing the edge f.*

(i) *Show that there exists an M_f-alternating cycle C_e containing the edge e. (Hint: See Exercise 11.3.3.)*
(ii) *If $f \notin E(C_e)$, show that there is a removable ear R of G such that either f does not belong to $E(R)$, or f belongs to $E(R)$ but both ends of f have degree two in G.*

   (Hint: Apply the result in Exercise 11.3.4 by taking C_e to be H, and the perfect matching M_f to be M.)
(iii) *Now use induction on the number of edges to show that e and f are contained together in a conformal cycle of G.* *(Little (1974, [53]))*

11.3.6 *Prove that every matching covered graph distinct from K_2 and C_{2n}, ($n \geq 2$) has two edge-disjoint removable ears. (We shall come across a much stronger statement (Theorem 16.12) in Part II. A proof of that statement appeared in [11].)*

Hint: Use induction on the number of edges. First deal with the simple cases in which G either has multiple edges or has vertices of degree two. Then show that the statement is valid if G is either a brick or a brace. What remains then is dealing with the case in which G has nontrivial tight cuts. By Theorem 5.6, any matching covered graph that has nontrivial tight cuts, either has a nontrivial barrier cut or 2-separation cut. Consider these two cases to complete the proof.

11.4 The Number of Ears in a Decomposition

In general, a matching covered graph may have many ear decompositions. This is the case even when the graph under consideration is bipartite. However, we noted in Chapter 3, the number r of graphs in an ear decomposition of a bipartite graph G is always $m-n+2$ and hence is an invariant of G. This is not the case with nonbipartite graphs, as the number of graphs in an ear decomposition $\mathcal{G}$ of G depends on the number of double ears used in the decomposition.

The following relationship may be established by a simple counting argument (Exercise 11.4.1).

Proposition 11.12 *Let $d(\mathcal{G})$ denote the number of double ear additions used in an ear decomposition $\mathcal{G}$, and let $r(\mathcal{G})$ be the number of graphs in $\mathcal{G}$. Then:*

$$r(\mathcal{G}) = m - n + 2 - d(\mathcal{G}). \tag{11.10}$$

Every ear decomposition of the Petersen graph uses exactly two double ears, but there are graphs which admit ear decompositions with different numbers of double ears. See Exercise 11.4.5.

11.4.1 Ear decompositions and perfect matchings

A list $(M_1, M_2, \ldots, M_r)$ of perfect matchings of a matching covered graph G is said to be in *stepwise descending order* if there are edges $e_1, e_2, ..., e_r$ such that, for $1 \leq i \leq r$, the edge e_i is in M_i but not in any M_j with $j < i$. The set $\{\chi^{M_1}, \chi^{M_2}, \ldots, \chi^{M_r}\}$ of incidence vectors of such a list of perfect matchings is a linearly independent subset of the matching space $\mathcal{L}in(G)$. (To see this, consider the $r \times m$ matrix whose i^{th}-row is χ^{M_i}. The $r \times r$ submatrix of this matrix corresponding to the edges $e_1, e_2, \ldots, e_r$ is a lower diagonal matrix with ones on its main diagonal.) If $\{\chi^{M_1}, \chi^{M_2}, \ldots, \chi^{M_r}\}$ also happens to generate $\mathcal{L}in(G)$, we shall refer to it as a *stepwise canonical basis* of $\mathcal{L}in(G)$.

We saw in Chapter 3 how an ear decomposition of a bipartite matching covered graph can be used to generate a set of perfect matchings whose incidence vectors are linearly independent (see Corollary 3.16 and Exercise 6.3.3). More generally, an ear decomposition of any matching covered graph G may be used to generate a list of perfect matchings which is in stepwise descending order, thereby yielding a linearly independent subset of $\mathcal{L}in(G)$. This assertion can be proved by induction on the number of subgraphs in the ear decomposition and relies on the following Proposition 11.13.

Recall that if R is a path of odd length, then $E_{\text{even}}(R)$ and $E_{\text{odd}}(R)$ denote the sets of even and odd numbered edges of R, respectively; if R is a union of two disjoint odd paths P_1 and P_2, then $E_{\text{even}}(R) = E_{\text{even}}(P_1) \cup E_{\text{even}}(P_2)$, and $E_{\text{odd}}(R) = E_{\text{odd}}(P_1) \cup E_{\text{odd}}(P_2)$.

Proposition 11.13 *Let G be a matching covered graph which is obtained from one of its matching covered subgraphs H by the addition of a (single or double) ear R. Then, given any list $(N_1, N_2, \ldots, N_k)$ of perfect matchings of H in stepwise descending order, there exists a list $(M_1, M_2, \ldots, M_k, M_{k+1})$ of perfect matchings of G in stepwise descending order such that $N_i = M_i \cap E(H)$, for $1 \leq i \leq k$.*

Proof Let $E_{\text{even}}(R)$ and $E_{\text{odd}}(R)$ be defined as above. Clearly, any perfect matching of G which contains one edge of $E_{\text{even}}(R)$ contains all edges of $E_{\text{even}}(R)$. Likewise, any perfect matching of G which contains one edge of $E_{\text{odd}}(R)$ contains all edges $E_{\text{odd}}(R)$. Furthermore, $M_i := N_i \cup E_{\text{even}}(R)$ is a perfect matching of G for $1 \leq i \leq k$. Let M_{k+1} be a perfect matching of G which contains an edge of $E_{\text{odd}}(R)$. Then M_{k+1} includes all the edges in $E_{\text{odd}}(R)$.

It is easy to see that $(M_1, M_2, \ldots, M_k, M_{k+1})$ is a list of perfect matchings of G which is in stepwise descending order (Exercise 11.4.2). □

The following theorem can be easily derived by applying the above proposition recursively (see Exercise 11.4.3).

Theorem 11.14 *Given a matching covered graph G, and an ear decomposition $\mathcal{G} := (G_1, G_2, \ldots, G_r)$ of G, there exists an associated list $(M_1, M_2, \ldots, M_r)$ of perfect matchings of G which is in stepwise descending order such that, for $1 \leq i \leq r$, $(M_1 \cap E(G_i), M_2 \cap E(G_i), \ldots, M_i \cap E(G_i))$ is a list of perfect matchings of G_i which is in stepwise descending order.* □

The stepwise descending property of the r perfect matchings $(M_1, M_2, \ldots, M_r)$ of G, as described in the above theorem, implies that their incidence vectors $\{\chi^{M_1}, \chi^{M_2}, \ldots, \chi^{M_r}\}$ are linearly independent over any field $\mathbb{F}$. This observation, together with Equation 11.10, implies the following corollary.

Corollary 11.15 *If a matching covered graph G has an ear decomposition $\mathcal{G}$ consisting of r subgraphs, then the dimension of the matching space over any field $\mathbb{F}$ is at least r. Furthermore, since, $r(\mathcal{G}) = m - n + 2 - d(\mathcal{G})$ by Equation 11.10, where $d(\mathcal{G})$ is the number of double ear additions used obtaining $\mathcal{G}$, it follows that*

$$d(\mathcal{G}) \geq m - n + 2 - \dim(\mathcal{L}in(G, \mathbb{F})), \tag{11.11}$$

where $\mathcal{L}in(G, \mathbb{F})$ is the matching space of G over the field $\mathbb{F}$ and $\dim(\mathcal{L}in(G, \mathbb{F}))$ *is its dimension.* □

Example 11.16 Figure 11.1 shows an ear decomposition of the Möbius ladder $\mathbb{M}_8$. Using this decomposition, one can obtain the following five perfect matchings of the brick $\mathbb{M}_8$:

$$\left.\begin{array}{l} M_1 := \{12, 57, 68, 34\} \\ M_2 := \{15, 78, 26, 34\} \\ M_3 := \{13, 24, 57, 68\} \\ M_4 := \{38, 47, 15, 26\} \\ M_5 := \{56, 78, 12, 34\} \end{array}\right\} \tag{11.12}$$

If we let $e_1 := 12$, $e_2 := 15$, $e_3 := 13$, $e_4 := 38$, and $e_5 := 56$, it can be verified that the edge e_i appears in M_i but not in M_j for $j < i$. Thus $(M_1, M_2, M_3, M_4, M_5)$ is in stepwise descending order. The incidence vectors of these five perfect matchings are the rows of the matrix shown in Table 11.1.

	12	15	13	38	56	57	68	34	78	26	24	47
χ^{M_1} :	1	0	0	0	0	1	1	1	0	0	0	0
χ^{M_2} :	0	1	0	0	0	0	0	1	1	1	0	0
χ^{M_3} :	0	0	1	0	0	1	1	0	0	0	1	0
χ^{M_4} :	0	1	0	1	0	0	0	0	0	1	0	1
χ^{M_5} :	1	0	0	0	1	0	0	1	1	0	0	0

The first five columns of the above matrix form a lower triangular matrix with 1's on the main diagonal. It follows that the rows of this matrix are linearly independent. Since the dimension of $\mathcal{L}in(\mathbb{M}_8)$ is five, this matrix is thus a basis matrix of $\mathcal{L}in(\mathbb{M}_8)$.

Table 11.1 The incidence vectors of the five perfect matchings

Example 11.17 Figure 11.4 shows an ear decomposition $\mathbb{P} := (G_1, G_2, \ldots, G_5)$ of the Petersen graph $\mathbb{P}$. This decomposition may be used to obtain five perfect matchings of $\mathbb{P}$ whose incidence vectors are linearly independent over any field. Thus the dimension of the matching space of $\mathbb{P}$ over any field $\mathbb{F}$ is at least five. This lower bound falls short of the actual dimension over the reals by one, but over the field $\mathbb{Z}_2$, it is tight. To see this, note that the Petersen graph has, altogether, six perfect matchings. As each edge is in precisely two of those six matchings, their incidence vectors are linearly dependent over $\mathbb{Z}_2$.

By Theorem 11.14, given any ear decomposition $\mathcal{G}$ of a matching covered graph G with $r := r(\mathcal{G})$ subgraphs, one can find a list of r perfect matchings which is in stepwise descending order. The set of incidence vectors in that list is always linearly independent. If one's aim is to find a basis for $\mathcal{L}in(G)$ by this means, one must clearly employ an ear decomposition of G with the largest possible number r of subgraphs. Since $r = m - n + 2 - d$, where $d := d(\mathcal{G})$ is the number of double ear additions used in $\mathcal{G}$, it would thus be of advantage to find an ear decomposition $\mathcal{G}$ which involves as few double ear additions as possible.

All ear decompositions of bipartite graphs utilize only single ear additions. Any ear decomposition of a nonbipartite matching covered graph, in particular that of a brick, must use at least one double ear addition; and any ear decomposition of a Petersen brick requires two double ear additions. The following theorem establishes a lower bound on the number of double ear additions in an ear decomposition of a matching covered graph.

Theorem 11.18 *Let G be a matching covered graph. For any ear decomposition $\mathcal{G} := (K_2 = G_1 \subset G_2 \cdots \subset G_r = G)$ of G,*

$$d(\mathcal{G}) \geq (b + p)(G)$$

Proof By induction on the number of edges. If $r = 1$, then $G = K_2$, and the statement is clearly true. So, suppose that $r \geq 2$, and let R be the ear deleted from $G = G_r$ to obtain G_{r-1}. Clearly, $\mathcal{G}' := (K_2 = G_1 \subset G_2 \cdots \subset G_{r-1})$, the subsequence of $\mathcal{G}$ consisting of its first $r - 1$ terms, is an ear decomposition of G_{r-1}. By the induction hypothesis,

$$d(\mathcal{G}') \geq (b + p)(G_{r-1}) = (b + p)(G - R)$$

Note that $d(\mathcal{G}) = d(\mathcal{G}')$ if R is a single ear, and $d(\mathcal{G}) = d(\mathcal{G}') + 1$ if R is a double ear. The assertion now follows from (11.3) and (11.4) in Exercise 11.2.2. □

We shall prove in a later chapter that the above bound is tight in the sense that, given any matching covered graph G, there exists an *optimal ear decomposition* of G, namely an ear decomposition of which G involves precisely $(b + p)(G)$ double ear additions. (In particular, we shall show that any brick which is not a Petersen brick has an ear decomposition which uses just one double ear addition.)

Returning to the example of the Möbius ladder $\mathbb{M}_8$, we note that, the ear decomposition we found contains just one double ear addition, and the number r of graphs is equal to $m - n + 2 - d = 12 - 8 + 2 - 1 = 5$. This number coincides with the the dimension of the matching space of $\mathbb{M}_8$. Lo and behold! we have a stepwise canonical basis for $\mathcal{L}in(\mathbb{M}_8)$.

Unfortunately, things don't work so well for the Petersen graph. Every ear decomposition of this brick uses two double ear additions, and hence the above method yields only $r = m-n+2-d = 15-10+2-2 = 5$ perfect matchings whose incidence vectors are linearly independent. This is one short of the required number because the dimension of the matching space of the Petersen graph is six. But all is not lost because the Petersen graph has just six perfect matchings, and their incidence vectors are linearly independent!

Exercises

11.4.1 *Give a proof of Proposition 11.12.*

▷**11.4.2** *Show that the list $(M_1, M_2, \ldots, M_k, M_{k+1})$ of perfect matchings defined in the proof of Proposition 11.13 is in stepwise descending order.*

▷**11.4.3** *Give a proof of Theorem 11.14.*

11.4.4

(i) *Show that the basis of $\mathcal{L}in(K_6)$ defined in Exercise 6.3.7 is not a stepwise canonical basis;*
(ii) *Find a stepwise canonical basis of $\mathcal{L}in(K_6)$.*

▷**11.4.5** *Consider the graph G by adding an edge to the Petersen graph (see Figure 11.5(a)).*

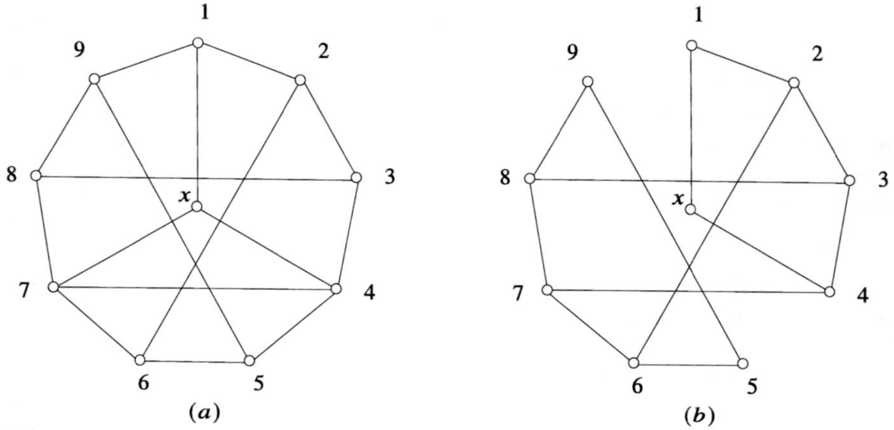

Fig. 11.5 (a) $G := \mathbb{P} + 47$; (b) H, a bisubdivision of $\overline{C_6}$ in G

(i) *Using the fact that the Petersen graph is a conformal subgraph of G, find an ear decomposition of G which involves two double ear additions.*
(ii) *Find an ear decomposition of the graph H shown in Figure 11.5(b) which uses just one double ear.*
(iii) *Complete the ear decomposition of H to an ear decomposition of G by adding the three 'missing edges' as single ears.*
(iv) *Now use the ear decomposition you have found to obtain a basis for the matching space of G.*

11.4.6 *Consider the solid brick Σ_8 shown in Figure 7.2.*

(i) *Find an ear decomposition of Σ_8 using just one double ear deletion; and*
(ii) *using it, find a stepwise canonical basis for $\mathcal{L}in(\Sigma_8)$, and justify your answer.*

▷**11.4.7 (Ear Decompositions of Solid Matching Covered Graphs)** *Let G be a solid matching covered graph. Using the identities in Exercises 11.2.2 and 11.2.4 prove that if $\mathcal{G}$ is any ear-decomposition of G, then:*

$$d(\mathcal{G}) = (b + p)(G) = b(G).$$

Deduce that any ear decomposition of a solid matching covered graph is an optimal ear decomposition.

11.5 Retract of a Graph

Given any matching covered graph G, there exists a related matching covered graph $\widehat{G}$ which can be obtained from G by means of special types of tight cut contractions known as bicontractions. The significance of this graph $\widehat{G}$, called the retract of G, is that, when G is not an even cycle, there exists a one-to-one correspondence between the set of all removable ears of G and the set of all removable classes of $\widehat{G}$.

We shall first define bicontractions, explain their relevance, and then describe how the retract of a matching covered graph can be obtained using bicontractions.

Bicontraction of a vertex of degree two

Let v_0 be a vertex of degree two in a matching covered graph G of order four or more, and let v_1 and v_2 be its two neighbours. Then $\partial(X)$, where $X := \{v_0, v_1, v_2\}$, is a tight cut of G, and the matching covered graph G/X is said to be obtained from G by the *bicontraction of* v_0. See Figure 11.6 for an illustration.

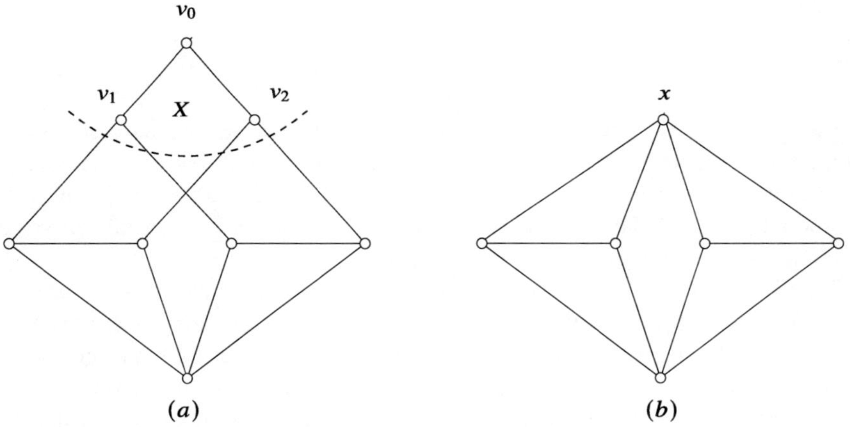

Fig. 11.6 Bicontraction of vertex v_0: (a) G, (b) $G/(X \to x)$

Motivated by the above considerations, we now define the notion of the retract of a matching covered graph.

The retract of a matching covered graph

The *retract* of a matching covered graph G, denoted by $\widehat{G}$, is a matching covered graph obtained from G by the following recursive procedure: If G has order two, or if G has no vertices of degree two, then $\widehat{G} := G$. Otherwise, $\widehat{G} := \widehat{G_v}$, where G_v is the graph obtained by bicontracting a vertex v of degree two in G. Clearly, if G has more than one vertex of degree two, this recursive procedure is not uniquely determined. However, it can be shown that, up to isomorphism, the retract $\widehat{G}$ does not depend on the order in which the bicontractions are performed. (An inductive proof of this result first appeared in CLM (2005, [13]). Theorem 11.19 and Corollary 11.20 provide a constructive proof.) Since each bicontraction preserves the property of being matching covered, we note that $\widehat{G}$ is matching covered. See Figure 11.7 for an illustration.

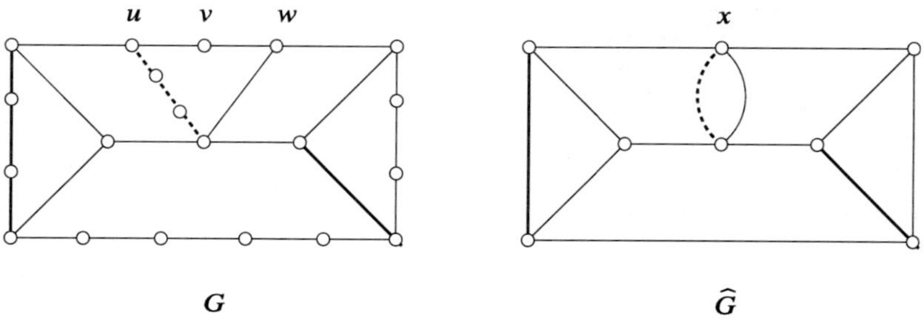

Fig. 11.7 The retract $\widehat{G}$ of a matching covered graph G

Let us extend the notion of retract to paths of a matching covered graph. Let P be a path of a matching covered graph G joining vertices of degree three or more whose internal vertices, if any, have degree two in G. The path P has an image in the retract $\widehat{G}$ of G. That image is obtained by bicontracting the internal vertices of P, is denoted $\widehat{P}$ and is called *the retract of P in G*.

Let G be a matching covered graph of order at least four which is not an even cycle. We shall refer to vertices of degree at least three as the *branch vertices* of G, and those of degree two, if any, as the *subdivision vertices* of G. Every edge of G clearly lies on a path whose ends are branch vertices but whose internal vertices, if any, are not. We shall refer to such paths as *lobes* in G, and note that all internal vertices of a lobe are subdivision vertices. A lobe P is an *even lobe* or *odd lobe* according to the parity of its length. It should be noted that odd lobes are the same as ears whose ends are branch vertices, and also that any two lobes of G are internally disjoint. The matching covered graph G shown in Figure 11.7 has one even lobe, namely the path uvw connecting the two branch vertices u and w. All the edges other than uv and vw belong to odd lobes (or ears) in G.

The following theorem elucidates the interaction between even and odd lobes in a matching covered graph G.

Theorem 11.19 *Let G be a matching covered graph of order at least four which is not an even cycle. Let G_{even} denote the subgraph of G induced by the set of edges that belong to even lobes of G. Then the set of branch vertices of G in any component of G_{even} is a barrier of G. Furthermore,*

(i) *the graph G_{even} is acyclic, and*
(ii) *no odd lobe connects two branch vertices of G that belong to the same component of G_{even}.*

Proof Let H be a component of G_{even}. By definition, H is the union of a subset of the set of even lobes of G. Let $B(H)$ denote the set of vertices of H which are branch vertices of G. Since H is connected, it follows that there are at least $|B(H)| - 1$ even lobes that connect pairs of vertices in $B(H)$. Since each even lobe has an odd number of internal vertices, it follows that $G - B(H)$ has at least $|B(H)| - 1$ odd components. Using Exercise 1.2.1 we now conclude that $B(H)$ is a barrier of G.

Suppose that a connected component H of G_{even} has a cycle. Then, there are at least $|B(H)|$ even lobes that connect pairs of branch vertices of G in H, and hence $G - B(H)$ has at least $|B(H)|$ odd components. It must in fact have precisely $|B(H)|$ odd components because G is matching covered, implying that H is a cycle. Since each vertex in $B(H)$ is a branch vertex of G, and since G is not a cycle itself, it follows that the graph $G - B(H)$ would have components other than the $|B(H)|$ odd components of $H - B(H)$. This is impossible because G is matching covered and $B(H)$ is a barrier of G. We conclude that each component of G_{even} is a tree, and hence that G_{even} is acyclic. This proves statement (i).

Let H be a component of G_{even}. As already noted, the set $B(H)$ of branch vertices of G in H is a barrier of G. Suppose that there is an an odd lobe P which connects two vertices u and v belonging to $B(H)$. If the length of P is one, then uv would be an edge which joins two vertices belonging to the barrier $B(H)$ of G. And, if P has length three or more, then $P - \{u, v\}$ would be an even component of $G - B(H)$. Both these are impossible because $B(H)$ is a barrier of G. This proves statement (ii). □

Readers may use the graph shown in Figure 11.8, on page 248, to identify its even lobes, its odd lobes and its removable ears (Exercise 11.5.1).

Series reduction of an odd lobe (or ear) P connecting two branch vertices s and t has the effect of replacing the path P by its retract $\widehat{P}$, whose single edge connects the two ends s and t of P. (This does not affect the edge sets of the other lobes, odd or even.) And, the series reduction of an even lobe P has the effect of deleting the internal vertices of P and then identifying its ends s and t to form a single vertex, which is the only vertex of the retract $\widehat{P}$ of P. (This does not lead to the identification of the ends of any lobe (odd or even) other than P, and thus, the edge sets of all lobes other than P remain intact. This follows from Theorem 11.19.)

We leave the proof of the following corollary to the above theorem as Exercise 11.5.3.

Corollary 11.20 *Let G be a matching covered graph on at least four vertices which is not an even cycle. Then, for each component H of G_{even}, the cut $\partial(H)$ is a tight cut of G. The retract $\widehat{G}$ is, up to isomorphism, the graph obtained from G by first shrinking each component of G_{even} to a single vertex, and then replacing each odd lobe P by a single edge, which is the only edge of the retract $\widehat{P}$ of P in G.* □

Clearly the graph K_2 has no removable ears. In the even cycle C_{2n} there are $2n$ removable single ears, and the retract of C_{2n} is C_2. Suppose now that G is a matching covered graph of order at least four which is not an even cycle. And, suppose that R is a removable ear (single or double) of G. It follows from Proposition 11.6 that each constituent single ear of R is an odd lobe whose ends are branch vertices. For a single or double ear R of G, denote by $G(\widehat{R})$ the graph obtained from G by replacing each constituent path P of R by its retract $\widehat{P}$. The following conclusion can be drawn from the preceding analysis:

Theorem 11.21 *Let G be a matching covered graph which has at least four vertices and is not an even cycle. Then*

(i) *an ear P in G is a removable single ear of G if and only if $\widehat{P}$ is a removable edge of $\widehat{G}$, and*
(ii) *a double ear $\{P, Q\}$ in G is a removable double ear of G if and only if $\{\widehat{P}, \widehat{Q}\}$ is a removable doubleton of $\widehat{G}$.* □

Note that the underlying simple graph of the graph $\widehat{G}$ shown in Figure 11.7 is the bicorn $\mathbb{H}_8$; $\widehat{G}$ is obtained by duplicating the only removable edge of $\mathbb{H}_8$. Thus $\widehat{G}$ has two removable edges, and two removable doubletons. Correspondingly, the graph G has two removable single ears, and two removable double ears.

Exercises

11.5.1 *Identify (i) even lobes, (ii) odd lobes, and (iii) removable ears in the graph shown in Figure 11.8.*

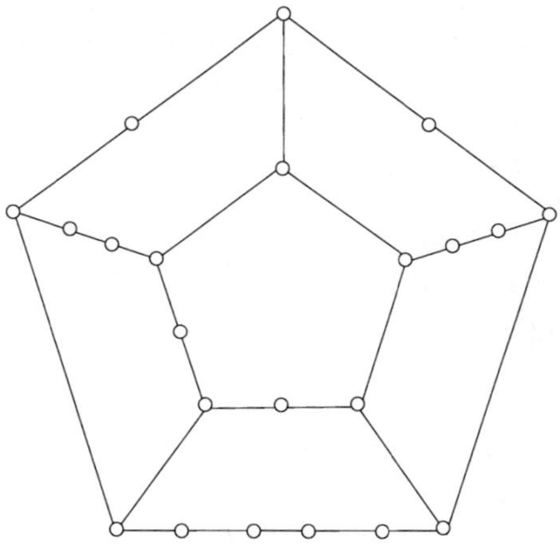

Fig. 11.8 Graph for Exercise 11.5.1

▷**11.5.2** *Show that the retract of a near-bipartite graph is also near-bipartite.*

▷**11.5.3** *Give a proof of Corollary 11.20.*

▷**11.5.4** *Let H be any cubic matching covered graph. Show that any matching covered graph G whose retract is H is a bisubdivision of H.*

▷**11.5.5** *Let G be a matching covered graph such that the maximum degree Δ of its vertices is three or more.*

(i) *Prove that G has at least Δ edge-disjoint removable ears.*
(ii) *Let M be a perfect matching of G. Prove that G has at least $\Delta - 1$ edge-disjoint removable ears R such that $M - R$ is a perfect matching of $G - R$.*

Hint: apply Theorem 11.21 and Exercise 8.3.8.

▷**11.5.6** *Let $\widehat{G}$ be the retract of a matching covered graph G. Show that*

$$b(G) = b(\widehat{G}), \text{ and } (b+p)(G) = (b+p)(\widehat{G})$$

∗**11.5.7 (Number of removable ears of solid graphs)** *Let G be a solid matching covered graph. Prove each of the two statements below.*

(i) *If G is neither K_2 nor an even cycle then G has at least $m-n+2-b$ edge-disjoint removable ears. In particular, if G is bipartite and is neither K_2 nor an even cycle then G has at least $m - n + 2$ edge-disjoint removable single ears.*
(ii) *If G is a solid near-brick which is not a bisubdivision of K_4 then G has at least $m - n - 1$ edge-disjoint removable single ears.*

11.6 Ear Decompositions of Critical Graphs ♯

Recall that a graph G is critical if, for any vertex v of G, the subgraph $G - v$ has a perfect matching. It turns out that critical graphs afford decompositions which are somewhat analogous to ear decompositions of matching covered graphs. In this section we present a brief discussion of these decompositions.

It is easy to see that all critical graphs are connected, and that every nontrivial critical graph is nonbipartite. A critical graph may have cut vertices; however, all blocks of a critical graph are also critical (Exercise 11.6.1).

Proposition 11.22 *Suppose that G is a critical graph, and H is a proper critical subgraph of G.*

(i) *For any (odd) ear path P of H in G, the graph $H + P := H \cup P$ is also critical.*
(ii) *For any odd cycle C in G which has just one vertex in common with H, the graph $H + C := H \cup C$ is also critical.* □

We leave the proof of above assertion as Exercise 11.6.2. In the latter case of the statement, we refer to C as a *closed ear* and, in the former case, for contrast, we refer to P as an *open ear*.

With the exception that, in dealing with decompositions of critical graphs, we permit closed ears, the terminology we use here is analogous to the one introduced in the context of ear decompositions of matching covered graphs. Thus, a critical subgraph H of a critical graph G is *conformal* if the graph $G - V(H)$ has a perfect matching. A *near-perfect matching* of a graph G is a matching M of G which covers all but one of the vertices of G. So, a subgraph H of G is conformal, in the above-defined sense, if there is a near-perfect matching M of G such that $M \cap E(H)$ is a near-perfect matching of H, and $M - E(H)$ is a perfect matching of $G - V(H)$.

An *ear decomposition* of a critical graph G is a sequence

$$\mathcal{G} := (K_1 = G_1, G_2, \ldots, G_r = G) \tag{11.13}$$

of conformal critical subgraphs of G such that for $1 \leq i < r$, the graph G_{i+1} is obtained from G_i by adding to it either an open ear or a closed ear. (Note that if G is nontrivial, the second graph G_2 is an odd cycle and is necessarily obtained from G_1 by the addition of a closed ear.) For an example, see Figure 11.9.

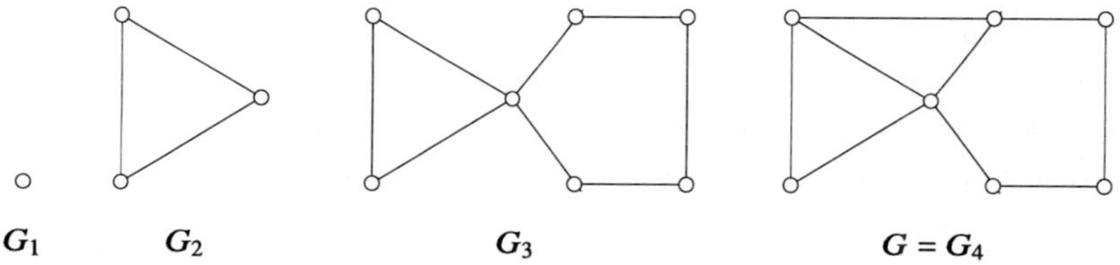

Fig. 11.9 An ear decomposition of a critical graph

In the above example, the first two ear additions are those of closed ears. Clearly, it is possible to find an ear decomposition of this graph G which involves just one closed ear addition (Exercise 11.6.3).

Although a critical graph may have different ear decompositions, the number of graphs in the sequence is simply a function of the order and size of the graph.

Proposition 11.23 *Let $\mathcal{G} := (K_1 = G_1, G_2, \ldots, G_r = G)$ be an ear decomposition of a critical graph G. Then,*

$$i = |E(G_i)| - |V(G_i)| + 2, \text{ for } 1 \leq i \leq r.$$

In particular, the number r of graphs in $\mathcal{G}$ is equal to $|E| - |V| + 2$ (see Exercise 11.6.4).

□

Theorem 11.24 *Given any conformal critical subgraph H of a critical graph G, there exists an ear decomposition of G starting with H. That is, there exists a sequence $(H = G_i, G_{i+1}, \ldots, G_r = G)$, where $i = |E(H)| - |V(H)| + 2$, such that, for $i \le j < r$, the graph G_{j+1} is obtained from G_j by adding either an open ear or a closed ear.*

Proof If H is a spanning subgraph of G, the edges in $E(G) - E(H)$ can be added, one by one, as open ears. So, we may assume that $V(H) \subset V(G)$. Since G is connected, there must be a vertex v of H adjacent to a vertex u in $V(G) - V(H)$. Let M_v be a near-perfect matching of G such that $M_v \cap E(H)$ is a near-perfect matching of H which covers all vertices of H except v. Such a perfect matching must exist because H is a conformal critical subgraph of G. In addition, since G is critical, there is a perfect matching M_u of $G - u$. Clearly, there must exist an (M_v, M_u)-alternating path P_{uv} in G connecting u and v. Let w be the first vertex of P_{uv} in $V(H)$, and let P_{uw} be the uw-segment of P_{uv}. If $w = v$, then vuP_{uv} is a cycle of odd length which can be added to H as a closed ear to obtain a conformal critical subgraph of G which has more edges than H. On the other hand, if $w \ne v$, then vuP_{uw} is an ear path of H and it can be added to H as an open ear to the same effect. This procedure can be clearly carried out recursively to obtain the desired type of ear decomposition of G. □

For any vertex v of a critical graph G, the subgraph induced by $\{v\}$ is clearly a conformal subgraph of G. Thus, the above theorem implies the existence of an ear decomposition of G starting with K_1. Lovász and Plummer (see [59, page 197]) established the following strengthening of the above theorem for 2-connected critical graphs.

Theorem 11.25 *Every 2-connected critical graph G admits an ear decomposition in which all ear additions, subsequent to the first one, are those of open ears.*

Proof By induction on $|E| - |V|$. If $|E| - |V| = 0$, then G is an odd cycle, and the assertion clearly holds in this case. So, suppose that $|E| - |V| \ge 1$. Since G is 2-connected, in any ear decomposition of G, the last ear added is necessarily an open ear. Thus, if there is a decomposition of G in which the penultimate graph is 2-connected, then the induction hypothesis may be applied to that graph and the validity of the assertion may be deduced. With this in view, choose an ear decomposition $\mathcal{G} := (G_1, G_2, \ldots, G_r)$ of G such that the block of the largest size of G_{r-1} has the maximum possible number of edges.

If G_{r-1} is 2-connected, there is nothing more to prove. Suppose that G_{r-1} has more than one block, and let P_{uv} be the ear added to G_{r-1} to obtain $G = G_r$. Since G is 2-connected, it follows that G_{r-1} has precisely two end blocks and that P_{uv} connects a vertex u in one end block to a vertex v in the other end block. Now let M_u be a perfect matching of $G_{r-1} - u$, let M_v be a perfect matching of $G_{r-1} - v$, and let Q_{vu} be the (M_u, M_v)-alternating vu-path in G_{r-1}. Since Q_{vu} connects vertices belonging to the two end blocks of G_{r-1}, it must traverse through each block of G_{r-1} exactly once. For each block B the first edge of $Q_{vu} \cap E(B)$ is an edge of M_v and the last is an edge of M_u. Let L denote a block of G_{r-1} with the largest possible number of edges, and let s and t be the first and last vertices of Q_{vu} in $V(L)$. See Figure 11.10 for an illustration in which G_{r-1} has three blocks with L being the middle block (the edges indicated by solid lines belong to M_u, and those indicated by dashed lines belong to M_v).

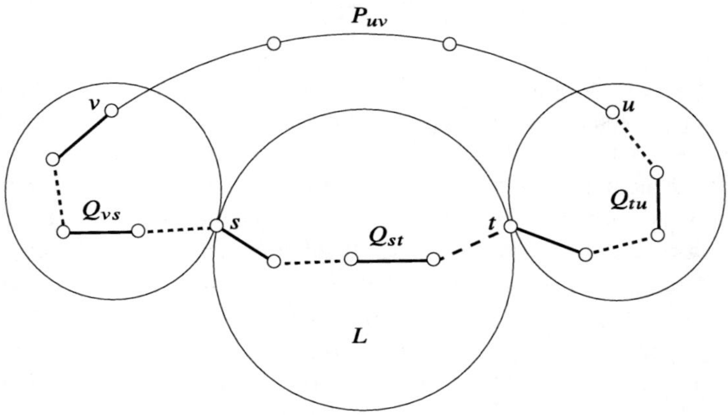

Fig. 11.10 The graph G_{r-1}, largest block L of G_{r-1}, and the last ear P_{uv}

Now let R_{ts} denote the path which is composed of the tv-segment Q_{tu} of Q_{vu}, the path P_{uv}, and the vs-segment Q_{vs} of Q_{vu}. It can be verified that $L \cup R_{ts}$ is a 2-connected conformal subgraph of G (Exercise 11.6.5). If $\mathcal{G}'$ is an ear decomposition of G starting with $L \cup R_{ts}$, the largest block of the penultimate member of $\mathcal{G}'$ has more edges than L. This contradicts the choice of $\mathcal{G}$. □

Although the above proof is not constructive, it is possible to recast it as a constructive procedure for finding an ear decomposition of a 2-connected critical graph of the required type. We leave it as an exercise to the readers to describe such a procedure.

Exercises

11.6.1 *Show that all blocks of a critical graph are critical.*

11.6.2 *Give a proof of Proposition 11.22.*

11.6.3 *Find an ear decomposition of the critical graph G in Figure 11.9 which involves a single closed ear addition.*

11.6.4 *Give a proof of Proposition 11.23.*

11.6.5 *Supply the missing details in the proof of Theorem 11.25.*

11.7 Notes

The idea of ear decompositions, though not the terminology, originated in the work of Whitney (1932). He showed that any 2-connected graph may be obtained from a cycle by adding a sequence of ears (see Chapter 5 of Bondy and Murty [3]). The theory of ear decompositions of matching covered graphs seems to have had its origins in the works of Kotzig published mostly in Slovenian journals in the nineteen-fifties, and of Hetyei published in 1964 in a Hungarian journal. (References to the works of these two authors can be found in the treatises by Lovász and Plummer [59] and Schrijver [84]). Many significant advances were made by Lovász and Plummer and a cohesive account was presented for the first time in their book. But some of their proofs were rather involved. Szigeti [86] gave a simpler proof of the Two Ear Theorem 11.7.

Our own approach to this theory has been in the general framework of the dependence relation and removable classes described in Chapters 8 and 9. This approach has helped us in finding simpler proofs of several basic results in the theory, and also led us to generalizations of some. For example, CLM [8] proved that every matching covered graph G which is different from K_2 and C_{2n} has at least $\Delta(G)$ removable ears, which implies that G has $\Delta(G)!$ ear decompositions. In the next chapter we shall present a very simple proof of a result of Lovász which states that every non-bipartite matching covered graph has an ear decomposition in which either the third graph is a bisubdivision of K_4 or the fourth graph is a bisubdivision of $\overline{C_6}$. One of the main results in Part II is that every brick which is not a Petersen brick has an ear decomposition which involves just one double ear addition.

Chapter 12
Matching Minors

Contents

12.1 Matching Minors

In this final chapter of Part I, we bring together ideas of bicontractions of vertices of degree two and deletions of removable classes described in the preceding chapters to introduce two closely related types of minors that are specific to the theory of matching covered graphs. They play a role in this subject that is similar to that of the familiar type of minors in graph theory. We consider first the more general of these two types of minors which was introduced by Norine and Thomas (2007, [78]).

MATCHING MINORS

A matching covered graph J is a *matching minor* of another matching covered graph G if there is some conformal subgraph F of G such that J is obtainable from F by means of bicontractions.

C. L. Lucchesi, U. S. R. Murty, *Perfect Matchings*, Algorithms and Computation in Mathematics 31, https://doi.org/10.1007/978-3-031-47504-7_12

As an example, consider the tricorn $\mathbb{H}_{10}$ shown in Figure 12.1(a). The wheel W_5 can be obtained from the conformal subgraph of $\mathbb{H}_{10}$ depicted in Figure 12.1(b) by means of bicontractions. Likewise, K_4 can be obtained from the conformal subgraph depicted in Figure 12.1(c) by means of the same operations. Thus W_5 and K_4 are matching minors of $\mathbb{H}_{10}$.

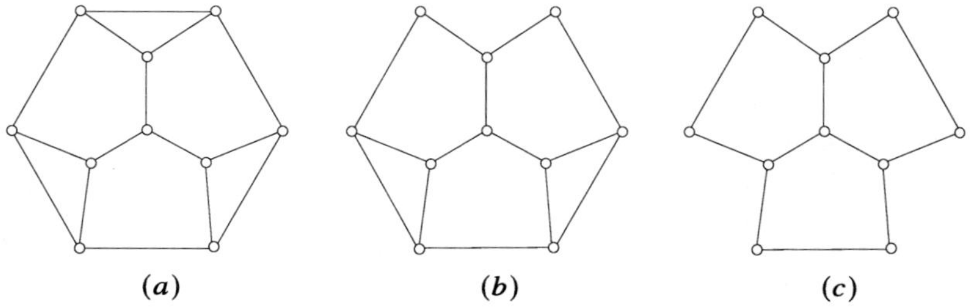

Fig. 12.1 The wheels W_5 and $W_3 = K_4$ are matching minors of $\mathbb{H}_{10}$

The following observation is an obvious consequence of the definition of a matching minor:

Proposition 12.1 *Any matching minor of a conformal subgraph of a matching covered graph G is also a matching minor of G.* □

Theorem 11.11 implies that a matching covered subgraph F of a matching covered graph G is a conformal subgraph of G if and only if it can be obtained from G by a sequence of deletions of removable ears. But the deletion of a removable ear R from G (or, from any conformal subgraph of G) can be effected by first applying suitable bicontractions to reduce R to a removable class in the resulting graph and then deleting it from that graph. Thus any conformal subgraph of G can be obtained from it by means of a sequence of bicontractions of vertices of degree two and deletions of removable classes. It follows that if J is a matching minor of G, then it can be obtained from G by a sequence of bicontractions and deletions of removable classes. We shall see that, conversely, any graph J which can be obtained from G by a sequence of such operations is a matching minor of G. We need the operation of bisplitting a vertex defined below for establishing this converse.

BISPLITTING A VERTEX

Let x be a vertex of degree at least two in a matching covered graph G. The operation of *bisplitting* the vertex x consists of first replacing x by two nonadjacent vertices v_1 and v_2, distributing the edges in $\partial(x)$ between v_1 and v_2 so that each of these two vertices is incident with at least one edge in $\partial(x)$, then adding a new vertex v_0 and joining it to both v_1 and v_2. See Figure 12.2.

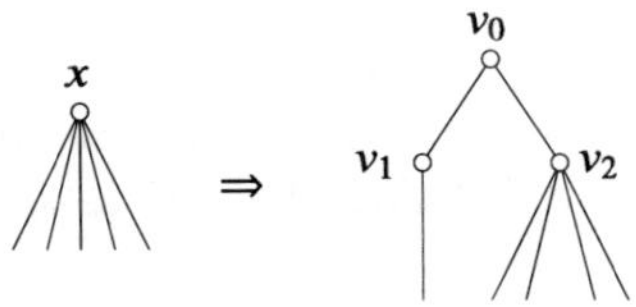

Fig. 12.2 Bisplitting a vertex x

It is easy to check that the resulting graph G' is matching covered in which the vertex v_0 has degree two and that G can be recovered from G' by bicontracting v_0. (In this context, we refer to the new vertex v_0 as the *inner vertex* and the two new vertices v_1 and v_2 as *the outer vertices*.)

Theorem 12.2 *A matching covered graph J is a matching minor of another matching covered graph G if and only if there exists a sequence*

$$G = G_1 \supset G_2 \supset \cdots \supset G_r = J \tag{12.1}$$

of matching covered graphs such that, for $2 \le i \le r$, the graph G_i is obtained from the graph G_{i-1} by either deleting a removable class or by bicontracting a vertex of degree two.

Proof We have already seen that any matching minor of G can be obtained from it by means of a sequence of bicontractions and deletions of removable classes. This proves the 'only if' part of the assertion.

To prove the converse, suppose that there exists a sequence of matching covered graphs as in (12.1) satisfying the stated properties. We shall prove by induction on the number r of graphs in (12.1) that J is a matching minor of G. This is obvious if $J = G$. So, we may assume that $J \neq G$.

By hypothesis, ($G_2 \supset \cdots \supset G_r = J$) is a sequence of matching covered subgraphs such that, for $3 \le i \le r$, the graph G_i is obtained from the graph G_{i-1} by either deleting a removable class from G_{i-1}, or by bicontracting a vertex of degree two in G_{i-1}. By the induction hypothesis, it follows that J is a matching minor of G_2. Thus there exists a conformal subgraph F of G_2 such that J is obtainable from F by means of bicontractions alone.

Case 1 *Graph G_2 is obtained from $G_1 = G$ by the deletion of a removable class R.*

Clearly, any conformal subgraph of $G_2 = G - R$ is also a conformal subgraph of G. Thus, F is a conformal subgraph of G. We conclude that J is a matching minor of G.

Case 2 *Graph G_2 is obtained from $G_1 = G$ by bicontracting a vertex, say v_0, of degree two.*

If the contraction vertex x resulting from the bicontraction of v_0 is not a vertex of F, then F is a subgraph of G. It can be shown easily that it is also a conformal subgraph of G. On the other hand, if x is a vertex of F, then F is not a conformal subgraph of G. However, by a suitable bisplitting of x, one can obtain a graph F' which is a conformal subgraph of G. Furthermore, J can be obtained from F' by a sequence of bicontractions, showing that J is a matching minor of G. We leave the details as Exercise 12.1.1(i). □

The following result is a consequence of Theorem 12.2. We leave its proof as Exercise 12.1.1(ii).

Corollary 12.3 *A matching covered graph J is a matching minor of a matching covered graph $G \neq J$ if and only if*

(i) *either there is a removable class R of G such that J is a matching minor of $H := G - R$;*
(ii) *or there is a vertex v of degree two in G such that J is a matching minor of the graph H obtained from G by the bicontraction of v.* □

12.1.1 Matching-minor-closed classes of graphs

A class $\mathcal{G}$ of matching covered graphs is *matching-minor-closed* if all matching minors of any member of $\mathcal{G}$ are also members of $\mathcal{G}$. An obvious example of such a class is the one consisting of all planar matching covered graphs. A less obvious example is the class of all solid matching covered graphs. (It follows from Theorem 7.1 that a graph obtained from the bicontraction of a vertex of degree two in a solid matching covered graph is solid. And, it follows from Theorems 10.14 and 10.15 that the deletion of a removable class from a solid matching covered graph results in a solid matching covered graph. Thus, any matching minor of a solid matching covered graph is also solid.) The fact that the class of solid matching covered graphs is matching-minor-closed implies that no nonsolid matching covered graph can be a matching minor of a solid matching covered graph. This suggests the possibility of a forbidden matching minor characterization of solid matching covered graphs. One such forbidden matching minor is $\overline{C_6}$, which happens to be the smallest nonsolid matching covered graph. However, not all '$\overline{C_6}$-free' matching covered graphs are solid. For example, $\overline{C_6}$ is not a matching minor of $\mathbb{P}$, the Petersen graph, but $\mathbb{P}$ is not solid. An account of what we know about the relationship between matching covered graphs which are solid and those which are '$\overline{C_6}$-free' is described in Section 12.3.

We shall come across several important examples of matching-minor-closed classes of matching covered graphs and their forbidden matching minor characterizations in Part III. In such problems the graph or graphs which we wish to forbid tend to be either simple bricks or simple braces. The results described in the following subsection play a useful role in finding solutions to such problems.

12.1.2 Matching minors and tight cuts

We leave the proof of the following preliminary proposition as Exercise 12.1.2.

Proposition 12.4 *Let G be a matching covered graph. A simple graph J is a matching minor of G if and only if J is a matching minor of the underlying simple graph of G.* □

We first consider the case in which the matching minor of interest is a simple brick J. Our objective is to show that J is a matching minor of a matching covered graph G if and only if it is a matching minor of one of the bricks of G. We present the proof of this assertion through a sequence of lemmas. Before we do that, it is in order to clarify the terminology we use concerning bricks of a matching covered graph G. By a brick of G we shall mean a brick resulting from any tight cut decomposition of G and, by a *simple brick* of G we shall mean the underlying simple graph of a brick of G. By Theorem 4.17, the list of simple bricks of G does not depend on the choice of the tight cut decomposition employed.

Lemma 12.5 *Let $C := \partial(X)$ be a nontrivial tight cut of a matching covered graph G. If a simple brick J is a matching minor of a C-contraction of G then it is also a matching minor of G.*

Proof Assume that J is a matching minor of a C-contraction of G. We prove, by induction on $|E|$, that J is a matching minor of G. Let $G_1 := G/(\overline{X} \to \overline{x})$ and $G_2 := G/(X \to x)$ denote the two C-contractions of G. Adjust notation so that J is a matching minor of G_1.

Let Y be a maximal superset of X such that $D := \partial(Y)$ is a nontrivial tight cut of G. Then, D is a tight cut of G_2. Moreover, if D is a trivial cut of G_2 then $D = C$. Let $H_1 := G/(\overline{Y} \to \overline{y})$ and $H_2 := G/(Y \to y)$ be the two D-contractions of G.

12.5.1 *The brick J is a matching minor of H_1.*

Proof If $C = D$ then $H_1 = G_1$; in that case, J is a matching minor of H_1. Suppose thus that C and D are distinct. The cut C is a nontrivial tight cut of H_1. The brick J is a matching minor of G_1, in turn a C-contraction of H_1. By induction, with H_1 playing the role of G, we conclude that the brick J is a matching minor of H_1. □

Case 1 *The graph H_2 has a removable class R_2 disjoint with D.*

The set R_2 is then a removable class of G, the cut D is tight in $G - R_2$ and H_1 is intact as a D-contraction of $G - R_2$. By induction, J is a matching minor of $G - R_2$. By Corollary 12.3, J is a matching minor of G.

Case 2 *The graph H_2 has a vertex v_2 of degree two such that $\partial(v_2)$ and D are disjoint.*

Let G' be the graph obtained from G by bicontracting vertex v_2. The cut D is tight in G' and H_1 is intact as a D-contraction of G'. By induction, J is a matching minor of G'. By Corollary 12.3, J is a matching minor of G.

12.5.2 *If H_2 is a brace then either Case 1 or Case 2 applies.*

Proof If H_2 has a removable edge which is not incident with the contraction vertex y then Case 1 applies. Suppose then that H_2 does not have a removable edge which is not incident with the contraction vertex y. By Theorem 8.3, H_2 has only four vertices. Moreover, H_2 has a vertex v of degree two in the same part of the bipartition of H_2 that contains vertex y. Case 2 then applies. □

By the maximality of Y, the graph H_2 does not have nontrivial tight cuts. We may thus assume that H_2 is a brick.

12.5.3 *If the graph H_1 is also a brick then it has multiple edges.*

Proof Suppose that H_1 is also a brick. Then, G is the splicing of two bricks. By hypothesis, the cut D is tight and nontrivial; therefore G is not a brick. By Proposition 5.15, there exists a 2-separation $S := \{u, v\}$ of G, with $u \in Y$ and $v \in \overline{Y}$, such that every edge of the cut D is incident with a vertex in S. (See Figure 12.3.)

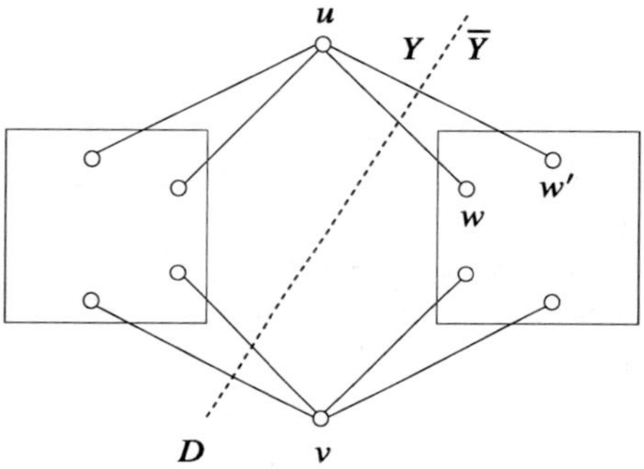

Fig. 12.3 The case in which G is not a brick but both H_1 and H_2 are

If u is adjacent to just one vertex $w \in \overline{Y} - v$, then $\{v, w\}$ would be a barrier of the brick H_2, which is absurd. Therefore, u is adjacent to at least two vertices in $\overline{Y} - v$. It follows that $H_1 = G/\overline{Y}$ has multiple edges incident with the contraction vertex $\overline{y}$. □

By hypothesis, J is a simple brick; hence J is not isomorphic to H_1. As J is a matching minor of H_1, it follows that a sequence such as (12.1), starting with H_1 and ending with J, has two or more graphs. We now consider two cases, depending on whether the second graph of the sequence is obtained by the deletion of a removable class or by the bicontraction of a vertex of degree two.

Case 3 *The graph H_1 has a removable class R_1 such that J is a matching minor of $H_1 - R_1$.*

Suppose that R_1 and D are disjoint. The set R_1 is then a removable class of G, the cut D is tight in $G - R_1$ and $H_1 - R_1$ is a D-contraction of $G - R_1$. By induction, J is a matching minor of $G - R_1$. By Corollary 12.3, J is a matching minor of G.

Suppose then that R_1 and D are not disjoint. In that case, R_1 has one edge, say e, in D. By Proposition 8.14, edge e induces a minimal class, say, R_2, of the dependence relation on the edges of H_2. Thus, only edges of R_2 depend on edges of R_2 and every edge of R_2 depends on edge e. As H_2, a brick, is 3-connected, R_2 is a removable class of H_2, by Corollary 9.2. If R_2 does not contain the edge e, then R_2 would be a subset of $E(H_2) - D$, $H_2 - R_2$ would be matching covered and Case 1 would apply. Therefore, we may assume that R_2 contains the edge e. In this case, $D - e$ is tight in $G - R_1 - R_2$, and $H_1 - R_1$ is a $(D - e)$-contraction of $G - R_1 - R_2$. By induction, J is a matching minor of $G - R_1 - R_2$. But $G - R_1 - R_2$ is a conformal subgraph of G with J as a matching minor, implying that J is a matching minor of G itself (see Proposition 12.1).

Case 4 *H_1 has a vertex v of degree two such that J is a matching minor of the graph H'_1 obtained by bicontracting v in H_1.*

Suppose that $\partial(v)$ and D are disjoint. Let G' be the graph obtained from G by bicontracting vertex v. The cut D is tight in G' and H'_1 is a D-contraction of G'. By induction, J is a matching minor of G'. By Corollary 12.3, J is a matching minor of G.

Suppose thus that $\partial(v)$ and D are not disjoint. Since H_2 is a brick, the cut D has at least three edges and, hence $v \neq \overline{y}$. Therefore, precisely one of the two neighbours of v, say w, is in $\overline{Y}$. See Figure 12.4.

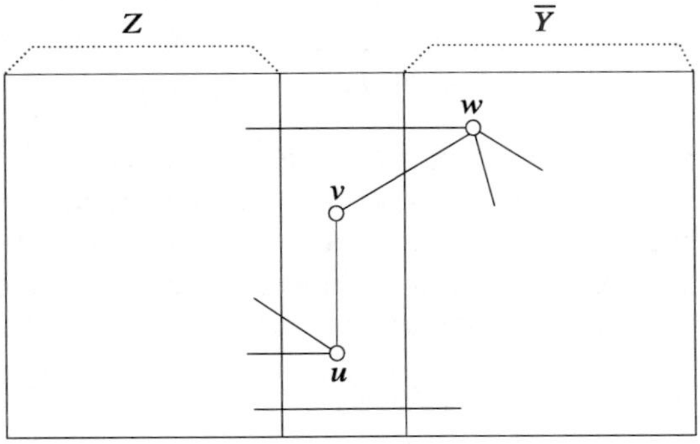

Fig. 12.4 The case in which H_1 has a vertex of degree two

Let u be the other neighbour of v, and let $Z := Y - \{u, v\}$. The cut $\partial(Z)$ is tight in G as well as in G', the graph obtained from G by bicontracting v. Furthermore, the graph $G'/(\overline{Z} \rightarrow \overline{z})$ is isomorphic to H'_1. By induction, J is a matching minor of G'. Hence J is a matching minor of G, by Corollary 12.3. □

Corollary 12.6 *Every simple brick of a matching covered graph G is a matching minor of G.* □

Theorem 12.7 *Let G be a matching covered graph. A simple brick J is a matching minor of G if and only if J is a matching minor of some simple brick of G.*

Proof If J is a matching minor of some simple brick of G then J is a matching minor of G, by Corollary 12.6. We prove the converse by induction on $|E|$.

Assume that J is a matching minor of G. If G is a brick, then J is a matching minor of the underlying simple graph of G, which is the simple brick of G. We may thus assume that G is not a brick. In particular, this implies that G and J are distinct. By Corollary 12.3, J is a matching minor of a graph G', obtained from G either by bicontracting a vertex of degree two or by the removal of a removable class of G.

Consider first the case in which G' is the result of the bicontraction of a vertex of degree two of G. By induction, J is a matching minor of a simple brick of G'. But, every simple brick of G' is a simple brick of G. Thus, J is a matching minor of a simple brick of G.

We may thus assume that $G' = G - R$, where R is a removable class of G. As G is not a brick, it follows that G has a nontrivial tight cut $C := \partial(X)$. Let G_1 and G_2 denote the two C-contractions of G. By induction, J is a matching minor of some simple brick of $G - R$, say H. The cut $C' := C - R$ is nontrivial and tight in $G - R$. Thus, H is a simple brick of some C'-contraction of $G - R$. The two C'-contractions of $G - R$ are $G_1 - R$ and $G_2 - R$. Adjust notation so that H is a simple brick of $G_1 - R$. By Corollary 12.6, J is a matching minor of $G_1 - R$. Thus, J is a matching minor of G_1. By induction, J is a matching minor of some simple brick L of G_1. The graph L is a simple brick of G. Thus, J is a matching minor of a simple brick of G. □

Corollary 12.8 *Let G be a matching covered graph and let v be a vertex of degree two of G. A simple brick J is a matching minor of G if and only if J is a matching minor of the graph obtained from G by the bicontraction of v.* □

Curiously, the analogue of Corollary 12.6 for braces is not valid: a simple brace J of a matching covered graph G need not be a matching minor of G. For example, $K_{3,3}$ is a simple brace of $K_{3,3} \odot K_4$, but it is not a matching minor of $K_{3,3} \odot K_4$ (Exercise 12.1.6). However, if G is a bipartite matching covered graph, then every simple brace of J of G is a matching minor of G (Exercise 12.1.8).

Exercises

12.1.1 *Provide the missing details in the proofs of:*

(i) *Theorem 12.2;*
(ii) *Corollary 12.3.*

12.1.2 *Give a proof of Proposition 12.4.*

▷**12.1.3** *Give examples of simple bricks J and separating cuts C of a brick G such that J is a matching minor of some C-contraction of G but it is not a matching minor of G.*

12.1.4 *Prove Corollary 12.8.*

▷**12.1.5** *Show that the Petersen graph is a matching minor of the Blanuša snark depicted in Figure 12.5.*

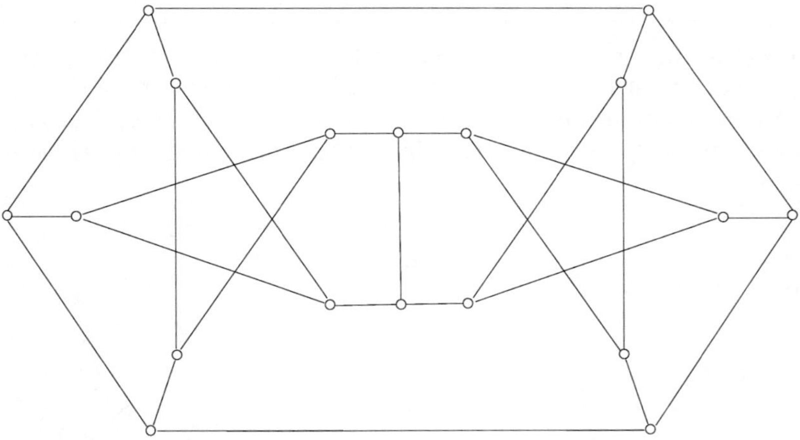

Fig. 12.5 The Blanuša snark

12.1.6 *Show that $K_{3,3}$ is not a matching minor of $K_{3,3} \odot K_4$.*

12.1.7 *Let G be a matching covered graph and let J be a simple brick that is a matching minor of G. Prove that if G and J are not isomorphic then one of the following properties holds:*

(i) *either G has vertices of degree two and, for every vertex v of degree two, J is a matching minor of the graph obtained from G by the bicontraction of v, or*
(ii) *G has minimum degree three or more and G has a removable class R such that J is a matching minor of $G - R$.*

▷**12.1.8** *Show that a simple brace J is a matching minor of a bipartite matching covered graph G if and only if it is a matching minor of a simple brace of G.*

12.2 Conformal Minors

We now turn to a special class of matching minors which we refer to as conformal minors that have been in vogue, though not under this name, since they were introduced by Little (1975, [54]). They play a role in this theory that is akin to that of topological minors in the theory of embeddings of graphs in surfaces. We recall the definition of conformal minor, originally given in Chapter 3.

CONFORMAL MINORS

A matching covered graph J is a *conformal minor* of another matching covered graph G if some bisubdivision of J is a conformal subgraph of G.

As Figure 12.6(a) shows, K_4 is a conformal minor of $\mathbb{P}$, the Petersen graph. But $\overline{C_6}$ is not a conformal minor of $\mathbb{P}$ (Exercise 12.2.3). However, if $\mathbb{P} + e$ is any graph obtained by adding an edge e to $\mathbb{P}$ joining two nonadjacent vertices, then $\overline{C_6}$ is a conformal minor of $\mathbb{P} + e$ as illustrated in 12.6(b), and also of $K_4 \odot \mathbb{P}$, as illustrated in 12.6(c).

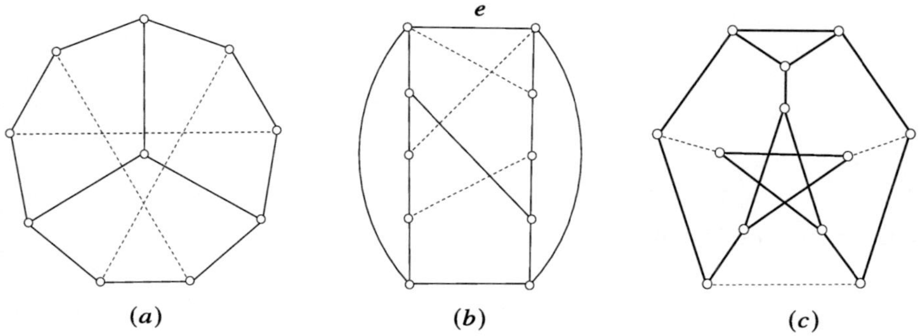

Fig. 12.6 Conformal minors: (a) K_4 of $\mathbb{P}$; (b) $\overline{C_6}$ of $\mathbb{P} + e$; (c) $\overline{C_6}$ of $K_4 \odot \mathbb{P}$

A *restricted bicontraction* is the bicontraction of a vertex of degree two that has at least one neighbour which is also of degree two. It follows from 11.11 that if H is a conformal matching covered subgraph of a matching covered graph G, then H can be obtained from G by a sequence of deletions of removable ears (single or double). But the deletion of an ear amounts to first reducing that ear to one of length one by means of restricted bicontractions, and then deleting the only edge of that ear. Likewise, if H is a bisubdivision of a graph J then J can in fact be obtained from H by means of restricted bicontractions. Thus any conformal minor J of a matching covered graph G can be obtained from it by means of a sequence of deletions of removable classes and restricted bicontractions. Conversely, it is not difficult to see that any graph J obtainable from G by a sequence of such operations is a conformal minor of G (Exercise 12.2.1). The following proposition, which is analogous to Theorem 12.2, records these observations.

Proposition 12.9 *A matching covered graph J is a conformal minor of another matching covered graph G if and only if there exists a sequence*

$$G = G_1 \supset G_2 \supset \cdots \supset G_r = J \qquad (12.2)$$

of matching covered graphs such that, for $2 \leq i \leq r$, the graph G_i is obtained from the graph G_{i-1} by either deleting a removable class from G_{i-1}, or by a restricted bicontraction of a vertex of degree two in G_{i-1}. □

As an immediate consequence of the above proposition, we have:

Corollary 12.10 *Every conformal minor of a matching covered graph G is also matching minor of G.* □

The converse is not true in general, due to the fact that unrestricted bicontractions are permissible in obtaining a matching minor of G. (For example, the wheel W_5 is a matching minor of the tricorn $\mathbb{H}_{10}$ as we have already observed (see Figure 12.1). But it is not a conformal minor of $\mathbb{H}_{10}$ for the simple reason that $\mathbb{H}_{10}$ has no vertices of degree greater than three and hence cannot have a bisubdivision of W_5 as a conformal subgraph.) However, it is easily seen that if a cubic matching covered graph J is obtained from a matching covered graph H by means of bicontractions, then H must be a bisubdivision of J. Consequently, we have the following:

Corollary 12.11 *A cubic matching covered graph J is a matching minor of a matching covered graph G if and only if J is a conformal minor of G.* □

12.2.1 Unavoidable conformal minors

The following fundamental theorem due to Lovász provides the motivation and also serves as a prototype for many of our investigations.

UNAVOIDABLE CONFORMAL MINORS: K_4, $\overline{C_6}$

Theorem 12.12 (Lovász (1983, [57]))

Every nonbipartite matching covered graph has a bisubdivision of either K_4 or of $\overline{C_6}$ as a conformal subgraph. □

In other words, every nonbipartite matching covered graph has at least one of K_4 and $\overline{C_6}$ as a conformal minor.

In view of Corollary 12.11, the following result implies Theorem 12.12:

Theorem 12.13 *Every nonbipartite matching covered has either K_4 or $\overline{C_6}$ as a matching minor.*

Proof By induction on $|E|$. Consider first the case in which G has a removable edge, e. By the monotonicity of function b, the graph $G - e$ is nonbipartite. By induction, $G - e$ has either K_4 or $\overline{C_6}$ as a matching minor. In that case, the assertion holds.

Consider next the case in which G has a nontrivial tight cut $C := \partial(X)$. At least one C-contraction of G, say, G_1, is nonbipartite. By induction, G_1 has either K_4 or $\overline{C_6}$ as a matching minor. By Lemma 12.5, G also has one of K_4 or $\overline{C_6}$ as a matching minor.

Suppose then that G is free of nontrivial tight cuts and also that it has no removable edges. As G is nonbipartite, we deduce that G is a brick. As it has no removable edges, it is simple. Moreover, by Theorem 9.14, G itself is either K_4 or $\overline{C_6}$. □

The original proof of the above theorem given by Lovász was somewhat involved. Carvalho and Lucchesi (1996, [7]) found a simpler proof based on the result of Little stated in Exercise 11.3.5. CLM (1999, [8]) deduced this result from Theorem 9.18 by showing that every near-bipartite brick G different from K_4 and $\overline{C_6}$ has an edge e such that $G - e$ is near-bipartite.

An ear decomposition $(K_2 = G_1 \subset G_2 \subset \cdots \subset G_r = G)$ of a nonbipartite matching covered graph G is *primary* if either the third graph G_3 is a bisubdivision of K_4 or the fourth graph G_4 is a bisubdivision of $\overline{C_6}$. Lovász [57] uses the term *canonical* to indicate that an ear decomposition is primary. Theorems 12.12 and 11.11 imply the following fundamental result concerning ear decompositions.

Theorem 12.14 (Primary Ear Decomposition Theorem) *Every nonbipartite matching covered graph has a primary ear decomposition.* □

For example, the ear decomposition of the Petersen graph $\mathbb{P}$ described in Figure 11.4 is a primary ear decomposition because the third graph is a bisubdivision of K_4; and the ear decomposition of $\mathbb{P} + 47$ described in Exercise 11.4.5(iii) is a primary ear decomposition because the fourth graph is a bisubdivision of $\overline{C_6}$. □

Exercises

12.2.1 *Show that a matching covered graph J is a conformal minor of another matching covered graph G if and only if J can be obtained from G by means of deletions of removable classes and restricted bicontractions.*

12.2.2 *Consider the graph G shown in Figure 12.7. Find an ear decomposition of G which is primary, and one which is not primary.*

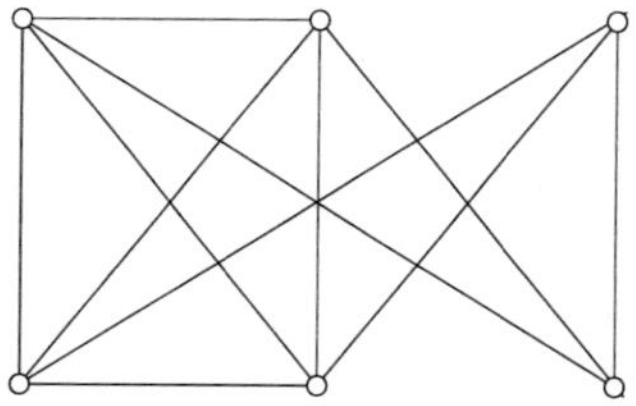

Fig. 12.7 Primary Ear Decomposition (Exercise 12.2.2)

12.2.3 *Show that the Petersen graph does not have a conformal subgraph that is a bisubdivision of $\overline{C_6}$. Hint: note that $\mathbb{P}$ is a transitive brick, that $\overline{C_6}$ is cubic, and use Corollaries 12.11 and 12.8.*

12.2.4 *Show that neither the bicorn $\mathbb{H}_8$ nor the tricorn $\mathbb{H}_{10}$ has $\overline{C_6}$ as a conformal minor.*

12.2.5 *Consider the family $\mathbb{S}_{2n}$, of staircases defined in Section 2.5. Show that $\mathbb{S}_{2n}$, $(n \geq 3)$, has $\overline{C_6}$ as a conformal minor if and only if n is odd.*

12.2.6 *Show that $\overline{C_6}$ is a conformal minor of the graph shown in Figure 12.8 which is obtained by adding an edge to the staircase $\mathbb{S}_{12}$.*

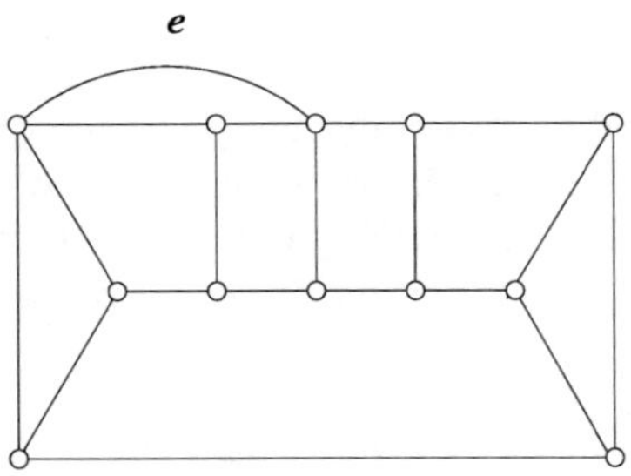

Fig. 12.8 $\overline{C_6}$ is a conformal minor of $\mathbb{S}_{12} + e$

∗ **12.2.7** *The objective of this exercise is to prove, by induction, that every brace of order six or more has either $K_{3,3}$ or the cube as a conformal minor.*

(i) *Prove that for every simple brace G of order ten or more and for every (removable) edge e of G, the underlying simple graph J of the retract H of $G - e$ has minimum degree three or more. Hint: consider the ways in which H may have multiple edges and prove that if J has minimum degree two then G has a nontrivial barrier containing fewer than five vertices.*

(ii) *Prove that the only simple brace of order six is $K_{3,3}$ and also prove that every brace of order eight has the cube as a spanning subgraph.*

(iii) *Prove the assertion by induction. If the brace has order ten or more, delete an edge e from G, use the first item to prove that $G - e$ has a brace of order six or more and then apply Corollary 12.11 and the result stated in Exercise 12.1.8.*

12.3 Graphs With Specified Conformal Minors

> J-BASED AND J-FREE GRAPHS
>
> Let J be any specific matching covered graph. We say that a matching covered graph G is *J-based* if there exists a conformal subgraph of G which is isomorphic to a bisubdivision of J; and is *J-free* if there is no such subgraph of G.
>
> In other words, G is J-based if J is a conformal minor of G; otherwise, it is J-free.

Using the above terminology, Theorem 12.12 may be restated as follows:

Theorem 12.15 *Every nonbipartite matching covered graph is either K_4-based or is $\overline{C_6}$-based (or both).* □

For example, the Petersen graph $\mathbb{P}$ is K_4-based, but is $\overline{C_6}$-free; the pentagonal prism $\mathbb{P}_{10}$ is $\overline{C_6}$-based, but K_4-free; and the graph $\mathbb{P} + 47$ shown in Figure 11.5 is both K_4-based and $\overline{C_6}$-based.

The problems of deciding which matching covered graphs are K_4-based and, similarly, which are $\overline{C_6}$-based are far from being solved. The following assertion, which is relevant to the solutions of these problems, first appeared in a paper by Kothari and Murty (2015, [50]). But it follows directly from Theorem 12.7 and the result in Exercise 11.5.4.

Theorem 12.16 ([50]) *Suppose that J is a cubic brick and that C is a tight cut of a matching covered graph G. Then G is J-free if and only if each C-contraction of G is J-free. (In particular, G is J-free if and only if each brick of G is J-free.)* □

The restriction that J be a cubic brick, hence a simple brick, is crucial for the validity of the above statement. (As already noted, it is not valid even for cubic braces.)

Two unsolved problems

Problem 12.17 *Characterize K_4-free bricks.*

Problem 12.18 *Characterize $\overline{C_6}$-free bricks.*

Since both K_4 and $\overline{C_6}$ are cubic bricks, it follows from Theorem 12.16 that solutions to the above problems would provide characterizations of K_4-free and $\overline{C_6}$-free matching covered graphs.

Using a brick generation theorem of Norine and Thomas, which will be described in Part-II, Kothari and Murty [50] were able to resolve them in the special case of planar bricks. In the remainder of this section we describe their results with some explanation but without proofs. We do this to give the reader an inkling of the topics to be studied in Parts II and III.

Adopting the terminology of Bondy and Murty [3], we shall refer to a planar embedding of a planar graph as a *plane graph*. A well-known theorem of Whitney (1933) asserts that every simple 3-connected planar graph has a unique embedding in the plane (see Theorem 10.28 in [3]). The import of this assertion is that if G' and G'' are two plane graphs obtained by embeddings of a simple 3-connected plane graph G, then a cycle of G is a facial cycle in G' if and only if it is also a facial cycle in G''. Clearly this statement is also true for subdivisions of simple 3-connected planar graphs.

12.3.1 K_4-free planar bricks

Suppose that G is a plane K_4-based graph. Then there exists a sequence $(G_1, G_2, \ldots, G_r)$ of plane graphs such that G_1 is a bisubdivision of K_4 and, for $1 \le i \le r-1$, the plane graph G_{i+1} is obtained by adding an ear to the plane graph G_i. Clearly, as G is a plane graph, any ear path of G_i is confined to a face of G_i. Thus adding a single ear P_i to G_i amounts to dividing a face of G_i into two faces by drawing ear path P_i joining two vertices on the boundary of that face. It is easy to verify that G_1 has 4 odd faces, and the number of odd faces of G_i is nondecreasing along the sequence.

Consequently, any plane matching covered graph that is K_4-based must have at least four odd faces (Exercise 12.3.1(ii)). But not every plane matching covered graph with four (or more) odd faces is K_4-based. For example, the graph depicted in Figure 12.9 has four odd faces but it is K_4-free (Exercise 12.3.2).

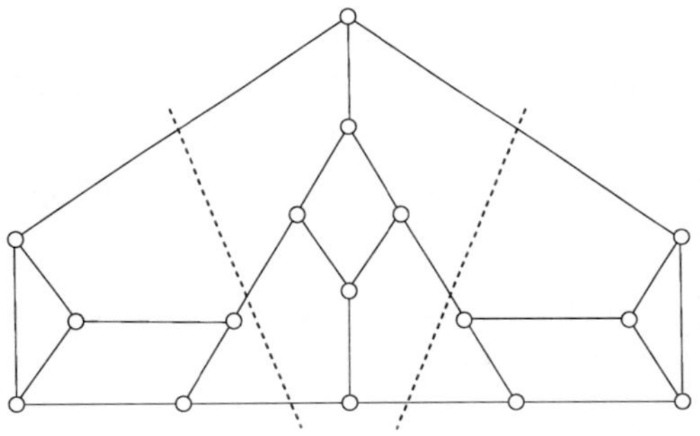

Fig. 12.9 A planar matching covered graph that is K_4-free

However, any plane brick with four or more odd faces is K_4-based. This follows from the following characterization of K_4-free plane bricks.

A CHARACTERIZATION OF PLANAR K_4-FREE BRICKS

Theorem 12.19 (Kothari and Murty [50]) *A simple planar brick is K_4-free if and only if its (unique) planar embedding has precisely two odd faces.*

A cubic nonplanar K_4-free graph of order 14, found by Nishad Kothari, is shown in Figure 12.10. It happens to be the smallest nonplanar K_4-free brick.

It is easy to see that the trellis shown in Figure 12.10 is a nonplanar brick (Exercise 12.3.3(i)). Apart from the obvious 'left-right' and 'top-bottom' reflectional symmetries, it also has the 'insider-out' symmetry which maps the quadrilateral $Q_1 := u_1u_2v_2v_1u_1$ onto $Q_2 := s_1s_2t_2t_1s_1$, and *vice-versa*.

We shall briefly sketch here a proof, by contradiction, to the effect that the trellis is K_4-free. Let G denote the trellis and assume that it is K_4-based. This assumption implies that brick G has a subgraph H of G such that (i) H is a bisubdivision of

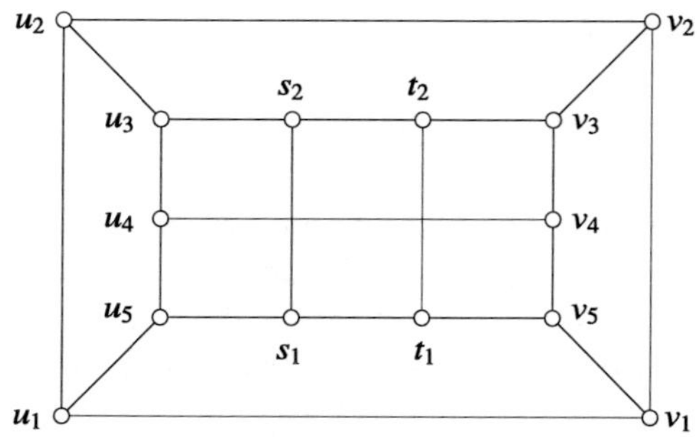

Fig. 12.10 Trellis, the smallest nonplanar K_4-free brick

K_4, and (ii) the graph $G - V(H)$ has a perfect matching, say M. Let H^* denote the spanning subgraph of G induced by the edge set $E(H) \cup M$. In H^*, the four branch vertices of H have degree three, all the subdivision vertices of H have degree two, and all the vertices which are incident with edges in the matching M have degree one. It is easy to establish the following statements (assuming that G is K_4-based) (Exercise 12.3.3(ii)).

Proposition 12.20 *Any two cycles of the graph H^* have at least one vertex in common.* □

Proposition 12.21 *If e is any removable edge of G that is not in $E(H^*)$, then the matching covered graph $G - e$ is also K_4-based.* □

Corollary 12.22 *Any removable edge e of G such that $G - e$ is K_4-free must be an edge of H^*.* □

Consider for example the two edges u_1u_2 and u_2v_2. The bricks of $G - e_1$ and $G - e_2$ happen to be just their retracts:

$$G_1 := \widehat{G - u_1u_2} = G - u_1u_2/(\{u_5, u_1, v_1\} \to x)/(\{u_3, u_2, v_2\} \to y) \text{ and}$$
$$G_2 := \widehat{G - u_2v_2} = G - u_2v_2/(\{u_3, u_2, u_1\} \to x)/(\{v_3, v_2, v_1\} \to y).$$

These bricks are shown in Figure 12.11

Note that the bricks of both $G - u_1u_2$ and $G - u_2v_2$ are planar bricks with two odd faces. Hence, by Theorem 12.19, they are K_4-free. Clearly, the edge v_1v_2 is similar to u_1u_2, and the edge u_1v_1 to u_2v_2. Hence $G - v_1v_2$ and $G - u_1v_1$ are also K_4-free. It now follows from Corollary 12.22 that the four edges u_1u_2, u_2v_2, v_2v_1 and v_1u_1 all belong to $E(H^*)$, and hence that the 4-cycle $Q_1 := u_1u_2v_2v_1u_1$ is a cycle in H^*.

The 'inside-out' symmetry of the trellis mentioned earlier implies that the 4-cycle $Q_2 := s_1s_2t_2t_1s_1$ is also a cycle in H^*. The two cycles Q_1 and Q_2 are vertex-disjoint cycles in H^*. This contradicts Proposition 12.20. We conclude that the trellis is K_4-free.

A description of an infinite family of K_4-free nonplanar bricks, which includes the trellis, is given in Exercise 12.3.4. Other constructions of K_4-free bricks and fascinating conjectures relating to them were given by Kothari (2019, [46]).

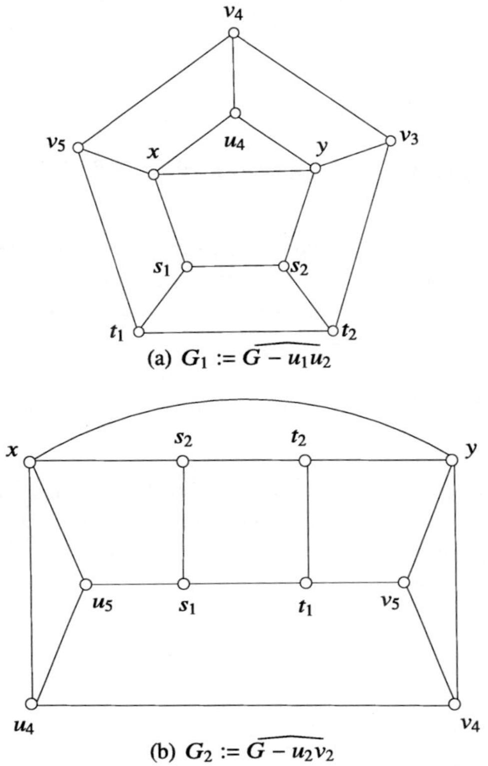

Fig. 12.11 The bricks of $G - u_1u_2$ and $G - u_2v_2$

12.3.2 $\overline{C_6}$-free planar bricks

Now let us turn to the problem of characterizing $\overline{C_6}$-free planar bricks. Odd wheels, being solid bricks, are $\overline{C_6}$-free. It can be verified that $\mathbb{H}_{10}$, the tricorn, is $\overline{C_6}$-free (Exercise 12.2.4), and also that any staircase $\mathbb{S}_{2n}$, where n is an even integer which is at least four, is $\overline{C_6}$-free (Exercise 12.2.5). Kothari and Murty showed that these are the only simple $\overline{C_6}$-free planar bricks:

THE LIST OF ALL SIMPLE $\overline{C_6}$-FREE PLANAR BRICKS

Theorem 12.23 (Kothari and Murty [50])
The only simple planar $\overline{C_6}$-free bricks are the odd wheels, the tricorn and staircases of order $4k$. □

Since every solid brick is $\overline{C_6}$-free and since the tricorn and staircases are not solid, we have the following characterization of simple planar solid bricks, originally proved by CLM (2006, [15]).

Corollary 12.24 (Kothari and Murty [50]) *The only simple planar solid bricks are the odd wheels.* □

12.3.3 A special property of the Petersen graph

The problem of characterizing nonplanar $\overline{C_6}$-free matching covered graphs and the seemingly unrelated problem of characterizing nonplanar solid bricks are now known to be equivalent. This is a consequence of the following theorem due to Lucchesi, Carvalho, Kothari and Murty [64]:

Theorem 12.25 *The Petersen graph is the only simple nonplanar $\overline{C_6}$-free brick that is not solid.* □

We conclude this part with the assertion above, which establishes a property of the Petersen graph that is not shared by any other simple brick. This graph will continue to play a prominent role in the chapters to come.

Exercises

12.3.1
Let G be a K_4-based plane matching covered graph. Let $(G_1, G_2, \ldots, G_r)$ be an ear decomposition of G which starts with a bisubdivision of K_4.

(i) *Verify that the number of odd faces of G_i is at least as large as the number of faces of G_{i-1}, for $i = 2, \ldots, r$.*
(ii) *Deduce that any plane matching covered graph that is K_4-based must have at least four odd faces.*

12.3.2 *The planar matching covered graph shown in Figure 12.9 has four odd faces. Using Theorem 12.16, show that this graph is K_4-free.*

12.3.3

(i) *Show that the trellis is nonplanar.*
(ii) *Give proofs of Propositions 12.20, 12.21, and Corollary 12.22.*

▷**12.3.4** *Let $k \geq 2$ be any positive integer, and consider the prism $\mathbb{P}_{4k+2}$ obtained from the two odd cycles $u_1u_2 \cdots u_{2k+1}u_1$ and $v_1v_2 \cdots v_{2k+1}v_1$ by adding an edge joining u_i to v_i, for $1 \leq i \leq 2k+1$. Now replace the edge $u_{2k+1}v_{2k+1}$ by the path $u_{2k+1}s_1t_1v_{2k+1}$, and the edge u_3v_3 by the path $u_3s_2t_2v_3$. Add the edges s_1s_2 and t_1t_2. We shall denote the resulting graph by Tr_{4k+6}. Note that Tr_{14} is the trellis shown in Figure 12.10.*

(i) *Draw the graph Tr_{18}.*
(ii) *Show that Tr_{4k+6} is a nonplanar K_4-free brick for each $k \geq 2$.*

▷**12.3.5** *The Heawood graph, shown in Figure 12.12, is the unique cubic graph of girth six with the fewest number of vertices. Show that it is $K_{3,3}$-free. (Hint: At least three edges of $K_{3,3}$ have to be bisubdivided to obtain a bipartite matching covered graph that is free of 4-cycles.)*

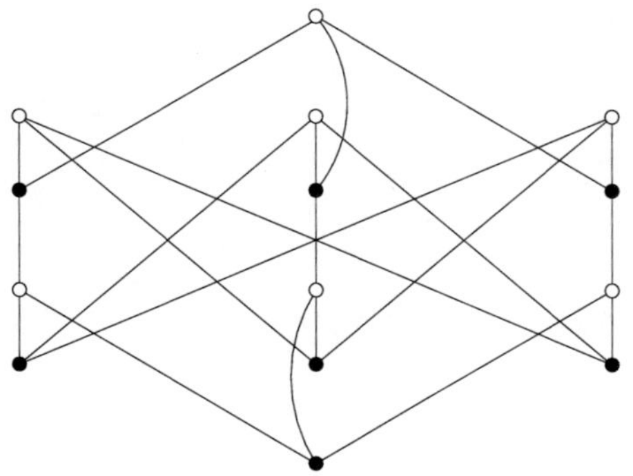

Fig. 12.12 The Heawood graph (a nonplanar $K_{3,3}$-free brace)

12.4 Notes

The first author to recognize the importance of the idea of conformal minors of matching covered graphs was Little (1972, [51]). Following the lead of Kasteleyn (1963, [44]), he considered graphs which admit a certain special type of orientation called a *Pfaffian orientation*. He showed that a bipartite matching covered graph admits a Pfaffian orientation if and only if it is $K_{3,3}$-free. (The Heawood graph shown in Figure 12.12 is an example of such a graph.) Later on, Robertson, Seymour and Thomas (1999, [82]) and, independently, McCuaig (2001, [68]), established a structural characterization of $K_{3,3}$-free bipartite matching covered graphs which led to a polynomial-time algorithm for recognizing such graphs. These results will be described in greater detail in Part III.

Norine and Thomas (2007, [78]) introduced the notion of matching minor. In that paper they established an important theorem concerning matching minors known as the 'splitter theorem'. This theorem was one of the tools used in the proof of Theorem 12.25.

Brick and brace generation procedures are a source for the inductive tools that are needed for analyzing properties of matching covered graphs with specified excluded matching minors. These procedures are the subject matter of Part II. Some of the future advances in this theory may well come from a deeper understanding of the structure of certain matching-minor-closed classes of matching covered graphs!

Part II
Brick and Brace Generation

Chapter 13
A Conjecture of Lovász Concerning Bricks

Contents

13.1 b-Invariant Edges in Bricks

The properties of matching covered graphs with nontrivial tight cuts may often be deduced from those of their bricks and braces. But to deal with bricks and braces themselves, which are free of such cuts, one requires subtler inductive tools. In case of bricks, for example, optimal ear decompositions, that is, ear decompositions with as few double ear additions as possible, turn out to be useful for analyzing their properties.

The seminal work of Lovász on the matching lattice (mentioned in Section 6.3.2), and a conjecture posed by him in that context, served as the motivation for much of the work of CLM (2002, [9], [10] and [11]). This work led to the development of the theory in several different directions. Independently, motivated by questions related to Pfaffian orientations of graphs, Little (1975, [54]), McCuaig (2001, [68]), Robertson, Seymour and Thomas (1999, [82]), and Norine and Thomas (2007,[78]) established deep results on bricks and braces. The objective of this part of the book is to describe some of these developments.

Recall that the matching lattice of a matching covered graph G, denoted by $\mathcal{Lat}(G)$, consists of all integer linear combinations of incidence vectors of perfect matchings of G. Clearly, $\mathcal{Lat}(G)$ is a subset of $\mathcal{Lin}(G) \cap \mathbb{Z}^E$, where $\mathcal{Lin}(G)$ is

C. L. Lucchesi, U. S. R. Murty, *Perfect Matchings*, Algorithms and Computation in Mathematics 31, https://doi.org/10.1007/978-3-031-47504-7_13

the matching space of G. Thus, every $\mathbf{x}$ in $\mathcal{L}at(G)$ is a regular integer vector (see Theorem 6.7). But, as we have noted, the converse does not always hold; the vector $\mathbf{1}$ of all 1's is in the matching space of the Petersen graph $\mathbb{P}$, but it does not belong to the matching lattice of $\mathbb{P}$. We shall refer to a brick whose underlying simple graph is $\mathbb{P}$ as a *Petersen brick*. The essential part of Lovász's characterization of the matching lattice is the following theorem which we mentioned in Chapter 6:

Theorem 13.1 (Lovász [58]) *For any matching covered graph G,*

$$\mathcal{L}at(G) = \mathcal{L}in(G) \cap \mathbb{Z}^E \tag{13.1}$$

if and only if no brick of G is a Petersen brick. □

The proof of the above theorem given by Lovász [58] was somewhat involved. His search for a simpler proof led him to surmise the existence of special types of removable edges in bricks. Recall that a removable edge e in a matching covered graph G is *b-invariant* if the equality $b(G - e) = b(G)$ holds. For brevity, we shall refer to a b-invariant removable edge simply as a b-invariant edge.

LOVÁSZ (1987, [58])

Conjecture 13.2 *Every brick different from K_4, $\overline{C_6}$, and $\mathbb{P}$ has a b-invariant edge.*

The above conjecture has several relevant and interesting ramifications involving special types of removable ears in matching covered graphs. One such type is described below.

13.1.1 *b*-Invariant ears in matching covered graphs

Suppose that R is a removable ear in a matching covered graph G. Recall that if R is a double ear, then $b(G - R) = b(G) - 1$, whereas if R is a single ear, then $b(G - R) \geq b(G)$ (see Exercise 11.2.2). The latter inequality suggests the following definition: a removable single ear R of a matching covered graph G is *b-invariant* if $b(G - R) = b(G)$.

A simple calculation shows that if R is either a removable double ear, or is a b-invariant single ear of a matching covered graph G, then the dimension of the matching space of $H := G - R$ is just one less than that of G (see Exercise 11.2.5), and a basis for $\mathcal{L}in(G)$ can be obtained from a basis for $\mathcal{L}in(H)$ by using the procedure described in Proposition 11.13.

Clearly, every removable ear of a bipartite matching covered graph is b-invariant. But, in general, an arbitrary removable single ear in a matching covered graph need not be b-invariant. The following statement may be regarded as a generalization of Lovász's Conjecture:

REMOVABLE-EAR-VERSION OF CONJECTURE 13.2

Conjecture 13.3 *Any nontrivial matching covered graph G, none of whose bricks is the Petersen graph, either has a removable double ear or has a b-invariant single ear.*

Apart from the above, we shall come across many other statements which are closely related to Lovász's Conjecture. One of the interesting facts that we shall learn is that this conjecture is equivalent to the statement that every brick that is not a Petersen brick admits an ear decomposition with just one double ear.

13.1.2 Our approach to Conjecture 13.2

How does one approach a conjecture such as this? Tight cut decompositions are not applicable because this is a statement about bricks. Thus it was necessary for us to look for new techniques and inductive tools for dealing with bricks.

Any brick different from K_4 and $\overline{C_6}$ has a removable edge. An examination of conditions under which an arbitrary removable edge in a brick is not b-invariant led us to the serendipitous discovery of solid bricks and the striking fact that every removable edge in such a brick is b-invariant (see Theorem 10.16). This left us with the problem of having to deal with bricks which have nontrivial separating cuts. A natural idea then was to take the cut-contractions with respect to a nontrivial separating cut and apply induction. Unfortunately there is a hitch; neither of the two cut-contractions with respect to an arbitrary separating cut of a brick need be a brick, or even a near-brick whose properties are closely akin to those of bricks. But, fortunately, we were able to show that every nonsolid brick G has a robust cut; that is nontrivial separating cut C such that both the C-contractions of G are near-bricks. This was a major step. Existence of robust cuts in nonsolid bricks is established in the next chapter.

The uniqueness of the role of the Petersen graph is perhaps the most mysterious and fascinating aspect of Lovász's characterization of the matching lattice. The notion of the characteristic of a separating cut introduced in Exercise 4.3.10 was helpful in resolving this mystery; we were able to show that every nonsolid brick different from a Petersen brick has a robust separating cut whose characteristic is three.

Further techniques and ideas were necessary to prove Conjecture 13.2. (The first proof of this conjecture appeared in the Ph.D. thesis (1996, [6]) of Marcelo de Carvalho). Article [10] has a proof of a strengthened form of this conjecture, and several consequences of this result were described in [11]. One of them, not unexpectedly, is a proof of Theorem 13.1. All these will be discussed in due course.

We shall present a proof of Conjecture 13.2 in Chapter 15 that does not rely on the notion of characteristic. This became possible after we discovered Lemma 10.10 which is a much stronger property of odd wheels used in earlier proofs of Lovász's conjecture.

Exercises

▷**13.1.1** *Verify that no edge of the Petersen graph $\mathbb{P}$ is b-invariant.*

▷**13.1.2** *List all the removable edges in the two bricks G and H shown in Figure 13.1, and the b-invariant edges among them.*

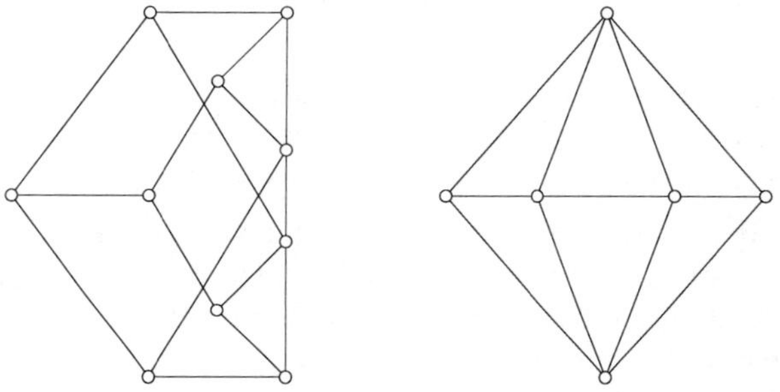

Fig. 13.1 b-Invariant edges, Exercise 13.1.2

13.2 $(b + p)$-Invariant Edges in Bricks

Recall that a Petersen brick is one whose underlying simple graph is the Petersen graph; and also that for a matching covered graph G, the symbol $p(G)$ denotes the number of its Petersen bricks.

A removable edge e of a brick G is *$(b + p)$-invariant* if

$$(b + p)(G - e) = (b + p)(G).$$

In a Petersen brick, every edge is $(b + p)$-invariant (Exercise 13.2.1). If G is not a Petersen brick, then $b(G) = 1$ and $p(G) = 0$. Thus, in such a brick G, a removable edge e is $(b + p)$-invariant if it is b-invariant and the brick of $G - e$ is not a Petersen brick. As an example, consider the brick $G := \mathbb{P} + e$ shown in Figure 13.2, with $b(G) = 1$ and $p(G) = 0$. The edge e is b-invariant because $b(G - e) = 1$; however, it is not $(b + p)$-invariant because $(b + p)(G - e) = 2$; but the edges

f_1 and f_2, indicated by dashed lines, are $(b + p)$-invariant because, for $i = 1, 2$, $(b + p)(G - f_i) = 1$ (Exercise 13.2.2(ii)). (We also note that, although $\mathbb{P}$ has no separating cuts of characteristic three, the graph $\mathbb{P} + e$ does; the cut $\partial(X)$ has characteristic three, Exercise 13.2.2(i).)

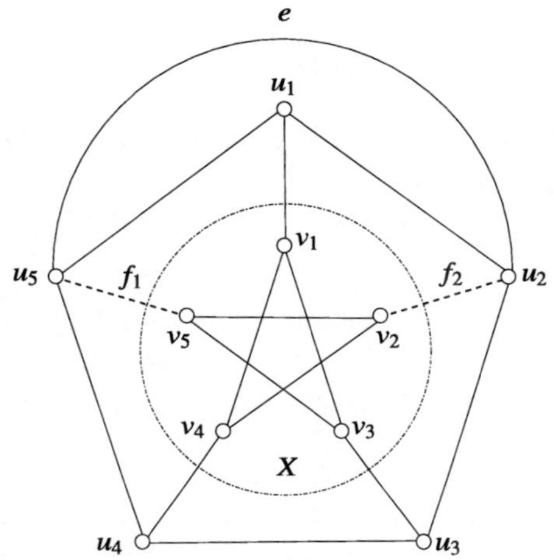

Fig. 13.2 The edges f_1 and f_2 are $(b + p)$-invariant in $\mathbb{P} + e$

In any brick which is not a Petersen brick, every $(b + p)$-invariant edge is also b-invariant. But, as we have noted above, a b-invariant edge in a brick need not be $(b + p)$-invariant. However, as we shall see later in this section, if a brick has a b-invariant edge then it also has a $(b + p)$-invariant edge. We first present a lemma which is crucial for establishing this assertion.

13.2.1 The three case lemma

Let G be a brick, and let e be a b-invariant edge of G. Then $G - e$ is a near-brick; and therefore, one of the two contractions with respect to any tight cut of $G - e$ is bipartite. It turns out that the brick of $G - e$ can be obtained from $G - e$ by contracting the bipartite shores of at most two nontrivial tight cuts.

Let $C := \partial(X)$ be a cut of G such that $C - e$ is a nontrivial tight cut of $G - e$, and suppose that $J := (G - e)/(\overline{X} \to \overline{x})$ is bipartite. We refer to the part to which $\overline{x}$ belongs as the *inner part* of J, and the other part of J as its *outer part*. The following property is easily proved:

Lemma 13.4 *Let G be a brick and let e be a b-invariant edge of G. Suppose that $C := \partial(X)$ is a cut of G such that $C - e$ is a nontrivial tight cut in $G - e$, and that $J := (G - e)/(\overline{X} \to \overline{x})$ is bipartite. Then, no end of e lies in the outer part of J and at least one end of e lies in the inner part of J.*

Proof By hypothesis, G is a brick, e is b-invariant and $C - e$ is a nontrivial tight cut of $G - e$. Thus, one of the $(C - e)$-contractions of $G - e$ is bipartite. Let B and I denote the outer and inner parts of J, respectively. If edge e has an end in B or if e has no end in I, then B would be a nontrivial barrier of G which is absurd because G is a brick. Thus, no end of e lies in B and at least one end of e lies in I. □

THE THREE CASE LEMMA

Lemma 13.5 *Let G be a brick. Suppose that e is a b-invariant edge of G such that $G - e$ is not a brick, and let H be the brick of $G - e$, obtained by a tight cut decomposition of $G - e$. Then, one of the following three alternatives holds (see Figure 13.3):*

(i) *either G has a cut $C_1 := \partial(X_1)$ such that $C_1 - e$ is a nontrivial tight cut of $G - e$ and $J_1 := (G - e)/(\overline{X_1} \to \overline{x_1})$ is bipartite, $H = (G - e)/(X_1 \to x_1)$ and the edge e has one end in the inner part of J_1, and the other end in $\overline{X_1} = V(H) - x_1$,*
(ii) *or G has two cuts cuts $C_1 := \partial(X_1)$ and $C_2 := \partial(X_2)$ such that X_1 and X_2 are disjoint, $C_1 - e$ and $C_2 - e$ are nontrivial tight cuts of $G - e$, and for $i = 1, 2$, $J_i := (G - e)/(\overline{X_i} \to \overline{x_i})$ is bipartite, $H = (G - e)/(X_1 \to x_1)/(X_2 \to x_2)$ and the edge e has one end in the inner part of J_1, and the other end in the inner part of J_2,*
(iii) *or G has a cut $C_1 := \partial(X_1)$ such that $C_1 - e$ is a nontrivial tight cut of $G - e$ and $J_1 := (G - e)/(\overline{X_1} \to \overline{x_1})$ is bipartite, $H = (G - e)/(X_1 \to x_1)$ and the edge e has both ends in the inner part of J_1.*

Proof Assume that $G - e$ is not a brick. Let $C_1 := \partial(X_1)$ be a nontrivial cut of G such that $C_1 - e$ is tight and one of the contraction vertices of H is x_1, obtained by contracting X_1 to x_1. As e is b-invariant, J_1 is bipartite. Moreover, either e has both ends in the inner part of J_1, or e has one end in the inner part of J_1, the other end in $\overline{X_1}$. In the former case, we have the last of the three asserted cases. Assume thus that e has one end in the inner part of J_1, the other end in $\overline{X_1}$. If x_1 is the only contraction vertex of H then the first of the three cases holds. We may thus assume that H has more than one contraction vertex. Let x_2 be another contraction vertex of H, distinct from x_1. Let $J_2 := (G - e)/(\overline{X_2} \to \overline{x_2})$. Then, J_2 is bipartite. Moreover, edge e has one end in the inner part of J_2. This conclusion holds for each contraction vertex x_2 of H distinct from x_1. We deduce that H has precisely two contraction vertices. Moreover, the second of the three cases holds. □

We now introduce the notion of the index of a b-invariant edge e in a brick G which depends on the pertinent case of Lemma 13.5. The relevance of this notion, which may not be evident at this stage, will hopefully become clear in Chapter 17 where special types of b-invariant edges called thin edges are discussed.

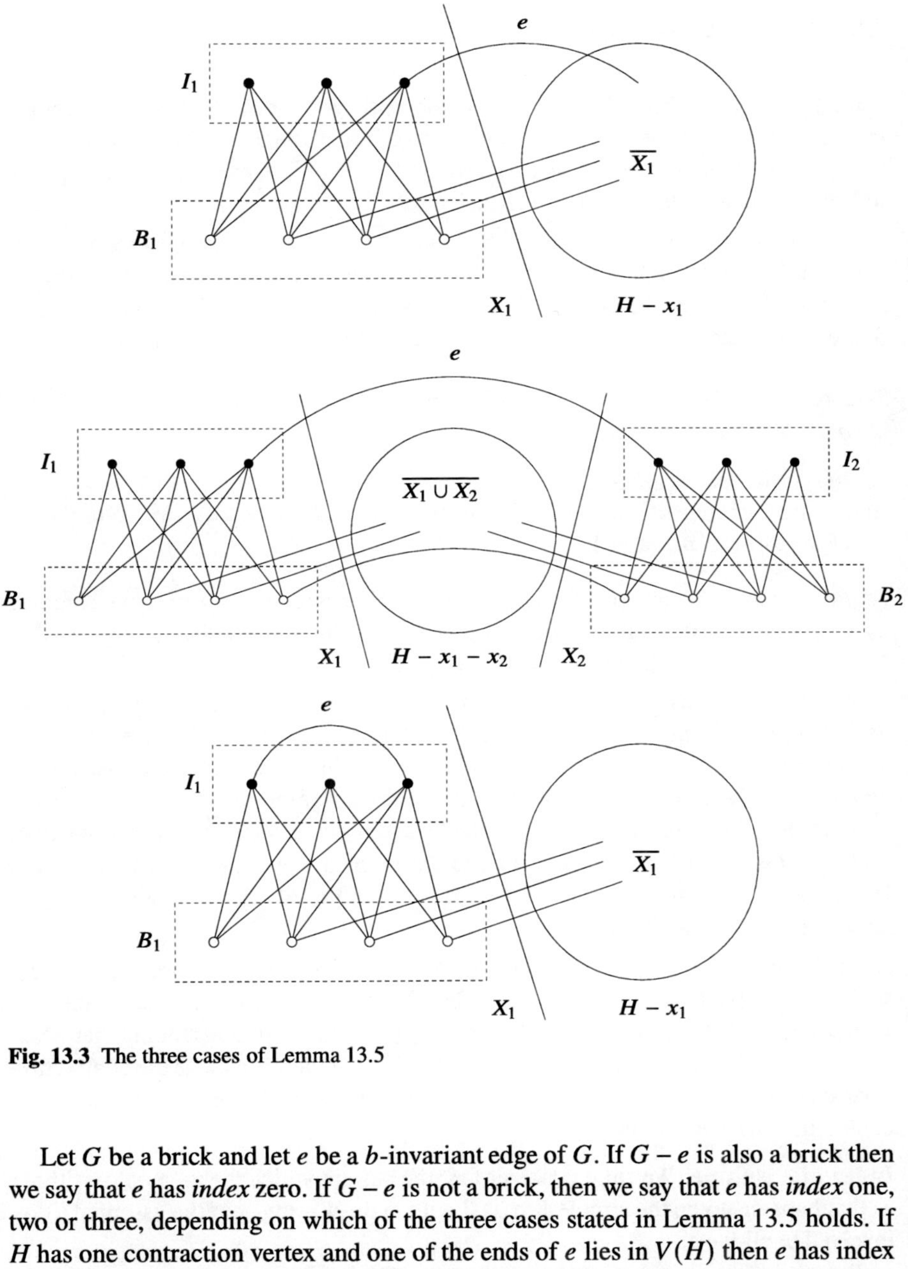

Fig. 13.3 The three cases of Lemma 13.5

Let G be a brick and let e be a b-invariant edge of G. If $G - e$ is also a brick then we say that e has *index* zero. If $G - e$ is not a brick, then we say that e has *index* one, two or three, depending on which of the three cases stated in Lemma 13.5 holds. If H has one contraction vertex and one of the ends of e lies in $V(H)$ then e has index one. If H has two contraction vertices then e has index two. Finally, if the last of the three cases holds then e has index three.

Whereas the index is at the moment just a pointer into the cases of the Three Case Lemma, this particular numbering will turn out to be mathematically useful in later chapters.

13.2.2 The Petersen graph

With the aid of the Three Case Lemma, we now proceed to prove the main theorem of this section. In its proof, we shall implicitly make use of the fact that the diameter of $\mathbb{P}$ is two and that, given any two paths $s_0s_1s_2$ and $t_0t_1t_2$ of $\mathbb{P}$, there is an automorphism of $\mathbb{P}$ which maps s_i to t_i, for $0 \le i \le 2$.

The interested reader will note that the proof of the following theorem uses many basic results such as the subadditivity of the function b (Theorem 4.20), the monotonicity of the function b (Theorem 9.8) and the notion of the characteristic. Some of the ideas used in this proof will recur in the next two chapters.

Theorem 13.6 *Let G be a brick and let e be a b-invariant edge of G. If the edge e is not $(b+p)$-invariant, then:*

(i) *the characteristic of G is three, and*
(ii) *some other edge of G is $(b+p)$-invariant (in fact G has at least two nonadjacent $(b+p)$-invariant edges).*

Proof Since e is b-invariant, $b(G-e) = 1$. Let H denote the brick of $G-e$ obtained by a tight cut decomposition. Suppose that e is not $(b+p)$-invariant, implying that G itself is not a Petersen brick, but H is.

By the Three Case Lemma, the index of e is either zero, one, two, or three. Let us first suppose that the index of e is zero. In this case $H = G - e$. If the edge e were a multiple edge of G, then it would itself be $(b+p)$-invariant. Thus, e joins vertices which are not adjacent in H. The underlying simple graph of H is $\mathbb{P}$, the Petersen graph. Considering the symmetries of the Petersen graph, we may thus assume that, up to multiple edges, G is isomorphic to the graph $\mathbb{P}+e$ depicted in Figure 13.2. It has $(b+p)$-invariant edges different from e and is of characteristic three (see Exercise 13.2.2(i) and (ii)).

In dealing with the remaining cases, we shall adopt the notation used in the statement of the Three Case Lemma. Also, in our illustrations, we shall label the vertices of the Petersen brick H as in Figure 13.2, with the understanding that when the index is either one or three, one vertex of H is to be identified with the contraction vertex x_1, and that when the index is two, then two vertices of H would serve as contraction vertices x_1 and x_2.

Index of e is three: We shall consider first the case in which the index of e is three as this happens to be the simplest to deal with and yet brings out the essential ideas involved in all the cases.

By the third item in the Three Case Lemma, the graph $G-e$ has a nontrivial tight cut $\partial(X_1)$ such that

$$J_1 := (G-e)/(\overline{X_1} \to \overline{x_1}) \text{ is bipartite, } H = (G-e)/(X_1 \to x_1),$$

and the edge e has both ends in I_1. See Figure 13.4. Note that in the figure I_1 is shown to have just two vertices, but I_1 may be larger. Also, note that a neighbour of the contraction vertex $x_1 = u_1$ in H, may be joined to more than one vertex in B_1, although our figures indicate just one edge between each of u_5, v_1 and u_2 and B_1. However, since $\partial(X_1)$ is not tight in G, there is some perfect matching which meets this cut in at least three edges. This implies that there is a matching which matches the three vertices u_5, v_1 and u_2 with distinct vertices in B_1.

We shall establish the first part of the assertion by showing that $C := \partial(X)$, where $X := \{v_1, v_2, v_3, v_4, v_5\}$, is a separating cut of G whose characteristic is three. The two C-contractions of $G - e$ are near-bricks with 5-wheels as their bricks. Therefore, C is a separating cut in $G - e$. To show that C is a separating cut in G, we must show that there is a perfect matching M_e of G containing e that meets C in precisely one edge. The existence of such a perfect matching is obvious and is shown in Figure 13.4(a).

Now to show that C has characteristic three, one must find a perfect matching which meets C in three edges. Such a perfect matching M'_e is shown in Figure 13.4(b).

Finally, let us turn to showing that f_1 is a $(b + p)$-invariant edge of G. The graph $(G - e)/\overline{X}$ is a 5-wheel and f_1 is a spoke of that wheel. Therefore $(G - e - f_1)/\overline{X}$ is matching covered (see Exercise 2.2.4), and in fact a near-brick. Also, the brick of $(G - e)/X$ is a 5-wheel and f_1 is a spoke of that wheel, and is not in the tight cut $\partial(I_1 \cup B_1)$. Thus $(G - e - f_1)/X$ is a near-brick. The cut $C - f_1$ is a robust cut of $G - e - f_1$ because, being of characteristic three, it is not tight, and both the $(C - f_1)$-contractions of $G - e - f_1$ are near-bricks. By the subadditivity of the function b (Theorem 4.20),

$$b(G - e - f_1) < b((G - e - f_1)/X) + b((G - e - f_1)/\overline{X})$$

Since, as observed above both $(G - e - f_1)/X$ and $(G - e - f_1)/\overline{X}$ are near-bricks, it follows that $G - e - f_1$ is a near-brick. The edge e is matchable in $G - f_1$ because the perfect matching M_e, which contains e, does not contain f_1. By the monotonicity of the function b (Theorem 9.8) it now follows that $G - f_1$ is a near-brick. The characteristic of $G - f_1$ is three because the cut $C - f_1$ has characteristic three. Consequently, the only brick of $G - f_1$ also has characteristic three, implying that it is not a Petersen brick. We conclude that f_1 is a $(b+p)$-invariant edge of G. Similarly, the edge f_2 is also a $(b + p)$-invariant edge of G.

Analyzing the cases involving other indices is very similar except that, in some cases, asserting the existence of perfect matchings M_e and M'_e is not so straightforward as in the case discussed above. For example, if the index of the edge e is one, the edge e joins a vertex in I_1 to a vertex in $V(H) - x_1$, but that vertex may or not be adjacent to the contraction vertex x_1 of H. In the former case, finding the appropriate perfect matchings M_e and M'_e is quite easy, but in the latter case, it is not straightforward. To illustrate the sort of arguments needed we shall discuss this case.

Index of e is one and e has an end in $V(H) - x_1$ that is not adjacent to x_1. As we did in the previous case, let us identify the contraction vertex x_1 with the vertex u_1 of H and take w_1 belonging to I_1 and u_3 as the ends of the edge e. To find the perfect matching M_e such that (i) $e \in M_e$, (ii) $f_1 \notin M_e$, and $|M_e \cap C| = 1$, consider first the graph $G - v_1 - w_1$. Since G is a brick, $G - v_1 - w_1$ has a perfect matching, say F. It matches u_2 and u_5 with distinct vertices, say w_2 and w_5 in B_1, the vertices

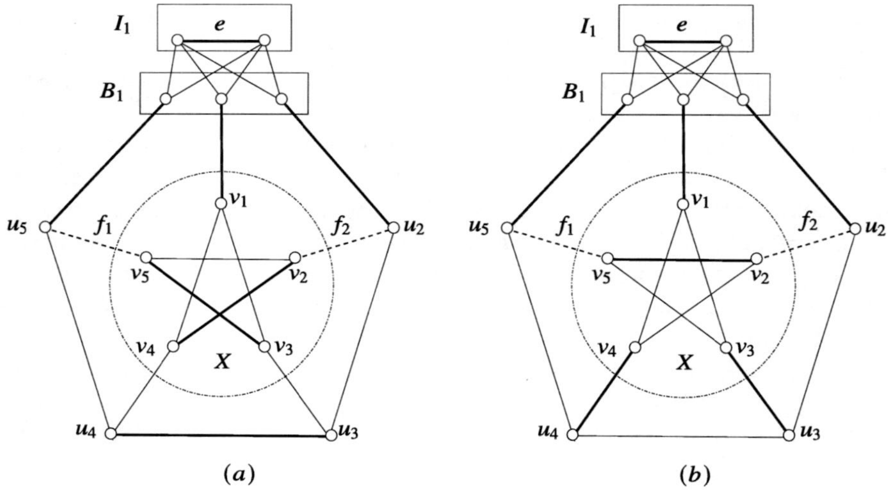

Fig. 13.4 Index of e is three; (a) $|M_e \cap C| = 1$; (b) $|M'_e \cap C| = 3$

in $I_1 - w_1$ with vertices in $B_1 - w_2 - w_5$, and includes a perfect matching, say F' of the 6-cycle $v_2v_5v_3u_3u_4v_4v_2$. Now define M_e as follows:

$$M_e = (F - F') \cup \{e, v_1v_3, v_2v_5, u_4v_4\}$$

It can be checked that M_e satisfies the three properties stated above. An analogous argument starting with a perfect matching of $G - u_2 - w_1$ yields the matching M'_e which meets the cut $\partial(X)$ in three edges. Deducing that G has characteristic three and that f_1 and f_2 are $(b + p)$-invariant is very similar to the previous case.

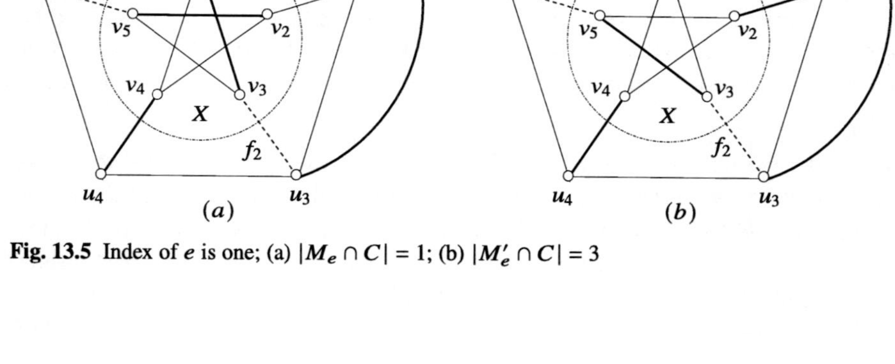

Fig. 13.5 Index of e is one; (a) $|M_e \cap C| = 1$; (b) $|M'_e \cap C| = 3$

We refer to paper [9] for full details concerning the cases not discussed here. □

Exercises

13.2.1 *Show that in a Petersen brick every edge is $(b + p)$-invariant.*

∗**13.2.2** *Let $G := \mathbb{P} + e$ be the brick depicted in Figure 13.2. Show that:*

(i) *the cut $C := \partial(X)$ is a separating cut of characteristic three in G; and*
(ii) $(b + p)(G - e) = 2$, *whereas* $(b + p)(G - f_1) = (b + p)(G - f_2) = 1$.
Hint: Find the retracts of $G - e$, $G - f_1$ and $G - f_2$ and use the fact that a matching covered graph and its retract have the same number of bricks, and the same number of Petersen bricks; see Exercise 11.5.6.
(iii) *Find all the $(b + p)$-invariant edges in the brick $\mathbb{P} + e$.*

13.3 Notes

Laci Lovász mentioned Conjecture 13.2 to the second author in a private communication around 1988. In a talk given at a workshop in Ardèche, France, in 1994, Santosh Vempala stated that he and Laci had established the validity of this conjecture and sketched some ideas. At this stage, the work on Marcelo Carvalho's thesis, under the supervision of the first author, had also progressed towards a solution. The second author joined the team some time later, and the three authors, CLM, produced a preliminary report with the title "A Proof of the Lovász-Vempala Theorem" in May 1997. We sent Laci a copy of that report and inquired if his work with Santosh on this conjecture had been written up. With his characteristic kindness and generosity, he wrote to us in a letter dated 11 June, 1998: "we have not finished the writing up, and have no claim on the result". (With that letter he enclosed two short notes. One of them is [61], whose contents have been described in Section 9.3.) It appears that both Laci and Santosh became interested in problems in other areas and abandoned this project. This subject is poorer for it!

Chapter 14
Robust Cuts in Bricks

Contents

14.1 Robust Cuts in Nonsolid Bricks ♯

ROBUST CUTS

Recall that a separating cut C in a matching covered graph G is *robust* if both the C-contractions of G are near-bricks. The purpose of this section is to establish the existence of robust cuts in nonsolid bricks.

In the brick G shown in Figure 14.1, both C and D are nontrivial separating cuts. It can be seen that both C-contractions of G are bricks, but one of the two D-contractions of G has two bricks. Therefore C is a robust cut of G, but D is not.

Let G be a nonsolid brick. By definition, G has nontrivial separating cuts. Consider any nontrivial separating cut D of G, and let M_0 be a perfect matching in G such that $|M_0 \cap D| > 1$. Now let $\mathcal{C}$ be the collection of cuts C of G such that (i) $|M_0 \cap C| > 1$ and (ii) $C \preceq D$, where $\preceq$ is the precedence relation on cuts defined in Section 7.2.2. We leave as Exercise 14.1.1 the proof that every cut in $\mathcal{C}$ is separating.

C. L. Lucchesi, U. S. R. Murty, *Perfect Matchings*, Algorithms and Computation in Mathematics 31, https://doi.org/10.1007/978-3-031-47504-7_14

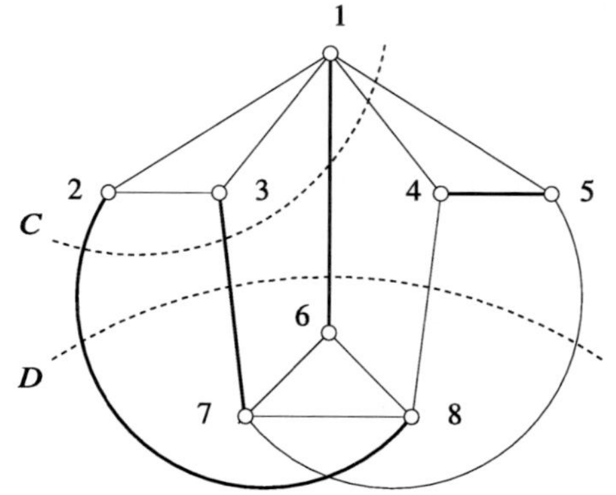

Fig. 14.1 Cut C is robust but D is not

Let $\mathcal{C}_*$ be the subcollection of $\mathcal{C}$ consisting of those cuts that are minimal with respect to the precedence relation $\preceq$. We refer to a member of $\mathcal{C}_*$ as a cut which is *M_0-induced by D*. (See Exercise 14.1.4 for an example.) We shall show that every such cut is a robust cut of G. We start with two preliminary lemmas.

Lemma 14.1 *Let $C := \partial(X)$ be a cut in $\mathcal{C}_*$. Then every 2-vertex cut is a barrier of the C-contraction $G_1 := G/(\overline{X} \rightarrow \overline{x})$ of G. (Equivalently, no tight cut of the graph G_1 is a 2-separation cut.)*

Proof No 2-vertex cut of G_1 can be a vertex cut of G, as G, a brick, is 3-connected; it follows that every 2-vertex cut of G_1 contains $\overline{x}$. Suppose now that one such cut, $\{\overline{x}, v\}$, is not a barrier of G_1.

Let D_1 and D_2 denote two crossing 2-separation tight cuts of G_1 associated with the 2-separation $\{\overline{x}, v\}$. See Figure 14.2.

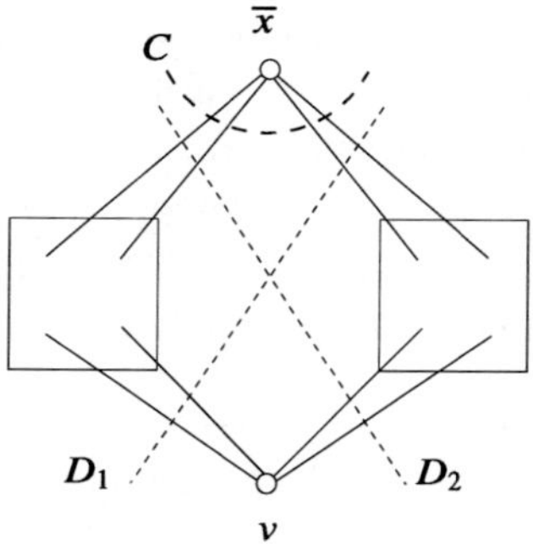

Fig. 14.2 2-separation cuts D_1 and D_2 in G_1

Recall that $\mathcal{M}$ denotes the collection of perfect matchings of G. For any perfect matching $M \in \mathcal{M}$, we clearly have:

$$|M \cap D_1| + |M \cap D_2| = |M \cap \partial(\overline{x})| + |M \cap \partial(v)| = |M \cap C| + 1. \tag{14.1}$$

Since $|M_0 \cap C|$ is greater than one, at least one of $|M_0 \cap D_1|$ and $|M_0 \cap D_2|$ is greater than one. Adjust notation so that $|M_0 \cap D_1| > 1$.

It follows immediately from equation (14.1) that $|M \cap D_1| \leq |M \cap C|$ for all $M \in \mathcal{M}$, implying that $D_1 \preceq C$. Since G is a brick, D_2 is not a tight cut of G. Therefore, there must exist a perfect matching M_2 of G such that $|M_2 \cap D_2| > 1$. Hence, by equation (14.1), $|M_2 \cap D_1| < |M_2 \cap C|$, which implies that D_1 strictly precedes C. This contradicts the minimality of C. □

Lemma 14.2 *Let $C := \partial(X)$ be a cut in $\mathcal{C}_*$. Suppose that the C-contraction $G_1 := G/(\overline{X} \to \overline{x})$ of G is not a brick. Then, G_1 is not bicritical and, for any nontrivial barrier B_1 in G_1:*

(i) *the contraction vertex $\overline{x}$ of G_1 is in B_1, and*
(ii) *$G_1 - B_1$ has precisely one nontrivial odd component, H_1, and the cuts C and $C_1 := \partial(H_1)$ are matching-equivalent.*

Proof The graph G_1 is matching covered because C, a cut in $\mathcal{C}$, is a separating cut of G and G_1 is a C-contraction of G. Suppose that G_1 is not a brick. By Lemma 14.1, G_1 is not bicritical. Let B_1 denote a nontrivial barrier in G_1.

(i) The contraction vertex $\overline{x}$ has to be in B_1; because G, a brick, is bicritical.

(ii) As G_1 is matching covered, B_1, a barrier of G_1, cannot contain both ends of any edge of G_1. Therefore, if $H_1, H_2, \ldots, H_t$, $t = |B_1|$, denote the odd components of $G_1 - B_1$, then

$$\sum_{i=1}^{t} |M \cap \partial(H_i)| = |M \cap C| + (t-1), \tag{14.2}$$

for any perfect matching M of G (see Exercise 14.1.2). The above equation implies that $\partial(H_i) \preceq C$, for $1 \leq i \leq t$, and also that $|M_0 \cap \partial(H_i)| > 1$ for at least one value of the index i. Adjust notation so that $|M_0 \cap \partial(H_1)| > 1$, and let $\partial(H_1)$ be the cut C_1. The fact that $|M_0 \cap C_1| > 1$ implies that $|V(H_1)| > 1$, and also that $C_1 \in \mathcal{C}$. (All cuts $\partial(H_j)$, $j \neq 1$, turn out to be trivial.) See Figure 14.3.

If possible, suppose that H_j is nontrivial for some $j \neq 1$. Since G is a brick, the cut $\partial(H_j)$ cannot be tight in G. Thus, there exists a perfect matching M_j of G such that $|M_j \cap \partial(H_j)| > 1$. It now follows from equation (14.2) that $|M_j \cap C_1| < |M_j \cap C|$. This implies that C_1 strictly precedes the cut C, which contradicts the minimality of C. We conclude that H_1 is the unique nontrivial component of $G_1 - B$. From equation (14.2) we infer that C and C_1 are matching equivalent. □

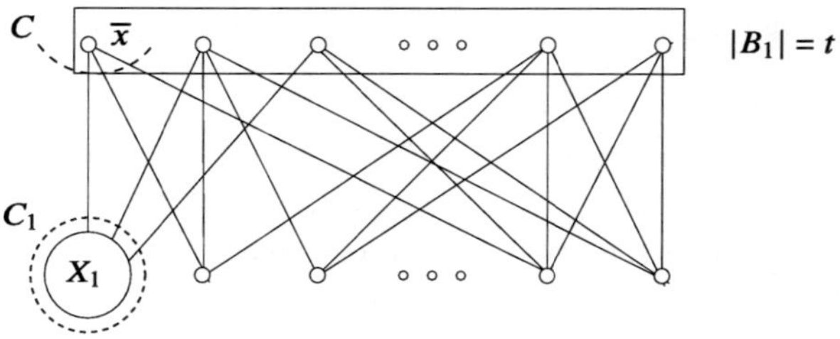

Fig. 14.3 $G_1 := G/(\overline{X} \to \overline{x})$ (Lemma 14.2)

Theorem 14.3 *Let G be a nonsolid brick. Let $\mathcal{C}$ be the collection of nontrivial separating cuts of G. Let $\mathcal{C}_*$ be the subcollection of $\mathcal{C}$ which are minimal with respect to the precedence relation $\preceq$ on the cuts of G. Let $C := \partial(X)$ be a cut in $\mathcal{C}_*$ and let $G_1 := G/(\overline{X} \to \overline{x})$. Then, there exists a subset X' of X (possibly equal to X) such that:*

(i) *the cuts C and $C' := \partial(X')$ are matching-equivalent cuts of G,*
(ii) *the graph $G' := G/(\overline{X'} \to \overline{x'})$ is a brick, and*
(iii) *the C'-contraction $H' := G_1/(X' \to x')$ of G_1 is bipartite and matching covered, and its two contraction vertices, $\overline{x}$ and x', are in distinct parts of its bipartition.*

Consequently, G_1 is a near-brick.

Proof If G_1 is a brick then the assertion holds immediately, with $X' := X$. We may thus assume that G_1 is not a brick. To prove part (i) we now apply Lemma 14.2. The graph G_1 is not bicritical. Let B_1 denote a maximal nontrivial barrier of G_1. The contraction vertex $\overline{x}$ of G_1 is in B_1. The graph $G_1 - B_1$ has precisely one nontrivial component, say, K'. Let $X' := V(K')$. The cut $C' := \partial(X')$ is matching-equivalent to C. This proves part (i).

Let us now prove part (ii), which asserts that G' is a brick. For this, assume the contrary. As C and C' are matching-equivalent, the cut C' is also in $\mathcal{C}_*$. By Lemma 14.2, G' has a nontrivial barrier, B'_1, which contains the contraction vertex $\overline{x'}$ of G'. For any vertex v of $B'_1 - \overline{x'}$, the graph $G' - \overline{x'} - v$ does not have a perfect matching. This implies that the graph $K' - v$ does not have a perfect matching. Thus, K' is not critical. By Theorem 3.2, B_1 is not maximal, a contradiction. This proves part (ii).

The validity of part (iii) follows immediately from (i) $\overline{x} \in B_1$, (ii) K' is the unique nontrivial component of $G_1 - B_1$ and (iii) $X' = V(K')$.

Finally, as the cuts C and C' are matching-equivalent in G, it follows that C' is tight in G_1. One of the C'-contractions of G_1 is the graph G', a brick. The other C'-contraction of G_1 is the bipartite matching covered graph H'. We conclude that G_1 is a near-brick. □

Note that the theorem implies, in particular, the following:

Corollary 14.4 *Let G be a nonsolid brick. Let $\mathcal{C}$ be the collection of nontrivial separating cuts of G. Let $\mathcal{C}_*$ be the subcollection of $\mathcal{C}$ which are minimal with respect to the precedence relation $\preceq$ on the cuts of G. Every cut in $\mathcal{C}_*$ is robust. Consequently, every nonsolid brick has a robust cut.* □

Underlying the above proof, there is a polynomial-time algorithm for finding a robust cut in a nonsolid brick G which accepts any nontrivial separating cut D of G as input and returns a robust cut of G, induced by D, as output (Exercise 14.1.3). However, there is a catch, we do not know if the problem of deciding whether or not a given brick has a nontrivial separating cut is in the complexity class P.

14.1.1 Peripheral robust cuts

If C is a robust cut of a brick G, then the two C-contractions of G are near-bricks, but, in general, need not be bricks. In proving properties of bricks, we cannot directly apply the induction hypothesis to the C-contractions if they were not bricks. Fortunately, one can infer from Theorem 14.3 a certain structural result that applies to all nonsolid bricks.

Let $C := \partial(X)$ be a robust cut of a brick G. Let X', C', G' and H' be as in the statement of Theorem 14.3. If we apply Theorem 14.3, with $\overline{X}$ playing the role of X and $G_2 := G/(X \to x)$ playing the role of G_1, we deduce that (i) there exists a subset X'' of $\overline{X}$ such that $C'' := \partial(X'')$ is matching-equivalent to C, (ii) the graph $G'' := G/(\overline{X''} \to \overline{x''})$ is a brick and (iii) the C''-contraction $H'' := G_2/(X'' \to x'')$ of G_2 is bipartite and matching covered, and its two contraction vertices, x and x'', are in distinct parts of its bipartition. Thus, the graph $H := G/(X' \to x')/(X'' \to x'')$, which is the splicing of H' and H'', is bipartite and matching covered, and its two contraction vertices, x' and x'', are in distinct parts of the bipartition of H. Moreover, the brick G is the result of the splicing of the bipartite matching covered graph H and the two bricks G' and G''. The following assertion summarizes these observations.

Corollary 14.5 *Let G be a brick, and let $C := \partial(X)$ be a robust cut of G. Then, there exist a subset X' of X and a subset X'' of $\overline{X}$ such that:*

(i) *the three cuts C, $C' := \partial(X')$ and $C'' := \partial(X'')$ are matching-equivalent robust cuts of G,*
(ii) *the graphs $G/\overline{X'}$ and $G/\overline{X''}$ are both bricks, and*
(iii) *the graph $G/X'/X''$ is bipartite and matching covered.* □

We refer to $\partial(X')$ and $\partial(X'')$ as *peripheral robust cuts*, and to the bricks $G/\overline{X'}$ and $G/\overline{X''}$ as the corresponding *peripheral bricks*. See Figure 14.4 for an illustration.

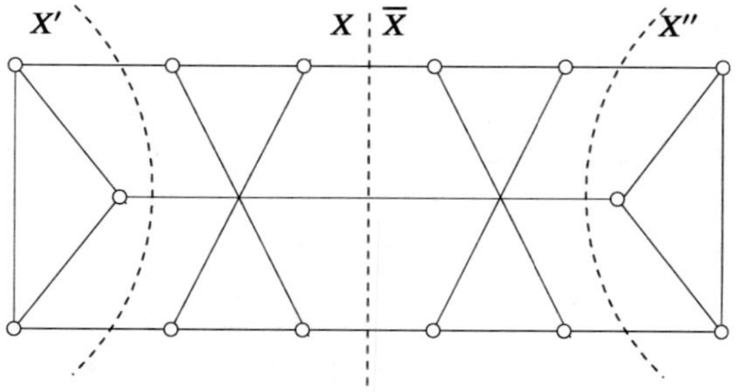

Fig. 14.4 Peripheral robust cuts (Corollary 14.5)

Exercises

▷**14.1.1** *Prove that every cut in C is separating. Hint: apply Theorem 4.2.*

14.1.2 *Verify the validity of equation (14.2). (Compare this equation with equation (10.2).)*

▷**14.1.3** *Describe a polynomial-time algorithm which accepts a brick G and a nontrivial separating cut D and returns a robust cut in G. Hint: there is no need to have the resulting robust cut to be M_0-induced, but it is possible to determine M_0, because there exists a pair $\{u_1v_1, u_2v_2\}$ of nonadjacent edges of C, such that the graph $G - u_1 - v_1 - u_2 - v_2$ is matchable.*

14.1.4 *Consider the brick G shown in Figure 14.1, with the cuts C and D indicated by broken lines, and the perfect matching M_0, indicated by thick lines, such that $|M_0 \cap D| = 3$. Show that the cut C is M_0-induced by D, and verify that C is a robust cut.*

▷**14.1.5** *Give an example to show that the peripheral bricks with respect to a robust cut of a nonsolid brick need not be solid bricks.*

A subset X of the vertex set V of a matching covered graph G is *critical* if X is nontrivial and the subgraph $G[X]$ of G induced by X is critical.

∗**14.1.6**

(i) *Show that a brick G is nonsolid if and only if there are two disjoint critical subsets X and Y of V such that the graph $G - (X \cup Y)$ is matchable.* CLM *(2004, [13]).*
(ii) *Deduce the Reed-Wakabayashi Theorem 7.8 from the above statement.*

▷**14.1.7** *Let $C := \partial(X)$ be a robust cut of a brick G such that X is minimal. Show that the near-brick $G_1 := G/(\overline{X} \to \overline{x})$ is in fact a brick.*

14.1.8 *Let $C = \partial(X)$ be a robust cut of a brick G such that $G_1 := G/(\overline{X} \to \overline{x})$ is a Petersen brick, and let Q be the vertex set of a pentagon of G_1 which does not contain the contraction vertex $\overline{x}$. Show that $\partial(Q)$ is a robust cut of G whose characteristic is three.*

14.2 Robust Cuts with a Shore of Minimum Size

By definition, if C is a robust cut of a nonsolid brick G, then both C-contractions of G are near-bricks. By Corollary 14.4, every nonsolid brick has robust cuts. As indicated in the statement of Exercise 14.1.7, it is always possible to find a robust cut C such that one of the C-contractions of C is a brick. However, there is no guarantee that such a C-contraction be a solid brick, as indicated in the statement of Exercise 14.1.5.

The objective of this subsection is to prove the stronger statement that every nonsolid brick G has a robust cut C such that one of the two C-contractions is a solid brick.

ROBUST CUTS HAVING MINIMUM SHORES

Lemma 14.6 (CLM (2005, [13])) *Let G be a nonsolid brick and let X be a subset of V of minimum possible cardinality such that $C := \partial(X)$ is a robust cut of G. Then the C-contraction $G_1 := G/(\overline{X} \to \overline{x})$ is a solid brick.*

Proof It follows from the minimality of $|X|$ that G_1 is a brick (see Exercise 14.1.7).

Suppose that G_1 is a nonsolid brick, and let D_0 be a nontrivial separating cut of G_1. By Exercise 4.2.4 the pair $\{D_0, C\}$ is cohesive in G.

If D_0 is a robust cut of G, we have an immediate contradiction as to the minimality of $|X|$. If this is not the case, we shall show that there is a robust cut of G with a shore with fewer than $|X|$ vertices. Towards this end, let M' be a perfect matching of G_1 such that $|M' \cap D_0| \geq 3$. (Such a matching exists because D_0 is a nontrivial separating cut of brick G_1.) Extend M' to a perfect matching M of G. (Such a matching exists because C is separating in G.) Let D be a cut M-induced by D_0 in G. By Corollary 14.4, cut D is robust in G. Note that M has only one edge in C, yet it has three or more edges in D. Consequently, C and D are not matching-equivalent. We now analyse three cases, depending on where D is located in relation to C.

Suppose first that cut D is a cut of G_1. Cuts C and D are not matching-equivalent; thus, they are distinct. Then, D is a robust cut of G and a nontrivial cut of G_1, in contradiction to the minimality of $|X|$.

Now suppose that D is a cut of $G_2 := G/(X \to x)$. The perfect matching M of G has three or more edges in D and only one edge in C. Thus, $M \cap E(G_2)$ is a perfect matching of G_2 that may be extended to a perfect matching of G that contains just one edge in D_0. This is a contradiction, as $D \preceq D_0$.

Finally, suppose that C and D cross, and let Y be the shore of D such that $|X \cap Y|$ is odd. Let $I := \partial(X \cap Y)$ and let $U := \partial(\overline{X} \cap \overline{Y})$. The pair $\{C, D\}$ is cohesive. By Theorem 4.14,

(i) no edge of G joins a vertex in $X \cap \overline{Y}$ to a vertex in $\overline{X} \cap Y$,
(ii) for each perfect matching M of G,

$$|M \cap C| + |M \cap D| = |M \cap I| + |M \cap U|, \tag{14.3}$$

(iii) the collection $\{C, D, I, U\}$ is cohesive, and
(iv) the cuts I and U are separating.

14.6.1 *Cut U is trivial in G. In fact, the shore $\overline{X} \cap \overline{Y}$ is a singleton.*

Proof Let us first show that U is tight in G_2. So, suppose that the cut U is not tight in G_2. Then, a perfect matching of G_2 containing three or more edges in U may be extended to a perfect matching M'' of G containing just one edge in cut C and just one edge in cut D_0. From (14.3) we deduce that M'' contains three or more edges in D, contradicting the fact that D precedes D_0. We deduce that U is tight in G_2.

The cut C is robust in G and G_2 is a C-contraction of G. Thus, the graph G_2 is a near-brick. As U is tight in G_2, it follows that some U-contraction of G_2 is bipartite. Thus, U has a shore in G_2, say, Z, such that the subgraph of G_2 induced by Z is bipartite. Moreover, the majority part Z_+ of Z is a barrier of G_2, and U is the only nontrivial barrier cut associated with Z_+. We shall now prove that $Z = \overline{X} \cap \overline{Y}$ and then deduce that Z is a singleton.

Suppose first that the contraction vertex x of G_2 is in Z_+. In that case, the cuts U and C are matching-equivalent. From (14.3), we infer that cuts D and I are also matching-equivalent. By Corollary 14.4, the cut I is robust in G. The shore $X \cap Y$ of I is a proper subset of X. Thus, we have a contradiction to the minimality of $|X|$.

We may thus assume that x is not in Z_+. In that case, Z_+ is a barrier of G. As G, a brick, is bicritical, it follows that Z_+ is a singleton. The cut U is a barrier cut associated with Z_+. As Z_+ is a singleton, Z_+ is a shore of U in G_2. The shore $(Y - X) + x$ of U in G_2 is not a singleton, because $Y - X$ is nonempty. Thus, Z_+ must be the shore $\overline{X} \cap \overline{Y}$ of U. Indeed, $\overline{X} \cap \overline{Y}$ is a singleton and U is trivial. □

No edge of G joins a vertex of $X \cap \overline{Y}$ to a vertex of $\overline{X} \cap Y$. Thus, $X \cap Y$ is not a singleton, else C and D would both be nontrivial tight cuts of the brick G, which is absurd. Hence, $|\overline{Y}| = |\overline{X} \cap \overline{Y}| + |X \cap \overline{Y}| < |X \cap Y| + |X \cap \overline{Y}| = |X|$, implying that $|\overline{Y}| < |X|$. This contradicts the minimality of $|X|$. We conclude that G_1 is a solid brick. □

Exercises

14.2.1 *Consider the cut C in the brick G shown in Figure 14.5. Show that C is a robust cut of G and that neither C-contraction of G is solid. (This example shows that minimality of $|X|$ in the proof of Lemma 14.6 cannot be replaced by minimality of X as a set.)*

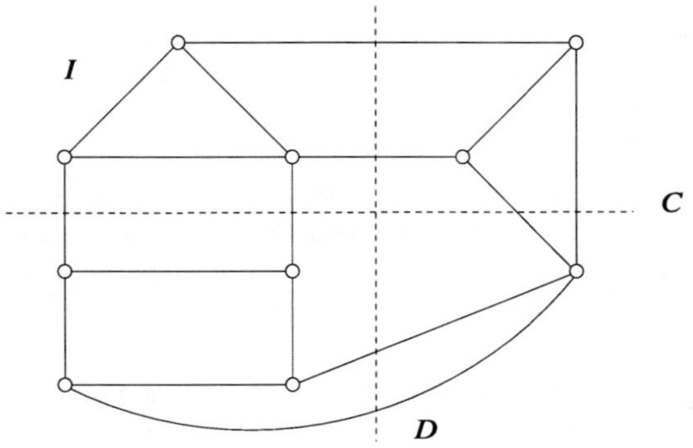

Fig. 14.5 Minimality of $|X|$ in Lemma 14.6 cannot be replaced by minimality of X as a set.

14.3 Quasi b-Invariant Edges

QUASI b-INVARIANT EDGES

A removable edge e in a nonsolid brick G is *quasi b-invariant* if $b(G - e) = 2$.

By Corollary 14.4, every nonsolid brick has robust cuts. As has already been mentioned, the strategy we use to prove the existence of b-invariant edges in a nonsolid brick G, distinct from K_4, $\overline{C_6}$ and $\mathbb{P}$, is to consider a robust cut C of G and use inductive arguments on the two C-contractions G_1 and G_2 of G. If e is an edge that is b-invariant in both G_1 and G_2, then both $G_1 - e$ and $G_2 - e$ are near-bricks. Thus, e is a removable edge of G and $C - e$ is a separating cut of $G - e$. By the subadditivity of function b (Theorem 4.20),

$$b(G - e) \leq b(G_1 - e) + b(G_2 - e) = 2, \tag{14.4}$$

where equality holds if and only if the cut $C - e$ is tight in $G - e$. By the monotonicity of function b (Theorem 9.8), $b(G) \leq b(G - e)$. Consequently, edge e is b-invariant in G if and only if $C - e$ is not a tight cut of $G - e$.

The following lemma, which plays a crucial role in several steps of our proof of Conjecture 13.2, records this crucial observation.

Lemma 14.7 *Let G be a brick, let C be a robust cut of G and let e be an edge that is b-invariant in both C-contractions of G. Then edge e is b-invariant in G if and only if $C - e$ is not tight in $G - e$. Equivalently, e is b-invariant in G if and only if there is a perfect matching M_0 of G such that (i) $e \notin M_0$, and (ii) $|M_0 \cap C| > 1$.* □

The following lemma and its corollary state easily proved sufficient conditions under which a cut D of G with the property that $D - e$ is tight in $G - e$ is a robust cut of G (Exercise 14.3.1).

Lemma 14.8 *Let G be a brick, let e be a quasi b-invariant edge of G, and let C be a nontrivial cut of G such that $C - e$ is a tight cut of $G - e$. Then C is a robust cut of G if and only if it is a separating cut of G.* □

By Theorem 4.2, a cut D of a matching covered graph G is a separating cut if and only if, given any edge of G, there is a perfect matching of G containing that edge which meets D in precisely one edge. The following corollary is an obvious consequence of the hypothesis that $D - e$ is tight in $G - e$ and Theorem 4.2.

Corollary 14.9 *Let e be a removable edge of a brick G and let C be a nontrivial cut of G such that $C - e$ is tight in $G - e$. If G has a perfect matching which contains the edge e and meets C in precisely one edge then C is robust in G .* □

Recall that the characteristic of a separating cut C in a matching covered graph G, denoted by $\lambda(C)$, is ∞ if C is tight, and is the smallest integer $k > 1$ such that there is a perfect matching of G which meets C in precisely k edges if C is not tight.

Lemma 14.10 *Let G be a brick, let C and D be nontrivial odd cuts of G, and let e be a quasi b-invariant edge of G such that the equality*

$$|M \cap C| + |M \cap D| = 2 + 2|M \cap \{e\}| \tag{14.5}$$

holds for each perfect matching M of G. Then, the cuts C and D are robust in G and have characteristic three.

Proof Since both C and D are odd cuts of G, any perfect matching of G has at least one edge in each of these cuts. For each perfect matching M of $G - e$, the equality (14.5) implies that $|M \cap (C - e)| = 1 = |M \cap (D - e)|$; hence $C - e$ and $D - e$ are both tight in $G - e$.

Let us now prove that C and D are both separating cuts in G and have characteristic three. By Theorem 4.2, a cut of G is a separating cut of G if and only if, given any edge f of G, there is a perfect matching M_f of G containing f and meeting that cut in exactly one edge. Since $C - e$ and $D - e$ are tight cuts of $G - e$, for any edge $f \neq e$, this condition is clearly satisfied. Since G is a brick, neither C nor D is a tight cut of G. Thus, there is a perfect matching M_C containing the edge e which meets C in at at least three edges. By equation (14.5), it follows that $|M_C \cap C| = 3$ and $|M_C \cap D| = 1$, and hence that D is a separating cut of G. Likewise, there is a perfect matching M_D containing the edge e which meets D in at at least three edges. By equation (14.5), it follows that $|M_D \cap D| = 3$ and $|M_D \cap C| = 1$, and hence that C is a separating cut of G. In sum, both cuts C and D are separating in G. Furthermore, as $|M_C \cap C| = 3 = |M_D \cap D|$, we deduce that $\lambda(C) = 3 = \lambda(D)$.

The cut $C - e$ is tight in $G - e$ and the perfect matching M_D of G contains the edge e and has just one edge in C. By Corollary 14.9, the cut C is robust in G. Likewise, the cut D is also robust in G. □

14.3.1 Barrier cut pairs and essentially 2-separation cut pairs

For the convenience of the readers we recall here the definitions of these two types of tight cuts.

Barrier cut pairs: Let G be a nonsolid brick, let e be a quasi b-invariant edge of G, and let B be a barrier of $G - e$ such that $G - e - B$ has precisely two nontrivial odd components, K and L. Then, the pair $(\partial(K) - e, \partial(L) - e)$ is a *barrier cut pair* of $G - e$. It should be noted that, since G itself is a brick, the edge e must have its ends in two different odd components of $G - e - B$. See Figure 14.6. (In this figure the edge e has no end in either $V(K)$ or $V(L)$. But these possibilities are not precluded.)

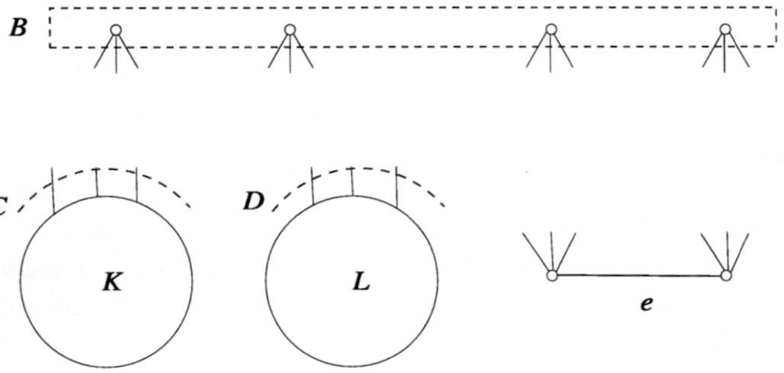

Fig. 14.6 The case in which $(C - e, D - e)$ is a barrier cut pair of $G - e$

Essentially 2-separation cut pairs: Let e be a quasi b-invariant edge of a nonsolid brick G and let $\{S, T\}$ be an essential 2-separation of $G - e$. Thus, (i) the two subgraphs $(G - e)[S]$ and $(G - e)[T]$ of $G - e$ induced by S and T, respectively, are bipartite subgraphs of $G - e$ such that their their majority parts S_+ and T_+ are special barriers of $G - e$, (ii) the cut $\partial(S)$ is the special barrier cut associated with the barrier S_+ and the cut $\partial(T)$ is the special barrier cut associated with the barrier T_+, and (iii) the pair $\{s, t\}$ is a 2-separation of the graph $(G - e)/(S \to s)/(T \to t)$. See Figure 14.7.

Let e be a quasi b-invariant edge of a nonsolid brick G and let C and D be two cuts of G such that $C - e$ and $D - e$ are either a barrier cut pair or an essentially 2-separation cut pair. We now establish fundamental properties of the cuts C and D.

Lemma 14.11 *Let e be a quasi b-invariant edge of a brick G, let B be a barrier of $G - e$ such that $G - e - B$ has precisely two nontrivial (odd) components, K and L. Then, the cuts $C := \partial_G(K)$ and $D := \partial_G(L)$ are both robust and have characteristic three.*

Proof Let M be a perfect matching of G. A simple counting argument shows that

$$|M \cap C| + |M \cap D| + (|B| - 2) = 2|M \cap \{e\}| + |B|;$$

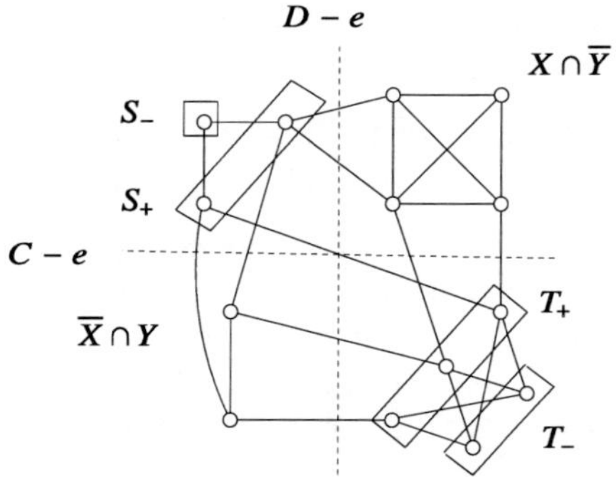

Fig. 14.7 Essentially 2-separation cuts $C - e$ and $D - e$ of $G - e$: $C = \partial(X), D = \partial(Y),\ \ I = \partial(X \cap Y)$, and $U = \partial(\overline{X} \cap \overline{Y})$.

hence equation (14.5) holds. By Lemma 14.10, the cuts C and D are robust and have characteristic three. □

Lemma 14.12 *Let e be a quasi b-invariant edge of a brick G and let $(C - e, D - e)$ be a pair of tight cuts of G associated with an essentially 2-separation $\{S, T\}$ of $G - e$. If either S or T has just one vertex, then both C and D are robust cuts of characteristic three.*

Proof Let X be a shore of C and Y be a shore of D such that $X \cap Y = S$. Assume that one of S and T has just one vertex. Adjust notation so that T is a singleton. If S is also a singleton then e joins a vertex in $X \cap \overline{Y}$ to a vertex in $\overline{X} \cap Y$. In that case, equality (14.5) holds, for each perfect matching M of G.

Alternatively, suppose that S is not a singleton; let M be any perfect matching of G. The edge e must have an end in the minority part S_- of S; hence $|M \cap \partial(S)| = 1 + 2|M \cap \{e\}|$. Moreover, no edge of G joins a vertex in $X \cap \overline{Y}$ to a vertex in $\overline{X} \cap Y$. Thus,

$$|M \cap C| + |M \cap D| = |M \cap \partial(S)| + |M \cap \partial(T)| = (1 + 2|M \cap \{e\}|) + 1;$$

hence equality (14.5) holds.

In both alternatives, equality (14.5) holds for each perfect matching M of G. By Lemma 14.10, the cuts C and D are robust and have characteristic three. □

Unlike the case with barrier cut pairs of tight cuts of $G - e$, the fact that $C - e$ and $D - e$ are an essentially 2-separation cut pair does not ensure that C and D are both separating cuts of G; see Exercise 14.3.2.

The fact below plays an important role in our proof of Lovász's conjecture which is presented in the next chapter.

A CONSEQUENCE OF COROLLARY 5.23

Suppose that e is a quasi-b-invariant edge in a nonsolid brick G, and let $C - e$ be a nontrivial tight cut of $G - e$ such that both $(C - e)$-contractions of $G - e$ are near-bricks. Then it follows from Corollary 5.23, that $C - e$ is either a member of a barrier cut pair or is a member of an essentially 2-separation cut pair.

Exercises

∗ **14.3.1** *Give a proof of Lemma 14.8.*

14.3.2 *Consider the brick G, the quasi b-invariant edge e in G, and the two cuts C and D shown in Figure 14.8. Show that (i) $(C - e, D - e)$ is an essentially 2-separation pair of tight cuts of $G - e$, (ii) D is not a separating cut of G although C is.*

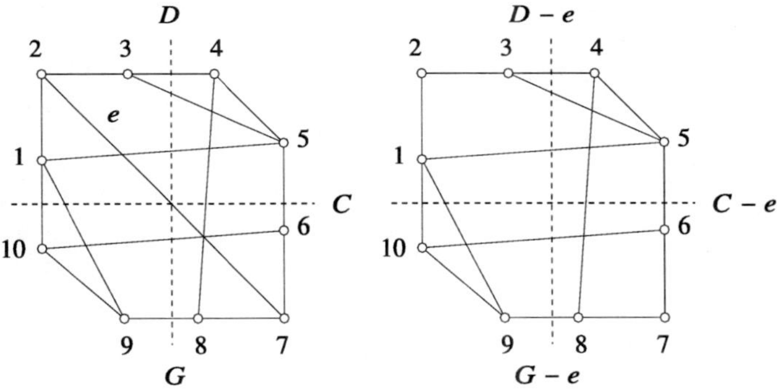

Fig. 14.8 The pair $(C - e), D - e)$ is an essentially 2-separation cut pair, but D is not separating in G.

14.4 Notes

The first published proof of the validity of Lovász's conjecture (Theorem 15.1) appeared in Carvalho's Ph. D. Thesis (1996, [6]), written under the supervision of Lucchesi and cosupervision of Murty. Later, it was published in a paper, by CLM (2002, [10]). That proof relied on a result concerning characteristics of bricks. It was shown (Theorem 4.1 in [10]) that if G is any nonsolid brick that is not a Petersen brick, then $\lambda(G) = 3$. In other words, every such brick has a separating cut C, and a perfect matching M_0, such that $|M_0 \cap C| = 3$. The inductive proof of that result made crucial use of the fact that if R is any removable class of a matching covered graph G, then $\lambda(G) \leq \lambda(G - R)$ (Exercise 10.2.6). Thus, to prove the required result by induction, it sufficed to show that any nonsolid brick G distinct from the Petersen graph either has a removable doubleton or has a $(b + p)$-invariant edge e such that $G - e$ is nonsolid. An unpublished report due to Campos and Lucchesi (2000, [5]) established the stronger result that if G is any nonsolid brick that is not a Petersen brick, and C is any nontrivial separating cut of G, then there exists a perfect matching M_0 of G such that $|M_0 \cap C| = 3$.

Chapter 15
b-Invariant Edges in Bricks

Contents

15.1 Existence of b-Invariant Edges

Using the results on robust cuts in nonsolid bricks established in the previous chapter, we now proceed to present a proof of Lovász's Conjecture 13.2:

C. L. Lucchesi, U. S. R. Murty, *Perfect Matchings*, Algorithms and Computation in Mathematics 31, https://doi.org/10.1007/978-3-031-47504-7_15

THE MAIN THEOREM

Theorem 15.1 *Every brick distinct from K_4, $\overline{C_6}$ and the Petersen graph has a b-invariant edge.*

For convenience we shall refer to K_4, $\overline{C_6}$, and $\mathbb{P}$ as *special* bricks.

The evolution of the approach to the Conjecture 13.2, as explained in Chapter 13, started with the observation that if a removable edge in a brick is not b-invariant, then that brick must have a nontrivial separating cut or, equivalently, that any removable edge in a solid brick is b-invariant. This led to the notion of robust cuts and to the consideration of the use of contractions with respect to a robust cut as an inductive tool. All this has been described in Chapter 13.

The proof we present here relies on many of the ideas developed in Carvalho's Ph. D. Thesis (1996, [6]), and in subsequently published, in streamlined and refined form, in papers by CLM (2002, [9, 10]). It also relies on results published some years later in CLM (2005, [13]), and in Lucchesi, Carvalho, Kothari and Murty (2018,[64]).

However, the very elegant climactic ending is new.

15.2 Proof of Theorem 15.1 ♯

The idea of the proof is to assume that (i) G is a brick distinct from K_4 and $\overline{C_6}$ and (ii) G has no b-invariant edge and deduce that G is the Petersen graph.

INDUCTION HYPOTHESIS

We shall prove Theorem 15.1 by assuming, inductively, that if G' is any brick that is not one of the three special bricks, with $|V(G')| < |V(G)|$, then G' has a b-invariant edge.

15.2.1 Reduction to simple nonsolid bricks

Every solid brick G distinct from K_4 has removable edges by Theorem 9.14, and by Theorem 10.16, every removable edge of such a brick G is b-invariant. Also, if G is a brick with multiple edges then clearly any member of a pair of parallel edges is b-invariant in G. We may thus suppose that G is a simple nonsolid brick distinct from $\overline{C_6}$.

THE IDEA OF REDUCTIONS

Our proof of Theorem 15.1 involves series of steps which we refer to as reductions. In each of these steps, we show that unless G has a certain property that the Petersen graph has, there is some edge of G that is b-invariant in G. The property that G is a simple nonsolid brick distinct from $\overline{C_6}$ is an example of reduction.

Let G be a simple nonsolid brick distinct from $\overline{C_6}$. By Corollary 14.4, G has robust cuts. We describe below how we choose a robust cut and also the notation that we use throughout the rest of the proof.

15.2.2 Definitions of X, C, M_0, G_1 and G_2

- Let X be a set of vertices of G having minimum cardinality such that $C := \partial(X)$ is robust.
- Let M_0 be a perfect matching of G such that $|M_0| = \lambda(C)$. (Thus, among the perfect matchings of G that meet C in more than one edge, M_0 has the fewest number of edges in common with C.)
- Let

$$G_1 := G/(\overline{X} \to \overline{x}) \quad \text{and} \quad G_2 := G/(X \to x)$$

denote the two C-contractions of G.

The following assertion records an immediate consequence of the minimality of $|X|$ and Lemma 14.6:

15.1.1 *The graph G_1 is a solid brick.* □

We shall prove that if G has no b-invariant edges and is distinct from $\overline{C_6}$ then $|M_0 \cap C| = 5$ and G is the Petersen graph. The following proposition is crucial to the inductive proof of this statement.

Proposition 15.2 *Let e be an edge of G such that both $G_1 - e$ and $G_2 - e$ are near-bricks. If $e \notin M_0$ then e is a b-invariant edge of G.*

Proof By hypothesis, e is not in M_0. Thus, M_0 is a perfect matching of $G - e$. As M_0 has three or more edges in C, it follows that $C - e$ is not a tight cut of $G - e$. By hypothesis, both $(C-e)$-contractions of $G-e$ are near-bricks. Hence, by subadditivity (Theorem 4.20), we deduce that $G - e$ is also a near-brick. □

15.2.3 Reduction to the case in which G_1 is an odd wheel

By statement 15.1.1, the C-contraction G_1 is a solid brick. In establishing the following assertion, we use the results concerning removable edges in solid bricks which are not odd wheels (up to multiple edges), described in Chapter 10, specifically Lemma 10.10 on Wheels.

15.2.1 *Suppose that G_1 is a solid brick that is not an odd wheel, up to multiple edges. Then some edge e of G_1 is b-invariant in G.*

Proof The graph G is simple. Thus, any multiple edges of G_1 must be in C. We may then obtain an underlying simple graph, G_1', of G_1, by removing multiple edges of G_1 in $C - M_0$. By Lemma 10.10 applied to G_1', the graph G_1' has a removable edge, say e, which is not in $M_0 \cup C$. The edge e is removable in G_1, and since G_1 is a solid brick, it follows from Theorem 10.16 that e is b-invariant in G_1. The graph $G_2 - e$ is also a near-brick, because $e \notin E(G_2)$. Since $e \notin M_0$, it follows from Proposition 15.2 that e is b-invariant in G. □

We may thus assume that G_1 is an odd wheel, up to multiple edges. Since G itself is simple, any multiple edges that G_1 has must be incident with its hub $\overline{x}$. If $|X| \geq 5$, then every edge of G_1 incident with $\overline{x}$ is b-invariant in G_1. On the other hand, if $|X| = 3$, an edge incident with $\overline{x}$ is b-invariant in G_1 if and only if it is a multiple edge. The following statement analyses the case in which an edge e incident with $\overline{x}$ is b-invariant in G_2 but not b-invariant in G_1.

Lemma 15.3 *Let e be an edge of G_2. If e is b-invariant in G_2 but is not b-invariant in G_1 then $|X| = 3$, $e \in M_0 \cap C$ and G_1 has an edge that is b-invariant in G.*

Proof Suppose that e is b-invariant in G_2 but not b-invariant in G_1. Every edge of G_2 that is not in C is b-invariant in G_1. Thus, e is in C. The edge e is in C and is not b-invariant in G_1. As G_1 is a wheel, this implies that the underlying simple graph of G_1 is K_4 and also that e is not a multiple edge of G_1. It follows that $|X| = 3$. Let u, v and w denote the three vertices of X and adjust notation so that e is incident with vertex u. See Figure 15.1.

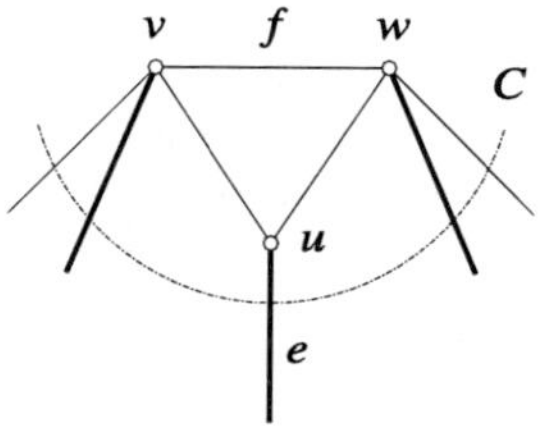

Fig. 15.1 Edge e is in C and not b-invariant in G_1. Solid lines indicate the edges of $M_0 \cap C$

The graph G_1 has precisely one edge, f, that joins the vertices v and w. Let $R := \{e, f\}$. One of the $(C - R)$-contractions of $G - R$ is $G_1 - R$, which is C_4, up to multiple edges. The other $(C - R)$-contraction of $G - R$ is $G_2 - e$, a near-brick. Thus, the graph $G - R$ is a near-brick. Moreover, the cut $C - e$ is tight in $G - R$.

As M_0 has more than one edge in C, and since e is not a multiple edge in C, it follows that $|M_0 \cap C| = 3$ and $e \in M_0$. Moreover, $f \notin M_0$. Thus, M_0 is a perfect matching of $G - f$ that contains edge e. As $G - R$ is matching covered, it follows that $G - f$ is matching covered. Thus, f is removable in G. By the monotonicity of function b, $1 \le b(G - f) \le b(G - R) = 1$; hence f is b-invariant in G. □

Corollary 15.4 *Let e be an edge of G_2. If e is b-invariant in G_2 and is not in M_0 then e is b-invariant in G_1 and in G.*

Proof Assume that e is b-invariant in G_2 and is not in M_0. As e is not in M_0, then e is b-invariant in G_1, by Lemma 15.3. By Proposition 15.2, e is b-invariant in G.□

The following results rely on the previous two statements.

CONDITIONS UNDER WHICH THE EXISTENCE OF A b-INVARIANT EDGE IN G_2 IMPLIES THE EXISTENCE OF A b-INVARIANT EDGE IN G

Lemma 15.5 *Let e be a b-invariant edge of G_2. The following properties hold:*

(i) *if e is not b-invariant in G_1 then e is not a multiple edge of G_1, $|X| = 3$ and G_1 has an edge that is b-invariant in G; and*
(ii) *if e is not in M_0 then e is b-invariant in G_1 and in G.* □

The graph G_2, being a C-contraction of G with respect to a robust cut, is a near-brick. We would like to show that, unless G has a b-invariant edge, G_2 is also an odd wheel. We are able to achieve this only by gradually unravelling the structure of G.

15.2.4 Reduction to the case in which G_2 is a brick

15.5.1 *If G_2 is not a brick then some edge of G is b-invariant in G.*

Proof Suppose that G_2 is not a brick. Then, it has nontrivial tight cuts. Let Y be a minimal proper superset of X such that $D := \partial(Y)$ is a nontrivial tight cut of G_2. The two shores of the cut D in G_2 are $Y' := (Y - X) + x$ and $\overline{Y}$, and the two D-contractions of G_2 are $H := G_2/(\overline{Y} \to \overline{y})$ and $H' := G_2/(Y' \to y')$. Since G_2 is a near-brick and D is a tight cut of G_2, one of H and H' is bipartite. But H' cannot be bipartite because, if it were, one of its parts would be free of contraction vertices, and that part would be a nontrivial barrier of G. This is impossible because G is a brick. Therefore Y' is the bipartite shore of D, and H is bipartite with bipartition (A, B), where A is the majority part Y'_+ of Y', and B is the minority part Y'_- of Y' together with $\overline{y}$. Thus

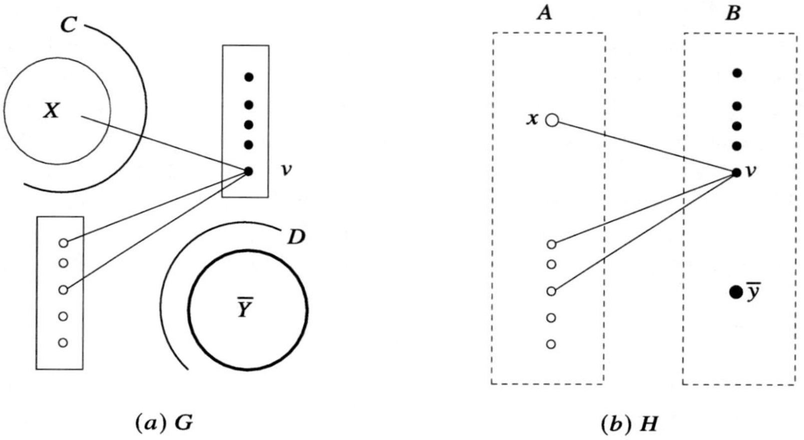

Fig. 15.2 (a) Cuts C and D in G; (b) $H = G_2/\overline{Y}$. (If $v \neq \overline{y}$ is any vertex in B, then some edge incident with v is b-invariant in both G_2 and G.)

the contraction vertex x belongs to A and the contraction vertex $\overline{y}$ belongs to B; all other vertices of H are vertices of G. See Figure 15.2.

The minimality of Y implies that the bipartite graph $H = G_2/(\overline{Y} \rightarrow \overline{y})$ is a brace. This can be proved quite easily by showing that if H is not a brace, then there is some set Z , with $X \subset Z \subset Y$ such that $\partial(Z)$ is a nontrivial tight cut of G_2 (Exercise 15.2.1).

Let $v \neq \overline{y}$ be a vertex of G that belongs to the same part of the bipartition of H to which $\overline{y}$ belongs. Suppose that, in G, two or more edges join v to vertices of X. Let e and f be two such edges incident with v. Since e and f are adjacent, not both can belong to the perfect matching M_0. Adjust notation so that $e \notin M_0$. In G_2, the edges e and f are multiple edges. Thus, e is b-invariant in G_2. Moreover, edge e does not belong to M_0. By Lemma 15.5, edge e is b-invariant in G.

So, suppose that, in G, vertex v is adjacent to at most one vertex of X. As G is a brick, the vertex v is adjacent to three or more vertices in G. Thus, in H, vertex v is adjacent to two or more vertices of $A - x$. It follows that H has six or more vertices. As has already been noted, H is a brace. Thus, every edge incident with v is removable in H. This clearly implies that there is an edge e which is incident with v that does not belong to $C \cup D \cup M_0$. It is now easy to see that e is b-invariant in both G_2 and G. (Exercise 15.2.2). □

We may thus assume that G_2 is a brick.

15.2.5 Reduction to the case in which G_2 is not a special brick

15.5.2 *Suppose that brick G_2 is a special brick. Then G_2 is not K_4; it must be either $\overline{C_6}$ or the Petersen graph. In either case, G has a b-invariant edge.*

Proof All the three special bricks are cubic. Thus, if G_2 is a special brick, then $|C| = 3$ implying that $G_1 = K_4$. If G_2 is K_4 then G is $\overline{C_6}$, a contradiction. If G_2 is $\overline{C_6}$ then G is the bicorn (Figure 15.3) and the only removable edge of G is b-invariant in G.

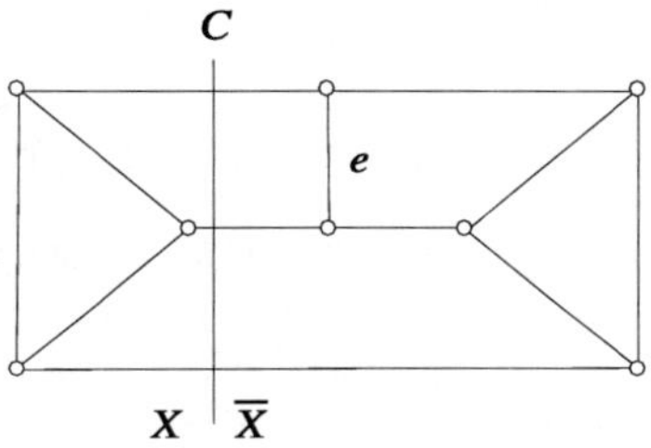

Fig. 15.3 Edge e is the only b-invariant edge of the bicorn

Suppose that G_2 is the Petersen graph (Figure 15.4).

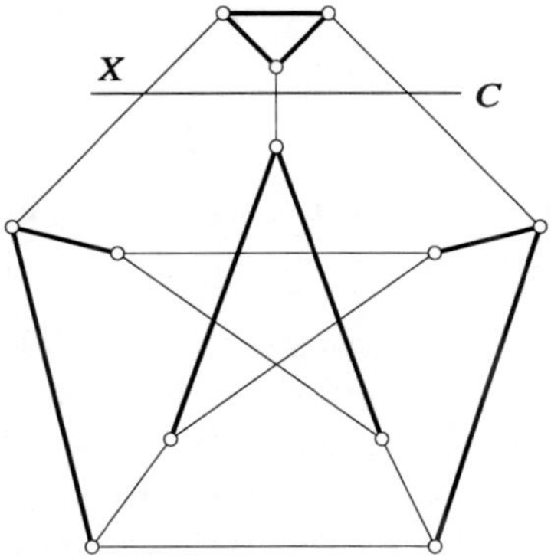

Fig. 15.4 Graph G in the case $G_2 = \mathbb{P}$. Solid lines indicate b-invariant edges of G

It is easy to prove that G has nine b-invariant edges (Exercise 15.2.3). □

15.2.6 Choosing a b-invariant edge e of G_2

The reductions we have described so far allow us to assume that G_1 is an odd wheel (up to multiple edges in C) and that G_2 is a brick, but not one of the three special bricks. By induction, G_2 has a b-invariant edge. We select a b-invariant edge e of G_2 according to the following criterion:

RULE FOR CHOOSING b-INVARIANT EDGE e OF G_2

If G_2 has b-invariant edges in C, then choose e to be an edge in C.

At this stage, we bring in a second perfect matching of G into play.

PERFECT MATCHING M_1

Let M_1 be a perfect matching of G that contains edge e and just one edge in C. (Since C is a separating cut of G, such a perfect matching must exist by Theorem 4.2.)

We have assumed that G does not contain b-invariant edges. In particular, e is not b-invariant in G. Lemma 15.5 immediately implies the following consequence:

Corollary 15.6 $e \in M_0 \cap M_1$. □

As the chosen b-invariant edge e of G_2 is not b-invariant in G, it does not necessarily mean that G has no b-invariant edges. Thus it helps to have some flexibility in selecting a b-invariant edge of G_2. The following result is relevant in this context.

Lemma 15.7 *Any edge of $G_2 - e$ which does not belong to M_1 and is b-invariant in $G_2 - e$ is also b-invariant in G_2.*

Proof Let $f \notin M_1$ be an edge of G_2 that is b-invariant in $G_2 - e$. Edge e is b-invariant in G_2; thus, $G_2 - e$ is a near-brick. Edge f is b-invariant in $G_2 - e$; consequently, $G_2 - e - f$ is a near-brick. As matching M_1 does not contain edge f, M_1 is a perfect matching of $G_2 - f$ that contains edge e. We deduce that graph $G_2 - f$ is matching covered. Moreover, by the monotonicity of function b, we have that

$$1 = b(G_2) \leq b(G_2 - f) \leq b(G_2 - e - f) = 1.$$

Hence, f is b-invariant in G_2. This conclusion holds for each edge f of G_2 that is b-invariant in $G_2 - e$ and is not in M_1. □

15.2.7 Reduction to the case in which e is a quasi b-invariant edge of G

Suppose that the b-invariant edge e of G_2 is not b-invariant in the wheel G_1. By Lemma 15.5, the graph G_1 contains an edge that is b-invariant in G.

We may thus assume that edge e is b-invariant in G_1.

The edge e, by our choice, is b-invariant in G_2. Thus, $G_1 - e$ and $G_2 - e$ are both near-bricks. If the cut $C - e$ is not tight in $G - e$ then, by the subadditivity of function b, the edge e is b-invariant in G.

Hence we may assume that $C - e$ is tight in $G - e$. In this case, e is not b-invariant in G. By Lemma 15.5, edge e lies in M_0. Moreover,

$$b(G - e) = b(G_1 - e) + b(G_2 - e) = 2.$$

This implies that e is a quasi b-invariant edge of G.

The above analysis shows that if G has no b-invariant edges, then the edge e, chosen to be a b-invariant edge of G_2, has to be a quasi b-invariant edge of G. So, we may suppose that $b(G - e) = 2$. In this case, as explained in Section 14.3, there exists a cut $D := \partial(Y)$ of G such that $\{C - e, D - e\}$ is either a barrier cut pair or an essentially 2-separation cut pair of $G - e$. The reduction described below eliminates the first possibility. In this reduction, and the rest of the proof of the Main Theorem, Lemma 8.19 plays a crucial role. The readers should review that lemma carefully before proceeding further.

15.2.8 Reduction to case in which $\{C - e, D - e\}$ is an essentially 2-separation cut pair of tight cuts of $G - e$

Suppose that $\{C - e, D - e\}$ is a barrier cut pair of tight cuts of $G - e$ associated with a barrier B of $G - e$. By Lemma 14.11, the cut D is robust, and both cuts C and D have characteristic three in G. The cut D is a nontrivial cut of G in one of the C-contractions of G. As the brick G_1 is solid, it must be the case that D is a separating cut of G_2. Adjust notation so that Y is the shore of D that is a subset of $\overline{X}$. (See Figure 15.5.) Let H denote the bipartite graph $(G - e)/(X \rightarrow x)/(Y \rightarrow y)$. Clearly, $H = (G_2 - e)/(Y \rightarrow y)$.

Proposition 15.8 $|C - e| \geq 3$.

Proof If $|C| \geq 4$ then the inequality holds. Assume thus that $|C| = 3$. In that case, $|X| = 3$, $G_1 = K_4$ and no edge of C is removable in G_1. The cut $C - e$ is tight in $G - e$; thus $e \notin C$. We deduce that $C - e = C$; hence the inequality holds with equality. □

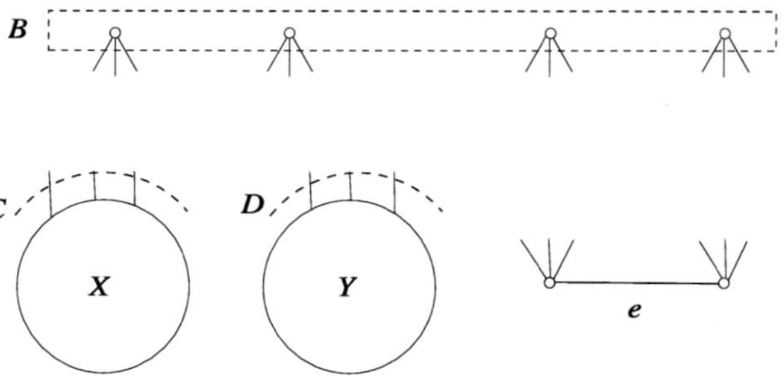

Fig. 15.5 The case in which $(C - e, D - e)$ is a barrier cut pair of $G - e$. The graph $H = (G - e)/(X \to x)/(Y \to y) = (G_2 - e)/(Y \to y)$ is bipartite.

Lemma 15.9 *At most one edge of $C - M_1$ is not b-invariant in G_2.*

Proof By Proposition 15.8, $|C - e| \geq 3$. We may thus suppose that the degree of x in H, which is equal to $|C - e|$, is at least three. By Lemma 8.19 (with the brick G_2 and the cut D, respectively, playing the roles of G and C in the statement of that lemma), at most one edge of $C - e$ is not removable in H. By definition, M_1 is a perfect matching containing e and meeting C in exactly one edge. If $e \in C$ then $C - M_1 = C - e$. Alternatively, if $e \notin C$ then at most one edge of C is not removable in H. In both alternatives, at most one edge of $C - M_1$ is not removable in H. Let f be any edge of $C - M_1$ that is removable in H. Then, as $C - e$ and $D - e$ are disjoint, it follows that f does not belong to cut D, and therefore that f is b-invariant in $G_2 - e$. Moreover, as f does not belong to M_1, it follows from Lemma 15.7 that edge f is b-invariant in G_2. This conclusion holds for each edge of $C - M_1$ that is removable in H. □

We first observe that, by the criterion used for choosing e, the above conclusion implies that edge e belongs to C. Thus, e is the edge of $C \cap M_1$; hence, again by the above conclusion, at most one edge of $C - e$ is not b-invariant in G_2. Moreover, as $e \in M_0$, we have that e is the edge of $M_0 \cap M_1 \cap C$; hence $C - M_0 - M_1 = C - M_0$.

Suppose that $|C| \geq 5$. In that case, $C - M_0$ consists of two or more edges, neither of which is e. Thus, $C - M_0$ contains an edge, e', which is b-invariant in G_2. By Lemma 15.5, e' is b-invariant in G_1 and in G.

Consider now the case in which $|C| \leq 4$. Hence $|X| = 3$ and G_1 is K_4, up to multiple edges in C. The edge e is in C and the cut $C - e$ is tight in $G - e$; hence e is a multiple edge in G_1. It follows that $|C| = 4$. The doubleton $M_0 \cap C - e$ contains an edge, e'', which is b-invariant in G_2. Moreover, e'' is not removable in G_1. By Lemma 15.5, G_1 contains an edge which is b-invariant in G.

We may thus assume that e is a quasi b-invariant edge of G, and also assume that the cut $C - e$ is tight in $G - e$ but is not a member of a barrier cut pair of $G - e$. Therefore, the brick G has a cut D such that $(C - e, D - e)$ is an essentially 2-separation cut pair of $G - e$.

15.2.9 Working with the hypothesis that $(C - e, D - e)$ is an essentially 2-separation cut pair of tight cuts of $G - e$

Let $\{S, T\}$ be an essential 2-separation of $G - e$ such that $C - e$ and $D - e$ are the two tight cuts of $G - e$ associated with $\{S, T\}$. Adjust notation so that $S = X \cap Y$ and $T = \overline{X} \cap \overline{Y}$ and, as usual, let I and U denote, respectively, $\partial(S)$ and $\partial(T)$.

SUBGRAPHS $G[X \cap Y]$ AND $G[X \cap \overline{Y}]$

As $G_1 := G/(\overline{X} \to \overline{x})$ is an odd wheel, it follows that: (i) the subgraph induced by $X \cap Y$ is a path of even length and (ii) the subgraph induced by $X \cap \overline{Y}$ is one of odd length. Furthermore, (iii) the cardinality of $X \cap Y$ is equal to either one or three because in $G - e$ the vertices in S_- have no neighbours outside S_+.

At this stage, all that we can say is that $|X \cap \overline{Y}|$ is a positive even integer. But assuming that G has no b-invariant edges, we will show that $|X \cap Y|$ and $|X \cap \overline{Y}|$ are, respectively 3 and 2 and thereby deducing that $G[X]$ is a pentagon. (This requires several intermediate steps.)

THE SUBSET F OF THE CUT C

We denote by F the set of edges in the cut C which have both their ends in $\overline{Y}$. In other words, $F = \partial(X \cap \overline{Y}) \cap U$. See Figures 15.6(a) and 15.6(b) where the edges in F are indicated by dotted lines.

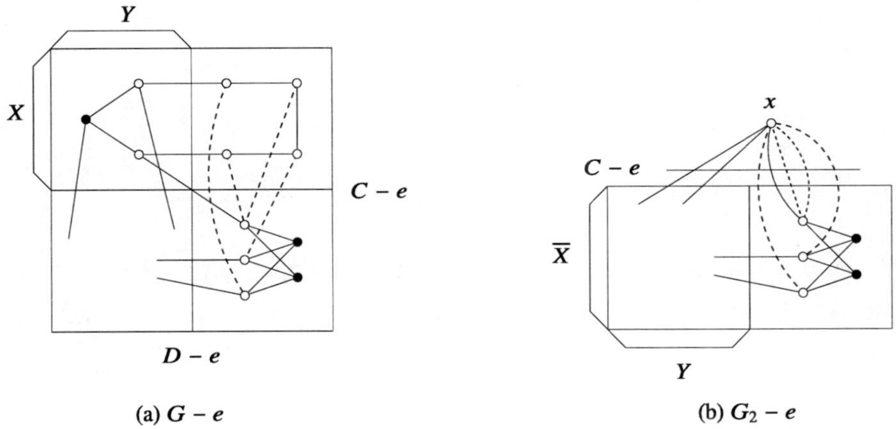

Fig. 15.6 The graphs $G - e$ and $G_2 - e$

Lemma 15.10 $|F - e| \geq |X \cap \overline{Y}| - 1$, *with equality only if* $|X \cap \overline{Y}| \geq 4$. *Consequently,* $|F - e| \geq 2$.

Proof Up to multiple edges in C, the graph G_1 is an odd wheel having the contraction vertex $\overline{x}$ as a hub. Thus, every vertex of $X \cap \overline{Y}$ is incident with an edge of C. Since $C - e$ and $D - e$ are two tight cuts of $G - e$ that cross, by Theorem 4.12 it follows that there is no edge of $G - e$ with one end in $X \cap \overline{Y}$ and one end in $\overline{X} \cap Y$.

If $|X \cap \overline{Y}| \geq 4$ then $|F - e| \geq |X \cap \overline{Y}| - 1$. If no end of e is in $X \cap \overline{Y}$ then $|F - e| \geq |X \cap \overline{Y}|$.

We may thus assume that $|X \cap \overline{Y}| = 2$ and e is incident with a vertex in $X \cap \overline{Y}$. In that case, as (i) $e \in C$, (ii) the cut $I = \partial(S)$ is tight in $G - e$ and (iii) $G[S] - e$ is bipartite, it follows that S is a singleton. Thus, $|X| = 3$. The graph G_1 is then K_4, up to multiple edges in C. The edge e is b-invariant in G_1; hence it is a multiple edge in G_1. We deduce that $|F - e| \geq 2 = |X \cap \overline{Y}|$. □

THE TWO $(U - e)$-CONTRACTIONS OF $G_2 - e$

Now we consider the two $(U - e)$-contractions

$$H' := (G_2 - e)/(\overline{X} \cap \overline{Y}) \text{ and } H'' := (G_2 - e)/((\overline{X} \cap Y) + x)$$

of the graph $G_2 - e$. See Figure 15.7.

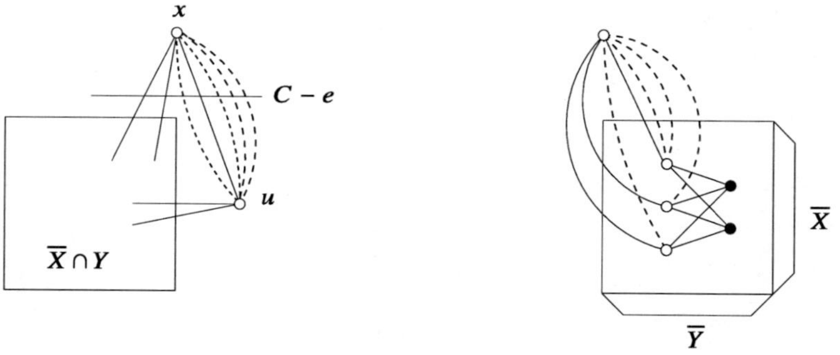

(a) $H' = (G_2 - e)/(\overline{X} \cap \overline{Y})$ (b) $H'' = (G_2 - e)/((\overline{X} \cap Y) + x)$

Fig. 15.7 The two $(U - e)$-contractions of $G_2 - e$

The following statement is an obvious observation (Exercise 15.2.4).

Lemma 15.11 *The bipartite graph* $H'' = (G_2 - e)/((\overline{X} \cap Y) + x)$ *is isomorphic to the* $(U - e)$*-contraction* $(G - e)/(X \cup Y)$ *of* $G - e$. □

Recall that M_0 is a perfect matching of G such that $|M_0 \cap C| = \lambda(C) \geq 3$ and that M_1 is a perfect matching of G such that $e \in M_1$ and $|M_1 \cap C| = 1$.

The following result is analogous to Lemma 15.9.

Lemma 15.12 *Every edge of $F - M_1$ is b-invariant in G_2.*

Proof Let f be any edge of F that does not belong to M_1. As the edge e does belong to M_1, it follows that f belongs to $F - e$. By Lemma 8.19 (with brick G_2 playing the role of G and U the role of C), every edge of $U - e$ is removable in the (bipartite) $(U - e)$-contraction $H'' = (G_2 - e)/((\overline{X} \cap Y) + x)$ of $G_2 - e$ (Figure 15.7(b)). In the other $(U - e)$-contraction $H' := (G_2 - e)/(\overline{X} \cap \overline{Y})$ of $G_2 - e$ (Figure 15.7(a)), from Lemma 15.10 it follows that all edges in $F - e$ are multiple edges and hence are removable. As $f \in F - e$, it follows that the edge f is removable in H' as well. In sum, f is removable in both $(U - e)$-contractions of $G_2 - e$. Thus, f is removable in $G_2 - e$.

The set $\overline{X} \cap \overline{Y}$ is the bipartite shore of the tight cut $U - e$ of $G_2 - e$. The graph $G_2 - e$ is a near-brick and H' is the $(U - e)$-contraction $(G_2 - e)/(\overline{X} \cap \overline{Y})$ of $G_2 - e$. Thus, H' is a near-brick. The edge f is a multiple edge of H'; hence $H' - f$ is a near-brick. That is, the $(U - e - f)$-contraction $H' - f$ of $G_2 - e - f$ is a near-brick. The other $(U - e - f)$-contraction of $G_2 - e - f$ is the bipartite matching covered graph $H'' - f$. We conclude that $G_2 - e - f$ is a near-brick. That is, f is b-invariant in $G_2 - e$. By hypothesis, f does not belong to M_1. Therefore, by Lemma 15.7, edge f is b-invariant in G_2. This conclusion holds for each edge f in $F - M_1$. □

Corollary 15.13 $e \in C$.

Proof Suppose that $e \notin C$. By Lemma 15.10, $|F| \geq 2$; hence $F - M_1$ is nonempty. By Lemma 15.12 we have that F contains an edge, f, that is b-invariant in G_2. The criterion adopted for choosing the edge e would have given preference to f; a contradiction. □

Reduction to the case in which $F \subseteq M_0$

Lemma 15.14 *Any edge in $F - M_0$ is b-invariant in G.*

Proof Let f be an edge in $F - M_0$. The edge e is in M_0 and is the only edge of $M_1 \cap C$. Thus, $f \in F - M_1$. By Lemma 15.12, f is b-invariant in G_2. As $f \notin M_0$ we conclude that f is b-invariant in G_1 and also b-invariant in G, by Lemma 15.5. This conclusion holds for each edge $f \in F - M_0$. □

We may thus assume that every edge of F is an edge in the perfect matching M_0. This clearly implies that F is a matching!

15.2.10 Reduction to the case in which $G[X]$ and $G[\overline{Y}]$ are pentagons

In this final reductive step we shall show, with the proviso that G has no b-invariant edges, that both the subgraphs $G[X]$ and $G[\overline{Y}]$ of G, induced by X and $\overline{Y}$, respectively, are pentagons. For this, it is convenient to establish the following somewhat technical result.

Lemma 15.15 *The cut U is nontrivial, $|X \cap \overline{Y}| = 2$ and the edge e has one end in the minority part of the shore T of U and one end in $X \cap Y$.*

Proof Clearly, $F \subseteq U$. By Lemma 15.14, we have that $F \subseteq M_0$. Hence

$$F - e \subseteq M_0 \cap U - e. \tag{15.1}$$

By Lemma 15.10, $|F - e| \geq 2$. It follows that

$$2 \leq |F - e| \leq |M_0 \cap U - e|. \tag{15.2}$$

Hence U is nontrivial and the shore T of U has three or more vertices.

The subgraph $G[T]$ of G induced by the shore T of U is bipartite; hence its majority part T_+ is a nontrivial barrier of $G - e$. As G is a brick, the edge e has an end in T_-. As $e \in C$, the edge e has its other end in X; hence the perfect matching M_0, which contains e, must match each vertex in T_-, other than the end of e, with a vertex in T_+. This implies that M_0 has precisely three edges in the cut U; one of these edges is e. Thus, equality holds throughout in equation (15.2); hence $|F - e| = |(M_0 \cap U) - e| = 2$. Moreover, from equation (15.1), we deduce that $F - e = (M_0 \cap U) - e$. As $|F - e| = 2$, it follows that $|X \cap \overline{Y}| = 2$, by Lemma 15.10.

The two edges of $F - e$ are in M_0 and are incident with the two vertices of $X \cap \overline{Y}$; hence $e \notin F$. In sum, e is incident with a vertex in T_- and is in $C - F$. Thus, e has an end in $X \cap Y$. □

Lemma 15.16 *The graph $G[X]$ is a pentagon.*

Proof We already know that G_1 is an odd wheel and hence that $G[X]$ is an odd polygon. Thus, to show that $G[X]$ is a pentagon, it suffices to show that $|X \cap Y| = 3$ and $|X \cap \overline{Y}| = 2$. By Lemma 15.15, we already know that $|X \cap \overline{Y}| = 2$.

Let us now prove that $|X \cap Y| = 3$. Suppose that $|X \cap Y| = 1$. Then, $|X| = |X \cap Y| + |X \cap \overline{Y}| = 3$ and both the edges of the matching F are b-invariant in G_2 and are not multiple edges. In this case, two edges of G_1 are b-invariant in G by Lemma 15.5.

Therefore $|X \cap Y| \geq 3$ and each minority vertex of the bipartite graph $G[X \cap Y]$ has to be incident with e. As we have already noted, this is possible only if there is just one minority vertex; equivalently, only if $|X \cap Y| = 3$. We conclude that $|X| = 5$ and hence that $G[X]$ is a pentagon. □

Lemma 15.17 *The cut D is robust and for every perfect matching M of G, $|M \cap C|, |M \cap D| \in \{1, 5\}$.*

Proof The two edges of F and the edge e are edges of $M_0 \cap C$. As M_0 is a perfect matching, it follows that the two vertices of S_+ are also incident with edges in $M_0 \cap C$. Thus, $|M_0 \cap C| = 5$. By definition of M_0, the characteristic of C is $|M_0 \cap C|$. Thus, C has characteristic five. As $|X| = 5$, $|M \cap C| \in \{1, 5\}$, for every perfect matching M of G. Neither S nor T is trivial. Thus, for each perfect matching M of G,

$$|M \cap I| = 1 + 2|M \cap \{e\}| \text{ and } |M \cap U| = 1 + 2|M \cap \{e\}|.$$

Moreover, no edge of G joins a vertex in $X \cap \overline{Y}$ to a vertex in $\overline{X} \cap Y$. Thus,

$$|M \cap C| + |M \cap D| = |M \cap I| + |M \cap U| = 2 + 4|M \cap \{e\}|.$$

As $|M \cap C| \in \{1, 5\}$, it follows that $|M \cap D| \in \{1, 5\}$. This conclusion holds for every perfect matching of G. In particular, M_0 contains the edge e and has just one edge in D. Moreover, $D - e$ is tight in $G - e$. By Corollary 14.9, D is robust. □

Lemma 15.18 *The graph $G[\overline{Y}]$ is also a pentagon.*

Proof We have already seen that the cuts I and U are nontrivial and e has one end in the minority part of S and the other end in the minority part of T. Thus, only one minority vertex of T is incident with e.

If possible, suppose that the shore T of U has five or more vertices. In this case, the minority part of T has two or more vertices; hence the minority part of T contains a vertex, v, not incident with edge e. Vertex v has degree at least three in $G_2 - e$. By Lemma 8.19, $\partial(v)$ contains two or more edges that are removable in the bipartite graph $(G - e)/(X \cup Y)$. At least one of these edges, say, f, does not lie in the perfect matching M_1. Edge f has both ends in T; hence $G_2 - e - f$ is a near-brick. As $f \notin M_1$, it follows from Lemma 15.7 that $G_2 - f$ is a near-brick. As f has both ends in $\overline{X}$, it is b-invariant in G_1. In sum, both C-contractions of $G - f$ are near-bricks. If C is not tight in $G - f$ then f is b-invariant in G, a contradiction.

We may thus assume that C is tight in $G - f$; hence f is quasi b-invariant in G and C is tight in $G - f$. Then, G has a cut D' such that $(C, D' - f)$ is either a barrier cut pair of $G - f$, or $(C, D' - f)$ is an essentially 2-separation cut pair of $G - f$. If $(C, D - f)$ is a barrier cut pair then C has characteristic three, by Lemma 14.11; this is a contradiction. So, suppose that $(C, D - f)$ is associated with an essentially 2-separation pair $\{S', T'\}$. Adjust notation so that $S' \subset X$ and $T' \subset \overline{X}$. The edge f has both ends in $\overline{X}$; hence S' is a singleton. By Lemma 14.12, C has characteristic three, a contradiction.

We deduce that T consists of precisely three vertices. The edge e is not in F; hence $G[\overline{Y}]$ is a pentagon. □

15.2.11 The conclusion: G is the Petersen graph

Making use of the various reductions described in the previous sections, we are now in a position to wrap up the proof of Lovász's Conjecture by showing that the only brick that has no b-invariant edges and is different from K_4 and $\overline{C_6}$ is the Petersen graph.

By Lemmas 15.16 and 15.18, the subgraphs $G[X]$ and $G[\overline{Y}]$ of G induced by X and $\overline{Y}$ are pentagons. Let $v_0v_1v_2v_3v_4v_0$ denote the pentagon $G[X]$, where v_0 is one of the ends of e, the two vertices v_1 and v_4 are the two majority vertices of $X \cap Y$, and v_2 and v_3 are the two vertices in $X \cap \overline{Y}$. Also, let $v'_0v'_1v_3v_2v'_4v'_0$ denote the pentagon $G[\overline{Y}]$, where v'_0 is the minority vertex of $\overline{X} \cap \overline{Y}$ and v'_1 and v'_4 are its two majority vertices. See Figure 15.8.

The two edges v_3v_1' and v_2v_4' are the edges in F. The perfect matching M_0 contains the edge $e = v_0v_0'$, the two edges of F, and it matches v_1 and v_4 with two distinct vertices in $\overline{X} \cap Y$. See Figure 15.8(a). Likewise, the perfect matching M_1 contains the edges e, v_1v_2, v_3v_4 and it matches v_1' and v_4' with two distinct vertices in $\overline{X} \cap Y$. See Figure 15.8(b).

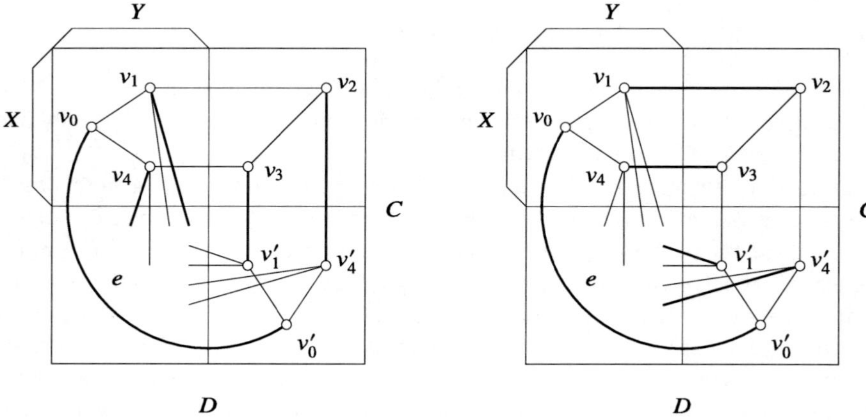

(a) M_0 (b) M_1

Fig. 15.8 The perfect matchings M_0 and M_1.

At this stage we can infer that G has at least ten vertices, and that $n - 8$ of them are in the quadrant $\overline{X} \cap Y$. We shall now proceed to show that this quadrant has precisely two vertices and that both $G[\overline{X}]$ and $G[Y]$ are also pentagons by using the fact that the two edges v_3v_1' and v_2v_4' in F are b-invariant in both G_1 and G_2 but not b-invariant in G (Exercise 15.2.5). Making use of this fact to deduce that G is indeed the Petersen graph involves a subtle use of symmetry as explained below.

FINAL STEP: CYCLIC ROTATIONS

Let v_2' and v_3' denote two vertices of $\overline{X} \cap Y$ that are respectively adjacent to vertex v_1 and vertex v_4. Likewise, let v_2'' and v_3'' denote two vertices of $\overline{X} \cap Y$ that are respectively adjacent to vertex v_1' and vertex v_4'. See Figure 15.9(a).

We first observe that the four vertices v_0, v_0', v_2 and v_3 have degree three. As noted above, edge $e_1 := v_3v_1'$ is also b-invariant in G_2 and in G_1, but not in G. Hence G has a robust cut $D_1 := \partial(Y_1)$ such that the pair $\{C - e_1, D_1 - e_1\}$ is associated with an essential 2-separation $\{S_1, T_1\}$ of $G - e_1$. Moreover, neither S_1 nor T_1 is trivial, and the graph $G[\overline{Y_1}]$ is a pentagon.

The minority vertex of $X \cap Y_1$ is the end v_3 of e_1 in X. Thus, the two majority vertices of $X \cap Y_1$ are the vertices v_2 and v_4. Consequently, the two vertices of $X \cap \overline{Y_1}$ are v_0 and v_1. This corresponds to the *rotation* of the pentagon $G[X]$ in which every vertex v_i is now mapped to the vertex v_{i+3}, where the indices are taken modulo 5. See Figure 15.9(b).

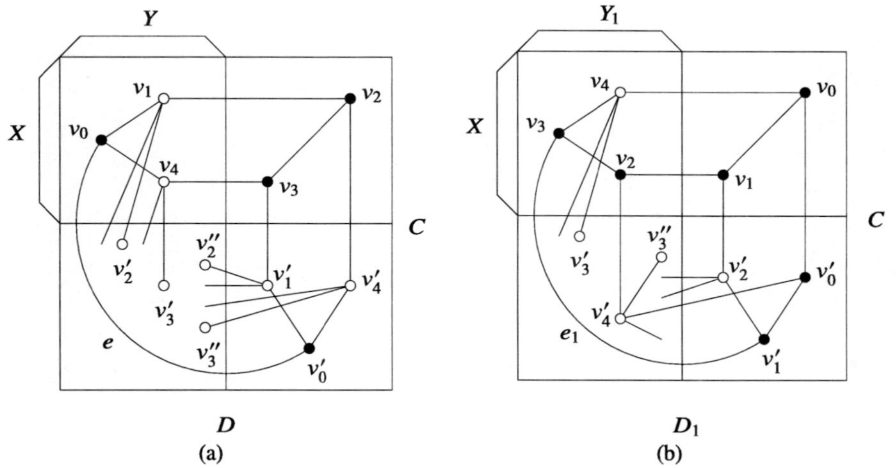

Fig. 15.9 Solid dots represent vertices known to have degree three. (a) The vertices v'_2, v'_3, v''_2 and v''_3. (b) the vertices v'_2 and v''_2 coincide.

We have seen that the vertices v_0, v'_0, v_2 and v_3 have degree three. There is a correspondence between the edges of the pentagons $G[\overline{Y}]$ and $G[\overline{Y_1}]$. The vertices v'_1 and v_1 correspond, with edge e_1 playing the role of e, respectively to the vertices v'_0 and v_3 in the case of edge e. Thus, the vertices v_1 and v'_1 also have degree three in G. Moreover, the vertex v'_2, which is adjacent to the vertex v_1, coincides with the vertex v''_2, which is adjacent to v'_1. Thus, the vertices v_1 and v'_1 are adjacent to precisely one and the same vertex v'_2 of $\overline{X} \cap Y$.

A similar reasoning may be applied with $e_2 := v_2v'_4$ playing the role of e, thereby concluding that both vertices v_4 and v'_4 have degree three and have a common neighbour v'_3 in $\overline{X} \cap Y$. In sum, all the eight vertices of the graph $G[X \cup \overline{Y}]$ have degree three. Moreover, the ends of the edges of $\partial(X \cup \overline{Y})$ in $\overline{X} \cap Y$ are in the set $R := \{v'_2, v'_3\}$. See Figure 15.10(a).

As G is a brick, it is 3-connected; thus, $\overline{X} \cap Y = R$. As $|\overline{X} \cap Y|$ is even, the vertices v'_2 and v'_3 are distinct. Moreover, they must have degree three or more; hence they have degree three and are adjacent. We conclude that G is the Petersen graph. See Figure 15.10(b).

This completes the proof of the validity of Lovász's Conjecture. □

A removable class R of a matching covered graph G is *optimal* if either R is a removable doubleton or R is a $(b + p)$-invariant singleton. The following corollary is an easy consequence of Theorems 15.1 and 13.6.

OPTIMAL REMOVABLE CLASSES

Corollary 15.19 *Every brick has an optimal removable class.* □

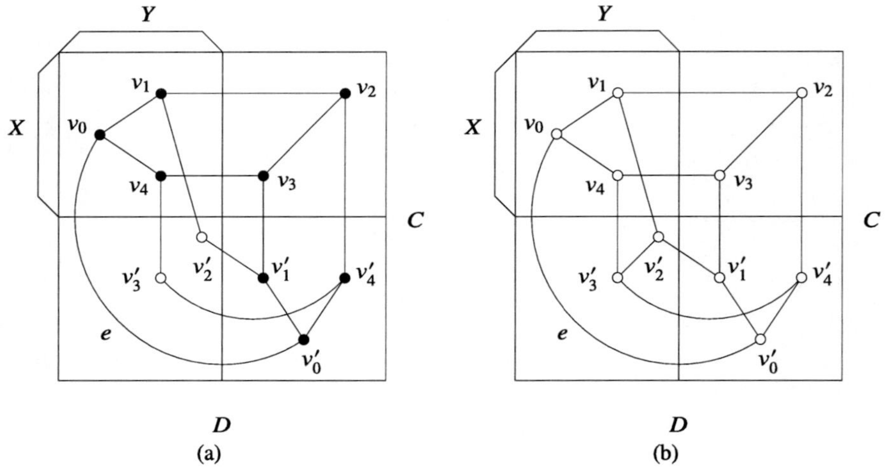

Fig. 15.10 (a) The vertices v_2' and v_3', where solid dots represent vertices known to have degree three; (b) the Petersen graph.

Exercises

15.2.1 *Show that the bipartite graph H defined in the proof of statement 15.5.1 is a brace.*

15.2.2 *In the proof of statement 15.5.1 explain why there is an edge e incident with v that is b-invariant in G.*

15.2.3 *Show that the nine edges indicated by solid lines in Figure 15.4 are the b-invariant edges in G.*

15.2.4 *Prove Lemma 15.11.*

15.2.5 *Show that the two edges v_3v_1' and v_2v_4' are b-invariant in both G_1 and G_2 (Hint: Use Lemma 15.12).*

Lemma 15.20 (Inheritance of removable classes) *Let G be a brick, let e be a b-invariant edge of G and let H be a brick obtained by a tight cut decomposition of $G - e$. Every removable class of H is a removable class of $G - e$. Moreover, if G is regular then every removable doubleton of H is a removable doubleton of G.*

∗ **15.2.6** *Prove Lemma 15.20.*

An edge of a matching covered graph G is *solitary* if it lies in precisely one perfect matching of G.

▷**15.2.7** *Prove that a brick $G \notin \{K_4, \overline{C_6}, \mathbb{P}\}$ is extremal if and only if (i) every b-invariant edge of G is solitary and (ii) for any b-invariant edge of G, the graph $G - e$ is extremal.*

15.3 Notes

As we have already mentioned, the first proof of Lovász's conjecture appeared in Carvalho's Ph.D. thesis [6]. As we shall see in the next chapter, using Theorem 15.1, it is fairly easy to show that every near-brick with $p = 0$ has an optimal ear decomposition (as defined in Section 11.4). To show that every matching covered graph has an optimal ear decomposition we needed to show that every brick G different from K_4, $\overline{C_6}$, the bicorn and the Petersen graph has two b-invariant edges. The proof of this stronger result appeared in [10]. We have chosen not to include that proof here. In Chapters 17 and 18 we shall present ramifications of Theorem 15.1 which lead to generation procedures for bricks. We shall also describe there analogous generation procedures for braces.

Chapter 16
The Matching Lattice and Optimal Ear Decompositions

Contents

16.1 The Matching Lattice and its Bases

The objective of this chapter is to present a characterization of the matching lattice of a matching covered graph. Our approach to this characterization, envisaged by Lovász as explained in Section 13.1, makes essential use of Theorem 15.1. This approach also enables us to answer two related questions; one concerning bases of the matching lattices, and the other concerning optimal ear decompositions of matching covered graphs. In the last section we give a description of the 'dual' approach adopted by Lovász [58].

C. L. Lucchesi, U. S. R. Murty, *Perfect Matchings*, Algorithms and Computation in Mathematics 31, https://doi.org/10.1007/978-3-031-47504-7_16

16.1.1 Lattices, generating sets, and bases

For the convenience of the reader, we start by recalling some basic definitions and results from the theory of lattices in Euclidean spaces. These are not necessary for understanding the material of this chapter, but we are including them here to give the readers an appreciation of the special nature of the matching lattice of a matching covered graph.

A subset $\mathbf{L}$ of $\mathbb{R}^m$ is a *lattice* if there exists a finite set $\mathbf{S}$ of rational vectors in $\mathbb{R}^m$, called a *generating set* of $\mathbf{L}$, such that $\mathbf{L}$ consists of all integer linear combinations of vectors in $\mathbf{S}$. (The matching lattice $\mathcal{L}at(G)$ of a matching covered graph G is the lattice generated by $\{\chi^M : M \in \mathcal{M}\}$.)

The theory of lattices is a part of geometric theory of numbers, and there is an extensive literature which deals with it. In recent years, combinatorial optimizers have been interested in lattices because of their connection with integer programming. The book by Schrijver [83] has a good introduction to aspects of lattices that are relevant for the material in this section.

Clearly, the same lattice $\mathbf{L}$ in $\mathbb{R}^m$ may have more than one generating set. But it is not obvious that every lattice must have generating sets that are linearly independent. The following fundamental theorem (Corollary 4.1b in [83]) asserts this to be true:

Theorem 16.1 *Every lattice has linearly independent generating sets, called its* bases. □

In the above respect, lattices are analogous to subspaces. However, there is one essential difference. Any generating set of a subspace contains a subset that is a basis of that subspace, but a generating set of a lattice may not contain a basis of that lattice. For example, $\{2, 3\}$ generates the lattice $\mathbb{Z}$ in $\mathbb{R}^1$, but it does not contain a subset that is a basis of $\mathbb{Z}$.

16.1.2 Reduction to bricks and braces

Now we return to the specific question of characterizing matching lattices of matching covered graphs. It would be helpful to the reader to review Section 6.2 and 6.3 before reading this section.

Recall that a vector $\mathbf{x}$ in the edge space $\mathbb{R}^E$ of a matching covered graph G is r-regular if $\mathbf{x}(C) = r$ for any tight cut C of G. We showed in Chapter 6 that a vector $\mathbf{x}$ belongs to the matching space $\mathcal{L}in(G)$ if and only if it is regular.

As the matching lattice $\mathcal{L}at(G)$ of a matching covered graph G is the set of integer linear combinations of incidence vectors of perfect matchings of G, it clearly is a subset of the matching space $\mathcal{L}in(G)$ which consists of all (real) linear combinations of the same set of vectors. Thus,

$$\mathcal{L}at(G) \subseteq \mathcal{L}in(G) \cap \mathbb{Z}^E. \tag{16.1}$$

Furthermore, as every vector in $\mathcal{L}in(G)$ is regular, the above inequality implies that every vector in $\mathcal{L}at(G)$ is a regular integer vector. However, as has been noted before, the converse is not always true; there are graphs, such as the Petersen graph, whose matching lattices do not include all the regular integer vectors.

Given any regular integer vector $\mathbf{x}$ in $\mathbb{Z}^E$, how do we decide whether or not it belongs to $\mathcal{L}at(G)$? The following observation is the first step in trying to find an answer to this question; we leave its proof, which is a straightforward adaptation of the proof of Theorem 6.11, as Exercise 16.1.2.

Lemma 16.2 *Let C be a tight cut of a matching covered graph G, and let G_1 and G_2 be the two C-contractions of G. A vector $\mathbf{x} \in \mathbb{Z}^E$ belongs to $\mathcal{L}at(G)$ if and only if the restriction of $\mathbf{x}$ to $E(G_i)$ belongs to $\mathcal{L}at(G_i)$, for $i = 1, 2$.* □

Consequently, given any tight cut decomposition $\mathcal{G}$ of G, the problem of deciding whether or not a given vector $\mathbf{x}$ is in the matching lattice of G reduces to deciding whether or not the restrictions of $\mathbf{x}$ to the edge sets of the bricks and braces in $\mathcal{G}$ are in their respective matching lattices.

Our solution to the problem of characterizing $\mathcal{L}at(G)$ goes hand-in-hand with an answer to a related question concerning the existence of bases of matching lattices.

16.1.3 The question of existence of bases

Although $\mathcal{L}at(G)$ is a proper subset of $\mathcal{L}in(G)$, the dimensions of these two sets are clearly the same, which, as we saw in Section 6.3, is equal to $m - n + 2 - b$. Obviously, $\mathcal{L}in(G)$ has bases which are subsets of $\{\chi^M : M \in \mathcal{M}\}$. However, *a priori*, it is not clear that $\mathcal{L}at(G)$ has a basis consisting solely of incidence vectors of perfect matchings. As a counterpart to Conjecture 13.2, we were led to the premise that $\mathcal{L}at(G)$ does indeed have such bases. (As observed in the previous subsection, in general, a generating set of a lattice need not contain a basis of the lattice! See Exercise 16.1.1 for an instructive example.)

As shown in Section 6.3, a basis for the matching space of a matching covered graph G can be obtained from bases of the matching spaces of its bricks and braces by using the merger operation. Finding a basis of the matching space of a brace is easy enough. In fact, more generally, using the fact that every bipartite matching covered graph distinct from K_2 has a removable ear, one may easily deduce the following result (see Exercise 6.3.3):

Lemma 16.3 *In any bipartite matching covered graph G one can find a list $\mathcal{B}$ of $m - n + 2$ perfect matchings in stepwise canonical order such that any regular integer vector in $\mathbb{Z}^E$ may be expressed as an integer linear combination of vectors in $\{\chi^M : M \in \mathcal{B}\}$.* □

We shall now prove Theorem 13.1 which says that if G is any matching covered graph none of whose bricks is a Petersen brick, then the equality $\mathcal{L}at(G) = \mathcal{L}in(G) \cap \mathbb{Z}^E$ holds.

MATCHING LATTICE OF GRAPHS WITH $p = 0$

Theorem 16.4 *Let G be a matching covered graph none of whose bricks is a Petersen brick. Then, G has a basis for the matching lattice consisting of $m - n + 2 - b$ perfect matchings. Moreover, a vector* $\mathbf{x}$ *in the edge space* $\mathbb{Z}^E$ *of G belongs to* $\mathcal{L}at(G)$ *if and only if it is regular.*

Proof Clearly, every $\mathbf{x}$ in $\mathcal{L}at(G)$ is regular. Our task is thus to show that every regular integer vector $\mathbf{x}$ belongs to $\mathcal{L}at(G)$, and also to establish a basis for $\mathcal{L}at(G)$ with the asserted properties. We do this by induction on $|E|$.

Case 1 *Graph G has a nontrivial tight cut C.*

Let G_1 and G_2 denote the two C-contractions of G. As G is free of Petersen bricks so are both G_1 and G_2. Furthermore, for $i = 1, 2$, the trivial tight cut of G_i associated with the contraction vertex is the same as C, whereas all the other tight cuts of G_i are also tight cuts of G. It follows that the restriction of $\mathbf{x}$ to $E(G_i)$ is regular. By the induction hypothesis, it follows that the restrictions of $\mathbf{x}$ to $E(G_1)$ and $E(G_2)$ are in $\mathcal{L}at(G_1)$ and $\mathcal{L}at(G_2)$, respectively. Using Lemma 16.2 we now deduce that $\mathbf{x}$ is in $\mathcal{L}at(G)$.

Turning to the question of the existence of a required type of basis of $\mathcal{L}at(G)$, we first note that for $i = 1, 2$, by induction, there exists a basis $\mathcal{B}_i$ of $\mathcal{L}at(G_i)$ consisting of $m_i - n_i + 2 - b_i$ perfect matchings of G_i. It is now a straightforward matter to verify that $\mathcal{B}_1 \vee \mathcal{B}_2$ consists of $m - n + 2 - b$ perfect matchings of G and that it is a basis of $\mathcal{L}at(G)$. See Exercise 16.1.4.

Case 2 *Graph G is a brace.*

In this case, the assertion follows from Lemma 16.3.

Case 3 *Graph G is a brick which is not a Petersen brick.*

If G is either K_4 or $\overline{C_6}$, it is easy to see that every regular vector in $\mathbb{Z}^E$ is in $\mathcal{L}at(G)$ and that $\{\chi^M : M \in \mathcal{M}\}$ is a basis of $\mathcal{L}at(G)$. We may therefore assume that G is not one of the three special bricks K_4, $\overline{C_6}$, and $\mathbb{P}$. By Theorem 15.1 G has a b-invariant edge, and hence, by Theorem 13.6, has a $(b + p)$-invariant edge, say e. Thus, the graph $G' = G - e$ is a near-brick whose brick is not a Petersen brick.

Let M_e be any perfect matching of G containing the edge e, and let $\mathbf{x}'$ denote the restriction of $\mathbf{x} - \mathbf{x}(e)\chi^{M_e}$ to $E - e$.

16.4.1 $\mathbf{x}'$ *is a regular integer vector in the edge space of* G'.

Proof Certainly $\mathbf{x}'$ is an integer vector in the edge space of G'. We must thus prove that $\mathbf{x}'$ is regular. For this, let r be the common value of $\mathbf{x}(\partial(v))$ for all $v \in V$ and let $r' := r - \mathbf{x}(e)$. Clearly, for each $v \in V$, $\mathbf{x}'(\partial(v)) = r - \mathbf{x}(e) = r'$. Let C be any nontrivial tight cut of G'. The graph G' is a near-brick; hence one C-contraction of G', say, H, is bipartite. Let (A, B) denote the bipartition of H, let y denote the contraction vertex of H, and adjust notation so that $y \in A$. Then, $\mathbf{x}'(C) = |B|r' - (|A| - 1)r' = r'$. We conclude that $\mathbf{x}'(C) = r'$, for each tight cut of G'; hence $\mathbf{x}'$ is regular. □

By induction, $\mathbf{x}'$ is in $\mathcal{L}at(G')$ and $\mathcal{L}at(G')$ has a basis $\mathcal{B}'$ consisting of incidence vectors of perfect matchings of G'. It is now easy to see that $\mathbf{x}$ is in $\mathcal{L}at(G)$ and that $\mathcal{B}' \cup \{\chi^{M_e}\}$ is a basis for $\mathcal{L}at(G)$. □

Lovász's own proof of Theorem 13.1, given in [58], is via a characterization of the 'dual' of the matching lattice. He established a clear correspondence between the two points of view in that paper. It relies on some basic facts from the theory of lattices. We shall give a brief description of these ideas in Section 16.3.

To complete our characterization of the matching lattices of all matching covered graphs, we need to present a description of matching lattices of Petersen bricks. Before we do that, we record here an attractive consequence of Theorem 16.4.

Theorem 16.5 *The edge set of any cubic brick G different from $\mathbb{P}$ may be written as the symmetric difference of a collection of perfect matchings of G.*

Proof Consider the incidence vector $\mathbf{1}$ of the edge set E of G. This vector is clearly a 3-regular vector. By Theorem 16.4 it follows that $\mathbf{1}$ is in $\mathcal{L}at(G)$. Thus, there exist integers $\alpha_M : M \in \mathcal{M}$ such that

$$\mathbf{1} = \sum_{M \in \mathcal{M}} \alpha_M \chi^M.$$

Consider any such linear combination of $\mathbf{1}$, and let $\mathcal{M}_{\text{odd}}$ and $\mathcal{M}_{\text{even}}$ denote, respectively, the subsets of $\mathcal{M}$ such that α_M is odd for each $M \in \mathcal{M}_{\text{odd}}$ and α_M is even for each $M \in \mathcal{M}_{\text{even}}$. We thus have:

$$\mathbf{1} = \sum_{M \in \mathcal{M}_{\text{odd}}} \alpha_M \chi^M + \sum_{M \in \mathcal{M}_{\text{even}}} \alpha_M \chi^M,$$

implying that:

$$\mathbf{1} \equiv \sum_{M \in \mathcal{M}_{\text{odd}}} \alpha_M \chi^M \pmod 2.$$

The above congruence implies that E is the symmetric difference of the perfect matchings in the set $M \in \mathcal{M}_{\text{odd}}$. □

It is not difficult to see that the edge set of $\mathbb{P}$ cannot be written as the symmetric difference of a set of perfect matchings of $\mathbb{P}$ (Exercise 16.1.10).

16.1.4 The matching lattice of the Petersen graph

Not every regular integer vector $\mathbf{x}$ in the edge space of the Petersen graph $\mathbb{P}$ is in its matching lattice $\mathcal{L}at(\mathbb{P})$. However, a simple condition on $\mathbf{x}$, in addition to its regularity, ensures that it is in $\mathcal{L}at(\mathbb{P})$.

We start by noting that the Petersen graph has twelve pentagons. For any pentagon Q of $\mathbb{P}$, we shall refer to $N := \partial(V(Q))$ as a *central* cut of $\mathbb{P}$. There are six central cuts, and each of these cuts is also a perfect matching of $\mathbb{P}$. Indeed, a set of edges of $\mathbb{P}$ is a central cut if and only if it is a perfect matching. Moreover, (i) every edge is in precisely two of the six perfect matchings of $\mathbb{P}$, and (ii) the intersection of any two distinct perfect matchings consists of precisely one edge.

Suppose that $\mathbf{x}$ is an r-regular integer vector that belongs to $\mathcal{L}at(\mathbb{P})$. Then, by definition, corresponding to each perfect matching M there exists an integer α_M such that

$$\mathbf{x} = \sum \alpha_M \chi^M$$

By equation (6.8), $r = \sum \alpha_M$.

Given any fixed perfect matching N of $\mathbb{P}$, every edge of N is in precisely one perfect matching of $\mathbb{P}$ that is different from N. It follows that

$$\mathbf{x}(N) = \Big(\sum_{M \in \mathcal{M}-N} \alpha_M\Big) + 5\alpha_N = r + 4\alpha_N \equiv r \pmod 4.$$

Not all regular integer vectors in the edge space of $\mathbb{P}$ satisfy this condition. For example, the 3-regular vector of all 1's does not, because, on any central cut it takes the value 5 and $5 \not\equiv 3 \pmod 4$. But the 7-regular vector $\mathbf{x}$ displayed in Figure 16.1(a) does ($\mathbf{x}(N) = -1 \equiv 7 \pmod 4$). (One of the curious consequences of the proof of Lemma 16.6 is that if $\mathbf{x}(N) \equiv r \pmod 4$ for any one particular perfect matching N, then $\mathbf{x}(M) \equiv r \pmod 4$ for any perfect matching M of $\mathbb{P}$. See Exercise 16.1.5.)

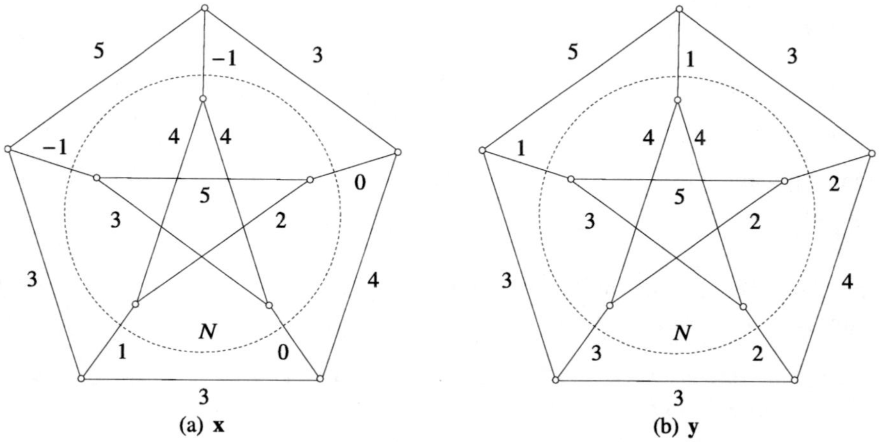

Fig. 16.1 (a) A 7-regular vector $\mathbf{x}$ in $\mathbb{Z}^E$; (b) the 9-regular vector $\mathbf{y} = \mathbf{x} + 2\chi^N$

Lemma 16.6 *An r-regular integer vector $\mathbf{x}$ in the edge space of $\mathbb{P}$, the Petersen graph, is in the matching lattice $\mathcal{L}at(\mathbb{P})$ if and only if $\mathbf{x}(N) \equiv r \pmod 4$, for each perfect matching N of $\mathbb{P}$. Furthermore, the set of the incidence vectors of the six perfect matchings of $\mathbb{P}$ is a basis for $\mathcal{L}at(\mathbb{P})$.*

Proof The only if part follows from the observations made above. We now proceed to prove the converse.

Let $\mathbf{x}$ be an r-regular vector on $\mathbb{Z}^E$, and let N be any fixed perfect matching (central cut) of $\mathbb{P}$. Assume that $\mathbf{x}(N) \equiv r \pmod 4$. This assumption implies that

$$d := (\mathbf{x}(N) - r)/4$$

is an integer.

Now consider the vector

$$\mathbf{y} := \mathbf{x} - d \cdot \chi^N.$$

Clearly $\mathbf{y}$ is an $(r - d)$-regular integer vector. Moreover,

$$\mathbf{y}(N) = \mathbf{x}(N) - 5d = (4d + r) - 5d = r - d.$$

Let $\mathbb{P}_1$ and $\mathbb{P}_2$ denote the two N-contractions of $\mathbb{P}$. The above property signifies that, not only is $\mathbf{y}$ an $(r-d)$-regular vector but that, for $i = 1, 2$, the restriction $\mathbf{y}_i$ of $\mathbf{y}$ to $E(\mathbb{P}_i)$ is also $(r-d)$-regular in $\mathbb{P}_i$. Both $\mathbb{P}_1$ and $\mathbb{P}_2$ are 5-wheels. By Theorem 16.4, $\mathbf{y}_i \in \mathcal{L}at(\mathbb{P}_i)$. Furthermore, for $i = 1, 2$, each perfect matching of $\mathbb{P}_i$ is the restriction of a perfect matching of $\mathbb{P}$ that has just one edge in common with N. It follows that $\mathbf{y}$ is an integer linear combination of the incidence vectors of the five perfect matchings of $\mathbb{P}$ that have just one edge in common with N. Since $\mathbf{x} = \mathbf{y} + d\chi^N$, it follows that $\mathbf{x}$ is in $\mathcal{L}at(\mathbb{P})$, and that $\{\chi^M : M \in \mathcal{M}\}$ is a basis of $\mathcal{L}at(\mathbb{P})$. □

In the example shown in Figure 16.1(a), the vector $\mathbf{x}$ is a 7-regular vector and $\mathbf{x}(N) = -1$. Thus, in this case $d = (-1 - 7)/4 = -2$. The vector $\mathbf{y} = \mathbf{x} + 2\chi^N$, which is displayed in Figure 16.1(b), is a 9-regular vector, with the additional property that $\mathbf{y}(N)$ is also equal to 9.

A Petersen brick G (a graph whose underlying simple graph is the Petersen graph $\mathbb{P}$) also has twelve central cuts that correspond to central cuts of $\mathbb{P}$. If G has multiple edges, then not all central cuts are perfect matchings of G. However, each central cut Q is a union of perfect matchings of G that meet it in five edges, and all other perfect matchings of G have precisely one edge in common with the cut Q.

If G is simple then it is the Petersen graph. If it is not simple, it has multiple edges and each multiple edge is clearly b-invariant. The following theorem may thus be proved by induction on the number of edges using the statement of Lemma 16.6 as the basis for induction (Exercise 16.1.6).

THE MATCHING LATTICE OF A PETERSEN BRICK

Theorem 16.7 *Let G be a Petersen brick. Then an r-regular integer vector $\mathbf{x}$ in the edge space of G is in $\mathcal{L}at(G)$ if and only if $\mathbf{x}(Q) \equiv r \pmod 4$, for each central cut Q of G. Furthermore, $\mathcal{L}at(G)$ has a basis consisting of incidence vectors of $m - n + 1$ perfect matchings of G.* □

As a corollary of Theorems 16.4 and 16.7 we have the following result which answers the question concerning the bases of matching lattices:

BASES OF $\mathcal{L}\mathrm{AT}(G)$

Corollary 16.8 *Every matching covered graph G has a basis for the matching lattice consisting of precisely* $m - n + 2 - b$ *perfect matchings.* □

It should be noted that Lex Schrijver has a proof of Theorem 16.4 in Chapter 38 of his treatise [84]. This proof is different from the one due to Lovász and relies on the study of properties of special barriers in a matching covered graph.

Exercises

16.1.1 *Consider the pentagonal prism* $G := \mathbb{P}_{10}$ *with its edges labelled as in Figure 16.2.*

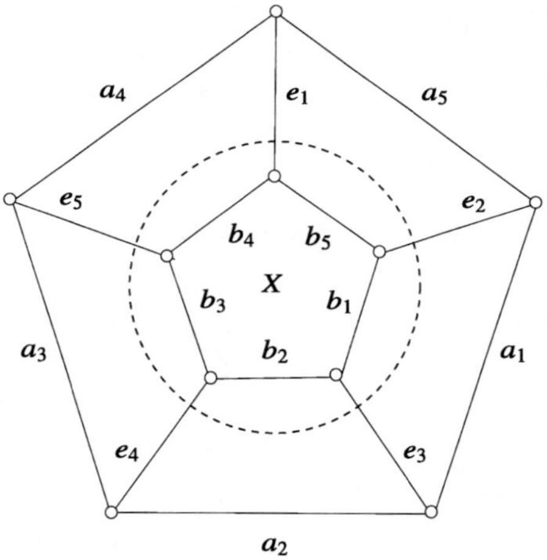

Fig. 16.2 The pentagonal prism $G := \mathbb{P}_{10}$.

Let $\mathcal{B}$ *denote the set of the incidence vectors of the five perfect matchings listed below:*

$$M_1 := \{e_1, a_1, a_3, b_1, b_3\}$$
$$M_2 := \{e_2, a_2, a_4, b_2, b_4\}$$
$$M_3 := \{e_3, a_3, a_5, b_3, b_5\}$$
$$M_4 := \{e_4, a_1, a_4, b_1, b_4\}$$
$$M_5 := \{e_5, a_2, a_5, b_2, b_5\}$$

(i) *Show that $\mathcal{B}$ is a basis of the sub-lattice of $\mathcal{Lat}(G)$ generated by the incidence vectors of perfect matchings that meet the robust cut $\partial(X)$ in precisely one edge.*
(ii) *Let M_6 denote the perfect matching $\{e_1, e_2, e_3, e_4, e_5\}$. Show that $\mathcal{B} \cup \{\chi^{M_6}\}$ is a basis of $\mathcal{Lin}(G)$.*
(iii) *Show that the vector* **1** *cannot be expressed as an integer linear combination of the the six vectors in $\mathcal{B} \cup \{\chi^{M_6}\}$.*
(iv) *Finally, conclude that $\mathcal{B} \cup \{\chi^{M_6}\}$ is a basis of $\mathcal{Lin}(G)$ but that it is not a basis of $\mathcal{Lat}(G)$.*
(v) *Find a basis of $\mathcal{Lat}(G)$ that includes $\mathcal{B}$.*

16.1.2 *Give a proof of Lemma 16.2.*

▷**16.1.3** *Figure 16.3 depicts a brick G, and an integral vector* $\mathbf{x}$*, where the unlabeled edges have weight one. Show that* $\mathbf{x}$ *lies in $\mathcal{Lat}(G)$, following the proof of Theorem 16.4.*

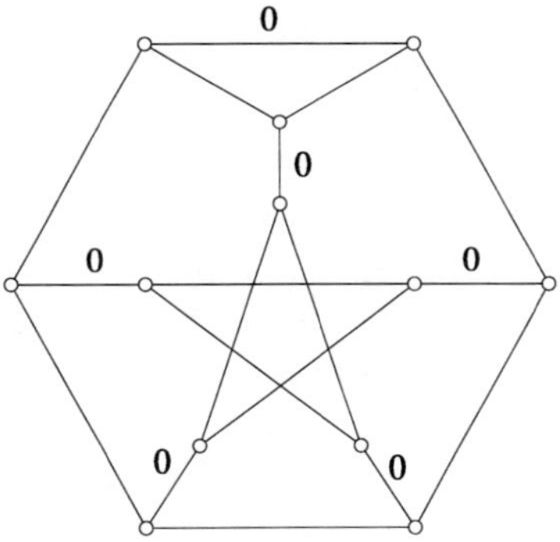

Fig. 16.3 The brick for Exercise 16.1.3

▷**16.1.4** *Adapt the proof of Theorem 6.11 to show that if C is a tight cut of G, and $\mathcal{B}_1$ and $\mathcal{B}_2$ are bases of $\mathcal{Lat}(G_1)$ and $\mathcal{Lat}(G_2)$, respectively, then $\mathcal{B}$ is a basis of $\mathcal{Lat}(G)$.*

▷**16.1.5** *Let* $\mathbf{x}$ *be an integral vector in the matching space of $\mathbb{P}$. Prove that if $\mathbf{x}(N) \equiv r \pmod 4$ for any one particular perfect matching N of $\mathbb{P}$, then $\mathbf{x}(M) \equiv r \pmod 4$ for any perfect matching M of $\mathbb{P}$. Hint: this statement is a consequence of the proof of Lemma 16.6.*

16.1.6 *Give a proof of Theorem 16.7.*

▷**16.1.7** *Prove Corollary 16.8.*

▷**16.1.8** *Give a polynomial-time algorithm which, given a matching covered graph G and a vector* $\mathbf{x}$ *in $\mathbb{Z}^E$, determines whether* $\mathbf{x}$ *lies in $\mathcal{Lat}(G)$.*

16.1.9 *Consider the cubic brick* $G = \mathbb{P} \odot K_4$*, depicted in Figure 16.3.*

(i) *Find an expression of* $\chi^E = \mathbf{1}$ *as an integer linear combination of the incidence vectors of perfect matchings of* G*.*
(ii) *Using the above linear combination, find a set of perfect matchings of* G *whose symmetric difference is the edge set* E*.*

16.1.10 *Show that* $E(\mathbb{P})$ *cannot be expressed as the symmetric difference of a collection of perfect matchings of* $\mathbb{P}$*.*

16.1.11 *Let* G *be a cubic matching covered graph. Show that the vector* $\mathbf{2} := (2, 2, \ldots, 2)$ *of all 2's in the edge space of* G *belongs to* $\mathcal{L}at(G)$*.*

16.2 Optimal Ear Decompositions

According to Theorem 11.18, the number of double ears in any ear decomposition of a matching covered graph G is at least $b(G)+p(G)$. This bound is always attainable, as shown by CLM (2002, [11]). In other words, any matching covered graph G has an ear decomposition with precisely $b + p$ double ears. We refer to such an ear decomposition as an *optimal ear decomposition* of G.

In seeking to find optimal ear decompositions of a matching covered graph G, it is convenient to think in terms of top-down decompositions in which we start with G and sequentially delete removable ears to obtain K_2. If R is any removable double ear of G, then $(b + p)(G - R) = (b + p)(G) - 1$. Thus, deleting R reduces the task of finding an optimal ear decomposition of G to one of finding such a decomposition of $G - R$. On the other hand, if R is an arbitrary removable single ear, then $(b+p)(G-R)$ may be even larger than $(b+p)(G)$ itself, and deleting R would be a setback to the task of finding an optimal ear decomposition of G. However, if R were a $(b + p)$-invariant single ear, then $(b + p)(G - R) = (b + p)(G)$. In this case, deletion of R reduces the task of finding an optimal ear decomposition of G to finding one of the graph $G - R$ with fewer edges than G. This leads us to the notion of an optimal ear as defined in the following inset:

OPTIMAL REMOVABLE EARS AND THEIR EXISTENCE

A removable ear R of a matching covered graph G is *optimal* if it is either a double ear or is a $(b + p)$-invariant single ear. (An *optimal removable class* of G is either a removable doubleton or is a $(b + p)$-invariant edge.). Our proof of the following theorem, which appeared in CLM [11], is along the lines of the hint given for establishing the weaker statement in Exercise 11.3.6.

Theorem 16.9 *Every matching covered graph* G *distinct from* K_2 *has an optimal removable ear.* □

From Theorem 16.9 it is easy to prove the following result, by induction on the number of edges of G (Exercise 16.2.1).

Corollary 16.10 *Every matching covered graph G has an optimal ear decomposition, which uses precisely $(b + p)(G)$ double ears.* □

As we shall see below, the validity of Theorem 16.9 for near-bricks is a consequence of Corollary 15.19.

16.2.1 The case of bipartite graphs and near-bricks

Let us prove Theorem 16.9 for matching covered graphs G that are either bipartite or near-bricks. That is, $b(G) \leq 1$. First note that the retract $\widehat{G}$ of G is also a matching covered graph with $b(\widehat{G}) = b(G) \leq 1$ and $(b + p)(\widehat{G}) = (b + p)(G)$. Observe that, by Theorem 11.21, there is a one-to-correspondence between removable ears of $\widehat{G}$ and those of G. Also note that if $\widehat{R}$ is an optimal removable class of $\widehat{G}$, then the corresponding ear R of G is an optimal removable ear of G (Exercise 16.2.2). In view of this, it is sufficient to prove the existence of optimal removable classes in bipartite graphs and near-bricks which are free of vertices of degree two. Furthermore, if G has multiple edges, then it clearly has a $(b+p)$-invariant edge. Thus, we may suppose that G is simple and that $\delta(G) \geq 3$. (Figure 16.4 shows a near-brick and its retract.)

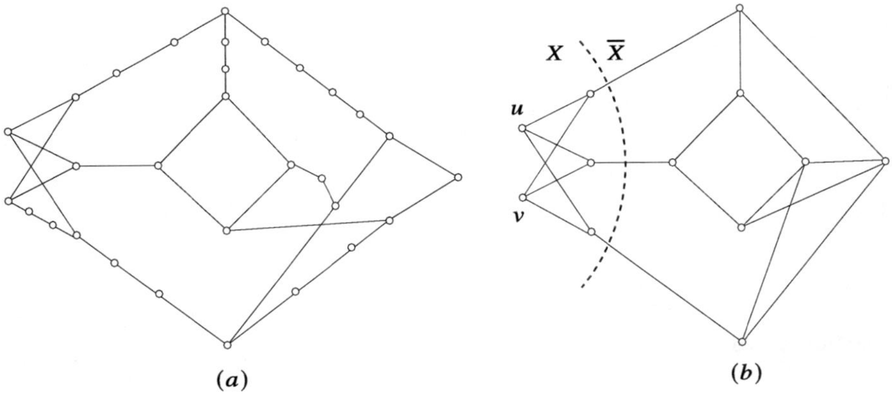

Fig. 16.4 (a) A near-brick; (b) its retract

If G has no nontrivial tight cuts, then G is a brace or a brick. If G is a brace, then as G is simple and $\delta(G) \geq 3$, it follows that G has order six or more and every edge of G is removable, hence $(b + p)$-invariant. If G is a brick then the validity of Theorem 16.9 follows from Corollary 15.19. So, suppose that G has nontrivial tight cuts. Then, since $b(G) \leq 1$, it follows that at least one of the shores of any tight cut is bipartite. Let $\partial(X)$ be a nontrivial tight cut of G such that: (i) $G_1 := G/(\overline{X} \to \overline{x})$ is bipartite and, (ii) subject to (i), the shore X is minimal. The minimality of X implies that G_1 is a brace, and the fact that G is a simple graph which is free of vertices of degree two implies that G_1 has order at least six. The majority part X_+ of X is a special barrier of G and $G[\overline{X}]$ is the only nontrivial component of $G - X_+$. Any edge

e incident with a vertex in X_- is a removable edge in the brace G_1 and is not an edge in the cut $\partial(X)$. Consequently, e is a removable edge of G. Also, clearly, $\partial(X)$ is a tight cut of $G - e$, implying that $(b + p)(G - e) = (b + p)(G)$ and hence that e is in fact a $(b + p)$-invariant edge of G. (In the near-brick shown in Figure 16.4(b), any edge incident with either u or with v is a $(b + p)$-invariant edge of that near-brick.)

The validity of Theorem 16.9 for bipartite graphs and near-bricks implies the existence of optimal ear decompositions for matching covered graphs that are bipartite or near-bricks. In fact, if R is an optimal removable ear of G and $b(G) \leq 1$ then $b(G - R) \leq 1$. Hence, an inductive argument based on the number of edges of G establishes the validity of Corollary 16.10 for bipartite matching covered graphs and for near-bricks.

16.2.2 The difficulty in dealing with graphs which have more than one brick

Suppose that G is a matching covered graph with two or more bricks. (See Figure 16.5 for an illustration.) Then, by the result established above, every brick of G has an optimal ear decomposition. It would be natural to somehow combine the optimal ear decompositions of the bricks of G to obtain an optimal ear decomposition of G.

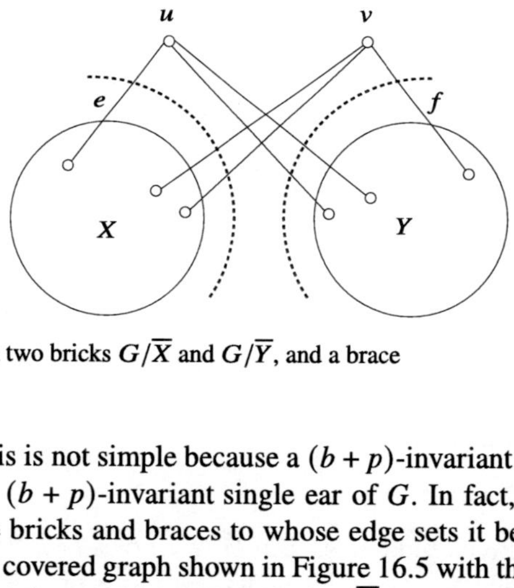

Fig. 16.5 A graph with two bricks $G/\overline{X}$ and $G/\overline{Y}$, and a brace

Unfortunately, this is not simple because a $(b + p)$-invariant single ear of a brick of G need not be a $(b + p)$-invariant single ear of G. In fact, it may not even be removable in all the bricks and braces to whose edge sets it belongs. Consider for example a matching covered graph shown in Figure 16.5 with the indicated tight cut decomposition. The edge e which belongs to $G/\overline{X}$ as also to the unique brace of order four is not removable in G as v is a cut vertex of $G - e$. Similarly, the edge f which is shared by $G/\overline{Y}$ and the brace is not removable in G. Thus, if e were the only $(b + p)$-invariant edge of $G/\overline{X}$, and f were the only $(b + p)$-invariant edge of $G/\overline{Y}$, we would be left with no obvious way of finding a $(b+p)$-invariant edge in G!

16.2.3 A strengthening of Theorem 15.1

To circumvent the difficulty pointed out above, it was necessary to prove a result which is stronger than Theorem 15.1. The following result was established by CLM (2002, [10]):

Theorem 16.11 *Every brick different from K_4, $\overline{C_6}$, $\mathbb{P}$, and the bicorn (Figure 15.3) has at least two $(b+p)$-invariant edges.* □

The proof of this theorem is rather involved and we have chosen to not include that proof here with the hope that keen students might be led to pursue the subject further.

The graphs K_4 and $\overline{C_6}$ have three removable doubletons. In the Petersen graph, each edge is removable and $(b+p)$-invariant. The bicorn has two removable doubletons, in addition to a $(b+p)$-invariant edge. These observations, together with Theorem 16.11, can be used to show the following strengthening of Theorem 16.9 (Exercise 16.2.7).

Theorem 16.12 *Every matching covered graph distinct from C_{2n} $(n \geq 2)$ and K_2 has two optimal removable ears.* □

16.2.4 Exchange property of optimal removable classes

Optimal removable ears in matching covered graphs satisfy an exchange property which will play a crucial role in the proofs of some important results in future chapters. The following lemma is a statement of that property.

Lemma 16.13 *Let G be a matching covered graph, let R be an optimal removable class of G and let S be an optimal removable class of $G - R$. If G has a perfect matching that contains the edges of R but is disjoint with S then $G - S$ is matching covered, S is an optimal removable class of G and R is an optimal removable class of $G - S$.*

Proof There are four cases to be considered depending whether R is a singleton or a doubleton, and whether S is a singleton or doubleton. Note that, in all these four cases, (i) $(G - R) - S = (G - S) - R$ is matching covered because, by hypothesis, S is removable in $G - R$, and (ii) there is a perfect matching of $G - S$ containing all edges of R. Therefore, in all cases, the graph $G - S$ is a matching covered graph. Thus, to show that S is an optimal removable class of G, it suffices to show that $(b+p)(G-S) = (b+p)(G)$ if S is a singleton, and $(b+p)(G-S) = (b+p)(G)-1$ if S is a doubleton. Similarly, to show that R is an optimal removable removable class of $G - S$, it suffices to that $(b+p)(G - S - R) = (b+p)(G - S)$ if R is a singleton, and $(b+p)(G - S - R) = (b+p)(G - S) - 1$ if R is a doubleton. Both these objectives can be achieved by using the monotonicity of the function $(b+p)$ established in Section 9.1.1.

We illustrate this idea by considering the special case in which $R = \{e\}$ and $S = \{f\}$ are singletons. Since $\{e\}$ is an optimal class in G, we have that $(b+p)(G-e) = (b+p)(G)$, and since $\{f\}$ is an optimal removable class in $G-e$, we have $(b+p)((G-e)-f) = (b+p)(G-e)$. As already noted, the fact that there is a perfect matching containing e but not f implies that $G-f$ is matching covered. Thus, to show that $\{f\}$ is an optimal removable class in G, it is necessary to show that $(b+p)(G) = (b+p)(G-f)$. Similarly, to show that $\{e\}$ is an optimal removable class of $G-f$, it is necessary to show that $(b+p)(G-f-e) = (b+p)(G-f)$.

By the monotonicity of the function $b+p$, we have:

$$(b+p)(G) \le (b+p)(G-f) \le (b+p)(G-e-f) = (b+p)(G-e) = (b+p)(G).$$

The two desired equalities follow.

We leave the proofs of the remaining three cases as Exercise 16.2.6. □

Exercises

16.2.1 *Prove that Theorem 16.9 implies Corollary 16.10.*

16.2.2 *Let $G \neq K_2$ be a matching covered graph. Show that any optimal removable class $\widehat{R}$ of the retract $\widehat{G}$ of G corresponds to an optimal removable ear R of G.*

16.2.3 *Find an ear decomposition with two double ears of the near-brick shown in Figure 16.6.*

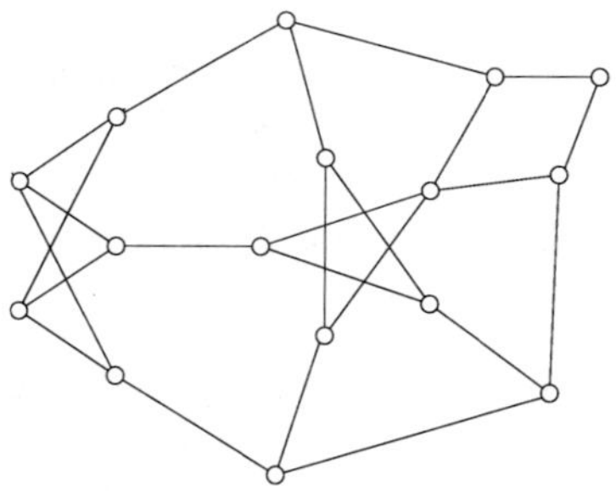

Fig. 16.6 A near-brick with $b + p = 2$

16.2.4 *Find an ear decomposition of $\mathbb{P} \oplus K_4$ with just one double ear.*

16.2.5 *Show that any near-brick G with $p = 1$ has an ear decomposition with two double ears.*

∗ **16.2.6** *Provide proofs of the remaining three cases of Lemma 16.13.*

▷**16.2.7** *Prove that Theorem 16.11 implies Theorem 16.12.*

∗ **16.2.8** *Give a polynomial-time algorithm which, given a matching covered graph G, produces an optimal ear decomposition of G and the corresponding sequence of optimal removable ears.*

16.3 Matching Integral Vectors ♯

The objective of this section is to provide a brief sketch of Lovász's original approach (in [58]) to the problem of characterizing matching lattices which did not use the idea of b-invariant edges. For this reason, in the ensuing discussion, we shall not make use of Corollary 16.8, whose proof depends on the existence of b-invariant edges in nonspecial bricks. To do so would be like *putting a horse before the cart.* But Theorem 16.1 which asserts that every lattice has a basis is a crucial ingredient of the proofs of the main results in this section.

BASIS MATRIX OF $\mathcal{L}at(G)$

Let G be any matching covered graph. By Theorem 6.13 the dimension of $\mathcal{L}in(G)$ is $d := m - n + 2 - b$. Thus, as a consequence of Theorem 16.1, there exists a $d \times m$ matrix $\mathbf{M}$ whose rows constitute a basis of $\mathcal{L}at(G)$. We shall refer to such a matrix as a *basis matrix* of $\mathcal{L}at(G)$.

For future reference we record below an observation that follows from the definition of a basis of a lattice.

Proposition 16.14 *Let* $\mathbf{M}$ *be a* $d \times m$ *basis matrix of* $\mathcal{L}at(G)$. *A vector* $\mathbf{x}$ *in* $\mathbb{Z}^E$ *is in* $\mathcal{L}at(G)$ *if and only if there is an integer vector* $\mathbf{a} = (a_1, a_2, \ldots, a_d)$ *such that* $\mathbf{aM} = \mathbf{x}$. □

MATCHING INTEGRAL VECTORS

A vector $\mathbf{y}$ in the edge space of a matching covered graph G is *matching integral* if $\mathbf{x}\mathbf{y}^t$ is an integer for each vector $\mathbf{x}$ in $\mathcal{L}at(G)$. Equivalently, $\mathbf{y}$ is matching integral if $\mathbf{y}(M) \in \mathbb{Z}$, for each $M \in \mathcal{M}$.

We shall denote the set of all matching integral vectors of G by $\mathcal{L}at^*(G)$. (In the theory of 'lattices', $\mathcal{L}at^*(G)$ is referred to as the *dual* of $\mathcal{L}at(G)$. See Schrijver [84].)

Recall that a vector $\mathbf{y_0}$ in the edge space of a matching covered graph G is matching orthogonal (Exercise 6.2.6) if $\mathbf{y}_0(M) = 0$, for each perfect matching M of G. Clearly each matching orthogonal vector is matching integral.

As specific examples of matching integral vectors, consider the three vectors $\mathbf{y}_0$, $\mathbf{z}$, and $\mathbf{y}_0 + \mathbf{z}$ in the edge space of $\overline{C_6}$ shown in Figure 16.7. (The coefficients corresponding to edges with no attached labels should be taken as zeroes). The first vector $\mathbf{y}_0$ is a matching orthogonal vector. The second vector $\mathbf{z}$, being an integer vector, is clearly matching integral. The third vector, which is the sum of the first two, is thus also a matching integral vector.

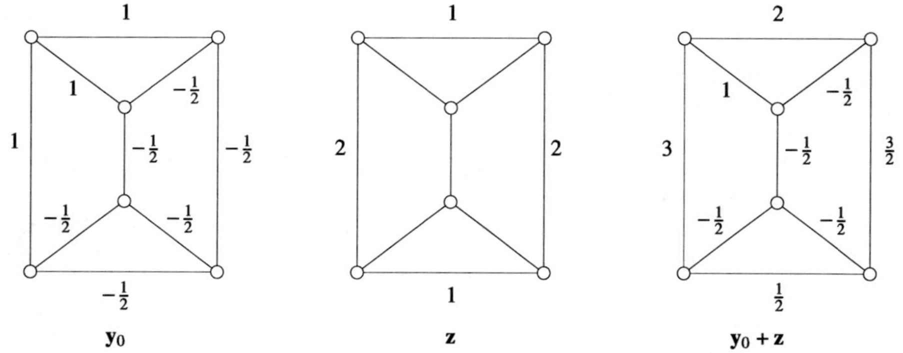

Fig. 16.7 Examples of matching integral vectors

The following proposition is just a restatement of what it means to say that a vector $\mathbf{y}$ in $\mathbb{R}^E$ is a matching integral vector.

Proposition 16.15 *Let* $\mathbf{M}$ *be a basis matrix of* $\mathcal{L}at(G)$. *A vector* $\mathbf{y}$ *in* $\mathbb{R}^E$ *is in* $\mathcal{L}at^\star(G)$ *if and only if* $\mathbf{M}\mathbf{y}^t$ *is an integer vector.* □

For any matching covered graph G, the set of *matching orthogonal vectors* is, by definition, equal to $\mathcal{L}in^\perp := (\mathcal{L}in(G))^\perp$, the orthogonal complement of the matching space $\mathcal{L}in := \mathcal{L}in(G)$. As noted above, the sum of a matching orthogonal vector and an integer vector is matching integral. Thus:

$$\mathcal{L}in^\perp + \mathbb{Z}^E \subseteq \mathcal{L}at^\star. \tag{16.2}$$

There are graphs for which the above inequality holds strictly. For example, the 5-cycle vector in the edge space of the Petersen graph $\mathbb{P}$ which is shown in Figure 16.8 is a matching integral vector, but it cannot be written as the sum of a matching orthogonal vector of $\mathbb{P}$ and an integer vector (Exercise 16.3.3).

However, Lovász [58] showed that the inequality (16.2) holds as an equation if and only if $\mathcal{L}at = \mathcal{L}in \cap \mathbb{Z}^E$. His proof of this assertion relies on three results from linear algebra. The first is an elementary fact concerning solution sets to systems of linear equations.

Lemma 16.16 *Let* S *be the set of all solutions to a system* $\mathbf{M}\mathbf{y}^\mathbf{t} = \mathbf{b}$ *of linear equations, let* S_0 *be the set of all solutions to the corresponding homogeneous system* $\mathbf{M}\mathbf{y}^\mathbf{t} = \mathbf{0}$, *and suppose that* $\mathbf{s}$ *is any vector in* S. *Then* $S = S_0 + \mathbf{s}$. □

The next two statements follow from a classical result concerning the existence of integer solutions to systems of linear equations with integer coefficients (Corollary 4.1a in [83]). We state them here in notation that is convenient for us:

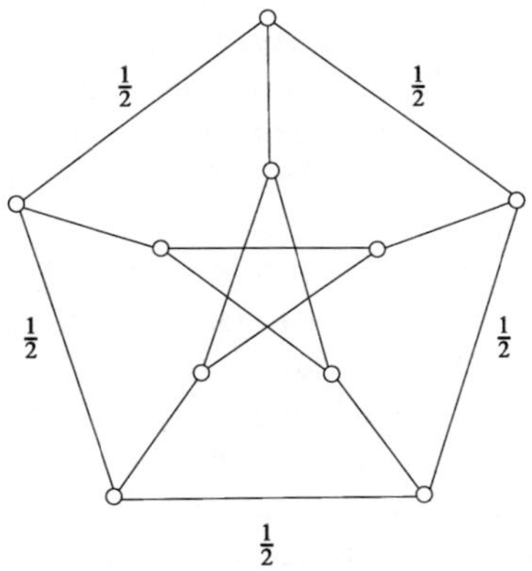

Fig. 16.8 A 5-cycle vector $\mathbf{w}$ in the edge space of $\mathbb{P}$.

Lemma 16.17 *Let* $\mathbf{M}$ *be a* $d \times m$ *integer matrix and let* $\mathbf{x}$ *be a* $1 \times m$ *integer vector. Then there exists a* $1 \times d$ *integer vector* $\mathbf{a}$ *satisfying the system* $\mathbf{aM} = \mathbf{x}$ *of linear equations if and only if there is* no $1 \times m$ *vector* $\mathbf{y}$ *such that* $\mathbf{My}^t$ *is an integer vector, but* $\mathbf{xy}^t$ *is not an integer.* □

Thus, to show that there is no integer vector $\mathbf{a}$ satisfying the equation $\mathbf{aM} = \mathbf{x}$, it suffices to display a vector $\mathbf{y}$ such $\mathbf{My}^t$ is an integer vector, but $\mathbf{xy}^t$ is not an integer. As an example, consider the following system of linear equations:

$$\begin{pmatrix} a_1 & a_2 \end{pmatrix} \begin{pmatrix} 1 & 1 \\ -1 & 1 \end{pmatrix} = \begin{pmatrix} 2 & 1 \end{pmatrix}.$$

This system has no integral solution because:

$$\begin{pmatrix} 1 & 1 \\ -1 & 1 \end{pmatrix} \begin{pmatrix} \frac{1}{2} \\ \frac{1}{2} \end{pmatrix} = \begin{pmatrix} 1 \\ 0 \end{pmatrix},$$

is an integer vector, whereas $(2, 1)(\frac{1}{2}, \frac{1}{2})^t = \frac{3}{2}$ is not an integer.

The next lemma is a variant of the one above.

Lemma 16.18 *Let* $\mathbf{M}$ *be a* $d \times m$ *integer matrix and let* $\mathbf{b}$ *be a* $d \times 1$ *integer vector. Then there exists a* $1 \times m$ *integer vector* $\mathbf{y}$ *satisfying the system* $\mathbf{My}^t = \mathbf{b}$ *of linear equations if and only if there is no* $1 \times d$ *vector* $\mathbf{a}$ *such that* $\mathbf{aM}$ *is an integer vector, but* $\mathbf{ab}$ *is not an integer.* □

We are now in a position to state and prove the main assertion of this section. In our proof $\mathbf{M}$ represents a basis matrix of $\mathcal{L}at(G)$.

Theorem 16.19 *For any matching covered graph* G*:*

$$\mathcal{L}at = \mathcal{L}in \cap \mathbb{Z}^E \quad \text{if and only if} \quad \mathcal{L}at^* = \mathcal{L}in^\perp + \mathbb{Z}^E.$$

Proof We have already noted that the inequality $\mathcal{L}in^{\perp} + \mathbb{Z}^E \subseteq \mathcal{L}at^{\star}$ holds regardless of whether or not $\mathcal{L}at = \mathcal{L}in \cap \mathbb{Z}^E$. Let us first suppose that there is an integer vector $\mathbf{x}$ in $\mathcal{L}in$ that is not in $\mathcal{L}at$. By Proposition 16.14, this is equivalent to saying that there is no integer vector $\mathbf{a}$ satisfying the system $\mathbf{aM} = \mathbf{x}$ of linear equations. By Lemma 16.17, it follows that there is a vector $\mathbf{y}$ such that $\mathbf{My}^t$ is an integer vector and $\mathbf{xy}^t$ is not an integer. The fact that $\mathbf{My}^t$ is an integer vector means that $\mathbf{y}$ is a matching integral vector. Since $\mathbf{x}$ is an integer vector in $\mathcal{L}in$, the fact that $\mathbf{xy}^t$ is not an integer clearly means that $\mathbf{y}$ is not in $\mathcal{L}in^{\perp} + \mathbb{Z}^E$ (Exercise 16.3.2). Thus, if $\mathcal{L}at \subset \mathcal{L}in \cap \mathbb{Z}^E$, then the strict inequality $\mathcal{L}in^{\perp} + \mathbb{Z}^E \subset \mathcal{L}at^{\star}$ holds. (Thus, a matching integral vector $\mathbf{y}$ such that $\mathbf{xy}^t$ is not an integer serves as a certificate for showing that $\mathbf{x}$ is not in $\mathcal{L}at(G)$. For example, the 5-cycle vector $\mathbf{w}$ shown in Figure 16.8 serves as a certificate to show that $\mathbf{1}$ is not in the matching lattice of $\mathbb{P}$.)

Turning to the converse, suppose that $\mathcal{L}at = \mathcal{L}in \cap \mathbb{Z}^E$. What we need to do is to show that this supposition implies that the equality $\mathcal{L}in^{\perp} + \mathbb{Z}^E = \mathcal{L}at^{\star}$ holds. Towards this end, let $\mathbf{z}$ be any matching integral vector. Then, by definition, $\mathbf{Mz}^t$ is an integer vector. Let $\mathbf{b} := \mathbf{Mz}^t$ and consider the equation:

$$\mathbf{My}^t = \mathbf{b}.$$

Using Lemma 16.18, we shall first show that the above equation has an integer solution. Thus, suppose that $\mathbf{a}$ is any vector such that $\mathbf{aM}$ is an integer vector. As $\mathbf{aM}$ is a linear combination of the rows of the basis matrix $\mathbf{M}$ of $\mathcal{L}in$, it belongs to $\mathcal{L}in$. Thus, being an integer vector it belongs to $\mathcal{L}in \cap \mathbb{Z}^E$ which, by our assumption, is equal to $\mathcal{L}at(G)$. Therefore, $\mathbf{aM} = \mathbf{cM}$, for some integer vector $\mathbf{c}$. Since the rows of $\mathbf{M}$ are linearly independent, it follows that $\mathbf{a} = \mathbf{c}$ is an integer vector. Consequently, $\mathbf{ab}$ is an integer. Therefore $\mathbf{My}^t = \mathbf{b}$ has an integer solution.

Suppose that $\mathbf{y} = \mathbf{s}$ is an integral solution to $\mathbf{My}^t = \mathbf{b}$. Noting that $\mathcal{L}in^{\perp}$ is the solution set of the homogeneous system $\mathbf{My}^t = \mathbf{0}$, we conclude, by Lemma 16.16, that the set of all solutions to the equation $\mathbf{My}^t = \mathbf{b}$ is a subset of $\mathcal{L}in^{\perp} + \mathbf{s}$. This means that if $\mathcal{L}at = \mathcal{L}in \cap \mathbb{Z}^E$ then any matching integral vector may be expressed as the sum of a matching orthogonal vector and an integer vector! □

The main theorem established by Lovász in [58] was the following:

Theorem 16.20 *If G is any brick which is not a Petersen brick then $\mathcal{L}at^{\star}(G) = (\mathcal{L}in(G))^{\perp} + \mathbb{Z}^E$. (See Exercise 16.3.1 for an illustration.)*

Using the above result and Theorem 16.19, Lovász deduced that $\mathcal{L}at(G) = \mathcal{L}in(G) \cap \mathbb{Z}^E$ for any such brick. He also showed how this theorem could be used to determine the dimension of the matching space of a matching covered graph over any finite field and, in particular, to derive an attractive statement concerning matching spaces over $\mathbb{Z}_2$. These results are described in the next section.

Exercises

16.3.1 *Consider the vector* $\mathbf{w}$ *in the edge space of the pentagonal prism shown in Figure 16.9. Show that* $\mathbf{w}$ *is a matching integral vector and express it as the sum of a matching orthogonal vector and an integer vector.*

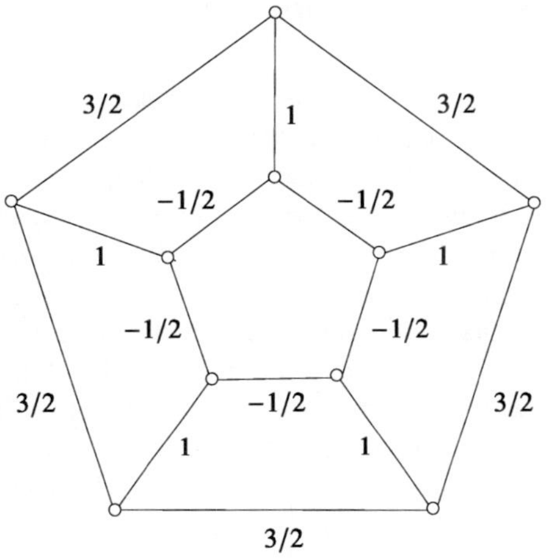

Fig. 16.9 A vector in the edge space of the pentagonal prism.

16.3.2 *Let* $\mathbf{x} \in \mathcal{L}in \cap \mathbb{Z}^E$ *and let* $\mathbf{y}$ *be a matching integral vector. Prove that if* $\mathbf{x}\mathbf{y}^t$ *is not an integer then* $\mathbf{y} \notin \mathcal{L}in^{\perp} + \mathbb{Z}^E$.

16.3.3 *Consider the vector in the edge space of* $\mathbb{P}$ *which is indicated by the edge labels in Figure 16.8 (only nonzero labels are shown). Show that this vector cannot be written as the sum of a matching orthogonal vector and an integer vector.*

16.4 Matching Spaces over $\mathbb{Z}_2$ ♯

Recall (from Section 6.4) that, given any field $\mathbb{F}$, the matching space of a matching covered graph G, denoted by $\mathcal{L}in(G, \mathbb{F})$, is the linear space over $\mathbb{F}$ generated by the set of incidence vectors of perfect matchings of G. There is a striking difference between matching spaces over fields of characteristic 2 and those over fields of odd characteristic. For example, if $\mathbb{F}$ is any field of characteristic 2, then $\dim(\mathcal{L}in(\mathbb{P}, \mathbb{F})) = 5$; on the other hand, if $\mathbb{F}$ is any field of odd characteristic, then $\dim(\mathcal{L}in(\mathbb{P}, \mathbb{F})) = 6$. In fact, by using the existence of optimal removable classes in bricks (Corollary 15.19), it may be shown that, for any matching covered graph G, $\dim(\mathcal{L}in(G, \mathbb{F})) = m - n + 2 - b$, if $\mathbb{F}$ has odd characteristic, whereas $\dim(\mathcal{L}in(G, \mathbb{Z}_2)) = m - n + 2 - (b + p)$. See Exercises 16.4.1 and 16.4.2.

In this section, for the sake of simplicity and clarity, we shall restrict ourselves to matching spaces over $\mathbb{Z}_2$. We shall describe how Lovász was able to use Theorem 16.20 to obtain an elegant characterization of the orthogonal complement of $\mathcal{L}in(G, \mathbb{Z}_2)$.

16.4.1 Matching orthogonal vectors

Let G be a matching covered graph, and let $\mathbb{F}$ be a field. A vector $\mathbf{y}$ in $\mathbb{F}^E$ is *matching orthogonal* over $\mathbb{F}$ if, for any perfect matching M of G, the sum $\mathbf{y}(M) = \sum_{e \in M} \mathbf{y}(e)$ is zero in that field. In other words, matching orthogonal vectors over $\mathbb{F}$ are the vectors in the orthogonal complement $\mathcal{L}in^{\perp}(G, \mathbb{F})$ of the matching space $\mathcal{L}in(G, \mathbb{F})$.

According to Theorem 16.20, any matching integral vector in the edge space $\mathbb{R}^E$ of a brick G is the sum of a matching orthogonal vector and an integer vector, provided that G is not a Petersen brick. Using this result which relates to vector spaces over $\mathbb{R}$, Lovász presented in [58] a simple characterization of matching orthogonal vectors over $\mathbb{Z}_2$. We describe below this ingenious application of Theorem 16.20.

We begin with the observation that a subset Q of the edge set of G is the support of a vector in $\mathbb{Z}_2^E$ which is matching orthogonal over $\mathbb{Z}_2$ if and only if $|Q \cap M|$ is even for every perfect matching M of G. (By the support of a vector $\mathbf{y} \in \mathbb{Z}_2^E$, denoted by $||\mathbf{y}||$, we mean the subset of the edge set E corresponding to nonzero coordinates of vector $\mathbf{y}$.)

Example 16.21 Supports of matching orthogonal vectors over $\mathbb{Z}_2$.

By Proposition 1.1, for any subset S of the vertex set of a matching covered graph G, and for any perfect matching M of G, the parity of $|\partial(S) \cap M|$ coincides with the parity of $|S|$. It follows that even cuts are supports of matching orthogonal vectors over $\mathbb{Z}_2$. In addition, if $n \equiv 0 \pmod 4$, all perfect matchings of G have even cardinality, implying that, in this case, complements of even cuts are also subsets of E of the required type. On the other hand, if $n \equiv 2 \pmod 4$, all perfect matchings of G have odd cardinality, implying that, in this case, it is the complement of odd cuts that fit the bill. In the Petersen graph, the edge set of any pentagon is the support of a matching orthogonal vector over $\mathbb{Z}_2$. This is neither a cut nor the complement of a cut!

Proposition 16.22 *Let* $\mathbf{y}$ *be a vector in* $\mathbb{Z}_2^E$*, and let* $\mathbf{w}$ *be the* $(0,1)$*-vector in* $\mathbb{R}^E$ *such that* $||\mathbf{w}|| = ||\mathbf{y}||$*. Vector* $\mathbf{y}$ *is a matching orthogonal vector over* $\mathbb{Z}_2$ *if and only if* $\frac{1}{2}\mathbf{w}$ *is a matching integral vector over* $\mathbb{R}$*.*

Proof By definition, $\mathbf{y}$ is matching orthogonal over $\mathbb{Z}_2$ if $\mathbf{y}(M)$ is zero in $\mathbb{Z}_2$, for any perfect matching M of G. Since $1 + 1 = 0$ in $\mathbb{Z}_2$, this amounts to saying that $\mathbf{y}$ is matching orthogonal over $\mathbb{Z}_2$ if and only if $||\mathbf{y}|| \cap M$ is an even subset of E. This is clearly equivalent to the statement that $\mathbf{w}(M)$ is an <u>even</u> integer, for any perfect matching M of G. The assertion follows. □

Theorem 16.23 *Let G be a brick which is not a Petersen brick, and let Q be a subset of E such that $|Q \cap M|$ is even for all $M \in \mathcal{M}$. Then either Q is an even cut, or it is the complement of an even cut when $n \equiv 0 \pmod 4$ and the complement of an odd cut when $n \equiv 2 \pmod 4$.*

Proof Let $\mathbf{w}$ denote the characteristic function χ^Q of the set Q. The first step in the proof is to deduce a property of $\mathbf{w}$ regarded as a vector over $\mathbb{R}$.

16.23.1 *Vector $\mathbf{w}$, regarded as a vector over $\mathbb{R}$, is the sum of a matching orthogonal vector and an integer vector all of whose coordinates are even integers.*

Proof The hypothesis implies that $\frac{1}{2}\mathbf{w}$, regarded as a vector over $\mathbb{R}$, is a matching integral vector. It now follows from Theorem 16.20 that $\frac{1}{2}\mathbf{w}$ is the sum of a matching orthogonal vector and an integer vector. As any scalar multiple of a matching orthogonal vector is also a matching orthogonal vector, the assertion follows. □

Let

$$\mathbf{w} = \mathbf{x} + \mathbf{z}, \tag{16.3}$$

where $\mathbf{x}$ is a matching orthogonal vector and z is a vector all of whose coordinates are even integers.

At this stage we invoke a result concerning matching orthogonal vectors stated in Exercise 6.2.6. According to that result, any matching orthogonal vector in the edge space of a brick is vertex induced, which means that there exists a real-valued function p on the vertex set of G such that $\mathbf{x}(uv) = p(u) + p(v)$, for any edge uv of G. (We shall refer to the values of p as vertex labels.) As each coordinate of $\mathbf{z}$ is an integer, and since each coordinate of $\mathbf{w}$ is 0 or 1, we deduce that each coordinate of $\mathbf{x}$ is also an integer. Thus, $p(u) + p(v)$ is an integer, for each edge uv of G. Moreover, as each coordinate of $\mathbf{z}$ is even, we conclude that

16.23.2 *There exists a real-valued function p on the vertex set of G such that, for any edge uv of G, we have:*

$$\mathbf{w}(uv) = p(u) + p(v) \pmod 2 \tag{16.4}$$

Furthermore, since $\mathbf{w}(uv) = 0$ or 1, it follows that $\mathbf{w}(uv) = 1$ if and only if $p(u)+p(v)$ is an odd integer. Equivalently, an edge uv belongs to the set Q if and only if $p(u) + p(v)$ is an odd integer.

Equation 16.4 does not imply that vertex labels themselves are integer-valued. However, we claim that all vertex labels are either integral, or all of them are half-integral. (Each real number r may be written as the sum of an integer z and a non-negative real number t less than 1. We shall refer to z and t, respectively, as the *integer and and fractional parts* of r. For example $-2.25 = -3 + .75$ and, therefore, the integer part of -2.25 is -3 and its fractional part is $.75$. A number is *half-integral* if its fractional part is a half.)

16.23.3 *If p is a vertex labelling of the brick G which satisfies equation* (16.4), *then either all labels are integers or all labels are half-integral.*

Proof If all labels are integral then the assertion holds. Suppose then that not all labels are integral. Let u and v be any two adjacent vertices in G. As noted before, $p(u)+p(v)$ must be an integer. Suppose that $p(u)$ is not integral; let t be its fractional part. Then, $0 < t < 1$. Moreover, $p(v)$ is not integral either and its fractional part is $1 - t$. It follows that if $t \neq \frac{1}{2}$ then G has to be bipartite: each edge joins a vertex having t as the fractional part of its label to a vertex having $1 - t$ as the fractional part of its label. As G, a brick, is not bipartite, we deduce that $t = \frac{1}{2}$. Thus, all the labels are half-integral. □

To obtain the desired conclusion, we now consider two cases depending on whether or not all vertex labels are integers.

Case 1 *All vertex labels are integers.*

Let S denote the set of all vertices whose labels are odd integers. Then the labels of all vertices in $\overline{S}$ are even integers. We then have that $p(u) + p(v)$ is an odd integer if and only if u and v belong to different shores of the cut $\partial(S)$. By Statement 16.23.2, this implies that $Q = \partial(S)$. And, by the hypothesis of the theorem, every perfect matching of G meets Q in an even number of edges, implying that Q is an even cut of G.

Case 2 *All vertex labels are half-integral.*

Let S denote the set of all vertices whose integer parts are odd integers. In this case, for any two adjacent vertices u and v of G, the sum $p(u) + p(v)$ is an odd integer if and only if u and v are in the same shore of $\partial(S)$. Thus, by Statement 16.23.2, it follows that Q is the complement of the cut $\partial(S)$. Let M be any perfect matching of G; then, clearly, $|M \cap \partial(S)| = |M| - |M \cap Q|$. By hypothesis, $|M \cap Q|$ is even. If $n \equiv 0 \pmod 4$, then, being a perfect matching of G, M has an even number of edges. Thus, in this case, $|M \cap \partial(S)|$ is even, implying that $\partial(S)$ is an even cut and that Q is the complement of $\partial(S)$. A similar argument shows that, when $n \equiv 2 \pmod 4$, the cut $\partial(S)$ is odd and Q is its complement. □

We leave the proof of the following corollary as Exercise 16.4.3.

Corollary 16.24 *If G is a brick which is not a Petersen brick, then:*

(i) *when $n \equiv 0 \pmod 4$, $\mathcal{L}in^{\perp}(G, \mathbb{Z}_2)$ is generated by $\chi^E = \mathbf{1}$ and the incidence vectors of all even cuts; and*
(ii) *when $n \equiv 0 \pmod 2$, $\mathcal{L}in^{\perp}(G, \mathbb{Z}_2)$ is generated by the incidence vector of the complement of a trivial cut and the incidence vectors of all even cuts.* □

The above corollary may now be used to derive a formula for the dimension of $\mathcal{L}in^{\perp}(G, \mathbb{Z}_2)$. For this purpose, we make use of some known facts concerning the row space (over $\mathbb{Z}_2$) of the incidence matrix $\mathbf{A}$ of a graph G. This space is known as the bond space of G; the support of any vector in it is a cut of G. If $\partial(X)$ and $\partial(Y)$ are two even cuts, then $\partial(X) \triangle \partial(Y)$ is also an even cut. This means that the (mod 2) sum of the incidence vectors of any two even cuts is also the incidence vector of some even cut of G. Thus the incidence vectors of even cuts of G is a subspace of the bond space of G. Furthermore, any basis of this subspace, together with the incidence vector any one odd cut generates the bond space (see Exercise 16.4.4). The dimension of the bond space of a connected graph G is known to be $n-1$ (Theorem 20.7 in [3]), and thus the dimension of its subspace generated by the incidence vectors of even cuts is $n-2$. This observation, together with Corollary 16.24 implies the following:

Corollary 16.25 *If G is any brick which is not a Petersen brick, then the dimension of $\mathcal{L}in^{\perp}(G, \mathbb{Z}_2)$ is $n-1$ and the dimension of $\mathcal{L}in(G, \mathbb{Z}_2)$ is $m-n+1$.*

Theorem 16.5 shows that if G is any cubic brick different from $\mathbb{P}$, then **1** is in the matching space $\mathcal{L}in(G, \mathbb{Z}_2)$. We shall not attempt here to present a characterization of matching spaces over $\mathbb{Z}_2$ of arbitrary matching covered graphs. This is somewhat more involved than the above mentioned characterization results concerning matching orthogonal vectors. We leave it as a special topic for interested readers to pursue.

Exercises

16.4.1 *Adapt the proof of Theorem 16.4 to prove that, for every matching covered graph G, $\dim(\mathcal{L}in(G, \mathbb{Z}_2)) = m - n + 2 - (b + p)$.*

▷**16.4.2** *Let $\mathbb{F}$ be a field of finite, odd characteristic. Prove that (i) the dimension of $\mathcal{L}at(G, \mathbb{F})$ is equal to $m - n + 2 - b$ and (ii) a vector $\mathbf{x} \in \mathbb{F}^E$ is in $\mathcal{L}at(G, \mathbb{F})$ if and only if it is regular. Hint: in the case $p = 0$, adapt the proof of Theorem 16.4 and in the case of Petersen bricks, adapt the proof of Lemma 16.6, as 4 has an inverse in $\mathbb{F}$.*

16.4.3 *Prove Corollary 16.24.*

16.4.4 *Prove that the incidence vectors of even cuts of a graph G is a subspace of the bond space of G. Furthermore, any basis of this subspace, together with the incidence vector any one odd cut generates the bond space of G.*

16.5 Notes

Chapter 38 of Volume A of Schrijver's three-volume treatise [84] contains a proof of Lovász's characterization of the matching lattice of a brick (Theorem 6.14) without reference to the dual of the matching lattice. It is based on a detailed study of properties of special barriers in matching covered graphs. The inductive nature of our proof, which relies on the existence of b-invariant edges in nonspecial bricks (Theorem 15.1), has the advantage that it can be iteratively used to obtain a basis for the matching lattice of any matching covered graph.

Recall the perfect matching integer cone $IntCone(G)$ of a matching covered graph G is the set of all non-negative integer linear combinations of incidence vectors of perfect matchings of G (see Section 6.3.2). One of the longstanding unresolved conjectures in graph theory is the following:

> FULKERSON'S CONJECTURE
>
> For any 2-connected cubic graph G, the vector $\mathbf{2} = (2, 2, \ldots, 2)$ of all 2's in the edge space of G belongs to its perfect matching integer cone $IntCone(G)$. (It is easy to show that $\mathbf{2}$ belongs $\mathcal{L}at(G)$ (see Exercise 16.1.11).)

The above conjecture amounts to saying that if G is any 2-connected cubic graph then there exist six perfect matchings of G such that each edge of G belongs to two of these six matchings.

The paper by Ma et al (2022, [65] has many references to results related to Fulkerson's Conjecture.

For any matching covered graph G, let us define the *Berge number*, denoted by $c(G)$, to be the minimum number of perfect matchings of G whose union is $E(G)$. An obvious lower bound for $c(G)$ is the maximum degree $\Delta(G)$. Clearly, $c(\mathbb{P}) = 5$.

It is not difficult to prove that $c(G) = \Delta(G)$ for bipartite graphs, but we do not know whether the same equality holds for solid graphs. Clearly, if G is an r-regular graph, then $c(G) = \Delta(G)$ if and only if G is r-edge colourable. Another interesting question: does the equality $c(G) = \Delta(G)$ hold, in general, for all planar matching covered graphs?

> BERGE'S CONJECTURE
>
> For every 2-connected cubic graph G, $c(G) \leq 5$.

Fulkerson's Conjecture clearly implies Berge's Conjecture. The converse is also true, as proved by Mazzuoccolo (2011, [66]).

Chapter 17
Thin Edges in Bricks and Braces

Contents

17.1 Thin Edges and Thin Barriers in Bricks

THIN EDGES IN BRICKS

An edge e of a brick G is *thin* if the retract of $G - e$ is a brick. Since $b(G-e) = b(\widehat{G-e})$, it follows that every thin edge of a brick G is a b-invariant edge of G. But, in general, not every b-invariant edge of a brick is thin (see Exercise 17.1.3). Figure 17.1 shows examples of thin edges, indicated by solid lines.

C. L. Lucchesi, U. S. R. Murty, *Perfect Matchings*, Algorithms and Computation in Mathematics 31, https://doi.org/10.1007/978-3-031-47504-7_17

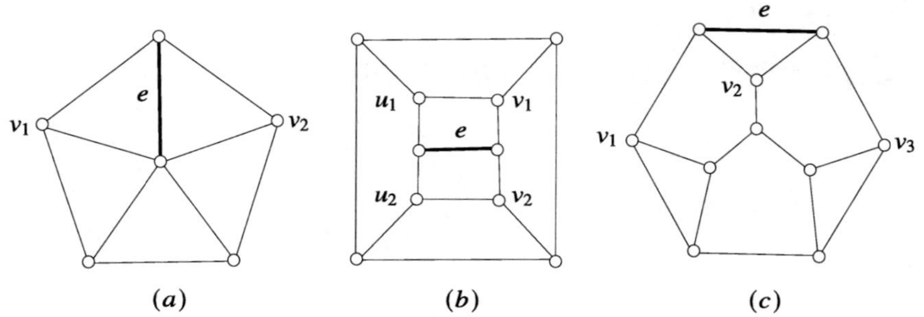

Fig. 17.1 Examples of thin edges in three different bricks

The notion of a thin edge in a brick was introduced by CLM (2006, [15]). However, the definition of a thin edge e in a brick G given there is in terms of 'sparsity' of barriers of $G - e$; and it was that definition that led to the choice of the adjective 'thin'.

Let G be a brick, and let e be a b-invariant edge of G. Every barrier of $G - e$ is special (Exercise 17.1.1). Thus, for any barrier B of $G - e$, the graph $G - e - B$ has just one nontrivial component and it follows that $|I(B)| = |B| - 1$, where $I := I(B)$ denotes the set of isolated vertices of $G - e - B$.

THIN BARRIERS

A barrier B of $G - e$ is *thin* if all vertices in I have degree two in $G - e$. In the brick G shown in Figure 17.1(a), the graph $G - e$ has a unique nontrivial barrier, namely $\{v_1, v_2\}$; in the brick G shown in Figure 17.1(b), the graph $G-e$ has two nontrivial barriers, namely $B_1 := \{u_1, u_2\}$ and $B_2 := \{v_1, v_2\}$; and in the brick G shown in Figure 17.1(c), the graph $G - e$ has a unique nontrivial maximal barrier, namely $\{v_1, v_2, v_3\}$. All these barriers are thin.

The following proposition is straightforward to establish (Exercise 17.1.4).

Proposition 17.1 *Let e be a b-invariant edge of a brick G, and let B be a thin barrier of $G - e$. Then $|B| \leq 3$. Furthermore:*

(i) *if $|B| = 2$ then, in G, one end of the edge e is the unique vertex in I and the other end of e is not in $B \cup I$, and*
(ii) *if $|B| = 3$ then, in G, the edge e joins the two vertices in I.* □

By the Three Case Lemma 13.5, when e is a b-invariant edge of a brick G, the brick of $G - e$ is obtained by at most two tight cut-contractions, each of which is a barrier cut associated with a special barrier. Based on this result and the above proposition, it is now easy to prove the following theorem. We leave its proof as Exercise 17.1.5.

Theorem 17.2 *A b-invariant edge e of a brick G is thin if and only if every barrier of $G - e$ is a thin barrier.* □

By Theorem 15.1, every brick which is not one of the three special bricks has a b-invariant edge. One of the principal results of this chapter is the following strengthening of that theorem:

THIN EDGE THEOREM FOR BRICKS

17.3 *Every brick distinct from K_4, $\overline{C_6}$ and the Petersen graph has a b-invariant edge that is thin.*

As a corollary of the above theorem, we have the interesting property of bricks stated below:

Corollary 17.4 *Every brick G of minimum degree greater than three has an edge e such that $G - e$ is a brick.* □

It follows from Theorem 17.3 that, given any brick G, there exists a sequence

$$(G_1, G_2, \ldots, G_k)$$

of bricks such that

(i) $G_1 = G$ and $G_k \in \{K_4, \overline{C_6}, \mathbb{P}\}$, and
(ii) for $1 \leq i < k$, the brick G_{i+1} is obtained from G_i by deleting an edge and then bicontracting the resulting vertices of degree two.

Conversely, as we shall see in Section 17.4, given any brick H, a larger brick can be obtained from H by means of one of four elementary operations involving bisplitting vertices (as defined in Chapter 12), and adding edges. This yields a simple procedure for generating bricks. In fact, some of these operations may also be used for generating braces.

17.1.1 Outline of the proof of the existence of thin edges in bricks

We now recall the definition of index of a b-invariant edge of a brick G, given in Chapter 13. Let G be a brick and let e be a b-invariant edge of G. If $G - e$ is also a brick then we say that e has *index* zero. If $G - e$ is not a brick, then we say that e has *index* one, two or three, depending on which of the three cases stated in Lemma 13.5 holds. Let H be the brick of $G - e$. If H has one contraction vertex and one of the ends of e lies in $V(H)$ then e has index one. If H has two contraction vertices then e has index two. Finally, if H has one contraction vertex and neither end of e lies in $V(H)$ then e has index three. If e is an edge of index zero, then $G - e$ is a brick and hence e is thin. If a b-invariant edge has index greater than zero and is not thin, we shall need to show that there is some other edge that is thin.

Let e be a b-invariant edge of a brick G that is not thin. Then $G - e$ must have a barrier B that is not thin, and thus I must have a vertex v of degree three or more in $G - e$. The set $\partial(v) - e$ can be shown to have at least one edge that is b-invariant in both G and $G - e$. For any such edge e', the brick of $G - e'$ would have at least as many vertices as the brick of $G - e$. Moreover, if equality holds then e has index two and in that case $\partial(v) - e$ would have a second edge, say e'', that is also b-invariant in both G and $G - e$. Finally, it turns out that equality of the numbers of vertices of the bricks of $G - e'$ and $G - e''$ is impossible, which implies that the brick of $G - e''$ has more vertices than the brick of $G - e$.

This leads us to define the *pre-rank* of a b-invariant edge e, denoted by $r_0(e)$, as the order of the brick of $G - e$. (For example, if e is of index zero, then $r_0(e) = |V|$.) The idea behind this notion is that the larger the pre-rank of a b-invariant edge is, the closer it is to being thin. Let e be a b-invariant edge that is not thin, let B be a maximal barrier of $G - e$ that is not thin and let s be a vertex in I whose degree in $G - e$ is at least three. Then, some edge that is incident with s is also b-invariant in G. As explained above, for any such e', we would have $r_0(e') \geq r_0(e)$. If perchance, the pre-ranks of e and e' happen to be equal, then the index of e would have to be two. In this case, it turns out some other edge e'' incident with s is b-invariant and its pre-rank is larger than that of e. (Two b-invariant edges e and f of G, of indices two and one, respectively, may have the same pre-rank, and it may be the case that e is thin while f is not. See Exercise 17.1.7.) Motivated by the above observation, we define the *rank* $r(e)$ of a b-invariant edge e as follows:

$$r(e) := \begin{cases} r_0(e) + 1, \text{ if index}(e) = 2 \\ \quad r_0(e), \text{ otherwise} \end{cases}$$

We prove Theorem 17.3 by showing that a b-invariant edge of a brick that has the maximum possible rank is thin!

As noted above, if a brick has a b-invariant edge that is not thin, then there is some other b-invariant edge of the brick whose rank is higher. In Section 17.2.2 we shall establish the tools necessary for finding such alternatives, and using these tools we shall present a proof of the existence of thin edges in bricks in Section 17.3.

Exercises

▷**17.1.1** *Let G be a brick, let e be a b-invariant edge of G and let B denote a barrier of $G - e$. The graph $G - e - B$ has precisely one nonbipartite component. Show that B is special. Hint: it suffices to show that all the bipartite components of $G - e - B$ are trivial. Assume, by way of contradiction, that $G - e - B$ has a nontrivial bipartite component $G[X]$. See Figure 17.2. Derive a contradiction by showing that one of X_+ and $B \cup X_-$ is a nontrivial barrier of G.*

17.1.2 *In each of the three bricks shown in Figure 17.1, show that the edge e is a thin edge.*

17.1.3 *Consider the brick G shown in Figure 17.3.*
Verify the following statements concerning the edges of G:

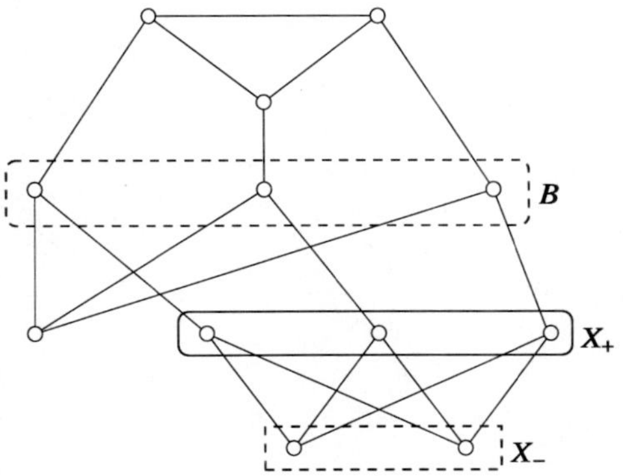

Fig. 17.2 Illustration for Exercise 17.1.1

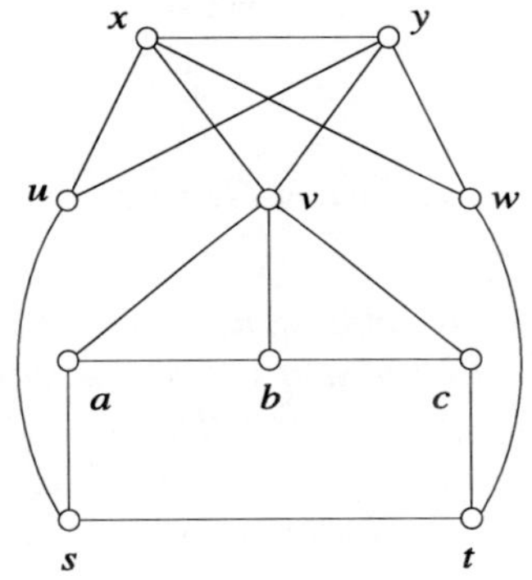

Fig. 17.3 The brick G in Exercise 17.1.3

(i) *ab is not removable;*
(ii) *xy is b-invariant but not thin;*
(iii) *bv and st are thin. (Altogether, there are ten thin edges in this brick.)*

17.1.4 *Give a proof of Proposition 17.1.*

17.1.5 *Give a proof of Theorem 17.2.*

17.1.6 *Let e be a thin edge of a brick G of order n. Verify the following statements:*

(i) *If e has index zero, then $r(e) = r_0(e) = n$;*
(ii) *if e has index one, then $r(e) = r_0(e) = n - 2$;*
(iii) *if e has index two, then $r(e) = n - 3$ whereas $r_0(e) = n - 4$; and*
(iv) *if e has index three, then $r(e) = r_0(e) = n - 4$.*

17.1.7 *Consider the brick G depicted in Figure 17.4. Show that $e = 48$ and $f = 13$ have the same pre-rank but that e (which is of index 2) is thin whereas the edge f (which is of index 1) is not.*

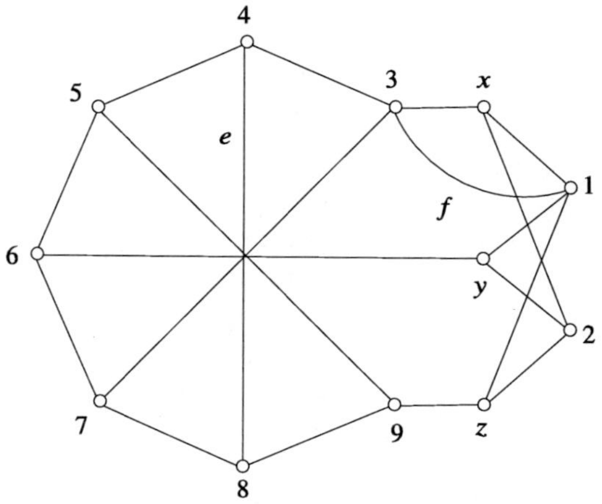

Fig. 17.4 Two edges of the same pre-rank, one thin and the other not

17.1.8 *Consider the four bricks shown in Figure 17.5. In each case show that edge f indicated by a solid line is thin.*

Hint: *(a) Observe that $\widehat{G-f}$ is obtained by splicing two bricks (one of order four and one of order six) and use Proposition 5.15; (b) observe that $\widehat{G-f}$ is isomorphic to the brick in Figure 17.5(a); (c) observe that $\widehat{G-f}$ is isomorphic to the graph obtained by the addition of an edge to the brick in Figure 17.5(a); (d) observe that $\widehat{G-f}$ is isomorphic to the brick shown in Figure 17.5(b).*

17.2 Barriers and Tight Cuts in Near-Bricks ♯

In this section we shall establish the relationship between barriers and tight cuts in near-bricks. It follows from Proposition 4.18 that for any tight cut C in a near-brick G, precisely one of the shores of C is bipartite and, furthermore, C is a barrier cut associated with a special barrier of G. It also follows that every bicritical near-brick is a brick. We have used these facts in many proofs in previous chapters, and shall use them here without explicit reference to it.

Let G be a near-brick, let C be a tight cut of G, and let X be the bipartite shore of C. We shall refer to the majority part B of $G[X]$ as the *external part* of X, and its minority part I as the *internal part*. See Figure 17.6 (the vertices of the barrier associated with the cut C are indicated by solid dots.)

Given any tight cut of a near-brick G, as noted at the beginning of this section, there is a unique special barrier that corresponds to it. Conversely, there is a unique nontrivial tight cut that corresponds to any special barrier of G. We shall use the following notation for nontrivial tight cuts associated with special barriers of near-bricks.

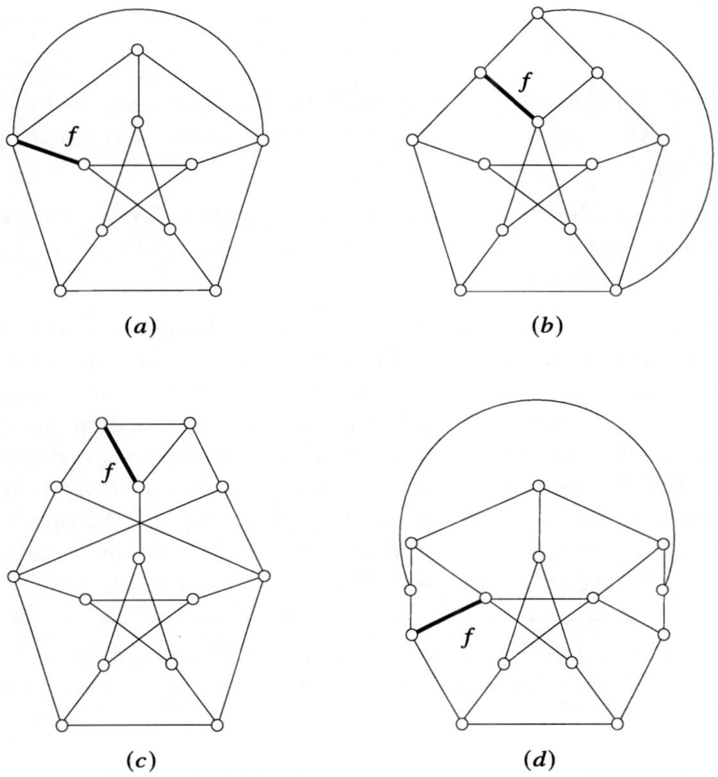

Fig. 17.5 Thin edges in four bricks related to the Petersen graph

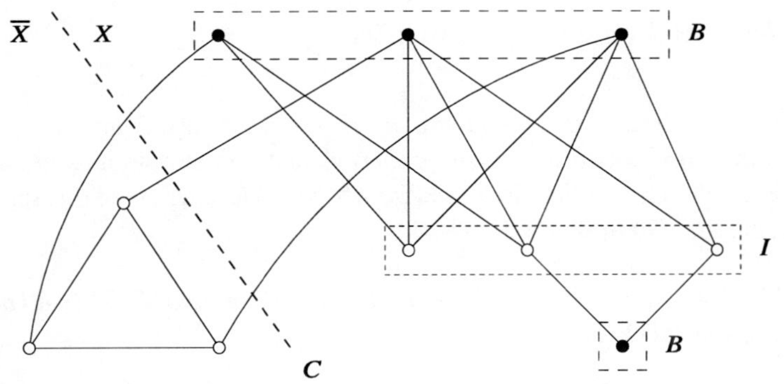

Fig. 17.6 Cuts and their shores in near-bricks

NOTATION: B, I, X, C

17.5 *Let C be a tight cut of a near-brick G. Then, we denote the bipartite shore of C by $X_G(C)$, and its majority and minority parts by $B_G(C)$ and $I_G(C)$, respectively.*

Correspondingly, let B be a special barrier of a graph G. Then we denote the set of isolated vertices of $G - B$ by $I_G(B)$; the set $B \cup I_G(B)$ by $X_G(B)$; and the cut $\partial_G(X_G(B))$ by $C_G(B)$.

(In both the above cases, we adopt obvious notational simplifications when the parameters involved are understood from context. For example, if the near-brick G and tight cut C are understood, then we simply write X, B and I, instead of $X_G(C)$, $B_G(C)$ and $I_G(C)$, respectively. Also, we adopt the usual notational conventions when some of the parameters under consideration are denoted by letters with either subscripts or superscripts. For example, if the tight cut of G under consideration is C_1, we denote $X(C_1)$, $B(C_1)$ and $I(C_1)$ simply by X_1, B_1, and I_1, respectively. Similarly, if B' is a special barrier of G, we denote $I(B')$, $X(B')$ and $C(B')$ simply by I', X' and C', respectively.)

It should be noted that, in general, not every barrier in a near-brick is special. However, as has been observed in Section 4.4, the following assertion holds (Exercise 4.4.5):

Proposition 17.6 *Every maximal barrier in a near-brick is special.* □

17.2.1 Near-bricks inherited from bricks

A near-brick obtained from the deletion of a b-invariant edge from a brick is said to be *inherited* from that brick. Many properties that do not hold in general for all near-bricks hold for near-bricks inherited from bricks. The purpose of this subsection is to describe some of those properties.

Lemma 17.7 *Let G be a brick and let e be a b-invariant edge of G. Then all barriers of $G - e$ are special (Exercise 17.1.1).* □

Lemma 17.8 *Let G be a brick, let e be a b-invariant edge of G and let B be a (special) maximal barrier of $G - e$. Then, graph $G/(X \to x)$ is a brick.*

Proof According to the notational conventions we have adopted, the cut $C := \partial(X)$ is a tight cut of $G - e$, the graph $G[X]$ induced by X is the bipartite shore of C, and B is its majority part. Let $G' := G/(X \to x)$. As every tight cut in a near-brick is a barrier cut, to establish that G' is a brick it suffices to show that (i) G' is a near-brick, and that (ii) G' has no nontrivial barriers.

Let us first show that G' is a near-brick. As the graph $G' - e$ is the nonbipartite $(C - e)$-contraction of the near-brick $G - e$, it follows that $G' - e$ is a near-brick. If $e \notin C$ then $G' = G' - e$, and G' is a near-brick. Suppose then that $e \in C$. As B is maximal, it follows, by Theorem 3.2, that the component $G[\overline{X}]$ of $G - e - B$ is critical. Thus, $\{x\}$ is a maximal barrier of $G' - e$. In G', edge e is incident with x. Thus, the graph G' is matching covered. As $G' - e$ is a near-brick, it follows, by the monotonicity of function b (Theorem 9.8), that G' is also a near-brick.

Now let us proceed to show that G' has no nontrivial barriers. Thus, let B' be any barrier of G'. Suppose first that $x \in B'$. In this case, as $G' - x$ is critical, it follows that $B' = \{x\}$. Alternatively, suppose that $x \notin B'$. In this case, B' is a barrier of G, where the set X is part of the set of vertices of some component of $G - B$ and its contraction produces the corresponding component of $G' - B$. As G is a brick, we deduce that B' is a singleton. In both alternatives, we deduced that B' is a singleton. As this conclusion holds for any barrier B' of G', it follows that G' is bicritical. Hence the assertion holds. □

Three Case Lemma–Revisited

In Section 13.2.1 we observed that if e is a b-invariant edge of a brick G, then e has to have at least one end in the minority part of the bipartite shore of any nontrivial tight cut of $G-e$. As a consequence, it follows that the brick of $G-e$ can be obtained from $G - e$ by at most two tight cut contractions. If $G - e$ is itself not a brick, this observation gives rise to the three cases to which the structure of $G-e$ must conform; that is, to the Three Case Lemma 13.5. In much the same way, if B is any special barrier of $G - e$, then e has to have at least one end in the set I of isolated vertices of $G-e-B$. As a consequence, it follows that $G-e$ has at most two maximal nontrivial special barriers. This observation also leads to the same conclusions concerning the structure of $G-e$. Since our emphasis here is on special barriers of $G-e$, we restate below that lemma using the notation introduced above. It will prove to be crucial for computing ranks of b-invariant edges in bricks.

Suppose that e is a b-invariant edge of a brick G and that $G - e$ is itself not a brick. In this case, $G - e$ must have nontrivial maximal special barriers. If there is only one such barrier, say B, then the brick of $G - e$ is $(G - e)/(X \to x)$ which is obtained from $G - e$ by contracting the bipartite shore of the tight cut of $G - e$ associated with B. However, if B_1 and B_2 are two nontrivial maximal special barriers of $G-e$, the tight cuts C_1 and C_2 associated with B_1 and B_2, respectively, may cross each other. In this case, it would not be right to conclude that the brick of $G - e$ is $(G - e)/(X_1 \to x_1)/(X_2 \to x_2)$. Clearly, being distinct maximal barriers of $G - e$, by Theorem 3.11, B_1 and B_2 are disjoint, but $X_1 := B_1 \cup I_1$ and $X_2 := B_2 \cup I_2$ need not be disjoint. This situation occurs when $B_2 \cap I_1$ is nonempty (which implies that $B_1 \cap I_2$ is also nonempty, see Exercise 17.2.1).

For example, consider the brick G shown in Figure 17.7. The edge e is b-invariant in G, and $G - e$ has two maximal nontrivial barriers which are indicated by large black and white discs, respectively. The bipartite shores of the tight cuts associated with these barriers are X_1 and X_2, and the cuts C_1 and C_2 cross each other.

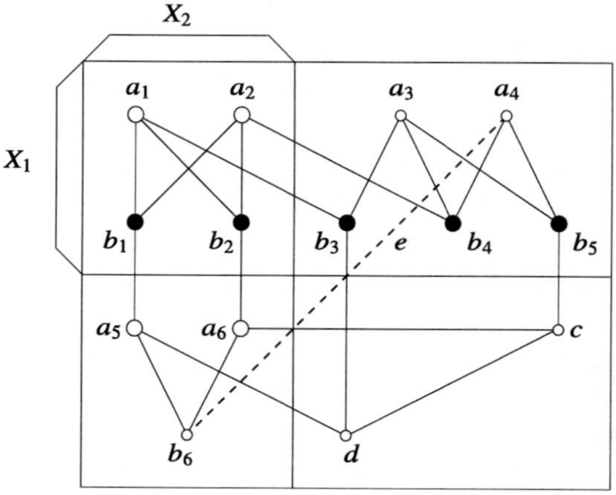

Fig. 17.7 The case in which the tight cuts associated with two maximal barriers B_1 and B_2 of $G - e$ cross ($B_1 = \{b_1, b_2, b_3, b_4, b_5\}$ with $I_1 = \{a_1, a_2, a_3, a_4\}$ and $B_2 = \{a_1, a_2, a_5, a_6\}$ with $I_2 = \{b_1, b_2, b_6\}$)

We now present the version of the Three Case Lemma, stated in terms of maximal nontrivial barriers of $G - e$, that we shall use in this chapter.

Lemma 17.9 (The Three Case Lemma) *Let G be a brick, and let e be a b-invariant edge of G. Then, one the following alternatives holds:*

(i) **index(e) = 0**: *graph $G - e$ is a brick, or*
(ii) **index(e) = 1**: *graph $G - e$ has precisely one maximal nontrivial barrier, B, the edge e has only one end in I and the graph $(G - e)/(X \to x)$ is a brick, or*
(iii) **index(e) = 2**: *graph $G - e$ has precisely two maximal nontrivial barriers, B_1 and B_2, and*

- *edge e has one end in $I_1 - X_2$ and the other in $I_2 - X_1$,*
- *the graph $(G - e)/(X_1 \to x_1)$ is a near-brick, $B_2 - X_1$ is its unique nontrivial maximal barrier and $X_2 - X_1$ is the bipartite shore of a nontrivial tight cut,*
- *the graph $(G - e)/(X_1 \to x_1)/((X_2 - X_1) \to x_2)$ is a brick (see Figure 17.7 — up to multiple edges, by Corollary 4.13, this is the same as the brick $(G - e)/((X_1 - X_2) \to x_1)/(X_2 \to x_2)$), or*

(iv) **index(e) = 3**: *graph $G - e$ has precisely one maximal nontrivial barrier, B, the edge e has both ends in I and the graph $(G - e)/(X \to x)$ is a brick.*

Rank Computations

If an edge e is of index zero, then $G - e$ is a brick. In this case, $r(e) = r_0(e) = |V|$. We describe below the procedure for finding the rank of e when the index of e is positive.

Suppose that a b-invariant edge e of a brick G has index one or three and suppose that B is the only nontrivial maximal barrier of $G - e$. Then the edge e might have both its ends in I (index$(e) = 3$) or one end in I and one end in $\overline{X}$ (index$(e) = 1$). In either case, the graph obtained from $G - e$ by contracting X to a single vertex is the brick of $G - e$. Thus, in this case:

$$r(e) = r_0(e) = |V| - |X| + 1 = |\overline{X}| + 1. \tag{17.1}$$

Now suppose that a b-invariant edge e of a brick G has index two and suppose that B_1 and B_2 are the two nontrivial maximal barriers of $G - e$. In this case, $B_2' := B_2 - X_1$ is the only maximal nontrivial barrier of $G' := (G - e)/(X_1 \to x_1)$ and $G'/((X_2 - X_1) \to x_2)$ is the brick of $G - e$. Thus, the pre-rank $r_0(e)$ of e is equal to $(|V| - |X_1|) + 1 - (|X_2 - X_1|) + 1 = |\overline{X_1}| - |X_2 - X_1| + 2$, implying the following formula for its rank:

$$r(e) = r_0(e) + 1 = |\overline{X_1}| - |X_2 - X_1| + 3 \leq |\overline{X_1}|. \tag{17.2}$$

For example, if e is the indicated b-invariant edge in the brick shown in Figure 17.7, $r(e) = |\overline{X_1}| - |X_2 - X_1| + 3 = 5 - 3 + 3 = 5$.

The next proposition is a simple consequence of (17.1) and (17.2).

Proposition 17.10 *Suppose that B is a maximal barrier of $G - e$ and that $\partial(X)$ is the tight cut of $G - e$ associated with B. Then $r(e) \leq |\overline{X}| + 1$, with equality only if the index of e is not equal to two.* □

Note that $r_0(e)$ is always even as it is the order of a brick. Thus, $r(e)$ is odd only when e is of index 2.

17.2.2 Removable edges in near-bricks inherited from bricks

Suppose that G is a brick and e is a b-invariant edge of G that is not thin. Our objective then is to show that there is some other b-invariant edge of G that is thin. The first step is to show that there exist b-invariant edges (thin or not) other than e. This section is devoted to the development of tools that are necessary for finding such edges. If B is a nontrivial maximal barrier of $G - e$, it turns out that one can always find a b-invariant edge of G incident with a vertex in I that is of degree greater than two in $G - e$. The main tool used in establishing this result is Lemma 8.19 proved in Chapter 8. We shall also use the following easily proved proposition.

Proposition 17.11 *Let G be a matching covered graph, let e be a removable edge of G, and let e' be an edge of $G - e$ that is removable in $G - e$. Edge e' is removable in G if and only if G has a perfect matching that contains edge e but does not contain edge e'.* □

ALTERNATIVE CHOICES FOR b-INVARIANT EDGES

Theorem 17.12 *Let G be a brick, let e be a b-invariant edge of G and let B be a maximal nontrivial (special) barrier of $G - e$. Let v be a vertex of I that has degree strictly greater than two in $G - e$. Then, the following properties hold:*

(i) *At most one edge of $\partial(v) - e$ is not b-invariant in $G - e$.*
(ii) *At most one edge of $\partial(v) - e$ is b-invariant in $G - e$ without being b-invariant in G.*
(iii) *If e is of index two, then every edge of $\partial(v) - e$ that is b-invariant in $G - e$ is also b-invariant in G.*

Proof Let $C := \partial(X)$, let $H := (G - e)/(\overline{X} \to \overline{x})$ and let $G_1 := G/(X \to x)$. Of these two, H is bipartite because X is the bipartite shore of $C - e$, where $I + \overline{x}$ and B are the two parts of its bipartition. The graph G_1 is a brick, by Lemma 17.8.

17.12.1 *Let e' be any edge in $\partial(v) - e$ that is removable in H. Then e' is a b-invariant edge of $G - e$. Furthermore, if the edge e does not depend on e' in G, then e' is a b-invariant edge of G itself.*

Proof The edge e' is in $H - C$. The two $(C - e)$-contractions of $G - e - e'$ are $H - e'$ and $G_1 - e$. The graph $H - e'$ is bipartite and matching covered, because H is the bipartite $(C - e)$-contraction of $G - e$ and e' is removable in H. The graph $G_1 - e$ is a near-brick, because e is b-invariant in G and $G_1 - e$ is the nonbipartite $(C - e)$-contraction of $G - e$. Thus, the graph $G - e - e'$ is the splicing of $H - e'$ and $G_1 - e$; hence $G - e - e'$ is matching covered. Consequently, e' is removable in $G - e$. Moreover, the cut $C - e$, a tight cut of $G - e$, is tight in $G - e - e'$. Thus, $G - e - e'$ is a near-brick. As $G - e$ is a near-brick, we conclude that e' is b-invariant in $G - e$.

Now suppose that e does not depend on e' in G. Then there is a perfect matching of G which contains e but not e'. It follows from Proposition 17.11 that e' is removable in G. As e' is b-invariant in $G - e$, by the monotonicity of the function b we deduce that

$$b(G) \leq b(G - e') \leq b(G - e - e') = b(G - e) = b(G).$$

Consequently, the edge e' is b-invariant in G itself. This proves the second part of the assertion. □

By Lemma 8.19, at most one edge of $\partial(v) - e$ is not removable in H, hence at most one edge of $\partial(v) - e$ is not b-invariant in $G - e$. This completes the proof of the first item of the assertion. Also, clearly, edge e could be dependent on at most one edge incident with v. Thus, at most one edge of $\partial(v) - e$ is b-invariant in $G - e$ but not b-invariant in G. This proves the second part of the assertion.

Suppose that e has index two. Assume, contrary to the third item of the assertion, that $\partial(v) - e$ contains an edge, say e', that is b-invariant in $G - e$ but not b-invariant in G. From statement 17.12.1 we infer that e depends on e' in G. That is, every perfect matching of G that contains edge e also contains edge e'. Therefore, graph

$G - e'$ has barriers that include both ends of edge e. Let B' be a maximal barrier of $G - e'$ that includes both ends of e. Graph $G - e - e'$ is matching covered; therefore e is the only edge of G that depends on e'. We conclude that in G no edge other than e has both its ends in B'.

As e has index two, one of the ends of e, say, y, is in $G_1 - x$. Thus, in G_1, the ends of e are y and the contraction vertex x.

17.12.2 *The vertex y is the only vertex of $G_1 - x$ in B'.*

Proof Certainly, y, an end of e, is in B'. Let z be any vertex of $G_1 - x - y$. The graph G_1 is a brick; hence $G_1 - y - z$ has a perfect matching, M_1. Let f be the edge of M_1 incident with the contraction vertex x of G_1. Thus, f is the edge of M_1 in C. As y is an end of e in G_1, the matching M_1 does not contain the edge e. Thus, f is an edge of $C - e$. As e' is not in C, the edges e' and f are distinct. As e' is removable in H, the graph $H - e'$ has a perfect matching, M_H, which contains the edge f. Thus, $M := M_1 \cup M_H$ is a perfect matching of $G - y - z - e'$. Consequently, $z \notin B'$. This conclusion holds for each vertex z of $G_1 - x - y$. We deduce that $V(G_1 - x) \cap B' = \{y\}$. □

All the vertices of $G_1 - x - y$ are in components of $G - e' - B'$. The graph G_1, a brick, is 3-connected. Thus, $G_1 - x - y$ is connected. We deduce that all the vertices of $G_1 - x - y$ are in the same component, K, of $G - e' - B'$.

17.12.3 *Every neighbour of y in $G - e$ is in $V(G_1 - x)$.*

Proof By the Three Case Lemma, y is an isolated vertex of $G_1 - e - B_2$, where B_2 is the maximal nontrivial barrier of $G_1 - e$. As G_1 is a brick, B_2 is not a barrier of G_1; hence the end x of e in G_1 is not in B_2. Thus, every neighbour of y in $G - e$ is in $V(G_1 - x)$. □

We conclude that all the neighbours of y in $G - e$ are in K. In that case, $B'' := B' - y$ is a nonempty barrier of $G - e'$ and $V(K) + y$ is the set of vertices of an even component of $G - e' - B''$. This is not possible (by Proposition 2.6). We deduce that e' is b-invariant in G. This conclusion holds for each edge $e' \in \partial(v) - e$ which is b-invariant in $G - e$, provided index$(e) = 2$. □

Figure 17.8 depicts a brick G, where vw is the only edge of $\partial(v)$ which is removable in both G and $G - e$, thereby illustrating the necessity of index$(e) = 2$ in Theorem 17.12.

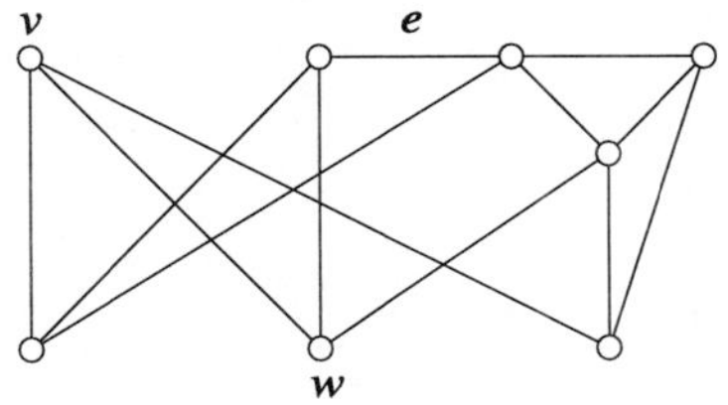

Fig. 17.8 Edge vw is the unique edge in $\partial(v)$ that is removable in both G and $G - e$.

Given any b-invariant edge e of a brick G, and a maximal barrier B of $G - e$ that is not thin, our objective is to show that there is some edge of G whose rank is higher than that of e. We shall in fact see in the next section that some edge incident with an isolated vertex of $G - e - B$ has higher rank than that of e. Our proof of this statement does not rely on the following proposition, whose proof is left as Exercise 17.2.6.

Proposition 17.13 *Let s be any isolated vertex of $G - e - B$ having degree at least three in $G - e$, and let e' be any edge incident with s that is b-invariant in both $G - e$ and G. Then, $r(e') \geq r(e)$.* □

Exercises

∗**17.2.1** *Let e be a b-invariant edge of a brick G and suppose that B_1 and B_2 are two maximal nontrivial barriers of $G - e$. Show that $|B_1 \cap I_2| = |B_2 \cap I_1|$.*

∗**17.2.2** *Let e be a b-invariant edge of a brick G of index two, and let B_1 and B_2 denote the two maximal nontrivial barriers of $G - e$. Suppose that the corresponding cuts C_1 and C_2 cross. Prove that*

(i) *the graph $G[X_1 \cap X_2]$ is equipartite and*
(ii) *the sets $X_1 - X_2$ and $X_2 - X_1$ are bipartite shores of nontrivial tight cuts of $G - e$.*

17.2.3 *Consider the brick G shown in Figure 17.7. Determine the two graphs $(G - e)/(X_1 \to x_1)/((X_2 - X_1) \to x_2)$ and $(G - e)/((X_1 - X_2) \to x_1)/(X_2 \to x_2)$, and verify that they are isomorphic, up to multiple edges.*

17.2.4 *Prove Proposition 17.11.*

17.2.5 *Verify that the graph G in Figure 17.8 is a brick and also that vw is the only edge incident with v which is removable in both G and $G - e$.*

▷**17.2.6** *Prove Proposition 17.13. Towards this goal, follow the two steps below:*

(i) *Prove that $G - e - e'$ has a tight cut decomposition which produces a brick which is equal to a brick of $(G - e)/(X \to x)$, having precisely the same set of contraction vertices.*
(ii) *Prove that every brick of $G - e'$ has at least as many vertices as a brick of $G - e - e'$. Hint: for every cut D of G, if $D - e'$ is tight in $G - e'$ then $D - e - e'$ is tight in $G - e - e'$.*

17.3 The Existence of Thin Edges in Bricks ♯

In this section we shall present a proof of Theorem 17.3. By Theorem 15.1, any brick G different from K_4, $\overline{C_6}$ and $\mathbb{P}$ has a b-invariant edge e. If that edge e is not itself thin, our strategy, as explained earlier, is to show that there exists another b-invariant edge f of G such that $r(f) > r(e)$. We start with a few pertinent observations before we turn to a proof of that statement.

Thus suppose that e is a b-invariant edge of G that is not thin. Then the near-brick $G - e$ has at least one nontrivial maximal barrier that is not thin. Let B be one such barrier of $G - e$ and let s be a vertex of I that has degree greater than two in $G - e$. We have seen in the last section that at least one edge of $G - e$ incident with s is b-invariant in both G and $G - e$. Let e' be such an edge. Let $G_1 := G/(\overline{X} \to \overline{x})$ and $G_2 := G/(X \to x)$ denote the two C-contractions of G. The following facts are easily observed:

17.13.1 *The edge e' has both its ends in X, whereas the edge e has either one (Figure 17.9(a)) or two ends in I (Figure 17.9(b)). (The former case must occur when* $\text{index}(e) \in \{1, 2\}$ *and the latter case must occur when* $\text{index}(e) = 3$.*)* □

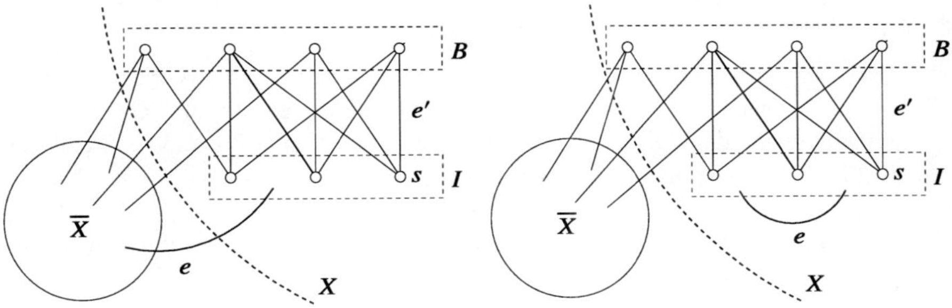

Fig. 17.9 The two b-invariant edges e and e'

As B is a barrier of $G - e$, then it certainly is a barrier of $G - e - e'$. Conversely, since the edge e' has an end in the barrier B of $G - e$, it follows that any barrier B' of $G - e - e'$ which contains B is also a barrier of $G - e$. Since B is a maximal barrier of $G - e$, it follows that B is a maximal barrier of $G - e - e'$. The following proposition records this observation:

Proposition 17.14 *The set B is a maximal barrier of $G - e - e'$.* □

The following lemma plays a crucial role in the proof of Theorem 17.3.

Lemma 17.15 *Let C' be a nontrivial cut of G such that $C' - e'$ is tight in $G - e'$. If C' does not cross C then X' is a proper subset of X.*

Proof The cut $C' - e'$ is tight in $G - e'$ but $C - e' = C$ is not. Thus, $C' \neq C$. As C' does not cross C, the possible containment relations between X and X' are: (i) $X' \subset \overline{X}$, (ii) $\overline{X} \subset X'$, (iii) $X \subset X'$, or (iv) $X' \subset X$. Our task is to show that (iv) holds. We do this by showing that the other possibilities cannot occur.

Edge e' has at least one end in X', but has no end in $\overline{X}$. Therefore (i) cannot hold. By Lemma 17.8, the graph $G_2 = G/(X \rightarrow x)$ is a brick, implying that $G[\overline{X}]$, which has neither e nor e' as an edge, is nonbipartite. But $G[X'] - e - e'$ is bipartite. Therefore (ii) does not hold either.

One end of the edge e' is in B and one end in I. Therefore, if X were a subset of X', an end of e' would be in B'. But this would mean that B', which is a nontrivial barrier of $G - e'$ is also a nontrivial barrier of the brick G itself. This is absurd. Thus, (iii) also cannot hold. □

We are now in position to prove the result concerning ranks that leads to a proof of Theorem 17.3.

THE RANK AUGMENTATION THEOREM FOR BRICKS

17.16 *Let e be a b-invariant edge in a brick G and suppose that e is not thin. Then G has a b-invariant edge f such that $r(f) > r(e)$.*

Proof As above, let s be a vertex in I which has degree at least three in $G - e$, and let e' be any edge incident with s which is b-invariant in both $G - e$ and G. We divide the proof of the assertion into two main cases.

Case 1 *For every maximal nontrivial barrier B' of $G - e'$, the corresponding set X' is a proper subset of X.*

By the Three Case Lemma, the brick of $G - e'$ is obtained from $G - e'$ by the contraction of proper subsets of X. As X is nontrivial and odd, the set of vertices of the brick of $G - e'$ includes not only $\overline{X}$, but also at least three more vertices. From this, by Proposition 17.10,

$$r(e') \geq r_0(e') \geq |\overline{X}| + 3 \geq r(e) + 2,$$

implying that $r(e') > r(e)$. Thus, in this case, the assertion of the Theorem holds, with $f := e'$. (See Example 17.17 for an illustration.)

Example 17.17 Consider the b-invariant edge $e = uv$ in the brick G shown in Figure 17.10. The index of e is one, $X = \{a_1, a_2, a_3, a_4, b_1, b_2, b_3, b_4, b_5\}$, and $r(e) = 4$. If e' is the edge a_4b_2, then its index is one, $\{a_1, a_2, a_3\}$ is the unique nontrivial maximal barrier of $G - e'$, and $X' = \{a_1, a_2, a_3, b_1, b_2\}$ is a proper subset of X. It can be verified that $r(e') = 8 > r(e)$. (Edge a_2b_2, a b-invariant edge of index two, is another edge which satisfies the hypothesis of Case 1.)

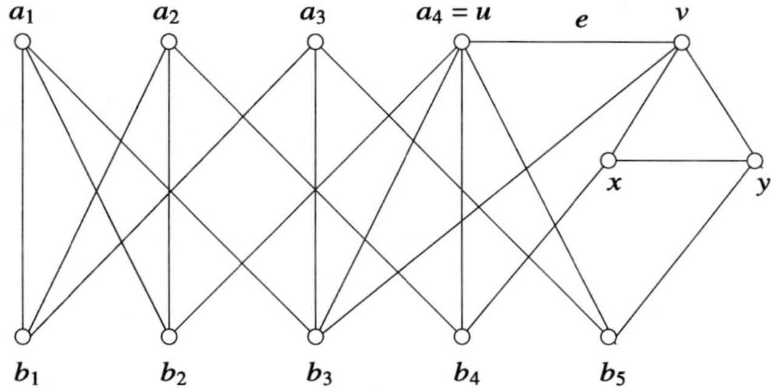

Fig. 17.10 Illustration for Case 1, Theorem 17.16

Case 2 *$G - e'$ has a maximal nontrivial barrier B' such that its corresponding set X' is not a proper subset of X.*

In this case, by Lemma 17.15, the corresponding cut C' crosses the cut C.

Lemma 17.18 *Let C' be a cut of G such that $C' - e'$ is tight in $G - e'$. Suppose that C and C' cross. Let $D := \partial(X - X')$ and let $D' := \partial(X' - X)$. Then, $X \cap X'$ and $\overline{X} \cap \overline{X'}$ are both even and the following properties hold:*

(i) *cut D is nontrivial and $D - e - e'$ is tight in $G - e - e'$, $X - X'$ is its bipartite shore, $B - X'$ the external part and $I - X'$ the internal part of $X - X'$,*
(ii) *cut $D' - e$ is tight in $G - e$, $X' - X$ is its bipartite shore, $B' - X$ the external part and $I' - X$ the internal part of $X' - X$, and*
(iii) *edge e' has one end in $I - X'$, the other in $B \cap I'$.*

(See Figures 17.12 and 17.13.)

Proof The cuts $C - e$ and $C' - e - e'$ are both tight in $G - e - e'$. Therefore the four subgraphs of $G - e - e'$ induced by $X, \overline{X}, X'$ and $\overline{X'}$ are connected. Thus, $G - e - e'$ has four edges $v_i w_i$, $1 \leq i \leq 4$, where $v_1 \in X \cap X'$, $w_1 \in X' - X$, $v_2 \in X - X'$, $w_2 \in \overline{X} \cap \overline{X'}$, $v_3 \in X \cap X'$, $w_3 \in X - X'$, $v_4 \in X' - X$ and $w_4 \in \overline{X} \cap \overline{X}'$. See Figure 17.11. The end in X of any edge in the cut $C - e$ is in B, and thus the two vertices v_1 and v_2 are in B. For a similar reason, the two vertices v_3 and v_4 are in B'.

17.18.1 *The two sets B and B' are disjoint.* □

Proof Suppose that $B \cap B' \neq \emptyset$. By Proposition 17.14, the set B is a maximal barrier of $G - e - e'$. Clearly, B' is a barrier of $G - e - e'$. Thus, our assumption implies that $B' \subseteq B$. This is a contradiction, because B is a subset of X, whereas B' contains the vertex v_4 in $X' - X$. □

Let us now prove that $X \cap X'$ is even. Suppose that $X \cap X'$ is odd. As $C - e$ and $C' - e - e'$ are tight cuts of $G - e - e'$, then, by Theorem 4.12, cuts $\partial(X \cap X') - e - e'$ and $\partial(\overline{X} \cap \overline{X'}) - e - e'$ are also tight cuts in $G - e - e'$, and each shore of these tight cuts induces a connected subgraph of $G - e - e'$. The subgraphs of $G - e - e'$ induced

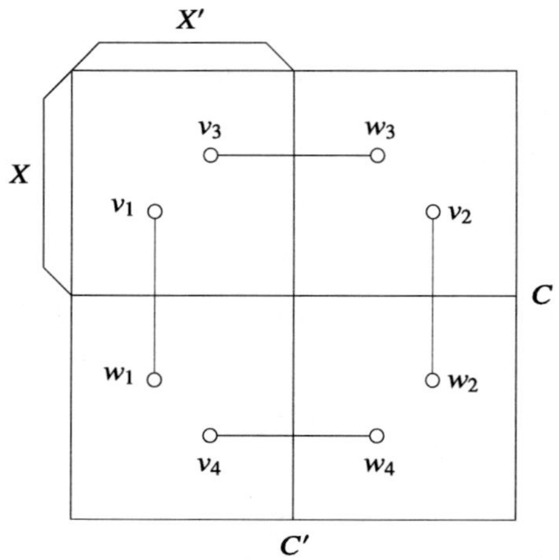

Fig. 17.11 The four edges $v_i w_i$, $1 \le i \le 4$

by X and by X' are both bipartite; so too is $(G-e-e')[X \cap X']$. Moreover, as $(G-e-e')[X \cap X']$ is connected, its bipartition is unique. Thus, $\{B \cap X', I \cap X'\} = \{B' \cap X, I' \cap X\}$.

The vertex v_1 is in B and also in the external part of $X \cap X'$. Thus, the external part of $X \cap X'$ is $B \cap X'$. Likewise, v_3 is in B' and also in the external part of $X \cap X'$. Thus, the external part of $X \cap X'$ is $B' \cap X$. We conclude that $B \cap X' = B' \cap X$. This is a contradiction, because B and B' are disjoint by statement 17.18.1. We conclude that $X \cap X'$ is even. It follows that $\overline{X} \cap \overline{X'}$ is also even. By Theorem 4.12, $D - e - e'$ and $D' - e$ are tight cuts of $G - e - e'$. See Figure 17.12.

Clearly, $X - X'$ induces a bipartite connected subgraph of $G - e - e'$. Thus, $\{B - X', I - X'\}$ is the bipartition of $(G - e - e')[X - X']$. The vertex v_2 is in B and in the external part of $X - X'$. Thus, $B - X'$ is the external part of $X - X'$, and $I - X'$ is its internal part. Likewise, $X' - X$ is the bipartite shore of $D' - e$. The set $B' - X$ is the external part of $X' - X$, and $I' - X$ is its internal part.

To complete the proof, recall that edge e' has its end s in I, and its other end, say t', in B. We wish to show that s is in $I - X'$. Suppose, to the contrary, that s is in X'. The cut C' is nontrivial and $C' - e'$ is tight in $G - e'$; hence $s \notin B'$; otherwise C' is tight in G. It follows that $s \in I'$. Then, $s \in I \cap I'$. Let f be any edge of $G - e - e'$ incident with s. The end of f distinct from s is in $B \cap B'$, a contradiction, as B and B' are disjoint by statement 17.18.1. It follows that $s \in X - X'$, and hence that $s \in I - X'$. As $I - X'$ is the internal part of $X - X'$, we deduce that D is nontrivial. The end t' of e' must be in B. Moreover, at least one end of e' is in I'. We conclude that t' is in $B \cap I'$. □

We have assumed that $G - e'$ has a maximal barrier B' whose correspoding cut C' crosses C. So, we shall adopt the notation introduced in the statement of Lemma 17.18. In particular, D' denotes the cut $\partial(X' - X)$.

Case 2.1 The cut D' is trivial. See Figure 17.12.

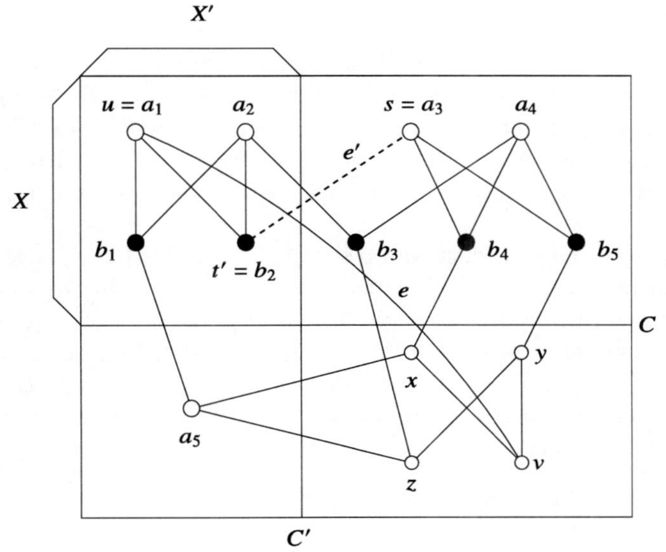

Fig. 17.12 Illustration for Case 2.1, Theorem 17.16.

As $\overline{X} \cap X'$ is trivial, it follows that $|\overline{X}| = |\overline{X} \cap \overline{X'}| + 1$. The set X corresponds to the maximal nontrivial barrier B of $G - e$. By Proposition 17.10,

$$r(e) \leq |\overline{X}| + 1 = |\overline{X} \cap \overline{X'}| + 2. \tag{17.3}$$

As $X \cap \overline{X'}$ is nontrivial, $|\overline{X'}| > |\overline{X} \cap \overline{X'}| + 1$. The set X' corresponds to the maximal nontrivial barrier B' of $G - e'$. Suppose that the index of e' is not equal to two. By Proposition 17.10,

$$r(e') = |\overline{X'}| + 1 > |\overline{X} \cap \overline{X'}| + 2.$$

It then follows, from (17.3), that $r(e') > r(e)$.

We may thus assume that e' has index two. Let B'' be the maximal nontrivial barrier of $G - e'$ distinct from B'. One of the ends of e' is in I', the other is in I''. The end t' of e' is in I'; hence the end s of e' is in I''. The vertex s is in I; hence it is in $I \cap I''$. It follows that the neighbours of s in $G - e - e'$ are in $B \cap B''$; hence $B'' \cap B$ is nonempty. By Proposition 17.14, B is a maximal barrier of $G - e - e'$. The set B'', a barrier of $G - e'$, is also a barrier of $G - e - e'$. As B and B'' are not disjoint, we conclude that $B'' \subseteq B$. The set of isolated vertices of $G - e - e' - B$ coincides with the set I of isolated vertices of $G - e - B$. Thus, the set of isolated vertices of $G - e - e' - B''$ is a subset of I; hence $I'' \subseteq I$. Consequently, $X'' \subseteq X$; hence the second contraction is of the set $X'' - X'$, a subset of $X - X'$. We deduce that $r_0(e') \geq |\overline{X} \cap \overline{X'}| + 2$. As e' has index two, we have that

$$r(e') > |\overline{X} \cap \overline{X'}| + 2. \quad (17.4)$$

From (17.4) and (17.3), we conclude that $r(e') > r(e)$. (See Example 17.19 for an illustration.)

Example 17.19 Consider the brick G depicted in Figure 17.12. The edge $e = uv$ is a b-invariant edge of index two of G and the sets $B_1 := \{b_1, b_2, b_3, b_4, b_5\}$ and $B_2 := \{a_3, a_4, x, y\}$ are the two nontrivial maximal barriers of $G - e$. The brick of $G - e$ is obtained by shrinking $X_1 = \{b_1, b_2, b_3, b_4, b_5, a_1, a_2, a_3, a_4\}$ to a single vertex, and the set $X_2 = \{x, y, v\}$ to a single vertex. Thus e is of index two, and $r(e) = 5$.

The edge $e' = a_3b_2$ incident with the vertex $s = a_3$ in I_1 is an edge which is b-invariant in both $G-e$ and G. The index of e' is two and $B' = \{a_1, a_2, a_5\}$ is one of the two maximal barriers of $G - e'$ with $I' = \{b_1, b_2\}$ and $X' = \{a_1, a_2, a_5, b_1, b_2\}$. The end $t' = b_2$ of e' is in $B \cap I'$. The second maximal barrier of $G - e'$ is $B'' = \{b_3, b_4, b_5\}$ with $I'' = \{a_3, a_4\}$ and $X'' = \{b_3, b_4, b_5, a_3, a_4\}$. The brick of $G - e'$ has two contraction vertices in addition to the four vertices x, y, z and v of the set $\overline{X} \cap \overline{X'}$. Thus $r(e') = 7$.

Case 2.2 The cut D' is nontrivial. See Figure 17.13.

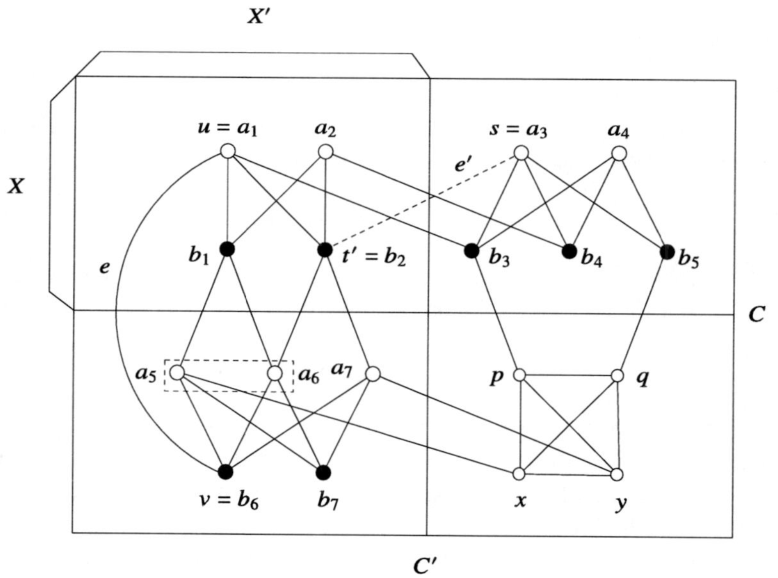

Fig. 17.13 Illustration for Case 2.2, Theorem 17.16. Both e and e' are b-invariant edges of index two with equal ranks. It can be verified that if M is any perfect matching of $G - \{a_5, a_6\}$, then $\{e, e'\}$ is a subset of M. (Thus a_5 and a_6 play the roles of w_1 and w_2 in the analysis of Case 2.2.)

17.19.1 *The edge e has an end, v, in the internal part $I' - X$ of $X' - X$. Consequently, e has index two.*

Proof The cut $D' - e$ is a nontrivial tight cut of $G - e$; hence e has an end, v, in the internal part $I' - X$ of $X' - X$. This implies that $B' - X$ is a nontrivial barrier of $G - e$ disjoint with B. Consequently, e has index two. □

All vertices adjacent to v in $G - e$ are in the external part $B' - X$ of $X' - X$. Since $B' - X$ is nontrivial, vertex v, which is in the internal part of $X' - X$, has at least two neighbours, say w_1 and w_2, in $B' - X$. Let M be a perfect matching of $G - \{w_1, w_2\}$. Such a matching exists because G is a brick. A simple counting argument implies the following statement:

17.19.2 $\{e, e'\} \subset M$. □

17.19.3 *There exist at least two edges incident with s which are b-invariant in both G and $G - e$.*

Proof By definition of s, its degree in $G - e$ is at least three. The assertion follows from the fact that the index of e is two, by Theorem 17.12. □

Let $e'' \neq e'$ be an edge incident with s which is b-invariant in both G and $G - e$. We may now repeat the preceding analysis with e'' playing the role of e'. If, for every nontrivial barrier B'' of $G - e''$ the corresponding set X'' is a proper subset of X then Case 1 applies and we would have $r(e'') > r(e)$. Thus, we may assume that $G - e''$ has a maximal nontrivial barrier B'' such that the corresponding set X'' is not a proper subset of X.

Thus, Case 2 applies and the corresponding cut C'' crosses the cut C. If the cut $D'' := \partial(X'' - X)$ is trivial, then Case 2.1 applies and we would again have $r(e'') > r(e)$. Thus, to conclude that $r(e'') > r(e)$, it suffices to show that the cut D'' of G is trivial.

Suppose that D'' is not trivial. In this case, the end v of e is in $I'' - X$; hence all the neighbours of v in $G - e$ are in $X'' - X$. In particular, the vertices w_1 and w_2 are in $X'' - X$ (as well as in $X' - X$). As above, let M be any perfect matching of $G - \{w_1, w_2\}$. The same counting argument which implies that $e' \in M$ also implies that $e'' \in M$. But this is absurd because e' and e'' are distinct edges which are both incident with s and M is a matching. It follows that in the case of edge e'', one of the first two alternatives must occur. The proof of the rank augmentation theorem for bricks is now complete. (See Example 17.20 for an illustration. In that illustration the two vertices a_5 and a_6 play the roles of w_1 and w_2.) □

Example 17.20 Let G be the brick shown in Figure 17.13. The edge $e = uv$ is a b-invariant edge of index two of G and $B = \{b_1, b_2, b_3, b_4, b_5\}$ is a maximal barrier of $G - e$, with $I = \{a_1, a_2, a_3, a_4\}$ and $X = \{b_1, b_2, b_3, b_4, b_5, a_1, a_2, a_3, a_4\}$. The edge $e' = a_3b_2$, incident with $a_3 \in I$ is an edge that is b-invariant in both G and $G - e$, and $B' = \{a_1, a_2, a_5, a_6, a_7\}$ is a maximal barrier of $G - e'$, with $I' = \{b_1, b_2, b_6, b_7\}$ and $X' = \{a_1, a_2, a_5, a_6, a_7, b_1, b_2, b_6, b_7\}$.

The cuts C and C' cross, and $B' - X = \{a_5, a_6, a_7\}$ is a maximal barrier of $(G - e)/X$. Thus the index of e is two and it is easy to see that $r(e) = 7$. The edge e' is also of index two because $\{b_3, b_4, b_5\}$, which is disjoint from B', is a maximal barrier of $(G - e')/X'$. Thus $r(e') = r(e) = 7$. It is not difficult to see that any perfect matching M of $G - \{a_5, a_6\}$ must contain both e and e'.

There are edges, distinct from e' which are also incident with a_3 and are b-invariant in both G and $G - e$. For example, a_3b_4 and a_3b_5 are two such edges. The graph $G - a_3b_4$ has one maximal nontrivial barrier, namely $\{a_2, a_4\}$. The bipartite shore $\{a_2, b_4, a_4\}$ of the tight cut associated with this barrier is a subset of X. So Case 1 applies, and it can be seen that $r(a_3b_4) = 16$. The graph $G - a_3b_5$ also has just one nontrivial barrier, namely $\{a_4, q\}$. The tight cut associated with this this barrier crosses C and the readers may verify that here Case 2.1 applies and that $r(a_3b_5) = 16$.

Proof (of Theorem 17.3) Let G be a brick distinct from the three special bricks. By Theorem 15.1, G has b-invariant edges. By Theorem 17.16, a b-invariant edge of G having maximum rank is thin. □

In fact, the above proof of the existence of thin edges in bricks may be adapted to actually prove the following:

Corollary 17.21 *Let G be a brick, let e be a b-invariant edge of G and let H be the underlying simple graph of any brick of $G - e$. If e is not thin then G has a b-invariant edge e' such that $r(e') > r(e)$, e' is b-invariant in $G - e$ and H is the underlying simple graph of any brick of $G - e - e'$.* □

Corollary 17.22 *Let G be a brick, let e be a b-invariant edge of G and let J be a simple brick. If J is a matching minor of $G - e$ then G has a thin edge f such that J is a matching minor of $G - f$.* □

Exercises

▷**17.3.1** *Prove Corollary 17.21.*

▷**17.3.2** *Prove Corollary 17.22.*

17.4 Building Bricks

In this section we give precise definitions of the four operations alluded to in the introduction. We shall refer to these four operations as *expansion operations*. Given any brick H, any one of these operations may be used to produce a new brick G from H such that $|E(G)| > |E(H)|$.

Three of the four expansion operations involve the device of bisplitting vertices which was introduced in Chapter 12. We restate the definition of bisplitting here and introduce the notation, terminology and properties that we shall be using.

BISPLITTING A VERTEX (REVISITED) AND CONSTRAINED BISPLITTING

Let G be a matching covered graph and let x be a vertex of G. The operation of *bisplitting* the vertex x consists of first replacing x by two vertices b_1, b_2 and distributing the edges in $\partial(x)$ between b_1 and b_2 so that each of these two vertices is incident with at least one edge in $\partial(x)$, and then adding a vertex a and joining a to b_1 and to b_2. We shall refer to b_1 and b_2 as the two *new outer vertices* and a as the *new inner vertex*.

In a special type of bisplitting, called a *constrained bisplitting*, we require that each outer vertex have at least three distinct neighbours.

We denote by $G\{x \to (a, b_1, b_2)\}$ the graph resulting from a constrained bisplitting of vertex x of a graph G, having a as the inner vertex, and b_1 and b_2 as the two outer vertices. As noted in Chapter 12, it is easy to check that $H := G\{x \to (a, b_1, b_2)\}$ is matching covered and that G can be recovered from H by bicontracting a. Furthermore, $b(H) = b(G)$.

We shall use the following property in the analysis of the four operations. Its proof is left as Exercise 17.4.1.

Proposition 17.23 *Let G be a matching covered graph, let x be a vertex of G, let $H := G\{x \to (a, b_1, b_2)\}$, let B denote a maximal barrier of H and let $X := \{a, b_1, b_2\}$. Then:*

(i) *if B and X are disjoint then B is a barrier of G,*
(ii) *if B and $\{b_1, b_2\}$ have a vertex in common then $\{b_1, b_2\} \subseteq B$ and the set $(B - b_1 - b_2) + x$ is a barrier of G, and*
(iii) *if $a \in B$ and $B \neq \{a\}$ then the vertices b_1 and b_2 are in distinct components of $H - B$ and $B - a$ is a barrier of G.* □

17.4.1 The four expansion operations

The simplest of the four operations is the one defined below:

1. Edge addition: Let u and v be two distinct vertices of a brick H. Obtain G from H by adding a new edge e joining u and v. It is easy to see that G is a brick.

The three remaining operations are more complicated and involve the device of constrained bisplitting vertices. Theorem 17.25 shows that an application of any one of these operations on a brick H results in a brick G of higher order.

2. Expansion of a vertex by a barrier of size two: Let H be a brick in which x is a vertex of degree four or more, and let

$$H_1 := H\{x \to (a, b_1, b_2)\}. \tag{17.5}$$

Now obtain G from H_1 by joining the new inner vertex a to a vertex w of $H - x$ by an edge labelled e (see Figure 17.14). We allow the possibility of w and x being adjacent in G. We shall refer to the graph G thus constructed as a graph obtained from H by *an expansion of x by a barrier of size two.*

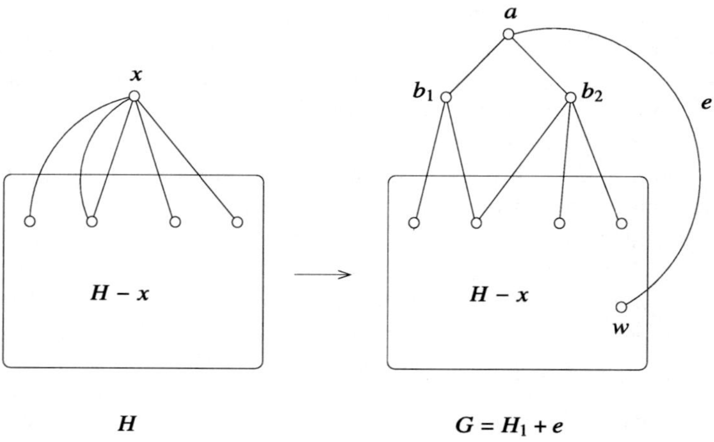

Fig. 17.14 Expansion of a vertex x by a barrier of size two

3. Expansion of two vertices by barriers of size two: Let H be a brick in which x and x' are two distinct (possibly adjacent) vertices of degree at least four, and let

$$H'_2 := H\{x \to (a, b_1, b_2)\} \quad \text{and} \quad H_2 := H'_2\{x' \to (a', b'_1, b'_2)\}. \tag{17.6}$$

We require that each of the vertices b_1, b_2, b'_1 and b'_2 must have at least one neighbour in $V(H) - x - x'$.

Now obtain G from H_2 by joining the two vertices a and a' by an edge labelled e (see Figure 17.15). We shall refer to the graph G thus constructed as a graph obtained from H by *expansions of x and x' by barriers of size two.*

4. Expansion of a vertex by a barrier of size three: Let H be a brick containing a vertex x of degree at least five, and let

$$H'_3 := H\{x \to (a_1, b_1, t)\} \quad \text{and} \quad H_3 := H'_3\{t \to (a_2, b_2, b_3)\}. \tag{17.7}$$

Naturally, we require t to be adjacent in H'_3 to three or more vertices in $V(H) - x$. We adjust notation so that b_2 is the vertex in $\{b_2, b_3\}$ which is adjacent to a_1.

Now obtain G from H_3 by joining a_1 and a_2 by an edge labelled e (see Figure 17.16). We shall refer to the graph G thus constructed as a graph obtained from H by *an expansion of x by a barrier of size three.*

We have already noted that adding any edge to a given brick H results in a brick G with one more edge. Now we turn to the task of showing that any application of the three above defined operations of expanding vertices of a given brick H results in a brick G of larger order.

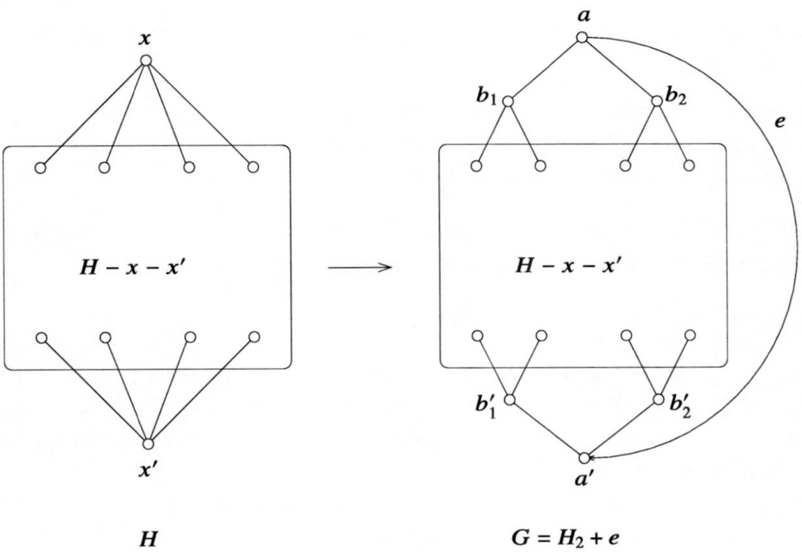

Fig. 17.15 Expansion of two vertices by barriers of size two

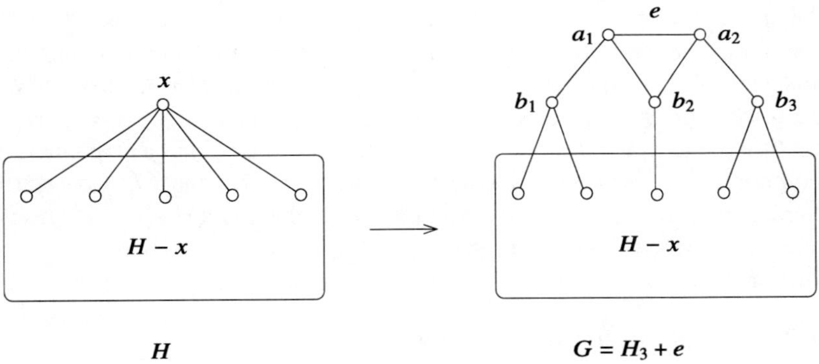

Fig. 17.16 Expansion of a vertex by a barrier of size three

In the ensuing discussion H_1, H_2 and H_3 denote, respectively, the graphs defined in equations (17.5), (17.6), and (17.7).

Our interest is in showing that, for $i = 1, 2, 3$, the graph $G = H_i + e$ is a brick. Towards this end, it is clearly necessary for us to be able to assert that edge e is matchable in G. By Corollary 2.2, e is matchable in G if and only if no barrier of H_i contains both the ends of e. With this in view, we proceed to determine the list of all maximal nontrivial barriers of H_i.

Lemma 17.24 *The following statements concerning barriers in the three graphs H_1, H_2 and H_3 hold:*

(i) *the only nontrivial barrier in H_1 is $\{b_1, b_2\}$;*
(ii) *the only two nontrivial barriers in H_2 are $\{b_1, b_2\}$ and $\{b'_1, b'_2\}$; and*
(iii) *the only nontrivial maximal barrier in H_3 is $\{b_1, b_2, b_3\}$.*

Proof Let B denote a maximal barrier of H_i, where $i \in \{1, 2, 3\}$. We consider the three cases separately, but their proofs have many common features.

Case 1 *The graph under consideration is H_1.*

Let $X_1 := \{a, b_1, b_2\}$ and let $B_1 := \{b_1, b_2\}$. We consider separately three cases, (i) B and X_1 are disjoint, (ii) B and B_1 are not disjoint and (iii) $a \in B$. In each of these three possibilities, we apply Proposition 17.23, **with H playing the role of G and H_1 playing the role of H.**

Suppose that B and X_1 are disjoint. Then, by Proposition 17.23(i), B is s a barrier of H. As H, a brick, is bicritical, B is trivial.

Suppose that B contains a vertex in B_1. Then, by Proposition 17.23(ii), $B_1 \subseteq B$ and $B' := (B - B_1) + x$ is a barrier of H. Again, as H is bicritical, we infer that $B' = \{x\}$; hence $B = \{b_1, b_2\}$.

Finally, suppose that $a \in B$. Let us prove that $B = \{a\}$. Suppose that $B'' := B - a$ is nonempty. Then, by Proposition 17.23(iii), the vertices b_1 and b_2 are in distinct components of $H_1 - B$, respectively, J_1 and J_2. Moreover, B'' is a barrier of H. As H is a brick, then B'' is a singleton; let w be its only vertex. Thus, $B = \{a, w\}$. The vertex b_1 is adjacent to three or more vertices of H_1. Thus, b_1 has a neighbour, c_1, in the component J_1. Likewise, b_2 has a neighbour, c_2, in component J_2. Consequently, the vertices c_1 and c_2 are in distinct components of the graph $H - w - x$. This is not possible, as H, a brick, is 3-connected.

We conclude that B_1 is the only nontrivial barrier of H_1.

Case 2 *The graph under consideration is H_2.*

Let $X_2 := \{a, b_1, b_2\}$, let $B_2 := \{b_1, b_2\}$, let $X'_2 := \{a', b'_1, b'_2\}$ and let $B'_2 := \{b'_1, b'_2\}$. By definition of H'_2 (equation (17.6)), the outer vertices b_1 and b_2 have degree three or more in H'_2. By Case 1, B_2 is the only nontrivial barrier of H'_2.

We consider separately three cases: (i) $B \cap X_2 = \emptyset$ or $B \cap X'_2 = \emptyset$, (ii) $B \cap (B_2 \cup B'_2) \neq \emptyset$, and (iii) $\{a, a'\} \subseteq B$. In each of these three possibilities, we apply Proposition 17.23, **with H'_2, X'_2 and H_2 playing respectively the role of G, X and H.**

Suppose that $B \cap X_2 = \emptyset$ or $B \cap X'_2 = \emptyset$. The definition of H_2 is symmetric; without loss of generality suppose that B and X'_2 are disjoint. By Proposition 17.23(i), B is a barrier of the graph H'_2. We have seen that B_2 is the only nontrivial barrier of H'_2. Thus, either B is a singleton or $B = B_2$.

Suppose now that B and $B_2 \cup B'_2$ are not disjoint. Adjust notation so that B and B'_2 are not disjoint. By Proposition 17.23(ii), $B'_2 \subseteq B$ and the set $B' := (B - B'_2) + x'$ is a barrier of H'_2. We have seen that B_2 is the only nontrivial barrier of H'_2. The vertex x' is in B' but is not in B_2; hence B' is trivial. That is, $B' = \{x'\}$, hence $B = B'_2$.

We may thus assume that B meets both X_2 and X_2', but it is disjoint with $B_2 \cup B_2'$. Thus, $\{a, a'\} \subseteq B$. We shall prove that this is not possible. By Proposition 17.23(iii) the set $B'' := B - a'$ is a barrier of H_2'. We have seen that B_2 is the only nontrivial barrier of H_2'. Thus, $B'' = \{a\}$; hence $B = \{a, a'\}$. Also by Proposition 17.23(iii), the vertices b_1' and b_2' are in distinct components of $H_2 - B$. For $k = 1, 2$, let J_k be the component of $H_2 - B$ that contains the vertex b_k'. By definition of the expansion, the vertex b_k' is adjacent in J_k to a vertex, c_k, which is not in $\{b_1, b_2\}$. Thus, the graph $H_2 - X_2 - X_2'$ is not connected. Consequently, the graph $H - x - x'$ is not connected. This is a contradiction, as H, a brick, is 3-connected.

We conclude that $\{b_1, b_2\}$ and $\{b_1', b_2'\}$ are the only two nontrivial barriers of H_2.

Case 3 *The graph under consideration is* H_3.

By definition of H_3' (equation (17.7)), the vertices b_1 and t have three or more neighbours in H_3'. Thus, Case 1 applies to H_3'; hence $\{b_1, t\}$ is the only nontrivial barrier of H_3'.

Let $B_3 := \{b_1, b_2, b_3\}$. In order to prove that B_3 is the only maximal nontrivial barrier of H_3, we shall consider three alternatives: (i) $B \cap B_3 \neq \emptyset$, (ii) either $a_1 \notin B$ or $a_2 \notin B$, and (iii) $\{a_1, a_2\} \subseteq B$. In possibilities (ii) and (iii) we shall apply Proposition 17.23, **with** H_3', $\{a_2, b_2, b_3\}$ **and** H_3 **playing respectively the role of** G, X **and** H.

Suppose that B and B_3 are not disjoint. By the maximality of B and the Canonical Partition Theorem, $B_3 \subseteq B$. The vertices a_1 and a_2 are isolated in $H_3 - B$. Thus, $B' := (B - B_3) + x$ is a barrier of H. As H is a brick, $B' = \{x\}$; hence $B = B_3$.

We may thus assume that B and B_3 are disjoint. Consider now the case in which at least one of the vertices a_1 and a_2 is not in B. By the symmetry of the definition of H_3, we may assume, without loss of generality, that $a_2 \notin B$. As B and B_3 are disjoint, it follows that B and $\{a_2, b_2, b_3\}$ are disjoint. By Proposition 17.23(i), B is a barrier of H_3'. We have seen that $\{b_1, t\}$ is the only nontrivial barrier of H_3'. We have assumed that B and B_3 are disjoint; thus the vertex b_1, a vertex of B_3, is not in B. We conclude that B is trivial in this case.

Finally, suppose that $\{a_1, a_2\} \subseteq B$. We shall prove that this is not possible. By Proposition 17.23(iii), $B - a_2$ is a barrier of H_3'. The only nontrivial barrier of H_3' is $\{b_1, t\}$. As $b_1 \notin B$, we have that $B - a_2 = \{a_1\}$, hence $B = \{a_1, a_2\}$. Also, by Proposition 17.23(iii), the vertices b_2 and b_3 are in distinct components of $H_3 - B$. By symmetry, the vertices b_1 and b_2 are in distinct components of $H_3 - B$. We deduce that one of the two components of $H_3 - B$ contains the vertices b_1 and b_3, whereas the other, say K, contains the vertex b_2. By definition of H_3, the vertex b_2 has three or more neighbours in H_3. Thus, K is nontrivial, hence b_2 is a cut vertex of H_3. This is not possible, as H_3 is matching covered.

We conclude that B_3 is the only maximal nontrivial barrier of H_3. □

Theorem 17.25 *Let H be a brick and let G be a graph obtained from H by one of the four operations of expansion. Then G is a brick.*

Proof As we have already noted, any graph G which is obtained from H by simply adding an edge is a brick. Let us suppose that G is obtained from H by applying one of the three remaining expansion operations. The retract of $G - e$ is the brick H; hence $G - e$ is a near-brick.

By Lemma 17.24, in each case, no end of e belongs to a barrier of $G - e$. Consequently, G is matching covered. Furthermore, by definition of $G - e$ and G, every vertex has at least three distinct neighbours in G.

By the monotonicity of function b (see Theorem 9.8), the graph G is a near-brick. We shall prove that G is bicritical. The required assertion would then follow because every bicritical near-brick is a brick (see Corollary 4.19).

If possible suppose that B is a maximal nontrivial barrier of G. As G is a near-brick, B is special, by Proposition 17.6. If $|B| = 2$ then, as every vertex of G is adjacent to three or more vertices, we deduce that B is not special, a contradiction. We conclude that $|B| \geq 3$.

Since $G-e$ is a spanning matching covered subgraph of G, it follows that B is also a nontrivial barrier of $G - e$. By Lemma 17.24, G must have been obtained from H by the expansion of a vertex by a barrier of size three and $B = \{b_1, b_2, b_3\}$. The graph G is matching covered, B is a barrier of G, yet $G - B$ has an even component, induced by the pair $\{a_1, a_2\}$.

In all cases considered, the assumption that B is a maximal nontrivial barrier of G leads to a contradiction. We conclude that G is bicritical. Being a bicritical near-brick, G is a brick. □

We remark that the above proof also shows that, in every case, the edge e is a thin edge of the brick G and that H is the brick of $G - e$. As a complement to the above result, we then state the following theorem, which is easily proved. (Exercise 17.4.3).

Theorem 17.26 *Every brick H different from the three basic bricks can be obtained from one of them by a sequence of applications of the four operations of expansion.* □

See Figure 17.17 for an illustration. The graph H is obtained from the pentagonal prism by adding an edge. It is a plane brick with two odd faces, and hence it is K_4-free. (See Theorem 12.19.) By expanding the two vertices x and y in H into barriers of size two, we obtain G. (Note that G is a redrawing of the trellis depicted in Figure 12.10.) Conversely, by deleting the thin edge u_1u_5 from G and shrinking $\{u_2, u_1, v_1\}$ to x and $\{u_4, u_5, v_5\}$ to y we recover H.

A natural question that arises at this stage is whether all the four operations are really required for building bricks. This is indeed the case. (Exercise 17.4.7.)

$(b + p)$-invariant thin edges

Let G be a brick that is not a Petersen brick. If G can be obtained from a Petersen brick by means of an expansion operation, then it can also be obtained from a brick that is not a Petersen brick by some expansion operation. This amounts to saying that if a brick G is not a Petersen brick and has a thin edge e such that $\widehat{G - e}$ is a Petersen

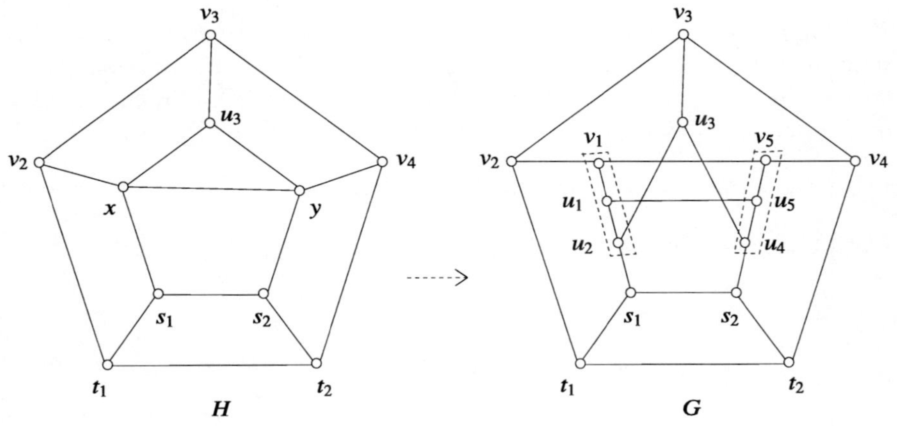

Fig. 17.17 Obtaining the trellis G from the graph H by expanding x and y into barriers of size two.

brick, then it also has a thin edge f such $\widehat{G-f}$ is not a Petersen brick. (Some cases of this assertion are illustrated in Figure 17.5.)

Theorem 17.27 *Every brick G that is not a Petersen brick and is distinct from K_4 and $\overline{C_6}$ has a thin edge e such that the brick of $G-e$ is not a Petersen brick.* □

This theorem can be verified directly by straightforward case analysis similar to the one used in the proof of Corollary 15.19. We leave it to the readers to work out the details (Exercise 17.4.6).

———≀≀———

In view of Theorem 17.27, every brick that is not a Petersen brick may be obtained from K_4 and $\overline{C_6}$ by a sequence of applications of the four operations of expansion. This is the sense in which any brick may be generated from one of the two bricks mentioned above, or, conversely, any brick may be reduced to one of them by deletions of thin edges and bicontractions!

CLM (2006, [15]) used Theorem 17.27 as an inductive tool to show that the only simple solid planar bricks are the wheels. A simpler proof of this result was established by Kothari and Murty (2015, [50]) using the idea of 'strictly thin edges' which will be introduced in the next chapter.

Exercises

▷**17.4.1** *Prove Proposition 17.23. Hint: in the proof of the last part of the assertion, if b_1 and b_2 are in the same component, K, of $H-B$, then the subgraph of G induced by $V(K)+a$ is an even component of $G-(B-a)$.*

▷**17.4.2** *In the definition of the operation of expansion of two vertices by barriers of size two, there is a requirement that "each of b_1, b_2, b'_1 and b'_2 must have at least three neighbours, and at least one neighbour in $V(H - x - x')$." Explain why this requirement is necessary for the resulting graph G to be a brick.*

17.4.3 *Prove Theorem 17.26. Hint: use induction on $|E|$.*

17.4.4 *Generate all the cubic bricks on four, six, and eight vertices.*

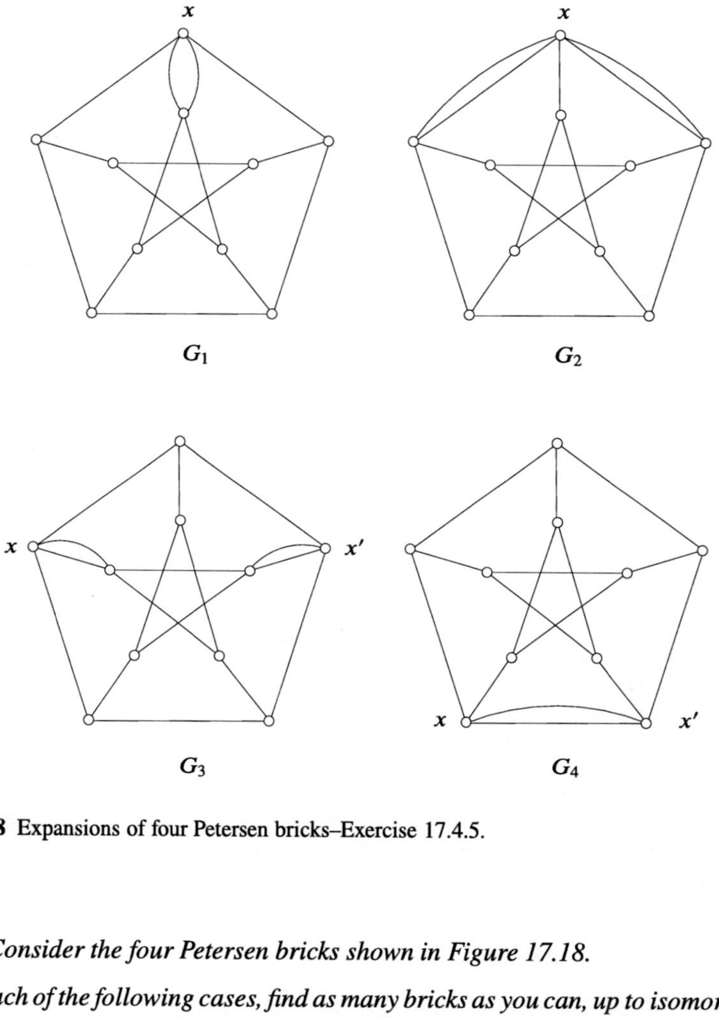

Fig. 17.18 Expansions of four Petersen bricks–Exercise 17.4.5.

▷**17.4.5** *Consider the four Petersen bricks shown in Figure 17.18.*

(i) *In each of the following cases, find as many bricks as you can, up to isomorphism, that can be obtained by means of the indicated expansion operations:*

(a) *expansion of vertex x in G_1 by a barrier of size two (four bricks);*
(b) *expansion of vertex x in G_2 by a barrier of size three (two bricks);*
(c) *expansion of vertices x and x′ in G_3 and G_4 by barriers of size two (respectively, one and two bricks).*

(ii) *For each brick G obtained by an expansion operation as in the first part, show that G has a thin edge f such that the retract of $G - f$ is a brick that is not a Petersen brick. (See Exercise 17.1.8.)*

▷**17.4.6** *Prove Theorem 17.27.*

▷**17.4.7** *Verify that all the four operations are necessary for the generation of all bricks starting with the three basic bricks. For each of the four operations we give below a hint:*

(i) *edge addition: consider any brick having minimum degree four or more;*
(ii) *expansion of a vertex by a barrier of size two: consider the odd wheels of order six or more;*
(iii) *expansion of a vertex by a barrier of size three: consider the tricorn $\mathbb{H}_{10}$, shown in Figure 17.1(c);*
(iv) *expansion of two vertices by barriers of size two: consider triangle-free cubic bricks distinct from the Petersen graph.*

17.5 Thin Edges in Braces

THIN EDGES IN BRACES AND THEIR INDICES

The *index* of a removable edge e in a brace G is the number of ends of e having degree two in $G - e$. If G has order at least six, it is easy to verify that the index of e is the number of contraction vertices in the retract $\widehat{G - e}$ of $G - e$.

A removable edge e in a brace G is *thin* if the retract of $G - e$ is also a brace. Any removable edge in a brace of order two or four is a multiple edge and is clearly thin. Every edge in a brace of order six or more is removable (Theorem 8.3), but not all edges are necessarily thin. For example, in the brace G of order 12 shown in Figure 17.19, edge f is one of the ten edges that are not thin; the edges e_0, e_1 and e_2 in this brace are three of the nine edges which are thin, and their indices are zero, one and two, respectively.

In this section we develop a theory pertaining to thin edges in braces similar to the one related to thin edges in bricks described in the previous sections. The principal result (Theorem 17.34) we establish states that every brace distinct from K_2 and C_4 has at least two thin edges. A consequence of this assertion is that any brace may be obtained from either K_2 or C_4 by sequentially applying suitable expansion operations much the same way any brick may be obtained from one of K_4, $\overline{C_6}$ and $\mathbb{P}$.

Most of the ideas needed for establishing the existence of thin edges in braces are similar to the ones that arose in connection with the proof the existence of thin edges in bricks. One crucial difference arises in connection with the selection of shores of tight cuts. For example, if e is a b-invariant edge of a brick G, the graph $G - e$ is a near-brick, and one of the shores of a nontrivial tight cut C of $G - e$ is bipartite and

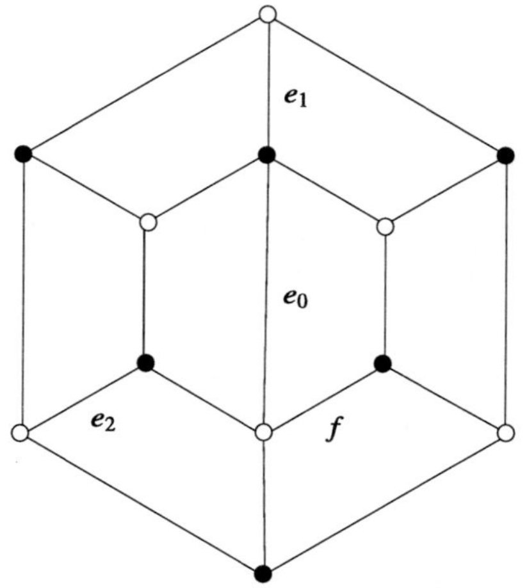

Fig. 17.19 Examples of thin and nonthin edges in a brace: edges e_0, e_1 and e_2 are thin; edge f is not thin.

the other is nonbipartite; hence the brick of $G - e$ may be obtained by a tight cut decomposition that starts by contracting the bipartite shore of C. However, when e is a removable edge of a brace G, the graph $G - e$ is a bipartite matching covered graph, and both shores of any nontrivial tight cut C of $G - e$ are bipartite, and it is not clear which C-contraction of $G - e$ yields the largest brace of $G - e$.

Another important distinction is in the structure of the bipartite matching covered graphs obtained from a brace by the deletion of an edge. We shall address these differences first before turning to the proof of the existence of thin edges in braces.

17.5.1 Graphs obtained by the deletion of an edge from a brace

Let G be a brace on six or more vertices, and let $e := uv$ be an edge of G. Then, by Theorem 8.3, $G - e$ is matching covered. We shall now establish the structure of a tight cut decomposition of $G - e$.

Tight cut decompositions of $G - e$

Suppose that $G - e$ is not a brace. Then, the ends u and v of e belong to different shores of any nontrivial tight cut of $G - e$. Let $\mathcal{C}$ be the (laminar) family of tight cuts used by a tight cut decomposition of $G - e$. As indicated in Exercise 8.1.10, the shores of the cuts in $\mathcal{C}$ containing u form a nested family $X_1 \subset X_2 \subset \cdots \subset X_k$ of proper subsets of V (see Figure 17.20), each containing u, such that

$(G-e)/\overline{X_1}$ is a brace,
$(G-e)/X_i/\overline{X_{i+1}}$ is a brace for $2 \le i < k$,
$(G-e)/X_k$ is a brace.

Suppose that $G-e$ is not a brace. Then, clearly the set X_1 is the minimal nontrivial subset of V such that $u \in X_1$ and $\partial(X_1)$ is a tight cut of G. Hence X_1 is unique. For a similar reason X_k is also unique. It follows that if $G-e$ has at most three braces, then it has a unique tight cut decomposition. However, in general, $G-e$ may have more that one tight cut decomposition, but all of them conform to the above pattern. See Exercise 17.5.3.

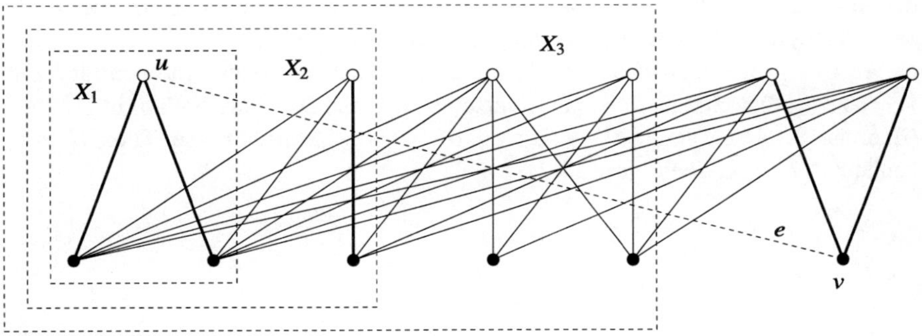

Fig. 17.20 The nested structure of shores of tight cuts of $G-e$.

Removable edges in $G-e$

The following lemma plays a role in the proof of the thin edge theorem for braces similar to the one that Lemma 8.19 played in the proof of the thin edge theorem for bricks.

Lemma 17.28 *Let G be brace of order at least six and let e be an edge of G. Then (see Exercise 8.4.1):*

(i) *for any vertex v that has degree at least three in $G-e$, at most one edge of $G-e$ incident with v is not removable in $G-e$.*
(ii) *if $C := \partial(X)-e$ is a nontrivial tight cut of $G-e$ then each edge of C is removable in $G-e$.* □

Maximal shores of tight cuts of $G-e$

Let H be the underlying simple graph of a brace of $G-e$. Let X be a maximal set of vertices of G such that $C := \partial(X)$ is nontrivial, $C-e$ is tight in $G-e$ and H is the underlying simple graph of a brace of $G'-e$, where $G' := G/X$.

Lemma 17.29 *If $G' - e$ is not a brace then $\overline{X}$ has a unique (proper) subset Y such that H is the underlying simple graph of $G'' - e$, where $G'' := G'/Y$.*

Proof Suppose that $G' - e$ is not a brace. Consider any tight cut decomposition of $G' - e$. The maximality of X implies the existence of a nontrivial cut $D := \partial(Y)$, where Y is a proper subset of $\overline{X}$ and $G'' := G'/Y$, such that H is the underlying simple graph of $G'' - e$. Our task is thus to prove that Y is unique. For this, we shall prove that every tight cut decomposition of $G' - e$ uses cut $D - e$. Note that as H is a brace, then $G'' - e$ is also a brace.

In fact, let Y' be a subset of $\overline{X}$, let $D' := \partial(Y')$ be nontrivial and suppose that $D' - e$ is tight in $G - e$. The cut $D' - e$ is tight in $G - e$; therefore Y' contains an end of e, say v. As $C - e$ is tight in $G - e$, the vertex v is the only end of e in $\overline{X}$. As $D - e$ is tight in $G - e$, it follows that v is also a vertex of Y.

We first prove that D' and D do not cross. For this, assume the contrary. Let $I := \partial(Y \cap Y')$ and $U := \partial(\overline{Y} \cap \overline{Y'})$. The cuts $I - e$ and $U - e$ are both tight in $G - e$ (Exercise 8.1.10). The cut $U - e$ is a cut of $G'' - e$, in turn a brace. Thus, $U = C$; hence $\overline{Y} \cap \overline{Y'} = \overline{X}$. See Figure 17.21.

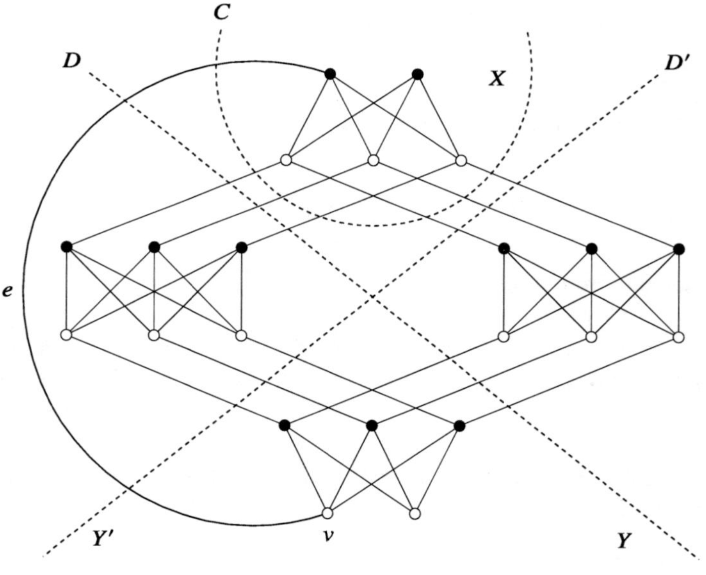

Fig. 17.21 The cuts C, D and D'.

It follows that $G/(X \cup (Y - Y'))/(Y \cap Y')$ is isomorphic, up to multiple edges, to G'', in contradiction to the maximality of X. We deduce that D and D' do not cross.

As $v \in Y \cap Y'$, it follows that one of Y and Y' is a subset of the other. The set Y is not a proper subset of Y'; else $D' - e$ would be a nontrivial tight cut of $G'' - e$, in turn a brace, a contradiction. The set Y' is then a subset of Y. This conclusion holds for each subset Y' of $\overline{X}$ such that $\partial(Y') - e$ is a nontrivial tight cut of $G - e$. It follows that every tight cut decomposition of $G' - e$ uses the cut Y. Indeed, Y is unique. □

Exercises

17.5.1 *Find an edge e in the brace G shown in Figure 17.22 that is not thin and display a tight cut decomposition of $G - e$.*

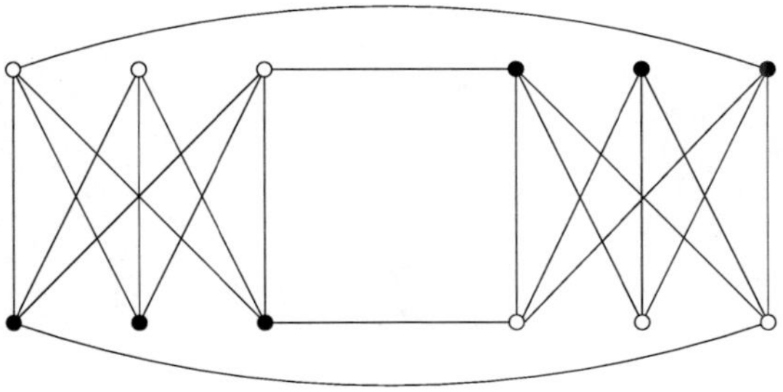

Fig. 17.22 Brace G for Exercise 17.5.1.

17.5.2 *Consider the brace G shown in Figure 17.30. The two edges $e = a_1b_5$ and $e' = a_3b_2$ happen to have the property that both $G - e$ and $G - e'$ are matching covered graphs with three braces. Find the unique tight cut decompositions of the two bipartite graphs $G - e$ and $G - e'$.*

17.5.3 *Find two different tight cut decompositions of the matching covered graph obtained by deleting the edge u_1v_6 from the brace G shown in Figure 17.23.*

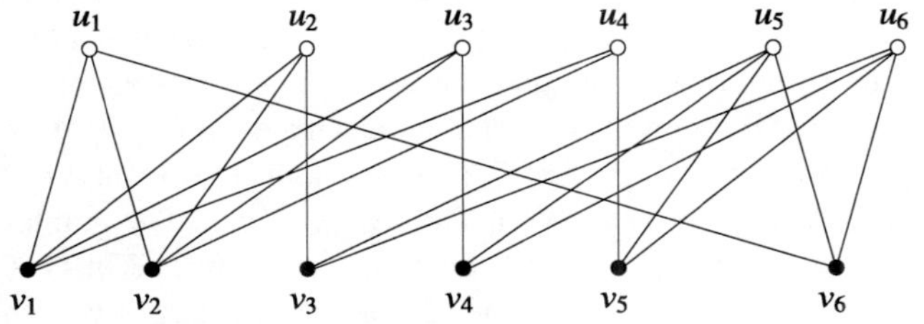

Fig. 17.23 $G - u_1v_6$ has two different tight cut decompositions.

17.5.4 *Consider the Heawood graph with the labelling of its vertices as shown in Figure 17.24, and let $e := a_0b_0$. Show that the retract of the graph obtained by deleting edge e from the Heawood graph is isomorphic to the biwheel $\mathbb{B}_{10}$. (Since the Heawood graph is edge transitive, it follows that all its edges are thin.)*

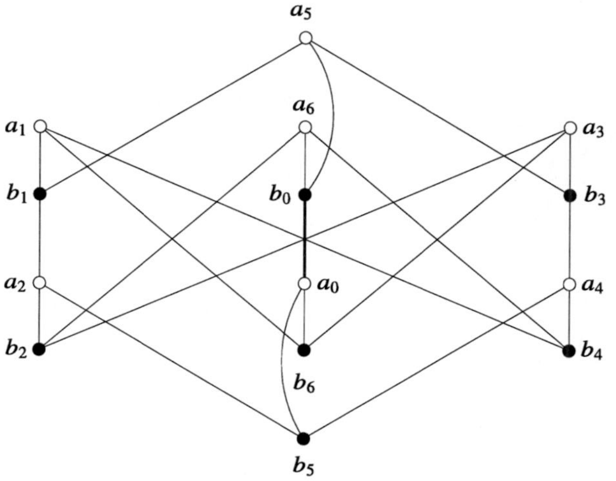

Fig. 17.24 Figure for Exercise 17.5.4.

17.6 The Existence of Thin Edges in Braces ♯

The rank of an edge

Our proof technique is constructive and is based on the notion of rank of an edge. Given an edge e of G which is not thin, we shall show that there exists an edge of higher rank than e. This leads us to the conclusion that an edge of G of maximum rank is a thin edge with the desired property. We used a similar technique for showing that every brick different from K_4, $\overline{C_6}$, and the Petersen graph has a thin edge.

Our objective is to define the notion of the rank of an edge and then show that an edge of maximum rank is a thin edge. But, first, we introduce a closely related function which we shall refer to as 'pre-rank'. The *pre-rank* of e, denoted by $r_0(e)$, is defined to be the maximum of the orders of all braces of $G - e$. As an example, consider the brace G shown in Figure 17.25. The graph $G - aw$ has two braces, one being the cube and the other being $K_{3,3}$. Thus, $r_0(aw) = 8$. The graph $G - ab$ has two braces: one has four vertices and the other has ten vertices. Thus $r_0(ab) = 10$.

In the brace shown in Figure 17.25, the edge ab has the largest possible pre-rank and is a thin edge. However, this is not in general true. It turns out that, in defining the rank of an edge, in addition to considering the number of vertices in a largest brace of $G - e$, we need also to consider the number of contraction vertices that brace has. To illustrate this point, consider the brace G in Figure 17.26.

Both graphs $G - e$ and $G - f$ have $K_{3,3}$ as a brace, up to multiple edges. The graph $G - e$ has two braces isomorphic to $K_{3,3}$, up to multiple edges (consider the tight cut $C - e$). Edge f is thin and the graph $G - f$ has only one brace isomorphic to $K_{3,3}$, up to multiple edges; it has two other braces, both of order four. Thus, both e and f have pre-rank six. But edge e is not thin, whereas f is thin. Note that the braces of $G - e$ have only one contraction vertex, whereas the brace of $G - f$ on six

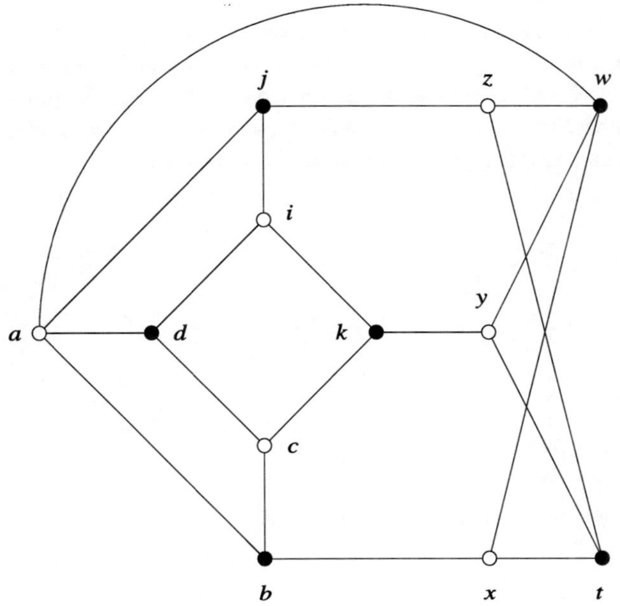

Fig. 17.25 The pre-rank of an edge, example: $r_0(aw) = 8$, $r_0(ab) = 10$.

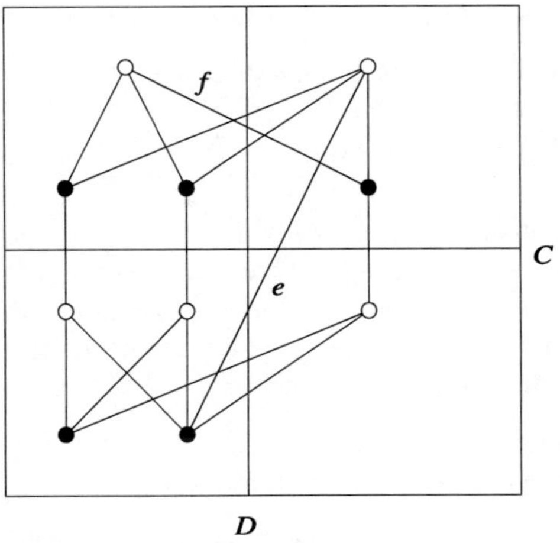

Fig. 17.26 $r_0(f) = r_0(e) = 6$, but $r(f) = 7$ and $r(e) = 6$.

vertices has two contraction vertices. So, in defining the rank of an edge e, we take into account the number of contraction vertices in a largest brace of $G - e$, giving preference to braces having two contraction vertices. We thus define the *rank* $r(e)$ of an edge e removable in G to be:

$$r(e) := \begin{cases} r_0(e) + 1, & \text{if } G - e \text{ has a brace of order } r_0(e) \\ & \text{with two contraction vertices} \\ r_0(e), & \text{otherwise.} \end{cases}$$

We remark that the same edge e may provide different choices for the brace of $G - e$. For example, in Figure 17.27, $G - e$ has two braces isomorphic to $K_{3,3}$ up to multiple edges. One brace contains the vertices in $\{1, 2, 7, 8, 9\}$, plus a contraction vertex. The other brace contains vertices in $\{3, 4, a, b\}$, plus two contraction vertices. Thus, the latter is responsible for the value seven for $r(e)$.

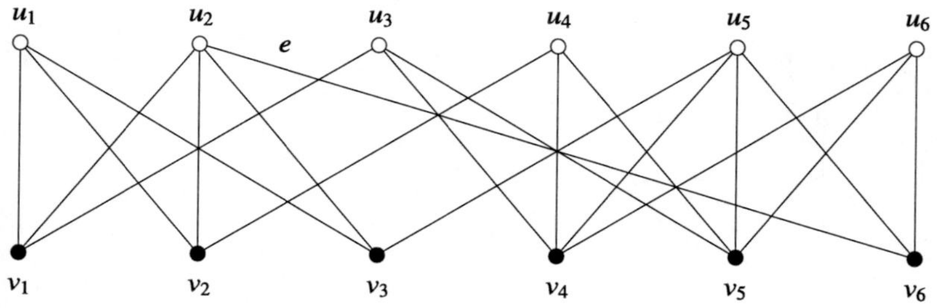

Fig. 17.27 The graph $G - e$ has two braces of order six

Theorem 17.30 (The Rank Augmentation Theorem for Braces) *Let $e := uv$ be an edge in a simple brace $G := [A, B]$ of order six or more and suppose that e is not thin. Then there exist two edges e' and e'' in G such that:*

(i) *e' and e'' are adjacent to each other, but not to e;*
(ii) *$r(e') \geq r(e)$, $r(e'') \geq r(e)$; and*
(iii) *either $r(e') > r(e)$ or $r(e'') > r(e)$.*

Proof By the definition of $r(e)$, there is a brace of $G - e$ of order $r_0(e)$. As e is not thin by hypothesis, $G - e$ has at least two braces, and $r(e) < |V|$. For every tight cut decomposition of $G - e$, every brace has at most two contraction vertices. Consider all the tight cut decompositions of $G - e$. Let $\mathcal{G}^\star$ denote the set of those braces $G^\star$ of $G - e$ that satisfy the following properties:

(i) $G^\star$ has order $r_0(e)$; and
(ii) if $r(e)$ is odd then $G^\star$ has two contraction vertices.

We note that $r_0(e)$ is at least four (all braces of any bipartite matching covered graph of order at least four have order at least four).

Let $\mathcal{G}$ be a tight cut decomposition of $G - e$ that contains braces in $\mathcal{G}^\star$. As e is not thin, every brace in $\mathcal{G}^*$ has a contraction vertex that is the result of the contraction of a set having five or more vertices. In particular, every brace in $\mathcal{G}$ that lies in $\mathcal{G}^\star$ has a contraction vertex that is the result of the contraction of a set containing five or more vertices.

Let $\mathcal{X}$ be the set of the shores X of tight cuts used by $\mathcal{G}$ such that $|X| \geq 5$ and $(G - e)/X$ contains a brace in $\mathcal{G}^\star$. Let X be a maximal set in $\mathcal{X}$, and let $C := \partial(X)$. The maximality of X implies that the graph $H := (G - e)/(X \to x)$ contains one brace in $\mathcal{G}^*$ and x is a contraction vertex of that brace.

We now consider two possibilities, depending on the parity of $r(e)$. If $r(e)$ is even then $H = G^*$ and $r(e) = r_0(e) = 1 + |\overline{X}|$. Alternatively, if $r(e)$ is odd then $\overline{X}$ has a proper subset Y such that (i) the cut $\partial(Y) - e$ is nontrivial and tight in $G - e$, (ii) the graph $G^\star := H/Y$ is a brace in $\mathcal{G}^*$ and (iii) no brace of $(G - e)/\overline{Y}$ is in $\mathcal{G}^*$; moreover, $r(e) = 1 + r_0(e) = 1 + (2 + |\overline{X} - Y|) \leq |\overline{X}|$, where the last inequality follows from the fact that $|Y| \geq 3$. In sum,

$$r(e) \leq 1 + |\overline{X}|, \tag{17.8}$$

with equality only if $G^\star$ has precisely one contraction vertex.

Note that edge e has its ends in X_- and $\overline{X}_-$. Let u denote the end of e in X_- and let v be its other end. Recall that (A, B) denotes the bipartition of G. Adjust notation so that u lies in A. (See Figure 17.28.)

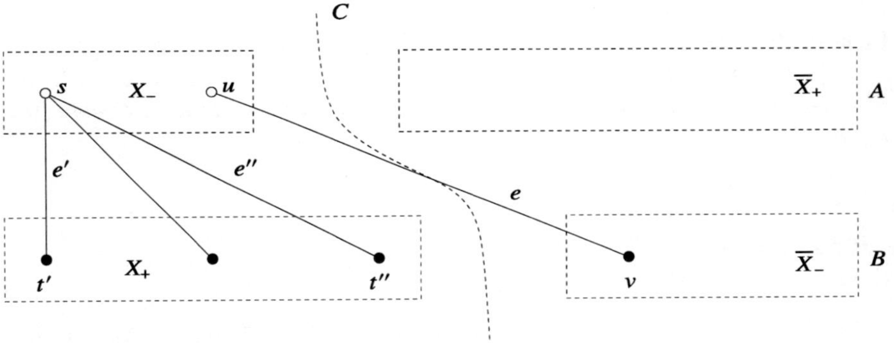

Fig. 17.28 Edges e, e' and e''.

As $|X| \geq 5$, it follows that X_- contains two or more vertices. Let $s \in X_-$ be a vertex distinct from u. Then, s has degree at least three in $G - e$. By Lemma 17.28, there are edges e' and e'', both incident with s and removable in $G-e$. Then $G-e-e'$ and $G - e - e''$ are matching covered subgraphs of $G - e$.

Let t' and t'' denote the ends of e' and e'' in B. The vertices of G adjacent to s in $G - e$ lie all in X_+. Thus, both t' and t'' lie in X_+ (Figure 17.28). This establishes that e' and e'' are adjacent to each other, but not to e. Since G is a brace, G has a perfect matching that contains edges e and e'', but not e', hence $G - e'$ is matching covered, by Proposition 17.11. Likewise, $G - e''$ is matching covered. We shall now prove that $r(e') \geq r(e)$, and characterize under what conditions equality holds. For this, we consider some cases. In all cases except the last, we show that $r(e') > r(e)$. In the last case we conclude that $r(e') = r(e)$. A similar reasoning holds for e''. We then finally show that equalities $r(e') = r(e)$ and $r(e'') = r(e)$ cannot both hold. This part of the proof is remarkably similar to the proof of the Rank Augmentation Theorem for Bricks (Theorem 17.16), but there are some differences, especially due to the fact that in a near-brick every tight cut has precisely one bipartite shore, whereas in a bipartite graph clearly both shores of a tight cut are bipartite.

The next statement is similar to a property established in the proof of Theorem 17.16 and its proof is left as Exercise 17.6.1.

17.30.1 *Let C' be a cut of G such that $C' - e'$ is tight in $G - e'$. If C and C' do not cross then $\overline{X}$ is a proper subset of a shore of C'.* □

Case 1 *No tight cut in $G - e'$ crosses C.*

In this case, by statement 17.30.1, the set $\overline{X}$ is properly contained in a shore of every tight cut in $G-e'$. Then $\overline{X}$ is properly contained in the vertex set of a brace H of $G-e'$. It follows that H has at least $|\overline{X}|+3$ vertices. Thus, $r(e') \geq |V(H)| \geq |\overline{X}|+3 > r(e)$, establishing that $r(e') > r(e)$.

Case 2 *$G - e'$ has tight cuts that cross C.*

The proof of the next result is somewhat similar to the proof of Lemma 17.18 (Exercise 17.6.2).

Lemma 17.31 *Let $C' := \partial(X')$ be a cut of G such that $C' - e'$ is tight in $G - e'$. Suppose that C and C' cross and also that the end t' of e' is in $X \cap X'$. Let $D := \partial(X - X')$ and let $D' := \partial(X' - X)$. Then, the following properties hold (see Figure 17.29):*

(i) *the end s of edge e' is in $X - X'$,*
(ii) *$X_+ \subset B$ and $X'_+ \subset A$,*
(iii) *the cut $D' - e$ is tight in $G - e$, and*
(iv) *the cut D is nontrivial and $D - e - e'$ is tight in $G - e - e'$.* □

Let X' be a maximal subset of V such that $t' \in X'$, $C' := \partial(X')$ is nontrivial, $C' - e'$ is tight in $G - e'$ and the cuts C and C' cross. By statement 17.30.1, the maximality of X' implies that every nontrivial tight cut of $(G - e')/X'$ has a shore that includes the set $\overline{X} \cap \overline{X'}$.

Case 2.1 Cut D' is trivial.

Recall that $r(e) \leq 1 + |\overline{X}| = 2 + |\overline{X} \cap \overline{X'}|$. Every nontrivial tight cut of $(G - e')/X'$ has a shore that includes the set $\overline{X} \cap \overline{X'}$, hence $r_0(e') \geq 2 + |\overline{X} \cap \overline{X'}| \geq r(e)$. If $D - e'$ is not tight then $r_0(e') > 2 + |\overline{X} \cap \overline{X'}| \geq r(e)$. Alternatively, if $D - e'$ is tight then $r_0(e') = 2 + |\overline{X} \cap \overline{X'}|$ and the corresponding brick of $G - e'$ has two contraction vertices, hence $r(e') > 2 + |\overline{X} \cap \overline{X'}| \geq r(e)$. In both alternatives, $r(e') > r(e)$. (See Example 17.32 for an illustration.)

Example 17.32 (Case 2.1, Theorem 17.30) Consider the brace G depicted in Figure 17.29. The edge $e = a_1b_5$ is a removable edge of G, the cut $C := \partial(X)$, where $X := \{a_1, a_2, a_3, b_1, b_2, b_3, b_4\}$ is a tight cut of $G - e$. Both C-contractions of $G - e$ happen to be braces of order eight, each with just one contraction vertex. Thus $r(e) = 8$. The edge $e' = a_3b_2$ is removable in both $G - e$ and G. The cut $C' := \partial(X')$, where $X' := \{a_1.a_2, a_4, b_1, b_2\}$, is a tight cut of $G - e'$. The two cuts C and C' cross, and $|\overline{X} \cap X'| = 1$. The graph $G - e'$ has three braces of orders six, eight and four. The largest brace of $G - e'$, which is of order eight, has two contraction vertices. Therefore, $r(e') = 9$.

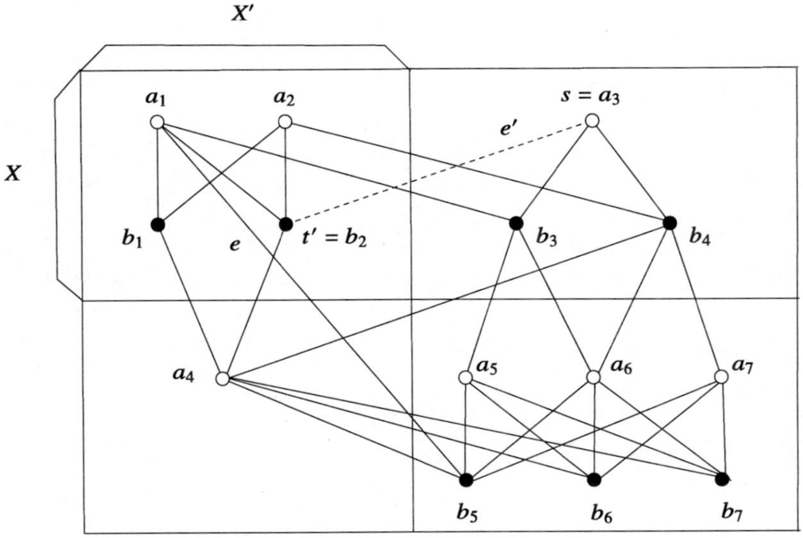

Fig. 17.29 Illustration for Case 2.1, Theorem 17.30

Case 2.2 Cut D' is nontrivial.

As $D' - e$ is tight in $G - e$, the edge e has an end in the minority part of the shore $X' - X$ of D'. The minority part of $X' - X$ is a subset of the part B of the bipartition of G. Let v be the end of e in $X' - X$. The other end u of e is in A and also in the minority part of X. As $C' - e'$ is tight in $G - e'$, we deduce that u is in $X \cap X'$. See Figure 17.30.

As every tight cut of $(G - e')/X'$ has a shore that includes $\overline{X} \cap \overline{X'}$, the graph $H := (G-e')/X'/(X-X')$ is a brace. This brace H is isomorphic to $G/X/(X' - X)$, up to multiple edges. The cut $D' - e$ is a nontrivial cut of $(G - e)/X$; therefore G^* has two contraction vertices. We deduce that $r(e) = r(e') = |\overline{X} \cap \overline{X'}| + 3$. Let $Y := X' - X$. Note that Y is a proper subset of $\overline{X}$, such that $(G - e)/X/Y$ is G^*. By Lemma 17.29, Y is unique.

A similar reasoning may then be applied to e''. The only case in which the inequality $r(e'') \geq r(e)$ holds with equality occurs when G has a cut $C'' := \partial(X'')$ that crosses C, $C'' - e''$ is tight and nontrivial in $G - e''$ and $Y = X'' - X$.

Thus, $X' - X = Y = X'' - X$; therefore there exist two vertices in $Y \cap A$, say a' and a'', both of which are adjacent to the end v of e. The vertex s has a neighbour, b', distinct from both t' and t''. The vertex b' is in $B - (X' \cup X'')$. Finally, in addition to its two contraction vertices, the brace G^* has at least two vertices in the equipartite set $\overline{X} \cap \overline{X'}$, one of which, say b'', is in B. In fact, $b'' \in B - (X' \cup X'')$. In sum, a' and a'' are two vertices in $A \cap (X' \cap X'')$ and the vertices b' and b'' are two vertices in $B - (X' \cup X'')$. The graph $G - a' - a'' - b' - b''$ has a perfect matching, M. It is necessary for M to contain both e' and e'', which is a contradiction, as e' and e'' are adjacent. The proof of the Rank Augmentation Theorem for Braces is complete. □

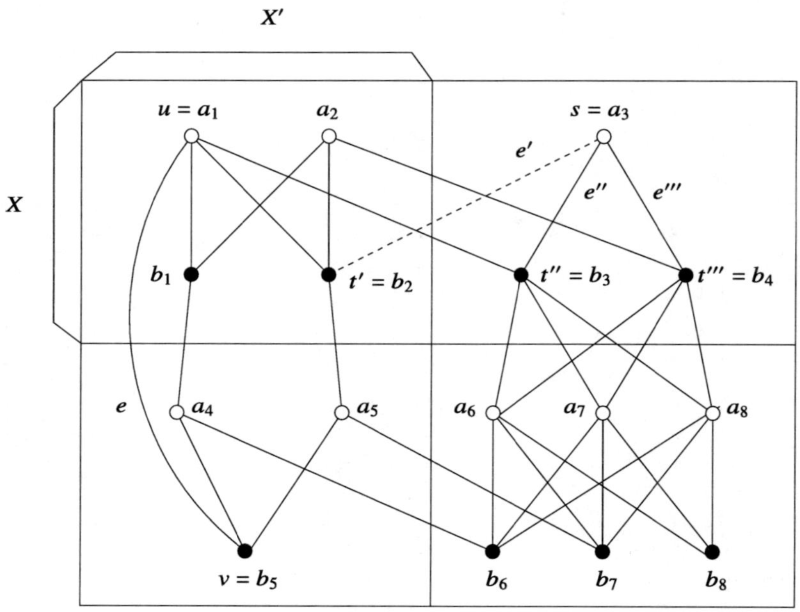

Fig. 17.30 Illustration for Case 2.2, Theorem 17.30

Example 17.33 (Case 2.2, Theorem 17.30) Consider the brace G depicted in Figure 17.30 with the edge $e = a_1b_5$ in it. The cut $C := \partial(X)$, where $X := \{a_1, a_2, a_3, b_1, b_2, b_3, b_4\}$ is a tight cut of $G - e$. The graph obtained from $G - e$ by contracting the set X to a single vertex x, and the set $\{a_4, a_5, b_5\}$ to a single vertex y is a brace of order eight of $G - e$ with two contraction vertices. It follows that $r(e) = 9$. (Note that the graph obtained from $G - e$ by contracting $\overline{X}$ is also a brace of order eight, but that brace has only one contraction vertex.)

The edge $e' = a_3b_2$ is removable in both $G - e$ and G. The cut $C' := \partial(X')$, where $X' := \{a_1.a_2, a_4, a_5, b_1, b_2, b_5\}$, is a tight cut of $G - e'$. The two cuts C and C' cross, and $|\overline{X} \cap X'| = 3$. The graph obtained from $G - e'$ by contracting X' to a single vertex x', and the set $\{a_3, b_3, b_4\}$ to a single vertex y' is a brace of order eight with two contraction vertices. Thus the rank $r(e')$ of e' is also 9. However, it can be verified that both the other edges $e'' = a_3b_3$ and $e''' = a_3b_4$ incident with the vertex a_3 have higher rank (Exercise 17.6.3).

If we set $a' := a_4$ and $a'' := a_5$, and $b' := b_4$ and $b'' = b_8$, then any perfect matching M of $G - a' - a'' - b' - b''$ must contain both the edges e and e'. It cannot contain any edge incident with s other than e'. (The contradiction that both e' and e'' belong to M was arrived at by assuming that $r(e'') = r(e') = r(e)$. But this situation does not arise.)

As a corollary of the above theorem we have:

THIN EDGE THEOREM FOR BRACES

17.34 *Every brace G distinct from K_2 and C_4 has at least two thin edges.*

Proof Each member of a pair of multiple edges of G is thin. We may thus assume that G is simple. The braces K_2 and C_4 are the only simple braces on fewer than six vertices. We may thus assume that G has six or more vertices, in which case every edge of G is removable. If every edge of G is thin then the assertion holds. We may thus assume that G has edges that are not thin.

Let r' be the maximum rank of edges of G that are not thin. Let e be an edge of G such that $r(e) = r'$. By the Rank Augmentation Theorem for Braces, G has two adjacent removable edges, e' and e'', such that at least one of e' and e'' is thin. If e' and e'' are both thin then the assertion holds. We may thus assume that precisely one of e' and e'', say, e', is thin. In that case, $r(e'') = r(e) = r'$. We then apply the Rank Augmentation Theorem for Braces with e'' playing the role of e, thereby obtaining a thin edge that is not adjacent to e'', hence distinct from e'. Indeed, G has at least two thin edges. □

Exercises

17.6.1 *Prove statement 17.30.1.*

$*$ **17.6.2** *Prove Lemma 17.31. Hint: see the proof of Lemma 17.18.*

17.6.3 *In the brace considered in Example 17.33, show that* $r(e'') = 12$ *and* $r(e''') = 14$.

$*$ **17.6.4** *Prove the following strengthening of Theorem 17.16 for solid bricks, whose statement is similar to that of Theorem 17.30:*

Theorem 17.35 *Let e be a b-invariant edge in a simple solid brick G and suppose that e is not thin. Then there exist two b-invariant edges e' and e'' in G such that:*

(i) *e' and e'' are adjacent to each other;*
(ii) $r(e') \geq r(e)$, $r(e'') \geq r(e)$;
(iii) *either* $r(e') > r(e)$ *or* $r(e'') > r(e)$; *and*
(iv) *if the common end of e' and e'' is also an end of e then* $r(e') > r(e)$ *and* $r(e'') > r(e)$.

Hint: use the exchange property of solid graphs (Theorem 10.23) to prove the existence of the two b-invariant edges e' and e''.

$*$ **17.6.5** *Use Theorem 17.35 to prove that every solid brick distinct from K_4 has two thin edges. Hint: apply Theorem 9.16 to obtain two removable edges and then imitate the proof of Theorem 17.34.*

17.7 Building Braces

In this section we give a precise definition of the three operations necessary for building all braces from K_2 and C_4. These operations are applicable to every brace H, producing a brace G. We refer to these three operations as *expansions* of a brace. Let H be a brace on which the operation will be applied and let (A, B) denote the bipartition of H. The resulting graph G will be a brace.

Edge addition: Let $x \in A$ and $y \in B$ be vertices of H. Obtain G from H by adding a new edge joining x and y. It is easy to see that G is a brace.

Expansion of a vertex by a barrier of size two: Let H be a brace and let x be a vertex of B whose degree is at least four. Let $H' := H\{x \to (a, b_1, b_2)\}$ be such that, in the underlying simple graph of H', b_1 and b_2 have degree at least three. Now obtain G from H' by joining a to a vertex w of $B - x$ by an edge labelled e. We shall refer to the graph G thus constructed as a graph obtained from H by *an expansion of x by a barrier of size two.*

Expansion of two vertices by barriers of size two: Let H be a brace and let $x \in B$ and $y \in A$ be two vertices of H whose degrees are at least four. Let $H'' := H\{x \to (a, b_1, b_2)\}\{y \to (b, a_1, a_2)\}$ be such that, in the underlying simple graph of H'', each of b_1, b_2, a_1 and a_2 has degree at least three and, if H has more than two vertices, each of b_1, b_2, a_1 and a_2 has at least one neighbour in $V(H - x - y)$. Now obtain G from H'' by joining a and b by an edge labelled e. We shall refer to the graph G thus constructed as a graph obtained from H by *an expansion of x and y by barriers of size two.* (Note that the order in which the two vertices x and y are split is immaterial.) Figure 17.31 shows how the cube can be obtained from a brace of order four by applying the above defined operation.

Theorem 17.36 *Let H be a brace and let G be a graph obtained from H by one of the three operations of expansion. Then G is a brace.* □

We remark that the above result also indicates that, in every expansion operation, the edge e is a thin edge of the brace G and that H is the brace of $G - e$. As a complement to the above result, we then state the following theorem, which is easily proved. (Exercise 17.7.3).

Theorem 17.37 *Every brace G different from the special braces K_2 and C_4 can be obtained from one of them by a sequence of applications of the three operations of expansion.* □

A natural question that arises at this stage is whether all the three operations are really required for building braces. This is indeed the case. (Exercise 17.7.5.)

CLM (2015, [18]) proved the following strengthening of Theorem 17.37:

Theorem 17.38 *Let G be a brace and let J be a simple brace that is a matching minor of G. If G and J are not isomorphic then G has a thin edge e such that J is a matching minor of $G - e$.* □

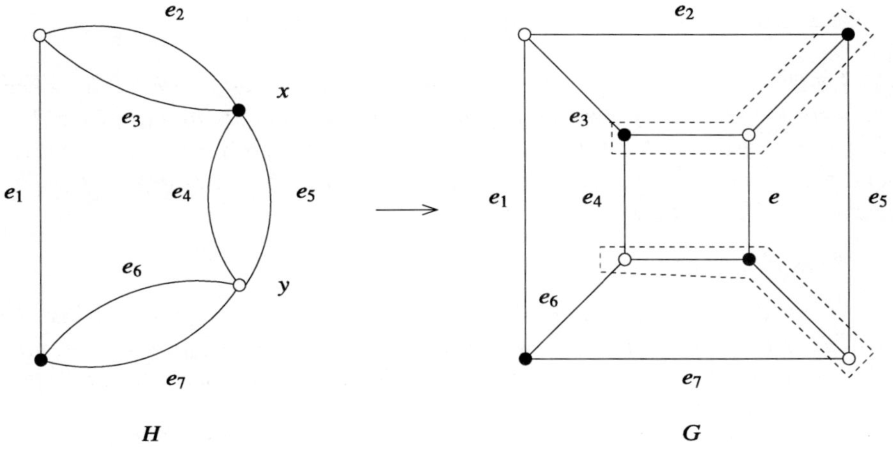

Fig. 17.31 Obtaining the cube (G) from a brace (H) of order four by expansion of two vertices by barriers of size two

Corollary 17.39 *Let G be a brace and let J be a simple brace that is a matching minor of G. Then G may be obtained from J by repeated applications of the three expansion operations.* □

Exercises

▷**17.7.1** *In the definition of the operation of expansion of two vertices by barriers of size two, there is a requirement that "if H has more than two vertices, each of b_1, b_2, a_1 and a_2 has at least one neighbour in $V(H-x-y)$." Explain why this requirement is necessary for the resulting graph G to be a brace if H has more than two vertices.*

▷**17.7.2** *Prove Theorem 17.36.*

17.7.3 *Prove Theorem 17.37. Hint: use induction on $|E|$.*

17.7.4 *Generate all the cubic braces on eight or fewer vertices.*

▷**17.7.5** *Verify that all the three operations are necessary for the generation of all braces starting with the two basic braces. For each of the three operations we give below a hint:*

(i) *edge addition: consider any brace having minimum degree four or more;*
(ii) *expansion of a vertex by a barrier of size two: consider the biwheels of order six or more;*
(iii) *expansion of two vertices by barriers of size two: consider cubic braces of order six or more.*

This exercise is relevant to the subject of the next chapter.

▷**17.7.6** *The families of graphs the following questions pertain to are all described in Section 2.5; some of these graphs are bricks and some are braces. (Hint: The statement in Exercise 8.2.4 is useful in finding the answers to some of the following questions. We recommend to the readers to first work with specific examples of graphs with around ten vertices.)*

(i) *Show that:*

(a) *the thin edges of the wheel* W_{2k+1}*,* $k \geq 2$*, are precisely its* $2k+1$ *spokes;*
(b) *each edge of the biwheel* $\mathbb{B}_8$ *(the cube) is a thin edge and, for* $n \geq 5$*, the thin edges of the biwheel* $\mathbb{B}_{2n}$ *are precisely the* $2n-2$ *edges incident with its two hubs* h_1 *and* h_2*;*
(c) *the thin edges of the truncated biwheel* $\mathbb{T}_{2n}$*,* $n \geq 4$ *are precisely the* $2n-4$ *edges incident with its two hubs* h_1 *and* h_2 *other than* h_1v_1*,* h_1v_{2n-2}*,* h_2v_1 *and* h_2v_{2n-2}*;*
(d) *the thin edges of the prism* $\mathbb{P}_{2n}$*,* $n \geq 5$ *are precisely its n rungs* u_1v_1*,* u_2v_2*,* $\ldots$*,* u_nv_n*;*
(e) *the thin edges of the Möbius ladder* $\mathbb{M}_{2n}$*,* $n \geq 5$*, are precisely its rungs* u_1v_1*,* u_2v_2*,* $\ldots$*,* u_nv_n*; and*
(f) *the thin edges of the staircase* $\mathbb{S}_{2n}$*,* $n \geq 4$*, are precisely its rungs* u_2v_2*,* $\ldots$*,* $u_{n-1}v_{n-1}$*.*

(ii) *In each of the above cases, show that the retract of the graph obtained by deleting a thin edge is not simple. (In the terminology to be introduced in the next chapter, such thin edges are not strictly thin.)*

17.8 Notes

Kothari (2016, [47]), (2019, [48]) proved that, with the exception of K_4 and $\overline{C_6}$, every near-bipartite brick G has a thin edge e such that the retract H of $G-e$ is also near-bipartite. In fact, he proved that one may fix the removable doubleton R of G so that R remains a removable doubleton of H.

Chapter 18
Strictly Thin Edges in Bricks and Braces

Contents

18.1 Strictly Thin Edges

A brick is solid if and only if its underlying simple graph is solid. A brick or brace has a specified simple matching covered graph J as a matching minor if and and only if its underlying simple graph has J as a matching minor. As these examples show, the properties of bricks and braces that are of interest to us are mostly those which are enjoyed by simple bricks and braces. This, as explained below, brings into question the efficacy of using thin edges as inductive tools for establishing properties of bricks and braces.

C. L. Lucchesi, U. S. R. Murty, *Perfect Matchings*, Algorithms and Computation in Mathematics 31, https://doi.org/10.1007/978-3-031-47504-7_18

Consider as an example the problem of characterizing simple planar solid bricks. It turns out that every such brick G is an odd wheel. How do we go about proving that this is indeed the case? If the order n of G is four, then it is clearly the complete graph K_4 which is the wheel W_3. So, suppose that $n \geq 6$ and assume inductively that every simple planar solid brick with fewer than m edges is a wheel. As both $\overline{C_6}$ and $\mathbb{P}$ are not solid, G is not a special brick. Thin Edge Theorem 17.3 implies that G has an edge e such that $G' := \widehat{G - e}$ is a brick. Using the fact that G is planar and solid, it is easy to deduce that G' is a planar solid brick (Exercise 18.1.1). As G' clearly has fewer edges than G, it follows from the induction hypothesis that the underlying simple graph of G' is a wheel and, furthermore, that G itself is obtained from such a graph by an expansion operation. There would be not too many cases to contend with if G' were simple. (See Exercise 18.1.2.) However, the Thin Edge Theorem does not guarantee that G' is simple. This necessitates the consideration of many more cases in order to be able to deduce that G is a wheel. (The authors confess to ploughing through the needed case analysis in the paper by CLM (2016, [15]). This was before they knew the contents of this chapter.)

The above example is just one instance where it would be advantageous to have recursive procedures for generating simple bricks and braces within the realm of simple graphs. (See Exercise 18.1.3 for a second example.) In order to establish the existence of such procedures, one needs the notion of a strictly thin edge in a matching covered graph which is either a brick or a brace.

STRICTLY THIN EDGES IN BRICKS AND BRACES

A thin edge e of a simple brick G is *strictly thin* if the retract of $G - e$ is a simple brick. Likewise, a thin edge of a simple brace is *strictly thin* if the retract of $G - e$ is a simple brace.

As we learnt in the last chapter, all bricks and braces, with a few isolated exceptions, have thin edges. But there exist infinite families of bricks and braces which do not have any strictly thin edges. However, we shall learn that these families of graphs can be identified and that they are precisely the ones described in Exercise 17.7.6. We list them below in the order that will be found to be convenient for later purposes:

(i) wheels,
(ii) biwheels,
(iii) truncated biwheels,
(iv) prisms,
(v) Möbius ladders and
(vi) staircases.

18.1.1 Families of graphs without strictly thin edges

McCuaig braces

In his pioneering paper entitled *Brace generation*, McCuaig (2001, [68]), identified the three families of braces that do not have strictly thin edges. They are: (i) biwheels, (ii) prisms $\mathbb{P}_n$, $n \equiv 0 \pmod 4$, and (iii) Möbius ladders $\mathbb{M}_n$, $n \equiv 2 \pmod 4$. We shall refer to the union of these families as the *McCuaig family of braces*. As noted before, no brace in this family has strictly thin edges. More significantly, the following assertion was established in [68]:

Theorem 18.1 (Strictly Thin Edge Theorem for Braces) *Every simple brace of order six or more that is not a member of the McCuaig family of braces has a strictly thin edge.*

Norine-Thomas bricks

In a paper entitled *Brick generation*, Norine and Thomas (2008, [79]) identified all the infinite families of simple bricks that are free of strictly thin edges. They are: (i) wheels, (ii) truncated biwheels, (iii) prisms $\mathbb{P}_n$, $n \equiv 2 \pmod 4$, (iv) Möbius ladders $\mathbb{M}_n$, $n \equiv 0 \pmod 4$, and (v) staircases. (Note that K_4 is the wheel W_3, and $\overline{C_6}$ is the prism $\mathbb{P}_6$.) We shall refer to the union of these families of bricks as the *Norine-Thomas family of bricks*. The theorem below follows from the results established in [79]:

Theorem 18.2 (Strictly Thin Edge Theorem for Bricks) *Every simple brick which is different from the Petersen graph $\mathbb{P}$ and which is not a member of the Norine-Thomas family of bricks has a strictly thin edge.*

It should be noted that the papers [68] and [78], mentioned in the definition of the families, contain more general results than the ones stated above. A description of those results will be given at the end of the chapter.

In the previous chapter, we presented separate proofs for Thin Edge Theorems (17.3 and 17.34) for bricks and braces. Relying on those two theorems, and the fact that bricks and braces are the only matching covered graphs that are free of nontrivial tight cuts, we are able to prove the two Strictly Thin Edge Theorems (18.1 and 18.2) together by considering the class of graphs consisting of both bricks and braces:

Definition 18.3 A matching covered graph is *indecomposable* if it is either a brick or a brace.

We shall present a characterization of indecomposable graphs which are free of strictly thin edges and deduce from it the above-mentioned theorems of McCuaig, and of Norine and Thomas.

Exercises

18.1.1 *Let e be a thin edge in a planar solid brick G. Show that* $\widehat{G-e}$ *is a planar solid brick. (Hint: Use Theorems 10.14 and 7.1.)*

18.1.2 *Find all the simple planar bricks that can be obtained from the wheels* W_5 *and* W_7 *by means of the four expansion operations, and identify those among the resulting bricks which are solid.*

18.1.3

(i) *Let G be a simple brick and let e be a strictly thin edge of G such that* $\widehat{G-e}$ *is the Petersen graph* $\mathbb{P}$*. Show that there exists another strictly thin edge f of G such that* $\widehat{G-f}$ *is not the Petersen graph.*

(ii) *Assuming the Strictly Thin Edge Theorem for Bricks, and using the result in the first part, show that every brick G that is not a Petersen brick contains a thin edge e such that* $\widehat{G-e}$ *is not a Petersen brick. (This provides a proof of Theorem 17.27.)*

18.2 Multiple Edges in Retracts

Let $\mathcal{G}$ denote the class of graphs consisting of K_2, C_4, the Petersen graph and the members of the six infinite families described in Section 18.1.1. Our principal objective is to to show that any simple indecomposable graph that is not a member of $\mathcal{G}$ has a strictly thin edge.

The two basic braces, K_2 and C_4, as well as the three special bricks, K_4, $\overline{C_6}$ and $\mathbb{P}$, are members of $\mathcal{G}$. Thus, by Theorems 17.3 and 17.34, every simple indecomposable graph that does not belong to $\mathcal{G}$ has at least one thin edge. In the remaining part of this section, we shall use the notation specified in the inset below:

> NOTATION: G, e, H, f AND g
>
> - G is a simple indecomposable graph which does not have any strictly thin edges, (our objective is to show that $G \in \mathcal{G}$);
> - e is a thin edge of G (which is not strictly thin);
> - $H := \widehat{G-e}$ is the retract of $G - e$; and
> - f and g are two edges which are parallel to each other in H.

Since G itself is simple, f and g must be incident in H with at least one contraction vertex (that is, one resulting from the bicontraction of an end of e).

Let us begin with a brief review of the conditions under which bicontractions of vertices of degree two in $G - e$ result in a graph with multiple edges. As e is thin but not strictly thin in G, we note that $H \neq G - e$. Thus, the index of e is either one, two or three, as defined in Chapter 13 (and reviewed in Chapter 17). In each of these cases, as described in the following inset, we can infer that either one or both ends of e have degree three in G.

> **VERTICES OF DEGREE THREE AND THE SOLID DOT CONVENTION**
>
> If the index of e is one, then exactly one end of e has degree three in G, and if the index of e is two, then both ends of e have degree three in G and they have no common neighbour. Finally, if the index of e is three, both ends of e have degree three in G and they have a common neighbour. This property holds for any thin edge of G.
>
> In proving that G must be a member of $\mathcal{G}$, we shall make repeated use of the above-stated property to deduce that G has at most two vertices whose degrees are greater than three. In our illustrations we shall indicate vertices whose degrees can be inferred to be three by solid dots.

18.2.1 How do multiple edges in the retract of $G - e$ arise? ♯

The index of the chosen thin edge e of G may be either one, two or three. In what follows we shall examine each of these cases separately to discern how parallel edges in $H = \widehat{G - e}$ might arise.

Case in which the index of e is one: Suppose that $e = uv$ is a thin edge of index one. Let v be the end of e that has degree three in G, and let v_1 and v_2 be its neighbours distinct from u. The retract of $G - e$ has only one contraction vertex and that vertex is the result of the bicontraction of v in $G - e$. Thus if f and g are parallel edges in the retract H of $G - e$, they must be adjacent edges in G such that one of them is incident with v_1 and the other with v_2. Without loss of generality, we may thus assume that $f := v_1w$ and $g := v_2w$ are incident with a common vertex w which is not in $\{v, v_1, v_2\}$. See Figure 18.1. The degree of w is at least four in G and it is possible for u and w to be the same vertex (in which case w would have degree five or more in G).

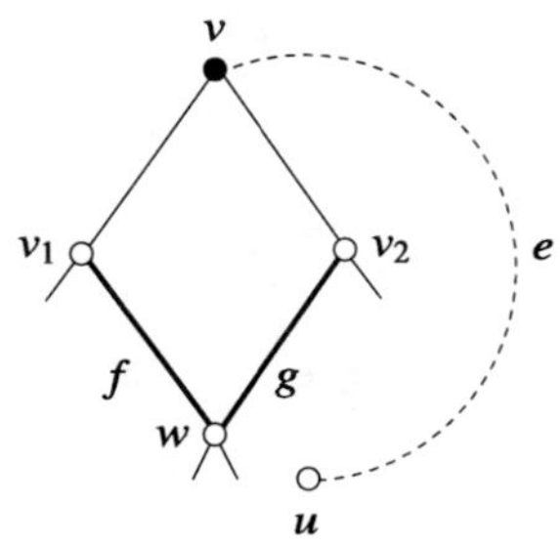

Fig. 18.1 Parallel edges in H, index$(e) = 1$

Case in which the index of e is two: Suppose that $e = uv$ is a thin edge of index two in G. The degrees of both u and v are three in G, and are two in $G - e$. Let u_1, u_2 and v be the three neighbours of u, and let v_1, v_2 and u be the three neighbours of v in G. As the index of e is two, u and v do not have a common neighbour, implying that $\{u_1, u_2\} \cap \{v_1, v_2\} = \emptyset$. There are essentially three possible situations under which f and g become parallel edges when u and v are bicontracted in $G - e$. These three possibilities are illustrated in Figure 18.2.

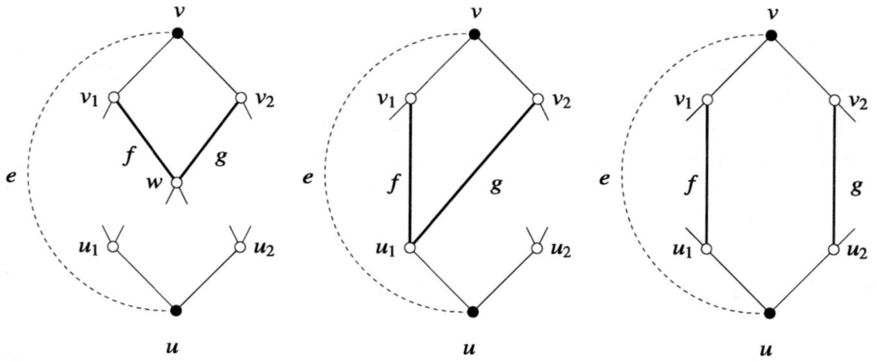

Fig. 18.2 Parallel edges in H, index$(e) = 2$

Case in which the index of e is three: Suppose now that $e = uv$ has index three. In this case, the ends u and v of e have a common neighbour in G. Let v_1 and v_2 be the neighbours of v in $G - e$, and let v_2 and v_3 be the neighbours of u in $G - e$. The bicontractions of u and v in $G - e$ result in a single contraction vertex. So, if f and g are parallel edges in the retract H of $G - e$, they must be incident with distinct vertices in $\{v_1, v_2, v_3\}$ and must have a common end, say w, which is not in $\{u, v, v_1, v_2, v_3\}$. The common end w of f and g has degree at least four in G. In Figure 18.3, which illustrates such an occurrence, edges f and g are indicated by solid lines.

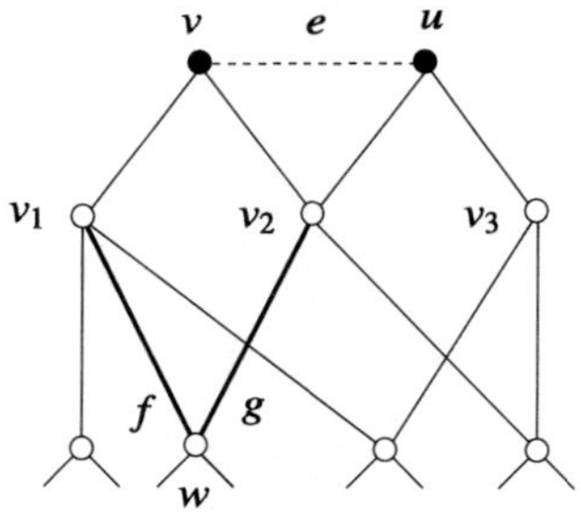

Fig. 18.3 Parallel edges in H, index$(e) = 3$

18.2.2 What is the status of those edges in G itself?

Recall that G is a simple indecomposable graph in which the edge e is thin, and that f and g are edges in the retract H of $G - e$ which are parallel to each other. Being multiple edges of H, both f and g are clearly removable in H. As H is the retract of $G - e$, it is obtained from $G - e$ by either one or two bicontractions. If C is the tight cut associated with a bicontraction, and if f belongs to C, it can be seen that f is a multiple edge (though not necessarily parallel to g) in both the C-contractions of $G - e$. Using this observation, it is easy to verify the following assertion. (We leave the details as Exercise 18.2.1).

Proposition 18.4 *Edge f (and similarly edge g) is removable in $G - e$.* □

If G is a brace (of order six or more), f and g are also removable in G, but they may or may not be thin (see Exercise 18.2.2). On the other hand, if G is a brick, neither f nor g needs to be removable in G. For example, in the bicorn $\mathbb{H}_8$ shown in Figure 18.4 the edge e is thin, the edges f and g are parallel edges in the retract of $\mathbb{H}_8 - e$, but neither of them is removable in $\mathbb{H}_8$.

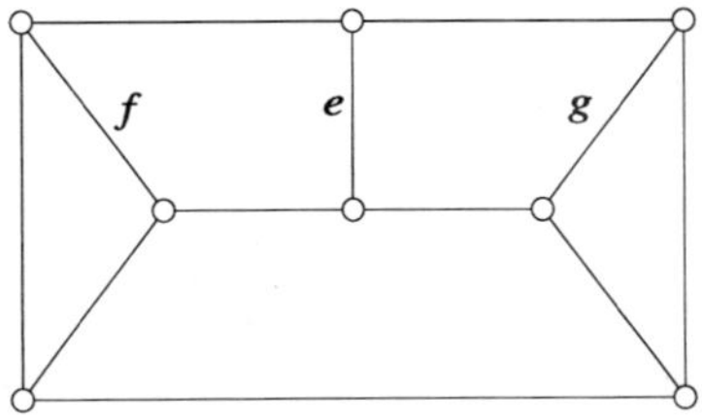

Fig. 18.4 Edges f and g are removable in $\mathbb{H}_8 - e$ but not in $\mathbb{H}_8$

The graph H is indecomposable and f and g are parallel edges in H; thus the graph $H - f$ is indecomposable. As f is a removable edge in $G - e$, the graph $G - e - f$ is matching covered; and in fact $H - f$ is the retract of $G - e - f$. The following proposition is thus obvious:

Proposition 18.5 *Up to multiple edges tight cut decompositions of $G - e - f$ and of $G - e$ produce the same list of indecomposable graphs.* □

18.2.3 Two key lemmas

Various possibilities occur depending on the status of f and g as to their removability or thinness in G. We shall examine all those possibilities and conclude that, in each case, the constraint that G has no strictly thin edges implies that G must contain certain configurations in the vicinity of the ends of e. In the next section we shall show that those local configurations must replicate themselves, and close up eventually, implying that G belongs to the family $\mathcal{G}$.

The configuration defined below features prominently in the first lemma.

HOUSES

A *house* with e as its *floor* and f as as its *ceiling* is an induced subgraph F of G consisting of a pentagon with precisely one chord where that chord is the edge f, and e is the edge of the pentagon disjoint from f, with the additional property that the ends of e and f have degree three in G. See Figure 18.5. (In $\mathbb{H}_8$ shown in Figure 18.4, there is a house with e as its floor and f as its ceiling, and one with e as its floor and g as its ceiling.)

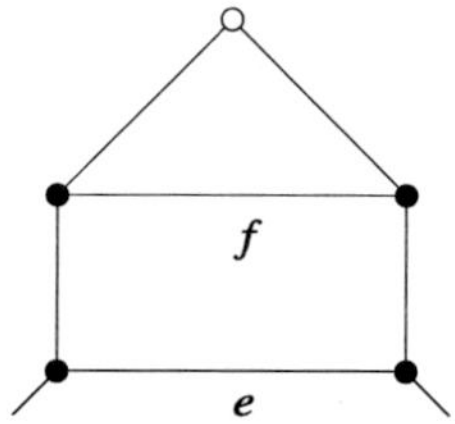

Fig. 18.5 A house with floor e and ceiling f

THE CASE IN WHICH f IS NOT REMOVABLE IN G

Lemma 18.6 *Let G be a simple brick, let $e := uv$ be a thin edge of G of index one or two and let f be a multiple edge in the retract H of $G - e$. If f is not removable in G then the following properties hold:*

(i) *if* $\text{index}(e) = 1$ *then f has an end of degree three which is adjacent to both ends of e, and*
(ii) *if* $\text{index}(e) = 2$ *then f is the ceiling of a house whose floor is e.*

See Figure 18.6.

Proof As G is a brick and e is thin, the retract H of $G - e$ is also a brick. Moreover, as f is a multiple edge of H, the graph $H - f$ is also a brick. We have seen that $G - e - f$ is matching covered. Moreover, the retract of $G - e - f$ is $H - f$, a brick. Thus, $G - e - f$ is a near-brick.

Assume that f is not removable in G, implying that $G - f$ is not matching covered. But, $G - e - f$ is matching covered; hence e is the only unmatchable edge in $G - f$ (equivalently, e is the only edge that depends on f). It follows that $G - e - f$ has a maximal barrier, say B, such that e is the only edge of G that has both ends in B. Edge f is the only edge which has ends in different components of $G - f - B$. As $G - e - f$ is a near-brick, the maximality of B implies that B is special, by Proposition 17.6.

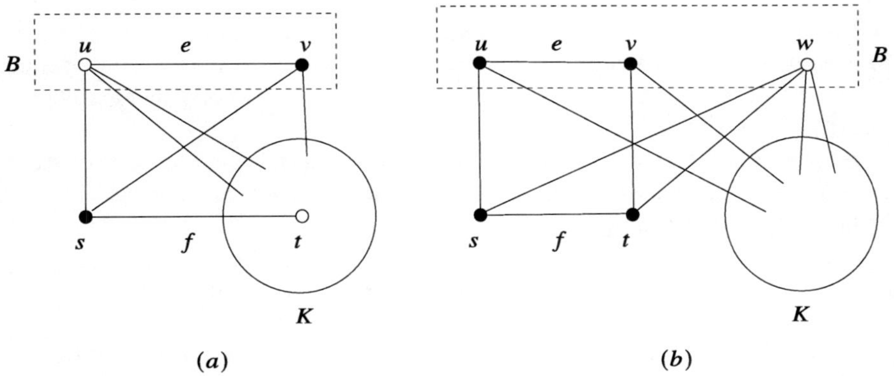

Fig. 18.6 Lemma 18.6: (a) $\text{index}(e) = 1$; (b) $\text{index}(e) = 2$

Suppose that $\text{index}(e) \in \{1, 2\}$. Let K denote the nontrivial component of $G - f - B$. As $H - f$, a brick, is the retract of $G - e - f$, up to multiple edges H is a brick of $G/\overline{V(K)}$; hence

$$|V(H)| \leq |V(K)| + 1.$$

As B is special,

$$2|B| - 1 = |V(G)| - |V(K)|.$$

As e is thin and $\text{index}(e) \in \{1, 2\}$,

$$|V(G)| - |V(H)| = 2 \cdot \text{index}(e).$$

Addition of the inequalities above and simplification yields

$$|B| \leq \text{index}(e) + 1. \tag{18.1}$$

Let I be the set of $|B| - 1$ isolated vertices of $G - f - B$. We shall now analyse two cases, which turn out to be correspond to the two items in the assertion of the lemma.

Case 1 *The set I contains a vertex, s, which is adjacent to both ends of e.*

Then, e is in a triangle of G. We have assumed that $\text{index}(e) \in \{1, 2\}$. If $\text{index}(e) = 2$ then the edge e is not in any triangle of G. Thus, $\text{index}(e) = 1$. In that case, from (18.1) we infer that $|B| = 2$; hence $B = \{u, v\}$. The vertex s is adjacent in G to three or more vertices. We conclude that s must then be an end of f and have degree three in G. Hence the assertion holds in this case.

Case 2 *No vertex of I is adjacent to both ends of e.*

Let s be a vertex in I. It has degree three or more in G. At least two of the neighbours of s are in B and one of them is not an end of e. Thus, $|B| \geq 3$. We have assumed that $\text{index}(e) \in \{1, 2\}$. From (18.1) we infer that $|B| = 3$ and $\text{index}(e) = 2$.

Let w be the vertex of $B - u - v$. Then, s is adjacent to one of the ends of e, and to w. Moreover, s is an end of f and has degree three in G. Likewise, the vertex $t \in I - s$ is also adjacent to an end of e and to w, it is an end of f and has degree three.

Finally, as $\text{index}(e) = 2$, no triangle of G contains an end of e. Thus, one of s and t is adjacent to u, the other is adjacent to v. We deduce that G has a house with e as its floor and f as its ceiling. □

THE CASE IN WHICH f IS REMOVABLE BUT NOT THIN IN G

Lemma 18.7 *Let G be a simple brace on ten or more vertices, or a simple brick, let $e := uv$ be a thin edge of G, let H be the retract of $G - e$ and let f and g be two parallel edges of H. Assume that f is removable in G but that it is not thin. Then, the following properties hold:*

(i) $\text{index}(e) = 2$ *and*
(ii) *edge g is thin in G and* $\text{index}(g) \leq 1$.

Proof Suppose that the edge f is removable in G. For any cut C of G, if $C - f$ is a tight cut of $G - f$ then $C - e - f$ is a tight cut of $G - e - f$. Thus by starting with a tight cut decomposition of $G - f$ and refining it we may obtain a tight cut decomposition of $G - e - f$. Let $\mathcal{F}$ be the set of indecomposable graphs obtained from a tight cut decomposition of $G - f$. We may refine $\mathcal{F}$ to a tight cut decomposition of $G - e - f$. The graph $H - f$ is the retract of $G - e - f$. Thus, $\mathcal{F}$ contains a graph F such that a tight cut decomposition of $F - e$ produces a graph isomorphic to H, up to multiple edges. Thus, $|V(F)| \geq |V(H)|$; hence

$$|V(G)| - 4 \geq |V(F)| \geq |V(H)| \geq |V(G)| - 4, \tag{18.2}$$

where the leftmost inequality follows from the fact that f is not thin, and the rightmost inequality comes from the fact that H is the retract of $G - e$ and e is thin. We conclude that $|V(F)| = |V(H)| = |V(G)| - 4$. Consequently, $\text{index}(e) \in \{2, 3\}$; hence both ends of e have degree three in G.

If F has two contraction vertices then the tight cut decomposition $\mathcal{F}$ of $G - f$ would contain, apart from F, two C_4's. If this were the case, f would be a thin edge of index two, a contradiction. We conclude that F has just one contraction vertex. Let $C := \partial(X)$ then be a nontrivial cut of G such that $C - f$ is tight in $G - f$, the graph $G_1 := (G - f)/(\overline{X} \to \overline{x})$ is bipartite and has order six, and $F = (G - f)/(X \to x)$. See Figure 18.7.

Let B denote the barrier of $G - f$ associated with C and let I be the set of isolated vertices of $G - f - B$. The edge e has no end in $\overline{X}$, because both ends of e have degree three and a tight cut decomposition of $F - e$ would produce an indecomposable graph of order $|V(H)| < |V(F)|$. Thus, e has both ends in X. Adjust notation so that $u \in I$ and $v \in B$. As $\text{index}(e) \in \{2, 3\}$, no edge of $G - e$ incident with an end of e is an edge of H, yet f is an edge of H. Thus, f is not incident with u. It follows that B consists of the three vertices of G adjacent to u. One of the vertices of B is v; the

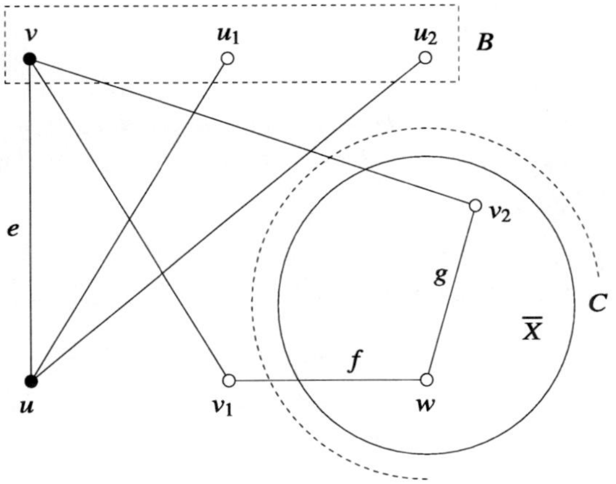

Fig. 18.7 Edge f is removable but not thin in G.

two vertices of $B - v$ are the two vertices of $G - e$ adjacent to u, say, u_1 and u_2. It follows that no triangle of G contains edge e; therefore $\text{index}(e) < 3$. We have seen that $\text{index}(e) \in \{2, 3\}$, hence $\text{index}(e) = 2$. This proves the first part of the assertion.

Let v_1 and v_2 be the two neighbours of v in $G - e$. Edge f is not incident with any vertex of B. In particular, f is not incident with u_1, nor with u_2. Thus, f and g have a common end, w, not in $\{u, v, u_1, u_2, v_1, v_2\}$ (see Figure 18.2). Moreover, one of f and g joins v_1 to w; the other joins v_2 to w. Adjust notation so that $f = v_1w$ and $g = v_2w$. The vertex w, the end of f distinct from v_1, is in $\overline{X}$. The edge g joins v_2 to w. Thus, both ends of g are in $\overline{X}$. In sum, the vertices v_2 and w are in $\overline{X}$ and v_1 is in I. Note that

$$N(v_1) \subseteq B + w.$$

As the edge g is not incident with any vertex in B, it follows that no end of g is adjacent to vertex u. We deduce that there cannot be a house with e as the floor and g as the ceiling. Therefore, by Lemma 18.6, edge g is removable in G. Assume, to the contrary, that g is not thin. We may then apply to g the argument that we applied to f, with v_2 playing the role of v_1 and conclude that $N(v_2)$, just like $N(v_1)$, is a subset of $B + w$. In $G - B - w$, the vertices u, v_1 and v_2 are isolated; therefore $B + w$ is a barrier of G. This is a contradiction, because, no brick, or a brace on ten or more vertices, can have a barrier of size four. Thus, g is thin.

If G is a brick then F is a brick and w is adjacent to three or more vertices in F. Alternatively, if G is a brace then by hypothesis it has order ten or more; hence F is a brace of order $|V(F)| = |V(G)| - 4 \geq 6$. In both alternatives, the end w of g is adjacent to three or more vertices in F; hence it is adjacent to four or more vertices in G. We conclude that $\text{index}(g) \leq 1$. □

Exercises

∗**18.2.1** *Let G be an indecomposable graph, let e be a thin edge of G, and let f and g be two parallel edges of $\widehat{G-e}$. Show that both f and g are removable in the matching covered graph $G-e$.*

18.2.2 *Consider the brace G shown in Figure 18.8. Verify the following: (i) e is a thin edge of G, (ii) f and g are multiple edges in $\widehat{G-e}$, and (iii) f is a thin edge of G, but g is not.*

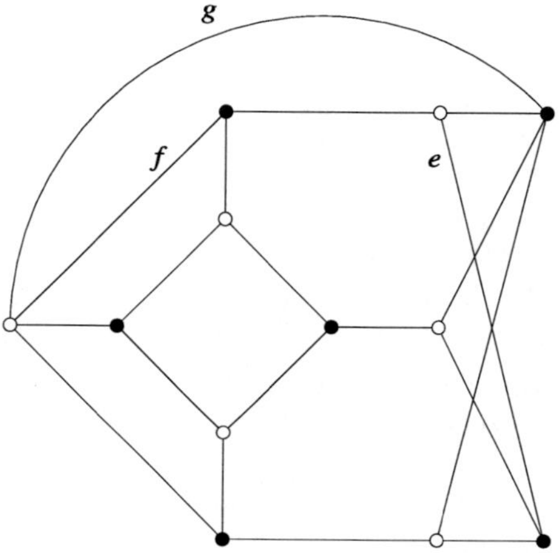

Fig. 18.8 Edge e is thin and edges f and g are multiple edges in $H = \widehat{G-e}$. Edge f is thin in G, but g is not. (This brace is the same as the one in Figure 17.25.)

18.3 Characterization of $\mathcal{G}^\#$

To establish the stated characterization of $\mathcal{G}$, we shall find it convenient to introduce a new parameter which is defined in the following inset.

> **DEFINITION: index(G)**
>
> For an indecomposable graph G which has thin edges, we define index(G) to be equal to the lowest index of a thin edge of G. For example, since all the thin edges of the tricorn $\mathbb{H}_{10}$ (depicted in Figure 17.1(c)) have index three, index($\mathbb{H}_{10}$) = 3. The brace G in Figure 18.8 has some thin edges of index one and some of index two, and none of index zero. Thus index(G) = 1.
>
> It is important to notice that every multiple edge of G is strictly thin. Thus, if index(G) > 0 or if G is free of strictly thin edges then G is simple.

18.3.1 Graphs G with index$(G) = 3$

It turns out that establishing the stated characterization of $\mathcal{G}$, thin edges of index three are easy to deal with. Let us first note that if index$(G) = 3$, then G is a brick.

Lemma 18.8 *Let G be a brick. If* index$(G) = 3$ *then every thin edge of G is strictly thin.*

Proof Let e be a thin edge of G. Let H be the retract of $G - e$. Assume, to the contrary, that H is not simple. Let f and g denote two multiple edges of H. All the ends of e and its neighbours are contracted to a single vertex in the case where index$(e) = 3$. Therefore, f and g have a common end, w, not adjacent to any end of e. See Figure 18.3. Graph H is a brick; therefore vertex w is adjacent in H to three or more vertices. We deduce that w has degree four or more in G.

Let M be a perfect matching of G that contains edge e. As f and g are adjacent in G, it follows that at most one of f and g lies in M. Adjust notation so that f does not lie in M. The edge f is removable in $G - e$ (Exercise 18.2.1). By Proposition 17.11, edge f is removable in G. By Lemma 18.7, edge f is thin in G. But the end w of f has degree four or more; therefore index$(f) < 2$. This is a contradiction to the hypothesis that index$(G) = 3$. We conclude that e is strictly thin. This conclusion holds for each thin edge e of G. □

In view of the above lemma, we may now restrict our attention to indecomposable graphs with index = 1 or 2 which have thin edges, but are free of strictly thin edges. The structure of such a graph neatly depends on the value of that parameter. We divide the proof accordingly.

18.3.2 Graphs G with index$(G) = 1$

In this case, odd wheels, biwheels and truncated biwheels emerge as the families of indecomposable graphs without strictly thin edges. The following lemma establishes the existence of a certain configuration in the vicinity of thin edges of index one in graphs with index = 1. It is a simple consequence of Lemmas 18.6 and 18.7. We shall state it in a notation that is convenient for describing how this configuration replicates itself.

Lemma 18.9 *Let G be a brace on ten or more vertices, or a brick. Assume that G is free of strictly thin edges. Let $e := hv_2$ be a thin edge of index one of G, where v_2 is the end of e with degree three. Let v_1 and v_3 denote the neighbours of v_2 distinct from h. Then, vertices v_1 and v_3 have both degree three in G and there is a vertex h' of degree four or more, possibly equal to h, that is adjacent to both v_1 and v_3 (Figure 18.9). Moreover:*

(i) *if $f := h'v_1$ is not removable in G, then h and h' are distinct and v_1 is adjacent to h (Figure 18.9(a)), and*

(ii) *if* $f := h'v_1$ *is removable in* G*, then it is thin of index one (Figures 18.9(b) and 18.9(c)).*

(Similar statements also apply to $g := h'v_3$*.)*

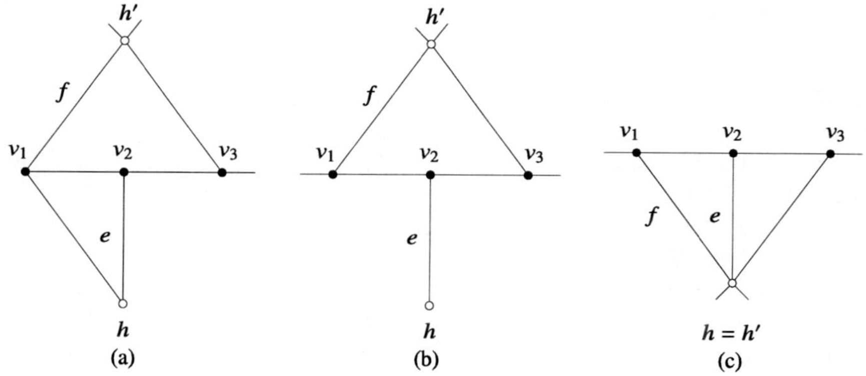

Fig. 18.9 Lemma 18.9: (a) f is not removable; (b) f is thin and $h' \neq h$; (c) f is thin and $h' = h$

Proof Let H denote the indecomposable graph obtained from $G - e$ by the bicontraction of v_2. By hypothesis, e is not strictly thin. Therefore, H has multiple edges. This implies that G has a vertex h', possibly equal to h, but distinct from v_2, that is adjacent to both v_1 and v_3.

Let us now prove that vertex h' has degree four or more in G. If G is a brick then h' is adjacent to three or more vertices in H. If G is a brace, then, by hypothesis, G has eight or more vertices; therefore H is a brace on six or more vertices. In both alternatives, h' is adjacent to three or more vertices in H. We conclude that h' has degree four or more in G.

To complete the proof, we consider separately two cases, depending on whether or not $f = h'v_1$ is removable in G. In the case in which f is not removable, then, by Lemma 18.6, the edges e and f are not adjacent; hence h and h' are distinct. Moreover, vertex v_1 has degree three and is adjacent to both h and v_2 (Figure 18.9(a)). The assertion holds in this case. Thus, suppose that f is removable in G. By Lemma 18.7, if the edge f were not thin, the index of e would have to be two. Thus, f is thin in G. By hypothesis, G is free of strictly thin edges. Therefore, $\text{index}(f) > 0$. One end of f, namely h', has degree four or more. As $\text{index}(f) \geq 1$, we deduce that the other end v_1 of f has degree three, and that edge f has index one.

By applying similar arguments to the edge $g := h'v_3$, we may conclude that v_3 also has degree three, and also similar properties for g. □

Theorem 18.10 *Let* G *be a brace on ten vertices or more, or a brick. Suppose that* G *is free of strictly thin edges. If* $\text{index}(G) = 1$*, then* G *is either an odd wheel, a biwheel or a truncated biwheel.*

Proof By hypothesis, G has a thin edge $e := hv_2$ of index one. Adjust notation so that v_2 has degree three. Then, h has degree four or more. Let v_1 and v_3 be the two neighbours of v_2 distinct from h. By Lemma 18.9, G has a vertex h' of degree four or more, possibly equal to h, that is adjacent to both vertices v_1 and v_3. Moreover, vertices v_1 and v_3 have both degree three.

Let thus $P := v_1v_2 \dots v_t$, $t \geq 3$, be a path of maximum length in $G - h - h'$ that has the following properties (see Figure 18.10):

(i) every vertex v_i of P has degree three, and is adjacent to h if i is even, and to h' if i is odd, and
(ii) for every internal vertex v_i of P, the edge of G that joins v_i to one of h and h' is thin of index one.

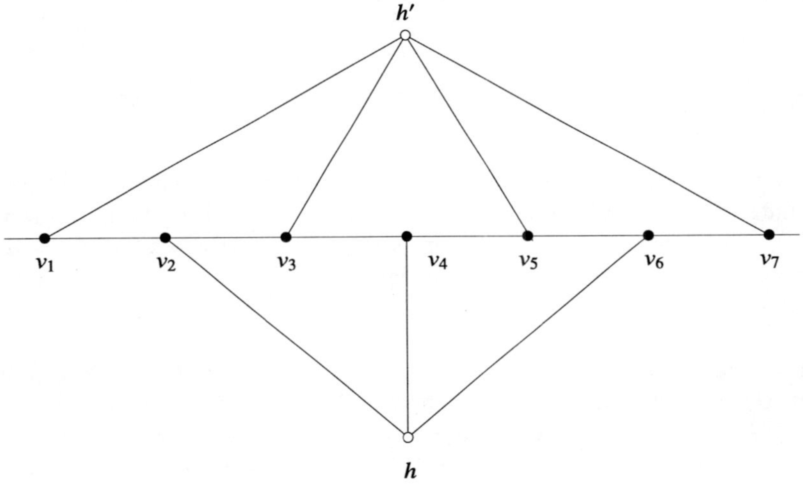

Fig. 18.10 An illustration for Theorem 18.10.

Let $f := h'v_1$, and $f' := xv_t$, where $x = h'$ if t is odd and $x = h$ if t is even. To draw the desired conclusion, we shall now consider the status of the edges f and f'.

Case 1 *One of the edges f and f' is removable in G.*

Adjust notation so that f is removable in G. By Lemma 18.9, we deduce that f is a thin edge of index one. Also, by Lemma 18.9, with f playing the role of e, we deduce that the neighbours of v_1, distinct from h', have degree three in G. Thus v_1 has a neighbour v_0 of degree three in G which is different from v_2. By the maximality of P, vertex v_0 lies in $V(P)$. All the vertices of P have degree three. Therefore, v_0 cannot be an internal vertex of P. We deduce that $v_0 = v_t$. Hence the subgraph of G induced by $V(P)$ is a cycle. Moreover, every vertex of P is adjacent only to vertices of $V(P) \cup \{h, h'\}$. Graph G, a brick or a brace on more than four vertices, is 3-connected. Therefore, $V(G) = V(P) \cup \{h, h'\}$. If t is odd then $h = h'$ and G is an odd wheel, on six or more vertices, with h as its hub. If t is even then h and h' are distinct and G is a biwheel, on ten or more vertices, with h and h' as its two hubs, In both alternatives, edge f' is also removable in G. (Figure 2.9 depicts the biwheel $\mathbb{B}_{10}$.)

Case 2 *Neither f nor f' is removable in G.*

By applying Lemma 18.9, and using the fact that f is not removable, we deduce that h and h' are distinct and v_1 is joined to both h and h'. Likewise, v_t is joined to both h and h'. The set $\{h, v_1, v_2\}$ induces a triangle; hence G is not bipartite. Thus, G is a brick. Moreover, every vertex of P is adjacent only to vertices of $V(P) \cup \{h, h'\}$. Graph G, a brick, is 3-connected. Therefore, $V(G) = V(P) \cup \{h, h'\}$. As h and h' are distinct, it follows that t is even. Then, as the degrees of h and h' are equal and greater than three, it follows that $t \geq 6$. We deduce that G is a truncated biwheel. (Figure 2.10 depicts the truncated biwheel $\mathbb{T}_{10}$.) □

18.3.3 Graphs G with index$(G) = 2$

Now we turn to indecomposable graphs G in which there are no strictly thin edges, and index$(G) = 2$. Every such graph turns out to be either a prism, or a Möbius ladder, or a staircase. As in the case index$(G) = 1$, we start with a lemma which establishes the existence of a certain configuration in the vicinity of every thin edge of index two.

Lemma 18.11 *Let G be a brace on ten vertices or more, or a brick. Suppose that* index$(G) = 2$ *and that G is free of strictly thin edges. Let $e := u_2v_2$ be a thin edge of index two. Let u_1 and u_3 be the two neighbours of u_2 in $G - e$ and let v_1 and v_3 be the two neighbours of v_2 in $G - e$. Then, the vertices u_1, u_3, v_1 and v_3 are distinct and have degree three. Moreover, there are precisely two edges joining vertices in $\{u_1, u_3\}$ to vertices in $\{v_1, v_3\}$ and those two edges are nonadjacent (see Figure 18.11(c)).*

Proof Let H be the retract of $G - e$. By hypothesis, e is not strictly thin. Therefore, H has multiple edges. Let f and g denote two multiple edges of H. Then, both edges have ends that are adjacent to the same end of e (see Figure 18.2). Adjust notation so that f is incident with vertex u_1, whereas g is incident with vertex u_3.

18.11.1 *Both ends of f have degree three in G. Moreover, if f is removable in G then f is thin. (A similar statement holds for g.)*

Proof If f is not removable in G then, by Lemma 18.6, edge f is the ceiling of a house whose floor is edge e; in that case, both ends of f have degree three. Suppose thus that f is removable in G. If f is not thin then, by Lemma 18.7, g is thin and has index one. This is a contradiction to the hypothesis that index$(G) = 2$. Thus, f is thin; hence index$(f) \in \{2, 3\}$. Consequently, both ends of f have degree three. A similar conclusion holds for edge g. □

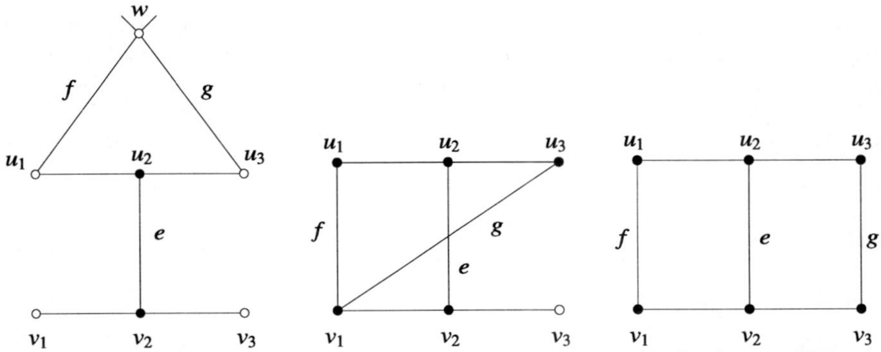

(a) w not adjacent to v_2 (b) w adjacent to v_2 (c) f and g nonadjacent

Fig. 18.11 Illustration for Lemma 18.11.

18.11.2 *The edges f and g are not adjacent.*

Proof Assume that f and g share a common vertex, w. Suppose that w is not adjacent to v_2 (see Figure 18.11(a)). Then, w is not adjacent to any end of e. By Lemma 18.6, edges f and g are both removable in G. As $\text{index}(G) = 2$, it follows, by Lemma 18.7, that f and g are both thin. In H, a brace on six or more vertices, or a brick, vertex w is adjacent to three or more vertices. Therefore, w has degree four or more in G. It follows that $\text{index}(f), \text{index}(g) \leq 1$, in contradiction to the hypothesis that $\text{index}(G) = 2$.

Suppose thus that $w \in \{v_1, v_3\}$. Adjust notation so that $w = v_1$ (see Figure 18.11(b)). The vertex v_1, an end of f, has degree three and it is adjacent to the three vertices of $B := \{v_2, u_1, u_3\}$. In $G - B$, the vertices v_1 and u_2 are both isolated. Hence B is a barrier of G. This conclusion implies that G is a brace on six vertices, a contradiction to the hypothesis that if G is a brace then it has ten or more vertices. Indeed, f and g are nonadjacent. □

It follows that both f and g join vertices in $\{u_1, u_3\}$ to vertices in $\{v_1, v_3\}$. Moreover, if we adjust notation so that $f = u_1v_1$, it follows that $g = u_3v_3$ (see Figure 18.11(c)). We conclude that the four vertices in $\{u_1, u_3, v_1, v_3\}$ have degree three. □

Theorem 18.12 *Let G be a brace on ten or more vertices, or a brick. Assume also that G is free of strictly thin edges. If* $\text{index}(G) = 2$ *then G is either a prism, a Möbius ladder or a staircase.*

Proof By hypothesis, G has a thin edge u_2v_2 of index two. Let u_1 and u_3 denote the two neighbours of u_2 distinct from v_2, let v_1 and v_3 denote the two neighbours of v_2 distinct from u_2. By hypothesis, e is not strictly thin. Therefore, by Lemma 18.11, the vertices in $\{u_1, u_3, v_1, v_3\}$ all have degree three. Moreover, there are precisely two edges that join u_1 and u_3 to v_1 and v_3, and the two edges are nonadjacent. Adjust notation so that u_1 is adjacent to v_1 and u_3 to v_3.

Let $P := u_1u_2\cdots u_t$ and $Q := v_1v_2\cdots v_t$, $t \geq 3$, be two disjoint paths in G of maximum length, such that for $i = 1, 2\ldots, t$, the vertices u_i and v_i are adjacent and both have degree three, and, if $1 < i < t$ then edge u_iv_i is thin of index two (see Figure 18.12.).

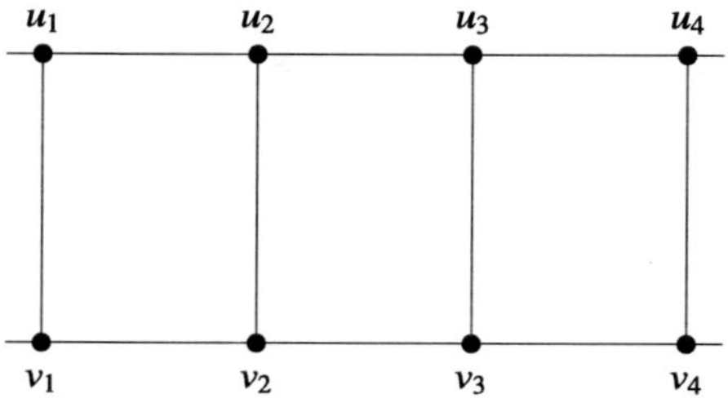

Fig. 18.12 Illustration for Theorem 18.12.

Case 1 *At least one of u_1v_1 and u_tv_t is removable in G.*

Adjust notation so that u_1v_1 is removable in G. By Lemma 18.7, u_1v_1 is thin in G. The index of u_1v_1 is two or three, because $\text{index}(G) = 2$. But the index of u_1v_1 cannot be three; otherwise there would be a vertex y adjacent to both u_1 and v_1, and consequently edge u_2v_2 would depend on edge u_1v_1, a contradiction. Indeed, u_1v_1 is thin of index two.

By Lemma 18.11, vertex u_1 has a neighbour u_0 distinct from u_2 and v_1, and v_1 has a neighbour v_0 distinct from u_1 and v_2, such that u_0 and v_0 are adjacent and have degree three. By the maximality of t, it follows that $\{u_0, v_0\} = \{u_t, v_t\}$. If $u_0 = u_t$ then $v_0 = v_t$ and G is a prism. If $u_0 = v_t$ then $v_0 = u_t$ and G is a Möbius ladder (Figure 2.7 depicts the prism $\mathbb{P}_{10}$ and the Möbius ladder $\mathbb{M}_{10}$). In both alternatives, u_tv_t is also removable in G.

Case 2 *Neither u_1v_1 nor u_tv_t is removable in G.*

By Lemma 18.6, with u_2v_2 playing the role of e, we have that G has a vertex, w_1, that, together with vertices u_1, u_2, v_1, v_2, span a house, where u_1v_1 is the ceiling and u_2v_2 the floor. Likewise, G has a vertex, w_t, which is adjacent to both u_t and v_t. Graph G, a brick or a brace, is 3-connected. Therefore, $V(G) = V(P) \cup V(Q) \cup \{w_1, w_t\}$. Moreover, as G has an even number of vertices, it follows that w_1 and w_t are distinct. Finally, w_1 and w_t must be adjacent, in order to have minimum degree three. We conclude that G is a staircase (Figure 2.8(b) depicts the staircase $\mathbb{S}_{10}$). □

18.3.4 The characterization of $\mathcal{G}^\sharp$

The following result combines the results of McCuaig [68] and Norine and Thomas [78].

Theorem 18.13 *Let G be a brace on six or more vertices or a brick. If G has no strictly thin edge then G is either the Petersen graph or is an odd wheel, a biwheel, a truncated biwheel, a prism, a Möbius ladder or a staircase.*

Proof Suppose that G has no thin edges. If G is a brace then G is either K_2 or C_4, both braces with fewer than six vertices. Thus, G is a brick, hence G is either K_4, $\overline{C_6}$ or the Petersen graph. The graph K_4 is an odd wheel and $\overline{C_6}$ a prism (and a truncated biwheel).

We may thus assume that G has thin edges. If $\text{index}(G) = 3$ then, by Lemma 18.8, every thin edge of G is strictly thin. Thus, $\text{index}(G) \leq 2$. As G has no strictly thin edges, its thin edges have positive index. Thus, $\text{index}(G) \in \{1, 2\}$.

The only simple brace on six vertices is $K_{3,3}$, a Möbius ladder. Therefore we may assume that if G is a brace then it has eight or more vertices. Every brace on eight vertices other than the cube has a thin edge of index zero. The cube is a prism. We may thus assume that if G is a brace then G has ten vertices or more.

If $\text{index}(G) = 1$ then, by Theorem 18.10, G is either an odd wheel, a biwheel or a truncated biwheel. If $\text{index}(G) = 2$ then, by Theorem 18.12, G is either a prism, a Möbius ladder or a staircase. □

Exercises

▷**18.3.1** *Why, in the statement of Lemma 18.7, it is required that the brace have ten or more vertices? And in Lemma 18.9? And in Theorem 18.10?*

18.4 Generating Simple Braces and Bricks

We conclude the chapter with a brief observation concerning the generation procedures for simple braces and bricks. These involve the four expansion operations described in Chapter 17. Suppose that G is a simple brick and e is a strictly thin edge of G, and H, the retract of $G - e$, is a simple brick. If $\text{index}(e) = 0$ then $G = H + e$. If $\text{index}(e) = 1$ then G is an expansion of H by a barrier of size two. If $\text{index}(e) = 2$ then G is an expansion of H by two barriers of size two. Finally, if $\text{index}(e) = 3$ then G is an expansion of H by a barrier of size three. The expansion operations for bricks were described in Section 17.4. Theorem 18.13 now implies the following result of Norine and Thomas (2007, [78]).

Theorem 18.14 *Given any simple brick G, there exists a sequence*

$$G_1, G_2, \ldots, G_k$$

of simple bricks such that: (i) G_1 is either an odd wheel, or a truncated biwheel, or a prism, or a Möbius ladder, or a staircase, or the Petersen graph, and $G_k = G$, and (ii) for $2 \leq i \leq k$, G_i is obtained from G_{i-1} by an expansion. □

With the exception of the expansion by a barrier of size three, all the other expansions have similar expansions for braces. The expansion operations for braces were described in Section 17.7. Theorem 18.13 also implies the following result of McCuaig (2001,[68]).

Theorem 18.15 *Given any simple brace G of order six or more, there exists a sequence*

$$G_1, G_2, \ldots, G_k$$

of simple braces such that: (i) G_1 is either a biwheel, or a prism, or a Möbius ladder, and $G_k = G$, and (ii) for $2 \leq i \leq k$, G_i is obtained from G_{i-1} by an expansion. □

Exercises

18.4.1 *Using Theorem 18.15, show that every brace of order six or more has a matching minor which, up to isomorphism, is either $K_{3,3}$ (which is a Möbius ladder) or the cube (which is both a prism and a biwheel).*

18.5 Splitter Versions

As noted earlier, Norine and Thomas [78] and McCuaig [68] proved generalizations of Theorems 18.14 and 18.15, respectively. They refer to those generalizations as 'splitter-versions', as they are motivated by a theorem on 3-connected graphs, due to P.D. Seymour, which is known as the splitter theorem. For the sake of clarity, we shall restrict ourselves here to bricks. Theorem 18.16, below, is the splitter-version of Theorem 18.14.

Recall (Chapter 12) that a matching covered graph H is a *matching minor* of another matching covered graph G if it can be obtained from a conformal subgraph of G by means of a sequence of bicontractions of vertices of degree two. For example, by Theorem 12.13, every brick contains either K_4 or $\overline{C_6}$ as a matching minor.

In order to be able to state the generalization of Theorem 18.14 proved by Norine and Thomas [78], we need to define a new class of graphs. A *prismoid* is either a truncated biwheel or one obtained from it by adding an edge joining its two hubs.

Theorem 18.16 *Let G be a brick which is not an odd wheel, a prismoid, a prism, a Möbius ladder, a staircase, or the Petersen graph, and let $H \notin \{K_4, \overline{C_6}\}$ be a simple brick that is a matching minor of G. Then there there exists a sequence $G_1, G_2, \ldots, G_k$ of simple bricks distinct from the Petersen graph such that: (i) $G_1 = H$, $G_k = G$, and (ii) for $2 \leq i \leq k$, G_i is obtained from G_{i-1} by an expansion.* □

An analogue of the above theorem for braces, due to McCuaig [68] is also described in [78]. Another proof of McCuaig's splitter theorem for braces was given by CLM [18]. Unfortunately, a proof of the more general theorem stated above is beyond the scope of this book.

18.6 Notes

Kothari [47] proved that every simple near-bipartite brick G, except for those in eight infinite families, has a strictly thin edge e such that the retract of $G - e$ is also near-bipartite. (If one wants to fix the removable doubleton and preserve it, then 11 infinite families are required.) This theorem and its proof also appear in Kothari and Carvalho (2020, [49]).

Part III
Pfaffian Orientations

Chapter 19
Pfaffian Orientations

Contents

19.1 The Pfaffian Orientation Problem

This chapter is dedicated mainly to some interesting questions arising from Tutte's original proof of his theorem, presented in Section 1.5, using Pfaffian identities. For the convenience of the readers, we recall briefly the relevant definitions.

According to Cayley's Theorem (1.9), the determinant of a skew-symmetric matrix $\mathbf{A} = (a_{ij})$ is the square of a polynomial in the entries of $\mathbf{A}$, called the Pfaffian of $\mathbf{A}$. As the determinant of a skew-symmetric matrix of odd order is always zero, we restrict our attention here to graphs of even order. Thus let G be a graph of order $2k$ and let D be an orientation of G. Then the adjacency matrix $\mathbf{A} = (a_{ij})$ of D is a skew-symmetric matrix. Let $M := \{e_1, e_2, \ldots, e_k\}$ be any perfect matching of G, and suppose that in the orientation D of G, the tail and the head of the edge e_i, for $1 \le i \le k$, are u_i and v_i, respectively. We may then associate with M the permutation $\pi(M)$ as defined below:

$$\pi(M) := \begin{pmatrix} 1 & 2 & 3 & 4 & \ldots & 2k-1 & 2k \\ u_1 & v_1 & u_2 & v_2 & \ldots & u_k & v_k \end{pmatrix}.$$

C. L. Lucchesi, U. S. R. Murty, *Perfect Matchings*, Algorithms and Computation in Mathematics 31, https://doi.org/10.1007/978-3-031-47504-7_19

The sign of $\pi(M)$ is equal to $(-1)^r$, where r is the number of inversions of $\pi(M)$. The sign of $\pi(M)$ does not depend on the order in which the edges of M are listed, but does depend on the orientation D of G. We define the *sign of M*, denoted $\text{sign}(M)$, to be the sign of $\pi(M)$. The Pfaffian $\text{Pf}(\mathbf{A})$ of the adjacency matrix $\mathbf{A}$ is then defined as the following polynomial in the entries of $\mathbf{A}$:

$$\text{Pf}(\mathbf{A}) := \sum \text{sign}(M)\, a_{u_1v_1}a_{u_2v_2}\dots a_{u_kv_k} \tag{19.1}$$

where the sum is taken over all M in $\mathcal{M}$, the set of perfect matchings of G.

Since, by the adopted convention that, for $1 \leq i \leq k$, vertex u_i is the tail and v_i is the head in D for any edge e_i of M, it follows that the product $a_{u_1v_1}a_{u_2v_2}\dots a_{u_kv_k}$ is always equal to one. Therefore, if the signs of all perfect matchings of G with respect to an orientation D were the same, then the absolute value of $\text{Pf}(\mathbf{A})$ would simply be the number of perfect matchings of D. In this case, that number could be computed efficiently by computing the determinant of $\mathbf{A}$ and taking its square root. Motivated by this observation, we now introduce the concept that is the object of our main focus in this part of the book.

PFAFFIAN DIGRAPHS AND PFAFFIAN GRAPHS

A digraph is *Pfaffian* if all its perfect matchings have the same sign. In the same vein, an (undirected) graph is *Pfaffian* if it admits an orientation that is Pfaffian.

We shall regard every digraph as an orientation of its underlying (undirected) graph. The same graph may admit one orientation that is Pfaffian and one which is not. For example, the second orientation of K_4 shown in Figure 1.8 is a Pfaffian orientation, but the first orientation is not.

Although our main interest is in orientations of matching covered graphs, we shall find it convenient to consider orientations of arbitrary undirected graphs which may not even be matchable. (Any orientation of an undirected graph with at most one perfect matching is a Pfaffian orientation!)

It should be noted that the signs of perfect matchings of a digraph do depend on the chosen enumeration of its vertices. However, the property of the digraph being Pfaffian or non-Pfaffian is independent of that enumeration. More specifically, two isomorphic digraphs are either both Pfaffian or both non-Pfaffian. (See Exercise 19.1.1.)

The above definitions lead to the following important decision problems:

Problem 19.1 (The Pfaffian recognition problem)

Given: A digraph D;
Decide: whether D is Pfaffian.

Problem 19.2 (The Pfaffian orientation problem)

Given: A graph G;
Decide: whether G admits a Pfaffian orientation.

Both these problems are known to be 'difficult' in general. However, we shall see that both of them are in co-NP and are polynomial-time equivalent. (The first proof of these results, due to Vazirani and Yanakakis, appeared in 1989 [91]. The CLM proof, using a different approach, appeared in 2005 [14]. This proof is presented in Section 20.4.) But in some special classes of graphs (such as planar graphs and bipartite graphs), polynomial-time algorithms have been devised to resolve these problems. These discoveries have some surprising and spectacular applications in the theory of algorithms.

As noted before, the idea of using Pfaffians in matching theory goes back to Tutte (1947, [89]). In his mathematical autobiography *Graph Theory as I have known it* (1998, [90]), he describes the thought process that led him to the discovery of the relationship between Pfaffians and perfect matchings in graphs. The notion of Pfaffian orientations is, however, due to Kasteleyn (1963, [44]). He established the fundamental result that every planar graph has a Pfaffian orientation and used it to count the number of perfect matchings in certain classes of planar graphs. His work was motivated by problems in statistical physics. We shall present a proof of Kasteleyn's Theorem in Section 19.3.

Exercises

19.1.1 *Let D and D' be two isomorphic digraphs on $\{1, 2, \ldots, 2n-1, 2n\}$, and let σ denote an isomorphism from D to D'. Show that there exists a constant $c \in \{1, -1\}$ such that for any perfect matching M of the underlying graph of D, $\mathrm{sign}(\pi_{D'}(\sigma(M))) = c \cdot \mathrm{sign}(\pi_D(M))$.*

19.1.2 *The adjacency matrix of the second orientation of K_4, shown in Figure 1.8 is:*

$$\begin{bmatrix} 0 & +1 & +1 & +1 \\ -1 & 0 & -1 & +1 \\ -1 & +1 & 0 & -1 \\ -1 & -1 & +1 & 0 \end{bmatrix}$$

Verify that the determinant of this matrix is nine, which is the square of the number of perfect matchings of K_4.

19.1.3 *Is the orientation of $K_{3,3}$ shown in Figure 19.1 a Pfaffian orientation? (Hint: Consider signs of the two perfect matchings $M_1 := \{12, 34, 56\}$ and $M_2 := \{12, 63, 45\}$.)*

▷**19.1.4** *Let R be a removable class of a matching covered graph G and let D be an orientation of G. Suppose that $D - R$ is Pfaffian, and all perfect matchings including R have equal sign. Prove that G is Pfaffian.*

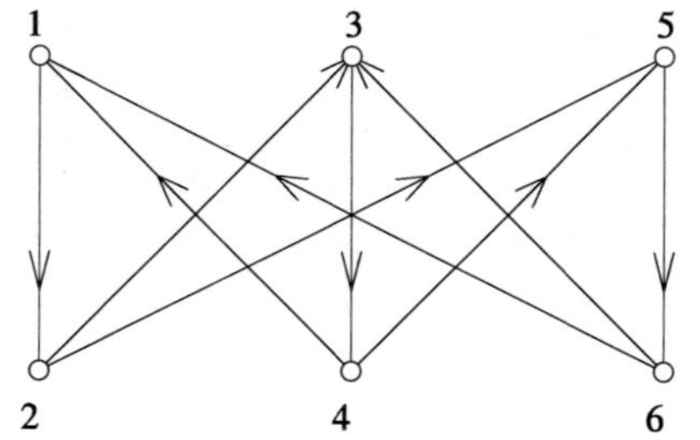

Fig. 19.1 An orientation of $K_{3,3}$

19.2 Equivalent Formulations

In this section we shall present several equivalent definitions of Pfaffian orientations in purely graph theoretical terms by using the terminology and notation described in the following inset:

> ODDLY AND EVENLY ORIENTED TRAILS
>
> Let D be an orientation of a graph G and let T be a trail of G. An arc of T is a *forward arc* of T if it is traversed from tail to head; and is a *reverse arc* of T if it is traversed from head to tail. The set of forward arcs of T is denoted $\text{fw}_D(T)$ and the set of reverse arcs of T is denoted $\text{rv}_D(T)$. (If D is understood, the subscript D is omitted.)
>
> We say that T is *oddly oriented* if $|\text{fw}(T)|$ is odd and is *evenly oriented* if $|\text{fw}(T)|$ is even.
>
> Note that $|\text{fw}(T)| + |\text{rv}(T)|$ is equal to the length of T. Consequently, if T has even length then $|\text{fw}(T)| \equiv |\text{rv}(T)| \pmod 2$. In that case, T is oddly oriented if and only its reverse is oddly oriented. In particular, this property holds for every even cycle and every path of even length of D.

In the first orientation of K_4 shown in Figure 1.8, the cycle $(1,3,4,2,1)$ is oddly oriented and the cycle $(1,3,2,4,1)$ is evenly oriented. The characterization of Pfaffian orientations of even cycles is a very important simple result, which has a fundamental, easily proved consequence (Exercises 19.2.1 and 19.2.2).

Proposition 19.3 *Let D be a digraph whose underlying graph is an even cycle, Q. The digraph D is Pfaffian if and only if Q is oddly oriented in D.* □

Corollary 19.4 *Let D be a digraph, let M_1 and M_2 be two perfect matchings of G and let k denote the number of evenly oriented (M_1, M_2)-alternating cycles in D. Then,* $\text{sign}(M_1) \cdot \text{sign}(M_2) = (-1)^k$. □

19.2.1 Oddly oriented cycles and conformal subgraphs

Recall that a matchable subgraph H of a graph G is conformal if $G - V(H)$ is matchable. For example, a 6-cycle in the Petersen graph is not conformal, but an 8-cycle is; see Figure 19.2.

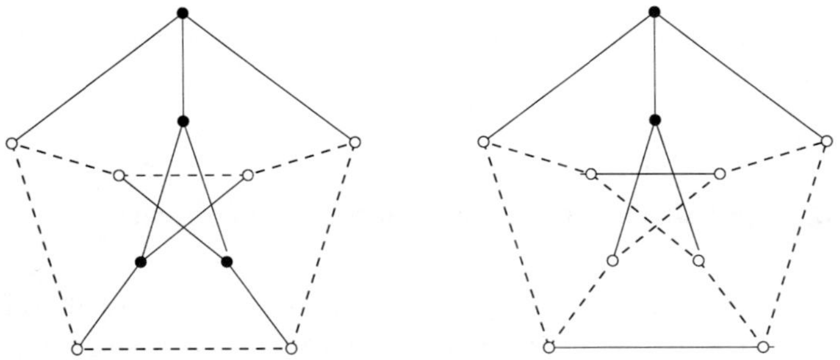

Fig. 19.2 Even cycles in the Petersen graph

Lemma 19.5 *Let G be a graph, let D be an orientation of G and let M be a perfect matching of G. The following assertions are equivalent:*

(i) *the orientation D of G is Pfaffian,*
(ii) *for every conformal subgraph H of G, the restriction D_H of D to H is Pfaffian,*
(iii) *every conformal cycle of D is oddly oriented,*
(iv) *every M-alternating cycle of D is oddly oriented.*

Proof (i)⇒(ii): Suppose that D is a Pfaffian orientation of G, let H be a conformal subgraph of G, let D_H be the restriction of D to H, and let N_1 and N_2 be two perfect matchings of H. As H is conformal, $G - V(H)$ has a perfect matching, N'. For $i = 1, 2$, let $M_i := N_i \cup N'$. Then, M_1 and M_2 are perfect matchings of G. As D is a Pfaffian orientation of G, $\text{sign}(M_1) = \text{sign}(M_2)$. We may relabel the vertices of G so that $V(H) = \{1, 2, \ldots, |V(H)|\}$ (see Exercise 19.1.1). It follows that N_1 and N_2 have the same sign in D_H. We deduce that any two perfect matchings of H have the same sign in D_H; hence D_H is a Pfaffian orientation of H.

(ii)⇒(iii): Assume that for every conformal subgraph H of G the restriction D_H of D to H is Pfaffian. In particular, for every conformal cycle C of G, D_C is Pfaffian. By Proposition 19.3, D_C is oddly oriented.

(iii)⇒(iv): Assume that every conformal cycle of D is oddly oriented. Every M-alternating cycle C of G is conformal, because $M - E(C)$ is a perfect matching of $G - V(C)$. Thus, every M-alternating cycle of D is oddly oriented.

(iv)⇒(i): Suppose that every M-alternating cycle of D is oddly oriented. Let N be a perfect matching of G. Every (M, N)-alternating cycle of D is an M-alternating cycle of D; hence it is oddly oriented. By Corollary 19.4, $\text{sign}(N) = \text{sign}(M)$. It follows that every perfect matching of D has the same sign that M has; hence D is a Pfaffian orientation of G. □

In light of the above lemma, one may deduce that the digraph D in Figure 19.1 is not Pfaffian simply from the fact that the cycle $(1, 2, 3, 4, 1)$ is conformal and evenly oriented. This, of course, does not immediately imply that $K_{3,3}$ is non-Pfaffian. However, $K_{3,3}$ is non-Pfaffian (see Exercise 19.2.11(iii)) and, indeed, it is the smallest non-Pfaffian graph. From Lemma 19.5 it is also easy to deduce the following consequence (Exercise 19.2.3).

Corollary 19.6 *A matching covered graph G is Pfaffian if and only if each of its conformal subgraphs is Pfaffian.* □

19.2.2 An application: the even directed cycle problem

A variety of algorithmic problems, not necessarily in graph theory and matrix theory, are reducible to the problem of recognizing whether or not a given directed bipartite graph is Pfaffian; see the paper by McCuaig (2004, [69]). One such problem is the following:

Problem 19.7 (The even directed cycle problem)
Given: a digraph D;
Decide: whether or not D has a directed cycle of even length.

The above problem may be reduced to the Pfaffian Recognition Problem as follows. Let D be any digraph. Obtain the bipartite digraph $B(D)$ from D as described below.

- First split each vertex v of D into two new vertices v^+ and v^-, so that all the arcs that leave v in D leave v^+ in $B(D)$, and that all the arcs that enter v in D enter v^- in $B(D)$. (Thus, if an arc a in D leaves a vertex x and enters a vertex y, then, in $B(D)$, arc a leaves x^+ and enters y^-.)
- Then, for each $v \in V(D)$, add an arc directed from v^+ to v^-.

The graph $B(D)$ is the *associated bipartite digraph* of D. See Figure 19.3. The following assertion is left as Exercise 19.2.8:

Theorem 19.8 *A digraph D has an even directed cycle if and only if the associated bipartite digraph is <u>not</u> Pfaffian.* □

The problem of deciding whether or not a given digraph has an odd directed cycle is easy (Exercise 19.2.13). For a long time, the complexity status of the even directed cycle problem remained unresolved. As we have seen above, that problem is equivalent to the problem of recognizing whether or not a given bipartite digraph is Pfaffian. In Chapter 21 we shall describe a polynomial-time algorithm to determine whether a bipartite graph is Pfaffian.

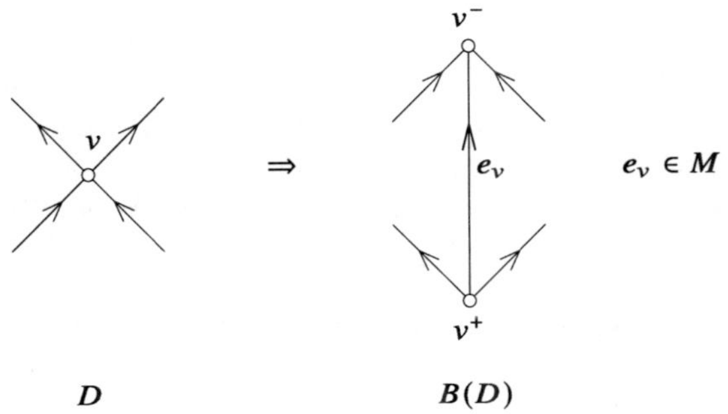

Fig. 19.3 The construction of $B(D)$

19.2.3 Non-Pfaffian graphs

How do we show that a given matching covered graph G does not admit a Pfaffian orientation, if that happens to be the case? A naïve approach would be to consider all possible orientations of G and in each of those orientations exhibit a conformal cycle that is not oddly oriented. This is clearly not feasible even for moderately large graphs. Little (1973, [52]) proposed an ingenious approach in which we choose an orientation D of G and try to show that no re-orientation of D could be a Pfaffian orientation of G. This may at first seem to be no easier than trying out all possible orientations. But Little showed that, when G is non-Pfaffian, there exists a collection of conformal cycles in G whose orientations in D satisfy certain parity conditions and, more significantly, such a collection of cycles serves as a certificate to the effect that no orientation of G could be a Pfaffian orientation.

INTRACTABLE COLLECTIONS OF CONFORMAL CYCLES

Let G be a matching covered graph. A collection $\mathcal{C}$ of conformal cycles of G is *intractable* if it satisfies the following two properties:

(i) every edge of G is in an even number of cycles of $\mathcal{C}$, and
(ii) an odd number of cycles in $\mathcal{C}$ are evenly oriented in some orientation D of G.

Proposition 19.9 *If a matching covered graph G has an intractable collection $\mathcal{C}$ of conformal cycles then, for every orientation D of G, an odd number of cycles in $\mathcal{C}$ is evenly oriented in D. Consequently, G is not Pfaffian.*

Proof Suppose that G has an intractable collection $\mathcal{C}$ of conformal cycles, and let D_0 be an orientation of G such that an odd number of cycles in $\mathcal{C}$ is evenly oriented in D_0. Let D be an orientation of G. Clearly, there is a sequence

$$D_0, D_1, \ldots, D_r = D, \quad r \geq 0$$

such that for every i, $0 \leq i < r$, D_{i+1} is obtained from D_i by reversing the orientation of an edge. By definition, an odd number of cycles in $\mathcal{C}$ is evenly oriented in D_0.

Also by definition, each edge of G is in an even number of cycles in $\mathcal{C}$. A simple inductive argument then shows that $\mathcal{C}$ contains an odd number of cycles which are evenly oriented in D_i, for $i = 0, 1, \ldots, r$. In particular, an odd number of cycles in $\mathcal{C}$ are evenly oriented in D. We deduce that D is not a Pfaffian orientation of G. This conclusion holds for each orientation D of G; hence G is not Pfaffian. □

We now illustrate the concept of intractable collections of conformal cycles with two examples, proving that neither $K_{3,3}$ nor the Petersen graph is Pfaffian.

Lemma 19.10 *The graph $K_{3,3}$ is not Pfaffian.*

Proof Consider the graph $K_{3,3}$, depicted in Figure 19.4(a), and the orientation D of $K_{3,3}$, shown in Figure 19.4(b).

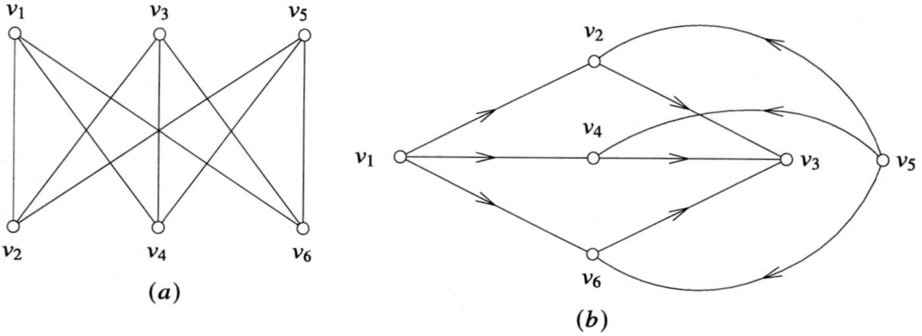

Fig. 19.4 The graph $K_{3,3}$ and the orientation D of $K_{3,3}$

The figure indicates a set $\mathcal{C} := \{C_1, C_2, C_3\}$ of three quadrilaterals, which are conformal cycles of $K_{3,3}$:

$$\begin{aligned} C_1 &:= (v_1, v_2, v_3, v_4, v_1) = K_{3,3} - v_5 - v_6 \\ C_2 &:= (v_1, v_4, v_3, v_6, v_1) = K_{3,3} - v_5 - v_2 \\ C_3 &:= (v_1, v_6, v_3, v_2, v_1) = K_{3,3} - v_5 - v_4 \end{aligned}$$

The three cycles of $\mathcal{C}$ are evenly oriented. Every edge of $K_{3,3}$ is in an even number of cycles in $\mathcal{C}$: the three edges incident with v_5 are not in any cycle in $\mathcal{C}$, each one of the other six edges is in precisely two of the three quadrilaterals. Thus, $\mathcal{C}$ is intractable; hence $K_{3,3}$ is not Pfaffian. □

Lemma 19.11 *The Petersen graph is not Pfaffian.*

Proof Figure 19.5 depicts the orientation D of the Petersen graph. Let $X := \{v_0, v_1, v_2, v_3, v_4\}$. The cut $\partial(X)$ is a directed cut in D; the pentagons $C' := (v_0, v_1, v_2, v_3, v_4, v_0)$ and $C'' := (v_5, v_6, v_7, v_8, v_9, v_5)$ are directed cycles in D. Figure 19.5 also shows the octagon $C_1 := (v_1, v_2, v_9, v_8, v_4, v_3, v_6, v_7, v_1)$, which is a conformal cycle of $\mathbb{P}$, it is equal to $\mathbb{P} - v_0 - v_5$.

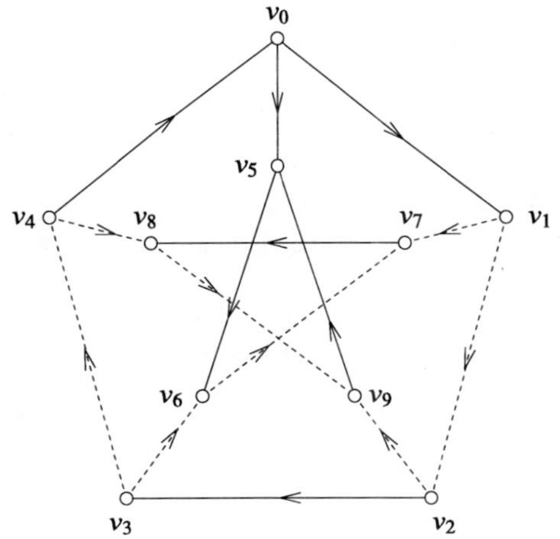

Fig. 19.5 The orientation D of the Petersen graph

The cycle C_1 has four edges in the directed cut $\partial(X)$; two of these edges are forward edges and two are reverse edges. The cycle C_1 has two edges in the cycle C'; one is a forward edge; the other is a reverse edge. Likewise, it has two edges in the cycle C''; one is a forward edge; one is a reverse edge. Thus, C_1 is evenly oriented in D. Consider now the automorphism

$$\varphi := (v_0, v_1, v_2, v_3, v_4)(v_5, v_7, v_9, v_6, v_8)$$

The orbit of C_1 under φ, denoted $\mathcal{C}$, consists of five octagons, listed below:

$$\begin{aligned}
C_1 &:= (v_1, v_2, v_9, v_8, v_4, v_3, v_6, v_7, v_1) = \mathbb{P} - v_0 - v_5 \\
C_2 &:= (v_2, v_3, v_6, v_5, v_0, v_4, v_8, v_9, v_2) = \mathbb{P} - v_1 - v_7 \\
C_3 &:= (v_3, v_4, v_8, v_7, v_1, v_0, v_5, v_6, v_3) = \mathbb{P} - v_2 - v_9 \\
C_4 &:= (v_4, v_0, v_5, v_9, v_2, v_1, v_7, v_8, v_4) = \mathbb{P} - v_3 - v_6 \\
C_5 &:= (v_0, v_1, v_7, v_6, v_3, v_2, v_9, v_5, v_0) = \mathbb{P} - v_4 - v_8
\end{aligned}$$

Each one of the five octagons in $\mathcal{C}$ is evenly oriented. Figure 19.6 shows views of C_1 and C_2.

Each edge of C' is in two octagons of $\mathcal{C}$. Likewise, every edge of C'' is in two octagons of $\mathcal{C}$. Finally, every edge of $\partial(X)$ is in four octagons of $\mathcal{C}$. In sum, every edge of $\mathbb{P}$ is in an even number of octagons of $\mathcal{C}$. We deduce that $\mathcal{C}$ is intractable; hence $\mathbb{P}$ is not Pfaffian. □

The fact that we were able to find intractable collections of conformal cycles for both $K_{3,3}$ and $\mathbb{P}$ is not a coincidence. The following theorem shows that this is always possible.

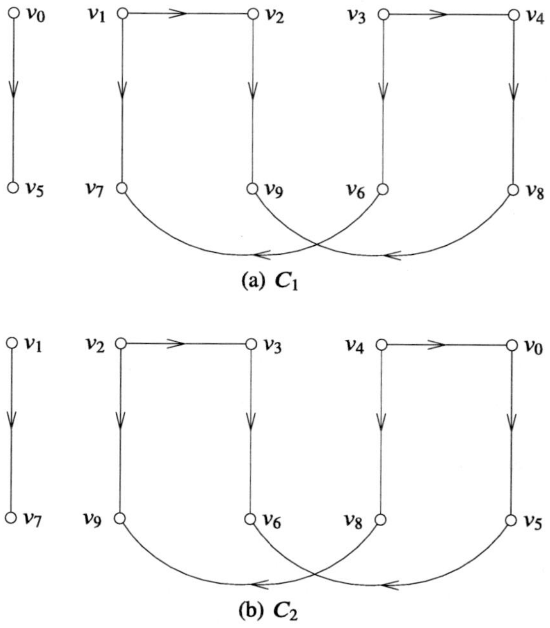

Fig. 19.6 Views of C_1 and C_2

Theorem 19.12 (Little (1973, [52])) *A matching covered graph G is non-Pfaffian if and only if it has an intractable collection of conformal cycles.*

Proof Let D be an orientation of G. We have seen in Proposition 19.9 that the intractability of a collection of conformal cycles does not depend on the orientation D.

For each conformal cycle C of G let

$$t(C) := \begin{cases} 0 \text{ if } C \text{ is oddly oriented in } D, \\ 1 \text{ if } C \text{ is evenly oriented in } D. \end{cases}$$

Consider the following system of linear equations over GF(2), where the set of variables is $\{x_e : e \in E(G)\}$:

$$\sum_{e \in E(C)} x_e = t(C) \quad \text{(one for each conformal cycle } C \text{ of } G). \tag{19.2}$$

19.12.1 *Suppose that the system* (19.2) *has a solution. Then, reversal in D of the edges in* $R := \{e : x_e \equiv 1 \pmod 2\}$ *yields a Pfaffian orientation F of G.*

Proof Let C be a conformal cycle of G. As the system of equations (19.2) has a solution, it follows that $|R \cap E(C)| \equiv t(C) \pmod 2$. That is, if C is oddly oriented in D then $|R \cap E(C)|$ is even and if C is evenly oriented in D then $|R \cap E(C)|$ is odd. We deduce that C is oddly oriented in F. This conclusion holds for each conformal cycle C of G; hence F is a Pfaffian orientation of G. □

19.12.2 *Suppose that G has a Pfaffian orientation, F, and let S be the set of edges of G whose orientations in F and in D are distinct. Then,*

$$x_e := \begin{cases} 1, \textit{ if } e \in S, \\ 0, \textit{ otherwise} \end{cases}$$

is a solution of the linear system (19.2).

Proof Let C be a conformal cycle of G. By hypothesis, F is a Pfaffian orientation of G; thus, C is oddly oriented in F. If C is oddly oriented in D then $t(C) = 0$ and $|S \cap E(C)|$ is even, whereas if C is evenly oriented in D then $t(C) = 1$ and $|S \cap E(C)|$ is odd. In both alternatives, $|S \cap E(C)| \equiv t(C) \pmod 2$. In other words, $\sum_{e \in E(C)} x_e \equiv t(C) \pmod 2$. Indeed x is a solution to the system (19.2). □

It follows that G is non-Pfaffian if and only if the system (19.2) has no solution. By the Theorem of Rouché-Capelli, the system has no solution if and only if the rank in GF(2) of the coefficient matrix is smaller than the rank in GF(2) of the augmented matrix. Thus, the system has no solution if and only if G has a collection $\mathcal{C}$ of conformal cycles such that $\triangle_{C \in \mathcal{C}} E(C) = \emptyset$ and $\sum_{C \in \mathcal{C}} t(C) \equiv 1 \pmod 2$. That is, the system has no solution if and only if G has an intractable collection of conformal cycles. In other words, G is not Pfaffian if and only if it has an intractable collection of conformal cycles. □

In Section 20.4 we shall show that given any matching covered graph G one can find an orientation D of G, which we refer to as a characteristic orientation of G, with the property that G is Pfaffian if and only if D is Pfaffian. Thus, if G happens to be non-Pfaffian, it suffices to show that D is non-Pfaffian. In order to do this, it suffices to point to a conformal cycle in G that is evenly oriented in D.

Exercises

∗**19.2.1** *Prove Proposition 19.3. Hint: (i) enumerate the* $2n$ *vertices of* Q *so that* $Q = 1, 2, \ldots, 2n, 1$ *and let* M_1 *and* M_2 *be the two perfect matchings of* Q*; (ii) verify that if* $\mathrm{rv}(Q)$ *is empty then* M_1 *and* M_2 *have distinct signs; (iii) prove that* $\mathrm{sign}(M_1) \cdot \mathrm{sign}(M_2) = (-1)^{|\mathrm{rv}(Q)|+1}$*, by induction on* $|\mathrm{rv}(Q)|$.

19.2.2 *Deduce Corollary 19.4 from Proposition 19.3.*

▷**19.2.3** *Deduce Corollary 19.6 from Lemma 19.5.*

▷**19.2.4** *Let G be a matchable, but not necessarily matching covered graph. Let N be the set of nonmatchable edges of G and let H := G − N. Prove the following properties:*

(i) *every component of H is matching covered,*
(ii) *an orientation D of G is Pfaffian if and only if its restriction to each component of H is Pfaffian, and*

(iii) *the graph G is Pfaffian if and only if each component of H is Pfaffian.*

Hint: use Lemma 19.5.

▷**19.2.5** *Figure 19.7 depicts an orientation of* $W_5 - e$ *obtained by deleting a spoke e from the 5-wheel* W_5*. By verifying that each conformal even cycle in* $W_5 - e$ *is oddly oriented, deduce that this orientation is a Pfaffian orientation of* $W_5 - e$*. Then show that there is precisely one way of extending the given orientation of* $W_5 - e$ *to a Pfaffian orientation of* W_5*.*

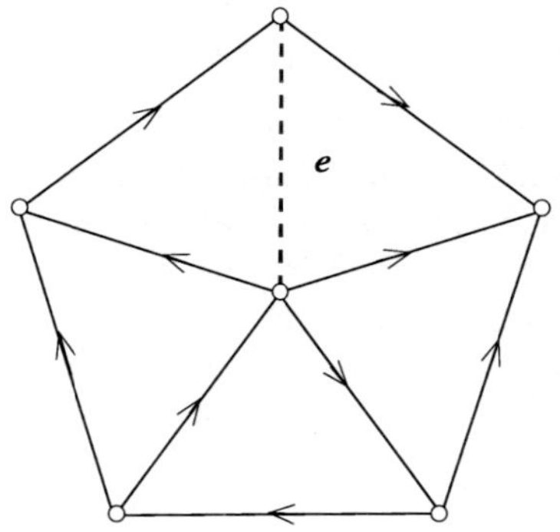

Fig. 19.7 Pfaffian orientations of W_5

▷**19.2.6** *Consider the conformal 4-cycle* $C_4 := (1, 2, 3, 4, 1)$ *in* K_4 *shown in Figure 19.8 with the indicated orientation which makes it an oddly oriented cycle. Find all the ways in which this orientation of* C_4 *may be extended to a Pfaffian orientation of* K_4*.*

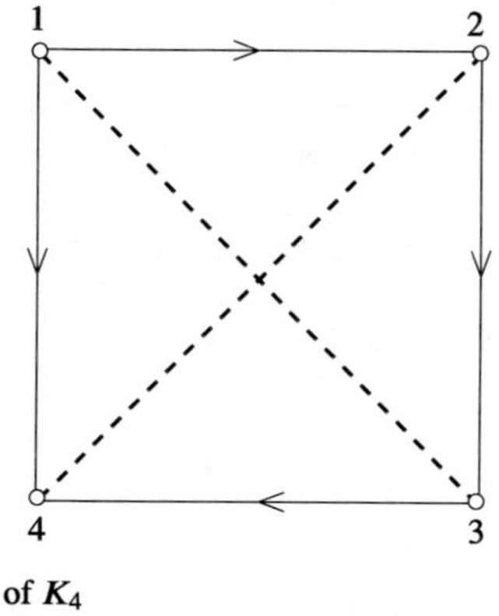

Fig. 19.8 Pfaffian orientations of K_4

▷**19.2.7** *Figure19.9 depicts $\overline{C_6}$ and two orientations, D_0 and D. Show that (i) both these are Pfaffian orientations of $\overline{C_6}$, (ii) both C-contractions of D are Pfaffian orientations of K_4 and that this is not the case with the orientation D_0.*

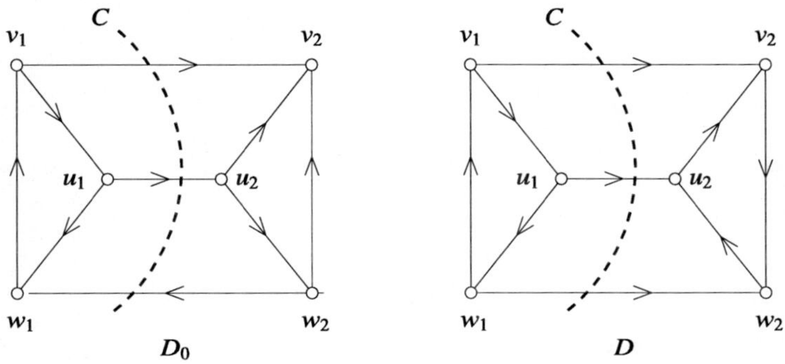

Fig. 19.9 Two Pfaffian orientations of $\overline{C_6}$

▷**19.2.8** *Prove Theorem 19.8. Hint: $M := \{v^+v^- : v \in V(D)\}$ is a perfect matching of $B(D)$ and every perfect matching of $B(D)$ has the same sign of M if and only if D has no directed cycle of even length.*

19.2.9 *The graph shown in Figure 19.10, known as the* Heawood *graph, is the smallest cubic graph of girth six. It has no conformal cycles of length* 0 (mod 4)*. Consider the orientation of the Heawood graph in which all edges are oriented downwards. Show that this orientation is Pfaffian.*

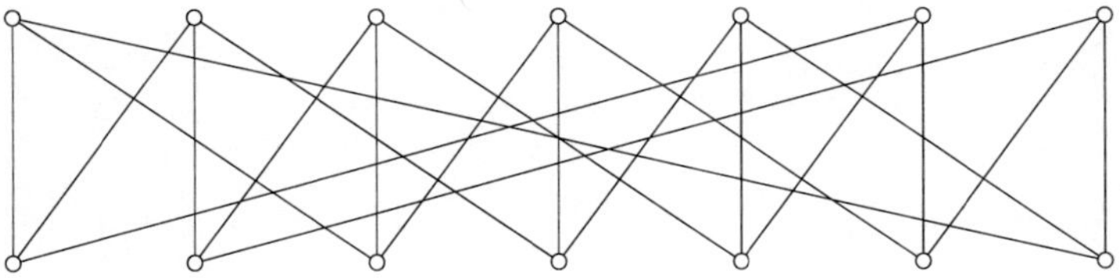

Fig. 19.10 The Heawood graph

19.2.10 *Consider the digraph D, known as the Koh-Tindell digraph, shown in Figure 19.11. Show that $B(D)$ is a Pfaffian orientation of the Heawood graph (see Exercise 19.2.9). Deduce that the Koh-Tindell digraph has no directed even cycles.*

▷**19.2.11 (C. H. C. Little)** *Consider the graph $G := K_{2,k}$, where k is an integer, $k \geq 3$. Let w_1 and w_2 be the two vertices of degree k, let $v_1, v_2, \ldots, v_k$ be the k vertices of degree two of G. For $i = 1, 2, \ldots, k$, let Q_i be the quadrilateral induced by the set $\{v_i, v_{i+1}, w_1, w_2\}$, where the indices are taken modulo k. Figure 19.12 depicts the particular case in which $k = 4$.*

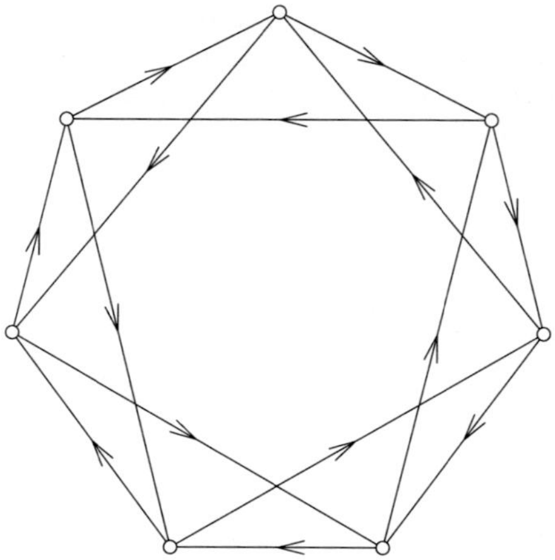

Fig. 19.11 The Koh-Tindell digraph

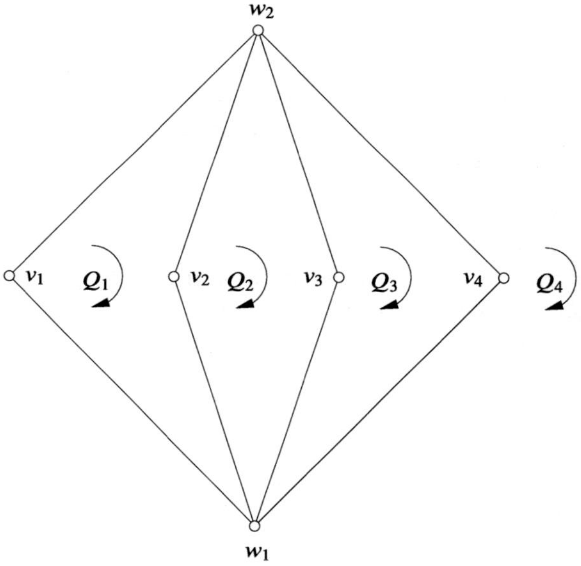

Fig. 19.12 Graph $K_{2,4}$

(i) *Prove that for every orientation of G the number r of oddly oriented quadrilaterals in* $\{Q_i : 1 \leq i \leq k\}$ *is even. Hint: observe that reversal of the orientation of any edge preserves the parity of r; hence the parity of r does not depend on the orientation of G; consider then the orientation in which* w_1 *is a source and* w_2 *is a sink.*

(ii) *In Chapter 3, we defined a bisubdivision of a graph G to be a graph obtained from G by replacing each edge by a path of odd length (possibly equal to one). Suppose that a graph G has a subgraph H which is a bisubdivision of* $K_{2,3}$. *Then, H has three quadrilaterals* Q_i, $i = 1, 2, 3$, *which are bisubdivisions of* C_2.

Show that if each of the three cycles is a conformal subgraph of G then G is not Pfaffian.

(iii) *Use part (ii) to prove that $K_{3,3}$ is not Pfaffian.*

▷**19.2.12 (C. H. C. Little (1973, [52]))** *Let G be a matching covered graph and let D be an orientation of G. A collection C of perfect matchings of G is D-intractable if it satisfies the following two properties: (i) every edge of G is in an even number of perfect matchings in C, and (ii) an odd number of perfect matchings in C has sign -1 in D (or sign 1). Property (i) implies that $|C|$ is even, because for any vertex v of G, $|C|$ is equal to the sum, over all edges e of $\partial(v)$, of the number of perfect matchings in C that contain edge e; property (i) also implies that property (ii) does not depend on the orientation D.*

For each perfect matching M of G, let

$$t(M) := \begin{cases} 0 \text{ if } \operatorname{sign}(M) = 1, \\ 1 \text{ if } \operatorname{sign}(M) = -1. \end{cases}$$

Consider the following system of linear equations over GF(2), *where the set of variables is* $\{x_e : e \in E(G)\}$*:*

$$\sum_{e \in M} x_e = t(M) \quad \textit{(one for each perfect matching M of G).} \tag{19.3}$$

(i) *Prove that the intractability of C does not depend on the orientation D, that is, prove that C is D-intractable if and only if it is F-intractable, for each orientation F of G.*

(ii) *Show that if the system* (19.3) *has a solution then reversal in D of the edges e such that* $x_e \equiv 1 \pmod 2$ *yields a Pfaffian orientation of G.*

(iii) *Prove that if G has a Pfaffian orientation, F, then*

$$x_e := \begin{cases} 1 \text{ if the orientations of } e \text{ in } D \text{ and } F \text{ are distinct,} \\ 0 \text{ otherwise} \end{cases}$$

is a solution of the linear system (19.3).

(iv) *Conclude that a graph is non-Pfaffian if and only if it has an intractable collection of perfect matchings. Hint: apply the criterion for the solvability of linear equations to show that the system has no solution if and only if G has a collection C of perfect matchings such that* $\triangle_{M \in C} M = \emptyset$ *and* $\sum_{M \in C} t(M) \equiv 1 \pmod 2$.

19.2.13

(i) *Show that a digraph has a directed cycle of odd length if and only if it has a closed directed walk of odd length.*

(ii) *Describe a polynomial-time algorithm for finding a directed closed walk of odd length, if it has one.*

19.3 Pfaffian Orientations of Planar Graphs

In this section we give a proof of Kasteleyn's Theorem which states that every planar graph is Pfaffian [44]. As our main interest is in Pfaffian orientations of matching covered graphs, we shall restrict ourselves to 2-connected graphs.

A *plane graph* is a planar graph that is embedded in the plane. We shall make use of the classical result of Whitney (1932, [93]) which says that in any 2-connected plane graph G all faces are bounded by cycles ([3, Theorem 10.7]). Any cycle in G which bounds a face is a *facial cycle*.

Given an embedding G of a planar graph, there is one face in that embedding which is unbounded. That face is referred to as the *outer face*, and all other faces as *inner faces*. In general, the bounding cycle of the outer face in one embedding G of a planar graph need not be the bounding cycle of the outer face in a different embedding of the same planar graph (see Exercise 19.3.1). However, given any planar embedding G of a planar graph and an inner face f of G, it is possible to find another planar embedding $\tilde{G}$ of that planar graph such that all facial cycles of G are also facial cycles of $\tilde{G}$ and, in addition, the image $\tilde{f}$ of f is the outer face of $\tilde{G}$. (See Theorem 9.3 in [2].)

In order to prove Kasteleyn's Theorem, we shall first establish that every plane graph admits a special type of orientation, which we refer as an odd orientation, and then show that any odd orientation of a matching covered plane graph is a Pfaffian orientation of that graph.

19.3.1 Odd orientations of plane graphs

Let G be a 2-connected plane graph. We shall adopt the convention that every cycle of G is traversed in the clockwise order, and an orientation of a cycle C of G is *odd* if there are an odd number of edges of the cycle which are oriented in that direction, and is *even* otherwise.

Odd orientations

An orientation D of a plane graph G is an *odd orientation* if all the facial cycles are odd, with the possible exception of the cycle bounding the outer face.

We shall need the following strengthening of the ear decomposition theorem [3, Theorem 5.8] to plane graphs.

Proposition 19.13
A 2-connected plane graph G has a sequence $(G_0, G_1, G_2, \ldots, G_r)$ of 2-connected plane subgraphs satisfying the following conditions:

(i) $G_r = G$ *and* G_0 *is a facial cycle of* G*;*
(ii) *for* $0 \le i < r$*, there is a path* P_i *which has its ends in* $V(G_i)$*, is internally-disjoint from* G_i*, and lies in the outer face of* G_i*, such that* $G_{i+1} = G_i \cup P_i$*.*

Lemma 19.14 *Every plane graph admits an odd orientation.*

Proof by induction on the number of edges. Let $(G_0, G_1, G_2, \ldots, G_r)$ be the sequence defined in the statement of Proposition 19.13. The assertion is obviously true if $G = G_0$ is itself a cycle. Let i be an integer such that $0 < i < r$, and assume inductively that G_i has an odd orientation D_i. We shall now show how to find an odd orientation D_{i+1} of G_{i+1}.

Note that G_{i+1} is obtained from G_i by adding a path P joining two vertices on the bounding cycle of the outer face F. The addition of P divides the face F into two regions which become faces of G_{i+1}. One of these faces is a face of G, bounded by a cycle, Q. The other face is the outer face of G_{i+1}. See Figure 19.13.

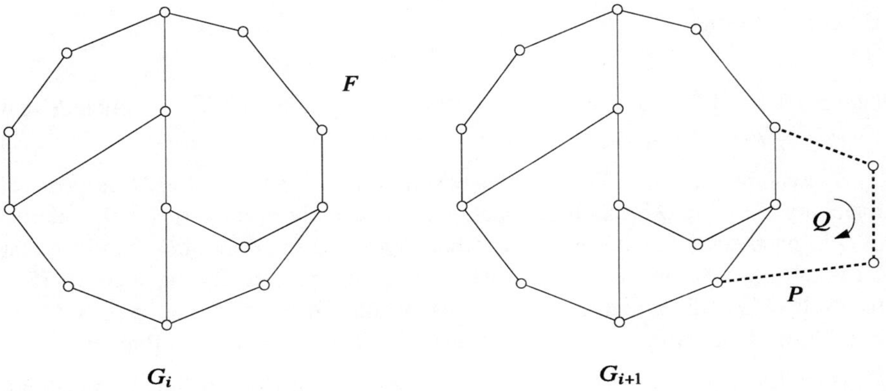

Fig. 19.13 The outer face F, the path P and the cycle Q

Extend D_i to an orientation D_{i+1} of G_{i+1} by orienting the edges of P such that Q is oddly oriented. Clearly, D_{i+1} is an odd orientation of G_{i+1}. The assertion is valid, by induction. □

Note that underlying the above inductive proof there is an algorithm for finding an odd orientation of any given 2-connected plane graph.

KASTELEYN'S THEOREM

Theorem 19.15 *Any odd orientation of a 2-connected matching covered plane graph is Pfaffian.*

Proof Let G be a matching covered plane graph and consider an odd orientation D of G (with the adopted convention that each cycle is traversed clockwise).

By the definition of odd orientation, if Q is any cycle of G, then the bounding cycles of all the faces in the interior of Q are oddly oriented in D.

19.15.1 *Let Q be any cycle of G and let p be the number of vertices in the interior of Q. Then, $|\mathrm{fw}(Q)| \equiv p + 1 \pmod 2$, where $\mathrm{fw}(Q)$ is the number of forward arcs of Q.*

Proof Let $\mathcal{F}$, q and ℓ denote, respectively, the collection of faces inside Q, the number of arcs in the interior of Q and the length of Q.

For each face F in $\mathcal{F}$, let Q_F denote the bounding cycle of F. Each one of the q arcs in the interior of Q is traversed twice by the bounding cycles of the faces in $\mathcal{F}$, once in each direction. Moreover, by hypothesis, $|\mathrm{fw}(Q_F)|$ is odd, for each face $F \in \mathcal{F}$. Thus,

$$|\mathrm{fw}(Q)| = \sum_{F \in \mathcal{F}} |\mathrm{fw}(Q_F)| - q \equiv |\mathcal{F}| - q \pmod 2. \tag{19.4}$$

By Euler's formula,

$$(|\mathcal{F}| + 1) + (p + \ell) - (q + \ell) = 2,$$

hence $p + 1 \equiv |\mathcal{F}| - q \pmod 2$. From this and equation (19.4), we conclude that $|\mathrm{fw}(Q)| \equiv p + 1 \pmod 2$, as asserted. □

By hypothesis, graph G is matching covered and 2-connected. Let M be a perfect matching of G. Let Q be an M-alternating cycle. Clearly, every vertex in the interior of Q is matched by an edge of M to another vertex in the interior of Q. It follows that the number p of vertices of G in the interior of Q must be even. By statement 19.15.1, the cycle Q is oddly oriented. This conclusion holds for each M-alternating cycle Q of G. From Lemma 19.5 we infer that the odd orientation D of G is Pfaffian. □

The following more general assertion may be easily derived from the above lemma (Exercise 19.3.3).

Corollary 19.16 *Every planar graph, regardless of whether it is matching covered or not, is Pfaffian.* □

Exercises

19.3.1 *Find a planar embedding of the pentagonal prism in which the outer face is a square and find a Pfaffian orientation of the pentagonal prism.*

19.3.2 *Supply the missing details in the proof of Lemma 19.14.*

19.3.3 *Give a proof of Corollary 19.16.*

19.4 Notes

There is a generalization of the notion of Pfaffian graphs, which is the notion of k-Pfaffian graphs. Let G be a graph, let $\mathcal{M}$ be the collection of perfect matchings of G.

A *k-orientation* of G is a sequence $\mathbf{D} := (D_1, D_2, \ldots, D_k)$ of k orientations of G. For each perfect matching $M \in \mathcal{M}$, the *sign of M in* $\mathbf{D}$ is the line vector

$$\text{sign}_{\mathbf{D}}(M) := (\text{sign}_{D_1}(M), \text{sign}_{D_2}(M), \ldots, \text{sign}_{D_k}(M)).$$

Let $\text{sign}_{\mathbf{D}}$ be the $|\mathcal{M}| \times k$ matrix, in which each line is the sign vector in $\mathbf{D}$ of a perfect matching of G. The k-orientation $\mathbf{D}$ is *Pfaffian* if the system

$$\text{sign}_{\mathbf{D}}\mathbf{x} = \mathbf{1}$$

has a solution, where $\mathbf{x}$ is a column vector of k rows and $\mathbf{1}$ is a column vector of $|\mathcal{M}|$ 1's. The graph G is *k-Pfaffian* if it has a Pfaffian k-orientation. Clearly, G is 1-Pfaffian if and only if G is Pfaffian. The graph $K_{3,3}$, for example, is 4-Pfaffian. The *Pfaffian number* of a graph G is the smallest integer k such that G is k-Pfaffian.

Galluccio and Loebel (1999, [37]) and, independently, Tesler (2000, [88]) proved that if a graph is embeddable in an orientable surface of genus g then its Pfaffian number is at most 4^g. Norine (1999, [77]) proved that a graph is 4-Pfaffian if and only if it has an immersion in the torus in which every perfect matching has an even number of crossings. In that same article, Norine proved that every 5-Pfaffian graph is 4-Pfaffian and every 3-Pfaffian graph is Pfaffian. Norine also conjectured that the Pfaffian number of every graph is a power of four. Miranda and Lucchesi (2011, [75]) found a counterexample, a graph whose Pfaffian number is six.

There are interesting open questions about this generalization. For example, we may think of the problems analogous to Problems 19.1 and 19.2. It is not even known whether these generalizations are polynomial-time equivalent.

Chapter 20
Similarity of Orientations and Characteristic Orientations

Contents

20.1 Similarity of Orientations

In this section we introduce an equivalence relation on the set of all orientations of a graph. The study of equivalence classes with respect to this relation turns out to play an important role in developing algorithms for finding Pfaffian orientations of matching graphs which are known to be Pfaffian and in showing that the Pfaffian Recognition Problem (19.1) and the Pfaffian Orientation Problem (19.2) are both in co-NP and are polynomial-time equivalent.

C. L. Lucchesi, U. S. R. Murty, *Perfect Matchings*, Algorithms and Computation in Mathematics 31, https://doi.org/10.1007/978-3-031-47504-7_20

SIMILAR ORIENTATIONS

Let D be a digraph, let S be a set of arcs of D. We denote by D rev S the digraph obtained from D by reversing the orientations of the edges of S.

Two orientations D_1 and D_2 of a graph G are said to be *similar* if G has a cut C such that $D_2 = D_1$ rev C.

We now record some basic properties of similarity of orientations. The proofs of some of these properties are left as Exercises 20.1.1 and 20.1.2.

Proposition 20.1

(i) *For any two sets X and Y of vertices of a graph, $\partial(X \triangle Y) = \partial(X) \triangle \partial(Y)$.*
(ii) *A set C of edges of a graph G is a cut if and only if $|C \cap E(Q)|$ is even, for every cycle Q of G.* □

Proposition 20.2 *Similarity is an equivalence relation on the collection of orientations of a graph.* □

Determining whether a set of edges of a matching covered graph is a cut

We now describe a linear-time algorithm to determine whether a set of edges of a graph is a cut.

DETERMINING WHETHER A SET IS A CUT

Algorithm 20.3
Input: a connected graph G and a set R of edges of G;
Output: either (i) a set $X \subseteq V$ such that $R = \partial(X)$ or (ii) a cycle Q of G that contains an odd number of edges in R.

The objective of the algorithm is to determine a 2-colouring of the vertices of G so that the ends of an edge have distinct colours if and only if the edge is in R.

To look for such a 2-colouring, we use a depth-first-search (DFS). Traverse the graph by a DFS, while 2-colouring the vertices, and constructing a tree, T.

- Arbitrarily choose an initial vertex as root and give to it a colour of your choice;
- when adding an edge to T, colour the new vertex so that the edge satisfies the requirement of the 2-colouring;
- when examining an edge e with both ends in T:
  - if the ends of e do not satisfy the requirement of the 2-colouring then return the cycle Q in $T + e$, which can be easily retrieved from the DFS structure; clearly, Q has an odd number of edges in R;
  - if the ends of e satisfy the requirement of the 2-colouring, continue.
- if all the edges satisfy the 2-colouring requirement then return X, the set of vertices coloured with one of the two colours: $R = \partial(X)$.

Proposition 20.4 *The Algorithm 20.3 runs in linear time.* □

20.1.1 Similarity and Pfaffian orientations

Proposition 20.5 *Let D_1 and D_2 be two similar orientations of a graph G. Then, D_1 is Pfaffian if and only if D_2 is Pfaffian.*

Proof Let C be the cut of G such that $D_2 = D_1$ rev C and let M be a perfect matching of G and let Q be an M-alternating cycle of G. Reversal of any edge of Q changes the parity of Q. By Proposition 20.1(ii), Q has an even number of edges in C. Thus, reversal of all the edges of $E(Q) \cap C$ preserves the parity of Q. We deduce that Q is oddly oriented in D_1 if and only if it is oddly oriented in D_2. This conclusion holds for each M-alternating cycle Q of G. By Lemma 19.5, D_1 is Pfaffian if and only if D_2 is Pfaffian. □

The above proposition implies that if an equivalence class of orientations of a graph G includes a Pfaffian orientation of G, then all members of that equivalence class are Pfaffian orientations of G. It turns out, as we shall see later, that if G is a Pfaffian bipartite graph then all its Pfaffian orientations belong to a single equivalence class. However, this is not true in general. See Exercise 20.1.3.

Proposition 20.6 *Let G be a matching covered graph and let v be a vertex of G. Each Pfaffian orientation of G has a similar orientation D such that v is a source (or a sink) of D.*

Proof Let D_0 be a Pfaffian orientation of G and let $\partial^-(v)$ denote the set of arcs of D_0 that enter vertex v. Consider a pair R of parallel edges of G. Clearly, R induces a conformal even cycle, Q, of G. As D_0 is Pfaffian, the cycle Q is oddly oriented in F; hence the two edges of R are parallel arcs in D_0. This conclusion holds for each pair of parallel edges of G. In particular, it holds for each pair of parallel edges incident with v. Consequently, if we let X be the set of tails of arcs in $\partial^-(v)$ and let $C := \partial(X)$, it follows that $C \cap \partial(v) = \partial^-(v)$. Thus, in the similar orientation $D := D_0$ rev C of D_0, the vertex v is a source. Moreover, in the similar orientation $D' := D$ rev $\partial(v)$ of D_0, the vertex v is a sink. □

20.1.2 Similarity and removable ears

The following property is easy to be verified. This result will play a crucial role in the proof of Lemmas 20.8 and 20.9.

Proposition 20.7 *Let P be a path of odd length and let D be an orientation of P such that P is oddly oriented in D. Then, there exists a set X of internal vertices of P such that P rev $\partial(X)$ is a directed path.*

Proof As P is oddly oriented in D, the number of forward arcs of P in D is odd. As P has odd length, the number of reverse arcs of P in D is even.

Let $P := v_1 v_2 \ldots v_k$. For $i = 1, 2, \ldots, k$, let r_i denote the number of reverse arcs of D in the traversal of the subpath $v_1 v_2 \ldots v_i$ of P. Let $X := \{v_i : r_i \text{ is odd}\}$. For each forward arc e, either both ends of e are in X or neither end of e is in X. For each reverse arc f, precisely one of the ends of f is in X. Thus, P rev $\partial(X)$ is a directed path from v_1 to v_k. Moreover, r_k is the number of reverse arcs of P, which is even, and r_1 is obviously even; hence each vertex in X is an internal vertex of P. □

Lemma 20.8 *Let R be a removable single ear of a matching covered graph G, let Q be a conformal cycle of G that contains all the edges of R, and let D_1 and D_2 be two orientations of G. If Q is oddly oriented in both D_1 and D_2 and if $D_1 - R$ and $D_2 - R$ are similar then D_1 and D_2 are also similar.*

Proof As $D_1 - R$ and $D_2 - R$ are similar, $V(G - R)$ has a subset X such that

$$D_2 - R = (D_1 - R) \text{ rev } \partial_{G-R}(X).$$

Let $C := \partial_G(X)$ and let $D_1' := D_1$ rev C. Then, $\partial_{G-R}(X) = C - R$; hence

$$D_2 - R = (D_1 - R) \text{ rev } (C - R) = (D_1 \text{ rev } C) - R = D_1' - R.$$

By Proposition 20.1(ii), Q has an even number of edges in C. As Q is oddly oriented in D_1, we infer that Q is oddly oriented in D_1'. Thus, Q is oddly oriented in both D_1' and D_2. Adjust the direction of the traversal of Q so that the segment P of Q induced by $E(R)$ is oddly oriented in D_1'.

We have seen that $D_1' - R = D_2 - R$. As Q is oddly oriented in D_2, it follows that P is also oddly oriented in D_2. By Proposition 20.7, there exists a set X_1 of internal vertices of P such that P is a directed path in $D_1'' := D_1'$ rev $\partial(X_1)$. Likewise, there exists a set X_2 of internal vertices of P such that P is a directed path in $D_2' := D_2$ rev $\partial(X_2)$. As X_1 and X_2 consist solely of internal vertices of P, it follows that

$$D_1'' - R = D_1' - R = D_2 - R = D_2' - R.$$

As P is a directed path in both D_1'' and D_2', we deduce that $D_1'' = D_2'$. The orientation D_1'' is similar to the orientation D_1', which in turn is similar to D_1. The orientation D_2' is similar to the orientation D_2. As $D_1'' = D_2'$, we conclude that D_1 and D_2 are similar. □

Lemma 20.9 *Let R be a removable double ear of a matching covered graph G, let Q be a conformal cycle of G that contains all the edges of R and let $C := \{D_1, D_2, D_3\}$ be a collection of three orientations of G. If Q is oddly oriented in each orientation in C and if the orientations in $\mathcal{D} := \{D - R : D \in C\}$ are pairwise similar then at least two of the orientations in C are similar.*

Proof As the orientations in $\mathcal{D}$ are similar, we infer that, for $i = 1, 2$, $V(G - R)$ has subsets X_i such that

$$D_3 - R = (D_i - R) \text{ rev } \partial_{G-R}(X_i).$$

Let $C_i := \partial_G(X_i)$ and let $D_i' := D_i$ rev C_i. In order to have a uniform notation, let $D_3' := D_3$. Then, $\partial_{G-R}(X_i) = C_i - R$; hence

$$D'_3 - R = D_3 - R = (D_i - R) \text{ rev } (C_i - R) = (D_i \text{ rev } C_i) - R = D'_i - R.$$

Let R_1 and R_2 be the two constituent paths of R. For $j = 1, 2$, let P_j be the subpath of Q induced by $E(R_j)$.

By Proposition 20.7, for $i = 1, 2, 3$ and for $j = 1, 2$ there exists a set Y_{ij} of internal vertices of P_j such that, in D'_i rev $\partial(Y_{ij})$, P_j is a directed path if P_j is oddly oriented in D'_i; otherwise the reverse of P_j is a directed path. For $i = 1, 2, 3$, let $D''_i := D'_i$ rev $\partial(Y_{i1} \cup Y_{i2})$.

We have seen that $D'_1 - R = D'_2 - R = D'_3 - R$. As each Y_{ij} is a set of internal vertices of P_j, we conclude that for $i = 1, 2$,

$$D''_3 - R = D'_3 - R = D_3 - R = D'_i - R = D''_i - R.$$

The orientations of P_1 in at least two of D''_1, D''_2 and D''_3 coincide. Thus, let $1 \leq i < j \leq 3$ be such that the orientations of P_1 in D''_i and in D''_j coincide. Then, the orientations of P_2 in D''_i and in D''_j also coincide. In that case, $D''_i = D''_j$. The orientation D''_i is similar to the orientation D'_i, which in turn is similar to the orientation D_i. Likewise, the orientation D''_j is similar to the orientation D'_j, which in turn is similar to the orientation D_j. We conclude that D_i and D_j are similar. □

Exercises

▷**20.1.1** *Prove Proposition 20.1.*

∗**20.1.2 (Similarity is an equivalence relation)**

(a) *Let C_1 and C_2 be two cuts of a digraph D. Show that $(D$ rev $C_1)$ rev $C_2 = D$ rev $(C_1 \triangle C_2)$.*
(b) *Using part (a) and Proposition 20.1(i), prove Proposition 20.2.*

20.1.3 *Find two dissimilar Pfaffian orientations D_1 and D_2 of K_4. Explain why D_1 and D_2 are dissimilar. (Hint: Let D_1 denote any Pfaffian orientation of K_4 and let D_2 be the converse of D_1, that is, one which is obtained from D_1 by reversing the orientations of all edges of K_4.)*

20.1.4 (Extensions of Orientations of Spanning Trees)

(a) *Suppose that G is a Pfaffian graph and that T is a spanning tree of G. Show that every orientation of T can be extended to a Pfaffian orientation of G. (Consider similar Pfaffian orientations of G and show that for each edge e of T, G has a cut C such that $C \cap E(T) = \{e\}$.)*
(b) *Suppose that (i) G is a Pfaffian graph, (ii) T_0 is a spanning tree of G and (iii) $T := T_0 + e$ is not bipartite. Show that every orientation D_T of T can be extended to a Pfaffian orientation of G. (Firstly, apply the reasoning of part (a) to obtain a Pfaffian orientation D of G which is an extension of $D_T - e$. If the orientations of e in D and in D_T do not coincide, let $D' := (D$ rev $E(G))$ rev $\partial(A)$, where (A, B) is the bipartition of T_0.)*

(c) *Use part (a) to show that $K_{3,3}$ is not Pfaffian.*

20.1.5 *Show that the Petersen graph is not Pfaffian. (This is not quite as straightforward as showing that $K_{3,3}$ is not Pfaffian.)*

20.2 Pfaffian Orientations and Tight Cuts

We shall now use similarity of orientations to prove the following fundamental result due to Little and Rendl (1991, [55]).

Theorem 20.10 *Let G be a matching covered graph and let $C := \partial(X)$ be a tight cut of G. The graph G is Pfaffian if and only if both C-contractions of G are Pfaffian.*

In what follows we shall first analyse a relationship between Pfaffian orientations of a graph and those of the contractions of that graph with respect to a separating cut, and then deduce the above assertion from that analysis.

20.2.1 Pfaffian orientations and separating cuts

We first proceed to show that if C is a tight cut of a matching covered graph G and D is an orientation of G such that C is a directed cut in D and both C-contractions of D are Pfaffian, then D is Pfaffian. One might think that given a tight cut C of G and an orientation D of G such that both C-contractions of D are Pfaffian then D would be Pfaffian. This is not the case. For a simple example, consider the orientation D of a hexagon depicted in Figure 20.1. The cut C is tight in G, both C-contractions of D are Pfaffian, yet D is not Pfaffian. In order for D to be Pfaffian, we require C to be a directed cut in D.

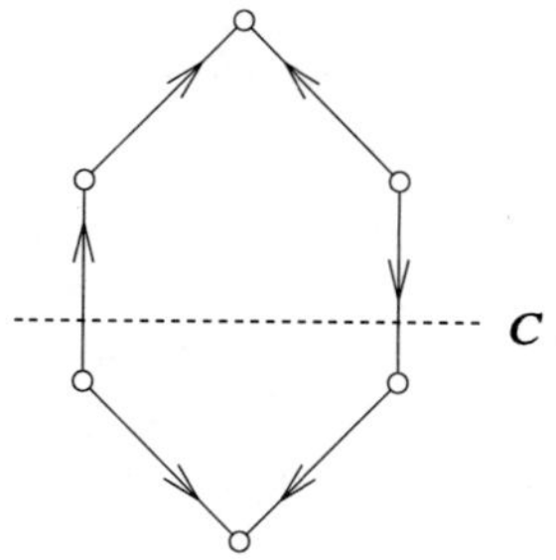

Fig. 20.1 The cut C is tight, both C-contractions of D are Pfaffian, yet D is not Pfaffian

We prove a similar statement, where C is a separating cut, which clearly implies the result mentioned above, if C is tight.

Lemma 20.11 *Let $C := \partial(X)$ be a separating cut of a matching covered graph G and let D be an orientation of G such that (i) C is a directed cut in D and (ii) both C-contractions of D are Pfaffian. Then, all the perfect matchings of G that contain only one edge in C have the same sign in D.*

Proof By hypothesis, the cut C is directed in D. Adjust notation so that X is the shore of C which contains the heads of all the arcs of C. Let $G_1 := G/(\overline{X} \to \overline{x})$ and let $G_2 := G/(X \to x)$ be the two C-contractions of G. Let $D_1 := D/(\overline{X} \to \overline{x})$ and let $D_2 := D/(X \to x)$ be the two C-contractions of D. By hypothesis, D_1 and D_2 are Pfaffian.

Let M and N be two perfect matchings of G and let e and f be two edges in C such that $M \cap C = \{e\}$ and $N \cap C = \{f\}$. We shall prove that M and N have the same sign by showing that every (M, N)-alternating cycle of G is oddly oriented. Thus, let Q be any (M, N)-alternating cycle of G. As M and N both have just one edge in C, it follows that for $i = 1, 2$, the sets $M_i := M \cap E(G_i)$ and $N_i := N \cap E(G_i)$ are perfect matchings of G_i.

If Q and C are disjoint then for some $i \in \{1, 2\}$, Q is an (M_i, N_i)-alternating cycle in G_i. As D_i is Pfaffian, Q is oddly oriented in D_i, hence it is oddly oriented in D. We may thus assume that Q and C have edges in common. Then, Q necessarily contains the edge e of $M \cap C$ and the edge f of $N \cap C$. Moreover, the edges e and f are the only two edges of Q in C. For $i = 1, 2$, the set $E(Q) \cap E(G_i)$ induces an M_i-alternating cycle, say Q_i, in G_i. As D_i is Pfaffian, Q_i is oddly oriented in D_i. Consider traversing Q, Q_1 and Q_2 so that e is a forward edge in each of the three traversals (and f is a reverse edge in each of the three traversals). Then, $\mathrm{fw}(Q_1) \cap \mathrm{fw}(Q_2) = \{e\}$. It follows that

$$|\mathrm{fw}(Q)| = |\mathrm{fw}(Q_1)| + |\mathrm{fw}(Q_2)| - 1 \equiv 1 \pmod{2}.$$

We conclude that Q is oddly oriented in D. Regardless of whether Q and C are disjoint, the cycle Q is oddly oriented in D. This conclusion holds for each (M, N)-alternating cycle Q of G. By Corollary 19.4, the perfect matchings M and N have the same sign. This conclusion holds for every pair $\{M, N\}$ of perfect matchings of G that contain just one edge in C. □

Our next step towards the proof of Theorem 20.10 is to consider a 'converse' of Lemma 20.11. We thus proceed to show that if G is a matching covered graph that admits a Pfaffian orientation and C is a separating cut of G, then both C-contractions of G also admit Pfaffian orientations. Before we present a proof of this lemma it is worth noting one subtle point. Suppose that G is a Pfaffian graph and C is a separating cut of G. The graph G, being Pfaffian, admits a Pfaffian orientation, say D. One might think that for any such orientation D of G, the two C-contractions of D are Pfaffian orientations of the corresponding C-contractions of G. But this is not the case. The Pfaffian orientation D_0 of $\overline{C_6}$ shown in Exercise 19.2.7 illustrates this possibility. Another illustration is given by Example 20.12.

Example 20.12 Let G be the graph obtained by splicing K_4 and the cube, and let D be the orientation of G depicted in Figure 20.2.

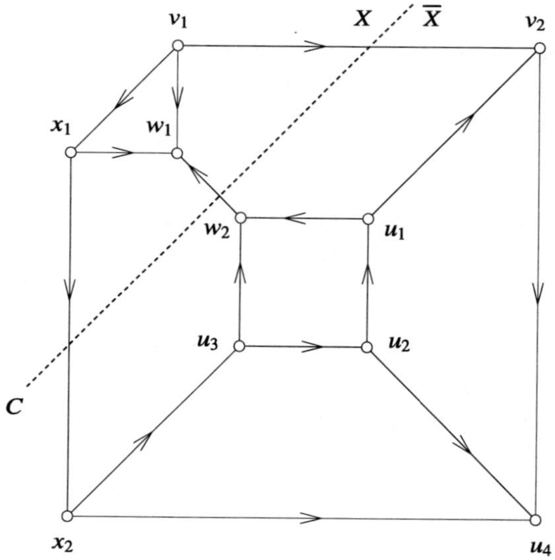

Fig. 20.2 A tight cut C of G and a Pfaffian orientation D of G where neither C-contraction of D is Pfaffian

The cut $C := \partial(X)$ is a tight cut of G. It may be verified that D is a Pfaffian orientation of G, but that neither C-contraction of D is Pfaffian. However, the orientation $D' := D$ rev $\partial(\{w_2, x_1, x_2\})$ is similar to D, hence it is Pfaffian and is such that C is a directed cut of D' and the two C-contractions of D' are Pfaffian. See Exercise 20.2.2.

Lemma 20.13 (CLM (2012, [16])) *Let $C := \partial(X)$ be a separating cut of a matching covered graph G and let D_0 be a Pfaffian orientation of G. Then G has a (Pfaffian) orientation D similar to D_0 such that C is directed in D and both C-contractions of D are Pfaffian.*

Proof We wish to show that there is an orientation D of G which is similar to D_0 such that C is a directed cut in D. From this, using Lemma 19.5 we shall deduce that the two C-contractions of D are Pfaffian orientations of the corresponding C-contractions of G.

With this as our objective, we now proceed to describe a procedure for determining a subset S of the ends of edges in C and show that $D := D_0$ rev $\partial(S)$ has the asserted properties. (Example 20.14, which is a continuation of Example 20.12, illustrates this step of the proof.) First note that D, being similar to D_0, is Pfaffian by Proposition 20.5.

Let $G_1 := G/(\overline{X} \to \overline{x})$ and $G_2 := G/(X \to x)$ be the two C-contractions of G. Also, let V_1 be the set of vertices of X that are ends of edges of C and let V_2 be the set of vertices of $\overline{X}$ that are ends of edges of C. Choose any edge $e := v_1v_2$ of cut C where $v_1 \in V_1$ and $v_2 \in V_2$. As C is a separating cut of G, there exists a perfect matching of G which has just the edge e in C. Let M be such a perfect matching of G. For $i = 1, 2$, let M_i be the perfect matching $M \cap E(G_i)$ of G_i that contains the edge e. See Figure 20.3.

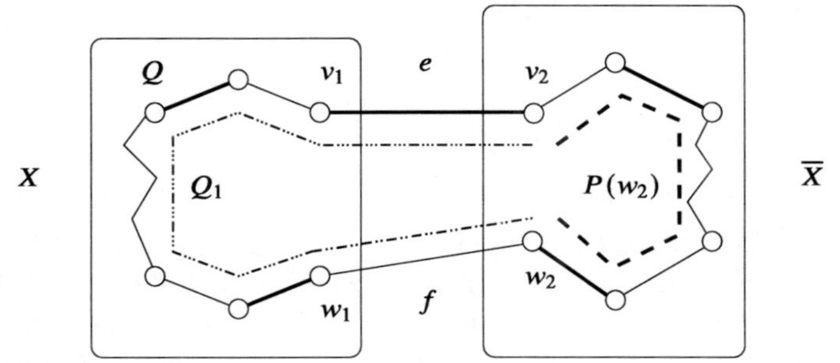

Fig. 20.3 Proof of Lemma 20.13

20.13.1 *For any vertex w_2 in V_2, the graph $G[\overline{X}]$ has an M_2-alternating path $P(w_2)$ of even length from v_2 to w_2. See Figure 20.3.*

Proof If $w_2 = v_2$ then let $P(w_2) := v_2$. We may thus assume that $w_2 \neq v_2$. As $w_2 \in V_2$, w_2 is incident with an edge in C, say f. As C is separating, G has a perfect matching, say N, such that $N \cap C = \{f\}$. Let Q be the (M, N)-alternating cycle in G that contains edge f. Then, this path Q meets C in precisely the two edges e and f. Let $P(w_2)$ denote the segment of Q in G_2 that joins v_2 and w_2. Clearly, $P(w_2)$ is an (M, N)-alternating path in $G_2 - x$, hence $P(w_2)$ is M_2-alternating. The edge of $P(w_2)$ incident with v_2 is in N, and the edge of $P(w_2)$ incident with w_2 is in M, hence $P(w_2)$ has even length. □

Likewise, for each vertex w_1 in V_1, define a path $P(w_1)$ to be an M_1-alternating path of even length in $G[X]$ that connects v_1 and w_1. Thus, for any vertex w in $V_1 \cup V_2$, there is an M-alternating path $P(w)$. We now define S as the subset of $V_1 \cup V_2$ consisting of those vertices w such that $P(w)$ is oddly oriented in D_0.

Recall that we have defined D to be equal to D_0 rev ∂S.

20.13.2 *For each vertex $w \in V_1 \cup V_2$, the path $P(w)$ is evenly oriented in D.*

Proof The ends v_1 and v_2 of e are not in S. Thus, reversal of the edges of $\partial(S)$ preserves the parity of $P(w)$ if and only if $w \notin S$. The vertex w is in S if and only if $P(w)$ is oddly oriented in D_0. We deduce that $P(w)$ is evenly oriented in D. This conclusion holds for each vertex $w \in V_1 \cup V_2$. □

For each w in $V_1 \cup V_2$, the path $P(w)$ has even length and is evenly oriented in D. Thus, the reverse of $P(w)$ is also evenly oriented in D.

20.13.3 *The cut C is directed in D.*

Proof Let $f := w_1w_2$ be any edge of $C - e$. The set $\{e, f\} \cup E(P(w_1)) \cup E(P(w_2))$ induces an M-alternating cycle, Q, of G. Let us traverse Q in D so that e is a forward edge of Q. The orientation D of G is similar to the Pfaffian orientation D_0 of G, hence it is Pfaffian. Thus, Q is oddly oriented. The paths $P(w_1)$ and $P(w_2)$ are evenly oriented. Thus, f is a reverse edge of Q in D. It follows that the heads of e and f in D are in the same shore of C. This conclusion holds for each edge f of $C - e$. Indeed, C is a directed cut in D. □

For $i = 1, 2$, let D_i be the restriction of D to G_i.

20.13.4 *The orientation D_1 of G_1 is Pfaffian.*

Proof Let Q_1 be an M_1-alternating cycle of G_1. To prove that D_1 is Pfaffian, we make use of Lemma 19.5, and show that Q_1 is oddly oriented in D_1. If Q_1 has no edge in C then Q_1 is an M-alternating cycle of G. As D is Pfaffian, Q_1 is oddly oriented in D, hence it is also oddly oriented in D_1. We may thus assume that Q_1 contains edges in C. Clearly, Q_1 contains precisely two edges in C, one of which is e. Let f be the other edge of Q_1 in C. Let w_2 be the end of f in $\overline{X}$. Let Q be the cycle of G induced by $E(Q_1) \cup E(P(w_2))$. The cycle Q is M-alternating in G. As D is Pfaffian, Q is oddly oriented in D. The path $P(w_2)$ is evenly oriented in D and the cut C is directed in D. Thus, Q_1 is oddly oriented in D_1. This conclusion holds for every M_1-alternating cycle Q_1 of G_1. It follows that D_1 is Pfaffian. □

A similar reasoning may be used to prove that D_2 is Pfaffian. The proof of Lemma 20.13 is complete. □

Example 20.14 In the digraph depicted in Figure 20.2, where the shore X of C is $\{v_1, w_1, x_1\}$,

$$
\begin{aligned}
M &:= \{v_1v_2, w_1x_1, u_1w_2, u_2u_3, u_4x_2\},\\
N &:= \{w_1w_2, v_1x_1, u_1v_2, u_2u_3, u_4x_2\}, \text{ and}\\
F &:= \{x_1x_2, v_1w_1, u_3w_2, u_1u_2, u_4v_2\}
\end{aligned}
$$

are perfect matchings such that the only edge in $C \cap M$ is $e := v_1v_2$, the only edge of $C \cap N$ is w_1w_2, and the only edge of $C \cap F$ is x_1x_2.

In the shore $\overline{X}$ of C, the (M_2, N_2)-alternating path from v_2 to w_2 is oddly oriented and the (M_2, F_2)-alternating path from v_2 to x_2 is also oddly oriented. In the shore X of C, the (M_1, F_1)-alternating path from v_1 to x_1 is oddly oriented, but the (M_1, N_1)-alternating path from v_1 to w_1 is not. Thus $S = \{w_2, x_1, x_2\}$, as indicated in Example 20.12.

Proof of Theorem 20.10

Let $G_1 = G/(\overline{X} \to \overline{x})$ and let $G_2 := G/(X \to x)$ be the two C-contractions of G. Suppose that G is Pfaffian. By Lemma 20.13, G has a Pfaffian orientation whose C-contractions are both Pfaffian. Thus, G_1 and G_2 are both Pfaffian.

Conversely, suppose that G_1 and G_2 are Pfaffian. By Proposition 20.6, G_1 has a Pfaffian orientation, say D_1, such that its contraction vertex $\overline{x}$ is a source in D_1. Likewise, G_2 has a Pfaffian orientation, say D_2, such that its contraction vertex x is a sink. Let D be the common extension of D_1 and D_2 to an orientation of G. Clearly, C is a directed cut of D. By Lemma 20.11, all the perfect matchings of G that contains just one edge in C have the same sign in D. As C is tight, every perfect matching of G contains just one edge in C. Thus, D is Pfaffian. We conclude that G is Pfaffian. □

It should be noted that the converse of Lemma 20.13 is false. That is, there are matching covered graphs that are not Pfaffian but have a separating cut C such that both its C-contractions are Pfaffian. For example, the Petersen graph is not Pfaffian but it is the splicing of two 5-wheels, which are Pfaffian. Another non-Pfaffian splicing of two 5-wheels is presented in Figure 20.4.

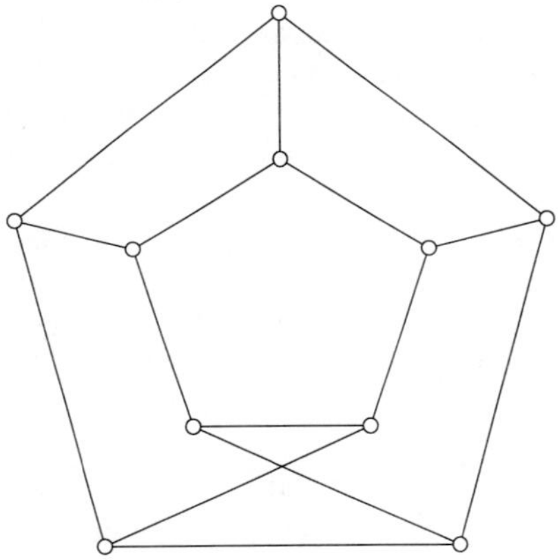

Fig. 20.4 A non-Pfaffian graph which is a splicing of two 5-wheels

In Chapter 21 we will exhibit other counterexamples of the converse of Lemma 20.13. In fact, we shall present an infinite family of such graphs.

Some immediate consequences of Theorem 20.10

Corollary 20.15 *A matching covered graph is Pfaffian if and only if all of its bricks and braces are Pfaffian.* □

> PFAFFIAN ORIENTATIONS AND MATCHING MINORS
>
> Recall that a graph H is a matching minor of graph G if H can be obtained from a conformal subgraph of G by means of a sequence of bicontractions. The following assertion may be easily deduced (Exercise 20.2.5) from Theorem 20.10.

Corollary 20.16 *A matching covered graph G is Pfaffian if and only if each of its matching minors (as defined in Chapter 12) is also Pfaffian.* □

A non-Pfaffian matching covered graph H is *matching-minor-minimal* if every proper matching-minor of H is Pfaffian. (For example, $K_{3,3}$ is such a graph.) This leads us to contemplate the possibility of 'excluded-matching-minor' characterisations of Pfaffian matching covered graphs which is analogous to Kuratowski's characterization of planar graphs. We shall explore this idea extensively in the next chapter by introducing the notion of a minor of a matching covered graph which is more general than that of a matching-minor.

Corollary 20.17 *Every optimal removable single ear of a Pfaffian matching covered graph is b-invariant.*

Proof Let G be a Pfaffian matching covered graph and let R be an optimal removable single ear of G. The graph G is Pfaffian and the Petersen graph is not Pfaffian. By Corollary 20.15, no brick of G is a Petersen brick, hence $p(G) = 0$. Likewise, $G-R$, a conformal subgraph of G, is Pfaffian, by Lemma 19.5(ii), hence $p(G - R) = 0$. By the monotonicity of function b, it follows that

$$(b + p)(G) = b(G) \leq b(G - R) \leq (b + p)(G - R) = (b + p)(G).$$

Equality holds throughout, hence $b(G - R) = b(G)$. Indeed, R is b-invariant. This conclusion holds for each optimal removable ear R of every Pfaffian matching covered graph. □

In Chapter 16 we have shown (Corollary 16.10) that every matching covered G has an ear decomposition which uses just $b + p$ double ears. Combined with the above result we thus have:

Corollary 20.18 *Every Pfaffian matching covered graph G has an ear decomposition which uses just b double ears.* □

Exercises

20.2.1 *Let G be a graph, let D be an orientation of G and let v be a vertex of degree two in G which is adjacent to two distinct vertices. Suppose that v is neither a source nor a sink in D. Prove that D is Pfaffian if and only if the digraph obtained from D by the bicontraction of v is Pfaffian.*

▷**20.2.2** *Show that:*

(i) *the directed graph D depicted in Figure 20.2 is a Pfaffian orientation of the graph G obtained by splicing* K_4 *and the cube;*
(ii) *neither C-contraction of D is Pfaffian,*

(iii) *the digraph* $D' := D$ rev $\partial(\{w_2, x_1, x_2\})$ *is also a Pfaffian orientation of* G*; and finally show that*

(iv) *both* C*-contractions of* D' *are Pfaffian.*

▷**20.2.3** *Prove that the graph depicted in Figure 20.4 is not Pfaffian.*

▷**20.2.4** *The objective of this exercise is to provide a proof of a weaker version of Lemma 20.13. Let* G *and* C *be as in the statement of Lemma 20.13. Let* G_1 *and* G_2 *be the two* C*-contractions of* G*, let* D *be a Pfaffian orientation of* G *and let* D_1 *and* D_2 *be the* C*-contractions of* D *which are orientations of* G_1 *and* G_2*, respectively. For each edge* e *of* C*, let* $\mathcal{M}_1(e)$ *denote the set of perfect matchings of* G_1 *that contain edge* e*.*

(i) *Prove that, for each edge* $e \in C$*, all the perfect matchings in* $\mathcal{M}_1(e)$ *have the same sign in* D_1*. Hint: consider a perfect matching* M_2 *of* G_2 *that contains edge* e *and the collection* $\mathcal{M}(e) := \{M_1 \cup M_2 : M_1 \in \mathcal{M}_1(e)\}$*.*

(ii) *From the previous part, conclude that* G_1 *is Pfaffian. Hint: define the set* R *to be* $\{e \in C : \mathrm{sign}(M_1) = -1 \quad \forall M_1 \in \mathcal{M}_1(e)\}$ *and deduce that* D_1 rev R *is Pfaffian. Thus,* G_1 *is Pfaffian. Likewise,* G_2 *is also Pfaffian.*

20.2.5 *Give a proof of Corollary 20.16.*

20.3 The Number of Similarity Classes ♯

Recall that if D is a Pfaffian orientation of G, then any orientation of G that is similar to D is also a Pfaffian orientation of G. Moreover, similarity is an equivalence relation.

> THE NUMBER $\tau(G)$ OF DISSIMILAR PFAFFIAN ORIENTATIONS
>
> Let $\tau(G)$ denote the number of equivalence classes of Pfaffian orientations of a graph G with respect to similarity.

In this section we shall prove that $\tau(G) = 2^{b(G)}$ for every Pfaffian matching covered graph G. (Thus, interestingly, $\tau(G)$ depends only on the number $b(G)$ of bricks of G.)

Theorem 20.19 *The equality* $\tau(G) = 2^{b(G)}$ *holds for every Pfaffian matching covered graph* G*.*

Proof The proof is divided in two parts, consisting of the proofs of Lemmas 20.20 and 20.23. □

20.3.1 An upper bound for $\tau(G)$

LEMMA: AN UPPER BOUND FOR THE NUMBER OF DISSIMILAR PFAFFIAN ORIENTATIONS

20.20 *Let G be a Pfaffian matching covered graph. Then, $\tau(G) \leq 2^{b(G)}$.*

Proof by induction on $|E|$. Adopt, as the induction hypothesis, that the inequality $\tau(H) \leq 2^{b(H)}$ holds for every Pfaffian matching covered graph H having fewer edges than G. Let C be a collection of $\tau(G)$ dissimilar Pfaffian orientations of G. If $G = K_2$ then the two orientations of G are Pfaffian and similar. In that case, $\tau(G) = |C| = 1$. We may thus assume that $G \neq K_2$.

By Theorem 16.9, G has an optimal removable ear, R. Let Q be a conformal cycle of G that contains an edge of R. Then, Q contains all the edges of R. The orientations in C are Pfaffian; hence Q is oddly oriented in each orientation in C. Let $\mathcal{D} := \{D - R : D \in C\}$. Each orientation in $\mathcal{D}$ is a Pfaffian orientation of $G - R$. By induction, $\tau(G - R) \leq 2^{b(G-R)}$.

Suppose that R is a single ear. By Corollary 20.17, R is b-invariant; hence $b(G - R) = b(G)$. By Lemma 20.8, the $\tau(G)$ orientations in $\mathcal{D}$ are pairwise dissimilar, because the orientations in C are pairwise dissimilar. Thus, $\tau(G) \leq \tau(G - R)$. In sum,

$$\tau(G) \leq \tau(G - R) \leq 2^{b(G-R)} = 2^{b(G)}.$$

The assertion holds if R is a single ear.

Suppose now that R is a double ear. Then, $b(G - R) = b(G) - 1$. The orientations in C are pairwise dissimilar. By Lemma 20.9, no three orientations in $\mathcal{D}$ are pairwise similar. Thus, $\mathcal{D}$ has a subcollection of at least $|\mathcal{D}|/2$ pairwise dissimilar Pfaffian orientations; hence $|\mathcal{D}|/2 \leq \tau(G - R)$. Thus, $|\mathcal{D}| \leq 2\tau(G - R)$. We deduce that $\tau(G) \leq 2\tau(G - R)$. In sum,

$$\tau(G) \leq 2\tau(G - R) \leq 2 \cdot 2^{b(G-R)} = 2 \cdot 2^{b(G)-1} = 2^{b(G)}.$$

The assertion also holds if R is a double ear. The proof of Lemma 20.20 is complete.□

Corollary 20.21 *Let G be a Pfaffian matching covered graph which is either bipartite or a near-brick. Then,*

$$\tau(G) = \begin{cases} 1, & \text{if } G \text{ is bipartite} \\ 2, & \text{if } G \text{ is a near-brick} \end{cases}$$

Proof Suppose that G is bipartite. As G is Pfaffian, $1 \leq \tau(G)$. By the Theorem, $\tau(G) \leq 2^{b(G)} = 1$. Thus, $\tau(G) = 1$.

Suppose that G is a near-brick. By the Theorem, $\tau(G) \leq 2^{b(G)} = 2$. As G is Pfaffian, let D be a Pfaffian orientation of G. The orientation D rev E is also a Pfaffian orientation of G. Moreover, as G is not bipartite, the set E is not a cut of G; hence D and D rev E are dissimilar. We deduce that $\tau(G) = 2$. □

20.3.2 A lower bound for $\tau(G)$

In order to determine a lower bound for $\tau(G)$, we need the following elementary property of matching covered graphs.

Proposition 20.22 *Let G be a matching covered graph and let v be a vertex of G. If $G - v$ is bipartite then G is also bipartite.*

Proof Suppose that $G - v$ is bipartite and let (A, B) denote a bipartition of $G - v$. Let $e := vw$ be an edge of G incident with v and let M be a perfect matching of G that contains the edge e. The vertex w is in $A \cup B$. Adjust notation so that $w \in B$. Every vertex of $A + v$ is matched with a vertex of B; thus $|B| = |A + v|$. It follows that $A + v$ is a barrier of G. As G is matching covered, $A + v$ is an independent set of vertices of G. Thus, $(A + v, B)$ is a bipartition of G. □

LEMMA: A LOWER BOUND FOR THE NUMBER OF DISSIMILAR PFAFFIAN ORIENTATIONS

20.23 *Let G be a Pfaffian matching covered graph. Then, $\tau(G) \geq 2^{b(G)}$.*

Proof by induction on $|E|$. Adopt as the inductive hypothesis that the inequality $\tau(H) \geq 2^{b(H)}$ holds for each matching covered graph H having fewer edges than G.

If G is a brace then, by Corollary 20.21, $\tau(G) = 1$ and if G is a brick then, again by Corollary 20.21, $\tau(G) = 2$. In both cases, the assertion holds.

We may thus assume that G is neither a brace nor a brick; hence G has nontrivial tight cuts. Let X be a minimal set of vertices of G such that $C := \partial(X)$ is a nontrivial tight cut.

Let $G_1 := G/(\overline{X} \to \overline{x})$ and $G_2 := G/(X \to x)$ be the two C-contractions of G. For $i = 1, 2$, let $b(G_i)$ be the number of bricks of G_i. Then, $b(G) = b(G_1) + b(G_2)$. The minimality of X implies that G_1 is either a brick or a brace.

The graph G is Pfaffian. By Theorem 20.10, G_1 and G_2 are Pfaffian. Let D_1 be a Pfaffian orientation of G_1. By induction and Lemma 20.20, $\tau(G_2) = 2^{b(G_2)}$. Let $\mathcal{D}_2$ be a collection of $2^{b(G_2)}$ pairwise dissimilar Pfaffian orientations of G_2. By Proposition 20.6, we may assume that the contraction vertex $\overline{x}$ of G_1 is a source in D_1 and the contraction vertex x of G_2 is a sink in each orientation in $\mathcal{D}_2$. For each orientation D_2 in $\mathcal{D}_2$, there is an orientation D of G which is the common extension of D_1 and D_2. We say that D is *generated by D_1 and D_2*. Let $\mathcal{D}$ be the collection of orientations of G generated by D_1 and the orientations in $\mathcal{D}_2$. Clearly, $|\mathcal{D}| = |\mathcal{D}_2| = 2^{b(G_2)}$.

20.23.1 *The collection $\mathcal{D}$ consists of $2^{b(G_2)}$ pairwise dissimilar Pfaffian orientations of G.*

Proof By Lemma 20.11, each orientation in $\mathcal{D}$ is a Pfaffian orientation of G. Let us now prove that the orientations in $\mathcal{D}$ are pairwise dissimilar.

Assume, to the contrary, that $\mathcal{D}$ contains two similar orientations, D and F, respectively generated by distinct orientations D_2 and F_2 in $\mathcal{D}_2$. Let R be the set of edges of G whose orientations in D and in F differ. We have assumed that D and F are similar; hence R is a cut of G, say, $\partial(Y)$. For each edge in $E(G_1)$, its orientations in D and in F coincide with its orientation in D_1. Thus, $R \subseteq E(G_2) - C$. We may then adjust notation, so that $Y \subseteq V(G_2 - x)$; hence R is a cut of G_2. Moreover, D_2 rev $R = F_2$; hence D_2 and F_2 are similar. This is a contradiction. We conclude that the $2^{b(G_2)}$ orientations in $\mathcal{D}$ are pairwise dissimilar Pfaffian orientations of G.□

If G_1 is a brace then $b(G) = b(G_2)$; hence $\mathcal{D}$ is a collection of $2^{b(G)}$ pairwise dissimilar Pfaffian orientations of G. The assertion holds in this case. We may thus assume that G_1 is a brick; hence $b(G) = b(G_2) + 1$. It follows that $|\mathcal{D}| = 2^{b(G)-1}$.

Let D_1' be the orientation D_1 rev $(E(G_1) - C)$ of G_1. Let $\mathcal{D}'$ be the collection of orientations of G which are generated by D_1' and the orientations in $\mathcal{D}_2$. Repeating the reasoning done above, we conclude that $\mathcal{D}'$ consists of $2^{b(G)-1}$ pairwise dissimilar Pfaffian orientations of G.

Let $D \in \mathcal{D}$ and $D' \in \mathcal{D}'$ and let R be the set of edges whose orientations in D and in D' differ. Clearly, $E(G_1) - C \subseteq R$. As G_1 is a brick, $G_1 - \overline{x}$ is nonbipartite, by Proposition 20.22. Thus, $G_1 - C$ has an odd cycle, Q. Every edge of Q is in R. By Proposition 20.1(ii), R is not a cut of G; hence D and D' are dissimilar. This conclusion holds for each $D \in \mathcal{D}$ and each $D' \in \mathcal{D}'$. We deduce that $\mathcal{D} \cup \mathcal{D}'$ is a collection of $2^{b(G)}$ pairwise dissimilar Pfaffian orientations of G.

In all alternatives considered, we proved that $\tau(G) \geq 2^{b(G)}$. The proof of Lemma 20.23 completes the proof of Theorem 20.19. □

Exercises

20.3.1 *Find two dissimilar Pfaffian orientations of* G, *for* $G = K_4, \overline{C_6}, \mathbb{M}_8$.

▷**20.3.2** *Consider the matching covered graph* G *with* $b(G) = 3$ *obtained by splicing a* K_4 *at three vertices which belong to one of the two colour classes of the cube. Find* $2^3 = 8$ *pairwise dissimilar Pfaffian orientations of* G.

20.4 Characteristic Orientations

Characteristic orientations
An orientation D of a (not necessarily matching covered) graph G is a *characteristic orientation* of G if it has the property that G is Pfaffian if and only if D is a Pfaffian orientation of G.

In this section, based on Theorem 20.19, we shall deduce the existence of a polynomial-time algorithm for finding a characteristic orientation of a graph. This striking result makes it possible for us to deduce the polynomial-time equivalence of the Pfaffian Recognition Problem 19.1 and the Pfaffian Orientation Problem 19.2.

20.4.1 Pfaffian extensions

We now present a fundamental result which is the basis of the polynomial-time algorithm that produces a characteristic orientation of a graph.

THE PFAFFIAN EXTENSION THEOREM

20.24 *Let G be a matching covered graph, let D be an orientation of G, let R be an optimal removable ear of G and let Q be a conformal cycle of G that contains all the edges of R. Suppose that (i) $D - R$ is Pfaffian and (ii) Q is oddly oriented in D. Then, G is Pfaffian if and only if D is Pfaffian.*

Proof If D is Pfaffian then certainly G is Pfaffian. To prove the converse, suppose that G is Pfaffian. Let $t := 2^{b(G)}$. By Theorem 20.19, the graph G has a collection $\mathcal{D} := \{D_1, D_2, \ldots, D_t\}$ of t dissimilar Pfaffian orientations. Let $D_0 := D$ and let $\mathcal{D}' := \{D_i - R : 0 \le i \le t\}$. For each ℓ such that $1 \le \ell \le t$, the orientation D_ℓ of G is Pfaffian; hence D'_ℓ is also a Pfaffian orientation of $G - R$. Moreover, the cycle Q is oddly oriented in D_ℓ, because Q is a conformal cycle of G. By definition, $D_0 - R$ is a Pfaffian orientation of $G - R$ and Q is also oddly oriented in D_0.

Case 1 *R is a single ear.*

By hypothesis, R is an optimal removable ear of G. As R is a single ear, it is $(b+p)$-invariant. By Corollary 20.17, R is b-invariant; hence $2^{b(G-R)} = 2^{b(G)} = t$. By Theorem 20.19, t is the number of dissimilar Pfaffian orientations of $G - R$. As $\mathcal{D}'$ consists of $t+1$ Pfaffian orientations of $G-R$, it follows that $\mathcal{D}'$ contains two similar orientations. Let D_i and D_j be orientations in $\mathcal{D} + D_0$, where $0 \le i < j \le t$ and $D_i - R$ and $D_j - R$ are similar. Moreover, Q is oddly oriented in both D_i and D_j. By Lemma 20.8, D_i and D_j are similar. The orientations in $\mathcal{D}$ are not similar; hence $i = 0$. We deduce that D is similar to some Pfaffian orientation in $\mathcal{D}$. Thus, D is Pfaffian.

Case 2 *R is a double ear.*

From the fact that R is a double ear we infer that $2^{b(G-R)} = 2^{b(G)-1} = t/2$. By Theorem 20.19, $t/2$ is the number of dissimilar Pfaffian orientations of $G-R$. As $\mathcal{D}'$ consists of $t+1$ Pfaffian orientations of $G - R$, it follows that $\mathcal{D} + D_0$ contains three orientations, D_i, D_j and D_k, where $0 \le i < j < k \le t$ and the three orientations $D_i - R$, $D_j - R$ and $D_k - R$ are similar. Moreover, Q is oddly oriented in D_i, D_j and D_k. By Lemma 20.9, two of the three orientations D_i, D_j and D_k are similar. The orientations in $\mathcal{D}$ are not similar; hence $i = 0$. We deduce that D is similar to some Pfaffian orientation in $\mathcal{D}$. Thus, D is Pfaffian. □

Corollary 20.25 *Let G be a Pfaffian matching covered graph and let R be an optimal removable ear of G. Every Pfaffian orientation of $G-R$ has an extension to a Pfaffian orientation of G.* □

20.4.2 Finding characteristic orientations

We are now in position to describe a polynomial-time algorithm alluded to at the beginning of this section. This algorithm finds a characteristic orientation of a graph in polynomial time. We first give a special case, in which the given graph is matching covered. This algorithm is then used in the full version, in which the given graph is not necessarily matching covered.

FINDING A CHARACTERISTIC ORIENTATION

Algorithm 20.26
Input: a (not necessarily matching covered) graph G;
Output: a characteristic orientation of G.

A fundamental case: G is matching covered.

We first apply a polynomial-time algorithm to obtain an ear decomposition

$$G_1 = K_2 \subset G_2 \subset \cdots \subset G_r = G$$

where, for $i = 2, 3, \ldots, r$, $G_{i-1} = G_i - R_i$ and R_i is an optimal removable ear of G_i. (See Exercise 16.2.8.)

We then proceed to obtain a sequence of orientations $(D_1, D_2, \ldots, D_r)$ such that D_i is an orientation of G_i, for $i = 1, 2, \ldots, r$. We arbitrarily orient G_1, thereby obtaining D_1.

To obtain D_i for $i = 2, 3, \ldots, r$, we first determine a conformal cycle Q_i of G_i that contains all the edges of R_i and then extend D_{i-1} to the orientation D_i of G_i by orienting the edges of R_i so that Q_i is oddly oriented in D_i.

To determine the cycle Q_i, we apply a polynomial-time algorithm to determine a perfect matching M_i of G_i that contains an edge e_i of R_i. (See Exercise 2.1.2.) Then, we determine the M_i-alternating cycle of G that contains an edge f_i adjacent to e_i. Clearly, Q_i is a conformal cycle of G_i that contains all the edges of R_i.

The orientation D_1 is a characteristic orientation of G_1, because G_1 is K_2. With the help of Theorem 20.24, it is easy to prove by induction on i that D_i is a characteristic orientation of G_i, for $i = 1, 2, \ldots, r$.

The generic case: G is not necessarily matching covered.

We first determine whether G is matchable, using any polynomial-time algorithm which finds a maximum matching of a graph. If G is not matchable then output an arbitrary orientation of G. We may thus assume that G is matchable. In polynomial time we then determine the set N of nonmatchable edges of G (see Exercise 2.1.2).

Let $H := G - N$. Each component of H is matching covered. Apply the algorithm to each (matching covered) component K of H, thereby obtaining a characteristic orientation D_K of K. Let $D_H := \cup_K D_K$ and let D be an arbitrary extension of D_H to an orientation of G.

Let us now see why D is a characteristic orientation of G. If G is not Pfaffian then no orientation of G is Pfaffian. In particular, D is not Pfaffian. We may thus assume that G is Pfaffian and we must show that D is Pfaffian. Let K be a component of H. As seen in Exercise 19.2.4, K is Pfaffian. As D_K is a characteristic orientation of K, we deduce that D_K is Pfaffian. This conclusion holds for each component K of H. Again, as seen in Exercise 19.2.4, D is Pfaffian.

Corollary 20.27 *Algorithm 20.26 determines a characteristic orientation of a graph in polynomial time.* □

20.4.3 The equivalence of Problems 19.1 and 19.2

Theorem 20.28 *The Problem 19.1, of determining whether a given graph is Pfaffian, is polynomial-time equivalent to the Problem 19.2, of deciding whether a given digraph is Pfaffian.*

Proof Let us first describe a polynomial-time reduction of Problem 19.1, of determining whether a given graph is Pfaffian, to the Problem 19.2, of deciding whether a given digraph is Pfaffian. For this, let G be a graph. Apply Algorithm 20.26 to graph G, thereby obtaining a characteristic orientation D of G. The graph G is Pfaffian if and only if D is Pfaffian.

Let us now describe a polynomial-time reduction of Problem 19.2, of determining whether a given digraph is Pfaffian, to the Problem 19.1, of deciding whether a given graph is Pfaffian.

We first give a particular case, in which the given graph is matching covered. This algorithm is then used in the full version, in which the given graph is not necessarily matching covered.

VERIFYING AN ORIENTATION

Algorithm 20.29
Input: an orientation D of a (not necessarily matching covered) graph G;
Output: it either determines that D may be generated by Algorithm 20.26 or it outputs a cycle which is conformal in G and evenly oriented in D.

A fundamental case: G is matching covered.

Our task is done by Algorithm 20.26, in the case that G is matching covered, by trying to keep each digraph in the sequence of orientations D_i of G as a subdigraph of D. If, at any stage, it is not possible to extend an orientation D_{i-1} to an orientation D_i, it follows that the cycle Q_i, a conformal cycle of G, is evenly oriented in D. In that case, output Q_i as a conformal cycle of G which is evenly oriented in D. Alternatively, we succeed in determining that D may be generated by Algorithm 20.26.

The generic case: G is not necessarily matching covered.

As in Algorithm 20.26, determine in polynomial time whether G is matchable and, if G is matchable, determine the set N of non matchable edges of G. If G is not matchable then Algorithm 20.26 outputs an arbitrary orientation of G. In particular, D may be generated by Algorithm 20.26.

We may thus assume that G is matchable. Let $H := G - N$. Each component of H is matching covered. For each component K of H, let D_K be the restriction of D to K.

Apply the algorithm to each component K of H, to verify whether D_K may be generated by Algorithm 20.26. Suppose that the algorithm determines that D_K may be generated by Algorithm 20.26, for each component K of H. Then, D is an extension of $D_H := \cup_K D_K$, and D may be generated by Algorithm 20.26.

Alternatively, suppose that, for some component K, the algorithm produces a cycle Q which is conformal in K and evenly oriented in D_K. The graph K is a conformal subgraph of H, which in turn is a conformal subgraph of G. Thus, Q is conformal in G. Moreover, Q is evenly oriented in D. The cycle Q is then the output of the algorithm in this case.

The polynomial-time reduction of Problem 19.2 to Problem 19.1 is done by Algorithm 20.29. Let D be an orientation of a graph G. If the algorithm produces a conformal cycle which is evenly oriented in D, then D is not Pfaffian. Alternatively, the algorithm determines that D may be generated by Algorithm 20.26. Then, by Corollary 20.27, D is a characteristic orientation of G; hence D is Pfaffian if and only if G is Pfaffian. □

We end this section with an important property of Pfaffian graphs, which will be used in Chapter 21, to justify a polynomial-time algorithm to determine whether a near-bipartite graph is Pfaffian.

THE INCLUSION-EXCLUSION PROPERTY OF PERFECT MATCHINGS

20.30 *Let G be a matching covered graph, let R be an equivalence class of the dependence relation on the edges of G, let D be an orientation of G and let Q be a conformal cycle of G that contains the edges of R and is oddly oriented in D. Then, D is Pfaffian if and only if the digraphs $D - R$ and $D - V(R)$ are both Pfaffian, where $V(R)$ is the set of the ends of the edges of R.*

Proof We may clearly suppose that G is not K_2; hence G contains an edge adjacent to an edge of R. The set R is an equivalence class of the dependence relation on the edges of G. Thus, if a perfect matching $M_{\overline{R}}$ contains an edge adjacent to an edge of R then it does not contain any edge of R and is thus a perfect matching of $G - R$. Alternatively, if a perfect matching M_R contains an edge of R then it contains all the edges of R. We deduce that the collection $\mathcal{M}$ of perfect matchings of G can be partitioned in two sets, $\mathcal{M}_{\overline{R}}$ and $\mathcal{M}_R$, respectively the collection of perfect matchings of G that contain no edge of R and the collection of those perfect matchings that contain all the edges of R.

The graph $G - R$ is thus a (spanning) matchable subgraph of G; hence $G - R$ is a conformal subgraph of G. Likewise, for any matching M in $\mathcal{M}_R$, the set $M - R$ is a perfect matching of $G - V(R)$; hence the graph $G - V(R)$ is a matchable subgraph of G. Moreover, the set R is a perfect matching of $G - (G - V(R))$. Thus, $G - V(R)$ is also a conformal subgraph of G. In sum, the graphs $G - R$ and $G - V(R)$ are both conformal subgraphs of G. If D is Pfaffian then $D - R$ and $D - V(R)$ are both Pfaffian.

To prove the converse, suppose that $D - R$ and $D - V(R)$ are both Pfaffian. Let M be a perfect matching of G. If M is a perfect matching of $G - R$, that is, if $M \in \mathcal{M}_{\overline{R}}$, then, as $D - R$ is Pfaffian, we infer that every M-alternating cycle of $G - R$ is oddly oriented in D; hence all the perfect matchings of $G - R$ have the same sign, say s, in D. Likewise, if $M \in \mathcal{M}_R$, then, as $M - R$ is a perfect matching of $G - V(R)$, it follows that every M-alternating cycle of $G - V(R)$ is oddly oriented in D; hence all the perfect matchings of $G - V(R)$ have the same sign, say t, in D.

Finally, the cycle Q is conformal in G, contains the edges of R and is oddly oriented in D. As Q is conformal, it is an M-alternating cycle for some perfect matching, M, of G. The set $N := M \triangle E(Q)$ is a perfect matching of G. As Q is oddly oriented in D, the matchings M and N have the same sign in D. One of M and N is in $\mathcal{M}_{\overline{R}}$ and has sign s in D, the other is in $\mathcal{M}_R$ and has sign t in D. We deduce that $s = t$. This conclusion implies that all the perfect matchings of G have the same sign in D; hence D is Pfaffian. □

Exercises

▷**20.4.1** *In the statement of Theorem 20.24, the ear R of G, if it is a single ear, is required to be $(b+p)$-invariant. In fact, if it is not $(b+p)$-invariant, then one of two things may happen: either $p(G-R) > 0$ or $p(G-R) = 0$ but $b(G-R) > b(G)$. In the first alternative, G is not Pfaffian, by Corollary 20.15, because Petersen bricks are not Pfaffian. In the second alternative, R is not b-invariant.*

(i) *Let G be the simple graph obtained from the Petersen graph by the addition of an edge, e. The edge e is b-invariant, but not $(b+p)$-invariant. Using Algorithm 20.26 , determine a characteristic orientation D' of $G - e$, and extend it to a characteristic orientation D of G. Verify that D is not Pfaffian.*

(ii) *The orientation D of the brick G depicted in Figure 20.5 is Pfaffian. The edge e is not b-invariant in G. The pair $R := \{f, g\}$ is a removable doubleton of the graph $G - e$. Thus, $G - e$ has two characteristic orientations: one is $D - e$ and the other is $(D - e)$* rev *R. Verify that the second characteristic orientation of $G - e$ is not extendable to a Pfaffian orientation of G. Hint: extend $(D - e)$* rev *R to an orientation D' of G by arbitrarily orienting edge e and then find two conformal cycles of G, both containing edge e and having distinct signs in D'.*

20.4.2 *Find characteristic orientations of $K_{3,3}$ and $\mathbb{P}$ and verify that they are not Pfaffian.*

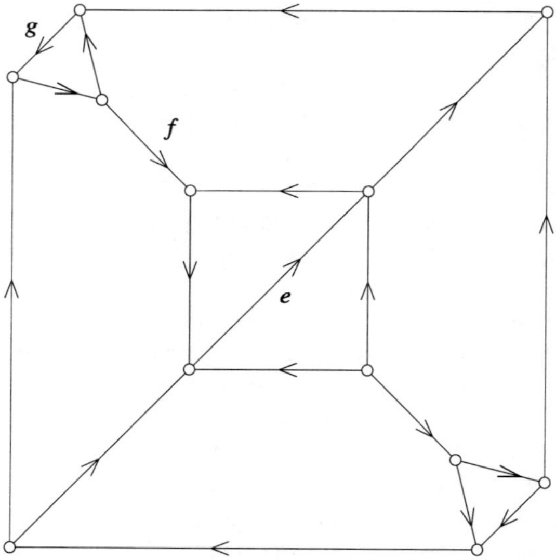

Fig. 20.5 The brick G and its orientation D

20.4.3 *Find maximal collections of dissimilar Pfaffian orientations of those graphs that represent the five Platonic polyhedra: the tetrahedron, the cube, the octahedron, the dodecahedron and the icosahedron.*

▷**20.4.4** *Let G be a matching covered graph G. Recall that we denote the numbers of vertices, edges, bricks, Petersen bricks, and the number of perfect matchings of G by n, m, b, p and Φ, respectively. We say that G is Pfaffian-extremal if $\Phi(G) = m - n + 2 - (b + p)$. For example, for each edge e, the graphs $K_{3,3} - e$ and $\mathbb{P} - e$ are both Pfaffian-extremal.*

(i) *Prove that if $\Phi(G) \leq m - n + 2 - (b + p)$ then in fact $p = 0$ and equality holds. (Recall, from Chapter 6, that $\Phi(G) \geq m - n + 2 - b$).*
(ii) *Prove that if G is Pfaffian-extremal and R is an optimal removable ear of G, then $G - R$ is Pfaffian-extremal.*
(iii) *Prove that every Pfaffian-extremal graph is Pfaffian and conclude that $K_{3,3} - e$ and $\mathbb{P} - e$ are Pfaffian, for each edge e. (Consider an optimal removable ear decomposition of G.)*

▷**20.4.5** *Prove that Problem 19.1 is in co-$\mathcal{NP}$. More precisely, show that given a non-Pfaffian digraph D, there is a succinct certificate to show that D is not Pfaffian, which can be checked in polynomial-time.*

▷**20.4.6** *Prove that Problem 19.2 is in co-$\mathcal{NP}$. More precisely, show that, given a non-Pfaffian matching covered graph G, there is a succinct certificate to show that G is not Pfaffian, which can be checked in polynomial-time. Hint: use the result of Exercise 20.4.5 and Algorithm 20.26.*

20.5 Notes

The reduction of Pfaffian matching covered graphs to Pfaffian bricks and braces was first proved by Little and Rendl (1991, [55]); see Theorem 20.10. The proof we give here is based on similarity of orientations and was published by CLM more than 20 years later (2012, [16]). The number of dissimilar Pfaffian orientations of a matching covered graph was published by CLM (2005, [14]). The equivalence of Problems 19.1 and 19.2 was published by Vazirani and Yannakakis (1989, [91]).

Chapter 21
Excluded-Minor Characterizations of Pfaffian Graphs

Contents

21.1 $\mathcal{S}$-Minors

As noted in Chapter 19, Kasteleyn (1963, [44]) not only pioneered the theory of Pfaffian orientations, but also established the first definitive result in this theory by showing that all planar graphs are Pfaffian.

C. L. Lucchesi, U. S. R. Murty, *Perfect Matchings*, Algorithms and Computation in Mathematics 31, https://doi.org/10.1007/978-3-031-47504-7_21

By Corollary 20.16, a matching covered graph is Pfaffian if and only if all its matching minors are Pfaffian. Thus any graph which admits a known non-Pfaffian graph (such as $K_{3,3}$) as a matching minor is also non-Pfaffian. Little (1975, [54]) advanced the theory of Pfaffian orientations by showing that a bipartite matching covered graph is Pfaffian if and only if it does not have a matching minor isomorphic to $K_{3,3}$.

Some years later, in collaboration with Fischer (2001, [34]), Little participated in another advance by characterizing near-bipartite graphs which admit a Pfaffian orientation. A proper description of this characterization requires the notion of a minor that is more general than that of a matching minor. This general notion, which plays a crucial role in this chapter, is defined in the inset below:

DEFINITION: S-MINOR

A *separation-deletion minor* of a matching covered graph G, or simply an *S-minor* of G, is a graph that is obtainable from G, up to isomorphism, by means of deletions of removable classes and contractions of shores of separating cuts. In other words, H is an S-minor of G if there exists a sequence

$$(G_1, G_2, \ldots, G_r) \quad (r \geq 1)$$

of graphs such that, $G_1 = G$, $G_r \cong H$ and, for $1 \leq i \leq r-1$, the graph G_{i+1} is obtained from G_i by either deleting a removable class or by contracting a shore of a separating cut to a single vertex. (We allow the possibility that $r = 1$; that is, we regard a graph G to be an S-minor of itself.)

As an example, consider the sequence (G_1, G_2, G_3, G_4) of graphs in Figure 21.1. The graph G_1 is a nonsolid brick and the cut C is a (nontight) separating cut of G_1. The graph G_2 is a C-contraction of G_1; it is a brick (which happens to be the solid brick Σ_8 shown in Figure 7.2), and e is a removable edge in it. The cut C in $G_3 := G_2 - e$ is a separating cut (in fact, a tight cut) and G_4 is obtained from G_3 by contracting one of the shores of C. Thus G_1, G_2, G_3 and G_4 are S-minors of G_1. (Note that K_4 is also an S-minor of G_1.)

It follows from the top-down version of the Ear Decomposition Theorem 11.3 that every conformal subgraph of a matching covered graph G is an S-minor of G. Consequently, every matching minor of G is also an S-minor of G. But not every S-minor of G is a matching minor of G. For example, $K_{3,3}$ is an S-minor of G_3, but it is not a matching minor of that graph (see Exercise 12.1.6).

From Corollary 19.6 and Lemma 20.13, we may now deduce the following important property of Pfaffian graphs, whose proof is left as Exercise 21.1.1.

Theorem 21.1 *A matching covered graph is Pfaffian if and only if all its S-minors are Pfaffian.* □

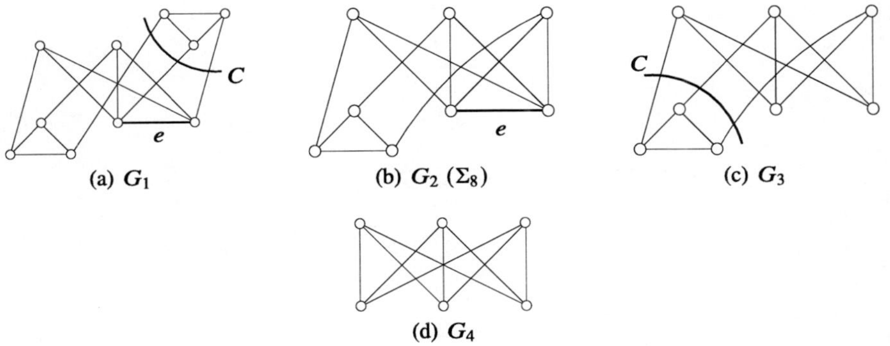

Fig. 21.1 S-minors

In light of the above theorem, to show that a given graph is non-Pfaffian, it suffices to produce an S-minor of that graph which is known to be non-Pfaffian. For example, since $K_{3,3}$ is known to be non-Pfaffian, and since it is a S-minor of each of the graphs in Figure 21.1, we may conclude that all of them are non-Pfaffian.

This leads us to the notion of an *S-minimal* non-Pfaffian graph. A non-Pfaffian matching covered is S-minimal if each of its proper S-minors is Pfaffian. In Section 21.2 we describe some properties of minimal Pfaffian graphs which are pertinent to characterizations of Pfaffian bipartite and Pfaffian near-bipartite graphs presented in Sections 21.3 and 21.4, respectively. In the latter section we show that the brace $K_{3,3}$ is the only S-minimal non-Pfaffian bipartite graph and the two bricks Γ_1 and Γ_2 depicted in Figures 21.2 and 21.3, respectively, are the only S-minimal non-Pfaffian near-bipartite graphs. This, in essence, is the theorem of Fischer and Little. (In their paper they characterize all matching minor-minimal non-Pfaffian near-bipartite graphs.)

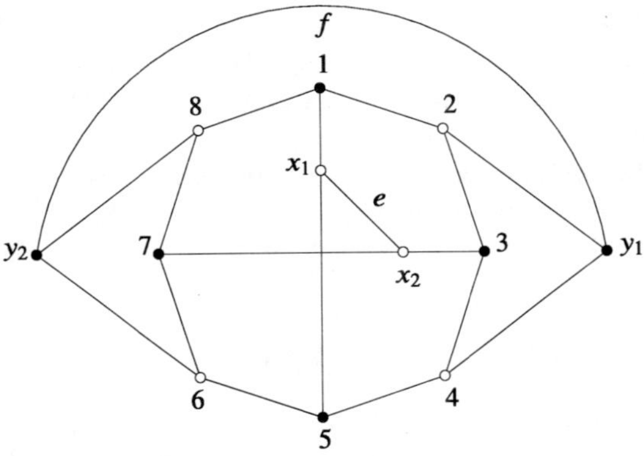

Fig. 21.2 The brick Γ_1

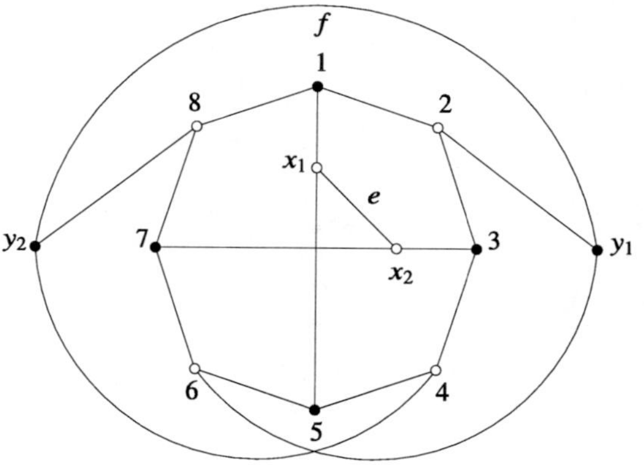

Fig. 21.3 The brick Γ_2

Exercises

21.1.1 *Prove Theorem 21.1.*

21.1.2 *Prove that every non-Pfaffian matching covered graph has order six or more.*

∗**21.1.3** *The objective of this exercise is to prove that the cube $\mathbb{P}_8$ is the only simple brace of order eight that is Pfaffian.*

(i) *Prove that the cube is the only cubic brace of order eight.*
(ii) *Prove that the cube is a subgraph of each brace of order eight.*
(iii) *Prove that every simple brace of order eight distinct from the cube is non-Pfaffian.*

▷**21.1.4** *Prove that the graphs Γ_1 and Γ_2 are non-Pfaffian. Hint: for $i = 1, 2$, find a characteristic orientation of $\Gamma_i - e - f$, using Algorithm 20.26 and extend it to a characteristic orientation D_i of Γ_i. Verify that the cycle $Q := (1, 2, \ldots, 8, 1)$ is conformal in Γ_i and evenly oriented in D_i. (For $i = 1, 2$: the graph $\Gamma_i - e - f$ has four vertices of degree two; the retract H_i of $\Gamma_i - e - f$ is C_4, up to multiple edges; extend any Pfaffian orientation of H_i to an orientation of $\Gamma_i - e - f$, by ensuring that each vertex of degree two is neither a source nor a sink.)*

21.2 S-Minimal non-Pfaffian Graphs

Motivated by Theorem 21.1, we now introduce the important notion of an S-minimal minor of a non-Pfaffian graph.

DEFINITION: S-MINIMAL NON-PFAFFIAN MINORS

A non-Pfaffian S-minor of given matching covered graph is *separation-deletion minimal*, or simply *S-minimal*, if all its proper S-minors are Pfaffian. For example, suppose that the non-Pfaffian brace $K_{3,3}$ is an S-minor of some graph G. Then $K_{3,3}$ is an S-minimal minor of G because all proper S-minors of $K_{3,3}$, being planar, are Pfaffian.

It follows from Corollary 20.15 that every S-minimal non-Pfaffian matching covered graph is either a brick or a brace. In addition, it is easy to see that such a graph is also simple. The only simple braces of order four or less are K_2 and C_4. The only simple brick of order four is K_4. It is easy to deduce that every non-Pfaffian graph has order at least six (Exercise 21.1.2). These observations imply:

Proposition 21.2 *All non-Pfaffian S-minimal minors of a non-Pfaffian graph are simple and are either bricks or braces of order six or more.* □

We shall assume the properties of S-minimal minors stated in the above proposition without explicitly stating them.

21.2.1 Properties of non-Pfaffian Graphs

The objective of this subsection is to establish a number of basic properties of non-Pfaffian matching covered graphs.

Theorem 21.3 *Let D be an orientation of a non-Pfaffian matching covered graph G and let R be a removable class of G. If $D - R$ is Pfaffian then the perfect matchings of G that contain R do not all have the same sign.*

Proof As $D - R$ is a Pfaffian orientation of $G - R$, all perfect matchings of $G - R$ have the same sign, say s, in D. If all the perfect matchings containing all the edges of R also have sign s in D, then D itself would be a Pfaffian orientation of G. And, if all the perfect matchings containing all the edges of R have sign $-s$ in D, the digraph obtained from D by reversing the orientation of one of the edges of R would be a Pfaffian orientation of G. Both these cases are impossible because, by hypothesis, G is non-Pfaffian. We conclude that the perfect matchings of G that contain R do not all have the same sign. □

In the following inset we introduce the notion of an R-triple which will play a crucial role in our proofs of the two theorems which provide characterizations of bipartite graphs and of near-bipartite graphs.

R-TRIPLES AND THEIR SIGNIFICANCE

Definition: Let R be a removable class of a matching covered graph G, let D be an orientation of G such that $D - R$ is Pfaffian and let M_1 and M_2 be perfect matchings of G both containing all the edges of R, and having distinct signs in D. Then, the orientation D of G and the perfect matchings M_1 and M_2 of G constitute, in that order, an *R-triple*. We denote such a triple by the ordered set (D, M_1, M_2). If R is a singleton, say, $\{e\}$, then we simply refer to the triple (D, M_1, M_2) as an *e-triple*.

The notion of an R-triple plays a significant role in our presentations of the proofs of the above mentioned theorems. This, as we shall see below, is due to the fact that if R happens to be an optimal removable class, then an associated R-triple serves as a certificate which demonstrates that G is non-Pfaffian.

Lemma 21.4 *Let G be a matching covered graph and let R be an optimal removable class of G. If G has an R-triple then G is not Pfaffian.*

Proof Suppose that G has an R-triple, (D, M_1, M_2). By definition, the orientation $D - R$ of $G - R$ is Pfaffian, the two perfect matchings M_1 and M_2 have distinct signs in D, and $R \subseteq M_1 \cap M_2$. Reversal of the orientation of any edge of R changes the sign of every perfect matching of G that contains that edge of R. It follows that M_1 and M_2 have distinct signs in every orientation of G that is an extension of the orientation $D - R$ of $G - R$, implying that no extension of the Pfaffian orientation $D - R$ of $G - R$ to an orientation of G could be Pfaffian. On the other hand, as R is an optimal class, by Corollary 20.25, if G were Pfaffian, it should be possible to extend any Pfaffian orientation of $G - R$ to a Pfaffian orientation of G. We conclude that G is not Pfaffian. □

The existence of an R-triple, without the requirement that R is an optimal removable class of G, does not necessarily imply that G is not Pfaffian. See Exercise 21.2.3.

Example 21.5 Consider the orientation D of $K_{3,3}$ shown in Figure 19.1. If we take R to consist of the edge $h := 56$, then $D - h$ is a Pfaffian orientation of $K_{3,3} - h$. The triple (D, M_1, M_2) is an h-triple, where $M_1 := \{12, 34, 56\}$ and $M_2 := \{14, 23, 56\}$, because $M_1 \triangle M_2$ induces an evenly oriented cycle. This is yet another proof that $K_{3,3}$ is not Pfaffian!

Example 21.6 As a second example, consider the brick Γ_1 with the orientation D depicted in Figure 21.4. The pair $R := \{e, f\}$ is a removable doubleton of Γ_1, where $e = x_1x_2$ and $f = y_1y_2$. The digraph $D - R$ is Pfaffian. (See Exercise 21.2.1.) The perfect matchings

$$M_1 := \{12, 34, 56, 78, x_1x_2, y_1y_2\} \text{ and}$$
$$M_2 := \{23, 45, 67, 81, x_1x_2, y_1y_2\}$$

have distinct signs in D, as the (M_1, M_2)-alternating cycle $(1, 2, 3, 4, 5, 6, 7, 8, 1)$ is evenly oriented. Thus, (D, M_1, M_2) is an R-triple corresponding to the optimal removable class R. It follows that Γ_1 is not Pfaffian.

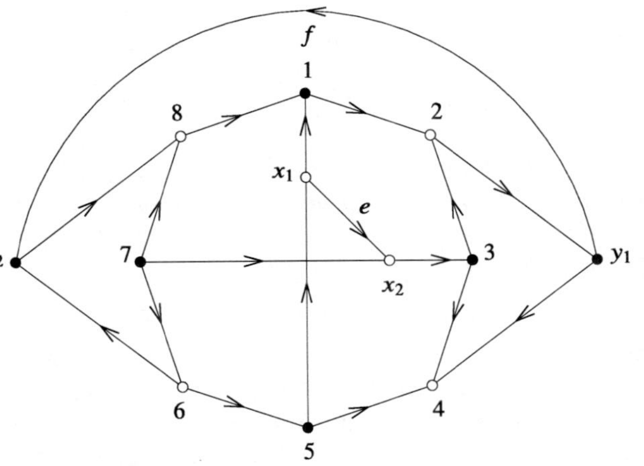

Fig. 21.4 The orientation D of Γ_1

Lemma 21.7 (Exchange Property: optimal classes) *Let G be a non-Pfaffian matching covered graph, let R be an optimal removable class of G and let (D, M_1, M_2) be an R-triple. Suppose that $G - R$ has an optimal removable class S which is disjoint with $M_1 \cup M_2$. Then, S is an optimal removable class of G, R is an optimal removable class of $G - S$ and the graph $G - S$ is not Pfaffian.*

Proof By definition of R-triple, $R \subseteq M_1 \cap M_2$. Thus, $R \subseteq M_1$ and M_1 is a perfect matching of $G - S$. From Lemma 16.13, with M_1 playing the role of M, we infer that S is an optimal removable class of G and R is an optimal removable class of $G - S$.

By definition of R-triple, the digraph $D - R$ is Pfaffian; hence $D - R - S$ is also Pfaffian. The sets M_1 and M_2 are perfect matchings of $G - S$ that have distinct signs in D. Thus, M_1 and M_2 are perfect matchings of $G - S$ that have distinct signs in $D - S$. Moreover, the optimal removable class R of $G - S$ is a subset of $M_1 \cap M_2$. In sum, $(D - S, M_1, M_2)$ is an R-triple of $G - S$. By Lemma 21.4, the graph $G - S$ is not Pfaffian. □

In a bipartite graph all removable classes are singletons and are optimal (b-invariant). Based on this observation, it is easy to deduce the following corollary of the above lemma. It plays an important role in our proof of Little's Theorem in the next section.

Corollary 21.8 (Exchange Property: bipartite graphs) *Let G be a bipartite non-Pfaffian matching covered graph, let e be a removable edge of G, and let (D, M_1, M_2) be an e-triple. Suppose that $G - e$ has a removable edge f which does not belong to $M_1 \cup M_2$. Then f is a removable edge of G and $G - f$ is not Pfaffian.* □

Suppose that G is a non-Pfaffian graph, R is a removable class of G, and that (D, M_1, M_2) is an R-triple of G. Then, by definition, the two perfect matchings M_1 and M_2 have different signs in D. It follows from Corollary 19.4 that there are an odd number of (M_1, M_2)-alternating cycles which are evenly oriented in D. In our approach to characterizations of S-minimal bipartite and near-bipartite non-Pfaffian graphs, it turns out to be convenient to ascertain that there is precisely one evenly oriented (M_1, M_2)-alternating cycle. The following lemma describes a condition under which this is the case.

Lemma 21.9 *Let G be a non-Pfaffian matching covered graph and let R be a removable class of G. Let (D, M_1, M_2) be an R-triple of G. If G has two or more (M_1, M_2)-alternating cycles then there exists an R-triple (D, M_1', M_2) such that G has an $(M_1' \cap M_2)$-alternating cycle.*

Proof Let C denote the collection of all (M_1, M_2)-alternating cycles of G. Suppose that C consists of two or more cycles. As (D, M_1, M_2) is an R-triple, the two perfect matchings M_1 and M_2 have distinct signs in D. By Corollary 19.4, it follows that the collection C includes an odd number of evenly oriented cycles (and it may also include some oddly oriented cycles). Let Q be an evenly oriented (M_1, M_2)-alternating cycle in C.

Let $T := \cup_{C \in C-Q} E(C)$ and let $M_1' := M_1 \triangle T$. It is easy to see that M_1' is a perfect matching of G and that Q is the only (M_1', M_2)-alternating cycle in G. Thus, once again, by Corollary 19.4 it follows that M_1' and M_2 have distinct signs in D. Furthermore those edges which belong to $M_1 \cap M_2$ also belong to $M_1' \cap M_2$. Thus, as $R \subseteq M_1 \cap M_2$, it follows that $R \subseteq M_1' \cap M_2$. In sum, D is a non-Pfaffian orientation of G, the set R is a removable class of G, the digraph $D - R$ is Pfaffian, the perfect matchings M_1' and M_2 have distinct signs in D and $R \subseteq M_1' \cap M_2$. Thus, (D, M_1', M_2) is also an R-triple of G. The assertion holds. □

It is important to understand the significance of the above lemma that does not depend on the choice of notation. This assertion relates two properties an R-triple (D, M_1, M_2) of a non-Pfaffian graph G may have: (i) the number of (M_1, M_2)-alternating cycles in G is just one; and (ii) there are $(M_1 \cap M_2)$-alternating cycles in G. What the lemma says is that if there is no R-triple of G with the second property, then all R-triples of graph G have the first property.

As an example, let $e = a_1 b_1$ be a removable edge in a minimal non-Pfaffian bipartite graph G. We shall show in the next section that, if (D, M_1, M_2) is any arbitrary e-triple of G, then $M_1 \cap M_2 = \{e\}$. This clearly implies that G has no $(M_1 \cap M_2)$-alternating cycles. By the above lemma, it follows that if (D, M_1, M_2) is any particular e-triple of G, then G has precisely one (M_1, M_2)-alternating cycle Q and that cycle is a spanning cycle of $G - a_1 - b_1$.

21.2.2 Monochromatic chords of even cycles

Let Q be an even cycle. A chord e of Q is *monochromatic* if both ends of e are in the same part of the bipartition of Q. In other words, e is monochromatic if and only if $Q + e$ is not bipartite. An 8-cycle in the Petersen graph, for example, has two chords and both of them are monochromatic. A Hamilton cycle in the pentagonal prism has five chords; two of them are monochromatic and the other three are not.

The following result establishes conditions under which all chords of an even cycle in a non-Pfaffian graph are monochromatic.

Lemma 21.10 *Let G be a non-Pfaffian simple matching covered graph and let R be a removable class of G with the property that if (D, M_1', M_2') is any R-triple of G, then $M_1' \cap M_2' = R$. Now let (D, M_1, M_2) be an R-triple of G, and let Q be any (M_1, M_2)-alternating cycle in G which is evenly oriented in D. Then, every chord of Q in G is monochromatic.*

Proof Assume, to the contrary, that Q has a chord, say uw, which is not monochromatic. See Figure 21.5.

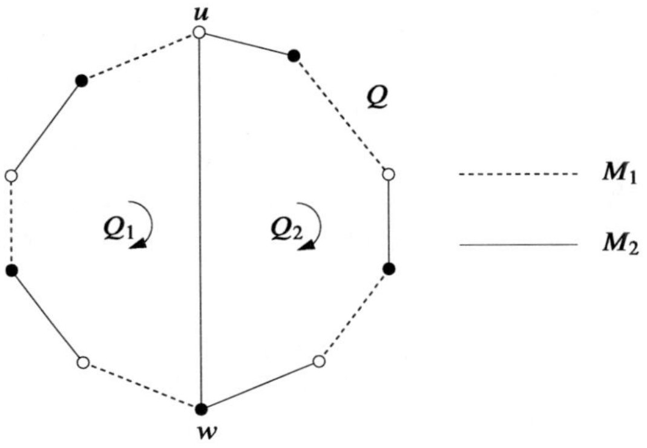

Fig. 21.5 The chord uw of cycle Q

As Q is bipartite, the vertices u and w define two segments of Q of odd length. And, as G is simple, each such segment has an odd length equal to three or more. These segments, plus the chord uw, define two cycles, Q_1 and Q_2. Using the fact that Q is evenly oriented in D, it is easy to deduce that one of Q_1 and Q_2 is oddly oriented and the other is evenly oriented (Exercise 21.2.4(i)). Moreover, one of Q_1 and Q_2 is M_1-alternating and the other is M_2-alternating. Adjust notation so that Q_1 is M_1-alternating and oddly oriented. Let $M_1' := M_1 \triangle E(Q_1)$. Then, $\text{sign}(M_1') = \text{sign}(M_1)$ (Exercise 21.2.4(ii)). Moreover, $R \subseteq M_1' \cap M_2$; hence (D, M_1', M_2) is an R-triple. As Q_2 has length four or more, we deduce that $(M_1' \cap M_2) - R$ is nonempty. This contradicts the hypothesis. It follows that every chord of Q is monochromatic. □

In the special case where G is a simple non-Pfaffian bipartite graph and $R = \{e\}$ is a removable singleton of G, the hypothesis of the above lemma implies that Q has no chords at all because the ends of any edge in G belong to different colour classes.

Exercises

▷**21.2.1** *The objective of this exercise is to prove that Γ_1 is an S-minimal non-Pfaffian graph. For the proof of some properties of Γ_1, it is convenient to note that the following permutations are automorphisms of Γ_1:*

$$(1\ 5)\ (2\ 4)\ (6\ 8)\ (3)\ (7)\ (x_1)\ (x_2)\ (y_1)\ (y_2)$$
$$(1)\ (5)\ (2\ 8)\ (3\ 7)\ (4\ 6)\ (x_1)\ (x_2)\ (y_1\ y_2)$$

(i) *Let J be the retract of the graph $H := \Gamma_1 - e - f$. The underlying simple graph of J is a quadrilateral. Find a Pfaffian orientation D_J of J.*
(ii) *Extend D_J to a Pfaffian orientation D_H of H. Hint: apply the result of Exercise 20.2.1 to each vertex of degree two of H, ensuring that each of the four vertices of degree two in H is neither a source nor a sink in D_H.*
(iii) *Prove that Γ_1 is not Pfaffian. (The cycle $(1, 2, \ldots, 8, 1)$ is evenly oriented and conformal in Γ_1).*
(iv) *Prove that for $g \in \{12, 23, 1x_1, 3x_2, 2y_1\}$ the graph $\Gamma_1 - g$ is Pfaffian. Taking into account the two automorphisms given above, conclude that $\Gamma_1 - g$ is Pfaffian, for each removable edge g of Γ_1.*
(v) *Let Q be a cycle in $\{(x_1, x_2, 3, 2, 1, x_1), (y_1, y_2, 8, 1, 2, y_1)\}$. Prove that $\Gamma_1/V(Q)$ is Pfaffian. Taking into account the two automorphisms given above, conclude that for each nontrivial separating cut C of Γ_1, both C-contractions of Γ_1 are Pfaffian.*

▷**21.2.2** *The objective of this exercise is to prove that Γ_2, depicted in Figure 21.3, is an S-minimal non-Pfaffian matching covered graph. Hint: it is convenient to notice that $(1\ 2\ 3\ 4\ 5\ 6\ 7\ 8)\ (x_1\ y_1\ x_2\ y_2)$ is an automorphism of Γ_2.*

(i) *Let J be the retract of the graph $H := \Gamma_2 - e - f$. The underlying simple graph of J is a quadrilateral. Find a Pfaffian orientation D_J of J.*
(ii) *Extend D_J to a Pfaffian orientation D_H of H. Hint: apply the result of Exercise 20.2.1 to each vertex of degree two of H, ensuring that each of the four vertices of degree two in H is neither a source nor a sink in D_H.*
(iii) *Prove that Γ_2 is not Pfaffian. (The cycle $(1, 2, \ldots, 8, 1)$ is evenly oriented and conformal in Γ_2).*
(iv) *Prove that for $g \in \{12, 1x_1\}$ the graph $\Gamma_2 - g$ is Pfaffian. Taking into account the automorphism given above, conclude that $\Gamma_2 - g$ is Pfaffian, for each removable edge g of Γ_2.*
(v) *Let Q be the cycle $(x_1, x_2, 3, 2, 1, x_1)$. Prove that $\Gamma_2/V(Q)$ is Pfaffian. Taking into account the automorphism given above, conclude that for each nontrivial separating cut C of Γ_2, both C-contractions of Γ_2 are Pfaffian.*

▷**21.2.3** *Lemma 21.4 requires that the removable class R be an optimal removable class. Give an example of a Pfaffian graph G, a removable class R of G and an R-triple of G.*

21.2.4 *Justify the following claims made in the proof of Lemma 21.10:*

(i) *the cycles Q_1 and Q_2 have distinct parities in D;*
(ii) *the signs of the two perfect matchings M_1' and M_1 are the same.*

21.3 Pfaffian Bipartite graphs

In this section we present a proof of the classical result of Little on non-Pfaffian bipartite graphs.

LITTLE'S THEOREM [54]

21.11 *The brace $K_{3,3}$ is the only S-minimal non-Pfaffian bipartite matching covered graph.*

Proof We have seen that $K_{3,3}$ is not Pfaffian (Lemma 19.10). Furthermore, all proper S-minors of $K_{3,3}$, being planar, are Pfaffian by Kasteleyn's Theorem (19.15). Thus $K_{3,3}$ is an S-minimal non-Pfaffian bipartite matching covered graph.

Now let G be any S-minimal non-Pfaffian bipartite matching covered graph. These properties of G, as we have already noted (right after the definition of S-minimal non-Pfaffian minors), imply that G is a simple brace of order six or more. To establish the desired assertion, we must prove that G is, in fact, the brace $K_{3,3}$. Towards this end, we first note that, by Theorem 8.3, every edge of G is removable. The following assertion may be deduced using Theorem 8.4 (see Exercise 8.4.1).

21.11.1 *For every (removable) edge e of G and for every vertex a of degree three or more in $G - e$, at most one edge of $G - e$ incident with a is not removable in $G - e$.* □

Let e be a (removable) edge of G. The minimality of G implies that $G - e$ is Pfaffian. By Theorem 21.3, G has e-triples. Let (D, M_1, M_2) be an e-triple of G.

21.11.2 *Every removable edge of $G - e$ is in $M_1 \cup M_2$.*

Proof Assume that $G - e$ has a removable edge, say f, which is not in $M_1 \cup M_2$. By Corollary 21.8, edge f is removable in G and $G - f$ is non-Pfaffian. This is in contradiction to the assumption that G is a minimal non-Pfaffian graph. □

21.11.3 *The graph G is cubic. Moreover, $M_1 \cap M_2 = \{e\}$.*

Proof Let v be any vertex of G. As G is a brace of order six or more, vertex v has degree three or more. We consider two cases depending whether or not v is incident with the edge e.

Case 1 *Vertex v is incident with e.*

In this case, as $e \in M_1 \cap M_2$, we infer that $\partial(v) - e$ and $M_1 \cup M_2$ are disjoint. By statement 21.11.2, no edge of $\partial(v) - e$ is removable in $G - e$. By statement 21.11.1, v has degree two in $G - e$; hence v has degree three in G.

Case 2 *Vertex v is not incident with e.*

In this case, v has degree three or more in $G - e$. By statement 21.11.1, at most one edge of $\partial(v)$ is not removable in $G - e$; all the other edges of $\partial(v)$ are removable in $G - e$. By statement 21.11.2, all the edges of $\partial(v)$ removable in $G - e$ are in $M_1 \cup M_2$. We deduce that v has degree three in G. Moreover, v is incident with two edges in $M_1 \cup M_2$; hence v is not incident with an edge in $M_1 \cap M_2$.

In both cases, these conclusions hold for each vertex v of G not incident with e. Indeed, G is cubic and e is the only edge in $M_1 \cap M_2$. □

Let a_1 and b_1 be the ends of the edge e.

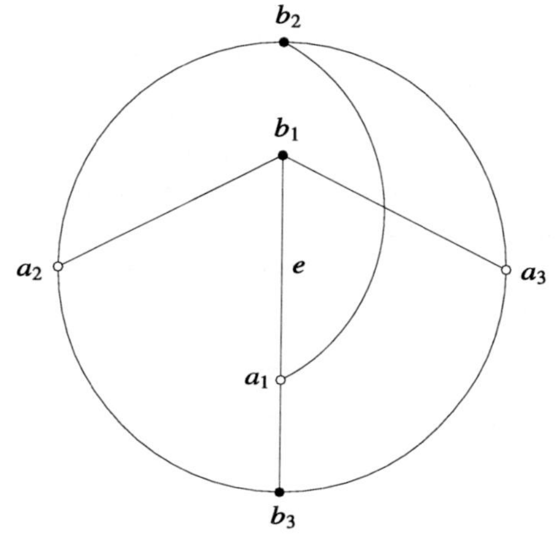

Fig. 21.6 The graph $K_{3,3}$

21.11.4 *The set $M_1 \triangle M_2$ induces a Hamilton cycle Q in the graph $G - a_1 - b_1$. Moreover, Q has no chords.*

Proof By statement 21.11.3, $M_1 \cap M_2 = \{e\}$. This equality holds for each e-triple (D, M_1, M_2) of G. Thus, for any e-triple (D, M_1, M_2), graph G is free of $(\overline{M_1 \cap M_2})$-alternating cycles. By Lemma 21.9, G has precisely one (M_1, M_2)-alternating cycle, say Q. Every vertex of $G - a_1 - b_1$ is incident with two edges in $M_1 \cup M_2$, hence it is a vertex of Q. This conclusion holds for each vertex of $G - a_1 - a_2$. It thus follows that Q is a Hamilton cycle of $G - a_1 - b_1$. By Lemma 21.10, every chord of Q is monochromatic. However, as G is bipartite, no edge joins two vertices belonging to the same colour class. Hence, Q has no chords. □

The brace G has order six or more, hence Q has length four or more. Every vertex of Q has degree three and has only two neighbours in $V(Q)$. Thus, every vertex of Q is adjacent to (precisely) one of the ends of e. See Figure 21.6.

As G is cubic and bipartite, it follows that the cycle Q has precisely four vertices. The two vertices of one part of Q are adjacent to a_1, the two vertices of the other part are adjacent to b_1. Indeed, as asserted, graph G is $K_{3,3}$. □

Pfaffian solid bricks

Every bipartite matching covered graph is solid, and many properties of bipartite graphs are valid more generally for solid graphs. This also happens to be the case in the context of Pfaffian orientations. In a paper [16] entitled *A Generalization of Little's Theorem on Pfaffian Orientations* CLM showed that every minimal non-Pfaffian brick must have a separating cut, and deduced that $K_{3,3}$ is the only minimal non-Pfaffian solid matching covered graph.

21.4 Pfaffian Near-Bipartite Graphs ♯

The Fischer-Little Theorem [34] may be restated as follows:

THE FISCHER-LITTLE THEOREM

21.12 *A near-bipartite graph is non-Pfaffian if and only if it contains one of the graphs $K_{3,3}$, Γ_1 and Γ_2 as an S-minor.*

Proof The graphs $K_{3,3}$, Γ_1 and Γ_2 are S-minimal non-Pfaffian (Theorem 21.11, Exercise 21.2.1 and Exercise 21.2.2). Let G be a non-Pfaffian near-bipartite graph. Assuming inductively that the statement of the theorem is valid for all near-bipartite graphs with fewer than $|E(G)|$ edges, we shall proceed to prove that G has one of $K_{3,3}$, Γ_1 and Γ_2 as an S-minor.

21.4.1 Some immediate implications of the induction hypothesis

If G has an S-minor G' that is bipartite and non-Pfaffian then, by Theorem 21.11, G' has $K_{3,3}$ as an S-minor and, hence, so does G. On the other hand, if G has a proper S-minor G'' which is near-bipartite and non-Pfaffian then, by induction, one of the three graphs $K_{3,3}$, Γ_1 and Γ_2 is an S-minor of G'' and, hence, of G.

We may thus assume that (i) every proper S-minor of G which is near-bipartite is Pfaffian and we may also assume that (ii) every bipartite S-minor of G is Pfaffian. Relying on these assumptions, we shall derive certain properties of G and deduce that G is either Γ_1 or Γ_2. We shall adopt the notation defined below in our proof:

NOTATION

- As G is near-bipartite, it has removable doubletons. Let $R := \{e, f\}$, where $e := x_1x_2$ and $f := y_1y_2$ be a removable doubleton of G.
- Let $Z := \{x_1, x_2, y_1, y_2\}$ be the set of the ends of edges e and f.

- Let H be the bipartite graph $G - R$ and let (A, B) denote the bipartition of H. Adjust notation so that the ends x_1 and x_2 of e are in A, and the ends y_1 and y_2 of f are in B.
- By Theorem 21.3 and Lemma 21.13, G has an R-triple. Let (D, M_1, M_2) be an R-triple.

Our approach to proving that G is either Γ_1 or Γ_2 has many features in common with the proof of Little's Theorem presented in the previous section. For example, analogous to Statement 21.11.2 we shall prove that every removable edge of H is in $M_1 \cup M_2$; and analogous to Statement 21.11.4 we shall prove that there is a unique (M_1, M_2)-alternating cycle, that it is a Hamilton cycle of $G - Z$, and that it is chordless. But establishing these results in the present context is technically much more involved. We begin with two simple observations.

Lemma 21.13 *The graph $H := G - R$ is Pfaffian.*

Proof The graph $G - R$ is a bipartite S-minor of G, and we have assumed that bipartite S-minors of G are Pfaffian. □

Lemma 21.14 *The graph G is a brick.*

Proof As $G-R$ is bipartite $b(G-R) = 0$. It follows from the monotonicity property (Theorem 9.9) that $b(G) = 1$ (that is, G is a near-brick). Suppose that G is not a brick. Then, G has nontrivial tight cuts. Let C be a nontrivial tight cut of G and let G_1 and G_2 denote the two C-contractions of G. As G is a near-brick, one of the C-contractions of G is bipartite and the other is a near-brick. Adjust notation so that G_1 is bipartite and G_2 is a near-brick.

21.14.1 *The graph G_2 is near-bipartite.*

Proof Every perfect matching of $G - R$ is a perfect matching of G. The cut C is tight in G; hence the cut $C - R$ is tight in $G - R$. As $G - R$ is bipartite, it follows that both $(C - R)$-contractions of $G - R$ are bipartite. As G_2 is a near-brick and since $G_2 - R$ is bipartite, it follows that R is a removable doubleton of G_2; hence G_2 is near-bipartite. □

The graph G_1 is a bipartite S-minor of G. We have assumed that bipartite S-minors of G are Pfaffian; hence G_1 is Pfaffian. The graph G_2 is a proper S-minor of G, because C is nontrivial and separating. Moreover, the graph G_2 is near-bipartite. We have assumed that near-bipartite proper S-minors of G are Pfaffian; hence G_2 is Pfaffian. In sum, both C-contractions of G are Pfaffian. From Theorem 20.10 we infer that G is Pfaffian, a contradiction. We deduce that G is a brick. □

21.4.2 Removable edges of H

Lemma 21.15 *Every removable edge of H is in $M_1 \cup M_2$.*

Proof Suppose that H has a removable edge h which is not in $M_1 \cup M_2$. The doubleton R is an optimal removable class of G. As $G - R$ is bipartite, the edge h is b-invariant in $G - R$; hence h is an optimal removable edge of $G - R$. By Lemma 21.7 (Exchange property: optimal classes), h is an optimal removable edge of G, R is an optimal removable class of $G - h$ and $G - h$ is not Pfaffian. However, G is a brick, h is a b-invariant edge of G and R is a removable doubleton of $G - h$. Thus, $G - h$ is a near-brick and R is a removable doubleton of $G - h$; hence $G - h - R$ is bipartite and matching covered. We deduce that $G - h$ is near-bipartite. Moreover, $G - h$ is not Pfaffian. We have assumed that every near-bipartite proper S-minor of G is Pfaffian; hence $G - h$ is Pfaffian, a contradiction! □

21.4.3 Degree of vertices

Lemma 21.16 *Let v be a vertex of G and let d be its degree in G. The following properties hold:*

(i) *If v is incident with two edges in $M_1 \cup M_2$ then $3 \leq d \leq 4$.*
(ii) *If v is incident in G with an edge h of $M_1 \cap M_2$, then $d = 3$. (In particular, all the four vertices in Z have degree three.) Moreover, if $h \notin R$ then h is the only edge of $\partial(v)$ removable in H.*

Proof By Lemma 8.6, at most two edges of $\partial_H(v)$ are not removable in H. By Lemma 21.15, every removable edge of H is in $M_1 \cup M_2$. If v is incident with two edges in $M_1 \cup M_2$ then $d \leq 4$. Alternatively, suppose that v is incident in G with an edge h in $M_1 \cap M_2$. Then, $d = 3$. Moreover, if $h \notin R$ then v is incident in H with a removable edge, which must necessarily be edge h. □

21.4.4 There is a unique (M_1, M_2)-alternating cycle in G

Lemma 21.17 *The graph G has no $(M_1 \cap M_2)$-alternating cycle.*

Proof Assume, to the contrary, that G has an $(M_1 \cap M_2)$-alternating cycle C. For $i = 1, 2$, let $M_i' := M_i \triangle E(C)$. It follows that, for $i = 1, 2$, the two perfect matchings M_i and M_i' have exactly the same edges in the set $E(G) - E(C)$, and the cycle C is the only (M_i, M_i')-alternating cycle in G. Thus, if C is oddly oriented in D then $\text{sign}(M_i') = \text{sign}(M_i)$, whereas if C is evenly oriented then $\text{sign}(M_i') = -\text{sign}(M_i)$. As M_1 and M_2 have distinct signs in D we conclude that in both alternatives the perfect matchings M_1' and M_2' of G have distinct signs in D. As $D - R$ is Pfaffian, it follows that M_1' and M_2' both contain an edge of R. As R is a removable doubleton of G, we infer that $R \subset M_1' \cap M_2'$. Hence (D, M_1', M_2') is an R-triple of G. The doubleton R is also a subset of $M_1 \cap M_2$; hence R and $E(C)$ are disjoint.

By Lemma 21.16, we have two contradictory conclusions. For suppose that $(e_1, e_2, \ldots, e_{2k})$, in that cyclic order, are the edges of the cycle C. Since this cycle is assumed to be $(M_1 \cap M_2)$-alternating, we may suppose that the edges of C in $M_1 \cap M_2$ are $e_1, e_3, \ldots, e_{2k-1}$. Then by Lemma 21.16 it follows that the only edges of C that are removable in H are $e_1, e_3, \ldots, e_{2k-1}$. On the other hand, by the definition of M_1' and M_2', it follows that C is an $(M_1' \cap M_2')$-alternating cycle and the edges of C that belong to $M_1' \cap M_2'$ are $e_2, e_4, \ldots, e_{2k}$. In this case, by Lemma 21.16, with (D, M_1', M_2') playing the role of (D, M_1, M_2), it follows that the only edges of C that are removable in H are $e_2, e_4, \ldots, e_{2k}$. This conclusion is a clear contradiction. □

The above result, together with Lemma 21.9, implies the following corollary which plays an important role in the proof of the Fischer-Little Theorem.

Corollary 21.18 *The graph G has precisely one (M_1, M_2)-alternating cycle.*

Proof By Lemma 21.17, G has no $(M_1 \cap M_2)$-alternating cycles. This assertion holds for each R-triple (D, M_1, M_2). By Lemma 21.9, G has precisely one (M_1, M_2)-alternating cycle. □

THE CYCLE Q

Let $Q := (v_1, v_2, \ldots, v_\ell, v_1)$ be the only (M_1, M_2)-alternating cycle of G. Since both edges e and f of R belong to $M_1 \cap M_2$, none of the ends of these two edges, namely the vertices in Z, lie on this cycle. Also, as G is simple, $\ell \geq 4$. Via a series of intermediate steps we shall show that Q is a Hamilton cycle of the graph $G - Z$, and that Q has no chords.

The first crucial step towards showing that Q is a Hamilton cycle of $G - Z$ is to show that $M_1 \cap M_2 = R$. The following two subsections establish results concerning short cycles in G which are required for achieving this objective.

21.4.5 Quadrilaterals in H

Lemma 21.19 *No quadrilateral of H contains two or more vertices in the set $Z := \{x_1, x_2, y_1, y_2\}$.*

Proof Assume, to the contrary, that H has a quadrilateral Q_4 that contains two or more vertices in the set Z.

Let us first consider the possibility that Q_4 contains both the ends y_1 and y_2 of edge f. In this case, since the degrees of all vertices in Z are three by Lemma 21.16, and since G has at least six vertices, it follows that the set $V(Q_4) - \{y_1, y_2\}$ is a 2-separation of G, in contradiction to the hypothesis that G is a brick. See Figure 21.7(a). Likewise, Q_4 does not contain both the ends x_1 and x_2 of edge e. Thus, Q_4

contains just one end of e and just one end of f. Adjust notation so that x_1 and y_1 are the vertices of Q_4 in Z. Clearly, x_1 and y_1 are adjacent in Q_4. Let $u \in A$ and $w \in B$ be the other two vertices of Q_4. That is, $Q_4 = (u, w, x_1, y_1, u)$. Let $e' := uy_1$ and $f' := wx_1$. (See Figure 21.7(b).) Now we proceed to show that $R' = \{e', f'\}$ is also a removable doubleton of G, and use it to derive a contradiction.

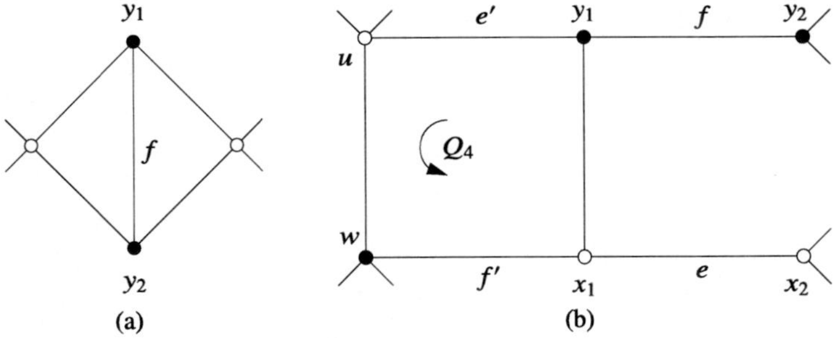

Fig. 21.7 A quadrilateral of H that contains two vertices in $\{x_1, x_2, y_1, uy_2\}$

In the bipartite graph $H = G - \{e, f\}$, the ears (u, w) and $S := (u, y_1, x_1, w)$ are parallel (that is, have the same ends); hence they are both removable in H. Thus, $H' := H - S$ is matching covered. Moreover, H' is the same as the graph $H - x_1 - y_1$ obtained from H by deleting x_1 and y_1. It follows that $(A - x_1, B - y_1)$ is the bipartition of H'. We may then add to H' the ear (x_2, x_1, y_1, y_2), thereby obtaining the bipartite matching covered graph H''. The graph H'' is equal to $G - e' - f'$; hence $R' := \{e', f'\}$ is a removable doubleton of G. The graph $G - R'$ is a bipartite S-minor of G. We have assumed that bipartite S-minors of G are Pfaffian. Thus, $G - R'$ is Pfaffian. By Theorem 21.3, G has two perfect matchings, M_1' and M_2', and an orientation D', such that (D', M_1', M_2') is an R'-triple of G. The edges of the doubleton R' belong to both M_1' and M_2' and, thus, the cycle Q_4 is an $(M_1' \cap M_2')$-alternating cycle of G. This is a contradiction to Lemma 21.17 (with the R'-triple (D', M_1', M_2') playing the role of the R-triple (D, M_1, M_2)). Indeed, no quadrilateral of H contains two or more vertices in the set Z. □

21.4.6 Triangles in G

Lemma 21.20 *No end of an edge of $M_1 \cap M_2$ is adjacent to both ends of an edge of R.*

Proof Suppose that $M_1 \cap M_2$ contains an edge h such that an end of h, say w, is adjacent to both ends of an edge of R. By the hypothesis and Lemma 21.16, the degree of w is three. Adjust notation so that w is adjacent to the ends x_1 and x_2 of e. The subgraph of G induced by the vertices of the set $T := \{x_1, x_2, w\}$ is a triangle in G. Let $C := \partial(T)$. We consider two cases depending on whether or not $h = f$. See Figure 21.8.

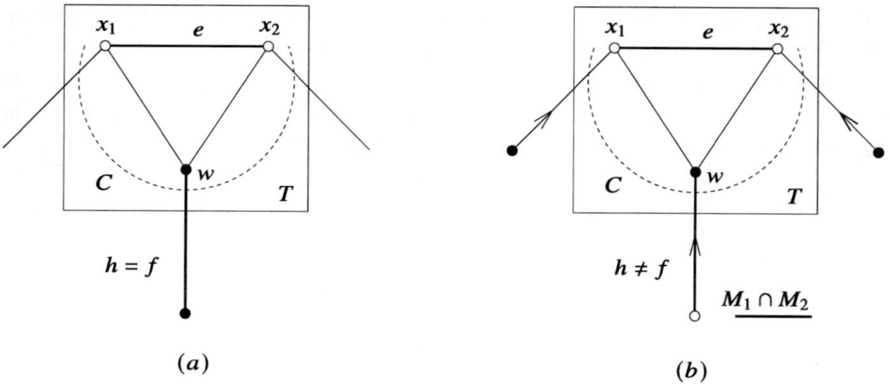

Fig. 21.8 The triangle T

Case 1 *The two edges h and f are the same (Figure 21.8(a)).*

The cut $C - f$ is tight in H, because w has degree two in H. Moreover, every perfect matching of G that contains an edge in R in fact contains precisely the edge f in C. Thus, C is tight in G. This is a contradiction, as G is a brick having more than four vertices.

Case 2 *The edges h and f are different. (Figure 21.8(b)).*

Let $G_1 := G/(T \to t)$ and let $G_2 := G/\overline{T}$. We shall prove that C is separating in G, and that the C-contraction G_1 of G is near-bipartite but is not Pfaffian.

By Theorem 4.2, in order to show that C is a separating cut of G we must show that, given any edge of G, there is a perfect matching of G containing that edge which meets C in precisely one edge. Firstly, note that edge h is removable in H by Lemma 21.16. Thus, the cut $C - h$ is tight in $H - h$ because w has degree two in $H - h$. So, every edge of $H - h$ is in a perfect matching that meets C in just one edge. On the other hand, every perfect matching that contains an edge of R contains the edge e and consequently the edge h. Thus, every edge of G is in a perfect matching of G that contains only one edge in C. We deduce that C is a separating cut of G.

The graph $G_1 - f - h$ is the result of a bicontraction of vertex w in $H - h$; hence $G_1 - f - h$ is bipartite. As G is a brick having more than four vertices, the nontrivial separating cut C is not tight, therefore G_1 is not bipartite. We conclude that $\{f, h\}$ is a removable doubleton of G_1; hence G_1 is near-bipartite.

Let X be a subset of T such that, in the digraph $D' = D$ rev $\partial(X)$, the cut C is a directed cut in D'. As D' is similar to D which is non-Pfaffian, it follows that D' is also non-Pfaffian. Cycle Q is an (M_1, M_2)-alternating cycle, and, by definition of an R-triple, both e and f belong to $M_1 \cap M_2$. Hence $\partial(X)$ and $E(Q)$ are disjoint. It follows that the orientation of the edges of the cycle Q are not changed. Thus, Q is evenly oriented in D' and hence M_1 and M_2 have distinct signs in D'. Moreover, h is the only edge of M_1 and of M_2 in the cut C. By Lemma 20.11, one of the C-contractions of G is not Pfaffian. As G_2, being K_4, is Pfaffian, we conclude that G_1, the other C-contraction of G, is not Pfaffian. In sum, G_1 is near-bipartite, non-

Pfaffian and a proper S-minor of G. This is a contradiction, as we have assumed that every near-bipartite proper S-minor of G is Pfaffian. □

21.4.7 The equality $M_1 \cap M_2 = R = \{e, f\}$

Our goal is to show that Q is a Hamilton cycle of the graph $G - Z$ and also that Q has no chords. Then, we will be able to prove that the length ℓ of the cycle Q is eight and deduce that G is one of Γ_1 and Γ_2, which is our ultimate goal. But, before proving that $\ell = 8$, there is one more hurdle to cross, namely proving the following lemma, which relies on the Zigzag Lemma 8.20, and then deriving its consequences.

Lemma 21.21 *Let $P := (u_1, u_2, u_3, u_4)$ be a path in H such that the vertices u_2 and u_3 are adjacent to three or more vertices of H. If no edge of P is removable in H then $u_2u_3 \in M_1 \cup M_2$.*

Proof Suppose that no edge of P is removable in H. By Lemma 8.20, there are partitions (A_1, A_2, A_3, A_4) of A and (B_1, B_2, B_3, B_4) of B such that

(i) $u_1 \in A_1, u_2 \in B_2, u_3 \in A_3, u_4 \in B_4$,
(ii) for $i = 1, 2, 3, 4$, $|A_i| = |B_i|$,
(iii) u_1u_2 is the only edge that joins a vertex in A_1 to a vertex of $B - B_1$,
(iv) u_3u_4 is the only edge that joins a vertex of $A - A_4$ to a vertex of B_4, and
(v) u_2u_3 is the only edge that joins a vertex of $B_1 \cup B_2$ to a vertex of $A_3 \cup A_4$.

See Figure 21.9.

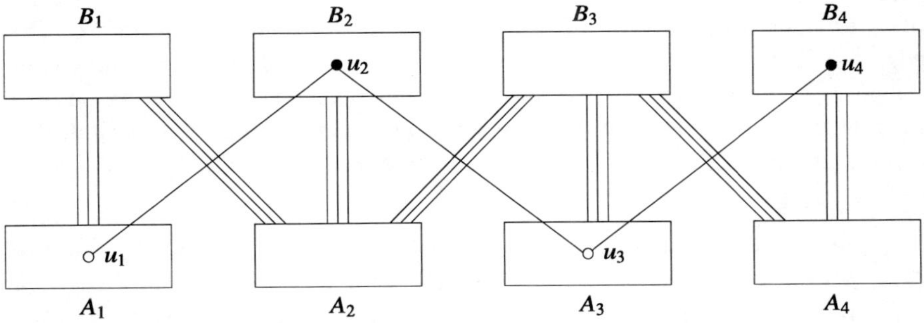

Fig. 21.9 The path P and the partitions of A and B

If we remove the $|A_1| + 1$ vertices of the set $B_1 + u_2$ from H, the vertices of A_1 become isolated. Thus, the cut $\partial_H(A_1 \cup B_1 \cup \{u_2\})$ is nontrivial and tight in the graph $H = G - R$. But, as G is a brick, this cut cannot be tight in G, and so the edge e must have an end in A_1 (Exercise 21.4.2). Let x_1 be the end of e in A_1. Similarly, if we remove the $|A_3| + |A_4| + 1$ vertices of the set $B_3 \cup B_4 \cup \{u_2\}$ from H, then the vertices of $A_3 \cup A_4$ become isolated. Thus, the end x_2 of e has to be in $A_3 \cup A_4$ (Exercise 21.4.2).

In sum, the end x_1 of e is in A_1 and the end x_2 of e is in $A_3 \cup A_4$. Likewise, f has an end, say y_1, in $B_1 \cup B_2$ and the other end y_2 in B_4.

21.21.1 *Let $X := A_1 \cup A_2 \cup B_1 \cup B_2$, and let C denote the cut $\partial(X)$ in G (see Figure 21.10). If $u_2u_3 \notin M_1 \cup M_2$, then (i) e and f are the only two edges of $M_1 \cap M_2$ in C and (ii) all the vertices of the cycle Q belong to one of the two shores of the cut C.*

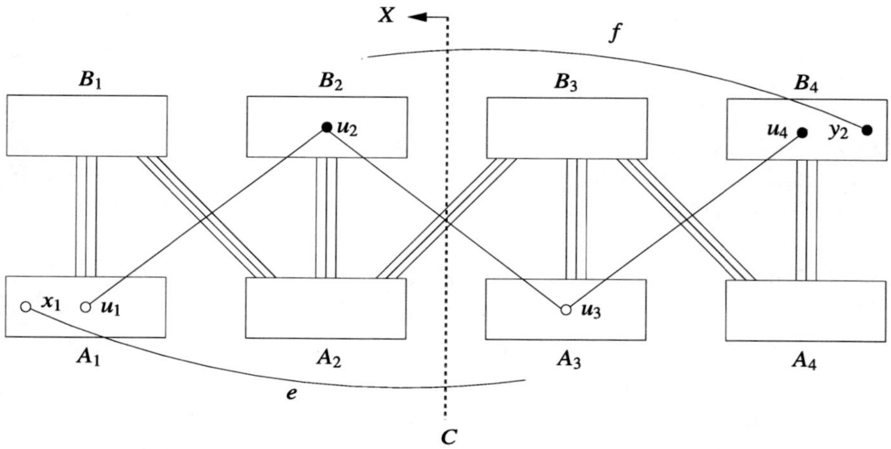

Fig. 21.10 The set X and the cut C

Proof As the cut C is even, every perfect matching of G meets C in an even number of edges; in particular, for $i = 1, 2$, $|M_i \cap C|$ is even. The edges e and f are in C. The edge u_2u_3 is the only edge of C that joins a vertex in $X \cap B$ to a vertex in $\overline{X} \cap A$. Suppose that $u_2u_3 \notin M_1 \cup M_2$. A simple counting argument then shows that no edge in $M_1 \cup M_2$ joins a vertex in A_2 to a vertex in B_3 (Exercise 21.4.2). We conclude that e and f are the only edges of $M_1 \cup M_2$ in C. It follows that $V(Q)$ is part of a shore of C. □

Without loss of generality, assume that $V(Q) \subseteq X$. In this case, the edges of $M_1 \cup M_2$ in the graph $G[\overline{X}]$ are edges in $M_1 \cap M_2$. Let N be the set of edges of $M_1 \cup M_2$ that have both ends in $\overline{X}$. All the edges of N are in fact edges of $M_1 \cap M_2$. Note that N is a perfect matching of the bipartite graph $J := G[\overline{X}] - x_2 - y_2$. We shall in fact show that J is empty, that is, we shall prove that $\overline{X} = \{x_2, y_2\}$.

By Lemma 21.17, G is free of $(M_1 \cap M_2)$-alternating cycles; hence N is the unique perfect matching of J (Exercise 21.4.2). The characterization of bipartite graphs with a unique perfect matching presented in Exercise 3.2.2 implies the following statement:

21.21.2 *There is an enumeration $v_1, v_2, \ldots, v_k$ of the vertices of $A \cap V(J)$ and an enumeration $w_1, w_2, \ldots, w_k$ of the vertices of $B \cap V(J)$ such that (i) $v_iw_i \in N$, for $i = 1, 2, \ldots, k$ and (ii) $v_iw_j \notin E(J)$ for $1 \le i < j \le k$. (See Figure 21.11.)* □

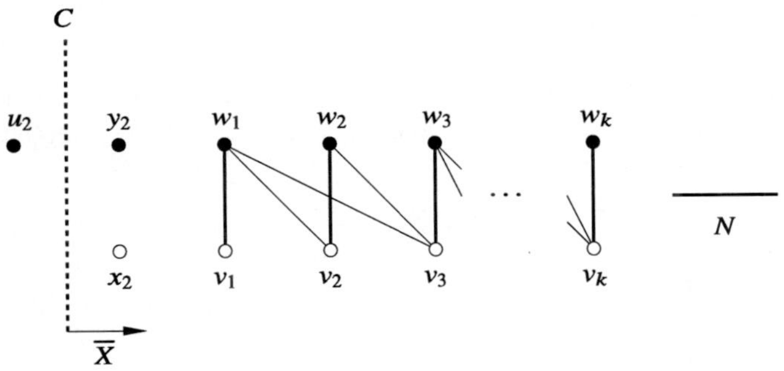

Fig. 21.11 The enumeration of $V(J)$

The vertex v_1, being an end of an edge in $M_1 \cap M_2$, has degree three in G by Lemma 21.16. The only neighbour of v_1 in J is the vertex w_1. Thus, $N(v_1) = \{w_1, y_2, u_2\}$. Consequently, $v_1 = u_3$. See Figure 21.12.

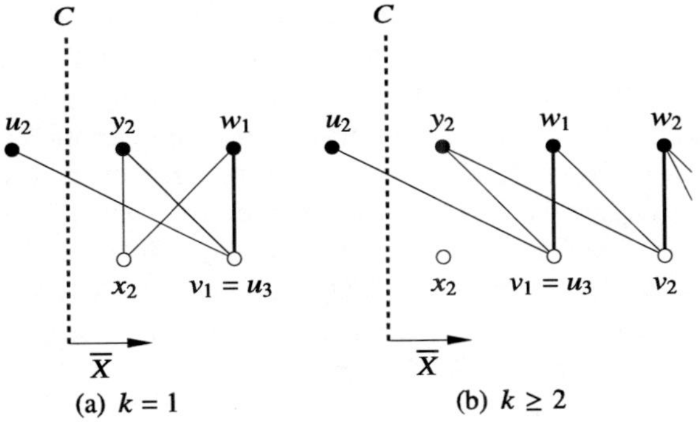

Fig. 21.12 The adjacencies of vertices in $A \cap V(J)$

Suppose that $k = 1$. Then, the vertex x_2, which has degree two in H, must be adjacent to w_1 and to y_2. See Figure 21.12(a). In this case $(v_1, w_1, x_2, y_2, v_1)$ is a quadrilateral in G that contains two vertices that are ends of edges e and f. This is a contradiction to Lemma 21.19.

Alternatively, suppose that $k \geq 2$. Then, $N(v_2) = \{w_2, w_1, y_2\}$, because v_2 is not adjacent to any vertex w_j, $j > 2$. Let $Y := \{v_1, y_2, v_2\}$. The cut $\partial(Y)$ is tight in H. The edges v_2w_1 and v_1w_1 are parallel in H/Y and v_2w_1 is not an edge of $H/\overline{Y}$. Hence edge v_2w_1 is removable in H. This is in contradiction to Lemma 21.15 because v_2w_1 does not belong to $M_1 \cup M_2$.

In both alternatives, we derived a contradiction. Thus, J is the empty graph; hence $\overline{X} = \{x_2, y_2\}$. This is not possible, because $\overline{X}$, which is $A_3 \cup A_4 \cup B_3 \cup B_4$, has four or more vertices. Indeed, if no edge of the path $P = (u_1, u_2, u_3, u_4)$ is removable in H then u_2u_3 is in $M_1 \cup M_2$. □

Lemma 21.22 $M_1 \cap M_2 = R$.

Proof Suppose that $(M_1 \cap M_2) - R$ contains an edge, u_2v_2. Let u_1 and u_3 be the two neighbours of u_2 distinct from v_2, and let v_1 and v_3 be the two neighbours of v_2 distinct from u_2. See Figure 21.13.

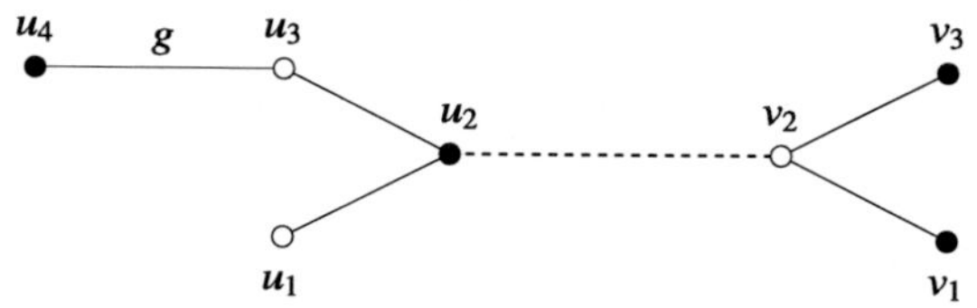

Fig. 21.13 The edge u_2v_2 of $M_1 \cap M_2$

By Lemma 21.20, at least one of u_1 and u_3 is not an end of an edge of R. Adjust notation so that u_3 is not an end of an edge of R. Then, u_3 has degree three or more in H. Let g be an edge of $\partial(u_3) - u_2u_3$. Then, $P := (u_1, u_2, u_3, u_4)$ is a path of H, where $g = u_3u_4$. The vertices u_2 and u_3 both have degree three or more in H. The edges u_1u_2 and u_2u_3 are not in $M_1 \cup M_2$; hence they are not removable in H. In particular, u_2u_3 is not in $M_1 \cup M_2$. By Lemma 21.21, the edge g is removable in H. This conclusion holds for each edge g in $\partial(u_3) - u_2u_3$. By Lemma 21.16, every edge in $\partial(u_3) - u_2u_3$ is in $M_1 \cup M_2$; hence u_3 has degree three and is in $V(Q)$.

In sum, u_3 has degree three, is in $V(Q)$ and the two edges of Q incident with u_3 are removable in H. Likewise, adjust notation so that v_3 is not an end of an edge of R. The vertex v_3 has degree three, is in $V(Q)$ and the two edges of Q incident with v_3 are removable in H. The *chordal path* (u_3, u_2, v_2, v_3) defines, together with the cycle Q, two cycles, Q_1 and Q_2. See Figure 21.14.

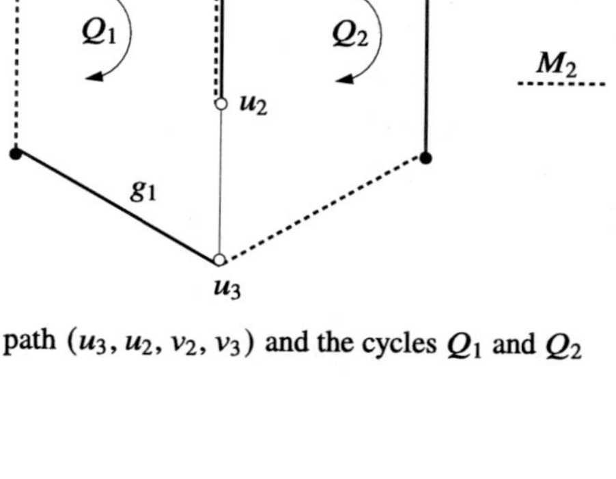

Fig. 21.14 The chordal path (u_3, u_2, v_2, v_3) and the cycles Q_1 and Q_2

One of Q_1 and Q_2 is M_1-alternating and the other is M_2-alternating. One of Q_1 and Q_2 is oddly oriented and the other is evenly oriented. Adjust notation so that Q_1 is M_1-alternating and oddly oriented. Let g_1 be the edge of Q_1 incident with u_3 and let $M_1' := M_1 \triangle E(Q_1)$. Then, $\text{sign}(M_1') = \text{sign}(M_1)$ and $R \subset M_1' \cap M_2$; hence (D, M_1', M_2) is an R-triple. The edge g_1, a removable edge of H, is not in $M_1' \cup M_2$. This is a contradiction to Lemma 21.15, with M_1' playing the role of M_1. □

21.4.8 The graphs Γ_1 and Γ_2

In order to prove that G is one of Γ_1 and Γ_2, we need yet another auxiliary result, to establish that $\ell = 8$.

Lemma 21.23 *The following properties hold:*

(i) *the cycle Q has no chords, and*
(ii) *no edge of G joins an end of e to an end of f.*

Proof (i): For each R-triple (D, M_1, M_2) of G, $R = M_1 \cap M_2$, by Lemma 21.22. By Lemma 21.10, every chord of Q is monochromatic. The graph $G - Z$ is bipartite. Thus, Q has no chords.

(ii): Assume, to the contrary, that an edge of G joins an end of e to an end of f. Adjust notation so that x_1y_1 is an edge of G. By Lemma 21.17, x_2 and y_2 are not adjacent. By Lemma 21.20, the vertices x_1 and y_2 are not adjacent. Likewise, x_2 and y_1 are not adjacent either. We shall now obtain a contradiction.

The vertex x_1 has degree three in G; hence $\partial(x_1)$ has precisely one edge that joins x_1 to a vertex of the cycle Q. Adjust notation so that x_1 is adjacent to vertex v_2 of Q. By Lemma 21.19, no quadrilateral contains the vertices x_1 and y_1. Thus, y_1 is not adjacent to the vertices v_1 and v_3. It follows that the length ℓ of Q is six or more. However, at most six edges join ends of edges e and f to vertices of Q. As Q has no chords, it follows that Q has length six. Moreover, y_1 is adjacent to v_5, y_2 is adjacent to v_1 and to v_3, and x_2 is adjacent to v_4 and to v_6. See Figure 21.15.

The graph $H' := H + x_2y_2$ is bipartite, matching covered and planar. By Corollary 19.16, H' is Pfaffian. The orientation $D - R$ of H is Pfaffian and x_2y_2 is an optimal removable edge of H'. By Corollary 20.25, $D - R$ has a Pfaffian extension to a Pfaffian orientation of H', say, F'. The cycle Q is conformal in H'. Moreover, Q is evenly oriented in D; hence it is evenly oriented in F'. This is a contradiction to Lemma 19.5(iii). We conclude that no edge of G joins an end of e to an end of f. □

Corollary 21.24 *The cycle Q is a chordless Hamilton cycle of the graph $G - Z$. Moreover, $|\partial(Z)| = 8$.* □

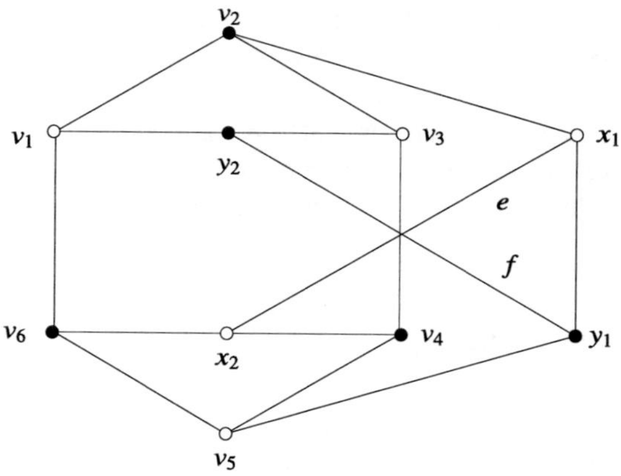

Fig. 21.15 The graph G, where x_1 and y_1 are adjacent

The finish line

Finally, we are now in a position to prove that the graph G is one of the graphs Γ_1 and Γ_2. Let $C := \partial(V(Q))$. By Corollary 21.24, $C = \partial(Z)$ and $|C| = 8$. Thus the length ℓ of Q is at most eight. As G is simple, $\ell \geq 4$; hence $\ell \in \{4, 6, 8\}$. Let us now examine the possibilities. Let $H_e := G - y_1 - y_2$ and let $H_f := G - x_1 - x_2$.

Suppose that $\ell \in \{4, 6\}$. Then, H_e is planar, and it has an embedding in the plane in which the cycle Q bounds the infinite face. See Figures 21.16(a) and 21.16(b). Likewise, H_f is planar and has an embedding in the plane such that Q bounds the infinite face. Thus, G is planar. In that case, G is Pfaffian, a contradiction.

We deduce that in fact $\ell = 8$. Thus, there are two possibilities for H_e, up to isomorphism (as well as for H_f). See Figures 21.16(c) and 21.16(d).

In the first possibility, denoted H_e' and depicted in Figure 21.16(c), the graph H_e' is planar and has an embedding in which Q bounds its infinite face. Thus, one of H_e and H_f must be isomorphic to the graph H_e'', depicted in Figure 21.16(d). If H_e is H_e' then H_f is isomorphic to H_e'' and the graph G is Γ_1. Alternatively, both H_e and H_f may be isomorphic to graph H_e'', in which case the graph G is Γ_2. Indeed, G is one of the graphs Γ_1 and Γ_2. □

Exercises

21.4.1 *Find a Pfaffian orientation of the graph $G' := G + x_2y_2$, where G is the graph depicted in Figure 21.15. Hint: extend a Pfaffian orientation of the planar bipartite matching covered graph $G' - R$ to a Pfaffian orientation of G' making the R-alternating quadrilateral oddly oriented.*

21.4.2 *Supply the missing details in the proof of Lemma 21.21.*

▷**21.4.3** *Prove that, as defined above, if $H_e = H_e'$ then H_f is isomorphic to H_e'' and G is Γ_1.*

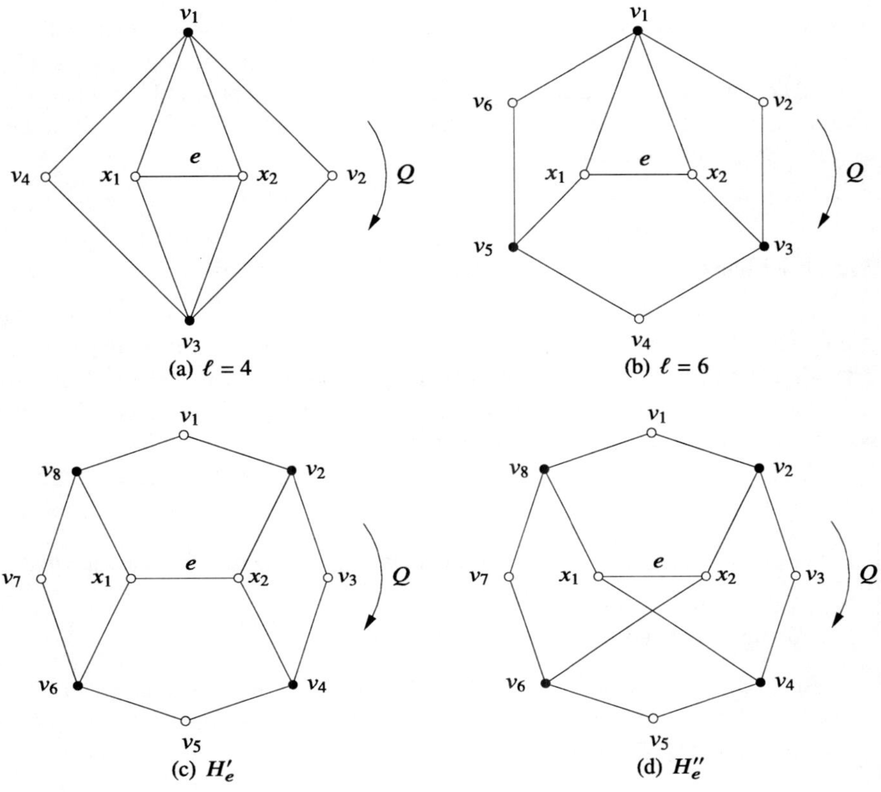

Fig. 21.16 The possibilities of graph H_e

⊳**21.4.4** *Prove that if neither H_e nor H_f is isomorphic to H'_e then G is Γ_2.*

21.4.5 *Let us refer to a near-bipartite graph G with a removable doubleton R as a* brace-based near-bipartite graph *if $G - R$ is a brace. Characterize brace-based non-Pfaffian near-bipartite graphs.*

21.5 Polynomial-time Algorithms

Little's excluded minor characterization of Pfaffian bipartite graphs provides yet another way of showing the Pfaffian Orientation Problem for bipartite graphs is in co-NP. (To show that a bipartite graph is not Pfaffian, one just has to exhibit a matching minor of that graph which is isomorphic to $K_{3,3}$.) Nearly 25 years later, Robertson, Seymour and Thomas [82] and, independently, McCuaig [68], succeeded in devising a polynomial-time algorithm which accepts any bipartite graph as input and, as output, it either displays a Pfaffian orientation of that graph or a matching minor isomorphic to $K_{3,3}$. This important algorithm is rather hard and a full description of it is beyond the scope of this book. However, we do provide in this section a sketch of the main ideas involved in it. Some years later Miranda and Lucchesi [74] showed how the polynomial-time algorithm for bipartite graphs could be adapted to obtain polynomial-time algorithms for near-bipartite graphs. Their idea is described towards the end of this section.

21.5.1 Efficient recognition of Pfaffian bipartite graphs

A polynomial-time algorithm for determining whether a bipartite matching covered graph is Pfaffian was discovered by McCuaig (2004, [69]) and, independently, by Robertson, Seymour and Thomas (1999, [82]). The concept of 4-cycle-sum, defined below, plays a central role in their algorithm.

> 4-CYCLE SUMS
>
> Let $G_1, G_2, \ldots, G_n$ $(n \geq 2)$ be bipartite graphs and let Q be a 4-cycle such that
>
> (i) each graph G_i has order six or more, and
> (ii) for any two distinct i and j, $1 \leq i < j \leq n$, $G_i \cap G_j = Q$.
>
> For each subset R of $E(Q)$, the (Q, R)-*sum* of the n *summands* G_i is the graph $G := (\bigcup_{i=1}^{n} G_i) - R$. If Q and R are understood, we simply say *4-cycle sum* instead of (Q, R)-sum.

Figure 21.17 depicts the rotunda, a 4-cycle sum of three cubes, where R is the set of the four edges of the quadrilateral Q, common to the three summands.

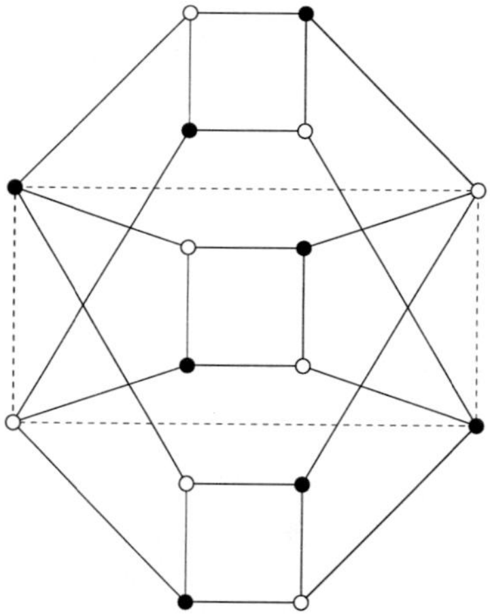

Fig. 21.17 The rotunda is a 4-cycle sum of three cubes - the edges of Q are represented by dashed lines.

The importance of 4-cycle sums is highlighted by the following three results, whose proofs are left as Exercises 21.5.1, 21.5.2 and 21.5.3.

Lemma 21.25 *If a 4-cycle sum of $n \geq 2$ bipartite graphs is a brace then each summand is a brace.* □

Lemma 21.26 *If a 4-cycle sum of $n \geq 3$ bipartite graphs is a Pfaffian brace then each summand is a Pfaffian brace.* □

Lemma 21.27 *Every brace which is a 4-cycle sum of $n \geq 2$ Pfaffian braces is Pfaffian.* □

4-CYCLE DECOMPOSITIONS AND THE HEAWOOD GRAPH

A bipartite graph G is *reducible* if it is a 4-cycle sum of three or more bipartite graphs. The next simple result is the basis of a polynomial-time algorithm to determine whether a bipartite graph is reducible.

Proposition 21.28 *A bipartite graph $G[A, B]$ is reducible if and only there exists a set X of vertices of G such that $|X \cap A| = 2 = |X \cap B|$ and $G - X$ has three or more components.* □

In view of Lemma 21.25, we may apply to a given brace G the following procedure, which produces a collection C of braces, called a *4-cycle decomposition of G*: if G is irreducible then let $C := \{G\}$; otherwise determine a 4-cycle sum of G of $n \geq 3$ summands G_i, $i = 1, 2, \ldots, n$, and recursively apply the procedure to each summand G_i, thereby obtaining a collection C_i, and then let $C := \cup_{i=1}^{n} C_i$.

The Heawood graph is Pfaffian and has a property, which does not seem to have a simple proof, that plays a crucial role in the algorithm:

Theorem 21.29 *The Heawood graph is the only simple Pfaffian irreducible nonplanar brace.* □

Description of the algorithm

The Heawood graph has girth six, and each summand of a 4-cycle sum has a quadrilateral. Thus, the Heawood graph is not a summand of any 4-cycle sum. From Theorem 21.29 and Lemmas 21.26 and 21.27 we then infer the following consequence, which is the basis of the polynomial-time recognition of Pfaffian bipartite graphs.

Corollary 21.30 *Let G be a brace whose underlying simple graph is not the Heawood graph and let C be a 4-cycle decomposition of G. The brace G is Pfaffian if and only if each brace in C is planar.*

Proof by induction on $|V|$. Let C be the 4-cycle decomposition of G. Consider first the case in which G is irreducible. This case is the basis of the inductive proof. In that case, $C = \{G\}$. By hypothesis, the underlying simple graph of G is not the Heawood graph. By Theorem 21.29, G is Pfaffian if and only if it is planar. The assertion holds in this case.

We may thus assume that G is reducible. Let $\mathcal{G}$ be a collection of $n \geq 3$ braces G_i, $i = 1, 2, \ldots, n$ such that G is the 4-cycle sum of the n graphs G_i. By Lemmas 21.26 and 21.27, G is Pfaffian if and only if each G_i is Pfaffian. For $i = 1, 2, \ldots, n$, let C_i be a 4-cycle decomposition of G_i. Then, $C = \cup_i C_i$. Each G_i has a quadrilateral. The girth of the Heawood graph is six. Thus, the underlying simple graph of G_i is not the Heawood graph. By induction, G_i is Pfaffian if and only if the collection C_i consists solely of planar braces. We deduce that G is Pfaffian if and only if C consists solely of planar braces. □

We now describe a polynomial-time algorithm to determine whether a given bipartite graph G is Pfaffian.

VERIFYING WHETHER A BIPARTITE GRAPH IS PFAFFIAN

Algorithm 21.31
Input: a (not necessarily matching covered) bipartite graph G;
Output: it checks whether G is Pfaffian.

A fundamental case: G is matching covered.

First we apply a tight cut decomposition procedure to G, thereby obtaining a collection $\mathcal{B}$ of braces (see Exercise 5.4.5). In view of Theorem 20.10, G is Pfaffian if and only if each brace in $\mathcal{B}$ is Pfaffian. These observations reduce the problem to the recognition of Pfaffian braces.

So, suppose that G is a brace. We first check whether the underlying simple graph H of G is the Heawood graph $\mathbb{H}_{14}$. If $H = \mathbb{H}_{14}$ then G is Pfaffian. If $H \neq \mathbb{H}_{14}$ then we apply a 4-cycle decomposition procedure to G, thereby obtaining a collection C of braces. We then determine, in polynomial time, whether each graph in C is planar. The brace G is Pfaffian if and only each graph in C is planar, by Corollary 21.30.

The generic case: G may not be matching covered.

First we determine whether G is matchable, using any polynomial-time algorithm which determines a maximum matching of a graph. If G is not matchable then G is Pfaffian. We may thus assume that G is matchable. Then, in polynomial time, we determine the set N of nonmatchable edges of G (Exercise 2.1.2). Let $H := G - N$. Each component of H is matching covered. The graph G is Pfaffian if and only if each component of H is Pfaffian (Exercise 19.2.4). So, the algorithm checks whether each component of H, a matching covered graph, is Pfaffian.

21.5.2 Efficient recognition of Pfaffian near-bipartite graphs

Miranda, in his Ph. D. thesis (2009, [72]), written under the supervision of Lucchesi, discovered a polynomial-time algorithm for the recognition of Pfaffian near-bipartite graphs. This result was also published in a paper, by Miranda and Lucchesi (2008, [74]). We give here a description of that algorithm. Theorem 20.30 plays a crucial role in the analysis of the algorithm.

Description of the algorithm

EFFICIENT RECOGNITION OF PFAFFIAN NEAR-BIPARTITE GRAPHS

Algorithm 21.32
Input: a near-bipartite graph G;
Output: it checks whether G is Pfaffian.

Let R be a removable doubleton of G and let $V(R)$ denote the ends of the two edges of R. We first use Algorithm 21.31 to determine whether the (bipartite) graphs $G-R$ and $G-V(R)$ are Pfaffian. (Note that $G-V(R)$ is matchable but may not be matching covered.) Both these graphs are conformal subgraphs of G. If one of them is not Pfaffian then G is not Pfaffian.

We may thus assume that the graphs $G-R$ and $G-V(R)$ are Pfaffian. The set R is a removable doubleton of G; hence R is an optimal removable class of G. We then apply Algorithm 20.26 to determine a characteristic orientation of $G-R$ and extend it to a characteristic orientation D of G, by choosing a conformal cycle Q of G that contains the edges of R and orienting the edges of R so that Q is oddly oriented in D.

Finally, we apply Algorithm 20.29 to $D-V(R)$. Consider first the case in which Algorithm 20.29 outputs a conformal cycle C of $G-V(R)$ that is evenly oriented in D. As $G-V(R)$ is a conformal subgraph of G, we infer that C is a conformal cycle of G; hence D is not Pfaffian. As D is a characteristic orientation of G we conclude that G is not Pfaffian.

Alternatively, suppose that $D-V(R)$ may be generated by Algorithm 20.26. Then, $D-V(R)$ is a characteristic orientation of $G-V(R)$, in turn a Pfaffian graph; hence $D-V(R)$ is Pfaffian. The graph $D-R$ is a characteristic orientation of $G-R$, in turn a Pfaffian graph; hence $D-R$ is Pfaffian. In sum, the digraphs $D-R$ and $D-V(R)$ are Pfaffian and the cycle Q contains both edges of R and is oddly oriented in D. By Theorem 20.30, D is Pfaffian. We conclude in this case that G is Pfaffian.

Exercises

$*$**21.5.1** *Let G be a brace which is a (Q, R)-sum of $n \geq 2$ summands G_i, $1 \leq i \leq n$. Prove that each summand is a brace.*

Hint: clearly, $G + R$ is a brace (Theorem 5.17), hence 2-extendable (Theorem 5.18).

(i) *Prove that each summand G_i is connected. Hint: apply the result proved in Exercise 5.4.6 to the graph $G - X$, where $X := V(Q) - v$ and v is a vertex of Q.*
(ii) *Prove that each summand G_i is matchable. Hint: extend a perfect matching of Q to a perfect matching of $G + R$.*
(iii) *Prove that each summand G_i is a brace. Hint: let e_1 and e_2 be any two nonadjacent edges of G_i and let M be a perfect matching of $G + R$ that contains both e_1 and e_2; use part (ii) to prove that the set $M \cap E(G_i)$ may be extended to a perfect matching of G_i by the addition of a subset of $E(Q)$.*

$*$**21.5.2** *Suppose that a Pfaffian brace G is a (Q, R)-sum of $n \geq 3$ summands $G_1, G_2, \ldots, G_n$. Prove that each summand G_i is Pfaffian.*

Hint: prove, by induction on $|R|$, that the brace $G + R$ is Pfaffian. Then, deduce that G_i is Pfaffian, because it is a conformal subgraph of $G + R$. Suppose that, for some edge $e \in R$, the brace $G + e$ is not Pfaffian; let D be an arbitrary extension of a Pfaffian orientation of G to an orientation of $G + e$.

(i) *Prove that the brace $G+e$ has a perfect matching M and an M-alternating cycle C such that $e \in M$ and C is evenly oriented in D but does not contain the edge e.*
(ii) *Prove that the set $E(C) - E(Q)$ has edges in at most two summands of $G+e$, and use a third summand to show that in fact G has a perfect matching N such that C is N-alternating in G. (That is the reason for requiring three or more summands in the 4-cycle sum!)*

$*$**21.5.3** *Suppose that G is a brace which is a (Q, R)-sum of $n \geq 2$ summands $G_1, G_2, \ldots, G_n$ which are all Pfaffian braces. Prove that G is Pfaffian.*

Hint: prove that $G + R$ is Pfaffian. Then, G, a conformal subgraph of $G + R$, is also Pfaffian.

(i) *Prove that all the summands G_i have Pfaffian orientations whose restrictions to Q coincide and are oddly oriented. Denote by D_i such a Pfaffian orientation of G_i. Let $D := \cup_i D_i$.*
(ii) *Let M be a perfect matching of $G + R$ that contains two edges of Q and let C be an M-alternating cycle of $G + R$. If the set $E(C) - E(Q)$ has edges in only one summand then certainly C is oddly oriented. Alternatively $E(C) - E(Q)$ has edges in two summands and contains both edges of $M \cap E(Q)$. In that case, the set $N := M \triangle E(Q)$ is a perfect matching of $G+R$ and the edges of $E(C) - E(Q)$ are partitioned into two N-alternating paths, each of which is in precisely one summand. A simple counting argument of forward edges then shows that C is oddly oriented in D.*

21.5.4 *Prove Proposition 21.28.*

▷**21.5.5** *Simulate the execution of Algorithm 21.32 having graphs Γ_1 and Γ_2 as input.*

∗**21.5.6 (Generalization of Algorithm 21.32 – Miranda [72])**

Let $H[A, B]$ be a matching covered bipartite graph; let G be a matching covered graph obtained from H by the addition of an edge e having both ends in A and the addition of $r \geq 1$ edges $f_1, f_2, \ldots, f_r$ having both ends in B. Describe a polynomial-time algorithm to determine whether G is Pfaffian.

Hint: for $i = 1, 2, \ldots, r$, the graph $H_i := H + e + f_i$ is near-bipartite, and the set $R_i := \{e, f_i\}$ is a removable doubleton of H_i.

21.6 An Infinite Family of Minimal Non-Pfaffian Graphs ♯

In this section we describe $\mathcal{NT}$, an infinite family of minimal non-Pfaffian graphs, discovered by Norine and Thomas (2008,[79]), who conjectured that the family is complete, in the sense that every minimal non-Pfaffian graph is in that family.

21.6.1 NT-minors

Although in this book we are concerned mostly with matching covered graphs, some authors develop theorems without restricting themselves to matching covered graphs. For example, in the definition of a Pfaffian digraph, we do not require the digraph to be matching covered. As such, Lemma 19.5 and Proposition 20.5 hold even if the graph is not matching covered. Indeed, we do not require the graph to be either matchable or connected, for that matter. In fact, if a digraph is not matchable then it is vacuously Pfaffian, because a non-Pfaffian digraph must have two perfect matchings having distinct signs. If a matchable graph is not connected then in order for it to be Pfaffian necessarily each of its components must be Pfaffian. These observations are summarized in the following result, and its proof is left as Exercise 21.6.1.

Proposition 21.33 *A graph G is non-Pfaffian if and only if it is matchable and at least one of its components is non-Pfaffian.* □

Norine and Thomas [79] define a concept of minor, which we will shall refer to as an NT-minor. A graph H is an *NT-minor* of a graph G if H is obtained from G by repeated applications of three operations: (i) removal of an edge, (ii) bicontraction of a vertex, and (iii) contractions of the edges of an odd cycle. In other words, H is an NT-minor of G if there exists a sequence

$$(G_1, G_2, \ldots, G_r) \quad (r \geq 1)$$

of graphs such that, $G_1 = G$, $G_r \cong H$ and, for $1 \leq i \leq r-1$, G_{i+1} is obtained from G_i by the application of one of the three operations mentioned above. (The Norine and Thomas graphs are simple, free of loops and multiple edges, so in their definition of bicontraction and contraction of an odd cycle they make sure to say that resulting loops and multiple edges are removed.) It can be shown that if G is Pfaffian then every NT-minor of G is Pfaffian (Exercise 21.6.2). Thus,

Proposition 21.34 *A graph is Pfaffian if and only if each of its NT-minors is Pfaffian.* □

21.6.2 Non-Pfaffian NT-minimal graphs

A non-Pfaffian graph G is *NT-minimal* (with respect to being non-Pfaffian) if application to G of any of the three operations mentioned earlier yields a Pfaffian graph. The notions of S-minimality and NT-minimality of connected non-Pfaffian graphs are equivalent (Exercise 21.6.4).

THE EQUIVALENCE OF S-MINIMALITY AND NT-MINIMALITY

Theorem 21.35 (Miranda (2008, [73]))

A connected graph is NT-minimal non-Pfaffian if and only if it is matching covered and S-minimal non-Pfaffian. □

21.6.3 The Norine and Thomas family $\mathcal{NT}$

We begin with the definition of a subcollection of $\mathcal{NT}$.

THE GRAPHS $G(k, \emptyset)$

For a positive integer k, let G_1 be the wheel W_{2k+1} whose rim is $(s_1, s_2, \ldots, s_{2k+1}, s_1)$, and let G_2 be the graph $K_{2k+1,3}$ with bipartition (T, U), where $T := \{t_1, t_2, \ldots, t_{2k+1}\}$ and $U := \{u_1, u_2, u_3\}$. The graph $G(k, \emptyset)$ is defined to be the result of the splicing of G_1 at its hub with G_2 at vertex u_3. See Figures 21.18 and 21.19.

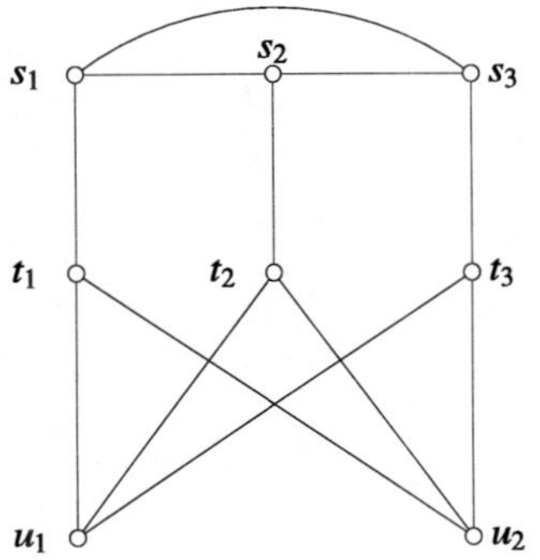

Fig. 21.18 The graph $G(1, \emptyset)$

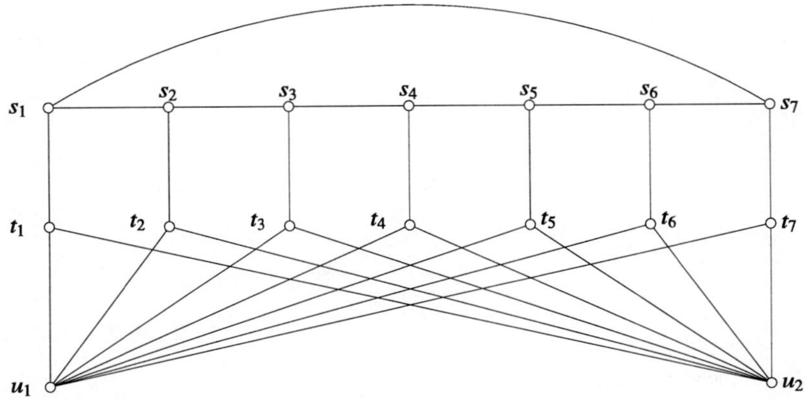

Fig. 21.19 The graph $G(3, \emptyset)$

THE GRAPHS $G(k, M)$

In the definition of $G(k, M)$ we use an auxiliary odd cycle $C_{2k+1} := (r_1, r_2, \ldots, r_{2k+1}, r_1)$. For every integer $k \geq 2$ and every matching M of the odd cycle C_{2k+1}, the graph $G(k, M)$ is obtained from $G(k, \emptyset)$ by the following procedure:

For each edge $r_i r_{i+1} \in M$, remove the edges $s_i s_{i+1}$, $t_i u_2$, $t_{i+1} u_1$ and add the edges $s_i t_{i+1}$ and $s_{i+1} t_i$.

See Figures 21.20 and 21.21.

We are now able to define the family $\mathcal{NT}$.

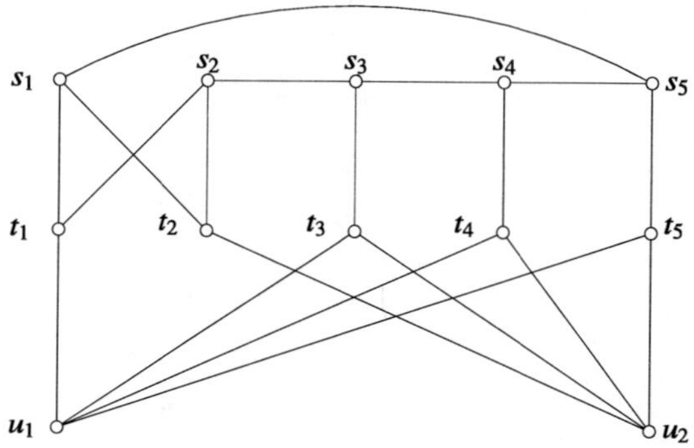

Fig. 21.20 The graph $G(2, \{r_1r_2\})$

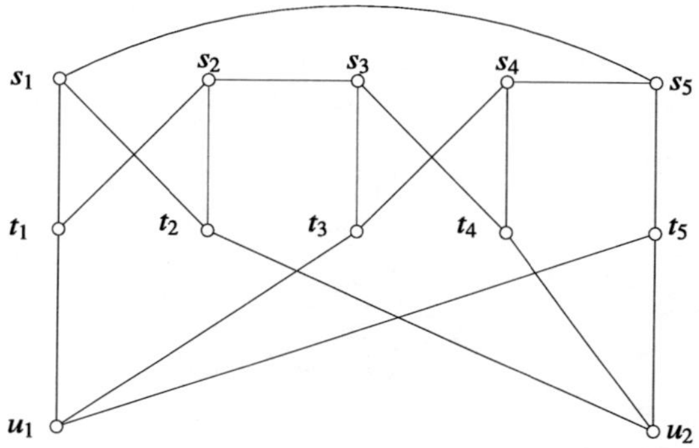

Fig. 21.21 The graph $G(2, \{r_1r_2, r_3r_4\})$

THE FAMILY $\mathcal{NT}$

Let $\mathcal{M}_k$ denote the set of (possibly empty) matchings of the odd cycle $C_{2k+1} := (r_1, r_2, \ldots, r_{2k+1}, r_1)$. Let

$$\mathcal{NT} := \{K_{3,3}, \mathbb{P}, \Gamma_2, G(1, \emptyset)\} \cup \{G(k, M) : k \geq 2, M \in \mathcal{M}_k\}$$

The reader may notice that Γ_2 is explicitly added to the family but Γ_1 is not. The reason for this discrepancy is that Γ_1 is isomorphic to $G(2, \{r_1r_2, r_3r_4\})$, depicted in Figure 21.21. (Exercise 21.6.5). The following result states that each graph in $\mathcal{NT}$ is a minimal non-Pfaffian graph, with one exception. Its proof is quite full of details; we refer the interested reader to Norine and Thomas' paper [79].

Theorem 21.36 (Norine and Thomas [79]) *Every graph in the family $\mathcal{NT}$ is non-Pfaffian and, with the exception of $G(1, \emptyset)$, is minimal.* □

Conjecture 21.37 (Norine and Thomas [79]) *Every connected minimal non-Pfaffian graph is isomorphic to a graph in $\mathcal{NT}$.*

Exercises

21.6.1

(i) *Prove Proposition 21.33.*
(ii) *Give an example of a Pfaffian graph which has a non-Pfaffian component.*

▷**21.6.2** *Prove that if G is Pfaffian then every NT-minor of G is Pfaffian.*

▷**21.6.3** *Prove that if G is a connected, non-Pfaffian graph which is not matching covered then one of the following alternatives holds:*

(i) *either G has an edge whose removal yields a connected, non-Pfaffian graph,*
(ii) *or G has a vertex of degree two.*

∗ **21.6.4** *Prove Theorem 21.35.*

21.6.5 *Prove that* $G := G(2, \{r_1r_2, r_3, r_4\})$*, depicted in Figure 21.21, is isomorphic to* Γ_1.

21.7 Notes

Miranda [71] observed that a partial result towards the proof of Conjecture 21.37 would be to prove that the graphs $K_{3,3}$, Γ_1 and Γ_2 are the only cubic minimal non-Pfaffian graphs.

Norine and Thomas [79] also consider two conjectures, in addition to Conjecture 21.37. For the first conjecture they define three operations, in addition to edge removal, vertex bicontraction and odd cycle contraction, which they call *compressions of type 1,2,3*. They then conjecture that a graph is non-Pfaffian if and only if it is either Γ_2 or it may be reduced to $K_{3,3}$ or the Petersen graph by repeated applications of the above operations.

For the second conjecture, they define an operation on a graph to be *valid* if the application of the operation to a Pfaffian graph yields a Pfaffian graph. They then conjecture that there exists a finite collection of valid operations such that every non-Pfaffian graph may be reduced to $K_{3,3}$ by repeated applications of operations in that collection.

Robertson, Seymour and Thomas, in their paper describing the polynomial-time algorithm to recognize Pfaffian bipartite graphs [82], proved that every nonplanar brace contains an odd subdivision of either $K_{3,3}$, the Heawood graph or the rotunda as a conformal subgraph. Another proof of this result and also the proof of the polynomial-time algorithm for the recognition of Pfaffian bipartite graphs, was given by P. Whalen in his Ph.D. Thesis, written under the supervision of Robin Thomas (2014, [92]).

Unsolved Problems

1. Characterize bricks, distinct from K_4, $\overline{C_6}$ and the Petersen graph, in which every b-invariant edge is solitary (page 320).
2. Characterize planar extremal bricks (page 136).
3. Every nonsolid extremal brick distinct from the Petersen brick is the result of a splicing of K_4, possibly with multiple edges, and an extremal brick (page 136).
4. There exists a positive integer N such that for every $n \geq N$, every brace of order n has at least $(n-2)^2/4$ perfect matchings (page 137).
5. There exists a positive integer N such that for every $n \geq N$, every brick of order n has at least $n-1$ perfect matchings (page 137).
6. Is the problem of recognizing whether a given brick is solid in the complexity class NP? Is it in P? (page 142).
7. Do there exist solid bricks which are 4-connected? (page 153).
8. There exists a positive integer N such that every simple solid matching covered graph of order $2n$ has at most n^2 edges, for all $n \geq N$ (page 153).
9. There exist a positive real number c and an integer N such that any brick G of order $n \geq N$ has at least cn removable edges (page 206).
10. Characterize wheel-like bricks as a splicing of two bricks (page 217).
11. Every planar wheel-like brick is a wheel (page 217).
12. Characterize K_4-free bricks (page 268).
13. Characterize $\overline{C_6}$-free bricks (page 268).
14. Every brace of order six or more has at least two nonadjacent thin edges (page 389).
15. There exists a positive constant c such that every brace of order n has at least cn thin edges (CLM, page 389).
16. Are the problems of recognizing a k-Pfaffian graph and a Pfaffian k-orientation polynomial-time equivalent? (page 435).
17. Every S-minimal non-Pfaffian graph is isomorphic to a graph in $\mathcal{NT}$ (Norine and Thomas, page 494).
18. The graphs $K_{3,3}$, Γ_1 and Γ_2 are the only cubic S-minimal non-Pfaffian graphs (Miranda, page 495).

C. L. Lucchesi, U. S. R. Murty, *Perfect Matchings*, Algorithms and Computation in Mathematics 31, https://doi.org/10.1007/978-3-031-47504-7

Some Classical Problems

1. Every planar r-graph is r-edge-colourable (Seymour, solved for $r \leq 8$, page 136).
2. Every r-graph is $(r + 1)$-edge-colourable. (Seymour, page 136).
3. For any 2-connected cubic graph G, the vector $\mathbf{2} = (2, 2, \ldots, 2)$ is in its perfect matching integer cone $IntCone(G)$ (Fulkerson, page 346).
4. For any matching covered graph G, denote by $c(G)$ the minimum number of perfect matchings of G whose union is $E(G)$. For every 2-connected cubic graph G, $c(G) \leq 5$ (Berge, page 346).

Appendix A
Solutions to Selected Exercises

Contents

C. L. Lucchesi, U. S. R. Murty, *Perfect Matchings*, Algorithms and Computation in Mathematics 31, https://doi.org/10.1007/978-3-031-47504-7

A.1 Chapter 1

Exercise 1.2.2

(i) Let M be a matching and let S be any subset of V. Then at least $o(G - S) - |S|$ odd components of $G - S$ have one or more vertices not covered by M. We conclude that

$$\mathrm{def}(M) \geq o(G - S) - |S|.$$

(ii) Suppose that equality holds for some set S. Let N be any matching of G. Then,

$$\mathrm{def}(N) \geq o(G - S) - |S| = \mathrm{def}(M),$$

hence $\mathrm{def}(N) \geq \mathrm{def}(M)$. This conclusion holds for each matching N of G, and so M is maximum.

Exercise 1.3.3

Let M be a matching of G and denote by $\mathcal{Q}$ the set of (M_0, M)-alternating paths and cycles of G. For each cycle C in $\mathcal{Q}$, we deduce that $|M \cap E(C)| = |M_0 \cap E(C)| \geq 1$. Let P be any path in $\mathcal{Q}$. As P is (M_0, M)-alternating, $|M \cap E(P)| \leq |M_0 \cap E(P)| + 1$. Moreover, the matching M_0 is maximal, thus $M_0 \cap E(P)$ is nonempty. Consequently,

$$\forall\, Q \in \mathcal{Q}, \quad |M \cap E(Q)| \leq |M_0 \cap E(Q)| + 1 \leq 2|M_0 \cap E(Q)|.$$

It follows that

$$\begin{aligned} |M| &= |M \cap M_0| + \textstyle\sum_{Q \in \mathcal{Q}} |M \cap E(Q)| \\ &\leq 2|M \cap M_0| + \textstyle\sum_{Q \in \mathcal{Q}} 2|M_0 \cap E(Q)| = 2|M_0|. \end{aligned}$$

We conclude that $|M| \leq 2|M_0|$. This inequality holds for every matching M. In particular, $|M^*| \leq 2|M_0|$.

Exercise 1.3.5

Start with a cycle C of length $r - 1$. Let A be a set of $r - 2$ new vertices (distinct from the vertices of C) and join each vertex in A with all vertices in C. The resulting graph, say K, has an odd number of vertices. Every vertex of A has degree $r - 1$ and every vertex of C has degree r. Graph K plays the role of an odd component in this construction.

Let S be a set of $r - 2$ new vertices (distinct from $A \cup C$) and join each vertex of S with one vertex in A forming a perfect matching in $S \cup A$. Now, take $r - 1$ copies of K and do the same links with the vertices in S. Figure A.1 illustrates this construction for $r = 5$. The resulting graph G is an $(r - 2)$-edge connected r-regular graph on an even number of vertices. Moreover, $G - S$ has $|S| + 2$ odd components. By Tutte's Theorem, G is not matchable.

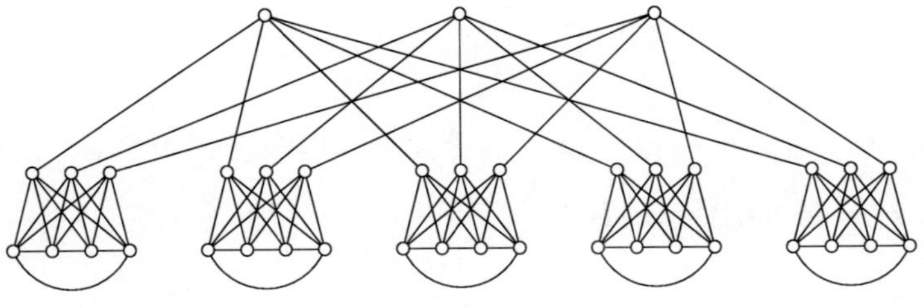

Fig. A.1 A 3-edge connected 5-regular graph which is not matchable

Exercise 1.5.6

Using Jacobi's identity, we shall now derive Tutte's identity. We shall do this taking $\mathbf{A}$ to be the matrix of indeterminates associated with an orientation of the complete graph. That would be an identity in $\binom{n}{2}$ variables. If G is any subgraph of K_n, by setting the variables corresponding to 'non-edges' to be zero, we would obtain the identity that corresponds to G.

If $\mathbf{A}$ is nonsingular, as would be the case when $G = K_n$, (1.14) reduces to:

$$\det(\mathbf{C}_{\overline{R}}) = (\det(\mathbf{A}))^{r-1} \det(\mathbf{A}_R) \tag{A.1}$$

We have proved this identity taking R to be $\{1, 2, ..., r\}$, but the identity is clearly valid for an r-subset of $\{1, 2, ..., n\}$.

Derivation of Tutte's Pfaffian identity requires two applications of Jacobi's identity, once with $r = 2$, and once with $r = 4$.

Let us first take R to be $\{i, j\}$. In this case, (A.1) reduces to:

$$\det\begin{pmatrix} 0 & c_{ij} \\ -c_{ij} & 0 \end{pmatrix} = \det(\mathbf{A}) \det(\mathbf{A}_{ij}). \tag{A.2}$$

Thus, $(c_{ij})^2 = \det(\mathbf{A}) \det(\mathbf{A}_{ij})$. Taking square roots, and noting that the Pfaffian of a skew-symmetric matrix is one of the square roots of its determinant, we have:

$$c_{ij} = \pm\, P\, P_{ij}. \tag{A.3}$$

Now let us take R to be $\{x, y, z, w\}$. In this case, (A.1) reduces to:

$$\det\begin{pmatrix} 0 & c_{xy} & c_{xz} & c_{xw} \\ -c_{xy} & 0 & c_{yz} & c_{yw} \\ -c_{xz} & -c_{yz} & 0 & c_{zw} \\ -c_{xw} & -c_{yw} & -c_{zw} & 0 \end{pmatrix} = (\det(\mathbf{A}))^3 \det(\mathbf{A}_{xyzw}). \tag{A.4}$$

This yields:

$$(c_{xy}c_{zw} - c_{xz}c_{yw} + c_{xw}c_{yz})^2 = (\det(\mathbf{A}))^3 \det(\mathbf{A}_{xyzw}). \tag{A.5}$$

Using (A.3), this reduces to:

$$P^4(\pm P_{xy}\, P_{zw} \ \pm\ P_{xz}\, P_{yw} \ \pm\ P_{xw}\, P_{yz})^2 = (\det(\mathbf{A}))^3 \det(\mathbf{A}_{xyzw}). \tag{A.6}$$

Since $P^4 = (\det(\mathbf{A}))^2$, we get:

$$(\pm P_{xy}\, P_{zw} \ \pm\ P_{xz}\, P_{yw} \ \pm\ P_{xw}\, P_{yz})^2 = (\det(\mathbf{A}))\det(\mathbf{A}_{xyzw}). \tag{A.7}$$

Now, upon taking square roots, we have Tutte's Pfaffian identity:

$$\pm P_{xy}\, P_{zw} \ \pm\ P_{xz}\, P_{yw} \ \pm\ P_{xw}\, P_{yz} \ =\ P\, P_{xyzw}.$$

A.2 Chapter 2

Exercise 2.2.5

Clearly, G_1 is connected; we must therefore prove that every edge of G_1 is matchable. Assume, to the contrary, that G_1 has an edge, $e := vw$, that is not matchable in G_1. Then, G_1 has a set S of vertices such that e has both ends in S and $o(G_1 - S) \geq |S|$.

If the contraction vertex $\overline{x}$ of G_1 is not in S then $o(G - S) = o(G_1 - S) \geq |S|$. In that case, as e has both ends in S, it follows that e is not matchable in G. This is not possible, as G is matching covered, by hypothesis.

We may thus assume that $\overline{x} \in S$. Let $T := S - \overline{x}$ and let $H := G[X] = G_1 - \overline{x}$. Clearly, $o(H - T) = o(G_1 - S) \geq |S| > |T|$. As S contains both ends of e, the set T is nonempty; hence the graph H is not critical (Exercise 1.3.9). This is not possible, as H, a graph that has an odd cycle as a hamiltonian cycle, is critical.

Exercise 2.2.6

(i) Suppose that D is strongly connected. Clearly, H is connected; hence G is connected. Let us show that every edge of G is matchable. Certainly, the edges of M are matchable. Let ab denote an edge of $E(G)-M$, where $a \in A$ and $b \in B$. As D is strongly connected, there is in D a directed path, P, from b to a. Thus, Pab is a directed cycle in D. The corresponding cycle in G is M-alternating and contains the edge ab, hence ab is matchable. This conclusion holds for each edge ab in $E(G) - M$. We deduce that every edge of G is matchable. As G is connected, in fact G is matching covered.

Conversely, suppose that G is matching covered. Clearly, G is connected; hence so too are H and D. To prove that D is strongly connected we must prove that every arc uv of D is in a directed cycle. If uv is in M or if uv is corresponds to an edge added to H as a parallel edge of an edge in M then uv is in a directed cycle of order two. Assume then that uv corresponds to an edge of $E(G) - M$. As G is matching covered, the edge uv is matchable, hence uv is in an M-alternating cycle, Q. The corresponding cycle in D is directed. Thus, uv is in a directed cycle of D. We conclude that every arc of D is in a directed cycle. As D is connected, we conclude that D is strongly connected.

(ii) Select an arbitrary node, v, of D and apply, for instance, two breadth-first searches [25], one determining the set S of nodes of D that are reachable from v by a directed path, the other determining the set T of nodes of D that may reach v by a directed path. The digraph D is strongly connected if and only if S and T are both equal to $V(D)$. It is also possible to determine whether D is strongly connected using only one depth-first search [25].

Exercise 2.3.1

If there exists a nonempty proper subset S of A such that $|N(S)| = |S|$ then every edge (if any) joining $N(S)$ with $A - S$ is not matchable. If there is no such an edge then G would not be connected, and consequently, not matching covered. In any case, G is not matching covered.

Conversely, suppose that G is not matching covered. By Theorem 2.9, A has a nonempty proper subset S such that $|N(S)| \leq |S|$. As G is matchable, $|N(S)| = |S|$, by Hall's Theorem.

Exercise 2.3.2

Let S be a minimal nonempty subset of A such that $|N(S)| = |S|$. It follows from Corollary 2.10 that the subgraph of G induced by $S \cup N(S)$ is matching covered.

Exercise 2.3.3

By induction on $|V(G)|$. By Proposition 2.11, G has a subset S of A such that the subgraph of G induced by $S \cup N(S)$ is matching covered. If $S = A$ then G is matching covered, and (A) and (B) constitute a Dulmage-Mendelsohn decomposition of G.

Otherwise, let $G' := G - (S \cup N(S))$. Then, G' is nonempty. Let $[A', B']$ denote the parts of G', where $A' \subset A$ and $B' \subset B$. For any perfect matching M of G, the restriction of M to G' is a perfect matching of G'. Then $G'[A', B']$ is matchable.

By induction, $G'[A', B']$ has a Dulmage-Mendelsohn decomposition, say, a partition $(A_1, \ldots, A_r)$ $(r \geq 1)$ of A' and a partition $(B_1, \ldots, B_r)$ of B'.

If we now set $A_0 := S$ and $B_0 = N(S)$, then partition $(A_0, \ldots, A_r)$ of A and partition $(B_0, \ldots, B_r)$ of B constitute a Dulmage-Mendelsohn decomposition of G.

Exercise 2.4.1

Let H be the result of the splicing of two matching covered graphs, G_1 and G_2, at vertices v_1 and v_2, respectively. Let C denote the cut of H that coincides with $\partial(v_i)$ in graph G_i, for $i = 1, 2$. If G_i has only two vertices then $G_i - v_i$ is connected. Alternatively, if G_i has order four or more then, by Corollary 2.7, G_i is 2-connected; hence $G_i - v_i$ is connected. In both alternatives, $G_i - v_i$ is connected; hence H is connected.

Let e be any edge of H. Adjust notation so that e is an edge of G_1. By hypothesis, G_1 has a perfect matching, M_1, that contains edge e. Let f denote the edge of M_1 in C. Let M_2 be a perfect matching of G_2 that contains edge f. Then, $M_1 \cup M_2$ is a perfect matching of G that contains edge e. Thus, edge e is matchable in H. That conclusion holds for each edge e of H. As H is connected, it follows that H is matching covered.

Exercise 2.4.2

Let v_i, for $1 \le i \le 5$, denote the vertices of the rim of the 5-wheel G, such that $v_1v_2v_3v_4v_5$ occur in that cyclic order in the rim. Order the vertices w_i of the rim of H, for $1 \le i \le 5$ analogously. The prism P_{10} may be obtained by defining $\theta(v_i) := w_i$, for $i = 1, 2, \ldots, 5$. The Petersen graph may be obtained by defining $\phi(v_i) := w_{2i-1}$ computed modulo 5.

Now consider any splicing $G \odot H$ that produces a graph distinct from the Petersen graph. Let C be the set of edges that join vertices of the rim of G to vertices of the rim of H.

We show that the hypothesis that $G \odot H$ is not the Petersen graph implies that C contains two edges that join two adjacent vertices of the rim of G to two adjacent vertices of the rim of H. To see this, assume the contrary. Then, given two adjacent vertices of the rim of G, their neighbours in the rim of H must be at distance two. Adjust notation so that v_1 and w_1 are adjacent. Then, v_2 is adjacent to one of w_3 and w_4. Without loss of generality, we may assume that v_2 is adjacent to w_3. It follows that v_3 is adjacent to w_5, v_4 is adjacent to w_2 and v_5 is adjacent to w_4. We deduce that $G \odot H$ is the Petersen graph, a contradiction.

Without loss of generality, we may assume that an edge e_i of C joins v_i to w_i, for $i = 1, 2$. Let M_i be the perfect matching of $G \odot H$ such that $M_i \cap C = \{e_i\}$. Those two perfect matchings are disjoint. As $G \odot H$ is cubic, it follows that it is 3-edge-colourable.

Assume, to the contrary, that the Petersen graph is 3-edge-colourable. As $\mathbb{P}$ is cubic, it follows that $E(\mathbb{P})$ has a partition in three perfect matchings. Every perfect matching of $\mathbb{P}$ must contain an odd number of edges in the cut C. Thus, one of the three disjoint perfect matchings of $\mathbb{P}$, say M, must contain precisely three edges in C. The ends of the edges of $M \cap C$ in the rim of G must be three cyclically consecutive vertices of the rim of G as a fourth edge of M joins the remaining two vertices of the rim of G. Likewise, the ends of the three edges of $M \cap C$ must be three consecutive vertices of the rim of H. The Petersen graph does not contain three edges in C with these properties.

Exercise 2.4.6

We now refer to the graphs depicted in Figure A.2. Clearly, $\mathcal{G}_4 = \{K_4\} = \{G_1\}$ and G_1 is vertex-transitive. Thus, $\mathcal{G}_6$ consists of just one graph, $\overline{C_6} = G_2$. Again, G_2 is vertex-transitive and $\mathcal{G}_8 = \{G_3\}$. The graph G_3 is not vertex-transitive. The orbit of every vertex contains one of the three black vertices. Thus, $\mathcal{G}_{10}$ consists of the three graphs G_4, G_5 and G_6.

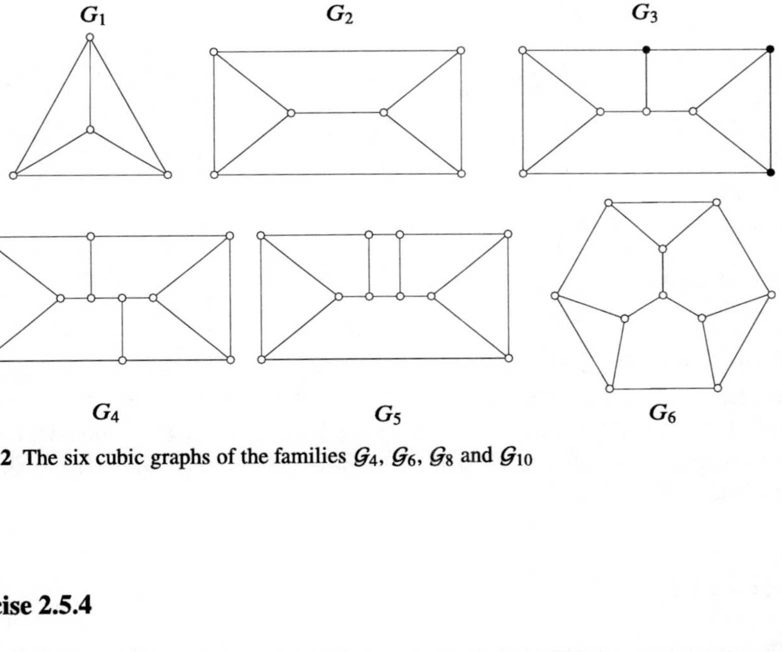

Fig. A.2 The six cubic graphs of the families $\mathcal{G}_4$, $\mathcal{G}_6$, $\mathcal{G}_8$ and $\mathcal{G}_{10}$

Exercise 2.5.4

The prism $\mathbb{P}_{2n}$ is obtained from the ladder $\mathbb{L}_{2n}$ by adding the edges u_1u_n and v_1v_n. Let us first compute $\Phi(\mathbb{P}_{2n})$ for $n \geq 3$ and odd. In this case, a perfect matching of $\mathbb{P}_{2n}$ either contains none of u_1u_n and v_1v_n or both of them. Using (2.4), we have

$$\Phi(\mathbb{P}_{2n}) = \Phi(\mathbb{L}_{2n}) + \Phi(\mathbb{L}_{2(n-2)}) = F(n+1) + F(n-1).$$

When $n \geq 4$ is even, a perfect matching of $\mathbb{P}_{2n}$ may contain any subset of $\{u_1u_n, v_1v_n\}$. Moreover, there are only two perfect matchings that contain precisely one of u_1u_n and v_1v_n. Using (2.4), we have

$$\Phi(\mathbb{P}_{2n}) = \Phi(\mathbb{L}_{2n}) + \Phi(\mathbb{L}_{2(n-2)}) + 2 = F(n+1) + F(n-1) + 2.$$

The formula for $\Phi(\mathbb{M}_{2n})$ can be deduced analogously.

Using the notation as in the definition of staircase, the formula for $\mathbb{S}_{2n}$ can be deduced from the fact that there are $F(n)$ perfect matchings containing the edge xy, and one perfect matching containing each one of the edges xu_1 and xv_1.

A.3 Chapter 3

Exercise 3.1.3

Figure A.3 depicts graph $G := K_{3,3} \odot K_{3,3}$. By Corollary 3.5, every barrier of G is

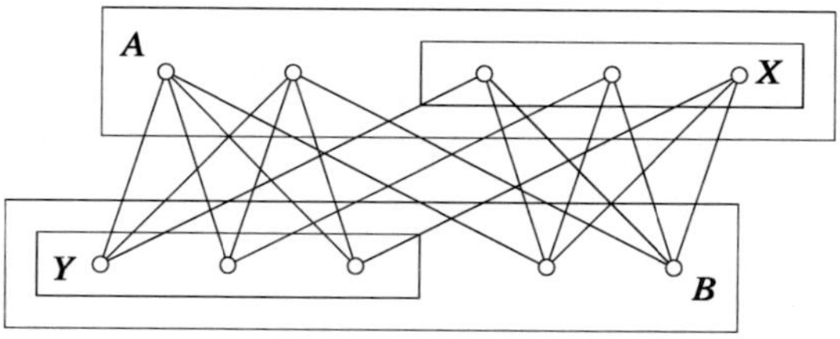

Fig. A.3 Graph $K_{3,3} \odot K_{3,3}$

a subset of one of A and B. Every vertex of G is a barrier. No two vertices of A constitute a barrier. No four vertices of A constitute a barrier. The set X is the only subset of A with just three vertices which is a barrier. Thus, the nontrivial barriers of G that are subsets of A are X and A. Likewise, the nontrivial barriers of G that are subsets of B are Y and B.

Exercise 3.1.4

Let B' be a proper superset of B, let us prove that B' is not a barrier of G. Each component of $G - B$ is critical, hence odd. Let $\mathcal{K}$ be the set of components of $G - B$ that contain vertices of B'. For each $K \in \mathcal{K}$, let $S_K := B' \cap V(K)$. As stated in Exercise 1.3.9, $o(K - S_K) \leq |S_K| - 1$ for each $K \in \mathcal{K}$. Thus,

$$\begin{aligned} o(G - B') &= o(G - B) - |\mathcal{K}| + \textstyle\sum_{K \in \mathcal{K}} o(K - S_K) \\ &\leq |B| - |\mathcal{K}| + \textstyle\sum_{K \in \mathcal{K}} [|S_K| - 1] = |B'| - 2|\mathcal{K}| < |B'|. \end{aligned}$$

Thus, B' is not a barrier of G. This conclusion holds for each proper superset B' of B. We deduce that B is a maximal barrier of G.

Exercise 3.1.7

If G is matching covered then $G - a - b$ is matchable, for each $a \in A$ and each $b \in B$, by Corollary 3.6.

Conversely, suppose that $G - a - b$ is matchable, for each $a \in A$ and each $b \in B$. By hypothesis, G has an edge, say, uv, $u \in A$, $b \in B$. The graph $G - u - v$ is matchable, hence G is also matchable.

Suppose that G is not matching covered. By Corollary 2.10, A has a nonempty proper subset S such that $|N(S)| = |S|$. Then, $A - S$ is nonempty, let $a \in A - S$. Also, $N(S)$ is nonempty, let $b \in N(S)$. By Hall's Theorem, the graph $G - a - b$ is not matchable, a contradiction. We conclude that G is matching covered.

Exercise 3.2.2

(i) Let P be a maximal M-alternating path in H and let u denote the origin of P. Suppose that the edge uv of M incident with u is not in P. By the maximality of P, the vertex v is in P. Thus, the edges of the subpath of P extending from u to v, plus the edge uv, induce an M-alternating cycle of H, in contradiction to the uniqueness of M. We deduce that the edge uv is in P. As P is a path, it follows that uv is the first edge of P. Thus, P starts with an edge in M. Suppose that u has degree greater than one, let uw be an edge of H incident with u but distinct from uv. The maximality of P implies that w is in P. As H is bipartite, the edges of the subpath of P extending from u to w, plus edge uw, induce an M-alternating cycle of H, again contradicting the uniqueness of M. We deduce that the origin of P has degree one and P starts with an edge in M. Likewise, the end of P has degree one and P ends with an edge in M. The length of P is odd, therefore it starts and ends in distinct parts of (A, B).

(ii) Let B denote a maximal barrier of G. Consider the core $\mathbb{H}(B)$ of G associated with B. The set A of contracted vertices of $\mathbb{H}$ constitutes, with B, a bipartition of $\mathbb{H}(B)$. By Theorem 3.2, every component of $G - B$ is odd and critical. It follows that every perfect matching of $\mathbb{H}(B)$ is extendable to a perfect matching of G. By hypothesis, G is 2-edge-connected. Thus, every vertex has degree two or more in $\mathbb{H}(B)$. It follows, from the first part, that $\mathbb{H}(B)$ has at least two perfect matchings. Thus, G has at least two perfect matchings.

(iii) By hypothesis, G has a (unique) perfect matching. Thus, G is matchable. Let us now consider a Dulmage-Mendelsohn decomposition of G. Let $(A_1, A_2, \ldots, A_r)$ be a partition of A and let $(B_1, B_2, \ldots, B_r)$ be a partition of B such that (i) for $i = 1, 2, \ldots, r$, the graph $H_i := G[A_i \cup B_i]$ is matching covered and (ii) for every edge $a_i b_j$ of G, with $a_i \in A_i$ and $b_j \in B_j$, $i \geq j$.
For $i = 1, 2, \ldots, r$ the graph H_i has a perfect matching, M_i. For any perfect matching N_i of H_i, the set $(M - M_i) \cup N_i$ is a perfect matching of G. By hypothesis, G has a unique perfect matching. It follows that H_i has a unique perfect matching. As H_i is matching covered, $H_i = K_2$. The assertion holds.

Exercise 3.3.1

By hypothesis, B is a maximal barrier, therefore, by Theorem 3.2, every component of $G - B$ is odd and critical. Thus, $K - a$ has a perfect matching, M_K. Likewise, for every component L of $G - B$ distinct from K, N has precisely one edge, e_L, incident with $V(L)$, and $L - v_L$ has a perfect matching, M_L, where v_L denotes the end of e_L in $V(L)$. Thus, $N \cup \left[\bigcup_{J \in \mathcal{O}(G-B)} M_J\right]$ is a perfect matching of $G - a - b$.

Exercise 3.3.5

Let G be a nonbipartite matching covered graph. Let B be a nontrivial maximal barrier of G. As G is nonbipartite, $G - B$ has a nontrivial component, K. By Theorem 3.2, K is critical. By Exercise 1.3.8, K has an odd cycle C. Recall that every edge of a matching covered graph joins vertices in distinct parts of the canonical partition. It follows that the vertices of C lie in at least three parts of the canonical partition. Thus, the canonical partition of G has at least four parts.

Let $r \geq 4$ be an integer. Let us now give an example of a matching covered graph G whose canonical partition has precisely r parts. If r is even then let $G := K_r$. If r is odd, let G be the graph obtained from $K_r - e$ by the addition of a vertex of degree two adjacent to both ends of e. If r is even then G is bicritical and has precisely r vertices. If r is odd then the two ends of e constitute a barrier, all the other barriers are singletons. In both cases, the canonical partition of G consists of precisely r barriers.

Exercise 3.3.10

For $i = 1, 2$, let B_i denote the barrier in the canonical decomposition of G that contains vertex u_i. The following three statements are equivalent:

1. The graph $G + e_i$ is not matching covered;
2. the graph $G - u_i - v_i$ is not matchable;
3. vertex v_i also lies in B_i.

Thus, the necessary and sufficient conditions are: (i) v_i lies in B_i, for $i = 1, 2$ and (ii) $G - u_1 - v_1 - u_2 - v_2$ is matchable.

Exercise 3.4.4

Consider first the case in which an ear decomposition of H may be extended to an ear decomposition of G. Let

$$G_1 = K_2, G_2, \ldots, G_q = H, \ldots, G_r = G,$$

be an ear decomposition of G which is an extension of an ear decomposition of H, where $1 \leq q \leq r$. By definition of ear decomposition, each G_i is matching covered, hence H is matchable. The set of even numbered edges of the ears P_i, $i = q+1, q+2, \ldots, r$, is a perfect matching of $G - V(H)$. Thus, H is a conformal subgraph of G.

To prove the converse, assume that H is a conformal subgraph of G. We prove that an ear decomposition of H may be extended to an ear decomposition of G by induction on $|E(G)| - |E(H)|$. If $H = G$ then the assertion holds trivially. We may thus assume that H is a proper subgraph of G.

As G, a matching covered graph, is connected, G has an edge f that is incident with a vertex in $V(H)$ but does not lie in H. Graph H is a conformal subgraph of G, let M be a perfect matching of G such that $M \cap E(H)$ is a perfect matching of H. Let M_f be a perfect matching of G that contains edge f. As in the proof of the Bipartite Ear Decomposition Theorem (3.14), consider an (M, M_f)-alternating cycle that contains edge f and let g denote the edge with which this cycle reenters $V(H)$ for the first time. Edge g also lies in M_f. Thus, the segment P of the cycle starting with f and ending with g has odd length and is internally disjoint with H. Moreover, $H' := H + P$ is a conformal subgraph of G, as $M - E(H')$ is a perfect matching of $G - V(H')$. Moreover, the ends of P lie in distinct parts of the bipartition of H, thus H' is matching covered and bipartite. The assertion holds, by induction.

A.4 Chapter 4

Exercise 4.2.2

Let M be any perfect matching of G. As $\{u, v\}$ is a 2-separation of G, every edge of C is incident with a vertex in $\{u, v\}$. Thus, $|M \cap C| \leq 2$. By definition, $\{u, v\}$ is not a barrier of G. Thus, every connected component of $G - u - v$ is even. This implies that cut C is odd. Thus, $|M \cap C|$ is odd. It follows that M contains precisely one edge in C. This conclusion holds for each perfect matching M of G. We deduce that C is tight. Likewise, cut D is also tight.

Exercise 4.2.4

(i) Let e be any edge of G. We shall prove that G has a perfect matching that contains e and just one edge in each of C and D. Consider first the case in which e is an edge of G/X. Let M_1 be a perfect matching of G/X that contains edge e and just one edge in D. Let f be the edge of M_1 in C. By definition, $G/\overline{X}$ is matching covered; let M_2 be a perfect matching of $G/\overline{X}$ that contains edge f. Then, $M_1 \cup M_2$ is a perfect matching of G that contains edge e and just one edge in each of C and D. Consider next the case in which e lies in $G/\overline{X}$. Let N_1 be a perfect matching of $G/\overline{X}$ that contains edge e. Let g be the edge of N_1 in C. Let N_2 be a perfect matching of G/X that contains edge g and just one edge in D. Then $N_1 \cup N_2$ is a perfect matching of G that contains edge e and precisely one edge in each of C and D.

(ii) Let M be any perfect matching of G. By hypothesis, C is tight in G; therefore M contains precisely one edge in C. It follows that the restriction of M to G/X is a perfect matching of G/X. By hypothesis, D is tight in G/X. Thus, M contains precisely one edge in D. This conclusion holds for each perfect matching M of G.

(iii) A trivial example is obtained by taking $D := C$ and C not tight in G. A less trivial example is shown in Figure A.4. Cut D is a 2-separation cut in G/X, but the perfect matching of G that contains all five edges in C contains three edges in D.

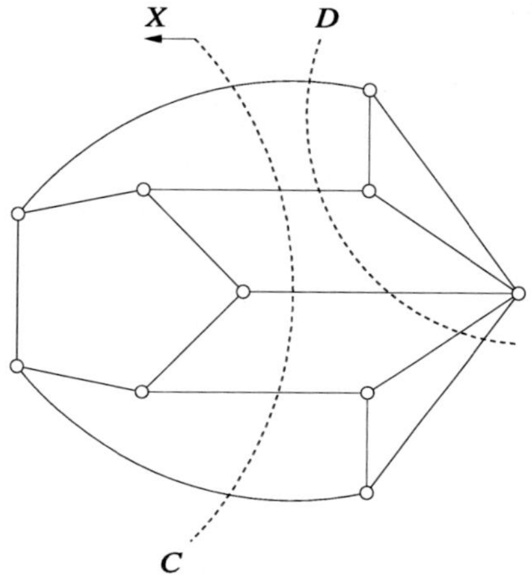

Fig. A.4 Example for the solution to Exercise 4.2.4 4.2.4

(iv) Let P be a pentagon of $\mathbb{P}$, the Petersen graph. Consider another pentagon, Q, which is neither P nor $\mathbb{P} - V(P)$. The cuts $C := \partial(P)$ and $D := \partial(Q)$ are both separating. Moreover, C and D have just one edge in common, and each perfect matching that contains that common edge has one edge in one of C and D and has five edges in the other.

Exercise 4.3.8

Suppose first that C and D do not cross and, without loss of generality, that $X \cap \overline{Y} = \emptyset$. In this case $X \subset Y$. Thus $I = \partial(X) = C$ and $U = \partial(Y) = D$, and hence the assertion holds.

So, we may assume that C and D do cross. Let M be any perfect matching of G. Denote by F the set of edges of G that join a vertex in $X \cap \overline{Y}$ and a vertex in $\overline{X} \cap Y$. Clearly,

$$|M \cap C| + |M \cap D| = |M \cap I| + |M \cap U| + 2|M \cap F|. \tag{A.8}$$

Let e be any edge of G. By hypothesis, $\{C, D\}$ is cohesive; thus G has a perfect matching, M_e, that contains edge e and just one edge in each of C and D. The cuts I and U are both odd; hence M_e contains at least one edge in each cut in $\{I, U\}$. From (A.8) we have that

$$2 = |M_e \cap C| + |M_e \cap D| = |M_e \cap I| + |M_e \cap U| + 2|M_e \cap F| \geq 2 + 2|M_e \cap F|. \quad \text{(A.9)}$$

We conclude that for each edge e of G there is a perfect matching M_e of G such that

(a) $e \in M_e$, $|M_e \cap F| = 0$,
(b) $|M_e \cap C| = 1$, $|M_e \cap D| = 1$, $|M_e \cap I| = 1$ and $|M_e \cap U| = 1$.

From (a) we deduce that $e \notin F$. This conclusion holds for each edge e of G. Thus, F is empty. That is, no edge of G joins a vertex in $X \cap \overline{Y}$ to a vertex in $\overline{X} \cap Y$.

As F is empty, from (A.8) we deduce the validity of equality (4.3).

For each edge e, G has a perfect matching M_e that satisfies (b). We deduce that $\{C, D, I, U\}$ is cohesive. In particular, the cuts I and U are both separating.

Exercise 4.3.9

The cut C is separating and the cut D is tight. Thus, the pair $\{C, D\}$ is cohesive (Exercise 4.2.3). By Theorem 4.14, no edge of G has an end in $X \cap \overline{Y}$ and an end in $\overline{X} \cap Y$ and the cuts I and U are separating.

Let M be a perfect matching of G. By Theorem 4.14, equality (4.3) holds. The cut D is tight; hence

$$|M \cap C| + 1 = |M \cap C| + |M \cap D| = |M \cap I| + |M \cap U| \geq 2.$$

If C is tight then $|M \cap I| + |M \cap U| = 2$; hence $|M \cap I| = 1 = |M \cap U|$; as this conclusion holds for each perfect matching M of G, we deduce that I and U are both tight.

Alternatively, if C is not tight then, for some perfect matching M of G, we have that $|M \cap C| > 1$. In that case, $|M \cap I| + |M \cap U| > 2$; hence at least one of the cuts I and U is not tight.

Exercise 4.3.11

If C is tight then then I and U are also tight. In that case, $\lambda(C) = \infty = \lambda(I) = \lambda(U)$. Thus, $\lambda(C) = \min\{\lambda(I), \lambda(U)\}$. We may thus assume that C is separating but not tight.

Let us first prove that $\lambda(C) \geq \min\{\lambda(I), \lambda(U)\}$. Let M be a perfect matching of G that contains precisely $\lambda(C)$ edges in C. As D is tight, we infer from (4.3) that

$$\lambda(C) + 1 = |M \cap I| + |M \cap U|.$$

As $\lambda(C) > 1$, at least one of I and U has more than one edge in M. Without loss of generality, suppose that $|M \cap I| > 1$. Then, $|M \cap I| \geq \lambda(I)$.

As the cuts I and U are both odd, we have that both I and U have at least one edge in M; hence neither I nor U has more than $\lambda(C)$ edges in M. In particular, $\lambda(C) \geq |M \cap I| \geq \lambda(I)$. We conclude that $\lambda(C) \geq \min\{\lambda(I), \lambda(U)\}$.

Let us now prove that $\lambda(C) \leq \min\{\lambda(I), \lambda(U)\}$. As C is not tight, at least one of I and U is not tight; hence $\min\{\lambda(I), \lambda(U)\} < \infty$. Without loss of generality, suppose that $\lambda(I) = \min\{\lambda(I), \lambda(U)\} < \infty$.

Let M_I be a perfect matching of G that contains $\lambda(I)$ edges in I. Cut D is tight; therefore M_I has just one edge in D, say f. As cut U is separating, G has a perfect matching, M_U, that contains f and just one edge in U. Then,

$$\left[M_I \cap E(G[Y])\right] \cup \left[M_U \cap E(G[\overline{Y}])\right] \cup \{f\}$$

is a perfect matching of G that contains precisely $\lambda(I)$ edges in C. It follows that $\lambda(C) \leq \lambda(I) = \min\{\lambda(I), \lambda(U)\}$.

Exercise 4.5.2

Let us define a family $\mathcal{G} := \{G_k : k \geq 2\}$ and prove that each G_k is a brick which has a separating cut whose C-contractions are K_4 and a graph H_k such that $b(H_k) = k - 1$. The graph G_k has precisely $2k + 2$ vertices:

$$V(G_k) := \{v_i, w_i : 0 \leq i \leq k\}.$$

Now let us add $4k + 1$ edges in order to obtain G_k (Figure A.5(a) depicts graph G_5):

1. Form $k - 1$ triangles by the addition of edges $v_i w_i$, $w_i v_{i+1}$ and $v_{i+1} v_i$, for $i = 1, 2, \ldots, k - 1$.
2. Form another triangle, by the addition of edges $v_0 w_0$, $w_0 w_k$ and $w_k v_0$.
3. Add edges $v_0 w_i$ for $i = 1, 2, \ldots, k - 1$.
4. Add edges $v_1 w_0$ and $v_k w_k$.

We remark that $G_2 = \overline{C_6}$. Let $X := \{v_0, w_0, w_k\}$ and let $C := \partial(X)$. The C-contraction $G' := G_k/\overline{X}$ is K_4, up to multiple edges. Let $G'' := G_k/(X \to x)$ be the other C-contraction of G_k. The set of 2-separations of G'' is $\{x, v_i\}$, $2 \leq i \leq k - 1$. The tight cut decomposition of G'' produces only K_4's, up to multiple edges, and $b(G'') = k - 1$. Thus, $b(G') + b(G'') = k$. Moreover, G_k is the splicing of G' and G''. Thus, G_k is matching covered.

Let us now prove that G_k is a brick, by induction on k. The basis of the induction is for $k = 2$, in which case $G_2 = \overline{C_6}$ is a brick.

Let us now assume that $k > 2$. Let $Y := \{v_{k-1}, v_k, w_{k-1}\}$. The set Y induces a triangle in G_k. Let $D := \partial(Y)$ and let $H := G_k/(Y \to y)$ (Figure A.5(b)). It is easy to see that H is isomorphic to $G_{k-1} + e$, where $e = v_0 w_{k-2}$.

Let us now prove that H is a brick. By induction, $H - e$ is a brick. Let F be any nontrivial cut of H. As $H - e$ is a brick, the graph $H - e$ has a perfect matching, M, that contains more than one edge in $F - e$. Thus, M is a perfect matching of H that contains more than one edge in F. We deduce that F is not tight. This conclusion

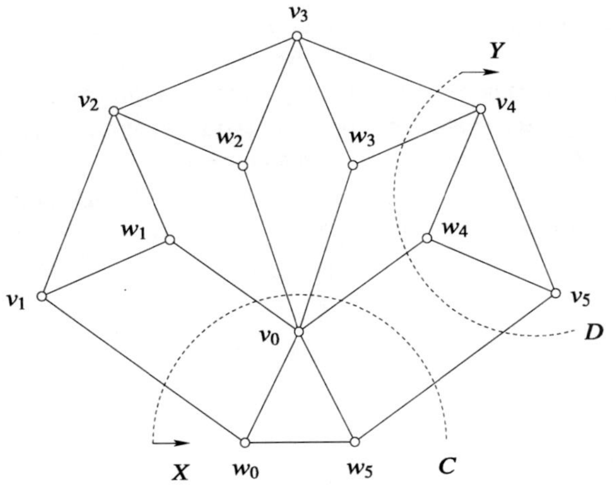

(a) G_5

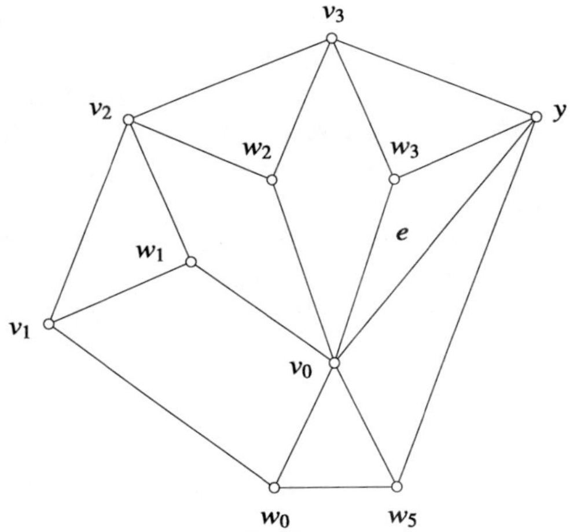

(b) $G_4 + e$

Fig. A.5 Graphs G_5 and $G_4 + e$

holds for each nontrivial cut F of H. Thus, H is either a brick or a brace. But H is not bipartite; hence H is a brick.

Let us now use the subadditivity of function b to prove that G_k is a near-brick. The graph H, a brick, is bicritical. Let M be a perfect matching of $H - v_0 - w_k$. Addition of the edges v_0w_{k-1} and v_kw_k to M yields a perfect matching of G that contains three edges in D. Thus, D is a separating cut of G which is not tight. Moreover, both D-contractions of G are bricks. By the subadditivity of function b, the graph G_k is a near-brick.

Finally, let us prove that G_k is free of nontrivial barrier cuts. We do this by showing that G_k is bicritical. Let $T := G_k[Y]$ be the triangle induced by the set Y. Let t_1 and t_2 denote two vertices of G_k. If neither t_1 nor t_2 lies in T then a perfect matching of $H - t_1 - t_2$ may be extended to a perfect matching of $G_k - t_1 - t_2$ by the addition of the appropriate edge of T. Alternatively, if t_1 and t_2 lie both in T then they are adjacent, and since G_k is matching covered it follows that t_1t_2 is matchable, hence $G_k - t_1 - t_2$ is matchable. We may thus assume that precisely one of t_1 and t_2, say, t_1, lies in T. In that case, any perfect matching of $H - y - t_2$ may be extended to a perfect matching of $G_k - t_1 - t_2$ by the addition of the appropriate edge of T. In sum, G_k is a bicritical near-brick, that is, G_k is a brick.

A.5 Chapter 5

Exercise 5.2.7

Suppose that $G[X]$ is bipartite. Let (A, B) denote its bipartition. Assume, to the contrary, that X contains two or more vertices. Graph G, a brick, is 3-connected. Thus, the ends of the three edges of C in X are distinct. Each end in X of an edge in C is adjacent to two or more vertices in X and a vertex of X not incident with any edge of C is adjacent to three or more vertices in X. We conclude that $|A|, |B| \geq 2$. Let a be the number of ends of edges of C in A. Let b denote the number of ends of edges of C in B. Then, $a + b = 3$.

Observe that, in $G - A$, $|B| - b$ vertices of B are isolated. As G is bicritical and $|A| \geq 2$, we have that $|B| - b \leq |A| - 2$. Likewise, $|A| - a \leq |B| - 2$. Adding those two inequalities, we have that $|A| + |B| - (a + b) \leq |A| + |B| - 4$, whence $a + b \geq 4$, a contradiction.

Exercise 5.2.8

First obtain a perfect matching M of G, using the algorithm of Micali and Vazirani [70] for finding a maximum matching in a graph in $O(m\sqrt{n})$-time.

Then, for each pair $\{u, v\}$ of vertices of G, verify whether $G-u-v$ is connected and matchable. Connectedness can be computed in linear time and checking if $G - u - v$ has a perfect matching can also be computed in time $O(m)$ by applying one iteration of Edmonds' maximum matching algorithm (see Gabow and Tarjan (1985, [36])), to get an augmenting path between the two M-exposed vertices of $G-u-v$. If $G-u-v$ is not matchable then a Tutte set S is obtained and $B := S + u + v$ is a (nontrivial) barrier of G. In that case, as G is nonbipartite, at least one (odd) component, K, of $G - B$, is nontrivial and the cut $\partial(K)$ is a nontrivial tight cut of G.

Alternatively, if the graph $G - u - v$ is matchable but not connected, it follows that each component of $G - u - v$ is even. Let L be one such component. The cut $\partial(V(L) + u)$ is a nontrivial tight cut of G.

Finally, if $G - u - v$ is matchable and connected for each pair $\{u, v\}$ of distinct vertices of G then G is a brick, by Theorem 5.7. The total running time of this algorithm is $O(m\sqrt{n}) + O(n^2) \times O(m) = O(mn^2)$.

Exercise 5.4.3

Suppose that G is a brace, let $e_1 := u_1v_1$ and $e_2 := u_2v_2$ be two nonadjacent edges of G. By Theorem 5.17, $G - u_1 - u_2 - v_1 - v_2$ has a perfect matching, M. Then, $M + e_1 + e_2$ is a perfect matching of G that contains both edges e_1 and e_2. We conclude that each pair $\{e_1, e_2\}$ of nonadjacent edges of G is in a perfect matching of G, hence G is 2-extendable.

Conversely, suppose that G is 2-extendable. Let us first prove that G is matching covered. By hypothesis, G is connected; therefore it remains to be shown that every edge of G is matchable. Let $e_1 := u_1v_1$, let us prove that e_1 is matchable. For this, we must prove that some edge of $G - e_1$ is nonadjacent to e_1. For this, let $u_2 \in A - u_1$ and $v_2 \in B - v_1$. If u_2 is not adjacent to v_1 then u_2 is adjacent to some vertex v_3 of $B - v_1$; in this case e_1 and u_2v_3 are nonadjacent edges of G. We may thus assume that u_2 is adjacent to v_1. Likewise, we may assume that v_2 is adjacent to u_1. By hypothesis, G has a perfect matching, M, that contains the edges u_1v_2 and u_2v_1. As G has order six or more, $A - u_1 - u_2$ contains a vertex, a, which is joined by an edge of M to a vertex $b \in B - v_1 - v_2$. The edges e_1 and ab of G are nonadjacent. In all cases considered, we conclude that G contains an edge nonadjacent to e_1. By hypothesis, G is 2-extendable, hence e_1 is matchable. This conclusion holds for each edge e_1 of G, hence every edge of G is matchable. As G is connected, we deduce that G is matching covered.

Let us now prove that G is a brace. For this, assume, to the contrary, that G has a nontrivial tight cut $C := \partial(X)$. Adjust notation so that $X_+ \subset B$ and $X_- \subset A$, where $|X_+| = |X_-| + 1$ (Theorem 4.8).

If all the edges of C are incident with the same vertex, u, in $A - X_-$, then $N(X_+) = X_- + u$ and $|N(X_+)| = |X_+|$. In that case, G is not matching covered (Exercise 2.3.1). Likewise, if all the edges of C are incident with the same vertex, v, in X_+, then $N(X_+ - v) = X_-$ and $|N(X_+ - v)| = |X_+ - v|$. We may thus assume that C contains two nonadjacent edges, f_1 and f_2. As C is tight, G has no perfect matching that contains both f_1 and f_2, a contradiction to the hypothesis that G is 2-extendable. Indeed, G is a brace.

Exercise 5.4.4

Let (A, B) be the bipartition of G.

Suppose that M has an edge, $e := uv$, $u \in A$, $v \in B$, such that $H := G - u - v$ is not matching covered. By hypothesis, the graph G has order six or more; hence H has order four or more. By Theorem 2.9, $A - u$ has a nonempty proper subset X such that

$$|N_H(X)| \leq |X|. \tag{A.10}$$

The set $M - e$ is a perfect matching of H; therefore, H is matchable. We deduce that equality holds in (A.10). Let $Y := X \cup N_H(X) \cup \{v\}$. The set $C := \partial(Y)$ is a tight cut of G. Moreover, X is a proper subset of $A - u$. Thus, $\overline{Y}$ is nontrivial. We conclude that C is a nontrivial tight cut of G. Hence, G is not a brace.

Conversely, suppose that $G - u - v$ is matching covered, for each $uv \in M$. To prove that G is a brace, let $X \subset A$ such that $0 < |X| < |A| - 1$. We want to prove that $|N(X)| \geq |X| + 2$. The graph G is matching covered, by hypothesis. From Theorem 2.9, we infer that $|N_G(X)| \geq |X| + 1$. Thus, M contains an edge ab such that $b \in N_G(X)$ and $a \in A - X$. By hypothesis, the graph $J := G - a - b$ is matching covered. As $0 < |X| < |A| - 1 = |A - a|$, we deduce that the set X is a nonnull proper subset of $A - a$. By Theorem 2.9, we infer that

$$|N_G(X)| = |N_J(X) + b| = |N_J(X)| + 1 \geq |X| + 2.$$

Thus, $|N_G(X) \geq |X| + 2$. This conclusion holds for each subset X of A such that $0 < |X| < |A| - 1$. By Theorem 5.17, G is a brace.

Let us now describe a polynomial-time algorithm that either determines that G is a brace or, alternatively, determines a nontrivial tight cut of G.

For each edge $e := uv$ of M, $u \in A$ and $v \in B$, let $H_e := G - u - v$ and let $\mathcal{H} := \{H_e : e \in M\}$.

For each H_e in $\mathcal{H}$, let D_e be the directed graph as described in Exercise 2.2.6, with H_e playing the role of G and D_e playing the role of D. In linear time it is possible to determine whether D_e is strongly connected. Thus, in linear time it is possible to determine whether H_e is matching covered.

Verification of whether each $H_e \in \mathcal{H}$ is matching covered can thus be done in time $O((m + n)n)$. Moreover, suppose that D_e is not strongly connected, for some $e \in M$. The linear time search determines a directed cut C in D_e. By definition of D_e, it follows that $C \subseteq E(H_e) - M$. Thus, C is a cut of H_e which has a shore X such that every edge of C is incident with a vertex in $X \cap B$. Moreover, as $M - e$ is a perfect matching of H_e, it follows that $|X \cap A| = |X \cap B|$. Consequently, the set $Y := X + v$ is the shore of a nontrivial tight cut of G.

Exercise 5.4.5

Let G be the given matching covered graph. In linear time, determine whether G is bipartite. If G is bipartite, use the $O(mn)$-time algorithm described in Exercise 5.4.4 to either attest that G is a brace, or alternatively, determine a nontrivial tight cut C of G. Likewise, if G is nonbipartite, use the $O(mn^2)$-time algorithm described in Exercise 5.2.8 to either attest that G is a brick or, alternatively, determine a nontrivial tight cut C of G. If G is a brick or a brace then the tight cut decomposition consists solely of G. Alternatively, if the nontrivial tight cut C is found, then recursively determine the tight cuts decompositions of both C-contractions of G and combine the results. The number r of nontrivial tight cuts of the tight cut decomposition is at most $n/2 - 2$ (Exercise 4.3.5). Thus, the two algorithms which are used to find nontrivial tight cuts are executed at most $2r + 1 \leq n - 3$ times. Thus, the algorithm runs in time $O(mn^3)$.

Exercise 5.4.6

Let S be a set of two or three vertices of G such that neither A nor B contains three vertices of S. Adjust notation so that $|S \cap A| \geq |S \cap B|$.

Case 1 *$S = \{u_1, v_1\}$, where $u_1 \in A$ and $v_1 \in B$.*

Let $H := G - S$ and let $a \in A - u_1$ and $b \in B - v_1$. By hypothesis, G is a brace. By Theorem 5.17, $H - a - b$ has a perfect matching, M. As G has order six or more, it follows that M has at least one edge. Thus, H has at least one edge and $H - a - b$ is matchable, for each $a \in A$ and $b \in B$. As stated in Exercise 3.1.7, H is matching covered; hence $G - S = H$ is 2-connected.

Case 2 *$S = \{u_1, u_2, v_1\}$, where $u_1, u_2 \in A$ and $v_1 \in B$.*

By Case 1, $G - u_1 - v_1$ is 2-connected, hence $G - S$ is connected.

Case 3 $S = \{u_1, u_2\} \subset A$.

Let v_1 be a vertex in B. By Case 2, $G - u_1 - u_2 - v_1$ is connected. As G, a brace, has order six or more, the vertex v_1 is adjacent to a vertex in $A - S$; otherwise $S + v_1$ is the shore of a nontrivial tight cut of G. It follows that $G - S$ is connected.

A.6 Chapter 6

Exercise 6.1.2

For every perfect matching M, clearly χ^M satisfies the non-negativity and degree constraints. For every odd set S of vertices, M contains an odd number of edges in $\partial(S)$. Thus, χ^M also satisfies the odd set constraints. Consider the linear combination

$$\mathbf{x} := \sum_{M \in \mathcal{M}} \alpha_M \chi^M, \quad 0 \leq \alpha_M.$$

Then, $\mathbf{x}$ satisfies the non-negativity constraints. Moreover, for every vertex v, we have that $\mathbf{x}(\partial(v)) = s$, and for every odd set S of vertices, we have that $\mathbf{x}(\partial(S)) \geq s$, where $s := \sum_{M \in \mathcal{M}} \alpha_M$. In particular, if the linear combination is convex then $s = 1$ and $\mathbf{x}$ satisfies the three constraints.

Exercise 6.1.4

Figure A.6 depicts $\mathbf{w}_1$ expressed as a convex linear combination of perfect matchings of G_1, Figure A.7 depicts $\mathbf{w}_2$ expressed as a convex linear combination of perfect matchings of G_2 and Figure A.8 depicts $\mathbf{w}$ expressed as a convex linear combination.

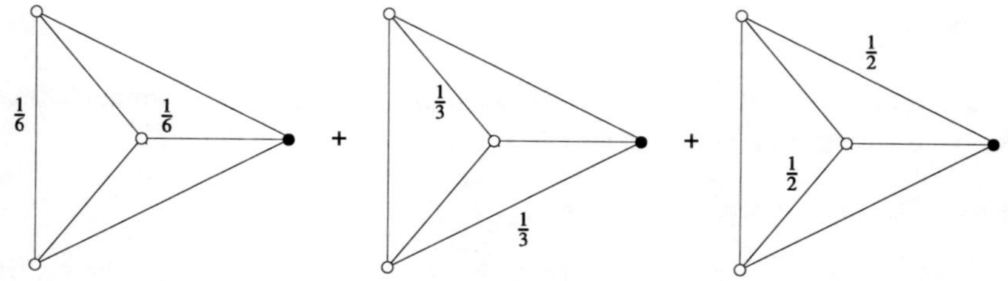

Fig. A.6 $\mathbf{w}_1$ expressed as a convex linear combination

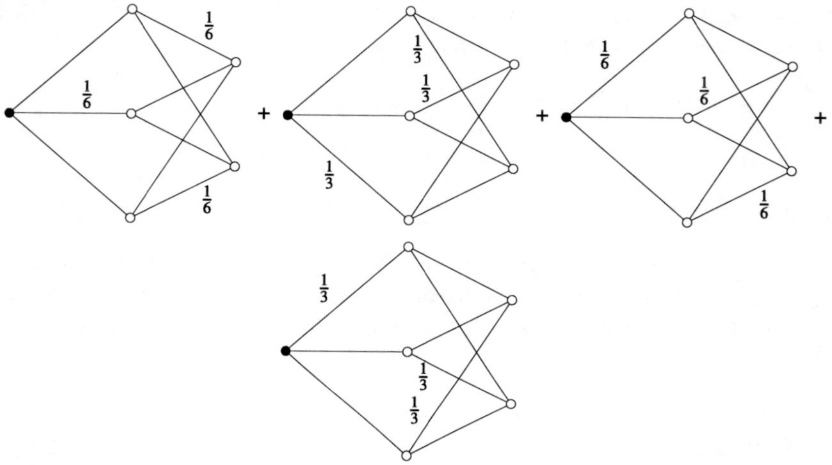

Fig. A.7 $\mathbf{w}_2$ expressed as a convex linear combination

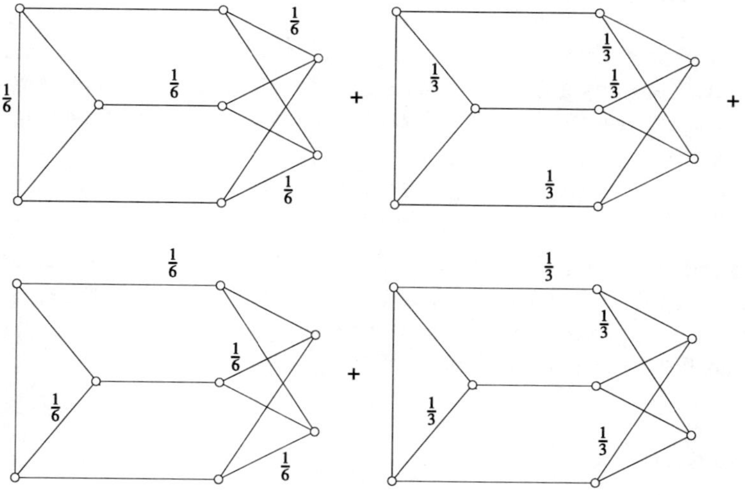

Fig. A.8 $\mathbf{w}$ expressed as a convex linear combination

Exercise 6.1.7

Let $\mathbf{x} := \sum_{M \in \mathcal{M}} \alpha_M \chi^M$. Our task is to prove that $\mathbf{w} = \mathbf{x}$. For this, we shall prove that $\mathbf{w}(f) = \mathbf{x}(f)$, for each edge f in E. Adjust notation so that $f \in E(G_1)$.

Let $\mathcal{M}_1(f)$ denote the set of perfect matchings of G_1 that contain edge f and let $\mathcal{N}(f)$ denote the set of those matchings in $\mathcal{N}$ that contain edge f. For each edge e of C, let $\mathcal{M}_1(e, f)$ denote the set of those matchings in $\mathcal{M}_1(f)$ that contain edge e, let $\mathcal{N}(e, f)$ denote the set of those matchings of $\mathcal{N}(f)$ that contain edge e and let $\mathcal{M}_2(e)$ denote the set of perfect matchings of G_2 that contain edge e.

Clearly, $\{\mathcal{M}_1(e, f) : e \in C\}$ is a partition of $\mathcal{M}_1(f)$ and $\{\mathcal{N}(e, f) : e \in C\}$ is a partition of $\mathcal{N}(f)$. Thus,

$$\mathbf{w}(f) = \mathbf{w}_1(f) = \sum_{M_1 \in \mathcal{M}_1(f)} \beta_{M_1} = \sum_{e \in C} \sum_{M_1 \in \mathcal{M}_1(e,f)} \beta_{M_1}, \tag{A.11}$$

$$\mathbf{w}(e) = \mathbf{w}_2(e) = \sum_{M_2 \in \mathcal{M}_2(e)} \gamma_{M_2} \quad (\forall e \in C), \tag{A.12}$$

$$\begin{aligned}\mathbf{x}(f) &= \sum_{M \in \mathcal{N}(f)} \alpha_M = \sum_{e \in C} \sum_{M \in \mathcal{N}(e,f)} \alpha_M \\ &= \sum_{e \in C} \sum_{M_1 \in \mathcal{M}_1(e,f)} \sum_{M_2 \in \mathcal{M}_2(e)} \alpha_{M_1 \cup M_2}.\end{aligned} \tag{A.13}$$

From (A.11), (A.12) and (A.13), we then deduce that

$$\begin{aligned}\mathbf{w}(f) &= \sum_{e \in C} \sum_{M_1 \in \mathcal{M}_1(e,f)} \frac{\beta_{M_1}}{\mathbf{w}(e)} \sum_{M_2 \in \mathcal{M}_2(e)} \gamma_{M_2} \\ &= \sum_{e \in C} \sum_{M_1 \in \mathcal{M}_1(e,f)} \sum_{M_2 \in \mathcal{M}_2(e)} \alpha_{M_1 \cup M_2} = \mathbf{x}(f).\end{aligned}$$

Exercise 6.1.12

If S is empty then $\mathbf{x} = \mathbf{0}$, a vector in $\mathit{Cone}(G)$. We may thus assume that S is nonempty. The non-negativity of $\mathbf{x}$ then implies that $r > 0$.

(i) Let $H := G[S]$ and let $X \subseteq V(H)$. For each $K \in \mathcal{O}(G - X)$, $\mathbf{x}(\partial(K)) \geq r$. Thus,

$$r \cdot o(G - X) \leq \sum_{K \in \mathcal{O}(G-X)} \mathbf{x}(\partial(K)) \leq \mathbf{x}(\partial(X)) \leq \sum_{v \in X} \mathbf{x}(v) = r \cdot |X|.$$

As $r > 0$, we deduce that $o(G - X) \leq |X|$. This conclusion holds for each subset X of $V(H)$; hence H is matchable, by Tutte's Theorem.

(ii) The proof in this case is like Case 1 of the proof of Theorem 6.1.

(iii) Let M be a perfect matching of $G[S]$. We prove that the inequalities (6.6a), (6.6b) and (6.6c) hold for $\mathbf{y}$ and $r - \epsilon$ playing respectively the roles of $\mathbf{x}$ and r. By definition, $\epsilon \le \mathbf{x}(e)$, for each edge e. Thus, $\mathbf{y}(e) = \mathbf{x}(e) - \epsilon \ge 0$. This establishes the validity of inequality (6.6a).
For each vertex v, $\mathbf{y}(\partial(v)) = \mathbf{x}(v) - \epsilon\chi^M(\partial(v)) = r - \epsilon$. The validity of inequality (6.6b) is established.
For each odd cut C such that $|M \cap C| = 1$, $\mathbf{y}(C) = \mathbf{x}(C) - \epsilon\chi^M(C) \ge r - \epsilon$.
For each odd cut C such that $|M \cap C| > 1$, by definition of ϵ,

$$\mathbf{y}(C) = \mathbf{x}(C) - \epsilon\chi^M(C) \ge \mathbf{x}(C) - \frac{\mathbf{x}(C) - r}{|M \cap C| - 1}|M \cap C| = \\ r - \frac{\mathbf{x}(C) - r}{|M \cap C| - 1} \ge r - \epsilon.$$

Note that if $\epsilon = \dfrac{\mathbf{x}(C) - r}{|M \cap C| - 1}$ then $\mathbf{y}(C) = r - \epsilon$. The validity of inequality (6.6c) is established.
By definition of ϵ, either $\epsilon = \mathbf{x}(e)$ for some edge e of S, or $\mathbf{y}(C) = r - \epsilon$, for some (nontrivial) odd cut C such that $|M \cap C| > 1$. In the former case, the support of $\mathbf{y}$ is a proper subset of S and the assertion follows by induction. In the latter case, the proof proceeds as in (ii).

Exercise 6.2.1

Let $\mathbf{x}$ be a 1-regular vector and let $\mathbf{y} := \sum_{M \in \mathcal{M}} |\mathcal{M}|^{-1}\chi^M$.

$$\text{Let } \mathcal{C} := \{C : C \text{ is an odd cut}, \mathbf{x}(C) < 1 < \mathbf{y}(C)\} \\ \text{and let } F := \{e \in E : \mathbf{x}(e) < 0\}.$$

If $F \cup \mathcal{C}$ is empty then $\mathbf{x} \in \mathcal{P}oly(G)$. We may thus assume that $F \cup \mathcal{C}$ is nonempty.

By definition of $\mathbf{y}$ and F, $\mathbf{x}(e) < 0 < \mathbf{y}(e)$, for each edge $e \in F$. By definition of $\mathbf{y}$ and $\mathcal{C}$, $\mathbf{x}(C) < 1 < \mathbf{y}(C)$, for each cut $C \in \mathcal{C}$. Thus,

$$\epsilon := \min\left[\left\{\frac{\mathbf{y}(e)}{\mathbf{y}(e) - \mathbf{x}(e)} : e \in F\right\} \cup \left\{\frac{\mathbf{y}(C) - 1}{\mathbf{y}(C) - \mathbf{x}(C)} : C \in \mathcal{C}\right\}\right]$$

satisfies the inequalities $0 < \epsilon < 1$. Let $\mathbf{z} := (1 - \epsilon)\mathbf{y} + \epsilon\mathbf{x}$. We must now prove that $\mathbf{z} \in \mathcal{P}oly(G)$.

Let e be an edge of G. By definition of $\mathbf{y}$, we have that $\mathbf{y}(e) > 0$. If $e \notin F$ then $\mathbf{x}(e) \ge 0$; hence $\mathbf{z}(e) = (1 - \epsilon)\mathbf{y}(e) + \epsilon\mathbf{x}(e) > 0$. Alternatively, if $e \in F$ then

$$0 < \epsilon \le \frac{\mathbf{y}(e)}{\mathbf{y}(e) - \mathbf{x}(e)}; \quad \text{hence} \quad \mathbf{z}(e) = \mathbf{y}(e) - \epsilon(\mathbf{y}(e) - \mathbf{x}(e)) \ge 0.$$

In both alternatives, $\mathbf{z}$ satisfies the inequalities (6.3a).

Let C be a tight cut of G. As $\mathbf{x}$ and $\mathbf{y}$ are both 1-regular, we deduce that $\mathbf{z}(C) = 1$; hence $\mathbf{z}$ is 1-regular. Thus, $\mathbf{x}$ satisfies the inequalities (6.3b).

Let C be an odd cut of G. If $C \notin \mathcal{C}$ then $\mathbf{x}(C), \mathbf{y}(C) \geq 1$; hence $\mathbf{z}(C) = (1-\epsilon)\mathbf{y}(C) + \epsilon\mathbf{x}(C) \geq 1$. Alternatively, if $C \in \mathcal{C}$ then

$$0 < \epsilon \leq \frac{\mathbf{y}(C) - 1}{\mathbf{y}(C) - \mathbf{x}(C)}; \quad \text{hence} \quad \mathbf{z}(C) = \mathbf{y}(C) - \epsilon(\mathbf{y}(C) - \mathbf{x}(C)) \geq 1.$$

In both alternatives, $\mathbf{z}$ satisfies the inequalities (6.3c). Indeed, $\mathbf{z} \in \mathcal{P}oly(G)$.

Exercise 6.2.5

Let $\mathcal{B}_0$ denote any basis of $\mathcal{L}in_0(G)$ and let $\mathbf{y}$ be any 1-regular vector of $\mathbb{R}^E$. Let us first show that $\mathcal{B} := \mathcal{B}_0 \cup \{\mathbf{y}\}$ is linearly independent. For this, consider a linear combination

$$\sum_{\mathbf{b} \in \mathcal{B}_0} \alpha_{\mathbf{b}}\mathbf{b} + \beta\mathbf{y} = \mathbf{0}.$$

Note first that $\beta = 0$; else $\mathbf{y}$ can be expressed as a linear combination of vectors in $\mathcal{B}_0$, a contradiction, as $\mathcal{B}_0$ is a basis of $\mathcal{L}in_0(G)$ and $\mathbf{y} \notin \mathcal{L}in_0(G)$. As $\beta = 0$, the linear independence of $\mathcal{B}_0$ then implies that $\alpha_{\mathbf{b}} = 0$, for each $\mathbf{b} \in \mathcal{B}_0$. We conclude that $\mathcal{B}$ is linearly independent. Let us now prove that $\mathcal{B}$ spans $\mathcal{L}in(G)$. For this, let $\mathbf{x} \in \mathcal{L}in(G)$. Then, $\mathbf{x}$ is r-regular, for some $r \in \mathbb{R}$. If $r = 0$ then $\mathbf{x} \in \mathcal{L}in_0(G)$ and $\mathbf{x}$ is equal to some linear combination of vectors in $\mathcal{B}_0$. Alternatively, if $r \neq 0$ then $\frac{1}{r}\mathbf{x} - \mathbf{y} \in \mathcal{L}in_0(G)$; therefore $\mathbf{x}$ is equal to some linear combination of vectors in $\mathcal{B}$. In both alternatives we deduce that $\mathcal{B}$ spans $\mathcal{L}in(G)$.

Exercise 6.2.6

(i) Let $\mathbf{x} \in \mathbb{R}^E$. By Theorem 6.7, $\mathbf{x} \in \mathcal{L}in(G)$ iff $\mathbf{x}$ is regular. Thus, for any specified vertex u, $\mathbf{x} \in \mathcal{L}in(G)$ iff $\chi^C\mathbf{x} - \chi^{\partial(u)}\mathbf{x} = 0$ for each $C \in \mathcal{T} - \{\partial(u)\}$. That is, $\mathbf{x} \in \mathcal{L}in(G)$ iff $\mathbf{B}\mathbf{x} = \mathbf{0}$. It follows that the rows of $\mathbf{B}$ span $(\mathcal{L}in(G))^\perp$.

(ii) Let $\mathbf{p} \in V^{\mathbb{R}}$ such that $\sum_{v \in V} \mathbf{p}(v) = 0$. Let $\mathbf{x} \in \mathbb{R}^E$ be defined by the equality $\mathbf{x}(uv) := \mathbf{p}(u) + \mathbf{p}(v)$ for each edge uv of G. Let M be any perfect matching of G. Then, $\mathbf{x}(\chi^M) = \sum_{uv \in M}[\mathbf{p}(u) + \mathbf{p}(v)] = \sum_{v \in V} \mathbf{p}(v) = 0$. Indeed, $\mathbf{x}$ is matching orthogonal.

(iii) Assume that G has at most one brick. From Exercise 6.2.3(ii), we know that $\mathbf{x}$ is regular if and only if $\mathbf{x}(u) = \mathbf{x}(v)$, for any two vertices u and v of G. In that case, we may use the $n \times m$ incidence matrix $\mathbf{A}$ of G, instead of $\mathbf{T}$ in the previous items. Let $\mathbf{x} \in (\mathcal{L}in(G))^\perp$. Then, $\mathbf{x}$ may be expressed as a linear combination of rows of $\mathbf{B}$:

$$\mathbf{x} = \sum_{v \in V-u} p(v)\left[\chi^{\partial(v)} - \chi^{\partial(u)}\right].$$

Let $p(u) := -\sum_{v \in V-u} p(v)$.

Then, $\sum_{v \in V} p(v) = 0$ and

$$\mathbf{x} = \sum_{v \in V-u} p(v)\chi^{\partial(v)} + p(u)\chi^{\partial(u)} = \sum_{v \in V} p(v)\chi^{\partial(v)}.$$

It follows that for every edge vw of G, $\mathbf{x}(vw) = p(v) + p(w)$.

Exercise 6.2.7

For each vertex v of G, let $\mathbf{A}_v$ denote the row of $\mathbf{A}$ corresponding to v. Likewise, for every edge e of G, let $\mathbf{A}_e^t$ denote the column of $\mathbf{A}$ that corresponds to edge e. Let us first prove that

$$\text{rank}(\mathbf{A}) \geq n - 1. \tag{A.14}$$

For this, we need the following auxiliary result.

A.1 *Let T be any nontrivial subtree of G. The set $S(T) := \{\mathbf{A}_e^t : e \in E(T)\}$ is linearly independent.*

Proof by induction on $|E(T)|$. As T is nontrivial, it has a vertex of degree one, say v. Let e be the edge of T incident with v. Consider any linear combination on the set $S(T)$ that produces the vector $\mathbf{0}$. The entry $\mathbf{A}_{ve}$ is the only nonnull entry in the row $\mathbf{A}_v$ of $\mathbf{A}$. Thus, the coefficient corresponding to e is zero. If e is the only edge of T then we are done. Otherwise, the restriction of the linear combination to $S(T - e)$ also produces $\mathbf{0}$; therefore all its coefficients are equal to zero, by induction. Indeed, the set $S(T)$ is linearly independent. □

The graph G is connected; therefore it has a spanning tree, T. The set $S(T)$ of the $n - 1$ columns corresponding to the edges of T is linearly independent. We then deduce the validity of (A.14).

Let us first consider the case in which G has a bipartition, (X, Y). Let us show that the n rows of $\mathbf{A}$ are not linearly independent. For this, consider the linear combination

$$\mathbf{y} := \sum_{v \in X} \mathbf{A}_v - \sum_{v \in Y} \mathbf{A}_v.$$

For each edge e of G, we have that $\mathbf{y}(e) = 0$. Thus, $\mathbf{y} = \mathbf{0}$. We conclude that $\text{rank}(\mathbf{A}) \leq n - 1$. From (A.14), we deduce that $\text{rank}(\mathbf{A}) = n - 1$, if G is bipartite.

Let us consider next the case in which G is not bipartite. For some odd positive integer r, let $v_1 v_2 \ldots v_r v_1$ be an odd cycle of G. For each vertex v of G, let $\alpha_v \in \mathbb{R}$; let

$$\mathbf{x} := \sum_{v \in V} \alpha_v \mathbf{A}_v. \tag{A.15}$$

Assume that $\mathbf{x} = \mathbf{0}$. For $i = 1, 2, \ldots, r-1$, we must have that $\mathbf{x}(v_i v_{i+1}) = 0$, hence $\alpha_{v_i} = -\alpha_{v_{i+1}}$. As the length r is odd, it follows that $\alpha_{v_i} = 0$, for $i = 1, 2, \ldots, r$. Thus, some coefficients α_v in (A.15) are equal to zero. From (A.14), we deduce that all coefficients are equal to zero. That is, the set of rows of $\mathbf{A}$ is linearly independent. We deduce that $\text{rank}(\mathbf{A}) = n$, if G is not bipartite.

Exercise 6.3.3

Consider any ear decomposition $(G_1, G_2, \ldots, G_r)$ of G, where $r \geq 1$, $G_1 = K_2$ and $G_r = G$. The proof is by induction on r. If $r = 1$ then the assertion holds immediately, with $M_1 := E(G_1)$. We may thus assume that $r > 1$. Let P_r denote the ear added to G_{r-1} to obtain G_r. By induction, G_{r-1} has $r-1$ perfect matchings N_i, $1 \leq i \leq r-1$, such that every integer vector in $\mathcal{Lin}(G_{r-1})$ is an integral linear combination of $\{\chi_i^N : 1 \leq i \leq r-1\}$. Each perfect matching N_i may be extended to a perfect matching M_i of G by adding the even numbered edges of P_r. In addition, the odd numbered edges of P_r are contained in some perfect matching M_r of G.

Let $\mathbf{x}$ be an integral vector in $\mathcal{Lin}(G)$. Let $d := \mathbf{x}(u)$ for some vertex u of G. Then, $\mathbf{x}(v) = d$, for each vertex v of G. In particular, as this equality applies to the internal vertices of P_r, it follows that for some integer d', (i) $\mathbf{x}(e) = d'$ if e is an odd numbered edge of P_r and (ii) $\mathbf{x}(e) = d - d'$, if e is an even numbered edge of P_r. Then, the restriction $\mathbf{y}$ of $\mathbf{x} - d'\chi^{M_r}$ to G_{r-1} is an integral vector of $\mathcal{Lin}(G_{r-1})$, because $\mathbf{y}$ is $(d-d')$-regular. By induction, $\mathbf{y}$ is an integral linear combination of the incidence vectors of N_i, $1 \leq i \leq r-1$. It follows that $\mathbf{x} - d'\chi^{M_r}$ is an integral linear combination of χ^{M_i}, for $1 \leq i \leq r-1$. Thus, $\mathbf{x}$ is an integral linear combination of χ^{M_i}, for $1 \leq i \leq r$.

A.7 Chapter 7

Exercise 7.2.3

(i) By hypothesis, the vector $\mathbf{x}$ is 1-regular and non-negative. Moreover, $\mathbf{x}(D) < 1$; thus

$$\sum_{K \in \mathcal{O}(G_1 - S)} \mathbf{x}(\partial(K)) \leq \mathbf{x}(\partial(S)) = |S| - 1 + \mathbf{x}(D) - 2\mathbf{x}(E(G_1[S])) < |S|.$$

As $o(G_1 - S) \geq |S|$, it follows that

$$\frac{\sum_{K \in \mathcal{O}(G_1 - S)} \mathbf{x}(\partial(K))}{o(G_1 - S)} < 1.$$

Hence $\mathbf{x}(\partial(K)) < 1$ for some $K \in \mathcal{O}(G_1 - S)$.

(ii) Let $F := \partial(K)$. Let us prove that $F \prec D$. For this, let M be a perfect matching of G. For each odd component J of $G_1 - S$, the cut $\partial(J)$ has at least one edge in M. Thus,

$$\begin{aligned} & |M \cap F| + o(G_1 - S) - 1 \\ \leq & \sum_{J \in \mathcal{O}(G_1 - S)} |M \cap \partial(J)| \leq |M \cap \partial(S)| \\ = & |M \cap D| + |S| - 1 - 2|M \cap E(G_1[S])| \end{aligned} .$$

As $|S| \leq o(G_1 - S)$, it follows that

$$|M \cap F| \leq |M \cap D|, \tag{A.16}$$

with equality only if $|S| = o(G_1 - S)$ and $M \cap E(G_1[S])$ is empty. If $|S| < o(G_1 - S)$ then $|M \cap F| < |M \cap D|$ and this inequality holds for each perfect matching M of G. In that case, $F \prec D$. We may thus assume that $|S| = o(G_1 - S)$. By hypothesis, G contains an edge, e, having both ends in S. From (A.16) we infer that $|M \cap F| \leq |M \cap D|$, with equality only if $e \notin M$. As G is matching covered, it has a perfect matching that contains edge e. We conclude that $F \prec D$.

Exercise 7.3.3

Let $\mathbf{w}$ be an extreme point of $\mathbb{P}$, let S be the support of $\mathbf{w}$, let $H := \mathbb{P}[S]$ and let K be a component of H. By Theorem 7.7, K is either K_2 or an odd cycle. If K contains only one edge, e, then $\mathbf{w}(e) = 1$, whereas if K is an odd cycle then $\mathbf{w}(e) = 1/2$, for each edge e of K. As $\mathbb{P}$ has an even number of vertices, an even number of odd cycles are components of H. The Petersen graph has girth five; therefore, either S is a perfect matching of $\mathbb{P}$ or H consists of two vertex-disjoint pentagons.

The set $\mathcal{M}$ of perfect matchings of $\mathbb{P}$ consists of precisely six perfect matchings. Each perfect matching M is a separating cut whose shores induce two edge-disjoint pentagons Q_1 and Q_2, where $E(Q_1) \cup E(Q_2) = E - M$. Thus, the set of extreme points of $\mathcal{Poly}(\mathbb{P})$ is a subset of

$$\mathcal{S} := \{\chi^M, 1/2 \cdot \chi^{E-M} : M \in \mathcal{M}\}.$$

Let $M \in \mathcal{M}$. The vector χ^M is the only vector in $\mathcal{S}$ whose support does not contain any edge in $E - M$ and the vector χ^{E-M} is the only vector in $\mathcal{S}$ whose support does not contain any edge in M. Thus, both χ^M and $1/2 \cdot \chi^{E-M}$ are in $\mathcal{S}$. This conclusion holds for each $M \in \mathcal{M}$. Thus, $\mathcal{S}$ is the set of extreme points of $\mathcal{Poly}(\mathbb{P})$.

Exercise 7.3.5

Let $\{C_1, C_2\}$ be any pair of vertex-disjoint disjoint odd cycles of G. We shall show that $G - (V(C_1) \cup V(C_2))$ is not matchable and then use the Reed-Wakabayashi Theorem 7.8 to deduce that G is solid.

Let $A := \{0, 1, 3, 5, h_2, v_2\}$ and let $B := \{2, 4, 6, h_1, h_3, v_1\}$. Note that (A, B) is an equipartition of the set $V(G)$.

The set $E_A := \{01, h_2v_2\}$ is the set of edges of G having both ends in A and the set $E_B := \{h_1v_1\}$ is the set of edges of G having both ends in B. Suppose that one of the cycles C_1 and C_2, say C_1, contains edge h_1v_1. Then, it also contains a vertex in $\{0, 1\}$ and a vertex in $\{h_2, v_2\}$, which implies that C_1 contains an end of each of the three edges of $E_A \cup E_B$. This implies that $G - V(C_1)$ is bipartite, a contradiction. Thus, no cycle in $\{C_1, C_2\}$ contains edge h_1v_1. That edge is the only edge of E_B. Moreover, E_A is a doubleton; thus each cycle in $\{C_1, C_2\}$ contains one of the edges of E_A. As $\{A, B\}$ is an equipartition of $V(G)$, the graph $G - (V(C_1) \cup V(C_2))$ is not matchable. It now follows from the Reed-Wakabayashi Theorem 7.8 that G is solid.

A.8 Chapter 8

Exercise 8.1.7(iii)

The vectors χ^e and χ^f are incidence vectors of the singletons $\{e\}$ and $\{f\}$. Thus the vector $\mathbf{x} = \chi^e - \chi^f$ has precisely two nonzero entries: $\mathbf{x}(e) = 1, \mathbf{x}(f) = -1$, and $\mathbf{x}(g) = 0$ for each edge $g \notin \{e, f\}$. Since every perfect matching of the brick G which contains the edge e also contains the edge f, it follows that $\mathbf{x}$ is a matching orthogonal vector of G. By part (ii), it follows that there is a real-valued function p on the vertex set of G (a labelling of the vertices of G) such that (i) $\sum_{v \in V} p(v) = 0$ and, (ii) $\mathbf{x}(st) = p(s) + p(t)$, for any edge st of G. Thus, in particular, if st is any edge different from e and f, then $p(s) = -p(t)$.

If possible, let $w_1w_2 \ldots w_{2k+1}w_1$ be an odd cycle in the graph $G-e-f$. Then, the above observation implies that $p(w_1) = -p(w_1)$. This is possible only if $p(w_i) = 0$ for every vertex on this odd cycle. Also since $G - e - f$ is matching covered, it is connected, and any two of its vertices are connected by a path. Thus, if $G - e - f$ has an odd cycle, it would follows that $p(v) = 0$ for each $v \in V$. But this impossible because, if $e = u_1u_2$, then $p(u_1)+p(u_2)$ has to be equal to one. We conclude $G-e-f$ is bipartite.

Exercise 8.1.10

Let (A, B) be the bipartition of G and adjust notation so that $a \in A$ and $b \in B$. Suppose that $G - e$ is not a brace; then a and b belong to the minority parts of different shores of any nontrivial tight cut of $G - e$.

(i) By hypothesis, the vertex a is in $X \cap Y$. As the cuts $C - e$ and $D - e$ are both nontrivial and tight in $G - e$, it follows that $a \in X_- \cap Y_-$ and b lies in $(\overline{X})_- \cap (\overline{Y})_-$. As $a \in A$ and $b \in B$, we deduce that X_+ and Y_+ are both subsets of B. Now consider the quadrant $X \cap \overline{Y}$. If it is empty then X is a subset of Y; in that case, $I = C$ and the assertion holds immediately. We may thus assume that $X \cap \overline{Y}$ is nonempty.
As $C - e$ is tight in $G - e$, the subgraph of $G - e$ induced by X is connected, therefore there is an edge in $G - e$ that joins a vertex $a_1 \in X \cap \overline{Y}$ to a vertex $b_1 \in X \cap Y$. See Figure A.9.

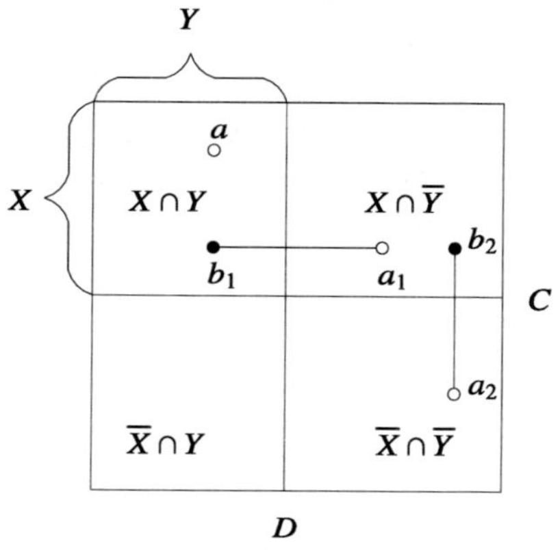

Fig. A.9 The four quadrants.

The vertex b_1 is in Y_+, in turn a subset of B; hence $a_1 \in A$. Likewise, the subgraph of $G - e$ induced by $\overline{Y}$ is connected; hence $G - e$ has an edge that joins a vertex b_2 in $X \cap \overline{Y}$ to a vertex a_2 in $\overline{X} \cap \overline{Y}$. The vertex b_2 is in X_+, in turn a subset of B. Thus, $\partial(X \cap \overline{Y}) - e$ contains an edge incident with a vertex in $A \cap X \cap \overline{Y}$ and an edge incident with a vertex in $B \cap X \cap \overline{Y}$. Therefore, $\partial(X \cap \overline{Y}) - e$ is not tight in $G - e$. By Theorem 4.12, one of the cuts $I - e$ and $\partial(X \cap \overline{Y}) - e$ is tight in $G - e$. We conclude that $I - e$ is tight in $G - e$.
The vertex a is in $X \cap Y \cap A$ and the vertex b_1 is in $X \cap Y \cap B$. Thus, $X \cap Y$ is nontrivial. Likewise, the vertex a_1 is in $X \cap \overline{Y} \cap A$ and the vertex b_2 is in $X \cap \overline{Y} \cap B$. Thus, both shores of I are nontrivial. Indeed, $I - e$ is a nontrivial tight cut of $G - e$. A similar reasoning may be used to prove that $U - e$ is a nontrivial tight cut of $G - e$.

(ii) Let $\mathcal{C}$ be the family of tight cuts corresponding to a tight cut decomposition of $G - e$. For each nontrivial cut $C := \partial(X)$ of G such that $C - e$ is tight in $G - e$, one of a and b is in X_-, the other is in $(\overline{X})_-$. Let $\mathcal{X}$ denote the collection of the shores X of cuts in $\mathcal{C}$ such that $a \in X$ (and $b \in \overline{X}$). For each $X \in \mathcal{X}$, $|X_+| = |X_-| + 1$ and ab is the only edge of G that joins a vertex in X_- to a vertex in $(\overline{X})_-$.

A.1.1 *For distinct X and Y in $\mathcal{X}$, one of X and Y is a (proper) subset of the other.*

Proof For distinct X and Y in $\mathcal{X}$, we have that $a \in X \cap Y$ and $b \in \overline{X} \cap \overline{Y}$. As $\partial(X)$ and $\partial(Y)$ do not cross, it follows that either $X \cap \overline{Y}$ is empty or $\overline{X} \cap Y$ is empty. Consequently, one of X and Y is a (proper) subset of the other. □

Let $k := |\mathcal{X}|$. Define now inductively the sequence $X_1, X_2, \ldots, X_k$ of the sets in $\mathcal{X}$ as follows: having defined $X_1, X_2, \ldots, X_{i-1}$, $1 \le i \le k$, let X_i be a minimal set in $\mathcal{X} - \{X_1, X_2, \ldots, X_{i-1}\}$. We shall now prove that $X_1 \subset X_2 \subset \cdots \subset X_i$, for $i = 1, 2, \ldots, k$, by induction on i. The basis of the inductive proof is for $i = 1$, and the assertion holds trivially. Let us now consider the inductive step, suppose that $1 < i$. By (A.1.1), one of X_{i-1} and X_i is a proper subset of the other. By the minimality of X_{i-1} in the collection $\{X_{i-1}, X_i \ldots, X_k\}$, we conclude that $X_{i-1} \subset X_i$. By induction, $X_1 \subset X_2 \subset \cdots \subset X_{i-1}$. Thus, $X_1 \subset X_2 \subset \ldots X_{i-1} \subset X_i$. We conclude that

$$X_1 \subset X_2 \subset \cdots \subset X_k. \tag{A.17}$$

See Figure A.10. Let us now prove (8.1). From (A.17), we infer that $\partial(X_1)$ is the

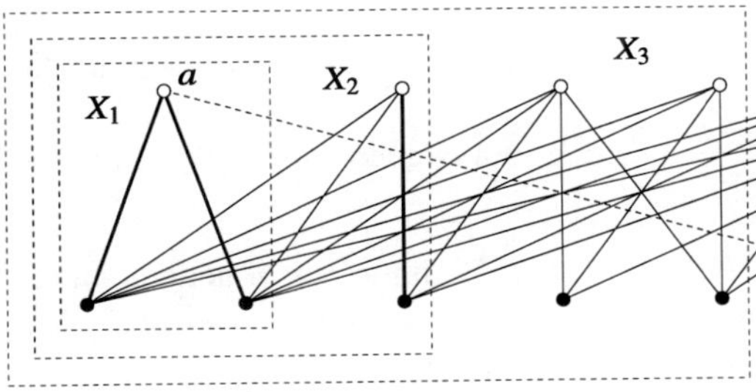

Fig. A.10 An illustration for Exercise 8.1.10, with $k = 3$

only cut of C in $H_1 := (G-e)/\overline{X_1}$, and it is trivial in H_1, hence H_1 is a brace, by definition of C. Likewise, for each i such that $2 \le i < k$, we infer from (A.17) that $\partial(X_i)$ and $\partial(X_{i+1})$ are the only two cuts of C in $H_i := (G-e)/X_i/\overline{X_{i+1}}$, and they are both trivial in H_i, hence H_i is a brace, by definition of C. Finally, from (A.17) we infer that $\partial(X_k)$ is the only cut of C in $H_k := (G-e)/X_k$, and it is trivial in H_k, hence H_k is a brace, by definition of C.

(iii) The graph H_1, which is $(G-e)/\overline{X_1}$, is a brace. Thus, it is free of nontrivial tight cuts. Let Y be a shore of a nontrivial tight cut of $G-e$ such that $a \in Y_-$. The cut $\partial(X_1 \cap Y) - e$ is nontrivial and tight in $G-e$, therefore it is a trivial cut of H_1. It follows that $X_1 \subseteq Y$. This conclusion holds for each shore Y of a nontrivial tight cut of $G-e$ that contains the vertex a. Thus, X_1 is unique. A similar reasoning may be used to prove the uniqueness of X_k. Consequently, if $G-e$ has three or less braces then $\mathcal{X}$ is unique.

Exercise 8.3.8

The proof is by induction on $|C|$. Let us first consider the basis case, in which C is empty. The maximality of C implies that G is free of even 2-cuts. Let v be a vertex of G having degree $\Delta(G)$ and let $\partial(v) := \{e_i : 1 \le i \le \Delta(G)\}$. For each e_i, let R_i denote a minimal dependence class induced by e_i. As G is free of even 2-cuts, it follows that $G - R_i$ is matching covered. That is, R_i is a removable class of G. Clearly, R_i and R_j are distinct, for $1 \le i < j \le \Delta(G)$. The assertion holds in this case.

Let us now consider the induction step. Let $\{e, f\} := \partial(X)$ be an even 2-cut in C. Let u_1 and u_2 denote the two ends of e and let v_1 and v_2 denote the two ends of f. Adjust notation so that u_1 and v_1 are both in X. Let G_1 denote the graph obtained from $G[X]$ by the addition of edge u_1v_1. Likewise, let G_2 denote the graph obtained from $G[\overline{X}]$ by the addition of edge u_2v_2. For $i = 1, 2$, note that G_i is isomorphic to a C_i-contraction of G, where C_i is one of the two barrier cuts associated with the barrier $\{u_1, v_2\}$ of G. Thus, G_1 and G_2 are matching covered. Moreover, for $i = 1, 2$, we have the following properties:

1. $\delta(G_i) \ge 3$.
2. The restriction C_i of $C - \{e, f\}$ to $E(G_i)$ is a maximal laminar collection of even 2-cuts of G_i.
3. Every removable class of G_i that does not contain edge u_iv_i is a removable class of G.

Moreover, $\Delta(G) = \max\{\Delta(G_i) : i = 1, 2\}$. Without loss of generality, assume that $\Delta(G) = \Delta(G_1)$. Let r_i denote the number of removable classes of G_i. Let r be the number of removable classes of G. By induction,

$$\begin{aligned} r_1 &\ge \Delta(G_1) + |C_1| \\ r_2 &\ge \Delta(G_2) + |C_2| \\ r &\ge (r_1 - 1) + (r_2 - 1). \end{aligned}$$

Addition of the three inequalities and simplification yields

$$r \ge \Delta(G) + \Delta(G_2) + \big[|C| - 1\big] - 2 \ge \Delta(G) + |C|.$$

Exercise 8.4.1

(i) Suppose that v is a vertex of degree three or more in $G - e$. Suppose, to the contrary, that two edges, e_1 and e_2, both incident with v, are not removable in $G - e$. Apply Theorem 8.4, with $r = 2$. Edge e is not incident with vertices in both B_1 and B_2. Adjust notation so that e is not incident with any vertex in B_1. As G is a brace of order six or more, vertex b_1 is adjacent to three or more vertices in G. As e is not incident with any vertex in B_1, in particular b_1 is adjacent to two or more vertices of A_1. Thus, $|A_1| \ge 2$. As e is not incident with any vertex of B_1, we conclude that $\partial(A_1 \cup (B_1 - b_1))$ is a nontrivial tight cut of G, a contradiction to the hypothesis that G is a brace.

(ii) Let H be a C-contraction of $G - e$, say, $H := (G - e)/(X \to x)$. Let (A, B) denote the bipartition of H, where the contraction vertex x of H is in A. As C is nontrivial, then the edge e joins a vertex in X_- to a vertex in $(\overline{X})_-$. Thus, edge e is not incident with a vertex in B. Let $f \in C$. Then, x is an end of f in H. Let b denote the other end of f in H. Assume, to the contrary, that f is not removable in H. By Lemma 8.2, A has a nonempty proper subset A_1 and B a nonempty proper subset B_1 such that $|A_1| = |B_1|$ and $f = bx$ is the only edge of H that joins a vertex of B_1 to a vertex not in A_1. The edge e is not incident with any vertex in B. In particular, b is not an end of e. Thus, b is adjacent in $G - e$ to three or more vertices. Consequently, b is adjacent to two or more vertices of A_1. Thus, $|A_1| \geq 2$. As e is not incident with any vertex in B, the set $A_1 \cup (B_1 - b)$ is the shore of a nontrivial tight cut of G, a contradiction.
We conclude that f is removable in H, a C-contraction of $G - e$. This conclusion holds for both C-contractions of $G - e$. Thus, f is removable in $G - e$.

A.9 Chapter 9

Exercise 9.1.6

(i) Assume that two cuts in C, say, $C := \partial(X)$ and $D := \partial(Y)$, cross. We now consider the four quadrants $X \cap Y$, $\overline{X} \cap Y$, $X \cap \overline{Y}$ and $\overline{X} \cap \overline{Y}$, which are, respectively, shores of the cuts $C_{XY}, C_{\overline{X}Y}, C_{X\overline{Y}}, C_{\overline{XY}}$. See Figure A.11.

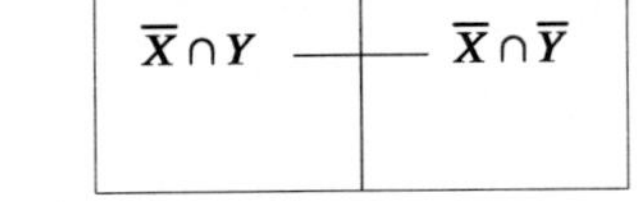

Fig. A.11 Two crossing cuts in the collection C

Let F be the set of edges that join a vertex in $X \cap Y$ to a vertex in $\overline{X} \cap \overline{Y}$ or join a vertex in $\overline{X} \cap Y$ to a vertex in $X \cap \overline{Y}$. By definition, $R_k = C \cap D$, hence $R_k = F$. Thus,

$$|C_{XY}| + |C_{\overline{X}Y}| + |C_{X\overline{Y}}| + |C_{\overline{XY}}| = 2|C| + 2|D| - 2|F| = 12.$$

As G, a brick, is 3-connected, we deduce that each of the four cuts is a 3-cut. No edge of R_k has both ends in any of the four quadrants; hence the subgraph of G induced by any one of the quadrants is bipartite. As seen in Exercise 5.2.7, each quadrant is a singleton; hence G has order four and is thus K_4. We conclude that if C is not laminar then G is K_4.

(ii) Suppose that G is not K_4. By the previous item, C is laminar. Let $C := \partial(X)$ and $D := \partial(Y)$ be two cuts in $\mathcal{C}$, where $X, Y \in \mathcal{X}$. By definition, $v_k \in X \cap Y$. As e_k, an edge of R_k, is in $C \cap D$, the vertex w_k is in $\overline{X} \cap \overline{Y}$. As C and D do not cross, we deduce that either $X \cap \overline{Y} = \emptyset$ or $\overline{X} \cap Y = \emptyset$. Thus, one of X and Y is a subset of the other. Moreover, C and D are distinct; hence the inclusion is proper. This conclusion holds for any two distinct cuts C and D of $\mathcal{C}$. Thus, the shores in $\mathcal{X}$ of the cuts in $\mathcal{C}$ form a nested collection.

(iii) For $i = 1, 2, \ldots, k-1$, let $C_i := \partial(X_i)$. Define the graphs H_i, $i = 1, 2, \ldots, k$ as follows:

a. let $H_1 := G[X_1]$,
b. for $i = 2, 3, \ldots, k-1$, let $H_i := G[X_i - X_{i-1}]$, and
c. let $H_k := G[\overline{X_{k-1}}]$. (See Figure 9.3.)

For $i = 1, 2, \ldots, k$, the cut $\partial(H_i)$ is equal to $R_{i-1} \cup R_i$; thus, it is a 4-cut. If H_i were not connected then $\partial(H_i)$ would not be a bond; hence G would not be 3-edge-connected. This is not possible. Thus, the k graphs H_i are the components of $G - R$.

Each H_i is a subgraph of $G - R_k$, in turn a bipartite (matching covered) graph. Thus, each H_i is bipartite. To prove that H_i is matchable, note that as R_{i-1} is a removable doubleton of G, the graph $G - R_{i-1}$ is a bipartite matching covered graph. Let v be the end in H_i of an edge of R_i. The vertex v has degree two or more in $G - R_{i-1}$; hence v is incident with an edge, e, which is not in R_i. Let M_e be a perfect matching of $G - R_{i-1}$ that contains the edge e. Then, $M_e \cap E(H_i)$ is a perfect matching of H_i. We conclude that H_i is matchable.

Exercise 9.2.3

The brick shown in Figure A.12 has two disjoint triangles. It is clearly not a prism or a Möbius ladder or a staircase. It has precisely two removable doubletons.

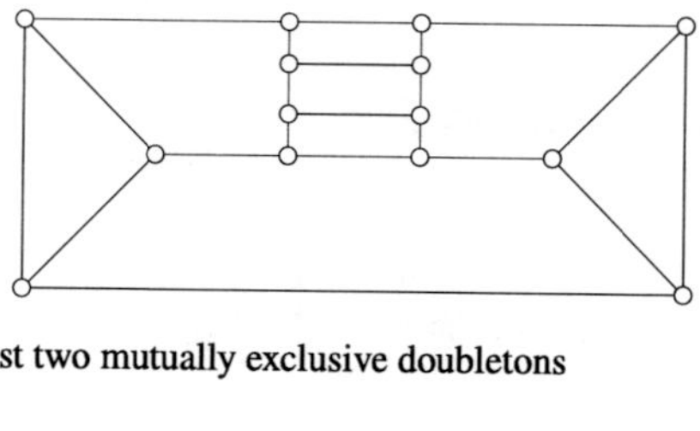

Fig. A.12 A brick with just two mutually exclusive doubletons

Exercise 9.2.4

(i) By Lemma 9.12, the doubletons R_1 and R_2 induce an aitch configuration. Adjust notation so that e_1 and e_2 share a common end, u_{12}, of degree three. Then, f_1 and f_2 also share a common end, v_{12}, also of degree three, and adjacent to u_{12}. Let h be the edge $u_{12}v_{12}$. See Figure 9.5.

(ii) Clearly, C contains also the edges e_2, f_1 and f_2, Assume, to the contrary, that C is not hamiltonian. The symmetric difference $M_1 \triangle M_2$ induces one or more edge-disjoint (M_1, M_2)-alternating cycles. It cannot induce two or more cycles; otherwise one of the cycles, say C', distinct from C, does not contain any of the four edges of $R_1 \cup R_2$; hence the set $M_1' := M_1 \triangle E(C')$ is a perfect matching of G that also contains the two edges of R_1. This is in contradiction to the uniqueness of M_1. Thus, C is the only cycle induced by $M_1 \triangle M_2$. If C is not hamiltonian then some vertex of G not in C is incident with an edge of $M_1 \cap M_2$. This is a contradiction to the fact that M_1 and M_2 are disjoint. Indeed, C is hamiltonian. The edge h is a chord of C.

(iii) Assume, to the contrary, that C has a chord h', distinct from h, that joins two vertices in P. The vertices u_{12} and v_{12} both have degree three and are joined by the chord h. Thus, h' joins internal vertices of P. Let w_i and w_j denote the ends of h', where $i < j$. Let P' denote the subpath of P from w_i to w_j. Then, $E(P')$ and $R_1 \cup R_2$ are disjoint. Adjust notation so that the first edge of P' is in M_1. As $G - R_1$ is bipartite, it follows that w_i and w_j lie in distinct parts of the bipartition of $G - R_1$. Thus, the last edge of P' also lies in M_1; hence the cycle C' obtained by the addition of the chord h' to the path P' is M_1-alternating. We conclude that $M_1 \triangle E(C')$ is a perfect matching of G that includes both edges of R_1, in contradiction to the uniqueness of M_1. We deduce that no chord of C, except h, has both ends in $V(P)$. Likewise, h is the only chord of C that has both ends in $V(Q)$.

(iv)

A.1.2 *Let w_i and x_j be internal vertices of P and Q, respectively and suppose that w_i and x_j are adjacent. The edge e_i of P that precedes w_i is in M_1 if and only if the edge f_j of Q that precedes x_j is in M_2.*

Proof We first recall that R_1 and R_2 are distinct equivalence classes of the dependence relation on the edges of G. Consequently, for $k \in \{1, 2\}$, an M_k-alternating cycle that contains an edge of R_k contains in fact both edges of R_k.
Let P' be the subpath of P from u_{12} to w_i and let Q' be the subpath of Q from u_{12} to x_j. Let C' be the cycle $P'w_ix_jR(Q')$, where $R(Q')$ is the reverse of Q'. If both edges e_i and f_j are in M_1 then C' is an M_1-alternating cycle that contains only one edge of R_1. Likewise, if e_i and f_j are both in M_2 then C' is an M_2-alternating cycle that contains only one edge of R_2. In both alternatives we derive a contradiction. □

We prove this part by induction on i. Adopt, as the inductive hypothesis, that for every j such that $1 \leq j < i$, the vertices w_j and x_j have degree three and are adjacent.

Assume, to the contrary, that either (i) w_i and x_i are not adjacent or (ii) w_i has degree four or more. By induction, for each r such that $1 \le r < i$, the vertices w_r and x_r have degree three and are adjacent. If w_i and x_i are not adjacent then w_i is adjacent to some vertex x_j, $j > i$. Alternatively, if w_i has degree four or more then it is also adjacent to some vertex x_j, $j > i$. In both alternatives, w_i is adjacent to some x_j, $j > i$. The subgraph $G - u_{12} - v_{12}$ is connected and bipartite. Thus, $j \ge i + 2$. Moreover, the vertex x_{j-1} is not adjacent to w_i. See Figure A.13.

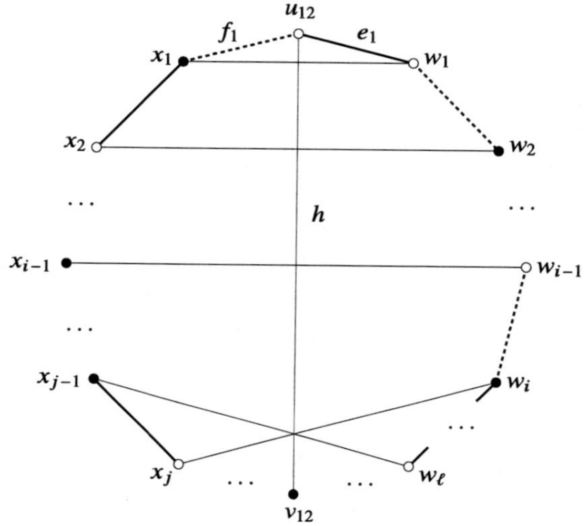

Fig. A.13 Illustration for item (iv)

In fact, x_{j-1} is not adjacent to any vertex of P that precedes w_i. Thus, x_{j-1} is adjacent to some vertex w_ℓ of P, where $\ell > i$, and i and ℓ have distinct parities. Let P' the subpath of P from w_i to w_ℓ and let $k \in \{1, 2\}$ be such that the first edge of P' is in M_k. Then, the last edge of P' also lies in M_k, because i and ℓ have distinct parities. The edge that precedes w_ℓ in P is in M_k. Thus, by statement (A.1.2), the edge $x_{j-1}x_j$ is also in M_k. We deduce that the cycle C'', obtained by adding the path $w_\ell x_{j-1} x_j w_i$ to P', is an M_k-alternating cycle that does not contain any edge in R_k. Thus, $M_k \triangle E(C'')$ is a perfect matching of G that contains both edges in R_k and is distinct from M_k. This is a contradiction to the uniqueness of M_k. The assertion holds, by induction.

5. We conclude that G is cubic and the paths P and Q have the same length. Moreover, each internal vertex w_i of P is joined to the corresponding internal vertex x_i of Q by a chord of C. If G has only four vertices then it is K_4. If G has order six or more, then it is a staircase.

Exercise 9.2.5

(i) Let v be a vertex of G. As G is a brick, then v has degree $d \geq 3$. Let $\partial(v) := \{e_i : 1 \leq i \leq d\}$. For $i = 1, 2, \ldots, d$, let R_i be a minimal equivalence class of the dependence relation induced by e_i. Those d classes are mutually exclusive and removable. If two or them are singletons then G has two removable edges. We conclude that at most one of those classes is a singleton. Then, at least two or these classes are removable doubletons, say, R_1 and R_2. Moreover, R_1 and R_2 are mutually exclusive.

(ii) Let e be a removable edge of H_i which is not in M_i. The graph $H_i - e$ is then matching covered and M_i is a perfect matching of $G - e$ that contains both edges of R_i. Thus, the edges of R_i are matchable in $G - e$; hence $G - e$ is matching covered. In other words, e is removable in G. This conclusion holds for each edge e which is removable in H_i but is not in M_i.

(iii) Suppose that $M_i - R_i$ contains an edge $g := vw$ which is not removable in H_i. The edge g is in $M_i - R_i$. Thus, v is not incident with any edge in R_i. It follows that v has degree three or more in H_i. By Lemma 8.6, at most two edges of $\partial(v)$ are not removable in H_i. One of these edges is g. As v has degree three or more in H_i, we deduce that $\partial(v) - g$ contains an edge, e, which is removable in H_i. Likewise, $\partial(w) - g$ contains an edge, f, which is removable in H_i. As G is simple, the edges e and f are distinct. In sum, e and f are distinct removable edges of H_i neither of which is in M_i. From part (ii) we infer that e and f are distinct removable edges of G, a contradiction.

(iv) Suppose that M_i is not unique and let N_i be a perfect matching of G distinct from M_i that contains the edges of R_i. Let C be an (M_i, N_i)-alternating cycle of G. As G is simple, the cycle Q has length four or more. Thus, $C \cap N_i$ contains two edges, e and f. Those two edges are in $N_i - M_i$. From parts (iii) and (ii), we infer that e and f are both removable in G.

(v) Suppose that $M_1 \cap M_2$ contains an edge, e. As $e \in M_1$, the edge e is removable in H_1. The graph $H_1 - e$ contains both edges of R_2. Let N_2 be a perfect matching of $H_1 - e$ that contains edge e_2. Clearly, N_2 is a perfect matching of G; hence it contains both edges of R_2. In sum, M_2 and N_2 are perfect matchings of G that contain both edges of R_2. But the edge e is in $M_2 - N_2$; hence M_2 and N_2 are distinct. This is a contradiction to the uniqueness of M_2.

(vi) By Theorem 9.15, the brick G is either K_4 or a staircase.

(vii) Let n denote the order of G. If $n = 4$ then G is K_4. We may thus assume that $n \geq 6$. Then, G is a staircase. The graph G has a hamiltonian cycle, C. As G is cubic, C has precisely $n/2$ chords. One of the chords joins the common vertices of both removable doubletons. Two other chords are edges of the two triangles of G. The remaining $n/2 - 3$ chords are removable. Thus, $n/2 - 3 \leq 1$; hence $n \leq 8$. We conclude that $n \in \{6, 8\}$. If $n = 6$ then $G = \overline{C_6}$, whereas if $n = 8$ then $G = \mathbb{H}_8$.

A.10 Chapter 10

Exercise 10.1.6

Proof (of Corollary 10.7) Let v be a vertex of G. Assume that for $i = 1, 2, 3$, edge vv_i is not removable in G. Then, for $i = 1, 2, 3$, V has an equipartition (B_i, I_i) such that $B_1 \cap B_2 = \{v_3\}$, $B_1 = I_3 - v + v_2$ and $B_2 = I_3 - v + v_1$. Then, $I_3 - v = B_1 \cap B_2 = \{v_3\}$, whence $I_3 = \{v, v_3\}$. In that case, as I_3 contains half the vertices of G, it follows that G contains precisely four vertices. □

Proof (of Corollary 10.8) Let v be a vertex of G. Suppose that neither vv_1 nor vv_2 is in a removable class of G. Then, in particular, neither vv_1 nor vv_2 is removable in G. By Theorem 10.5, v has precisely three neighbours, v_1, v_2 and v_3. For $i = 1, 2$, let B_i, I_i and H_i be as in the statement of Theorem 10.5. For $i = 1, 2$, let b_i denote the number of edges of G that have both ends in B_i. As $|B_i| = |I_i|$ and since vv_i has both ends in I_i, it follows that $b_i > 0$; otherwise vv_i would not be matchable in H_i. Moreover, all edges with both ends in B_1, and all edges with both ends in B_2, are incident with v_3. By hypothesis, the degree of v_3 is three or four. As v_3 is adjacent to v, which is a vertex of $I_1 \cap I_2$, it follows that $b_1 + b_2 \leq 3$; it follows that at least one of b_1 and b_2 is equal to one. Adjust notation so that $b_1 = 1$. Let f_1 denote the only edge of G having both ends in B_1. Then, $H_1 = G - vv_1 - f_1$. Neither vv_1 is matchable in $G - f_1$ nor f_1 is matchable in $G - vv_1$. We deduce that $\{vv_1, f_1\}$ is a removable doubleton of G. This contradicts the assumption that neither vv_1 nor vv_2 lies in a removable class of G. □

Exercise 10.2.2

Let G be a solid matching covered graph and let R be a set of edges of G such that $G - R$ is matching covered. We prove that $G - R$ is solid by induction on $|E|$. If R is empty then there is nothing to be proved. We may thus assume that R is nonempty.

If G is bipartite then $G - R$ is also bipartite, hence solid. Suppose that G is a brick. As $G - R$ is matching covered, it follows that R includes a removable class, S, of G. If S is a singleton then $G - S$ is solid, by Theorem 10.14. Alternatively, if S is a doubleton then $G - S$ is bipartite, hence solid. In both alternatives, $G - S$ is solid. By induction, $(G - S) - (R - S)$ is solid.

We may thus assume that G has a nontrivial tight cut, $C := \partial(X)$. Let G_1 and G_2 be the two C-contractions of G. The cut $C - R$ is tight in $G - R$. By Theorem 7.1, each brick of G is solid; hence each brick of G_i is solid, for $i = 1, 2$. Consequently, G_1 and G_2 are both solid. The graphs $G_i - R$ are matching covered. By induction, $G_1 - R$ and $G_2 - R$ are both solid. Thus, all the bricks of $G_1 - R$ and all the bricks of $G_2 - R$ are solid. Consequently, all the bricks of $G - R$ are solid; hence $G - R$ is solid.

Exercise 10.2.3

Let e be a removable edge of a solid matching covered graph G. We prove that $b(G - e) = b(G)$ by induction on $|E|$. If G is bipartite then $G - e$ is also bipartite and $b(G-e) = 0 = b(G)$, hence the assertion. If G is a brick then, by Theorem 10.16, $b(G - e) = 1 = b(G)$.

We may thus assume that G has a nontrivial tight cut, $C := \partial(X)$. Let G_1 and G_2 be the two C-contractions of G. By Theorem 7.1, each brick of G is solid; hence each brick of G_i is solid, for $i = 1, 2$. Consequently, G_1 and G_2 are both solid. By induction, $b(G_i - e) = b(G_i)$, for $i = 1, 2$. Clearly, $C - e$ is tight in $G - e$. Thus,

$$b(G - e) = b(G_1 - e) + b(G_2 - e) = b(G_1) + b(G_2) = b(G).$$

In all cases considered, e is b-invariant in G. This conclusion holds for each removable edge e of G.

By Theorem 7.1, each brick of a solid matching covered graph is solid. The Petersen graph is not solid; thus, no brick of a solid matching covered graph is a Petersen brick. Hence, $p(G) = 0$. Moreover, by Theorem 10.15, $G - e$ is also solid. Thus, $p(G - e) = 0$. We conclude that $(b + p)(G - e) = b(G - e) = b(G) = (b + p)(G)$.

Exercise 10.2.4

Let $C := \partial(X) \in \mathcal{C}$ and assume, to the contrary, that C is not separating in G. Then, a C-contraction of G, say, $G_1 := G/(\overline{X} \to \overline{x})$, is not matching covered. By definition of $\mathcal{C}$, the cut $C - R$ is separating in $G - R$. Hence, $G_1 - R$ is matching covered. Thus, only edges of R are not matchable in G_1.

Let e be an edge of R which is not matchable in G_1 and let B denote a maximal barrier of G_1 that contains both ends of e in G_1. As e is matchable in G, the set B is not a barrier of G. Thus, the contraction vertex $\overline{x}$ of G_1 is in B. Let E_B denote the set of edges of G_1 that contain both ends in B. Adopt the notation used in the proof of Lemma 10.13.

As $G_1 - R$ is matching covered, $E_B \subseteq R$. Let M be a perfect matching of G. As in equation 10.2, we have that

$$\sum_{K \in \mathcal{O}} |M \cap C_K| = |M \cap C| + (|\mathcal{O}| - 1) - 2|M \cap E_B|. \tag{A.18}$$

Let $J \in \mathcal{O}$. Then,

$$|M \cap C_J| + \sum_{K \in \mathcal{O}-J} |M \cap C_K| = \sum_{K \in \mathcal{O}} |M \cap C_K|, \tag{A.19}$$

$$|\mathcal{O}| - 1 \le \sum_{K \in \mathcal{O}-J} |M \cap C_K|. \tag{A.20}$$

Addition of equations (A.18)–(A.20) and simplification yields

$$|M \cap C_J| \leq |M \cap C| - 2|M \cap E_B|.$$

We conclude that $|M \cap C_J| \leq |M \cap C|$, with equality only if M and E_B are disjoint. We deduce that $C_J \prec C$.

Let us now prove that $C_J - R$ is separating in $G - R$. Let f be any edge of $G - R$. As $C - R$ is separating in $G - R$, there exists a perfect matching M_f of $G - R$ that contains edge f and just one edge in $C - R$. As $C_J \prec C$, we deduce that M_f contains just one edge in C_J. In sum, every edge of $G - R$ is in a perfect matching of $G - R$ that contains just one edge in $C_J - R$. We conclude that $C_J - R$ is separating in $G - R$.

So far we have proved that for every component J in O, $C_J \prec C$ and $C_J - R$ is separating in $G - R$. We now get a contradiction to the minimality of C, by showing that $\lambda_{G-R}(C_J - R) \leq \lambda_{G-R}(C - R)$ for some component J in O. If $C - R$ is tight in $G - R$ then $C_J - R$ is tight in $G - R$; in that case, $\lambda_{G-R}(C_J - R) = \infty = \lambda_{G-R}(C - R)$. We may thus assume that $C - R$ is not tight in $G - R$. Let M be a perfect matching of $G - R$ that contains precisely $\lambda_{G-R}(C - R) > 1$ edges in $C - R$. As $E_B \subseteq R$ and since M is a perfect matching of $G - R$, from (A.18) we infer that O contains a component J that satisfies the inequalities $1 < |M \cap C_J| \leq |M \cap C| = \lambda_{G-R}(C - R)$. We deduce that $\lambda_{G-R}(C_J - R) \leq \lambda_{G-R}(C - R)$.

We proved that O has a component J such that $C_J \prec C$, $C_J - R$ is separating in $G - R$ and $\lambda_{G-R}(C_J - R) \leq \lambda_{G-R}(C - R)$. This is a contradiction to the minimality of C in the relation $\prec$.

Exercise 10.2.5

(i) Suppose that R is a singleton, $\{e\}$. By hypothesis, $G - R - S$ is matching covered; hence the edge e is removable in $G - S$. The graph G is solid and e is removable in G; hence, by Theorem 10.17. $b(G - e) = b(G)$. By Theorems 9.8, and 9.9,

$$b(G - S) \leq b(G - e - S) = b(G - e) - 1 = b(G) - 1.$$

Thus, equality holds. By Corollary 10.18, the set S is a removable doubleton of G.

(ii) Consider next the case in which R is a doubleton, $\{e_1, e_2\}$. The graph $G - R - S$ is matching covered by hypothesis. Thus, either R is a removable doubleton of $G - S$ or one of the edges, say e_1, is removable in $G - S$ and the other edge of R is removable in $G - S - e_1$. If R is a removable doubleton in $G - S$ then $b(G - S) = b(G - R - S) + 1$, by Theorem 9.9. Alternatively, if e_1 is removable in $G - S$ and e_2 is removable in $G - e_1 - S$ then $b(G - S) < b(G - R - S) + 1$, by Theorem 9.8. In both alternatives, $b(G - S) \leq b(G - R - S) + 1$. By Theorems 9.8 and 9.9,

$$b(G - S) \leq b(G - R - S) + 1 = b(G - R) = b(G) - 1.$$

Thus, $b(G - S) = b(G) - 1$ and S is a removable doubleton of G.

Exercise 10.2.6

(a) If $G-R$ is solid then $\lambda(G) \leq \infty = \lambda(G-R)$ and the inequality holds immediately. We may thus assume that $G-R$ is not solid. Let D be a cut of G such that $D-R$ is separating in $G-R$ and $\lambda_{G-R}(D-R) = \lambda(G-R)$.
Using the result in Exercise 10.2.4, we infer that G has a separating cut C such that $C-R$ is separating in $G-R$ and $\lambda_{G-R}(C-R) \leq \lambda_{G-R}(D-R)$. By definition of D, $\lambda_{G-R}(D-R) = \lambda(G-R)$; hence $\lambda_{G-R}(C-R) = \lambda(G-R)$. Let M be a perfect matching of $G-R$ such that $|M \cap (C-R)| = \lambda(G-R)$. The set M is a perfect matching of G and C is separating in G; hence $\lambda(G) \leq |M \cap C| = |M \cap (C-R)| = \lambda(G-R)$, hence the inequality.
(b) If G is solid then $\lambda(G) = \infty$. By the first part, $\lambda(G-R) \geq \lambda(G)$. Thus, $\lambda(G-R) = \infty$; hence $G-R$ is solid.

A.11 Chapter 11

Exercise 11.5.7

Let us prove the two assertions by induction on m. In both (i) and (ii), suppose that the graph G has vertices of degree two. Consider the retract $\widehat{G}$ of G. Let $\widehat{m}$, $\widehat{n}$ and $\widehat{b}$ denote, respectively, the number of edges, vertices and bricks of $\widehat{G}$. Clearly, $\widehat{m} - \widehat{n} = m - n$ and $\widehat{b} = b$. If G is neither K_2 nor an even cycle then $\widehat{G}$ is neither K_2 nor an even cycle. If G is not a bisubdivision of K_4 then neither is $\widehat{G}$. Moreover, bicontractions preserve the solidity of G. In both assertions, the result follows by induction and Theorem 11.21. We may thus assume that the minimum degree δ of vertices of G is three or more, in both (i) and (ii).

(i): We assume that $\delta \geq 3$.

Case 1 *Graph G is a brace.*

If G has order two then, as G is not K_2, all its edges are removable. Likewise, if G has order six or more then all its edges are removable. The assertion thus holds if $n \neq 4$.

Suppose now that $n = 4$. As G is free of vertices of degree two, it has a perfect matching whose edges are both multiple edges of G. Thus, at most two edges of G are not removable; hence the number of removable edges of G is at least $m-2 = m-n+2$.

Case 2 *Graph G is a brick.*

If G is K_4 then it has three removable doubletons; hence the assertion holds, as $m - n + 2 - b = 3$. We may thus assume that G is not K_4. The only bricks that do not have removable edges are K_4 and $\overline{C_6}$. Moreover, $\overline{C_6}$ is not solid. Thus, G has a removable edge, e. By Theorem 10.14, the graph $G - e$ is solid. As G is solid, we infer from Theorem 10.16 that the edge e is b-invariant; hence the graph $G - e$ is a near-brick. By induction, $G - e$ has at least $m - n$ edge-disjoint removable ears. Let $P := v_0v_1 \dots v_\ell$, ℓ odd, be a constituent path of a removable ear of $G - e$. If $\ell \geq 3$ then, as $\delta \geq 3$, it follows that e joins the vertices v_1 and v_2. This is not possible, because in that case the set $\{v_0, v_1, v_2\}$ is the shore of a nontrivial tight cut of G. Thus, $G - e$ has at least $m - n$ edge-disjoint removable classes. Each removable class of $G - e$ includes a removable class of G, by Theorem 10.23. Consequently, in addition to edge e, G has at least $m - n$ edge-disjoint removable classes, none of which contains the edge e. The assertion holds.

Case 3 *Graph G has a nontrivial tight cut $C := \partial(X)$.*

Let $G_1 := G/(\overline{X} \to \overline{x})$ and $G_2 := G/(X \to x)$ be the two C-contractions of G. Every brick of G is solid; hence every brick of G_1 is also solid. Thus, G_1 is solid. Likewise, G_2 is solid. For $i = 1, 2$, let m_i, n_i and b_i denote respectively the number of edges, of vertices and of bricks of G_i. By induction, G_i has at least $r_i := m_i - n_i + 2 - b_i$ edge-disjoint removable ears.

Each vertex of G_1 distinct from the contraction vertex $\overline{x}$ has degree three or more in G_1. Thus, the r_1 edge-disjoint removable ears of G_1 are in fact removable classes of G_1. Likewise, the r_2 edge-disjoint removable ears of G_2 are removable classes of G_2. Each removable class of G_i disjoint with C is a removable class of G. For each edge e of C and for $i = 1, 2$, let

$$s_i(e) := \begin{cases} 0, \text{ if no removable class of } G_i \text{ contains the edge } e \\ 1, \text{ otherwise.} \end{cases}$$

A removable class R_1 of G_1 and a removable class R_2 of G_2 may share an edge in C. In that case, $G-(R_1 \cup R_2)$ is matching covered; hence $R_1 \cup R_2$ includes a removable class of G. Alternatively, $s_1(e)+s_2(e) = 1$ implies that the removable class of G_1 or of G_2 that contains the edge e is not removable in G. In sum, the number r of edge-disjoint removable classes of G is at least $r_1+r_2-\sum_{e \in C} \min\{1, (s_1(e)+s_2(e)\} \geq r_1+r_2-|C|$. Thus,

$$\begin{aligned} r &\geq r_1 + r_2 - |C| \\ &= (m_1 - n_1 + 2 - b_1) + (m_2 - n_2 + 2 - b_2) - |C| \\ &= (m_1 + m_2 - |C|) - (n_1 + n_2 - 2) + 2 - (b_1 + b_2) \\ &= m - n + 2 - b. \end{aligned}$$

In all cases considered, we deduce that G has at least $m - n + 2 - b$ edge-disjoint removable ears. In the particular case of bipartite graphs, each removable ear is a single ear and $b = 0$; hence in that case G has at least $m - n + 2$ edge-disjoint removable single ears.

(ii): We assume that $\delta \geq 3$ and also that the near-brick G is not K_4.

Case 1 *The graph G is a brick.*

As in the previous analysis, G has a removable edge, e and the graph $G - e$ is a solid near-brick. If $G - e$ is not a bisubdivision of K_4 then, by induction, $G - e$ has at least $m - n - 2$ edge-disjoint removable single ears. As in the analysis of (i), the $m - n - 2$ removable single ears of $G - e$ are in fact removable edges of $G - e$. Each removable edge of $G-e$ is a removable edge of G, by Theorem 10.23. Consequently, in addition to edge e, G has at least $m-n-2$ removable edges. The assertion holds.

Alternatively, the graph $G-e$ is a bisubdivision of K_4. As G is a brick, the edge e is a multiple edge of G and $G-e$ is K_4. Thus, G has two removable edges and $m-n-1=7-4-1=2$. The assertion holds.

Case 2 *The graph G has a nontrivial tight cut $C := \partial(X)$.*

As in the analysis of (i), define, for $i=1,2$, G_i, m_i and n_i. One of G_1 and G_2 is bipartite, the other is a near-brick. Adjust notation so that G_1 is bipartite. As in the analysis of (i), the removable ears of G_i are in fact removable classes of G_i. By part (i), G_1 has at least $r_1 = m_1 - n_1 + 2$ removable edges. If G_2 is not a bisubdivision of K_4 then, by induction, G_2 has at least $r_2 = m_2 - n_2 - 1$ removable edges. In that case, as in the analysis of (i), the number of removable edges of G is at least

$$r_1 + r_2 - |C| = m - n - 1.$$

Finally, suppose that G_2 is a bisubdivision of K_4. As every vertex of G_2 distinct from its contraction vertex x has degree three or more, it follows that G_2 is K_4. In that case, $r_2 = 0$ and $r_1 + r_2 - |C| = m_1 - n_1 + 2 - 3 = m_1 - n_1 - 1$. But $m - n = m_1 + 3 - (n_1 + 3) = m_1 - n_1$. The assertion holds.

A.12 Chapter 12

Exercise 12.2.7

(i) Let H be the retract of $G-e$ and let J be the underlying simple graph of H. Let v_0 and w_0 be the two ends of e. We have assumed that G is a simple brace of order ten or more. If the minimum degree $\delta(J)$ of the vertices of J is three or more, then, for every nontrivial subset X of V such that $\partial_J(X)$ is tight in G, a vertex of X_- has degree three or more; hence $|X| \geq 5$. It then follows that $G-e$ has a brace of order six or more.
Assume thus, to the contrary, that J has a vertex, x, of degree two. Necessarily, that vertex is a contraction vertex of H. Adjust notation so that x is the result of the contraction of the only two neighbours v_1 and v_2 of v_0 in $G-e$.
For $i=1,2$, let $N(v_i)$ denote the set of neighbours of v_i in G and let $B := N(v_1) \cup N(v_2)$. The vertices v_1 and v_2 are isolated in $G-B$. If $|B| \leq 3$ then B is a barrier of G, and we derive a contradiction, as G is a brace having more than six vertices. See Figures A.14(a) and A.14(b).
Another possibility for x to have degree two in J is the fact that w_0 also has degree three in G, and B consists of four vertices, two of which are the only two neighbours w_1 and w_2 of w_0 in $G-e$. Again, B is a barrier of G, because in this case the vertices v_1, v_2 and w_0 are isolated in $G-B$. We have a contradiction, as G is a brace of order ten or more.

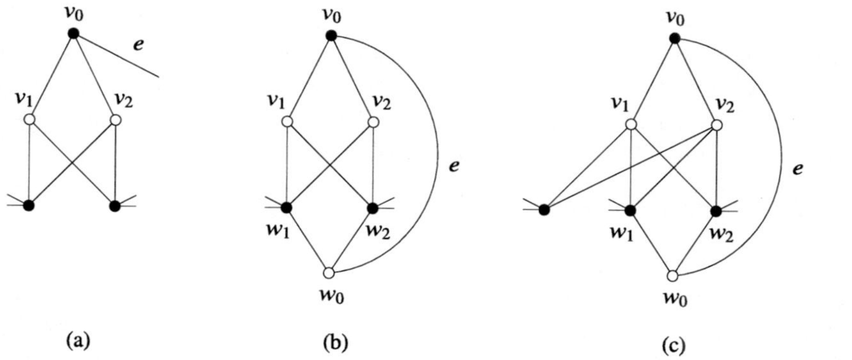

Fig. A.14 The possible causes for $\delta(J) = 2$

(ii) Every vertex of a brace of order six or more has degree three or more. Thus, $K_{3,3}$ is the only simple brace of order six. Let $G := G[A, B]$ be a simple cubic graph of order eight. No two vertices v and w of A have the same set of three neighbours; otherwise the vertex of B not adjacent to either v or w would have degree two, a contradiction. It follows that each vertex v of A corresponds to one of the four sets of three vertices of B. Consequently, the cube is the only simple cubic graph of order eight.
Let us now consider any simple graph $G := G[A, B]$ of order eight and minimum degree three or more. Let M be a maximal matching of G joining vertices of degree four of G. It is easy to see that the subgraph of G induced by M consists of all the vertices of degree four. Thus, $G - M$ is the cube. Addition of any edge to a brace yields another brace. Consequently, G is a brace. Moreover, the cube is a spanning subgraph of G.

(iii) Let us now prove the result by induction on $|E|$. If G has a pair of multiple edges then remove an edge from that pair and the assertion holds. We may thus assume that G is simple. If G has only six vertices then certainly it is $K_{3,3}$. Alternativehy, if G has order eight then it has the cube as a spanning subgraph; thus, the cube is a conformal subgraph of G.
Let us assume then that G has order ten or more. Let e be an edge of G. By the first part, $G - e$ has a brace, J, of order six or more. By induction, J has a conformal minor L which is either $K_{3,3}$ or the cube. Those two graphs are cubic; thus, L is a matching minor of J, by Corollary 12.11. By Exercise 12.1.8, L is a matching minor of G. Hence, again by Corollary 12.11, L is a conformal minor of G.

A.13 Chapter 13

Exercise 13.2.2

(i) The C-contraction $G/\overline{X}$ is the wheel W_5, whereas the C-contraction G/X is $W_5 + e$, a brick. Thus, C is separating. The set $\{u_1v_1, u_3v_3, u_4v_4, v_2v_5, e\}$ is a perfect matching of G that contains precisely three edges in C. Thus, $\lambda(C) = 3$.

(ii) The graph $G-e$ is $\mathbb{P}$, the Petersen graph, hence $(b+p)(G-e) = 2$. Figure A.15 depicts the retract H of $G - f_1 - f_2$, where x is the contraction vertex of H.

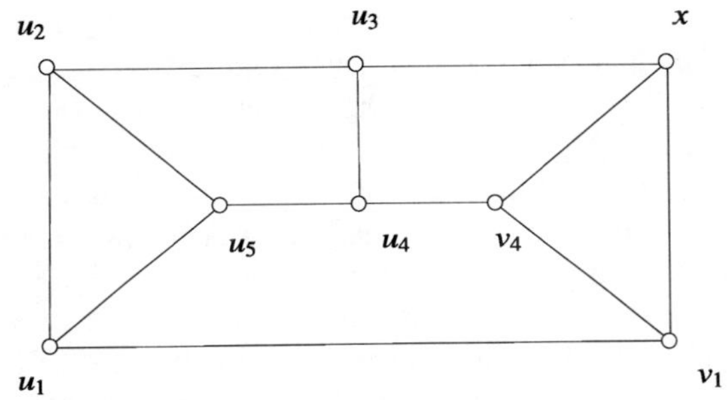

Fig. A.15 The retract of $G - f_1 - f_2$

The graph H is a brick, it is $\mathbb{H}_8$, the bicorn. The retract of $G - f_1$ is the graph $H + f_2$, hence also a brick. Likewise, the retract of $G - f_2$ is $H + f_1$, a brick. Moreover, $H+f_1$ and $H+f_2$ have only eight vertices, hence they are not Petersen bricks. We deduce that $(b + p)(G - f_1) = 1 = (b + p)(G - f_2)$.

(iii) There are eight $(b+p)$-invariant edges of G: $u_1u_2, u_1u_5, u_2u_3, u_4u_5, u_2v_2, u_5v_5, v_1v_3$, and v_1v_4. The edge u_1v_1 is nonremovable and the edge e is b-invariant, but not $(b + p)$-invariant. There are six edges which are not b-invariant : $u_3v_3, u_4v_4, u_3u_4, v_2v_4, v_3v_5$ and v_2v_5.

A.14 Chapter 14

Exercise 14.1.6

(i): Firstly, suppose that G has a nontrivial separating cut. Then, by Corollary 14.4, G has a robust cut, $C := \partial(X)$. By Corollary 14.5, X has a subset X' and $\overline{X}$ has a subset X'' such that the graph $H := G/(X' \to x')/(X'' \to x'')$ is bipartite and matching covered, with the contraction vertices x' and x'' in distinct parts of the bipartition of H. Furthermore, the graphs $G/\overline{X'}$ and $G/\overline{X''}$ are both bricks.

The sets X' and X'' are both critical subsets of V. The graph $G - (X' \cup X'')$ is equal to the graph $H - x' - x''$, which is matchable, by Corollary 3.6.

Conversely, suppose that G is a brick and it has disjoint critical sets X and Y such that $G - (X \cup Y)$ has a perfect matching, M. For each vertex v of X, let $M(v)$ denote a perfect matching of $G[X] - v$. Likewise, for each vertex v of Y, let $N(v)$ denote a perfect matching of $G[Y] - v$. Let

$$\mathbf{x} := \frac{1}{|X| - 1} \sum_{v \in X} \chi^{M(v)} + \frac{1}{|Y| - 1} \sum_{v \in Y} \chi^{N(v)} + \chi^{M}.$$

Note that $\mathbf{x}$ is nonnegative and 1-regular. Moreover, X is odd and $\mathbf{x}(\partial(X)) = 0$. By Theorem 6.1, $\mathbf{x} \notin \mathcal{Poly}(G)$. By Theorem 7.5, G is not solid.

(ii): Suppose that G has two disjoint odd cycles Q_1 and Q_2 such that $G' := G - (V(Q_1) \cup V(Q_1))$ is matchable. The sets $V(Q_1)$ and $V(Q_2)$ are both critical and disjoint. Moreover, G' is matchable. By the first part, G is nonsolid.

Conversely, suppose that G' is a nonsolid brick. By the first part, G has two disjoint critical sets, X and Y, such that $H := G - (X \cup Y)$ has a perfect matching, M_H. The graph $G[X]$ is critical; thus it has an ear decomposition $G_1, G_2, \ldots, G_r$, $r \geq 2$, such that each G_i is a conformal critical subgraph of $G[X]$. In particular, G_2 is a cycle, a conformal subgraph of $G[X]$ which is a cycle. Let $X' := V(G_2)$. The set X' is critical in $G[X]$; let M_X denote a perfect matching of $G[X] - X'$. Likewise, $G[Y]$ has a critical subgraph which is an odd cycle. Let Y' be the set of vertices of the odd cycle and let M_Y denote a perfect matching of $G[Y] - Y'$. The sets X' and Y' are both critical and $M_H \cup M_X \cup M_Y$ is a perfect matching of $G - (X' \cup Y')$.

Exercise 14.3.1

If C is robust then certainly it is separating. To prove the converse, suppose that C is separating. Let G_1 and G_2 denote the two C-contractions of G. As C is separating, the graphs G_1 and G_2 are matching covered. Moreover, as C is not tight in G, the graphs G_1 and G_2 are not bipartite. By the monotonicity of function b, $1 \leq b(G_1) \leq b(G_1 - e)$. Likewise, $1 \leq b(G_2) \leq b(G_2 - e)$. As e is quasi b-invariant and $C - e$ is tight in $G - e$, it follows that $b(G_1 - e) + b(G_2 - e) = b(G - e) = 2$. Consequently, the graphs G_1 and G_2 are near-bricks. In sum, both C-contractions of G are near-bricks. It follows that C is robust.

A.15 Chapter 15

Exercise 15.2.6

Let C be a tight cut of $G - e$. As e is b-invariant, one of the shores of C in $G - e$ is bipartite. By Lemma 8.19(i), every edge of C is removable in the bipartite C-

contraction of $G - e$. This implies that every edge of the brick H incident with a contraction vertex of H is in a tight cut C of $G - e$ and is removable in the bipartite C-contraction of $G - e$. It follows that every removable class of H is also removable in $G - e$.

Suppose that G is regular and R is a removable doubleton of H. By the previous part, R is a removable doubleton of $G - e$. As e is b-invariant, the graph $G - e$ is a near-brick. As R is a removable doubleton of $G - e$, the graph $G - e - R$ is bipartite and matching covered. Let (A, B) be the bipartition of $G - e - R$. One of the edges of R has both ends in A, the other has both ends in B. As G is regular, a simple counting argument then shows that e has one end in A an one end in B. By Corollary 3.7, the graph $G - R$ is also bipartite and matching covered. We conclude that R is a removable doubleton of G.

A.16 Chapter 16

Exercise 16.2.6

Let us assume that G has a perfect matching, M, which contains the edges of R and is disjoint with S. The set R is a removable class of G and the set S is a removable class of $G - R$; hence $G - R - S$ is matching covered. That is, every edge of $G - R - S$ is matchable and $G - R - S$ is connected. Clearly, $G - S$ is thus connected. Moreover, every edge of $G - R - S$ is matchable in $G - S$ and M is a perfect matching of $G - S$ that contains the edges of R; hence each edge of $G - S$ is matchable in $G - S$. We conclude that $G - S$ is matching covered.

Let H be a matching covered graph and let $T \subset E(H)$ such that $G - T$ is matching covered and T is either a singleton or a doubleton. If T is a singleton then the monotonicity of function b implies that $b(H - T) \geq b(H)$, with equality if and only if T is an optimal removable class of G. If T is a doubleton, the monotonicity of function b implies that $b(G - T) \geq b(G) - 1$, with equality if and only if T is a removable doubleton of G. In sum,

$$b(H - T) \geq b(H) - (|T| - 1), \tag{A.21}$$

with equality if and only if T is an optimal removable class of H.

We have seen that $G - S$ is matching covered and S is either a singleton or a doubleton. From equation (A.21) we infer that

$$b(G - S) \geq b(G) - (|S| - 1), \tag{A.22}$$

with equality only if S is an optimal removable class of G. Likewise,

$$b(G - R - S) \geq b(G - S) - (|R| - 1), \tag{A.23}$$

with equality only if R is an optimal removable class of $G-S$. From equations (A.23) and (A.22) we deduce that

$$b(G-R-S) \geq b(G) - (|R|-1) - (|S|-1), \tag{A.24}$$

with equality only if S is an optimal removable class of G and R is an optimal removable class of $G-S$.

As R is an optimal removable class of G and S is an optimal removable class of $G-R$, from equation (A.21) we infer that

$$b(G-R-S) = b(G-R) - (|S|-1) = b(G) - (|R|-1) - (|S|-1).$$

Thus, equality holds in equation (A.24). We conclude that S is an optimal removable class of G and R is an optimal removable class of $G-S$.

Exercise 16.2.8

Algorithm A.2 is a description of a polynomial-time algorithm to determine an optimal ear decomposition of a matching covered graph and the corresponding sequence of optimal removable ears.

Algorithm A.2 Given a matching covered graph G, determine an optimal ear decomposition of G and the corresponding sequence of optimal removable ears.

If $G = K_2$ then just return the sequence (G) as the ear decomposition and the empty sequence of removable ears. If $G \neq K_2$, obtain an optimal removable ear R of G using Algorithm A.3. Recursively determine an optimal ear decomposition $\mathcal{G}$ of $G-R$ and the corresponding sequence $\mathcal{R}$ of optimal removable ears. Add G as the last element of the sequence $\mathcal{G}$ and R as the last element of the sequence $\mathcal{R}$.

Algorithm A.3 Given a matching covered graph G distinct from K_2, determine an optimal removable ear of G.

First determine the retract H of G. This is obtained by contracting the set of vertices of each even lobe and by replacing each odd lobe by an edge. See Theorem 11.21. Then, determine the set R_1 of removable edges of H. To do this, for each edge e of H, determine whether $H-e$ is matching covered (See Exercise 2.2.2). Then, for each pair $\{e_1, e_2\}$ of edges of H not in R_1, determine whether $G - e_1 - e_2$ is matching covered.

If a removable doubleton $\{e_1, e_2\}$ is found, the odd lobes P_1 and P_2 of G which were replaced by e_1 and e_2 are the two constituent paths of a removable double ear of G.

So, suppose that H has no removable doubleton. By Theorem 16.9, R_1 must contain an edge that is $(b+p)$-invariant. Our task now is to find such an edge.

First obtain a tight cut decomposition C of H (See Exercise 5.4.5). Then, the parameter $(b+p)(H)$ is determined, by checking, for each graph $J \in C$ whether it is bipartite, and, if not bipartite, whether the underlying simple graph of J is the Petersen graph.

Likewise, for each removable edge e of H (an edge in R_1), determine the parameter $(b+p)(H-e)$ of $H-e$. There must be an edge e such that $(b+p)(H-e) = (b+p)(H)$. In other words, e is a $(b+p)$-invariant edge of H. The odd lobe of G which was replaced by e is thus a $(b+p)$-invariant ear of G.

A.17 Chapter 17

Exercise 17.2.1

By hypothesis, B_1 and B_2 are two maximal nontrivial barriers of $G - e$. By Theorem 3.11, B_1 and B_2 are disjoint. Thus, I_1 and I_2 are also disjoint; hence $X_1 \cap X_2 = (B_1 \cap I_2) \cup (B_2 \cap I_1)$. By the Three Case Lemma, index$(e) = 2$ and $X_2 - X_1$ is the bipartite shore of a nontrivial tight cut. Thus,

$$1 + |I_2 - B_1| = 1 + |I_2 - X_1| = |B_2 - X_1| = |B_2 - I_1|.$$

Furthermore, B_2 is a special barrier of $G - e$; hence

$$1 + |I_2| = |B_2|.$$

Subtraction of the first equality from the second one yields the asserted equality.

Exercise 17.2.2

We have seen, in the solution of Exercise 17.2.1, the equality $X_1 \cap X_2 = (B_1 \cap I_2) \cup (B_2 \cap I_1)$. We also saw that $|B_1 \cap I_2| = |B_2 \cap I_1|$. Thus, $G[X_1 \cap X_2]$ is equipartite. The second part of the assertion of the exercise comes immediately from the Three Case Lemma.

Exercise 17.6.2

By definition, both ends of e' are in X and its end t' is in X'. Thus, the end s of e' is in $X - X'$.

By definition, the end u of e in X is in A. Thus, $X_- \subset A$ and $X_+ \subset B$. Also by definition, the end t' of e' is in X' and in B. Thus, $X'_+ \subset A$.

Let $H := G - e - e'$ and let $F := \partial(X \cap X')$. Let us prove that $F - e - e'$ is not tight in H. For this, assume the contrary. The subset of H induced by $X \cap X'$ is then connected. Thus, $(A \cap X \cap X', B \cap X \cap X')$ is the only bipartition of the subgraph of H induced by $X \cap X'$.

The cuts $C - e$ and $C' - e - e'$ are both tight in H. Therefore the four subgraphs of H induced by X, $\overline{X}$, X' and $\overline{X'}$ are connected. Thus, H has four edges $v_i w_i$, $1 \le i \le 4$, where $v_1 \in X \cap X'$, $w_1 \in X' - X$, $v_2 \in X - X'$, $w_2 \in \overline{X} \cap \overline{X'}$, $v_3 \in X \cap X'$, $w_3 \in X - X'$, $v_4 \in X' - X$ and $w_4 \in \overline{X} \cap \overline{X}'$. See Figure 17.11.

As $X_+ \subset B$, the vertices v_1 and v_2 are in B. Likewise, $X'_+ \subset A$; thus, the vertices v_3 and v_4 are in A. Thus, $w_3 \in B$. In sum the cut $F - e - e'$ has two edges, one incident with v_1 and the other incident with v_3. Thus, $F - e - e'$ has two edges, one incident with a vertex in one of the parts of $H[X \cap X']$, the other incident with the other part of $H[X \cap X']$. This is a contradiction. We conclude that $F - e - e'$ is not tight in $G - e - e'$.

By Theorem 4.12, $D - e - e'$ and $D' - e - e'$ are tight cuts of $G - e - e'$. The edge e' has no end in $\overline{X}$. Thus, $D' - e$ is tight in $G - e$.

Let us now prove that D is nontrivial. The pair $(A \cap (X - X'), B \cap (X - X'))$ is the unique bipartition of $(G - e - e')[X - X']$. The edge $v_3 w_3$ is an edge of $D - e - e'$ which is incident with w_3, a vertex of $B \cap (X - X')$. Thus, $(X - X')_+ \subset B$. The end s of e' is in $A \cap (X - X')$, in turn a set equal to $(X - X')_-$. We deduce that D is nontrivial.

Exercise 17.6.4

By hypothesis, e is b-invariant but not thin. Thus, G has a nontrivial cut $C := \partial(X)$ such that $C - e$ is tight in $G - e$, X is the bipartite shore of $C - e$, B denotes the barrier of $G - e$ associated with $C - e$, and the set I of isolated vertices of $G - e - B$ has a vertex, s, which has degree three or more in $G - e$.

Let H denote the bipartite graph $(G - e)/\overline{X}$. At most one edge of $\partial(s) - e$ is not removable in H. Every edge of $\partial(s) - e$ that is removable in H is removable in $G - e$. By the exchange property of solid graphs (Theorem 10.23), every edge that is removable in $G - e$ is removable in G. Thus, at least two edges of $\partial(s) - e$, say, e' and e'', are removable in G. As G is solid, both e' and e'' are b-invariant in G.

By the proof of Theorem 17.16, $r(e') \ge r(e)$, $r(e'') \ge r(e)$ and at least one of these inequalities is strict. Suppose now that equality holds for one of these inequalities; suppose that $r(e') = r(e)$. By the proof of Theorem 17.16, Case 2.2 applies, G has a cut $C' := \partial(X')$ such that $C' - e'$ is tight in $G - e'$, C and C' cross, $s \in X - X'$ and e has an end in $X' - X$ and an end in $X \cap X'$. Thus, s is not an end of e.

Exercise 17.6.5

Let G be a solid brick distinct from K_4. Each member of a pair of multiple edges of G is thin. We may thus assume that G is simple. The bricks K_4, $\overline{C_6}$ and $\mathbb{H}_8$ are the only bricks with fewer than two removable edges (Theorem 9.16). The bricks $\overline{C_6}$ and $\mathbb{H}_8$ are not solid. Furthermore, by hypothesis, G is not K_4. Thus, G has at least two removable edges. If every removable edge of G is thin then the assertion holds. We may thus assume that G has removable edges that are not thin.

By Theorem 10.16, every removable edge of G is b-invariant. Thus, G has at least two b-invariant edges. Let r' be the maximum rank of b-invariant edges of G that are not thin. Let e be a b-invariant edge of G such that $r(e) = r'$. By Theorem 17.35, G has two adjacent b-invariant edges, e' and e'', such that at least one of e' and e'' is thin. Let s be the common end of e' and e''. If e' and e'' are both thin then the assertion holds. We may thus assume that precisely one of e' and e'', say, e', is thin. In that case, $r(e'') = r(e) = r'$. We then apply Theorem 17.35 with e'' playing the role of e, thereby obtaining a thin edge f. If f and e'' do not share the common end s of e' and e'' then f and e' are distinct; hence G has two thin edges. Alternatively, if f is incident with s then $\partial(s) - e''$ has two or more edges having maximum rank, hence thin.

A.18 Chapter 18

Exercise 18.2.1

If $\text{index}(e) = 0$ then $\widehat{G - e} = G - e$. In that case, the edges f and g are parallel in $G - e$, hence removable in $G - e$. We may thus assume that $\text{index}(e) > 0$.

Let $C := \partial_{G-e}(X)$ be a nontrivial tight cut of $G - e$ such that the set X is contracted to a single vertex which is a vertex of $\widehat{G - e}$. If $\text{index}(e) \in \{1, 2\}$ then $X := N_{G-e}(v) + v$, where v is an end of e. Alternatively, if $\text{index}(e) = 3$ then $X = N_{G-e}(\{v_1, v_2\}) + v_1 + v_2$, where v_1 and v_2 are the two ends of e. Let $H := (G - e)/\overline{X}$. The retract of H has just two vertices and C is its set of edges. Thus, each edge of C is removable in H. We deduce that every edge that is removable in $\widehat{G - e}$ is removable in $G - e$. In particular, the edges f and g are removable in $G - e$.

A.19 Chapter 19

Exercise 19.2.1

Let us fix a sense of traversal of D. Adjust notation so that the sequence of vertices of Q in this sense of traversal is $1, 2, 3, \ldots, 2n, 1$ (see Exercise 19.1.1). The edge set $E(Q)$ of Q has a partition into two perfect matchings, say M_1 and M_2, where $M_1 := \{\{1, 2\}, \{3, 4\}, \ldots, \{2n-1, 2n\}\}$, and $M_2 := \{\{2, 3\}, \{4, 5\}, \ldots, \{2n-2, 2n-1\}, \{2n, 1\}\}$.

Let us now prove that $\text{sign}(M_1) \cdot \text{sign}(M_2) = (-1)^{|\text{rv}(Q)|+1}$ by induction on $|\text{rv}(Q)|$. If $\text{rv}(Q) = \emptyset$, then Q is a directed cycle. It is easily seen that, in this case, $\text{sign}(M_1) = 1$ and $\text{sign}(M_2) = -1$, and thus that M_1 and M_2 have distinct signs. It follows that Q is evenly oriented and D is not Pfaffian. The assertion holds in this case.

Let us now consider the case in which $\text{rv}(Q)$ contains an edge, say e. Let D' be obtained from D by reversing the orientation of the edge e. For $i = 1, 2$, denote by $\text{sign}'(M_i)$ the sign of M_i in D'. Precisely one of M_1 or M_2, which contains the edge e, has its sign changed. Moreover, $|\text{rv}(Q')| = |\text{rv}(Q)| - 1$. By induction,

$$\text{sign}(M_1) \cdot \text{sign}(M_2) = -\,\text{sign}'(M_1) \cdot \text{sign}'(M_2) = -(-1)^{|\text{rv}(Q')|+1} = (-1)^{|\text{rv}(Q)|+1}.$$

The assertion holds, by induction. It follows that $\text{sign}(M_1) = \text{sign}(M_2)$ if and only if $|\text{rv}(Q)|$ is odd. That is, D is Pfaffian if and only if Q is oddly oriented.

A.20 Chapter 20

Exercise 20.1.2

(a) An arc e of D is reversed in $D' := (D \text{ rev } C_1) \text{ rev } C_2$ if and only if either (i) e is in C_1 but not in C_2 or (ii) e is in C_2 but not in C_1. Thus, the set of arcs that have distinct orientations in D and in D' is equal to $C_1 \triangle C_2$. That is, $(D \text{ rev } C_1) \text{ rev } C_2 = D \text{ rev } (C_1 \triangle C_2)$.

(b) The relation of similarity is certainly reflexive, because the empty set is a cut of any graph. The relation is also symmetric. For this, suppose that D_1 and D_2 are two orientations of G such that D_2 is similar to D_1. Then, G has a cut C such that $D_2 = D_1 \text{ rev } C$. In that case, $D_1 = (D_1 \text{ rev } C) \text{ rev } C = D_2 \text{ rev } C$; hence D_1 is similar to D_2.
Finally, let us prove that the relation of similarity is transitive. Let D_1, D_2 and D_3 be three orientations of a graph G, such that D_1 is similar to D_2 and D_2 is similar to D_3. Let C_1 be the cut of G such that $D_2 = D_1 \text{ rev } C_1$ and let C_2 be the cut of G such that $D_3 = D_2 \text{ rev } C_2$. Then,

$$D_3 = D_2 \text{ rev } C_2 = (D_1 \text{ rev } C_1) \text{ rev } C_2 = D_1 \text{ rev } (C_1 \triangle C_2),$$

where the last equality follows from the first part. By Proposition 20.1(i), $C_1 \triangle C_2$ is a cut of G; hence D_1 and D_3 are similar.

A.21 Chapter 21

Exercise 21.1.3

(i) Let G be a cubic brace of order eight and let (A, B) denote the bipartition of G. For every vertex v of A, the set $N(v)$ of neighbours of v consists of precisely three vertices of B; otherwise $N(v) + v$ would be the shore of a nontrivial tight cut of G. Thus, G is simple. Let f be a mapping from A to B such that

$f(v)$ is the vertex of $B - N(v)$, for each v in A. The mapping f is a bijection, for if $B - N(v) = B - N(w)$ for any two distinct vertices v and w of A then $N(v) \cup \{v, w\}$ is the shore of a nontrivial tight cut of G. As f is a bijection, the graph G is isomorphic to the cube.

(ii) Let G be a brace of order eight. We shall prove that the cube is a subgraph of G, by induction on $|E|$. If G has a multiple edge e, then, by induction, the cube is a subgraph of $G - e$, in turn a subgraph of G.
We may thus assume that G is simple; let (A, B) denote the bipartition of G. Every vertex of G has degree three or four. If G is cubic then, by the first part, G is the cube. We may thus assume that G has vertices of degree four. Clearly, the number of vertices of degree four in A is equal to the number of vertices of degree four in B. Let v be a vertex of degree four in A; let w be a vertex of degree four in B. Let $H := G - vw$.
Assume, to the contrary, that H is not a brace. Let X be a set of vertices of H which is a shore of a nontrivial tight cut of H. Adjust notation, by replacing X by the complement $\overline{X}$ if necessary, so that $X_+ = X \cap B$. The vertex v is in X_- and the vertex w is in $\overline{X}_-$. Thus, the vertex w is adjacent in G to each vertex of X_-. As vw is the only edge of G that joins a vertex in X_- to a vertex not in X, we deduce that X_- is the singleton, $\{v\}$. Then, $|X_+| = 2$; hence v has degree three in G, a contradiction. We conclude that H is a brace.
By induction, the cube is a subgraph of H, in turn a subgraph of G.

(iii) Let G be the simple brace obtained from the cube by the addition of an edge vw. Let D be a Pfaffian orientation of the cube. The orientation is unique, up to similarity. Let M be a perfect matching of G that contains the edge vw. The graph $G - v - w$ is a cycle which is evenly oriented (and M-alternating). Thus, G is not Pfaffian, by Lemma 21.4. Every simple brace of order eight distinct from the cube has the graph G as a subgraph; hence it is not Pfaffian.

Exercise 21.5.1

(i) To prove that each summand G_i is connected, note that the graph $G - X$ is connected (Exercise 5.4.6). Every path in $G - X$ which joins a vertex in $V(G_i) - X$ to vertex v is clearly a path in $G_i - X$. Thus, $G_i - X$ is connected; hence G_i is connected.

(ii) To prove that each summand G_i is matchable, let N be a perfect matching of Q; extend it to a perfect matching M of $G + R$. Then, $M \cap E(G_i)$ is a perfect matching of G_i.

(iii) To prove that each summand G_i is a brace, let e_1 and e_2 be any two nonadjacent edges of G_i. Let M be a perfect matching of $G + R$ that contains both edges e_1 and e_2. As G_i is matchable, $|V(G_i)|$ is even; hence the set M contains an even number of edges in $C := \partial_G(V(Q))$. Moreover, as G_i is bipartite, half of the edges of $M \cap C$ are incident with vertices of one part of the bipartition of G_i; the other half are incident with vertices of the other part. Thus, $M \cap E(G_i)$ is a matching of G_i which may be extended to a perfect matching of G_i by the addition of at most two edges of Q. The graph G_i is connected and 2-extendable; hence it is a brace.

Exercise 21.5.2

To prove that each summand G_i is Pfaffian, we prove that $G + R$ is Pfaffian, by induction on $|R|$. Then, G_i is a conformal subgraph of $G + R$; hence G_i is Pfaffian.

If R is empty then $G + R = G$ and $G + R$ is Pfaffian. We may thus assume that R is not empty. Let $Q := (q_1, q_2, q_3, q_4, q_1)$; let e be an edge of R and let $H := G + e$. Adjust notation so that $e = q_1q_2$. Extend a Pfaffian orientation of G to an orientation D of H by arbitrarily orienting edge e. If all the perfect matchings of H that contain the edge e have the same sign in D then either D is Pfaffian or the reversal of edge e in D yields a Pfaffian orientation of H. Assume, to the contrary, that H contains two perfect matchings that contain the edge e and have distinct signs in D. Then, H has a perfect matching M that contains the edge e and an M-alternating cycle C which is evenly oriented in D and does not contain the edge e. The set $E(C) - E(Q)$ contains edges in at most two summands. As $n \geq 3$, let G_j be a summand of G that does not contain any edge in $E(C) - E(Q)$. As G_j is a brace, it has a perfect matching, M_j, such that $M_j \cap E(Q) = q_3q_4$. Replace the edges of $M \cap (E(G_j) - q_3q_4)$ by the edges of $M_j - q_3q_4$. This operation yields a perfect matching N of G such that the cycle C is N-alternating and evenly oriented in $D - e$. Moreover, $e \notin N$. This is a contradiction, as $D - e$ is a Pfaffian orientation of G.

Exercise 21.5.3

Let $Q := (q_1, q_2, q_3, q_4, q_1)$. To prove that $G + R$ is Pfaffian, note that each summand G_i is Pfaffian and let D_i be a Pfaffian orientation of G_i. We may adjust notation, replacing D_i by a similar orientation if necessary, so that q_1 is a source of $D_i \cap E(Q)$ and q_4 is a sink of $D_i \cap E(Q)$. The cycle Q is a conformal subgraph of the brace G_i; hence Q is oddly oriented in D_i. We deduce that (q_1, q_4) and (q_1, q_2, q_3, q_4) are directed paths in D_i. Thus, $D := \cup_i D_i$ is an orientation of $G+R$. See Figure A.16(a).

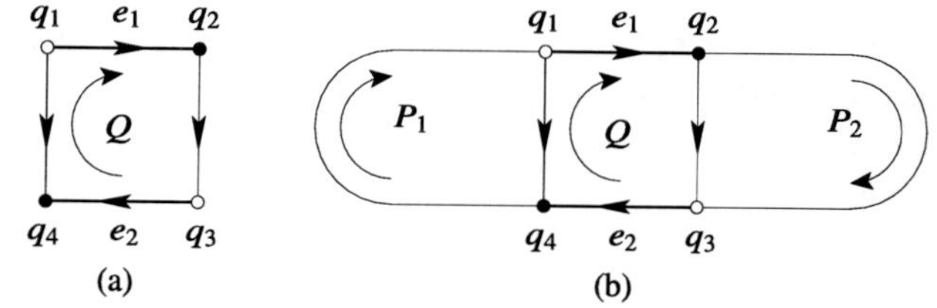

Fig. A.16 The orientation D and the cycle C

Let us now prove that D is Pfaffian. The graph G is a brace; therefore $G+R$ is also a brace. Let M be a perfect matching of $G+R$ that contains the edges $e_1 := q_1q_2$ and $e_2 := q_3q_4$. Let C be an M-alternating cycle of $G+R$. Suppose that all the edges of C are in the same summand of $G+R$, say, G_1. Then, as D_1 is a Pfaffian orientation of G_1, and since $M_1 := M \cap E(G_1)$ is a perfect matching of G_1, we deduce that C is oddly oriented in D_1. As D is an extension of D_1, we conclude that C is oddly oriented in D.

We may thus assume that $E(C) - E(Q)$ has edges in more than one summand of $G+R$. Then, both e_1 and e_2 are in C and $E(C) - E(Q)$ has edges in precisely two summands of $G+R$, say, G_1 and G_2. Then, C has two segments P_1 and P_2, where P_1 is a path in $G_1 - E(Q)$ from q_4 to q_1, P_2 is a path in $G_2 - E(Q)$ from q_2 to q_3, and $C = P_1 \cdot (q_1, q_2) \cdot P_2 \cdot (q_3, q_4)$. See Figure A.16(b). The set $N := M \triangle E(Q)$ is a perfect matching of $G+R$, $C_1 := P_1 \cdot (q_1, q_4)$ is an N-alternating cycle of G_1, and the cycle $C_2 := P_2 \cdot (q_3, q_2)$ is an N-alternating cycle of G_2. By the previous case, both C_1 and C_2 are oddly oriented in D. Moreover,

$$|\mathrm{fw}(C_1)| + |\mathrm{fw}(Q)| + |\mathrm{fw}(C_2)| \equiv |\mathrm{fw}(C)| \pmod 2.$$

We conclude that C is oddly oriented in D. This conclusion holds for each M-alternating cycle of $G+R$. Thus, $G+R$ is Pfaffian; therefore G is also Pfaffian.

Exercise 21.5.6

First apply Algorithm 21.31 to H and Algorithm 21.32 to each of the near-bipartite graphs $H_1, H_2, \ldots, H_r$. Each of these $r+1$ graphs is a conformal subgraph of G. If any one of these graphs is not Pfaffian then G is not Pfaffian.

We may thus assume that $H, H_1, H_2, \ldots, H_r$ are Pfaffian. We shall now produce a Pfaffian orientation of G. Apply Algorithm 20.26 to determine a characteristic (Pfaffian) orientation D_H of H and extend it to a characteristic (Pfaffian) orientation D_i of H_i, by making a conformal cycle of H_i that contains the edges of R_i to be oddly oriented in D_i. Note that reversal of both edges of R_i yields another Pfaffian orientation of H_i. We may thus adjust notation so that the orientation of edge e is the same in all the r orientations D_i. Let $D := \cup_i D_i$.

Let us now prove that D is Pfaffian. For this, let Q be a conformal cycle of G. If e is not in Q then Q is a conformal cycle of H, and Q is thus oddly oriented in D_H, hence oddly oriented in D. Alternatively, suppose that e is in Q. Then, precisely one of the edges $f_1, f_2, \ldots, f_r$ is in Q, say, f_i. Then, Q is a conformal cycle of H_i. Thus, Q is oddly oriented in D_i; hence Q is oddly oriented in D. In all cases considered, Q is oddly oriented in D. Thus, D is Pfaffian; hence G is Pfaffian.

Exercise 21.6.4

We must prove that a connected graph is NT-minimal non-Pfaffian if and only if it is matching covered and S-minimal non-Pfaffian.

if part: Suppose that a matching covered graph G is non-Pfaffian, but every proper S-minor of G is Pfaffian. Let us prove that every proper NT-minor of G is Pfaffian.

We begin by proving that G is free of nontrivial tight cuts. For each nontrivial tight cut C of G, both C-contractions of G are proper S-minors of G; hence Pfaffian. As G is non-Pfaffian we deduce, by Theorem 20.10, that G is free of nontrivial tight cuts. Consequently, every vertex of G is adjacent to three or more vertices.

Let H be a graph obtained from G by the application of either edge removal or contraction of edges of an odd cycle. We prove that in each case H is Pfaffian.

Let us first analyse the case in which $H = G - e$, for some edge e. Suppose that G is a brace. As G is not Pfaffian, it has six or more vertices; in that case, e is removable in G and H is matching covered. Thus, H is a proper S-minor of G; hence H is Pfaffian. We may thus assume that G is a brick. Let R be an equivalence class of the dependence relation induced by e. The edge e may be in R, it in fact may even be the only edge of R. By Corollary 9.3, R is a removable class of G. The graph $G - R$, a proper S-minor of G, is Pfaffian, hence $(G - R) - e$ is Pfaffian. The edges of $R - e$, if any, are not matchable in $G - e$, hence $G - e$ is Pfaffian. This conclusion holds for each edge e of G. In particular, we infer that G is free of multiple edges.

Suppose that H is obtained from G by the contraction of the edges of an odd cycle Q. Let $X := V(Q)$, and let $C := \partial(X)$. Then, $H := G/(X \to x)$. Let us prove that H is Pfaffian.

Suppose that H is not matching covered. It is certainly connected, as it is the result of the contraction of the edges of an odd cycle of the (connected) graph G. Thus, H has a nonmatchable edge, e. The graph $H - e$ is an NT-minor of $G - e$. We have seen above that $G - e$ is Pfaffian; hence $H - e$ is Pfaffian. As e is not matchable in H, we deduce that H is Pfaffian.

We may thus assume that H is matching covered. The graph $H' := G/(\overline{X} \to \overline{x})$ is also matching covered (Exercise 2.2.5). In sum, both C-contractions of G are matching covered and one of them, H, a proper S-minor of G, is Pfaffian. In the two possibilities analysed, the graph H is Pfaffian.

Indeed, the graph H, obtained from G by either removing an edge or by contracting the set of edges of an odd cycle, is Pfaffian. Moreover, G is free of vertices of degree two. We deduce that every proper NT-minor of G is Pfaffian.

only if part: Suppose that G is a connected non-Pfaffian graph such that every proper NT-minor of G is Pfaffian. We shall prove that (i) G is matching covered and (ii) every proper S-minor of G is Pfaffian.

Let us first prove that G is matching covered. Let e be any edge of G. The graph $G - e$, a proper NT-minor of G, is Pfaffian. As G is not Pfaffian, we deduce that e is matchable in G. This conclusion holds for each edge e of G. Moreover, by hypothesis, G is connected; hence G is matching covered. Let us now proceed to prove that every proper S-minor of G is Pfaffian.

Let R be a removable class of G; let us prove that $G - R$ is Pfaffian. For any edge $e \in R$, the graph $G - e$, a proper NT-minor of G, is Pfaffian; hence the graph $G - R$ is Pfaffian.

Let us now prove that for each nontrivial separating cut $C := \partial(X)$, the C-contraction $H := G/(X \to x)$ is Pfaffian. We do this by induction on $|X|$. Let $G_1 := G/(\overline{X} \to \overline{x})$.

Consider first the case in which G_1 has a nontrivial separating cut, say, $D := \partial_{G_1}(Y)$. Adjust notation, replacing Y by $V(G_1)-Y$ if necessary, so that the contraction vertex $\overline{x}$ of G_1 is not in Y. The collection $\{C, D\}$ is cohesive (Exercise 4.2.4). Thus, D is a separating cut of G. Moreover, C is a separating cut of the D-contraction $G' := G/(Y \rightarrow y)$ and H is a C-contraction of G'. By induction, G' is Pfaffian. By Lemma 20.13, H is Pfaffian.

Consider next the case in which G_1 has a removable class R disjoint from C. The cut C is nontrivial and separating in $G - R$. We have seen above that $G - R$ is Pfaffian. By Lemma 20.13, H is Pfaffian.

We may thus assume that G_1 is free of nontrivial separating cuts. In particular, this implies that G_1 is free of nontrivial tight cuts. Thus, G_1 is either a brace or a solid brick. We may also assume that every removable class of G_1 contains an edge in C.

Let us now prove that G_1 is not a brace. Assume the contrary, let (A, B) denote the bipartition of G_1, and adjust notation so that the contraction vertex $\overline{x}$ is in A. Let v be a vertex of $A - \overline{x}$. We have assumed that every removable class of G_1 contains an edge of C. This implies that no edge of G_1 incident with v is removable. By Theorem 8.3, G_1 has only four vertices. As G is free of multiple edges, v has degree two; hence H is the graph obtained from G by the bicontraction of v. Moreover, C is tight in G. Then, as G is not Pfaffian and G_1 is Pfaffian, we infer from Theorem 20.10 that H is not Pfaffian. This is a contradiction, as H is a proper NT-minor of G which is not Pfaffian.

We conclude that G_1 is a solid brick. Every removable class of G_1 contains an edge in C. By Corollary 10.11, G_1 is an odd wheel, up to multiple spokes, having $\overline{x}$ as its hub. In that case, the graph H is obtained by the contraction of the odd cycle which is the rim of H_1. Thus, H is a proper NT-minor of G, hence H is Pfaffian.

We conclude that G is matching covered and each proper S-minor of G is Pfaffian.

References

[1] E. Balas. Integer and fractional matchings. In P. Hansen, editor, *Studies on Graphs and Discrete Programming*, pages 1–13. North-Holland, 1981.

[2] J. A. Bondy and U. S. R. Murty. *Graph Theory with Applications*. Elsevier North Holland, 1976.

[3] J. A. Bondy and U. S. R. Murty. *Graph Theory*. Springer, 2008.

[4] R. L. Brooks, C. A. B. Smith, A. H. Stone, and W. T. Tutte. The dissection of rectangles into squares. *Duke Math. J.*, 7:312–340, 1940.

[5] C. N. Campos and C. L. Lucchesi. On the relation between the Petersen graph and the characteristic of separating cuts in matching covered graphs. Technical Report 22, Institute of Computing, University of Campinas, Brazil, 2000.

[6] M. H. Carvalho. *Decomposição Ótima em Orelhas para Grafos Matching Covered*. PhD thesis, Instituto de Computação – UNICAMP, 1996. in Portuguese.

[7] M. H. Carvalho and C. L. Lucchesi. Matching covered graphs and subdivisions of K_4 and $\overline{C}_6$. *J. Combin. Theory Ser. B*, 66:263–268, 1996.

[8] M. H. Carvalho, C. L. Lucchesi, and U. S. R. Murty. Ear decompositions of matching covered graphs. *Combinatorica*, 19:151–174, 1999.

[9] M. H. Carvalho, C. L. Lucchesi, and U. S. R. Murty. On a conjecture of Lovász concerning bricks. I. The characteristic of a matching covered graph. *J. Combin. Theory Ser. B*, 85:94–136, 2002.

[10] M. H. Carvalho, C. L. Lucchesi, and U. S. R. Murty. On a conjecture of Lovász concerning bricks. II. Bricks of finite characteristic. *J. Combin. Theory Ser. B*, 85:137–180, 2002.

[11] M. H. Carvalho, C. L. Lucchesi, and U. S. R. Murty. Optimal ear decompositions of matching covered graphs and bases for the matching lattice. *J. Combin. Theory Ser. B*, 85:59–93, 2002.

[12] M. H. Carvalho, C. L. Lucchesi, and U. S. R. Murty. The perfect matching polytope and solid bricks. *J. Combin. Theory Ser. B*, 92:319–324, 2004.

[13] M. H. Carvalho, C. L. Lucchesi, and U. S. R. Murty. Graphs with independent perfect matchings. *J. Graph Theory*, 48:19–50, 2005.

[14] M. H. Carvalho, C. L. Lucchesi, and U. S. R. Murty. On the number of dissimilar Pfaffian orientations of graphs. *RAIRO Theor. Inform. Appl.*, 39:93–113, 2005.

C. L. Lucchesi, U. S. R. Murty, *Perfect Matchings*, Algorithms and Computation in Mathematics 31, https://doi.org/10.1007/978-3-031-47504-7

[15] M. H. Carvalho, C. L. Lucchesi, and U. S. R. Murty. How to build a brick. *Discrete Math.*, 306:2383–2410, 2006.
[16] M. H. Carvalho, C. L. Lucchesi, and U. S. R. Murty. A generalization of Little's theorem on Pfaffian orientations. *J. Combin. Theory Ser. B*, 102:1241–1266, 2012.
[17] M. H. Carvalho, C. L. Lucchesi, and U. S. R. Murty. On the number of perfect matchings in a bipartite graph. *SIAM J. Discrete Math.*, 27:940–958, 2013.
[18] M. H. Carvalho, C. L. Lucchesi, and U. S. R. Murty. Thin edges in braces. *Electron. J.Combin.*, 22(#P4.14):38, 2015.
[19] M. H. Carvalho, C. L. Lucchesi, and U. S. R. Murty. On tight cuts in matching covered graphs. *J. Comb.*, 9:163–184, 2018.
[20] G. Chen, X. Feng, F. Lu, C. L. Lucchesi, and L. Zhang. Laminar tight cuts in matching covered graphs. *J. Combin. Theory Ser. B*, 150:177–194, 2021.
[21] G. Chen, X. Feng, F. Lu, and L. Zhang. Disjoint odd cycles in cubic solid bricks. *SIAM J. Discrete Math.*, 33:393–397, 2019.
[22] M. Chudnovsky, K. Edwards, K. Kawarabayashi, and P. D. Seymour. Edge-colouring seven-regular planar graphs. *J. Combin. Theory Ser. B*, 115:276–302, 2015.
[23] M. Chudnovsky, K. Edwards, and P. D. Seymour. Edge-colouring eight-regular planar graphs. *J. Combin. Theory Ser. B*, 115:303–338, 2016.
[24] A. Cobham. The intrinsic computational difficulty of functions. In Y. Bar-Hillel, editor, *Logic, Methodology and Phylosophy of Science, Proceeding of the 1964 International Congress*, Studies in Logic and the Foundation of Mathematics, pages 24–30. North-Holland, 1965.
[25] T. H. Cormen, C. E. Leiserson, R. L. Rivest, and C. Stein. *Introduction to Algorithms.* The MIT Press, England, 2009.
[26] M. H. de Carvalho, N. Kothari, C. L. Lucchesi, and U. S. R. Murty. Bricks with several removable doubletons. Unpublished, 2015.
[27] J. C. de Pina, C. G. Fernandes, B. A. Reed, and Y. Wakabayashi. A characterization of solid bricks. Personal Communication, 2003.
[28] A. L. Dulmage and N. S. Mendelsohn. Coverings of bipartite graphs. *Canad. J. Math.*, 10:517–534, 1958.
[29] Z. Dvořák, K. Kawarabayashi, and D. Král. Packing six T-joins in plane graphs. *J. Combin. Theory Ser. B*, 116:287–305, 2016.
[30] J. Edmonds. Maximum matchings and a polyhedron with (0,1) vertices. *J. Res. Nat. Bur. Standards Sect B*, 69B:125–130, 1965.
[31] J. Edmonds. Paths, trees, and flowers. *Canad. J. Math.*, 17:449–467, 1965.
[32] J. Edmonds, L. Lovász, and W. R. Pulleyblank. Brick decomposition and the matching rank of graphs. *Combinatorica*, 2:247–274, 1982.
[33] L. Esperet, F. Kardoš, A. D. King, D. Král', and S. Norine. Exponentially many perfect matchings in cubic graphs. *Adv. Math.*, 227:1646–1664, 2011.
[34] I. Fischer and C. H. C. Little. A characterisation of Pfaffian near bipartite graphs. *J. Combin. Theory Ser. B*, 82:175–222, 2001.
[35] G. Frobenius. Über zerlegbare Determinanten. *Sitzungber. König. Preuss. Akad. Wiss.*, XVIII:274–277, 1917.

[36] H. N. Gabow and R. E. Tarjan. A linear time algorithm for a special case of disjoint set union. *J. Comput. System Sci.*, 30:209–221, 1985.

[37] A. Galluccio and M. Loebl. On the theory of Pfaffian orientations. I. Perfect matchings and permanents. *Electron. J.Combin.*, 6, 1999.

[38] C. D. Godsil. *Algebraic Combinatorics*. Chapman and Hall, 1993.

[39] B. Guenin. Packing T-joins and edge colouring in planar graphs. Technical Report CORR 2003-09, Dept of Combinatorics and Optimization, University of Waterloo, 2003.

[40] P. Hall. On representatives of subsets. *J. London Math. Soc.*, 10:26–30, 1935.

[41] J. H. Halton. A combinatorial proof of Cayley's Theorem on Pfaffians. *J. Combin. Theory Ser. B*, 1:333–337, 1966.

[42] J. H. Halton. An identity of the Jacobi Type for Pfaffians. *J. Combin. Theory Ser. B*, 1:224–232, 1966.

[43] J. E. Hopcroft and R. M. Karp. An $n^{5/2}$ algorithm for maximum matchings in bipartite graphs. *SIAM J. Comput.*, 2:225–231, 1973.

[44] P. W. Kasteleyn. Dimer statistics and phase transitions. *J. Math. Phys.*, 4:287–293, 1963.

[45] K. Kawarabayashi and K. Ozeki. A simpler proof for the two disjoint odd cycles theorem. *J. Combin. Theory Ser. B*, 103:313–319, 2013.

[46] N. Kothari. Private communication. 2019.

[47] N. Kothari. *Brick Generation and Conformal Subgraphs*. PhD thesis, University of Waterloo, Waterloo, Ontario, Canada, 2016.

[48] N. Kothari. Generating near-bipartite bricks. *J. Graph Theory*, 90:565–590, 2019.

[49] N. Kothari and M. H. de Carvalho. Generating simple near-bipartite bricks. *J. Graph Theory*, 95:594–637, 2020.

[50] N. Kothari and U. S. R. Murty. K_4-free and $\overline{C_6}$-free planar matching covered graphs. *J. Graph Theory*, 82:5–32, 2015.

[51] C. H. C. Little. The parity of the number of 1-factors of a graph. *Discrete Math.*, 2:179–181, 1972.

[52] C. H. C. Little. Kasteleyn's theorem and arbitrary graphs. *Canad. J. Math.*, 25:758–764, 1973.

[53] C. H. C. Little. A theorem on connected graphs in which every edge belongs to a 1-factor. *J. Austral. Math. Soc.*, 18:450–452, 1974.

[54] C. H. C. Little. A characterization of convertible $(0,1)$-matrices. *J. Combin. Theory Ser. B*, 18:187–208, 1975.

[55] C. H. C. Little and F. Rendl. Operations preserving the Pfaffian property of a graph. *J. Aust. Math. Soc. (Series A)*, 50:248–257, 1991.

[56] L. Lovász. Three short proofs in graph theory. *J. Combin. Theory Ser. B*, 19:269–271, 1975.

[57] L. Lovász. Ear decompositions of matching-covered graphs. *Combinatorica*, 3:105–117, 1983.

[58] L. Lovász. Matching structure and the matching lattice. *J. Combin. Theory Ser. B*, 43:187–222, 1987.

[59] L. Lovász and M. D. Plummer. *Matching Theory*. Number 29 in Annals of Discrete Mathematics. Elsevier Science, 1986.

[60] L. Lovász and M. D. Plummer. *Matching Theory*. Reprint of the 1986 original. AMS Chelsea Publishing, 2009.

[61] L. Lovász and S. Vempala. On removable edges in matching covered graphs. Unpublished Manuscript.

[62] F. Lu, N. Kothari, X. Feng, and L. Zhang. Equivalence classes in matching covered graphs. *Discrete Math.*, 343, 2020.

[63] C. L. Lucchesi. On a theorem of Fischer and Little on Pfaffian orientations of near-bipartite graphs. Submitted for publication, 2023.

[64] C. L. Lucchesi, M. H. de Carvalho, N. Kothari, and U. S. R. Murty. On two unsolved problems concerning matching covered graphs. *SIAM J. Discrete Math.*, 32:1478–1504, 2018.

[65] Y. Ma, D. Mattiolo, E. Steffen, and I. H. Wolf. Pairwise disjoint perfect matchings in r-edge-connected r-regular graphs. *SIAM J. Discrete Math.*, 37:1548–1565, 2023.

[66] G. Mazzuoccolo. The equivalence of two conjectures of Berge and Fulkerson. *J. Graph Theory*, 68:125–128, 2011.

[67] D. McCarthy and R. G. Stanton, editors. *Selected Papers of W. T. Tutte*. Charles Babbage Research Centre, Manitoba, Canada, 1979. (in two volumes).

[68] W. McCuaig. Brace generation. *J. Graph Theory*, 38:124–169, 2001.

[69] W. McCuaig. Pólya's permanent problem. *Electron. J.Combin.*, 11, 2004.

[70] S. Micali and V. V. Vazirani. An $o(\sqrt{|V|}|e|)$ algorithm for finding maximum matching in general graphs. *Proc. 21st IEEE FOCS*, pages 17–27, 1980.

[71] A. A. A. Miranda. Private communication, 2008.

[72] A. A. A. Miranda. *Grafos Pfaffianos e Problemas Relacionados*. PhD thesis, Institute of Computing, University of Campinas, Brazil, 2009. In Portuguese.

[73] A. A. A. Miranda and C. L. Lucchesi. Private communication, 2008.

[74] A. A. A. Miranda and C. L. Lucchesi. A polynomial time algorithm for recognizing near-bipartite Pfaffian graphs. *Electron. Notes Discrete Math.*, 30:171–176, 2008. IV Latin-American Algorithms, Graphs and Optimization Symposium - LAGOS 2007, Chile.

[75] A. A. A. Miranda and C. L. Lucchesi. Matching signatures and pfaffian graphs. *Discrete Math.*, 311:289–294, 2011.

[76] D. Naddef. Rank of maximum matchings in a graph. *Math. Program.*, 22:52–70, 1982.

[77] S. Norine. Drawing 4-Pfaffian graphs on the torus. *Combinatorica*, 29:109–119, 2009.

[78] S. Norine and R. Thomas. Generating bricks. *J. Combin. Theory Ser. B*, 97:769–817, 2007.

[79] S. Norine and R. Thomas. Minimally non-Pfaffian graphs. *J. Combin. Theory Ser. B*, 98:1038–1055, 2008.

[80] J. Petersen. Die Theorie der regulären Graphs. *Acta Math.*, 15:193–220, 1891.

[81] R. Rizzi. Indecomposable r-graphs and some other counterexamples. *J. Graph Theory*, 32:1–15, 1999.

[82] N. Robertson, P. D. Seymour, and R. Thomas. Permanents, Pfaffian orientations and even directed circuits. *Ann. of Math.*, 150:929–975, 1999.

[83] A. Schrijver. *Theory of Linear and Integer Programming*. Wiley-Interscience Series in Discrete Mathematics and Optimization, 1998 (paperback edition).

[84] A. Schrijver. *Combinatorial Optimization: Polyhedra and Efficiency*. Springer, 2003.

[85] P. D. Seymour. On multi-colourings of cubic graphs and conjectures of Fulkerson and Tutte. *Proc. London Math. Soc*, 38:423–460, 1979.

[86] Z. Szigeti. The two ear theorem on matching-covered graphs. *J. Combin. Theory Ser. B*, 74:104–109, 1998.

[87] Z. Szigeti. Perfect matchings versus odd cuts. *Combinatorica*, 22:575–589, 2002.

[88] G. Tesler. Matchings in graphs on non-orientable surfaces. *J. Combin. Theory Ser. B*, 78:198–231, 2000.

[89] W. T. Tutte. The factorization of linear graphs. *J. London Math. Soc.*, 22:107–111, 1947.

[90] W. T. Tutte. *Graph Theory as I Have Known It*. Number 11 in Oxford Lecture Series in Mathematics and its Applications. Clarendon Press, Oxford, 1998.

[91] V. V. Vazirani and M. Yannakakis. Pfaffian orientation of graphs, 0-1 permanents, and even cycles in directed graphs. *Discrete Appl. Math.*, 25:179–190, 1989.

[92] P. Whalen. *Pfaffian Orientations, Flat Embeddings, and Steinberg's Conjecture*. PhD thesis, Georgia Institute of Technology, 2014.

[93] H. Whitney. Non-separable and planar graphs. *Trans. Amer. Math. Soc.*, 34:339–362, 1932.

[94] N. C. Wormald. Classifying K-connected cubic graphs. In A. F. Horadam and W. D. Wallis, editors, *Combinatorial Mathematics VI*, volume 748 of *Lecture Notes in Mathematics*, Berlin, Heidelberg, 1979. Springer.

[95] D. H. Younger. William Thomas Tutte. *Biogr. Mems Fell. R. Soc.*, 58:283–297, 2012.

List of Figures

C. L. Lucchesi, U. S. R. Murty, *Perfect Matchings*, Algorithms and Computation in Mathematics 31, https://doi.org/10.1007/978-3-031-47504-7

Glossary

$(G \odot H)_{u,v,\theta}$	the splicing of G at u and H at v with respect to θ	31
$(G \odot K_4)$	the splicing of vertex-transitive G and K_4	32
$(G \odot K_4)_u$	the splicing of G at u and K_4	32
$(b+p)(G)$	$b(G)+p(G)$	186
$A+x$	addition of element x to set A, commonly represented by $A \cup \{x\}$	xxi
$A-x$	deletion of element x from set A, commonly represented by $A \setminus \{x\}$	xxi
$B_G(C)$	the majority part of $X_G(C)$	354
$C \prec D$	cut C strictly precedes cut D	145
$C \preceq D$	cut C precedes cut D	145
$C_G(B)$	the cut $\partial_G(X_G(B))$	354
D rev S	the digraph obtained from D by reversing the orientations of the edges in S	438
E	the set of edges of a graph	xxii
$E_{\text{even}}(R)$	the set of even numbered edges of path(s) of R	231
$E_{\text{odd}}(R)$	the set of odd numbered edges of path(s) of R	231
$F(n)$	the nth Fibonacci number	38
$G-S$	the graph obtained from G by removing the set S (of edges or of vertices)	xxii
$G-e$	the graph obtained from G by the deletion of edge e	xxii
$G-v$	the graph obtained from G by the deletion of vertex v	xxii
$G[A,B]$	a graph G with bipartition (A,B)	13
$G \odot H$	the splicing of G and H	31
$I(G)$	the set of isolated vertices of a graph G	15
$I_G(B)$	the set of isolated vertices of $G-B$	354
$I_G(C)$	the minority part of $X_G(C)$	354
K_4	the complete graph on four vertices	27

C. L. Lucchesi, U. S. R. Murty, *Perfect Matchings*, Algorithms and Computation in Mathematics 31, https://doi.org/10.1007/978-3-031-47504-7

$N(S)$	the neighbourhood of a set S of vertices	xxii
$N(v)$	the neighbourhood of vertex v	xxii
$N_G(S)$	the neighbourhood of a set S of vertices in graph G	xxi
$N_G(v)$	the neighbourhood of vertex v in graph G	xxi
$S - T$	set difference, commonly represented by $S \setminus T$	xxi
V	the set of vertiocs of a graph	xxii
W_k	wheel whose rim has order k	28
X_+	the majority part of a bipartite graph, where X is its set of vertices	63
X_-	the minority part of a bipartite graph, where X is its set of vertices	63
$X_G(B)$	the set $B \cup I_G(B)$	354
$X_G(C)$	the bipartite shore of tight cut C of near-brick G	354
$IntCone(G)$	the integer cone of G	132
$Cone(G)$	the perfect matching cone of G	120
G/X	the graph obtained from G by contracting a nonempty subset X of V to a new single vertex and removing all resulting loops	xxiii
$G/(X \to x)$	the graph obtained from G by contracting a nonempty subset X of V to a new single vertex x and removing all resulting loops	xxiii
$\Delta(G)$	the maximum degree of the vertices of G	xxiii
Γ_1	the graph Γ_1, see Figure 21.2	463
Γ_2	the graph Γ_2, see Figure 21.3	464
$\mathcal{P}oly(G)$	the perfect matching polytope of G	115
$\mathcal{P}oly^*(G)$	the fractional perfect matching polytope of G	144
$\mathcal{P}oly^t(G)$	the tight-cut perfect matching polytope of G	144
$\mathrm{Pf}(\mathbf{A})$	Pfaffian of $\mathbf{A}$	17
$\Phi(G)$	the number of perfect matchings of G	38, 136
$\mathbb{R}^E$	the edge space of graph $G := (V, E)$	111
$G\{x \to (a, b_1, b_2)\}$	constrained bisplitting vertex x in graph G	369
$\mathbb{H}_8$	the bicorn	32
$\mathbb{B}_n$	the biwheel of order n	36
χ^F	the incidence vector of a set F of edges	112
$\delta(G)$	the minimum degree of the vertices of G	xxiii
$\det(\mathbf{A})$	the determinant of $\mathbf{A}$	16
$\mathrm{fw}(T)$	the set of forward edges of a trail T	420
$\kappa'(G)$	the edge-connectivity of graph G	xxii, xxiii
$\kappa(G)$	the connectivity of graph G	xxii
$\mathbb{L}_n$	the ladder of order n	35
$\lambda(C)$	the characteristic of cut C in a graph	73
$\lambda(G)$	the characteristic of graph G	73, 74
$\mathcal{L}at(G)$	the matching lattice of G	133
$\mathcal{L}at(G, \mathbb{F})$	the matching lattice of G over $\mathbb{F}$	135, 341

$\mathcal{L}at^*(G)$	the set of matching integral vectors of G	337
$\mathcal{L}in(G)$	the perfect matching space of G	121
$\mathcal{L}in(G, C)$	a subspace of $\mathcal{L}in(G)$	131
$\mathcal{L}in(G, \mathbb{F})$	the matching space of G over $\mathbb{F}$	135, 341
$\mathbb{H}(B)$	the core of a graph with respect to barrier B	44
$\mathbb{P}$	the Petersen graph	27
$\mathcal{B}$	the family of biwheels	36
$\mathcal{B}_1 \vee \mathcal{B}_2$	the merger operation of $\mathcal{B}_1$ and $\mathcal{B}_2$	128
$\mathcal{ML}$	the family of Möbius ladders	35
$\mathcal{M}$	the set of perfect matchings of a graph	35
$\mathcal{P}$	the family of prisms	34
$\mathcal{ST}$	the family of staircases	36
$\mathcal{TB}$	the family of truncated biwheels	37
$\mathcal{T}(G)$	the set of tight cuts of G	121
$\mathbb{M}_n$	the Möbius ladder of order n	34
$\overline{C_6}$	the triangular prism	27
$\overline{X}$	the complement $V - X$ of X	58
$\partial_G(H)$	the edge cut $\partial_G(V(H))$	xxii
$\partial_G(S)$	the edge cut of G associated with S	xxii
$\partial_G(v)$	the edge cut of G associated with $\{v\}$	xxii
$\mathbb{P}_n$	the prism of order n	34
$\text{rv}(T)$	the set of reverse edges of a trail T	420
$\text{sign}(M)$	the sign of a perfect matching M	16
$\mathbb{S}_n$	the staircase of order n	36
$\tau(G)$	the number of equivalence classes of Pfaffian orientations of G with respect to similarity	449
$\mathbb{H}_{10}$	the tricorn	32
$\mathbb{T}_n$	the truncated biwheel of order n	36
$\mathbf{x} \circ \mathbf{y}$	the scalar product of $\mathbf{x}$ and $\mathbf{y}$	163
$\mathbf{x}(F)$	$\sum_{e \in F} \mathbf{x}(e)$	112
$b(G)$	the number of bricks of G	76
$d(\mathcal{G})$	the number of double ears of ear decomposition $\mathcal{G}$	240
$e \Leftrightarrow f$	edges e and f are mutually dependent	164
$e \Rightarrow f$	edge e depends on (implies) edge f	164
m	the size of a graph	xxii
n	the order of a graph	xxii
$p(G)$	the number of Petersen bricks of G	186
$\mathcal{O}(G)$	the set of odd components of graph G	6
co-NP	the class of decision problems whose complements are solvable by nondeterministic algorithms in polynomial time	10
$\text{def}(M)$	the deficiency of a matching M	7
$o(G)$	the number of odd components of graph G	6

NP	the class of decision problems solvable by non-deterministic algorithms in polynomial time	10
P	the class of decision problems solvable in polynomial time	10

Index

C. L. Lucchesi, U. S. R. Murty, *Perfect Matchings*, Algorithms and Computation in Mathematics 31, https://doi.org/10.1007/978-3-031-47504-7